MECHANICAL ENGINEERING THEORY AND APPLICATIONS SERIES

THERMAL ENGINEERING RESEARCH DEVELOPMENTS

MECHANICAL ENGINEERING THEORY AND APPLICATIONS SERIES

Mechanical Vibrations of Discontinuous Systems
Juan Carlos Jauregui-Correa and Oscar M. Gonzalez Brambila
2009. ISBN: 978-1-60876-126-5

Thermal Engineering Research Developments
Jovan Evgova and Ognjan Kostadinov (Editors)
2010. ISBN: 978-1-60741-497-1

Mechanical Vibrations: Types, Testing and Analysis
Amy L. Galloway (Editor)
2010. ISBN: 978-1-61668-217-0

Mechanical Vibrations: Types, Testing and Analysis
Amy L. Galloway (Editor)
2010. ISBN: 978-1-61668-644-4 (Online Book)

Nondestructive Testing: Methods, Analyses and Applications
Earl N. Mallory (Editor)
2010. ISBN: 978-1-60876-157-9

Nondestructive Testing: Methods, Analyses and Applications
Earl N. Mallory (Editor)
2010. ISBN: 978-1-61668-422-6 (Online Book)

Mechatronic Design Automation: Emerging Research and Recent Advances
Zhun Fan
2010. ISBN: 978-1-61668-956-8

MECHANICAL ENGINEERING THEORY AND APPLICATIONS SERIES

THERMAL ENGINEERING RESEARCH DEVELOPMENTS

JOVAN EVGOVA

AND

OGNJAN KOSTADINOV

EDITORS

Nova Science Publishers, Inc.

New York

For permission to use material from this book please contact us:
Telephone 631-231-7269; Fax 631-231-8175
Web Site: http://www.novapublishers.com

LIBRARY OF CONGRESS CATALOGING-IN-PUBLICATION DATA

Thermal engineering research developments / editors, Jovan Evgova and Ognjan Kostadinov.
p. cm.
Includes index.
ISBN 978-1-60741-497-1 (hardcover)
1. Heat engineering. 2. Heat--Transmission. 3. Thermodynamics. I. Evgova, Jovan. II. Kostadinov, Ognjan.
TJ260.T4937 2009
621.402--dc22

2009035173

Published by Nova Science Publishers, Inc. ✦ New York

Contents

PREFACE

Different thermal engineering calculations can be useful tools to solve complex issues nowadays in thermal equipment design for several industrial applications. In this book, the Boltzmann Transport Equations (BTE) is examined to formulate the transport laws for equilibrium and irreversible thermodynamics. Particular attention is paid to the energetic processes in adsorption phenomena as well as in semiconductors from the view point of irreversible thermodynamics. The problems of heat exchange in laser medium and pump DLAs, as well as the associated thermal stress and thermal strain are also studied to seek the methods for enhancing heat transfer, lowering the device temperature and improving temperature uniformity and reducing thermal stress. The authors of this book also discuss the design, retrofit and the thermal performance analysis of industrial cooling systems. Other chapters include a proposal for a methodology to predict the heat transfer during the cutting process, an analysis of an optimization process for a reverse cycle machine, and numerical results concerning the simulation of semiconductors melts with free capillary surfaces.

Chapter 1 can be divided into four main parts, showing that different thermal engineering calculations can be useful tools to solve complex issues in nowadays in thermal equipment design for several industrial applications. The first one summarizes a new methodology for the design and optimization of a low NO_x-CO burner. To begin with, the working of an experimental 2 MW natural gas burner, in a superheated steam boiler was studied. A CFD model was built and validated with the data obtained from the above-mentioned burner, implementing new User Defined Functions. As a result of this, the first model was then modified in order to get a final CFD, low NO_x-CO emission model.

The second part is focused on a detailed study of the air pre-heaters in a thermoelectric power plant, where a preliminary large amount of ash deposition on one of the inlet conduits was observed, obstructing this way the incoming gas flow. Prior to the cleaning of the conduits, certain damage affected the boiler dome, and as a result of this, the pressure reached a very high value. All of this indicated that the cause of the clogging was not entirely physical. Comprehensive and detailed analytical studies, as well as the subsequent guideline for solving the situation were carried out.

The third part presents the complexity of a cogeneration process design for a power plant of reference. On one hand, a complete study of heat and electricity loads was carried out, also performing a comparative test between different alternatives, setting up selection criteria. On the other hand, the work goes through an economical analysis showing the viability of choosing natural gas engines instead of gas turbines. The investment-saving relationship was obviously considered.

Finally, a novel thermal optimization of the purging process for the bottles of CO_2 used in two big generators in a thermal power plant was described in deep. Because of both filling and emptying processes must be realized in an inert atmosphere, the experience demonstrated that the best fluid was the carbon dioxide. Almost the 90% of the content inside every bottle remains in liquid state, but as soon as the emptying process goes on, a quick and progressive cooling of this fluid occurs until the so called "carbonic ice" is finally formed, rendering useless this way the rest of the bottle.

In Chapter 2, the Boltzmann Transport Equations (BTE) is used to formulate the transport laws for equilibrium and irreversible thermodynamics and these BTE equations are suitable for analyzing system performance that are associated with systems ranging from macro to micro dimensions. In this regard, particular attention is paid to analyze the energetic processes in adsorption phenomena as well as in semiconductors from the view point of irreversible thermodynamics. The continuity equations for (i) gaseous flow at adsorption surface, and (ii) electrons, holes and phonons movements in the semiconductor structures are studied. The energy and entropy balances equations of (i) the adsorption system for macro cooling, and (ii) the thermoelectric device for micro cooling are derived that lead to expressions for entropy generation and system's bottlenecks. The BTE equation is applied to model the adsorption cooling processes for single-stage, multi-stage and multi-bed systems, and the simulated results are compared with experimental data. This chapter also presents a thermodynamic framework for the estimation of the minimum driving heat source temperature of an advanced adsorption cooling device from the rigor of Boltzmann distribution function. From this thermodynamic analysis, an interesting and useful finding has been established that it is possible to develop an adsorption cooling device as a green and sustainable technology that operates with a driving heat source temperature of near ambient. Moreover, the Onsager relations are applied to model the thermoelectric transport equations and, after coupling with Gibbs law and BTE, the temperature-entropy flux derivations are further developed and presented the energetic performances of thermoelectric cooling systems.

Thermal effects are among the major barriers obstructing the technological development in high-power solid-state laser. In Chapter 3, the problems of heat exchange in laser medium and pump DLAs, as well as the associated thermal stress and thermal strain are studied to seek the methods for enhancing the heat transfer, lowering the device temperature, improving the temperature uniformity and reducing the thermal stress, in order to promote the laser output and the beam quality.

For a side-pumped, side-cooled laser rod, mathematical models of the fluid flow, the heat transfer and the thermoelastic problems were developed. Both analytical and numerical methods were used to solve the 2-D axisymmetric fields of temperature, thermal stress and thermal strain. The various influencing factors are easily observed from the analytical expressions. The numerical results agree well with the analytical solution, while better exhibiting the conjugate heat transfer, the entrance effects and the end effects. Therefore the two methods should complement each other in engineering design.

For an orthotropic laser medium, mathematical models are presented for the heat conduction and the thermoelastic problems under different pumping schemes. The 3-D temperature, 2-D thermal stress and thermal strain fields are analytically derived. Full 3-D numerical simulations are performed on the coupled thermal-mechanical problem. Comparisons of temperature, thermal stress and thermal strain components are made among different pumping schemes. The maximum temperature and thermal stress under a Tophat pump beam are lower than under a Gaussian pump beam, whereas the thermal effects under Uniform pumping are the least significant of all; in the absence of external forces, the orthotropic thermal properties has more significant effects on the thermal strain than on the thermal stress.

For a pump DLA, three-dimensional conjugate numerical simulations using the inlet, average and variable thermal properties, respectively, were performed for the laminar water flow and heat transfer in rectangular microchannels with D_h of 0.333mm at *Re* of 101-1775. The calculated local and average characteristics of flow and heat transfer were compared among different methods, and also with the experiments, correlations and simplified theoretical solution data from published literatures. The results indicate that the "Variable property" method is superior in engineering application.

For a typical pump DLA, a liquid-flow microchannel heat sink was designed and fabricated by adopting dense, thin and short channels instead of the originally designed sparse, wide and long passages in order to increase the cooling capacity, improve the temperature uniformity, while achieving a small flow resistance. Under the maximum flow rate of 70ml/s in this experiment, the pressure drop across the channels is only 10.3kPa, the cooling capacity of the heat sink reaches 146.9W/cm^2.

In Chapter 4, the fluid dynamics of a free submerged jet is reviewed. Different conclusions are found in the literature for what concerns the lengths of the potential core region and the line source of the jet. The submerged slot jet of air is the research subject of the present authors to investigate its evolution with the distance from the slot exit. Hot anemometry is used to measure velocity and turbulence distributions on the exit as well as at several distances from the slot exit. The shadowgraph visualization technique is also employed to confirm the hot anemometry measurements.

The existence of an undisturbed region of flow, where the fluid dynamics characteristics are equal to those on the exit, was measured and visualized in a preliminary experiment with a small wind tunnel, named first, at the turbulent Reynolds number, based on the hydraulic diameter, equal to about Re=10,600-12,100. With a slot of height H=4.25 mm the dimensionless length of the undisturbed region, L_u, is equal to about L_u/H=1.6-2. The length of the undisturbed region decreases with the increase of the Reynolds number from the laminar to the turbulent regime. Further measurements of velocity and turbulence, carried on in the Reynolds number range from Re=13,500 to Re=22,100, confirmed that the dimensionless length of the undisturbed region span in the range L_u/H=1.6–2. The length of the undisturbed region of flow was visualized with the shadowgraph technique in the Reynolds range from Re=1,700 to Re=22,100. The length of the undisturbed region is L_u/H>3 for the laminar Reynolds number of Re=1,700 and is L_u/H=1 for the turbulent Reynolds number of Re=22,100.

The velocity and turbulence measurements were performed also in a different small wind tunnel, named second, with a slot height of 10 mm, for the range from Re=7,600 to Re=37,300. At Re=7,600 the undisturbed length is L_u/H=3–4, at Re=23,300 is L_u/H=1–2 and is zero at Re=37,300. The flow visualizations with the shadowgraph technique were performed in the same wind tunnel and slot, for the range from Re=7,100 to Re=33,400, confirming the dimensionless length of the undisturbed region of flow is L_u/H=3.8 for

Re=7,100, and $L_u/H=0$ for Re =33,400. The flow visualizations were performed also in the first wind tunnel, with a slot of height H=17 mm. The length of the undisturbed region is $L_u/H=2.35$ for Re=8,400, $L_u/H=1$ for Re=31,200 and $L_u/H=0$ for Re=52,900.

The introduction of a metallic grid inside the wind tunnel, between the converging duct and the slot, and on the slot exit, produces a shorter length of the undisturbed region, due to the higher turbulence. Hot anemometry was employed to measure velocity and turbulence in a free jet up to an impinged cylinder, set at two distances from the slot exit, equal to L/H=4 and 6, for the Reynolds number of Re=23,300. The undisturbed region is equal to $L_u/H=1$ when the cylinder is set at L/H=4 and $L_u/H=2$ when the cylinder is set at L/H=6.

The experimental results found by the present authors for the length of the undisturbed region of flow, obtained in turbulent as well as in laminar regime with hot anemometer and shadowgraph visualizations, are collected together with previous results of the literature, obtained with shadowgraph, Schlieren and PIV visualization in laminar regime, with a fairly good agreement.

The experimental heat transfer, or Nusselt number, measured by the present authors are reviewed in the present paper at the light of the evolution of a submerged jet, i.e. in order to study the influence of the length of undisturbed and potential core regions.

Preliminary experiments of heat transfer with the first small wind tunnel without the three metallic screens, i.e. at relatively high level of turbulence, were carried out to cool a cylinder of diameter D=10 mm with a slot of height H=5 mm. The experiments showed a qualitative increase of the average and local Nusselt number at Re=17,000. Heat transfer experiments were performed with the same wind tunnel, slot and circular cylinder for Reynolds number ranging from Re=4,000 to Re=20,000. The influence of the distance between the exit of the slot and the cylinder, X, was studied in the range of X/H=2–10. The local Nusselt number is a function of X/H, the angle from the impinging point and the Reynolds number.

Local heat transfer was investigated with the second wind tunnel at reduced turbulence and slot jets of different heights. The slot jet with the highest ratio between cylinder diameter and slot height, i.e., D/H, realizes the highest mean Nusselt number at the same Reynolds number. The distance from the slot exit where is present the highest mean Nusselt number is X/H=8 for D/H=4. The similar distance observed in a slot jet with D/H=2 is X/H=6.

Local heat transfer along the cylinder perimeter was measured in an upgraded version of the second wind tunnel, with lower turbulence level, for a large range of Reynolds numbers, i.e. from Re=7,400 to Re=37,300. Local Nusselt number, measured at different distances from the slot exit along the perimeter of the cylinder, presented interesting features from the point of view of the presence of the undisturbed region.

Mass conservation equation is employed to study the time evolution of the amount of non-renewable energy resources according to the extraction policy, while the energy conservation equation is transformed into an energy-capital conservation equation to study the time evolution of the price of non-extracted resources. The Hotelling rule for non-extracted resources, i.e. an exponential increase of the price of non-renewable resources at the rate of the current interest and time, is shown to be a special case of the general energy-capital conservation equation when the mass flow rate of extracted resources is unity. Mass and energy-capital conservation equations are coupled and solved jointly to investigated the time evolution of the extracted resources. The parameter PIFE, "Price Increase Factor of Extracted resources", is introduced as the difference between the interest rate of capital, typically the inflation rate, and the mass flow rate of extraction of non-renewable resources. The price of extracted resources increases if PIFE is greater than zero, i.e. if the mass flow rate of extraction is smaller than the inflation rate. The price is constant if PIFE is zero, i.e. if the mass flow rate of extraction is equal to the inflation rate. The price is decreasing with time if PIFE is smaller than zero, i.e. if the mass flow rate of extraction is greater than the inflation rate. The price evolution with time of the resources sold to the market is investigated in case of no-accumulation and no-depletion of the resources; i.e. when the resources are extracted and sold to the market at the same mass flow rate. The price of sold resources varies with time according to the relation between the parameters PIFE and PIFS, "Price Increase Factor of Sold resources", which is the difference between the extraction rate and the interest rates of sold resources, typically prime rate. The price of sold resources increases with time if the initial price is greater than CIPS, "Critical Initial Price of Sold resources", which depends on the initial price of extracted resources, the interest rate of non-extracted resources, and the difference between PIFS and PIFE or greater than CIPES, "Critical Initial Price Extreme of Sold resources", which depends on the initial price of extracted resources, the interest rate of non-extracted resources, and PIFS. The price of sold resources increases temporarily with time if the interest rates of non-extracted and extracted resources are equal, i.e. PIFE is equal to PIFS, and the initial price is greater than CIPES, "Critical Initial Price Extreme of Sold resources". Examples for oil and gas are discussed extensively in Chapter 5.

As presented in Chapter 6, laser chemical vapor deposition and infiltration (LCVD and LCVI) are based on the concept of building functional structures by deposition of solid materials from gas precursors by a laser beam in an environmentally controlled chamber. The chemical vapor deposition during these two processes can be based on reactions initiated pyrolytically, photolytically or a combination of both. The most important issue that the thermodynamic analysis can address is whether a given chemical reaction is feasible. For chemical reactions that are feasible, both composition plot and deposition maps can be obtained. The distinction between the conventional Chemical Vapor Deposition (CVD) and LCVD is that a laser beam heats a very small spot on the substrate and vapor deposition occurs only in the heated spot in the LCVD process. A series of thermal models of LCVD of Titanium Nitride on a finite slab under irradiation by a moving laser beam have been developed. Heat transfer in the substrate and gases, the chemical reaction on the substrate's top surface, and the mass transfer of gases in the chamber were taken into account. The effect of natural convection induced by the temperature and concentration gradients during LCVD was instigated. A parametric study on the LCVD by moving laser beam was carried out and an empirical correlation was proposed. To study the transport phenomena in LCVI, heat transfer in a gas-saturated powder layer subject to a short-pulsed volumetric heat source is modeled using a two-temperature model. The results show that the non-equilibrium is significant when the heat source pulse width is less than 1 microsecond and, therefore, heat transfer during nanosecond laser- powder interaction should be modeled using a non-equilibrium model. For the case with long pulse width, on the other hand, the equilibrium model is sufficient.

Chapter 7 looks at the design, retrofit and the thermal performance analysis of industrial cooling systems. In design, the effect of the arrangement of the exchanger network upon total surface area and cost is analyzed. In retrofit, a targeting methodology is presented that allows for the determination of the minimum water flow rate that will accomplish the required heat duty considering the actual installed surface area. Energy savings are achieved by reducing the actual pumping power to the minimum required by the minimum targeted water flow rate.

Machining remains one of the most important manufacturing process and the need for a protection of the machine tool against wear and breakage has become a priority over the past few years. The damage of the tool insert is obviously strongly correlated with the maximum temperature obtained close to the tool's tip. The goal of Chapter 8 is to propose an accurate methodology in two main steps in order to predict the heat transfer within the cutting tool.

The first part is dedicated to direct modelling of the cutting tool. After performing numerical simulations with the CFD code Fluent, an approach based on a quadrupole formulation is proposed in order to investigate thermal aspects of a cutting tool. The heat losses by convection and advection through the chip and the conduction in the tool holder are taken into account by heat transfer coefficients and a thermal contact resistance. The temperature evolution is obtained in the Laplace domain using Whittaker functions.

The second part concerns the inverse modelling in order to predict the temperature and the heat flux at the tool's tip. The model proposed which is not time consuming is implemented to inverse data. The difficulty in this problem comes from its ill-posed aspect: temperature and flux at the boundary (tool's tip) are determined through inner temperatures. If this model gives quite satisfactory results for temperature in the case of low noised data, this is not the case for heat flux. So the future time steps regularization technique is implemented to improve the estimations. The effects of particular in-time evolutions of the heat flux exhibiting step, ramp, slope and sinusoidal oscillations in the transient temperature determination are studied. The influence of the noise level, of the number of future time steps and also of the temperature profiles in the tool's tip vicinity (Cartesian, Cylindrical or Spherical) are also investigated.

Finally the conclusion presents some perspectives on a real-time tool's tip temperature and heat flux estimation adapted for turning process control.

The operating performance of a double-glass cover solar still with 10 mm spacing (i.e., solar still with cooling water flowing between a double-glass cover) was evaluated experimentally under the climatic conditions of Muscat, Oman. The base area of the still was 1 m^2 (i.e., 1 m length by 1 m width) and the cover of the still was made an angle of 23.4 degree with the horizontal optimized for 23.37 N latitude of Muscat. For comparison purposes, a conventional (single-glass cover) solar still with the same dimension was also designed and fabricated. The performances of the two stills were experimentally investigated under identical operational and weather conditions. The experimental study was conducted on some of the factors that affect the productivity of the solar stills. This included the water depth in the basin and the effect of the insulator. The tests were also performed using de-ionized water and sea water as feed. The productivities of the two stills and the temperatures of the water and the glass covers were determined. The quality of the water produced was tested. It was found that more yield of fresh water was obtained in case of a double-glass cover solar still as compared to the single-glass solar still. This is due to higher temperature difference between the water and inner glass cover. Chapter 9 has also revealed that the water depth is inversely proportional to the productivity of the stills. Hence, the daily production of the two stills was increased by reducing the depth of the water in the basin.

As presented in Chapter 10, the very first gas turbine was developed at Gottingen University in the last of 1930. After some arrangements, the engine was used after for planes replacing the comparatively heavy piston engines in this time. The first gas turbines efficiency was relatively low, surrounding 20%, and since the first prototype is designed, it was a standard for the engineers to improve their performance.

Currently the efficiency of the gas turbines is in the surroundings from 35 to 50%, it is a very significant progress since the thirties and which evolves until our days.

Reverse cycle machines are useful as refrigerating systems, heat pumps or air conditioning units. These machines generally imply phase change of working fluid in two heat exchangers (evaporator and condenser). Accordingly, the corresponding cycle is near of a reverse Carnot one. In Chapter 11, an optimization of such a machine is proposed and analysed by taking into account the internal irreversibility of the machine and generalized convective-like form of heat transfer laws, in case of limited size of heat exchangers through a parameter equivalent to a total NTU that is to be allocated to the two exchangers.

Various constraints are considered for the optimization; obtained results are discussed and compared. Perspectives are evoked for further developments.

In Chapter 12, the authors present numerical results concerning the simulation of semiconductor melts with free capillary surfaces, particularly silicon (Si) crystal growth by the floating zone method. The effect of fluid motion due to the gravity and surface tension is illustrated. The input power induced by the heat source is a Gaussian distribution and the surface tension is assumed dependent on temperature to models the heat experimental conditions. The mathematical model is a coupled system, which consists of axisymmetric heat equation and the Navier-Stokes equations in the melt with Marangoni boundary conditions and Boussinesq approximation. The partial differential equations obtained have been discretized by means of finite volume procedure, and the SIMPLER (Semi-Implicit Method for Pressure-Linked Equations Revised) algorithm is used to treat the coupled velocity-pressure equations. Different steady thermocapillary flows are obtained by varying the following significant parameters: temperature gradient conditions (Grashof number), surface tension gradient (Marangoni number), all of them relevant in practical crystal growth situations. Our results show that thermocapillary can have a quite significant effect on the stability of a primarily buoyancy-driven flow, as well as on the flow structures and dynamic behaviour.

In natural circulation boilers the downcomer tubes can be designed as heated or unheated tubes. The circulation of the working medium is produced by the difference in density between the water-steam mixture in the riser tubes and the water in the downcomer. The density of the water-steam mixture in the heated riser tubes is lower than that of the water in the less or unheated downcomer tubes, which results in a lifting force in the riser system and in a circulation of the working medium. The water circulation is controlled autonomously and is based on the heat flux. With an increase in the operation pressure, the density difference between the water-steam mixture in the riser tubes and the water in the downcomer tubes decreases. The physical boundary of natural circulation systems is the critical point (no separation of the steam and the water is possible). But a natural circulation is also possible under supercritical pressure. Nevertheless, the application of natural circulation systems is usually limited to a maximum drum pressure of approximately 180 bars. This pressure is based on the fact that the latent heat decreases with an increase in the operation pressure. This results in a decrease in the percentage of the evaporator heating surface on the overall heating surface of the boiler. If the water-steam separation in the drum is done with the help of cyclones, then the live steam pressure in front of the steam turbine decreases to approximately 160 bars. This lower pressure is the result of the higher pressure loss in the cyclones.

In Chapter 13 some modeling and stability aspects of natural circulation boilers will be discussed.

In: Thermal Engineering Research Developments
Editors: J. Evgova and O. Kostadinov, pp. 1-83

ISBN: 978-1-60741-497-1

Chapter 1

OPTIMIZING PRELIMINARY DESIGN OF INDUSTRIAL EQUIPMENT INVOLVING DIFFERENT THERMAL ENGINEERING CALCULATION PROCEDURES OVER A POWER PLANT

J.M. Blanco* and F. Peña

Dpto. Máquinas y Motores Térmicos, Escuela Técnica Superior de Ingeniería,
Universidad del País Vasco / E.H.U.
Alameda de Urquijo s/n (48013) Bilbao. España.

Abstract

This work can be divided into four main parts, showing that different thermal engineering calculations can be useful tools to solve complex issues in nowadays in thermal equipment design for several industrial applications. The first one summarizes a new methodology for the design and optimization of a low NO_x-CO burner. To begin with, the working of an experimental 2 MW natural gas burner, in a superheated steam boiler was studied. A CFD model was built and validated with the data obtained from the above-mentioned burner, implementing new User Defined Functions. As a result of this, the first model was then modified in order to get a final CFD, low NO_x-CO emission model.

The second part is focused on a detailed study of the air pre-heaters in a thermoelectric power plant, where a preliminary large amount of ash deposition on one of the inlet conduits was observed, obstructing this way the incoming gas flow. Prior to the cleaning of the conduits, certain damage affected the boiler dome, and as a result of this, the pressure reached a very high value. All of this indicated that the cause of the clogging was not entirely physical. Comprehensive and detailed analytical studies, as well as the subsequent guideline for solving the situation were carried out.

The third part presents the complexity of a cogeneration process design for a power plant of reference. On one hand, a complete study of heat and electricity loads was carried out, also performing a comparative test between different alternatives, setting up selection criteria. On the other hand, the work goes through an economical analysis showing the viability of

* E-mail address: jesusmaria.blanco@ehu.es. Phone: + 34 946014250; Fax: + 34 946014283 (Corresponding author)

choosing natural gas engines instead of gas turbines. The investment-saving relationship was obviously considered.

Finally, a novel thermal optimization of the purging process for the bottles of CO_2 used in two big generators in a thermal power plant was described in deep. Because of both filling and emptying processes must be realized in an inert atmosphere, the experience demonstrated that the best fluid was the carbon dioxide. Almost the 90% of the content inside every bottle remains in liquid state, but as soon as the emptying process goes on, a quick and progressive cooling of this fluid occurs until the so called "carbonic ice" is finally formed, rendering useless this way the rest of the bottle.

Keywords: Thermal efficiency, natural gas, CFD modelling, combustion, carbon dioxide, hydrogen, thermoelectric power plant, precipitators, cooling generators.

Nomenclature

a	=	Thermal diffusivity of the material on the metal plates (m^2/s).
a	=	Absorption coefficient.
a_o	=	Oxygen reaction order.
a_P	=	Equivalent absorption coefficient.
a_S	=	Characteristic strain rate.
A	=	Van Driest constant (= 26).
A	=	Wave amplitude (μm).
ADP	=	Acid dew point temperature (°C).
A_1	=	$1\ 10^{10}$ (s^{-1}).
A_2	=	$3\ 10^{12}$ (s^{-1}).
A_p	=	Characteristic particle surface (mm^2).
A_{pn}	=	Projected area of the particle *n*.
Aut	=	Automatic mode.
c_p	=	Specific heat of fluid (kJ/kg K).
c_{ps}	=	Specific heat at constant pressure of the heater's metal plates (kJ/kg K).
C	=	Linear-anisotropic phase function coefficient.
C	=	Lift constant.
C_s	=	Lift coefficient.
C_1, C_2	=	Integration constants for the CO_2 temperature equation.
$\overline{C}_1, \overline{C}_2$	=	Integration constants for the metallic cylinder temperature equation.
CEG	=	Combustion exhaust gases.
CFD	=	Computational Fluid Dynamics.
CO	=	Carbon monoxide.
CO_2	=	Carbon dioxide.
d	=	Internal diameter of the inlet duct (mm).
d_{pn}	=	Diameter of the particle *n*.
D	=	Diffusion coefficient.
$DCEG$	=	Dry combustion exhaust gases.
$erfc^{-1}$	=	Inverse complementary error function.
E	=	Electric power generated by the cogeneration installation *(MWh)*.
E	=	Energy term.

E = Elasticity constant.
E_a = Empirical constant (= 9,793).
E_1 = 28.045,95 (J $gmol^{-1}$).
E_2 = 251.151 (J $gmol^{-1}$).
E_P = Equivalent emission of the particles.
EEE = Equivalent Electrical Efficiency *(%).*
EU = European Union.
f = Mixture fraction.
f = Friction factor (Moody diagram).
f_{st} = Stoichiometric mixture fraction.
f_{pn} = Scattering factor associated with the particle *n*.
F = Phase.
F_a = Decay factor.
F_s = Lift force.
F_s = Fuel stream (generic).
$\vec{F}$ = External body forces.
g = Acceleration of gravity (m/s^2).
G = Incident radiation.
$G_{k,m}$ = Production of turbulence kinetic energy.
$G_{b,m}$ = Production of turbulence kinetic energy due to buoyancy.
GCV = Gross calorific value (kJ $[Nm^3\ fuel]^{-1}$).
GT = Gas Turbines.
h = Enthalpy for ideal gases (W/m^2 K).
h = Coefficient of heat transfer by convection or film coefficient (W/m^2 K).
h_f = Enthalpy of the liquid-phase fluid (kJ/kg).
h_{fg} = Vaporization latent heat (kJ/kg).
h_g = Enthalpy of the vapour-phase fluid (kJ/kg).
$[H_2O]$ = Water mass fraction.
HF = High flame mode.
I = Radiation intensity.
I_T = Turbulence intensity.
ICE = Internal Combustion Engine.
IE = Rate-of-rise for Primary Energy.
IRR = Internal Return Rate.
$\vec{J}_j$ = *Diffusion flux of species* j.
$J_{i,w}$ = *Diffusion flux of species* i *at the wall.*
k = Thermal conductivity of the metal plates on the regenerator heater (W/m K).
k_{fi} = *Rate coefficients used in the thermal* NO_x *model.*
k_r = *Reverse rate coefficients used in the thermal* NO_x *model.*
k_P = Turbulence kinetic energy at point *P*.
l = Litres.
L = Cylinder heat transfer length (m).
L = Length (m).
L = Heat transfer length (mm).

LAP	=	Ljungström Air Preheater.
LCV	=	Lower calorific value (kJ [Nm^3 fuel]$^{-1}$).
LF	=	Low flame mode.
m	=	Mass (kg).
$\dot{m}$	=	Mass flow (kg s^{-1}).
M	=	A general third body.
M_{CO2}	=	Molecular weight of CO_2 (= 44.01 g/mol).
$\overline{M}_{CO2}$	=	Molar volume of CO_2 (= 44.6189 mol/Nm^3).
M_υ	=	Correction coeficient for the viscosity.
M_λ	=	Correction coeficient for the thermal conductivity.
$M_{w,i}$	=	Molecular weight of *NO* (kg [gmol]$^{-1}$).
MCS	=	Measure and Control Station.
MP	=	Mechanical Precipitator.
n	=	Coefficient of excess of air (%).
N	=	Number of carbon atoms per molecule for the hydrocarbon fuel.
N	=	Number of bottles.
NAV	=	Net Actual Value.
N_i	=	Refractive index.
N_{NU}	=	Dimensionless Nusselt number.
N_{PR}	=	Dimensionless Prandtl number.
N_{RE}	=	Dimensionless Reynolds number.
NG	=	Natural gas.
NO_2	=	Nitrogen dioxide.
NO_x	=	Nitrogen oxides.
O_s	=	Oxidant stream (generic).
P	=	Pressure (mm. of water column).
P_s	=	Product stream (generic).
p	=	Static pressure (Pa).
Pr	=	Barometric pressure (Pa).
PC	=	Personal computer.
ppm	=	Particles per million.
PDF	=	Probability density function.
PR	=	Power plant of reference.
PVC	=	Polyvinyl chloride.
P_r	=	Molecular Prandtl number.
P_{rt}	=	Turbulent Prandtl number.
q	=	Heat (J).
$\dot{q}_v$	=	Volumetric thermal power (W/m^3).
$\dot{q}$	=	Wall heat flux (W/m^3).
$\dot{Q}$	=	Thermal power (W).
Q	=	Fuel consumption in the whole plant *(MWh)*.
q_r	=	Radiation flux.
r	=	Air-to-fuel ratio on a mass basis.
$\vec{r}$	=	Position vector.

r_B	=	Radius of the packing volume, by surface unit (mm).
r_0	=	Pipe radius (mm).
r_0	=	Internal radius of the bottle (m).
r_1	=	External radius of the bottle (m).
R	=	Overall efficiency of the cogeneration power plant.
R	=	The universal gas constant.
R_1, R_2	=	Conversion rates of *HCN* (s^{-1}).
$\overline{R}$	=	Hydraulic radius.
R_{cf}	=	Mean limiting reaction rate of fuel (kg m^{-3} s^{-1}).
s	=	Path length.
$\vec{s}$	=	Direction vector.
$\vec{s}\,'$	=	Scattering direction vector.
S	=	Heat transfer area (m^2).
S_c	=	Molecular Schmidt number.
S_{ct}	=	Turbulent Schmidt number.
S_G	=	User-defined radiation source.
S_h	=	Volumetric heat sources term (including heat of chemical reaction).
S_{HCN}	=	Consumption rates of *HCN* in reactions 1 and 2 respectively (kg m^{-3} s^{-1}).
S_m	=	Source term.
$S_{pl,HCN}$	=	Source of *HCN* (kg m^{-3} s^{-1}).
S_w	=	Swirl number.
SRT	=	Simple Recovery Term.
SSR	=	Spanish Statutory Regulation.
t	=	Time (s).
T	=	Local temperature (K).
T	=	Period (s^{-1}).
$\overline{T}$	=	Mean temperature (K).
T_P	=	Temperature at the cell adjacent to wall.
T_w	=	Temperature at the wall.
u'	=	Ratio of the root-mean-square of the velocity fluctuations.
u_{avg}	=	Mean flow velocity.
U	=	Overall heat transfer coefficient (conduction in plates and convection in air) (W/m^2 K).
U_P	=	Mean velocity of the fluid at point *P*.
U_c	=	Mean velocity magnitude at $y^* = y_T^*$.
UDF	=	User defined function.
URT	=	Updated Recovery Term.
v	=	Fluid velocity profile.
$\vec{v}$	=	Velocity (m s^{-1}).
V	=	Average fluid velocity (m/s).
V	=	Volume (Nm^3).
V	=	Useful thermal energy produced *(MWh)*.
V_g	=	Flow of gases inside the casing of the generator (Nm^3).
V_{CO2}	=	Flow of CO_2 (Nm^3).
W	=	Fluid weight (kg).

x	=	Distance to the surface (mm).
X_i	=	Mole fraction of specie $i = \frac{Y_i}{M_{w,i}}\left(\frac{\rho RT}{p}\right)$.
y	=	Depth of the boundary layer (μm).
y_P	=	Distance from point P to the wall.
y^*	=	Non-dimensional sub layer thickness.
y_T^*	=	Non-dimensional thermal sub layer thickness.
Y_j	=	Local species mass fraction.
$Y_{N,fuel}$	=	Mass fraction of nitrogen in the fuel.

Greek Symbols

Δ	=	*Increase.*
ΔE	=	Yearly saving of primary energy.
Λ	=	*Wave length (μm).*
$\Delta \theta m$	=	Average total logarithmic temperature of air and gas flows (°C).
ε	=	*Dissipation rate.*
$\in_{pn}$	=	*Emissivity of particle* n.
θ	=	*Temperature (°C).*
θ	=	*Temperature field for the metallic cylinder.*
$\overline{\theta}$	=	*Temperature field for the CO_2.*
θ_{w1}	=	*Experimental temperature of the inlet water (20°C).*
θ_{w1}	=	*Experimental temperature of the outlet water (15 °C).*
λ	=	*Thermal conductivity (W/m K).*
∂	=	*Partial derivative.*
ϕ	=	*Equivalence ratio.*
$\phi_{m,T}$	=	*Species mass fractions and temperature.*
$\overline{\phi}_{m,T}$	=	*Density-weighted mean species mass fractions and temperature.*
Φ	=	*Phase function.*
ρ_s	=	*Density of the metal plates on the heater (kg/m^3).*
φ	=	*Times of gas or air passage, through the rotary regenerative heat exchanger.*
ψ	=	*Number of moles.*
$\hat{\tau}$	=	*Stress tensor.*
ρ	=	*Density.*
ρ_m	=	*Mixture density.*
ρg	=	*Gravitational body force.*
μ	=	*Dynamic viscosity.*
υ	=	*Cinematic viscosity (m^2/s).*
μ_t	=	*Turbulent viscosity.*
ζ	=	*Unit tensor.*

κ	=	*Von Kármán constant (= 0.4187).*
k_{eff}	=	*Effective conductivity.*
σ	=	*Stefan-Boltzman constant (5.672 10^{-8} W m^{-2} K^{-4}).*
σ_P	=	*Equivalent particle scattering factor.*
σ_S	=	*Scattering coefficient.*
Γ	=	*Radiation quantity.*
χ	=	*Scalar dissipation rate.*
χ_{st}	=	*Stoichiometric scalar dissipation rate.*
ω	=	*Fluid specific weight.*
Ω'	=	*Solid angle.*

Subscripts

(a)	=	*Relative to the air.*
(b)	=	*Relative to the bottle.*
(e)	=	*Relative to the environment.*
(exp)	=	*Experimental.*
(g)	=	*Relative to the gas.*
(i)	=	*Inlet.*
(m)	=	*Average.*
(max)	=	*Maximum.*
(min)	=	*Minimum.*
(o)	=	*Outlet.*
(s)	=	*Relative to the surface.*
(w)	=	*Relative to the water.*

Superscripts

(b), (c)	=	*Lift coefficients.*
(i)	=	*Relative to the inlet.*
(o)	=	*Relative to the outlet.*
(F)	=	*Relative to the filling process.*
(D)	=	*Relative to the draining process.*

1. Introduction

Improving thermal engineering calculations have been revealed as a powerful tool in order to solve complex issues in nowadays thermal equipment design for several applications and processes in the industrial sector. This work has been focused in several of these applications, showing how different thermal engineering calculations can be used to solve problems related to heat transfer. Four are the examples that will be seen next.

The first part summarizes a new methodology for the design and optimization of a low NO_x-CO burner. To begin with, the working of an experimental *2 MW*, natural gas burner, in a superheated steam boiler was studied. A *CFD* model was built and validated with the data obtained from the above-mentioned experimental burner which included the input of working conditions, simplifications (if any), geometrical model, mesh refining process, setting up physical models, contour conditions, handling of algorithms of solution, and the incorporation of improved *UDF* (User Defined Functions). A new burner design evolved as a result of this and the first model was then modified to get a secondary *CFD,* low NO_x-*CO* emission model, which was called the "prototype burner".

To improve the design of the low NO_x emission burner, a Computational Fluid Dynamics (*CFD*) model was developed. *CFD* is the science of predicting fluid flow, heat transfer, mass transfer (as in perspiration or dissolution), phase change (as in freezing or boiling), chemical reaction (e.g., combustion), mechanical movement (e.g., fan rotation), stress or deformation of related solid structures (such as a mast bending in the wind), and related phenomena by solving, on a computer, the mathematical equations that govern these processes, by a numerical algorithm. The model includes the incorporation of improved *UDF* (user defined functions).

The second part provides a theoretical explanation of the causes that led to the clogging by ash, of one of the two groups of mechanical precipitators in a high-performance thermoelectric power plant, equipped with two "Ljungström" air preheaters. Following a detailed study of two of these mechanical precipitators, a large amount of ash deposited on one of the inlet conduits was observed, obstructing the incoming gas flow.

A comparison of the available data for the two most recent hopper cleaning operations revealed that, on the one hand, the amount of ash collected by the clogged precipitator was significantly less than that collected by the other one and, on the other hand, the temperature of the ash in the former was noticeably lower than in the latter. Prior to the cleaning of the conduits, a certain amount of damage was caused to the boiler dome, which meant that subsequent cleaning required the use of a hydro laser, where it was noted that inlet pressures were very high. All of this indicated that the cause of the clogging was not physical. This part provides a comprehensive analytical analysis that explains what happened inside this equipment.

The next part presents the complexity of a cogeneration process design for a power plant of reference and evaluates its performance efficiency. The performance of a cogeneration power plant is always a complex matter because such a plant supplies the amount of exhaust steam required by processes and simultaneously generates the electricity either being consumed in the factory or being exported to power supply when there is a production excess.

On the one hand, a complete study of heat and electricity loads has been carried out, also performing a comparison between different alternatives, setting up selection criteria and also offering the main characteristics of both equipment and auxiliary installations involved in.

On the other hand, an economical analysis showing the viability of the definitive model of cogeneration chosen by using natural gas engines instead of gas turbines has also been included, showing how their corresponding returns become more attractive. The investment-saving relationship has obviously been taken into account.

Finally the fourth part deals with the optimization process for both filling and emptying the H_2 in the alternators of big thermal power plants which must necessarily be performed in an inert atmosphere. Thus, operating experience has shown that carbon dioxide (CO_2) is the ideal fluid for this purpose, because its molecular weight is considerably higher than that of hydrogen. During the operations for sweeping the generators with air or H_2, cooling the CO_2 bottles in the natural flow of ambient air, it has repeatedly been shown that thermodynamic equilibrium is not attained, leading to a rapid cooling of the bottles to the freezing point of CO_2 (-56.61 °C), with the formation of what is referred to as “dry ice”, which obviously interrupts the flow.

This means that the available capacity of these bottles is clearly below the contents, which involves a significant loss of resources. This part describes the analytical study and implementation of a solution that optimises the emptying of bottles by solving the problem of CO_2 freezing.

All these parts are described next in depth with the same structure so the readers can easily find the most interesting one for them.

2. Part A: Design and Optimisation of a Burner through CFD Based on Experimental Data

2.1. Introduction

This study was carried out within the European Union (*EU*), where all the member states incorporate legislation about emissions. One of the primary areas of concern is the environment, especially NO_x emissions [1 - 3].

The discussion on the development of combustion systems in power generation ranges from the pre-environmental awareness era, which was characterized by its twin objectives of achieving complete combustion with minimum consumption of air and the capability to scale up to increased boiler unit performances, through the era of environmental awareness *(the 1970s),* which was characterized by the increasing focus on reduction of combustion generated pollution and finally to the present and immediate future when the focus is on a combination of clean combustion and high thermodynamic efficiency to meet the demand for mitigating CO_2 emissions.

From the 1970s, attention increasingly turned towards emission control technologies for the reduction of oxides of nitrogen and sulphur, the so-called acid rain precursors. By a better understanding of the NO_x formation and destruction mechanisms in flames, it became possible to reduce significantly their emission through combustion process modifications [4], for example, by maintaining sequentially fuel-rich and fuel-lean combustion zones in a burner flame or in the combustion chamber, or by injecting a hydrocarbon rich fuel into the NO_x bearing combustion products of a primary fuel, such as coal.

The European Union is expected to cover multiple steps towards the improvement of the environment. A first step is the fixing of limits for pollutants such as NO_x, based on technological advances with much focus on the extent of damages that they cause and on the technical possibilities to reduce their emission [5, 6].

The emission limit for combustors with less than *10 MW* power is $350\ mg/Nm^3$ according to the European Union Directives [7 - 9]. Nevertheless burners with lower emission rates are already available in the market, leading to the expectation that this set limit is clearly bound to drop even further. Following this logic, $150\ mg/Nm^3$ was set as the maximum emission limit for our new burner in order to guarantee that it works for as long a duration as possible. Technologies to diminish NO_x emission can be classified basically into three: The performance before combustion, during combustion, and after combustion.

There are several ways to diminish pollution [10 - 12], such as combustion with low input of air, combustion with attached eddies, staggered combustion, combustion with attached eddies combined with staggered combustion, combustion in porous regions, combustion in unstable regime, called "pulse combustion", and so on. There is also a special type of combustion called 'flameless oxidation'.

In contrast to combustion within stabilized flames, temperature peaks can be avoided in flameless oxidation. Due to this reason, the thermal *NO*-formation is largely suppressed; even at very high air preheat temperatures. The use of burners and boilers operated in flameless oxidation mode in continuous industrial combustion chambers, has proved to be reliable and is well-accepted by boiler users due to their uniform product quality [13].

Recent developments in heat recovery systems allow preheating of combustion air up to temperatures of *1,300 °C* and thus, fuel savings of up to *60 %* are achievable. The possibility of low NO_x emission from highly preheated air combustion is intensively discussed [14, 15]. Dilution of air with burned gases and combustion in low oxygen-concentration air are shown to be indispensable factors for low NO_x emission. Conventional burners will be either replaced with individual fuel and air injectors or substantially redesigned to facilitate uniformity of combustion conditions within the furnace [16, 17].

Nevertheless, in conventional burner designs, at higher combustion air temperatures, the NO_x emission is higher. Emission of NO_x and *CO* was lower when combustion air was preheated to high temperatures with low oxygen-concentration. The chemiluminescence intensity of *CH* and C_2 radicals is significantly affected by the preheated temperature of combustion air and the oxygen concentration in the oxidant [18].

Advanced reburning is another process that aids in NO_x emission reduction wherein injection of a hydrocarbon fuel such as natural gas downstream of the combustion zone is followed by injection of a nitrogen-containing species such as ammonia [19].

To better the design of the low NO_x emission burner, a Computational Fluid Dynamics (*CFD*) model was developed. *CFD* is the science of predicting fluid flow, heat transfer, mass transfer (as in perspiration or dissolution), phase change (as in freezing or boiling), chemical reaction (e.g., combustion), mechanical movement (e.g., fan rotation), stress or deformation of related solid structures (such as a mast bending in the wind), and related phenomena by solving, on a computer, the mathematical equations that govern these processes, by a numerical algorithm. The model includes the incorporation of improved *UDF* (user defined functions).

2.2. Aims and Methodology

A conventional *2 MW* burner of natural gas, working in a superheated water boiler was studied, both in experiment method and numerical method and the *CFD* model was validated with the data previously obtained from the experimental burner [20]. This included the input of working conditions, simplifications (if any), geometrical model, mesh refining, turbulence, combustion and radiation models, contour conditions, algorithms of solution including thermal conduction, fuel diffusion, viscosity, and chemical kinetics [21].Comprehensive measurements of velocities, temperature, gas composition (O_2, CO_2, CO, H_2, NO, CH_4), and radiation conditions inside the boiler were carried out [14]. The geometric configuration for the commercial unit will be modeled based on these, and hence the results obtained from the *CFD* model are expected to be sufficiently valid in making precise predictions in the near future. Numerical simulation of flames is a growing field, bringing significant improvements to our understanding of problem areas in combustion. Strong coupling between chemistry, transport, and fluid dynamics are involved [22].

Two important variables are required to define the appropriate contour conditions and to validate the results of the corresponding simulations. As mentioned before, it is imperative to measure the same variables that the simulation provides us, such as velocity, temperature fields, and so on, comparing both results under the same operating conditions. The two above-mentioned variables are:

1. Geometry: obtained from the planes supplied by the burner and boiler manufacturers, respectively.
2. Working conditions: obtained from the experimental unit in order to define the contour conditions, mainly for flow and energy, such as:

- *Burner*: gas consumption, temperature, composition, inlet pressure, air total pressure, and finally temperature of the air oxidizer.
- *Boiler*: flow of hot water, inlet and outlet pressure and temperature, temperature of the exhaust gases, both for the boiler and for the combustion chamber, temperature of the walls in the combustion chamber, flame aerodynamics, and finally composition of the exhaust gases.

Once this has been carried out, the new burner design entails the application of the same criteria that is mentioned above to the primary burner, so that the new model will be optimized in order to get another *CFD* model corresponding to a low NO_x-CO emission burner called the "prototype burner".

The objective of this study is to establish conditions that result in significantly reduced energy consumption, high efficiency and low pollution from the range of above-mentioned *2MW* power burners.

2.3. Experimental Procedure

Description of the Experimental Procedure

The installation consisted of three parts: Gas supply unit, burner, and finally the boiler. A monobloc type burner was chosen for the initial modelling, with two gears, flame arc ignition, and totally automatic. The combustion air is regulated by a servomotor and a corresponding valve. The main characteristics are shown in table 1.

Table 1. Main characteristics of the experimental burner

Power range (kW)	979 – 2,100
Natural gas consumption (Nm^3/h)	120 - 202
Drive motor (kW)	3
Voltage (V)	220/380
Frequency (Hz)	50
Weight (kg.)	125

Figure 1 shows the geometry of the above mentioned experimental burner. Only a part of it will be taken into account in order to get the physical domain of the *CFD* model, as shown in the figure. The air is supplied by a centrifugal fan and passes outside through *12* uniformly-placed gas nozzles, where the natural gas is finally injected, producing the required air-gas mixture while the initiating spark produces the flame.

The rest of the instrumentation used is described briefly below:

- *Temperature*: Temperature was measured at several points as described in table 2. The following equipments were used for the measurement of the maximum temperature of the flame and of the walls of the combustion chamber: a wet-bulb thermometer, a resistance thermometer, thermocouples, and a portable colorimetric pyrometer, model *QL2500*, with analogical *4-20 mA*. exit with data logger.

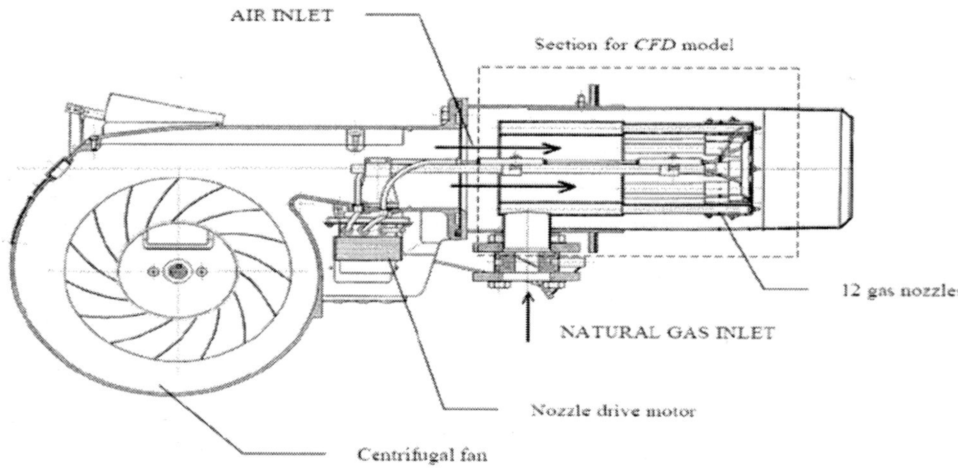

Figure 1. Main characteristics of the experimental burner.

Table 2. Description and objectives for the measurement points of reference

Measurement	Place	Objective
ENVIRONMENT:		
Temp. dry bulb, wet-bulb and barometric pressure.	Area surrounding the boiler.	External boundary conditions.
BOILER:		
Chimney: Gas analysis, outlet temperature	Gas evacuation channel of the boiler	Losses in exhaust gases.
Exit of the combustion chamber: Temperature and composition of gases.	Exhaust gases from the fireplace.	Heat transfer in the fireplace. Boundary conditions.
Inlet and outlet water: Flow, inlet temperature and pressure.	Pipeline of inlet water and boiler exit.	Heat consumption.
Combustion chamber: Maximum temperature of combustion. Temperature of walls.	Inside fireplace, across the peepholes of inspection.	Boundary conditions.
Fuel: Flow, inlet temperature and pressure.	Gas power supply.	Boundary conditions.

- *Pressure:* Bourdon gauge of graduated mask, pressure differential cut-out and differential gauges with *4-20 mA*. exit.
- *Barometric pressure*: Aneroid barometer.
- *Flow*: All the instruments incorporated in the installation were used, such as: a perforated plate (certified), for the water flow, a differential gauge of *4-20 mA*. exit, for the pressure drop, and a digital flow meter, model *RMC205* for the gas flow.
- *Gas composition*: Electronic gas analyzer, type *IMR 3000-P*, which enabled the determination of CO, CO_2, O_2, NO_x, and SO_x composition with the use of electrochemical cells. Additional information such as the environmental temperature, temperature of exhaust gases and so on was also provided. It was self-calibrated and allows a direct communication with a personal computer (*PC)* through a conventional cable-bus.

The boiler was loaded about one hour before the tests, to ensure stability, especially of the parameters of pressure, temperature, and flow of both gases and water [23]. The different instruments were read in order to verify the stability of the process. The parameters were considered to be stable when it is the variation between them did not exceed *5 %* in every measurement process [15]. Once stability was achieved, the input of information was commenced.

Nine points across the whole section, for different depths and widths were considered, for the measurements inside the chimney. The same procedure was used in the window of the combustion chamber. In all the cases only the average values were considered.

Experimental Results

All the tests and measurements were carried out with the experimental burner in automatic mode (working conditions of the burner were defined automatically, according to heat consumption) as well as manual alternative selection of low and high flame process monitoring respectively [24]. Table 3 summarizes the results obtained.

Table 3. Summary of the experimental measurements

PLACE	OPERATING MODE		
Measure value	Automatic	Low Flame	High Flame
ENVIRONMENT:			
Dry bulb temperature (K).	296	296	298
Barometric pressure (Pa).	101,329	101,328	101,330
Relative humidity (%).	78	52	63
CHIMNEY:			
CO (ppm)	3	1	4
CO_2 (%)	10.9	10.3	10.9
O_2 (%)	2.0	3.1	1.4
NO_2 (ppm)	20	17	18
NO (ppm)	40	49	51
Outlet temperature (K)	512	486	534
Coefficient of excess of air.	1.12	1.18	1.10

Table 3. Continued

PLACE Measure value	OPERATING MODE		
	Automatic	**Low Flame**	**High Flame**
COMBUSTIÓN CHAMBER:			
CO (ppm)	4	3	5
CO_2 (%)	10.6	10.1	10.8
O_2 (%)	2.1	3.2	1.7
NO_2 (ppm)	0	0	0
NO (ppm)	60	58	61
Outlet temperature (K).	1,360	1,350	1,362
Máx. temperature of combustion (K).	1,910	1,898	1,928
Walls temperature average (K).	1,005	1,002	1,009
WATER:			
Flow (kg/s).	12.2	12.2	12.2
Inlet temperature (K).	388.2	385.7	388.5
Outlet temperature (K)	417	410	418,5
Total pressure (Pa).	1,084.177	1,084.177	1,084.177
FUEL:			
Mass fuel flow (kg/s)	0.0352	0.0298	0.0377
Volumetric fuel flow ($Nm^3 h^{-1}$)	158.0	133.77	169.23
Inlet temperature (K).	295	295	296
Relative pressure (Pa).	6,754.1	7,453.32	6,864.9

The measurement process was considered as a decisive part of the design. Therefore, five reference areas were taken into account such as: environment, chimney, combustion chamber, water, and fuel. For each one of these areas, several items were measured in order to guarantee that the information needed for the *CFD* model would be precise enough. For this, a data-logger with a complex software was used in order to get the average value for each one of the measures.

These results combined with the physical properties of the materials and other radiation properties, [25 - 27], give enough information to define the boundary conditions and also to validate the next *CFD* model. Spectral absorption coefficients of the combustion products, including carbon dioxide, water vapour, and soot, are calculated based on the narrow-band model [28].

2.4. Numerical Simulations

This data was implemented in a compressible parallel finite volume-element solver called *FLUENT*, which is able to handle hybrid grids to simulate the whole combustion process [22]. In this way, the *CHEMKIN* code was implemented [29, 30], so the chemistry will be solved in the context of the burner in the *CFD* simulation.

At the outset a *CFD* model for the above-mentioned experimental burner was to be created or validated with the above-obtained measurements. Once this model was successfully completed, a second model corresponding to a low NO_x-CO burner (prototype burner) was to be created, comparing the results obtained from both models with reference to the NO_x and CO emissions.

The use of numerical simulations was shown to be a very important research instrument in combustion, [20, 31 and 32]. Working with *CFD* involves six fundamental phases:

1. Define the modeling goals.
2. Create the model geometry and grid.
3. Set up the solver and physical models.
4. Compute and monitor the solution.
5. Examine and save the results.
6. Consider revisions to the numerical or physical model parameters, if necessary,

Comprehensive modeling capabilities for any compressible and turbulent fluid flow problems were performed for steady-state analysis. A broad range of mathematical models for transport phenomena (heat transfer and chemical reactions) were combined with the ability to model complex geometries.

In this way, robust and accurate turbulence models are very important. The turbulence models have a broad range of applicability, and they include the effects of other physical phenomena, such as buoyancy and compressibility. Particular care has been devoted to addressing issues of near-wall accuracy through the use of extended wall functions.

Building a CFD Model for the Experimental Burner

Briefly, the hypotheses were related fundamentally to find some effective simplifications from the geometric point of view, which were necessary to manage a complex geometry such as the one showed in figure 1. After the geometry was successfully created by using a powerful pre/processor code, the next step was to realize the volume discretization (meshing).

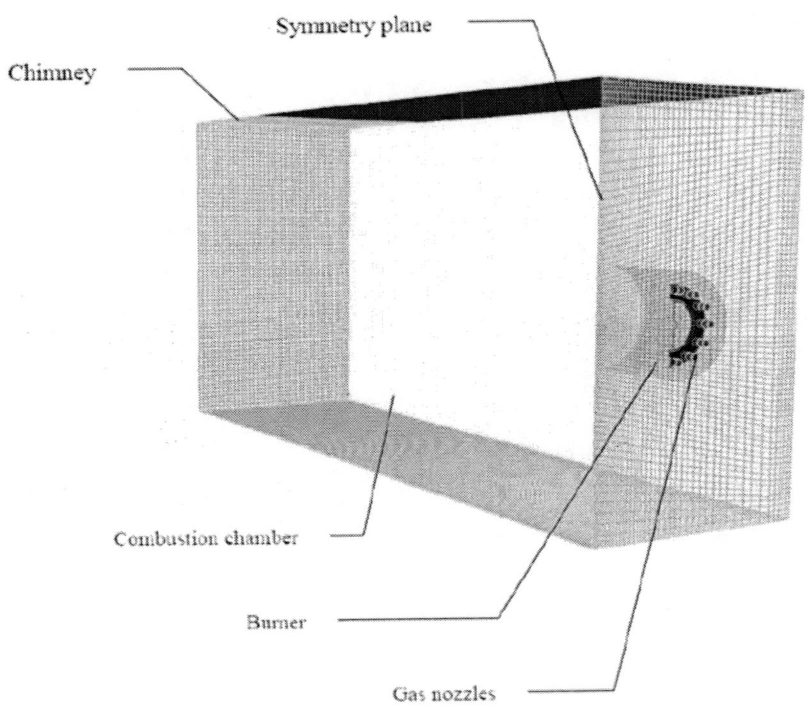

Figure 2. Discretised control volume over the symmetry plane.

Thus, the whole volume was discretized into finite elements called "cells" which will be the fundamental parts of the future calculations, because all the fluid equations will be solved for each one of these elements [33].

Hexahedral types of elements were used because they offered better results and also because the errors associated with numerical diffusion were lower [34]. Once the discretization of the geometric volume was completed, the mesh was to be totally exported, into the main code (processor), where usually the boundary conditions would be applied to complete the numerical model. It consisted of over 350,000 cells. Eventually, after the calculations, the post/processor allowed us to understand the results obtained, predicting the behaviour of the fluids inside the control volume.

Figure 2 shows part of the results corresponding to the above-mentioned meshing process, where the mesh of the burner and combustion chamber over the symmetry plane can be observed.

Defining the Basic Conservation Equations

Thus, the conservation equations for mass and momentum were solved for all flows. For flows involving heat transfer or compressibility, an additional equation for energy conservation was solved. For flows involving species mixing or reactions, a species conservation equation was solved. For the non-premixed combustion model, conservation equations for the mixture fraction and its variance were solved. Additional transport equations were also solved when the flow was turbulent [35].

The equation for conservation of mass, or continuity equation, can be written as follows [36]:

$$\frac{\partial \rho}{\partial t} + \nabla(\rho\, \vec{v}) = S_m \tag{1}$$

The source term is the mass added to the continuous phase from the dispersed second phase (e.g., due to vaporization of liquid droplets) and all the user-defined sources.

Conservation of momentum in an inertial (non-accelerating) reference frame is given by [37, 38]:

$$\frac{\partial}{\partial t}(\rho\, \vec{v}) + \nabla(\rho\, \vec{v}\, \vec{v}) = -\nabla p + \nabla(\bar{\tau}) + \rho\, \vec{g} + \vec{F} \tag{2}$$

The stress tensor is given by:

$$\bar{\tau} = \mu\left[\left(\nabla\vec{v} + \nabla\vec{v}^{\,T}\right) - \frac{2}{3}\nabla\vec{v}\varsigma\right] \tag{3}$$

where the second term on the right hand side is the effect of volume dilation. The energy equation is solved in the following form [39]:

$$\frac{\partial}{\partial t}(\rho\, E) + \nabla\left(\vec{v}\cdot(\rho\, E + \rho)\right) = \nabla\left(k_{eff}\cdot\nabla T - \sum_j h_j\cdot\vec{J}_j + \left(\bar{\tau}_{eff}\cdot\vec{v}\right)\right) + S_h \tag{4}$$

The first three terms on the right-hand side of equation (4) represent energy transfer due to conduction, species diffusion, and viscous dissipation, respectively, and the term E, is defined as,

$$E = h - \frac{p}{\rho} + \frac{v^2}{2} \tag{5}$$

where:

$$h = \sum_j Y_j \cdot h_j \text{ and } h_j = \int_{Tref}^{T} c_{p,j}\, dT \tag{6}$$

Defining Boundary Conditions

The boundary condition "mass flow inlet", for the fuel was chosen because the corresponding measurement value of natural gas flow was available. Thus, the consumption of air was obtained by calculating the ratio "kg air/kg fuel" for combustion with an excess air of *10 %*.

The condition of "outflow" for the outlet of the combustion chamber was adopted. Outflow boundary conditions were used as model flow exited where the details of the flow velocity and pressure are not known prior to solution of the flow problem. They are appropriate where the exit flow is close to a fully developed condition, as the outflow boundary condition assumes a zero, normal gradient for all flow variables, except pressure.

The external walls of the burner were considered to be adiabatic, considering that the gas nozzle was inside the refractory of the fireplace. With an optical pyrometer, the maximum temperature that was reached on the lateral walls was measured as described in table 3.

To evaluate correctly the temperature of the walls, an average temperature was considered and two criteria were used. For the first criterion, the temperature of the exhaust gases was determined from the combustion chamber. This temperature depends on the power of heat given out by the burner and the heat exchange across the walls in the fireplace [40]. For the second criterion, the heat across the walls for the thermal calculation was given by the manufacturer.

To fit this value, modelling was carried out several times, imposing a temperature to the walls verifying always the values of the average temperature for the *CEG* and the flow of total heat evacuated through the walls. The value reached for this temperature was *1,005.2 K* with a uniform distribution all over the side, top and rear walls. This value is very close to the maximum value of temperature measured up with the optical pyrometer also given in table 3.

We will next describe shortly the physical models adopted here:

- Turbulence model: k-ε standard,
- Radiation model: *P-1*.
- Combustion model: "Flamelet" model, (kinetic Gri-Mech 3.0 format Chemkin III).
- For the calculation of thermal properties of the fluids, such as conductivity, viscosity, and the NO_x, a user defined function (*UDF*) was implemented.
- The NO_x model includes thermal, prompt, fuel, and intermediate NO_2 routes.

Defining the Turbulence Model

Unfortunately no single turbulence model is universally accepted as being superior in dealing with all problem areas. The choice of turbulence model will depend on considerations such as the physics encompassed in the flow, the established practice for a specific class of problem, the level of accuracy required, the available computational resources, and the time available for the simulation [22, 28 and 31]. To make the most appropriate choice of model for our application, it was necessary to understand the capabilities and limitations of the various options.

The simplest "complete models" of turbulence were the two-equation models in which the solution of two separate transport equations allowed the turbulent velocity and length scales to be independently determined. The standard *k-ε* model in FLUENT fell within this class of turbulence model and became the workhorse of practical engineering flow calculations since the time it was proposed by Launder and Spalding [39]. Robustness, economy, and reasonable accuracy for a wide range of turbulent flows explained its popularity in industrial flow and heat transfer simulations. It is a semi-empirical model, and the derivation of the model equations relies on phenomenological considerations and empiricism.

The *k and ε* equations describing this model are as follows:

$$\frac{\partial}{\partial t}(\rho_m k)+\nabla(\rho_m \vec{v}_m k)=\nabla\left(\frac{\mu_t}{\sigma_k}\nabla k\right)+G_{k,m}+G_b-\rho_m \varepsilon \tag{7}$$

$$\frac{\partial}{\partial t}(\rho_m \varepsilon)+\nabla(\rho_m \vec{v}_m \varepsilon)=\nabla\left(\frac{\mu_{t,m}}{\sigma_\varepsilon}\nabla \varepsilon\right)+\frac{\varepsilon}{k}(C_{1\varepsilon}\cdot G_{k,m}-C_{2\varepsilon}\cdot\rho_m\cdot\varepsilon) \tag{8}$$

where:

$$G_{k,m}=\mu_{t,m}\left(\nabla\vec{v}_m+(\nabla\vec{v}_m)^T\right)\nabla\vec{v}_m \tag{9}$$

$$G_{b,m}=-g\cdot\frac{\mu_t}{\rho_m\cdot\sigma_h}\cdot\nabla\rho \tag{10}$$

and the rest of the experimental constants for our model are:

$$C_{1\varepsilon}=1.44;\quad C_{2\varepsilon}=1.92;\quad C_\mu=0.09;\quad \sigma_k=1.0;\quad \sigma_\varepsilon=1.3 \tag{11}$$

Turbulence intensity is defined as the ratio of the root-mean-square of the velocity fluctuations to the mean flow velocity such as:

$$I_T=\frac{u'}{u_{avg}} \tag{12}$$

For internal flows, the turbulence intensity at the inlets was totally dependent on the upstream history of the flow. If the flow upstream is under-developed and undisturbed, you can use low turbulence intensity [41, 42]. If the flow is fully developed, the turbulence intensity may be as high as a few percent.

Defining the Wall Functions

Wall functions are a collection of semi-empirical expressions, based on the proposal of [43] having the effect of a "bridge" or "link" for the solution variables at the near-wall cells and their corresponding quantities on the wall. They have been most widely used for industrial flows and comprise:

- Laws-of-the-wall for mean velocity and temperature (or other scalars).
- Formulae for near-wall turbulent quantities.

The standard wall functions work reasonably well for a broad range of wall-bound flows. However, they tend to become less reliable when the flow situations deviate too much from the ideal conditions that are assumed in their derivation. Among others, the constant-shear and local equilibrium hypotheses most restricted the universality of the standard wall functions. Accordingly, when the near-wall flows are subjected to severe pressure gradients, and when the flows are in strong non-equilibrium, the quality of the predictions is likely to be compromised [44].

The law-of-the-wall for mean velocity yields:

$$U^* = \frac{1}{\kappa} \cdot ln\left(E\, y^*\right) \tag{13}$$

where

$$U^* = \frac{U_P \cdot C_\mu^{1/4} \cdot k_P^{1/2}}{\tau_w / \rho} \quad \text{and} \quad y^* = \frac{\rho \cdot C_\mu^{1/4} \cdot k_P^{1/2} \cdot y_P}{\mu} \tag{14}$$

The law-of-the-wall for temperature included the contribution from the viscous heating [45]:

$$T^* = \frac{\left(T_w - T_P\right) \cdot \rho \cdot C_\mu^{1/4} \cdot k_P^{1/2}}{\dot{q}} = \begin{cases} Pr\, y^* + \frac{1}{2} \rho\, Pr \frac{C_\mu^{1/4} \cdot k_P^{1/2}}{\dot{q}} U_P^2 & \left(y^* \langle y_T^*\right) \\ Pr_t \left[\frac{1}{\kappa} \cdot ln\left(E\, y^*\right) + P\right] + \frac{1}{2} \rho \frac{C_\mu^{1/4} \cdot k_P^{1/2}}{\dot{q}} \left\{Pr_t \cdot U_P^2 + \left(Pr - Pr_t\right) \cdot U_c^2\right\} & \left(y^* \langle y_T^*\right) \end{cases} \tag{15}$$

where P is computed by [46] as:

$$P = 9.24 \cdot \left[\left(\frac{Pr}{Pr_t}\right)^{3/4} - 1\right] \cdot \left[1 + 0.28 \cdot exp\left(-0.007 \cdot \frac{Pr}{Pr_t}\right)\right] \tag{16}$$

and finally, the law-of-the-wall for species can be expressed for constant property flow with no viscous dissipation as:

$$Y^* = \frac{(Y_{i,w} - Y_i)\cdot \rho \cdot C_\mu^{1/4} \cdot k_P^{1/2}}{J_{i,w}} = \begin{cases} Sc\, y^* & (y^* \langle y_C^*) \\ Sc_t \left[\frac{1}{\kappa}\cdot ln(E\, y^*) + P_c\right] & (y^* \langle y_C^*) \end{cases} \tag{17}$$

where P_C and y_C^* are calculated in a similar way as P and y_T^*, the difference being that the Prandtl numbers are always replaced by the corresponding Schmidt numbers.

Defining the Radiation Model

Radiative heat transfer was included because the radiant heat flux, was high compared to the heat transfer rate due to convection or conduction. Typically this will occur at high temperatures where the fourth-order dependence of the radiative heat flux on temperature implies that radiation will dominate [47].

Heating or cooling of surfaces due to radiation and/or heat sources or sinks due to radiation within the fluid phase can be included in our model using a radiation model. The radiative transfer equation adopted for an absorbing, emitting, and scattering medium at position $\vec{r}$ in the direction $\vec{s}$, is [48]:

$$\frac{dI(\vec{r},\vec{s})}{ds} + (a+\sigma_s)\cdot I(\vec{r},\vec{s}) = a\cdot n^2 \cdot \frac{\sigma \cdot T^4}{\pi} + \frac{\sigma_s}{4\pi}\int_0^{4\pi} I(\vec{r},\vec{s})\cdot \Phi(\vec{s},\vec{s}\,')\, d\Omega \tag{18}$$

The *P-1* radiation model is based on the expansion of the radiation intensity into an orthogonal series of spherical harmonics [47]. The following equation is obtained for the radiation flux:

$$q_r = \frac{1}{3\cdot(a+\sigma_s) - C\cdot \sigma_s}\cdot \nabla G \tag{19}$$

The transport equation for G is:

$$\nabla\left(\left[\frac{1}{3\cdot(a+\sigma_s) - C\cdot\sigma_s}\right]\nabla G\right) - a\cdot G + 4\cdot a\cdot \sigma \cdot T^4 = S_G \tag{20}$$

The net radiative heat flux at flow inlets and outlets is computed in the same manner as at the walls, as described above. It has been assumed that the emissivity of all flow inlets and outlets is 1.0 (black body absorption).

For a gray, absorbing, emitting, and scattering medium containing absorbing, emitting, and scattering particles, the transport equation for the incident radiation can be written as [49]:

$$\nabla(\Gamma \nabla G) + 4\pi \cdot \left(a\cdot \frac{\sigma\cdot T^4}{\pi} + E_P\right) - (a + a_P)\cdot G = 0 \tag{21}$$

where:

$$\Gamma = \frac{1}{3(a + a_P + \sigma_P)}$$

$$\sigma_P = \lim_{V \to 0} \sum_{n=1}^{N} (1 - f_{pn}) \cdot (1 - \epsilon_{pn}) \cdot \frac{A_{pn}}{V}$$

$$E_P = \lim_{V \to 0} \sum_{n=1}^{N} \epsilon_{pn} \cdot A_{pn} \cdot \frac{\sigma \cdot T_{pn}^4}{\pi V} \tag{22}$$

$$a_P = \lim_{V \to 0} \sum_{n=1}^{N} \epsilon_{pn} \cdot \frac{A_{pn}}{V}$$

$$A_{pn} = \frac{\pi \cdot d_{pn}^2}{4}$$

So, finally, heat sources (sinks) due to particle radiation are included in the energy equation as follows:

$$-\nabla \cdot q_r = -4\pi \cdot \left(a \cdot \frac{\sigma \cdot T^4}{\pi} + E_P \right) + (a + a_P) \cdot G \tag{23}$$

Defining the Combustion Model

The laminar flamelet approach, modeled a turbulent flame brush as an ensemble of discrete, steady, laminar flames, called flamelets [40]. The individual flamelets were assumed to have the same structure as laminar flames in simple configurations, and were obtained by experiments or calculations. Using detailed chemical mechanisms, laminar opposed-flow diffusion flamelets for non-premixed combustion can be calculated. The laminar flamelets were then embedded in a turbulent flame using statistical methods (*PDF*).

The advantage of the laminar flamelet approach was that realistic chemical kinetic effects could be incorporated into turbulent flames [50, 51]. The chemistry could then be preprocessed and tabulated, minimizing computation substantially. However, the laminar flamelet model was limited to modeling combustion with relatively fast chemistry. The flame was assumed to respond instantaneously to the aerodynamic strain, and thus the model was unable to capture deep non-equilibrium effects such as ignition, extinction, and slow chemistry (similar to NO_x).

The conserved scalar model was interpreted as the most basic flamelet structure. The coupling between non-equilibrium chemistry and turbulence was achieved by the statistical description of two parameters: the mixture fraction (f) and the instantaneous scalar dissipation rate (χ). Under the assumption of equal diffusivities, a coordinate-free formulation of the flamelet structure was considered. The hypothesis of statistical independence of these two parameters was also discussed. It is shown, how local quenching of diffusion flamelets leads to a reduction of burnable flamelets [40].

This reduction of the complex chemistry to two variables allowed the calculations of flamelets to be preprocessed, and stored in look-up tables. By preprocessing the chemistry, computational costs were reduced considerably [52].

The mixture fraction definition can be understood in relation to common measures of reacting systems. Consider a simple combustion system involving a fuel stream, an oxidant stream, and a product stream symbolically represented at stoichiometric conditions as:

$$F_s + r \cdot O_s \rightarrow (1+r) \cdot P_s \tag{24}$$

Denoting the equivalence ratio in this way as:

$$\phi = \frac{(fuel/air)_{actual}}{(fuel/air)_{stoichiometric}} \tag{25}$$

So, the reaction in equation (24), under more general mixture conditions, can then be written as:

$$\phi \cdot F_s + r \cdot O_s \rightarrow (\phi + r) \cdot P_s \tag{26}$$

Looking at the left side of this equation, the mixture fraction (f) for the system as a whole can then be deduced as:

$$f = \frac{\phi}{\phi + r} \tag{27}$$

Equation (27) allows the computation of the mixture fraction at stoichiometric conditions (ϕ = 1) or at fuel-rich conditions ($\phi > 1$), or fuel-lean conditions ($\phi < 1$).

On the other hand, the scalar dissipation rate (χ) is defined as:

$$\chi = 2D \cdot |\nabla f|^2 \tag{28}$$

A turbulent flame brush was modeled as an ensemble of discrete laminar flamelets. For adiabatic systems, the density-weighted mean species mass fractions and temperature in the turbulent flame can be determined [40] as:

$$\overline{\phi}_{m,T} = \iint \phi_{m,T}(f, \chi_{st}) \cdot p(f, \chi_{st}) \cdot df \cdot d\chi_{st} \tag{29}$$

where:

$$\chi_{st} = \frac{a_s \cdot exp\left[-2 \cdot \left[erfc^{-1}(2 f_{st})\right]^2\right]}{\pi} \tag{30}$$

For our model where non-adiabatic laminar flamelets were considered, the additional parameter of enthalpy was required. However, the computational cost of modeling flamelets over a range of enthalpies is prohibitive, so some approximations were made. It was assumed that heat gain/loss to the system would have negligible effect on the species mass fractions, and adiabatic mass fractions were used [53, 54].

When introducing fuel to an oxidant, spontaneous ignition does not occur unless the temperature of the mixture exceeds the activation energy threshold required to maintain combustion. It is necessary to supply an ignition source to initiate combustion. This source may be a heated surface or inlet mass flow that heats the gas mixture above the required ignition temperature. Often, however, it is the equivalent of a spark: an initial solution state that causes combustion to proceed into a region of the *CFD* model that contains a sufficient fuel/air mixture for ignition to occur.

Defining the No_x Model

The post process NO_x model includes: thermal, prompt, fuel, and intermediate NO_2 routes.

A) Thermal Route

This is determined by a set of highly temperature-dependent chemical reactions known as the extended “Zeldovich mechanism”. The principal reactions governing the formation of thermal NO_x from molecular nitrogen are as follows:

$$
\begin{aligned}
O + N_2 &\Leftrightarrow N + NO \\
N + O_2 &\Leftrightarrow O + NO \\
N + OH &\Leftrightarrow H + NO
\end{aligned}
\tag{31}
$$

The rate constants for these reactions have been measured in numerous experimental studies [55, 56], and the data obtained from them have been critically evaluated by [57].

$$
\frac{d[NO]_T}{dt} = k_{f,1}[O]\cdot[N_2] + k_{f,2}[N]\cdot[O_2] + k_{f,3}[N]\cdot[OH] - k_{r,1}[NO]\cdot[N] - k_{r,2}[NO]\cdot[O] - k_{r,3}[NO]\cdot[H] \tag{32}
$$

The expressions for the rate coefficients (m^3 g^{-1} mol^{-1} s^{-1}) used in the thermal NO_x model are given below.

$$
\begin{aligned}
k_{f,1} &= 1.8 \cdot 10^8 \cdot exp(-38370/T) \\
k_{f,2} &= 1.8 \cdot 10^4 \cdot exp(-4680/T) \\
k_{f,3} &= 7.1 \cdot 10^7 \cdot exp(-450/T) \\
k_{r,1} &= 3.8 \cdot 10^7 \cdot exp(-425/T) \\
k_{r,2} &= 3.81 \cdot 10^3 \cdot exp(-20820/T) \\
k_{r,3} &= 1.7 \cdot 10^8 \cdot exp(-24560/T)
\end{aligned}
\tag{33}
$$

Hence, the NO source term due to thermal NO_x mechanisms is:

$$S_{Thermal,NO} = M_{w,NO} \cdot \frac{d[NO]_T}{dt} \tag{34}$$

B) Prompt Route

The presence of a second mechanism leading to NO_x formation was first identified by Fennimore [58] and was termed "prompt NO_x". There is good evidence that it can be significantly performed in some combustion environments, such as in low-temperature, fuel-rich conditions, and in situations where residence times are short. Surface burners or staged combustion systems, can create such conditions [59].

At present the prompt NO_x contribution to total NO_x from stationary combustors is small. However, as NO_x emissions are reduced to very low levels by employing new strategies (burner design or furnace geometry modification as it can be considered for this study), the relative importance of the prompt NO_x can be expected to increase. The route now accepted is as follows:

$$\begin{aligned}
&CH + N_2 \Leftrightarrow HCN + N \\
&N + O_2 \Leftrightarrow O + NO \\
&HCN + OH \Leftrightarrow CN + H_2O \\
&CN + O_2 \Leftrightarrow NO + CO \\
&CH_2 + N_2 \Leftrightarrow HCN + NH \\
&HCN + N \Leftrightarrow N_2 + \ldots
\end{aligned} \tag{35}$$

In other studies [60], comparison of probability density distributions for the location of the peak NO_x with those obtained for the peak CH have shown close correspondence, indicating that the majority of the NO_x at the flame base is prompt NO_x formed by the CH reaction [61].

The NO source term due to prompt NO_x mechanisms is:

$$S_{prompt,NO} = M_{w,NO} \cdot \frac{d[NO]_{Pr}}{dt} \tag{36}$$

and according to [62]:

$$\frac{d[NO]}{dt} = f \cdot k'_{pr} \cdot [O_2]^b \cdot [N_2] \cdot [FUEL] \cdot exp\left(-E'_a / RT\right) \tag{37}$$

where:

$$\begin{aligned}
&f = 4.75 + 0.0819\,n - 23.2\phi + 32\phi^2 - 12.2\phi^3 \\
&k'_{pr} = 6.4 \cdot 10^6 \left(RT/P\right)^{b+1} \\
&E'_a = 303474.125\ J/gmol
\end{aligned} \tag{38}$$

C) Fuel Route

It is well known that nitrogen-containing organic compounds present in liquid or solid fossil fuel can contribute to the total NO_x emissions during the combustion process. This fuel nitrogen is a particularly important source of nitrogen oxide emissions for residual fuel oil and coal, which typically contain *0.3-2 %* nitrogen by weight. Studies have shown that most of the nitrogen in heavy fuel oils is in the form of heterocycles and of significance are heterocyclic ring structures such as pyridine, quinoline, and amine type.

The extent of conversion of fuel nitrogen to NO_x was dependent on the local combustion characteristics and the initial concentration of nitrogen-bound compounds. Fuel-bound nitrogen-containing compounds were released into the gas phase when the fuel droplets or particles were heated during the devolatilization stage. From the thermal decomposition of these compounds in the reaction zone, radicals such as HCN, NH_3, N, CN, and NH were formed and converted to NO_x. These free radicals are subject to a double competitive reaction path [63, 64]. Other investigations have shown that hydrogen cyanide appears to be the principal product if fuel nitrogen is present in aromatic or cyclic form [65]. Although the route leading to fuel NO_x formation and destruction is still not completely understood, different investigators [66] seem to agree on a simplified model such as:

$$FUEL\ N_2 \rightarrow \begin{matrix} N_2\ intermediate\ or \\ HCN \end{matrix} \begin{cases} \xrightarrow{(1)\,O_2} NO \\ \xrightarrow{(2)\,NO} N_2 \end{cases} \tag{39}$$

The source terms in the transport equations can be written as follows:

$$\begin{aligned} S_{HCN} &= S_{pl,HCN} + S_{HCN-1} + S_{HCN-2} \\ S_{NO} &= S_{NO-1} + S_{NO-2} \end{aligned} \tag{40}$$

where:

$$\begin{aligned} S_{pl,HCN} &= \frac{R_{cf} \cdot Y_{N,fuel} \cdot M_{w,HCN}}{M_{w,N}} \\ S_{HCN-1} &= -R_1 \cdot \frac{M_{w,HCN} \cdot p}{R \cdot \overline{T}} \\ S_{HCN-2} &= -R_2 \cdot \frac{M_{w,HCN} \cdot p}{R \cdot \overline{T}} \\ S_{NO-1} &= -S_{HCN-1} \cdot \frac{M_{w,NO}}{M_{w,HCN}} = R_1 \cdot \frac{M_{w,NO} \cdot p}{R \cdot \overline{T}} \\ S_{NO-2} &= -S_{HCN-2} \cdot \frac{M_{w,NO}}{M_{w,HCN}} = R_2 \cdot \frac{M_{w,NO} \cdot p}{R \cdot \overline{T}} \end{aligned} \tag{41}$$

and finally:

$$\begin{aligned} R_1 &= A_1 \cdot X_{HCN} \cdot X_{O_2}^a \cdot exp(-E_1 / RT) \\ R_2 &= A_2 \cdot X_{HCN} \cdot X_{O_2}^a \cdot exp(-E_2 / RT) \end{aligned} \tag{42}$$

D) Intermediate N_2O Route

Nitrogen enters combustion systems mainly as a component of the combustion and dilution air. Under favorable conditions, which include elevated pressure and oxygen-rich conditions, this intermediate mechanism can contribute as much as *90 %* of the NO_x formed during combustion [67, 68]. This makes it particularly important in equipments such as gas turbines and compression-ignition engines. As these devices are operated at increasingly low temperatures to prevent NO_x formation through the thermal NO_x mechanism, the relative importance of the N_2O-intermediate mechanism is increasing [69].

The simplest form of the mechanism takes into account two reversible elementary reactions:

$$\begin{aligned} &N_2 + O + M \Leftrightarrow N_2O + M \\ &N_2O + O \Leftrightarrow 2NO \end{aligned} \tag{43}$$

Here, M is a general "third body". As the first reaction involves third bodies, the mechanism is favored at elevated pressures. Both reactions involve oxygen radical, which makes the mechanism favored in oxygen-rich conditions. Although it is not always justified, it is often assumed that the O atoms originate solely from the dissociation of molecular oxygen, [70, 71].

According to the kinetic rate laws, the rate of NO_x formation through the N_2O-intermediate mechanism is:

$$\frac{d[NO]}{dt} = \left(k_{f,2} \cdot [N_2O] \cdot [O] - k_{r,2}[NO]^2\right) \tag{44}$$

where:

$$[N_2O] = \frac{k_{f,1}[N_2] \cdot [O] \cdot [M] + k_{r2}[NO]^2}{k_{r,1}[M] + k_{f2}[O]} \tag{45}$$

In the above expressions, the forward rate constants and the corresponding reverse rate constants are:

$$\begin{aligned} &k_{f,1} = 4.44 \cdot 10^{32} \cdot T^{-8.358} \cdot exp(-28234/T) \\ &k_{f,2} = 2.90 \cdot 10^{7} \cdot exp(-11651/T) \\ &k_{r,1} = 44.00 \cdot 10^{8} \cdot exp(-28234/T) \\ &k_{r,2} = 1.45 \cdot 10^{-29} \cdot T^{9.259} \cdot exp(-11651/T) \end{aligned} \tag{46}$$

Validating the CFD Model

The experimental values measured with the above-mentioned pyrometer and the results obtained with the *CFD* model for the experimental burner, are shown in table 4. It shows that the calculation procedure was acceptable. A small difference in the maximum combustion temperature was expected, because it is well known that the combustion model used, tends to over-predict such values of temperature. As a consequence of this, the temperature of the *CEG* from the combustion chamber should be slightly lower than this given value.

Table 4. Validation of the *CFD* model for the experimental burner through the measurement procedure

Validation results	Experimental measurements	CFD model for the experimental burner
Heat transfer across the walls (W).	915,521 (*)	909,649.94
Outlet chamber temperature (K)	1,362.0	1,348.0
Maximum combustion temperature (K)	1,928	1,950
NOx (ppm)	61.04	64.72
CO (ppm)	4.6	3.2

(*) Estimated value from experimental calculations.

This also means an over-prediction of the NO_x, owed fundamentally to a major value of the combustion temperature corresponding to the *CFD* model. It is necessary to emphasize that the NO_x calculation process incorporated do not predict correctly the value of the emissions, so a correction was necessary. This is possible by introducing a link to a compiled library software solver called "User Defined Function" (*UDF*) as described below.

Further Improvements Introduced to the CFD Model

By analyzing the method for calculating the NO_x in FLUENT, it is observed that the models were obtained through tests in adiabatic reactors using techniques that have recently been perfected. The tests were carried out at temperatures higher than *2,200 K* and the mechanisms for NO_x formation were due to these high temperature values, so it is considered that the results obtained should be used for a qualitative analysis, rather than a quantitative one [72].

However, the situation changes when combustion takes place at lower temperatures, with other kinetic coefficients and mechanisms prevailing. It has also been clearly proved that in reactors of this type the temperature of the flame and more significantly the levels of NO_x can increase by as much as *2-5* times [73].

With regard to burners for boilers, as in this case where there are significant heat losses through radiation, combustion temperatures fall below *2,200 K* and therefore, the values of NO_x formation for the model used by FLUENT are much lower than the measurements performed, consistent with [74 - 77].

As far as the viability of the model is concerned, as seen in the previous table, the maximum combustion temperature, the outlet temperature in the combustion chamber (which governs the heat transfer in the chamber) and, the shape and aerodynamics of the flame both in the *CFD* model and in the experimental measurements, all coincide as can be seen in the above-mentioned table 4. This leads to the conclusion that it is the NO_x post-process model that is prone to errors and is unsuited to the case in question here.

It is imperative to determine the causes and mechanisms of NO_x formation in the initial burner in order to make the pertinent modifications to the prototype. Thus, it was decided to introduce a series of improvements for an accurate calculation of operating conditions of our burner.

Willian [78] studied a burner with combustion in the porous region and a low index of NO_x, with temperatures of around *1,800 - 1,850 K,* facilitating the correction and analysis of the prevailing mechanisms of NO_x formation under the above conditions. This temperature was attributed to strong component of radiant energy present, which resulted in a large amount of heat loss through radiation, precisely at the outlet of the porous region. Consequently, the temperature was lower than in a conventional burner and therefore, with lower thermal NO_x emission compared to all other previously explained mechanisms.

The experiments undertaken included measuring radiant flow, gas temperature and stable species concentration. Particular attention was paid to gas temperature profiles, in order to define the flame zone and compare this position to the predictions of theoretical models very similar to the one used in combustion models [79]. The result showed that the prevailing mechanism in NO_x formation was the "prompt route". Thus, the post-process calculation model is as follows:

- Thermal route: The reaction (32) finally accepted turns to:

$$\frac{d[NO]_T}{dt} = k_{178}[O]\cdot[N_2] \tag{47}$$

- Prompt route: the reaction (37) finally accepted turns to:

$$\frac{d[NO]_{Pr}}{dt} = k_{240}[CH]\cdot[N_2] \tag{48}$$

where the kinetic coefficients are given by [80 - 81]. By using the post-process model in FLUENT, the results obtained are in stationary condition, whereby it is proposed to modify the reaction order, using the experimental data provided by [78].

The reaction order was the gradient of a straight line plotted through each set of measurements, with the equation (48) being as follows:

$$\frac{d[NO]_{Pr}}{dt} = 1.2\times 10^7 \cdot k_{pr} \cdot f \cdot [O_2]^p \cdot [N_2]^l \cdot ([CH_4]+[C_2H_6])^{0.024} \cdot exp\left(\frac{-30303}{T}\right) \tag{49}$$

- Intermediate N_2O, The reaction finally used is:

$$\frac{d[NO]_{N_2O}}{d[NO]_T} = \frac{k_{185}[O]\cdot[N_2]\cdot[M]}{k_{178}[O]\cdot[N_2]} = \frac{k_{185}[M]}{k_{178}} \tag{50}$$

Using reasonable values for k_{185} [82 - 84], it is clearly important only at low temperatures. Accordingly, this modified and improved the calculation model used by FLUENT for the post-process determination of NO_x. This model was implemented as a *UDF*. These are programmed functions that can be dynamically linked to FLUENT's solver. They are written in the programming language "C" and are defined through macros, provided by the code. Thus, the values for the NO_x obtained at this instance (*61.80 ppm.*) represent a much better approach to the experimental results shown in the above-mentioned table 4. Accordingly, the corrected *CFD* model corresponding to the experimental burner could be considered as clearly validated.

Building a CFD Model for the Prototype Burner

Several models with the same geometric configuration but with different working conditions and "air/gas" ratios were studied for the new burner which had three air inlets and two gas inlets respectively.

They were placed with coaxial form and with the configuration of the previous burner. The primary air is supplied through the center nozzle while the secondary one goes out across a circumference section between an inside diameter matching the external wall of a nozzle (where the primary gas is supplied) and an outside diameter of *204 mm*.

The tertiary air is supplied by an external circumference of *326 mm*. diameter which matches the internal diameter of the burner. Finally the secondary gas is supplied by twelve nozzles with *18* holes each one of *2.5 mm*. diameter at the end, covered each one with a small bushing. The *CFD* layout is seen in figure 3.

This geometric configuration tries to get two combustion zones clearly identified, the primary one including the primary gas, primary and secondary air, and the secondary zone, including the secondary gas and the tertiary air.

The gas exit in both cases (primary and secondary) is normal to the axis of the burner since it was demonstrated that it offers better results for the gradual mixing process. The whole geometrical model consists of around *400,000* cells.

To complete the whole model the same physical models from the above-mentioned experimental burner *CFD* were implemented, as it was previously considered validated. These models are turbulence, radiation, combustion, and post-processing NO_x.

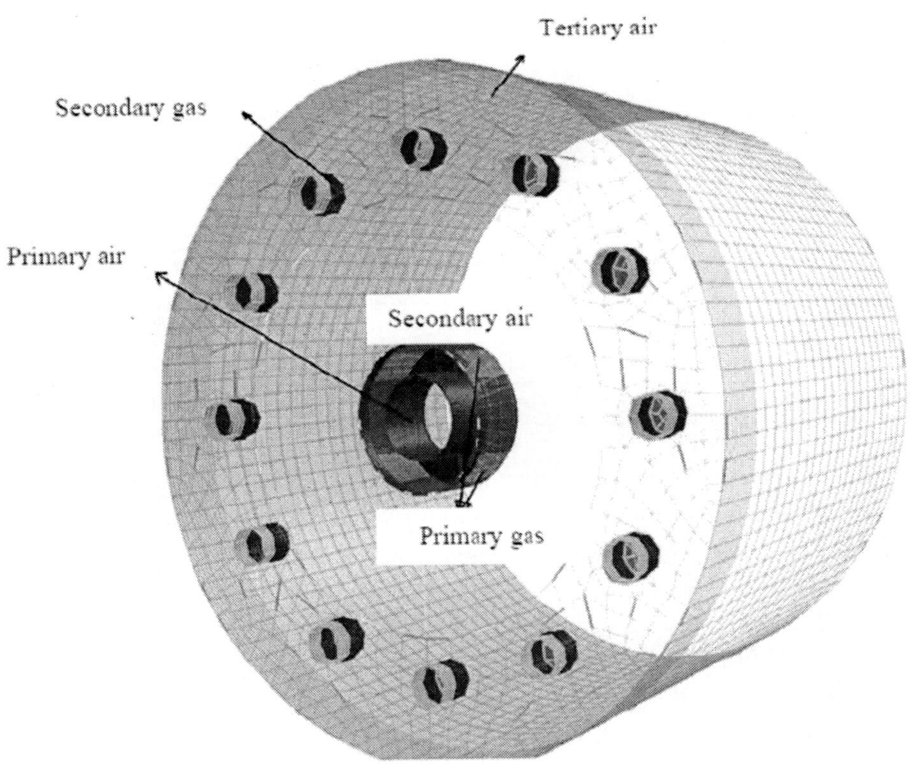

Figure 3. Detail of geometry and mesh of the prototype burner.

2.5. Results and Discussion

The results obtained for these models as a comparative test are shown as follows and one of the models will finally be selected.

Analysis of the Alternatives Studied for the Prototype Burner

The most relevant comparative results for the four alternatives corresponding to the new burner are summarized in table 5.

As seen in this table, all the models present a similar behavior, with the only exception being model "*D*". Nevertheless model "*C*" offers the best results both for combustion temperature as well as for NO_x and *CO* emissions. In this manner *CO* emission values are within the accepted levels for these types of burners. As a result the turbulence, kinetic chemistry, heat transfer, and NO_x emissions for model "*C*", which was the option finally chosen (prototype burner), will be closely analyzed in this work.

It is very difficult to identify the individual reactions and kinetic mechanisms under certain conditions in a *CFD*. In any case, a global analysis is possible. The kinetic mechanism used, called Gri-Mech 3.0 includes a wide documentation about its validation in [29], [30].

Table 5. Characteristics of the five *CFD* models for the prototype burner

Low NOx and CO burner *CFD*	**Model "A"**	**Model "B"**	**Model "C"**	**Model "D"**	**Model "E"**
Primary air (%).	2	3	3	2	4
Primary gas (%).	*60*	*50*	*75*	*25*	*90*
Secondary air (%).	38	47	37	68	44
Secondary gas (%).	*40*	*50*	*25*	*75*	*10*
Tertiary air (%)	60	50	60	30	52
ϕ primary	1.364	0.909	1.703	0.324	1.705
ϕ secondary	0.606	0.909	0.380	2.270	0.175
Maximum temperature (K)	1,870	1,890	1,830	2,018.3	1,846
Outlet chamber temperature (K)	1,369	1,358	1,384.6	1,387.7	1,381.6
NOx outlet average (ppm)	42.59	80.28	17.66	205.44	13.47
CO exit average (ppm)	3.0	3.8	2.9	3.1	3.4

In figure 4, flame speed opposed to equivalence ratio is seen in the form of a comparative study with works of other authors [66, 85 and 86]. This indicates that when working with values of the equivalence ratio $\phi < 0.9$ and $\phi > 1.2$ respectively, the flame speed clearly diminishes causing the flame front to advance slowly towards the mixture. Therefore the speed of the flow increases the main combustion zone, which leads to a stretching of the flame far from the exit of the burner.

For the model "*C*" in the primary zone the value of $\phi = 1.36$ corresponds to a speed of approximately *0.20 m/s* which is relatively low, while in the secondary zone $\phi = 0.66$ corresponds to a speed of *0.14 m/s*. This indicates that there are strong differences between both zones, which validate the structure of the temperature profiles adopted.

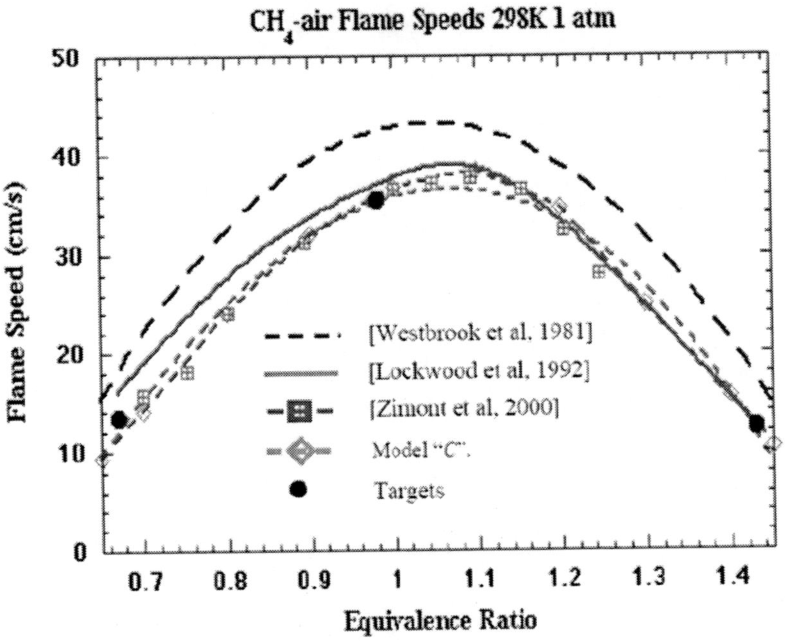

Figure 4. Comparative of the equivalence ratio and flame speed.

The NO_x emission in a combustion process, depends obviously on the conditions in which the combustion takes place [87]. Therefore if the species that have major mass fraction are known, it will be possible to identify the principal mechanisms for which they are formed. From the profiles of the species mass fraction menu, it can be appreciated that those of major mass fraction are: *HCN*, NH_3, and *HCNO*.

In figure 5, the NO_x emission obtained by the *UDF* through three ways (prompt, thermal, and fuel route associated) for model "*C*" are shown.

The high quantity of *HCN* in the position $x = 0.189$ *m.* (close to the thermocouple) with a temperature of *1,791.3 K,* favors *NO* production. This species is strongly associated with the "*Fennimore's mechanism*", also called "*prompt-NO*". In this matter the concentration of hydrocarbon radicals is quite relevant thus the production of *NO* begins by breaking the link "*N-N*". The most important reaction here is given by the first one of the equations (35), such as:

$$CH + N_2 \leftrightarrow HCN + N \tag{51}$$

On the other hand this atom of nitrogen produced in this reaction can form a molecule of *NO* through the "*Zeldovich's mechanism*". As is seen in the above-mentioned figure 5, the source of *NO* through the "thermal route" is bigger than the one formed through the "prompt route".

An increase of the N_2O mass fraction takes place for the same position ($x = 0.189$ *m*). This peak of N_2O also favours the formation of *NO*. Two peaks of maximum values are revealed for the total NO_x being situated at the positions: $x = 1$ *m* and $x = 2$ *m* respectively (scalar value: E-03).

The different terms having analyzed, it can be concluded that the NO_x emission by the "fuel route", represents the bigger contribution, followed by the "thermal" and the "prompt" routes respectively. The latter takes place in the post-combustion zone where the biggest mass fractions are the CO_2 and H_2O.

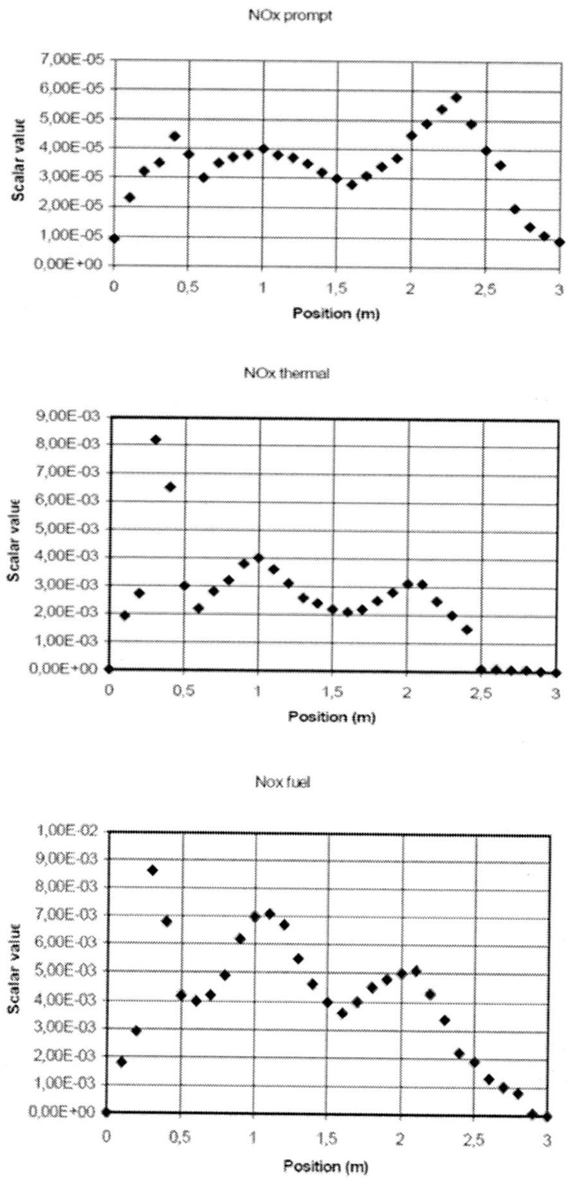

Figure 5. Mechanisms for the *NOx* for the *UDF*, over the symmetry axis of the prototype burner.

Comparative Results for Both Experimental (Corrected) and Prototype Burners

The NO_x emission in the prototype burner under standard working conditions in every stage, resulted in *17.66 ppm (36.11 mg/Nm*^{3}*)* as seen in table 6, which is lower than the emission level in the corrected experimental burner *(61.22 ppm)* and the emission limit recommended by the European regulation, as seen in [7 - 9].

Table 6. Summary of the comparative *CFD* results between experimental and prototype burners

Comparative results for both *CFD* burners	CFD model for the experimental burner (corrected)	Prototype *CFD* burner
Heat transfer across the walls (W).	909,649.94	867,716.60
Outlet chamber temperature (K)	1,348.0	1,384.6
Maximum combustion temperature (K)	1,950	1,830
NOx (ppm)	61.22	17.66
CO (ppm)	3.2	2.9

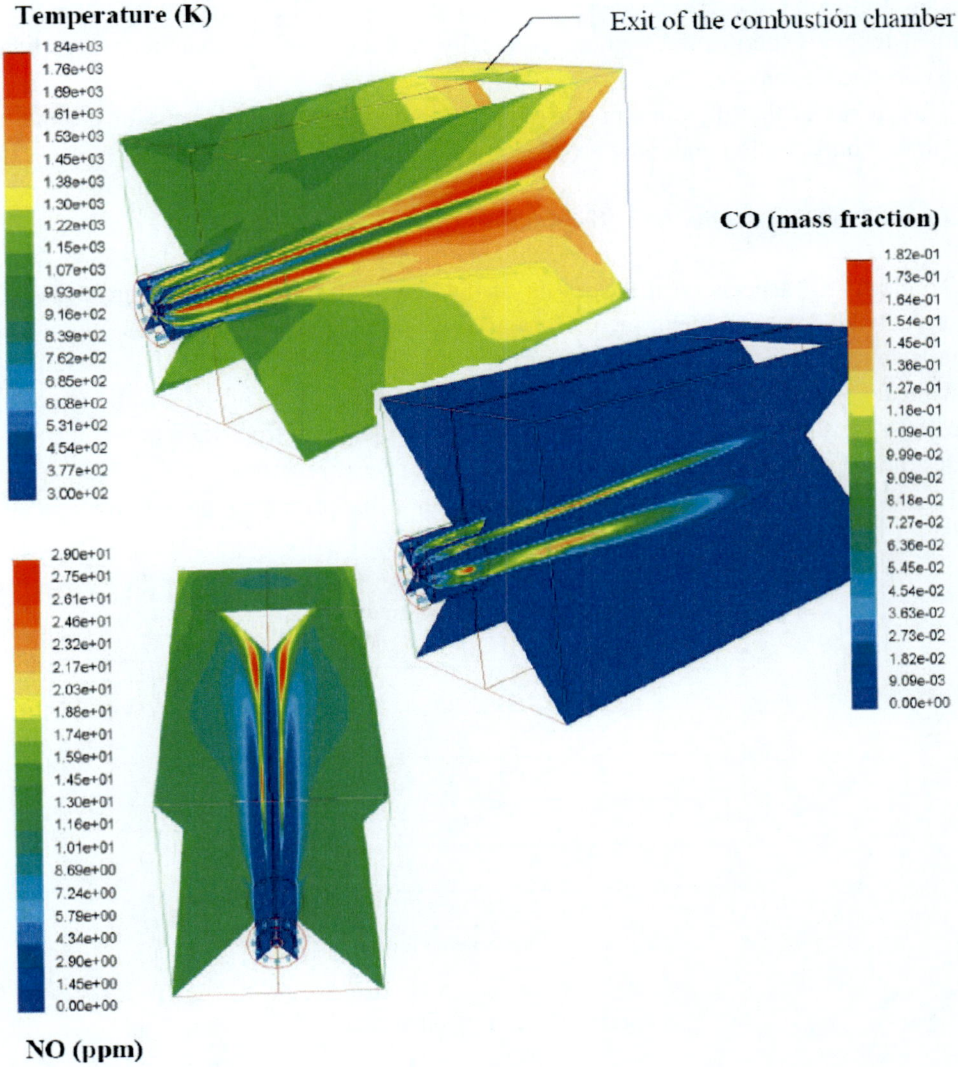

Figure 6. Contours of Temperature (K), CO (mass fraction) and NO (ppm) for the prototype burner.

According to this table, the maximum temperature of combustion at the exit of the combustion chamber decreases approximately to *120 K* and the NO_x emission decreases *to*

43.56 ppm. with relation to the new burner. The decrease of combustion temperature also decreases the heat transfer across the walls by about *4.6 %*, according to [88]. As a result, there is a temperature increase of about *36.6 K* at the outlet of the combustion chamber. Besides, a decrease in *CO* emission is achieved from *3.2* to *2.9 ppm*, as is seen in the same table.

The distribution of these values for the model "*C*" over the whole combustion chamber is shown in figure 6.

Contours of temperature show the effect of the conducting walls over the whole combustion chamber. In this way, these contours are finally detached towards the external walls, creating as uniform a temperature distribution in this area as possible. Anyway, a homogeneous recirculation region can be seen near the burner. Contours of *NO* are clearly influenced by the temperature as has been discussed above [87]. Finally, contours of *CO* are revealed following the flame shape, as expected, (but with lower values than the ones obtained for the experimental burner), especially at the end of the chamber where finally it almost becomes exhausted.

This indicates that the aims proposed in this paper have been successfully achieved by this burner configuration under the proposed working conditions.

Analysis of the Sensibility and Response of the Prototype Burner

To clarify all aspects on the working of the new burner and its sensibility to a relative response of the control system, the performance was analyzed under different conditions and air-gas relations, for each stage. The control system can be influenced by any mismatch of its mechanical components, referred to any irregular variations of the air-gas relationship in each stage [89], caused by the wear or disruptions of the elements from the control mechanisms besides the sensibility of the burner to the variation of these relations.

Figure 7 shows the sensitivity of the new burner in the primary zone with reference to the NO_x and *CO* emissions and temperature respectively.

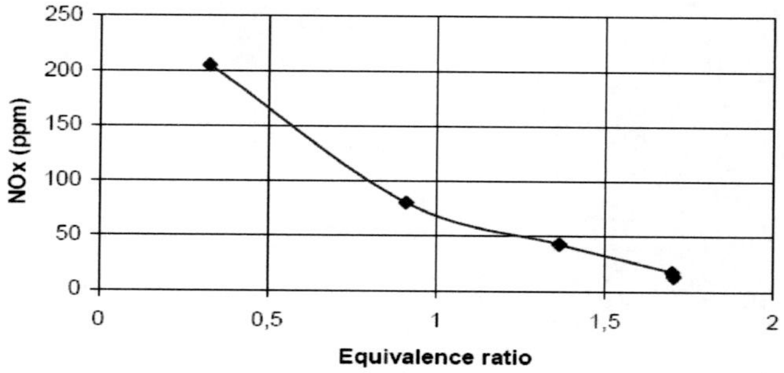

Figure 7. Continued on next page.

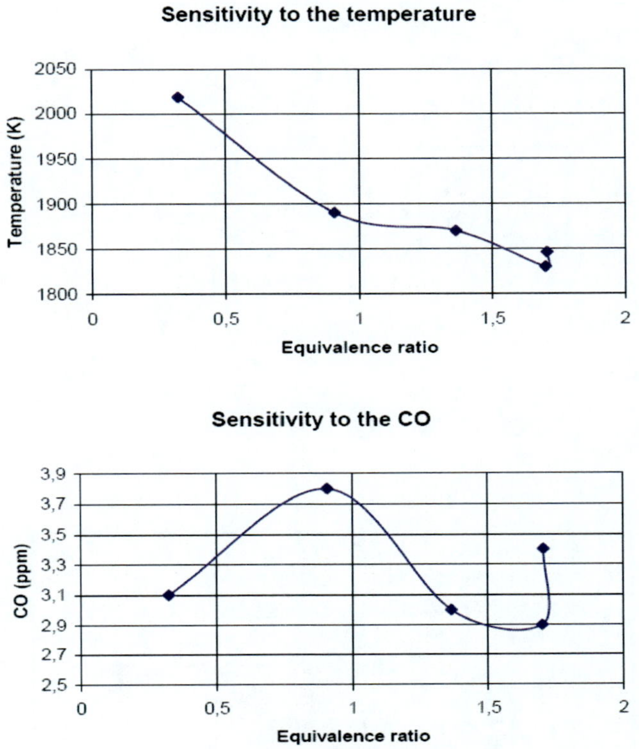

Figure 7. Prototype burner sensitivity in the primary zone with *NOx, CO* emissions, and Temperature.

According to this, the burner is slightly sensitive to the variation of NO_x emissions at the primary zone. When $\phi > 1.364$ this effect decreases and when $\phi > 1.3$ the effect over the temperature also decreases. The ideal relation for NO_x, *CO*, and temperature for the primary zone is reached at $\phi = 1.7$ and for the secondary zone $\phi = 0.38$.

Behaviour under a Combustion Based on Swirling Injections

One of the most generalized advantages in combustion associated with low NO_x emissions is the technology based on swirling injections [90, 91]. Therefore the behaviour of this geometric configuration has also been analyzed for our model under these working conditions.

The results corresponding to the temperature field obtained for a turbulent combustion based on swirling injections is shown in figure 8. The speed of the primary and secondary gas are: *15.89* and *24.67 m/s* respectively and the incidence angle for both cases are 79.64^o and 46.92^o respectively. Hence these components of the speed provoke a swirl number of *0.194* and *0.617* for both zones, primary and secondary respectively.

The swirl number is defined as the ratio of the axial flux of angular momentum to the axial flux of axial momentum as follows:

$$S_w = \frac{\int r \cdot w \cdot \vec{v} \cdot d\vec{A}}{\overline{R} \int u \cdot \vec{v} \cdot d\vec{A}} \tag{52}$$

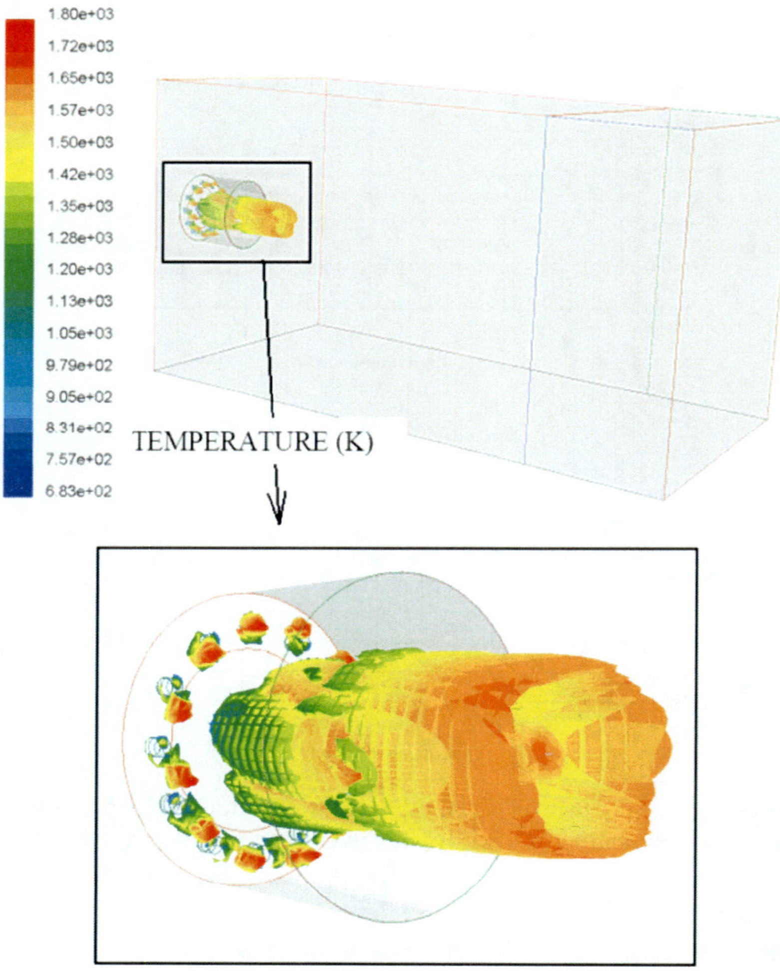

Figure 8. Contours of Temperature (K) under swirling injections.

For highly swirling flows ($S_w > 0.5$) the *k- ε* model yield appreciable improvements over other models. Taking all these in consideration the successful effects reached were an increase of maximum combustion temperature by about *1,890 K* and an average of *42.49 ppm.* of NO_x emission at the exit of the combustion chamber.

A comparison of this result with that of model *C*, with the same maximum combustion temperature, reveals that the NO_x*'s* emission is lower in this instance and they change the ϕ at every stage. Its effect is due to the shape of the temperature profiles and the mass fraction of species. This way, the temperature field inside the burner for this swirling injections based combustion is quite different from the traditional combustion at $x = 0.150$ *m*. (an intermediate position in the burner) and the two zones of high temperature values can be estimated (a primary zone with $\phi = 1.7$ and a secondary one with $\phi = 0.38$).

The primary zone is characterized by four jets of gas with angular distortion due to air rotation, which form a pseudo cylinder with the shape closer to a cone at the exit of the burner. As it passes through the combustion chamber, this cylinder increases its diameter keeping this behaviour until the end of the chamber. Traces of CH_4, C_2H_6, and CO are detected inside this cone, particularly increasing the concentration of the latter and reaching its maximum value between *1* to *1.5 m*. Finally, it becomes exhausted at the end of the chamber, where it remains together with the *OH*.

This type of jet configuration does not allow the gradual incorporation of the gases that usually cause recirculation in a chaotic way inside the chamber [92]. Hence the effect of a homogeneous recirculation found in cases of traditional combustion, is not observed here.

The control of the stoichiometry in every zone of combustion (staggered combustion) has been revealed as an effective technology for decreasing NO_x emissions. Nevertheless, staggered combustion is not the most suitable option here because of the geometry of the combustion chamber, where the effect of internal recirculation of gases is not negligible. This is an important aspect in the choice of low NO_x emission burner since it is necessary to design the burner and its combustion enclosure simultaneously.[93, 94].

From the kinetic point of view, large quantities of H_2O, CO, and H_2 are produced shortly after the methane is burned in this area. There is a reaction between methane and the radicals O, H, and OH to produce CH_3 so that it remains only in small traces of: O, H, and OH, due to the high affinity of the methane with these radicals.

The oxidation of the formaldehyde CH_2O to HCO and the subsequent formation of CO take place in areas of high concentrations of methyl and formaldehydes (rich zones $\phi > 1$). The oxidation of CO to CO_2 proceeds primarily of the route reaction:

$$CO + OH \rightarrow CO_2 + H \tag{53}$$

Therefore the rate of formation of CO_2 depends on the availability of OH radicals. Usually the origin of major quantity of species of hydrocarbon disables CO oxidation (it happens in the primary zone). This behaviour leads to the assumption that the rate of formation of reaction (54) is considerably smaller than the rate of formation of the reaction of H with species of hydrocarbon and that rate of formation of the reaction of oxidation of CO is also minor than the rate of formation of the species of hydrocarbon with the OH radical.

$$H + O_2 \rightarrow OH + O \tag{54}$$

As a result of this, it can be said that small quantities of hydrocarbon can restrict greatly the formation of CO_2 from CO and the level of OH radicals only rises when hydrocarbons and its fragments have already been burned, so CO_2 can be finally formed.

This can be the reason for the control of the flame temperature together with the kinetics of oxidation of N, [95 - 97], allowing this way the decreasing of the NO_x.

2.6. Conclusion

This study can be summarized as follows:

1. The scientific advances of the last twenty years have given to the international community, that is sensitive to environmental concerns, a number of protocols and agreements to limit NO_x emissions. Thus, the European Union will lower NO_x emission limits to a minimum, in its statutory regulations. This will compel all

European countries to comply with these limits, mainly by using natural gas with low NO_x emission technology.

2. The commonly known mechanisms for modelling nitrogen oxidation are still sensitive to special features and depend on the conditions in which the combustion takes place and the fuel burned. Multiple scientific publications concur on the identification of principal ways, but defer to the kinetic coefficients, as well as in the predominant mechanism under particular conditions, for every type of test and technology used. It is advisable to use a kinetic mechanism of combustion, which includes the mechanisms of nitrogen oxidation to enable treatment by *CFD*, and to subsequently calculate in post-process the aspects that the model is incapable of which include as a source term of NO_x in the equations of mass and species transport conservation.
3. It is demonstrated that the hypotheses of simplifications, geometric model, discretization, physical models, methodology and algorithms of solution, *UDF*, as well as its mathematical models, grouped as they were in this study, are correct and they are totally validated through experimental measurements also carried out here.
4. In accordance with the aim of this study, the design of a low NO_x-CO emission burner is obtained in the model C also called "prototype burner" and it satisfies the following characteristics:

    - It achieves maximum efficiency (over the *LCV*), with a concentration of oxygen in gases lower than *10 %* and minimizing the emission of unburned fuel, since *CO* content in the gases is lower than *0.1 %*.
    - The NO_x emissions in model C are well within the European statutory regulations, and comply with state-of-the-art technology for commercial low NO_x emission burners.
    - The flame provides maximum performance with a low combustion temperature for natural gas burners, and is totally stable without detaching from the flame pipe of the burner.
    - There is low pressure drop in the ducts of gas and air for the new model, allowing the use of same fans, placing a suitable coupling, with small modifications for the control of gas, at every stage.
    - The burner, allows a wide range of working conditions with a primary $\phi > 1.5$.

5. The use of numerical simulations has been revealed as a very important research instrument in combustion modelling, though there is scope for further improvements. The development of kinetic models for combustion, which involve the more complete and extensive kinetic of oxidation of *N*, is very difficult to be treated by a *CFD*, because the study involves managing multiple variables. Attempting simplifications could on occasion result in miscalculations. The calculation in post-process of NO_x emissions must be developed for each case of combustion and each kinetic model, by appropriately using available scientific publications and validating them it to whatever is necessary.
6. The model of NO_x developed in this work by *UDF*, can be used:

- To identify and calculate NO_x emissions when a combustion model is used, which includes gaseous fuel composed fundamentally of CH_4, together with a kinetic model with the oxidation of *N*.
- For temperatures lower than *1,950 K*, included by a strong radiation heat transfer.

7. Combustion processes used in each case of combustion must be carefully chosen, from the multiple technologies of decreasing NO_x emissions, taking into account both the burner and the combustion chamber. When the geometry of the chamber does not respond appropriately to a combustion based on swirling injections, with regard to internal and external recirculation of the gases, it can cause the opposite effect and thus increase both combustion temperature and emissions.
8. The staggered combustion achieving a geometric suitable configuration of the head gas of the burner is an effective technology for the decreasing of NO_x emissions, when the type of combustion enclosure does not respond appropriately to combustion based on swirling injections.
9. The most important and sensitive geometric aspects to the NO_x emissions are:

   - The exit of gas perpendicular to air flow.
   - The speed of the primary gas exit must be superior to that of secondary gas.
   - The number of gas nozzles for the primary combustion zone must be lower than 6 and always less to that of the secondary zone.
   - To establish a maximum of two combustion zones in the 2 MW power burner.
   - To pay special attention to the control of the air flow in the center of the flame, as well as its velocity. It is found that the geometric configurations are sensitive to the air in the central zone, since it defines the quantum of *CO* emission and with it the velocity control of reaction of some species, the dimensions and structure of flame and the temperature profiles.

10. The dimensions of the air and gas ducts are important variables, as also the length of the flame pipe, since these factors define the kinetic process of combustion, the flame shape, and the temperature and emission of unburned fuel and the NO_x emission.
11. To achieve a combustion temperature lower than *1,800 K* in any combustion equipment as the one treated here, it means an increase of the *CO* emissions and a decrease of the power of the burner.

3. Part B: Analytical Study of the Effects of the Clogging of a Mechanical Precipitator Unit

3.1. Introduction

This study provides a theoretical explanation of the causes that led to the clogging by ash, of one of the two groups of mechanical precipitators (*MP*) in a high-performance thermoelectric power plant, equipped with two Ljungström Air Preheaters (*LAP)*, arranged in the manner shown in figure 9.

The rotary regenerative heat exchanger was patented by the Swedish engineer Fredrik Ljungström in 1920. The basic component is a continuously rotating cylinder, called "the rotor", which is packed with thousands of square thin-walled steel plates, packed in frame baskets of heat transfer surface. During each revolution of the rotor, heat is absorbed by the heating surface passing through the hot gas stream and transferred to the combustion air stream. The basic operating scheme is shown in figure 10.

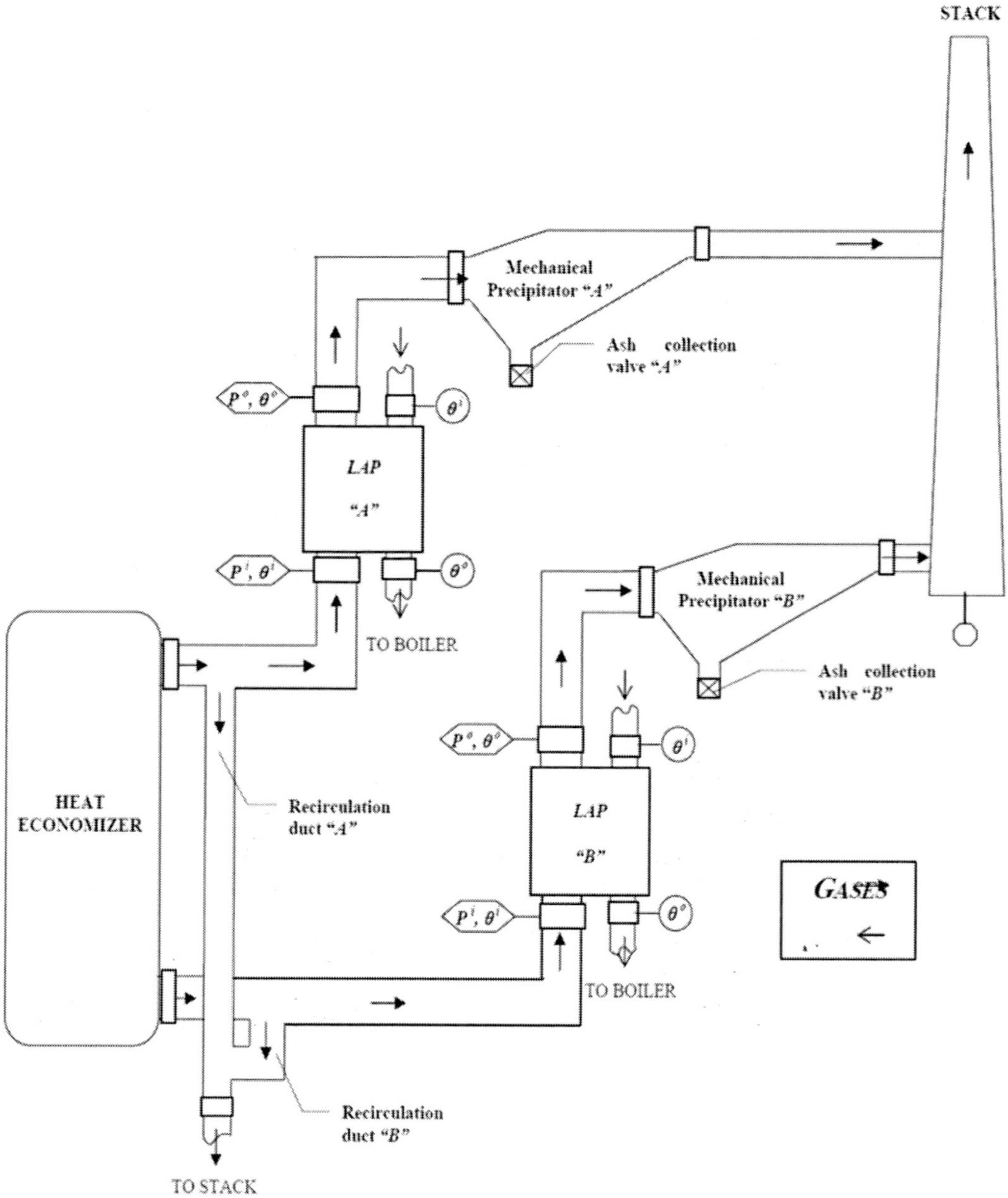

Figure 9. Simplified layout of the air heaters and mechanical precipitators in the power plant of reference.

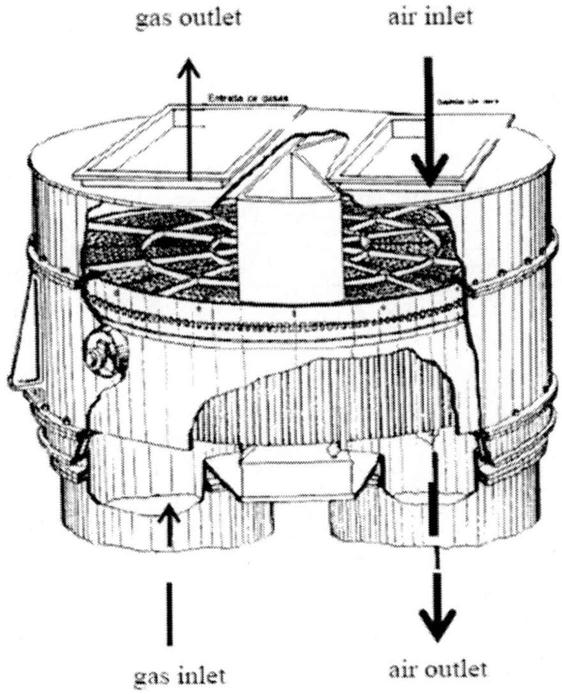

Figure 10. Description an air heater type "Ljungstrom" (ALSTOM Power Inc. Air Preheater Company).

This accumulated heat is released into the incoming air as the same surfaces pass through the other half of the structure. The heat transfer cycle is continuous in periods of about 60 seconds as the surfaces are alternately exposed to the outgoing gas and incoming air streams [98, 99]. Heat energy is captured and transferred to the incoming air for combustion before it is lost to the stack. The result is a substantial saving in fuel that would otherwise be required to bring the air up to combustion temperature [100].

Until now, such regenerators were designed to rotate steadily and continuously, being used mainly for gases near atmospheric pressure. They are used in boilers as air preheaters, in flue gas desulphurization plants as gas/gas heaters and in tail-end selective catalytic reduction NOx plants as gas pre-heaters. They are both designed and built to operate over extended periods with durable and uninterrupted service. Simplicity of design also makes them easy and economical to maintain while in operation and at scheduled outages [101, 102]. The majority of air pre-heaters in use in the power industry today are of the Ljungström type. In 1995, the American Society of Mechanical Engineers (ASME) declared them to be a "Mechanical Engineering Landmark".

When the exhaust gases leave each one of the air heaters in this installation, do not go directly to the stack, but first pass through the respective mechanical precipitators where the ash and cinders in the gases are retained by means of centrifugal forces created by settling cyclones, and collected later by a deposition tank through a deposition valve, as can be seen in figure 11a.

Following a thorough cleaning of the mechanical precipitators and a subsequent period of operation, in which a study was made of the pressure differences in the gas conduits operating under nominal load, a major obstruction was detected in just one of the precipitators. A subsequent inspection showed that a considerable amount of ash had been deposited on its inlet conduits, as can be seen in figure 11b, which was reducing the amount of incoming gas by 60 %. This is clearly an atypical circumstance, without precedent at all in the incident records of thermoelectric power plants.

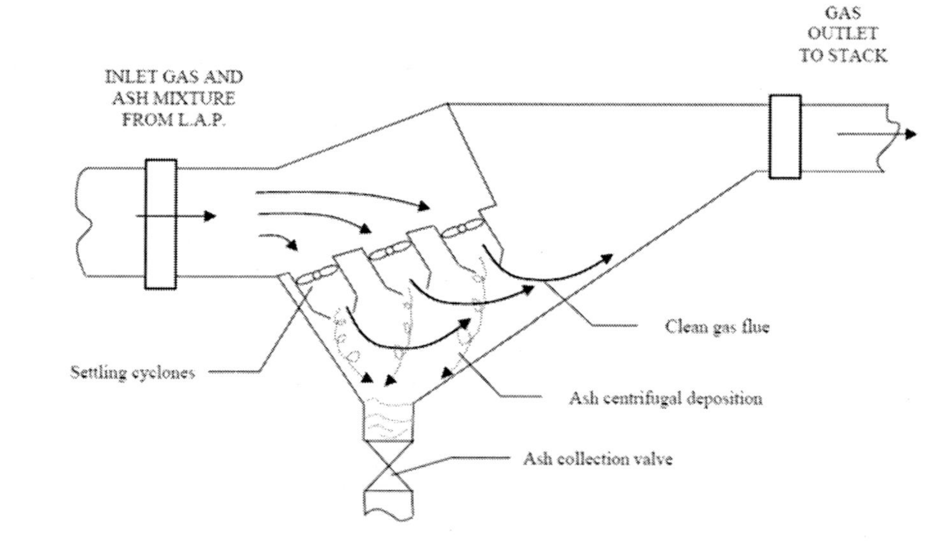

a) Nornal conditions L.A.P. (A)

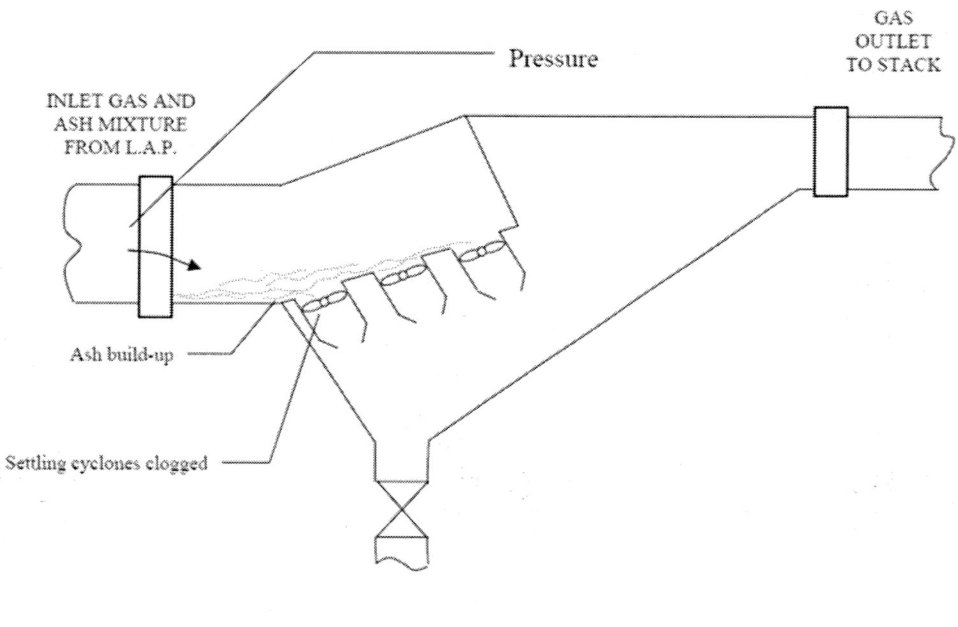

b) Clogging L.A.P. (A)

Figure 11. A two steps description of the clogging process for L.A.P. "A".

Furthermore, a comparison of the available data for the two most recent hopper cleaning operations showed that, on the one hand, the amount of ash collected by the precipitator that had become clogged was three-times less than that collected by the other and, on the other, the temperature of the ash held in the hoppers for the blocked one was significantly lower than the temperature of the ashes collected by the other one. This latter circumstance is directly related to the air and gas, inlet and outlet temperatures, as we shall see throughout this paper.

2.2. Aims and Methodology

Once the operating problem described in the preceding section has been roughly identified, the aim of this paper is to carry out a thorough study of the installation's operation in order to explain it in an effective way. Thus, there are sufficient background data to assess that, with the power plant operating at nominal load (480 MW) and the *LAP's* baskets (heat transfer surfaces) in perfect conditions, the relative pressure in the boiler furnace never should exceeds a value of 600 millimetres of water column, which in turn means heater inlet pressures of below 300 millimetres of water column, and never exceeding 350 millimetres of water column, with extremely dirty heaters due to the accumulation of waste from the exhaust gases.

It is important to note that immediately after the conduits were cleaned, normal values for pressure were reached. Anyway as a consequence of these cleaning operations, certain damage was caused to the boiler dome (top structure). This signalled a turning-point in the trends for the pressure differences in the gas conduits, as from that moment on, the pressure in the gas inlet on the (*LAP "A"*), reached a value of about 425 millimetres of water column, together with a furnace pressure of 730 millimetres of water column, so the power plant's effective output was reduced to 83 % of its rated figure.

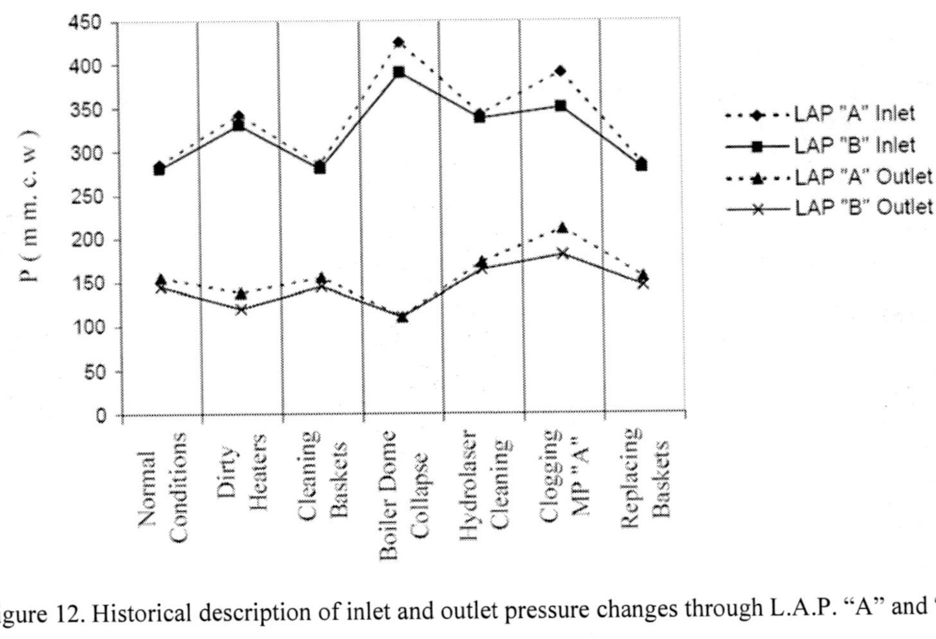

Figure 12. Historical description of inlet and outlet pressure changes through L.A.P. "A" and "B".

A more thorough cleaning was subsequently undertaken by means of a hydro laser system, involving a jet of water and perforation under rotation using bits that turn at between

1,500 and 2,000 rpm, whereby all deposits were removed without damaging the conduits at all. It was noted that the inlet pressures on both heaters were reduced significantly but clearly exceeded 300 millimetres of water column so it means that original values were not re-established in this way.

Even after a short operating period of time, a new and dangerous increasing of the inlet pressure was detected in (*LAP"A"*), as its mechanical precipitator becomes definitely clogged. A final decision of changing baskets was finally assumed, so as soon as the baskets were finally replaced, the normal values for pressure were again reached. All this time dependent process can be clearly seen in figure 12 where a historical description of inlet and outlet pressure changes through both units *"A"* and *"B"* was depicted.

This led us to suspect that the reason for the clogging was not physical, but due to some other causes. The first consideration related to the loss of power was to take into account the air flow inlet towards exit gases across the seal system of our *LAP*, which after measured only supposed less than 6 % of the total air flow, which was carefully certified for both units according to the historical available data from the power plant, so the conclusion was that it did not reveal much importance. A theoretical study was taken into account in order to explain the phenomena that gave rise to such anomalies above explained. A heat transfer survey, both in the heaters and conduits was carried out in order to ascertain the reasons for the loss of power already tested in one of them, as it was suspected that its origin might lie with the increasing of the heat transfer process inside the heater because of the turbulent flow; and a further result of it will address the accumulation of ash in the conduits of the mechanical precipitators.

3.3. Theoretical Time Dependent Study of the Heat Transfer Process

The heat transfer between the air and gas areas in the *LAP* responds to the case of "walls undergoing periodic changes in surface temperatures" [103]. The time dependent temperature variations on a surface inside the wall (metal plates) is calculated on the basis of the "equation on one-dimensional conduction in transient state" [104].

$$a \cdot \frac{\partial^2 \theta}{\partial x^2} = \frac{\partial \theta}{\partial t} \tag{55}$$

It is shown [105] that the solution to the expression is provided by:

$$(\theta - \theta_m) = (\theta_{máx} - \theta_m) \cdot e^{-x\sqrt{\frac{\pi}{aT}}} \cdot cos\left[x \cdot \sqrt{\frac{\pi}{a \cdot T}} - \frac{2\pi}{T} \cdot t\right] \tag{56}$$

This equation corresponds to a sinusoidal-type function with exponential decay, whereby we can define the following parameters:

$$A = (\theta_{máx} - \theta_m) \cdot e^{-x \cdot \sqrt{\frac{\pi}{aT}}} \tag{57}$$

$$F_a = e^{-x\cdot\sqrt{\frac{\pi}{aT}}} \tag{58}$$

$$F = \left[x\cdot\sqrt{\frac{\pi}{a\cdot T}} - \frac{2\pi}{T}\cdot t\right] \tag{59}$$

$$\Lambda = 2\cdot\sqrt{\pi\cdot a\cdot T} \tag{60}$$

The function decays, around an average temperature, as x increases, whereby its time dependent graphic representation is shown in figure 13.

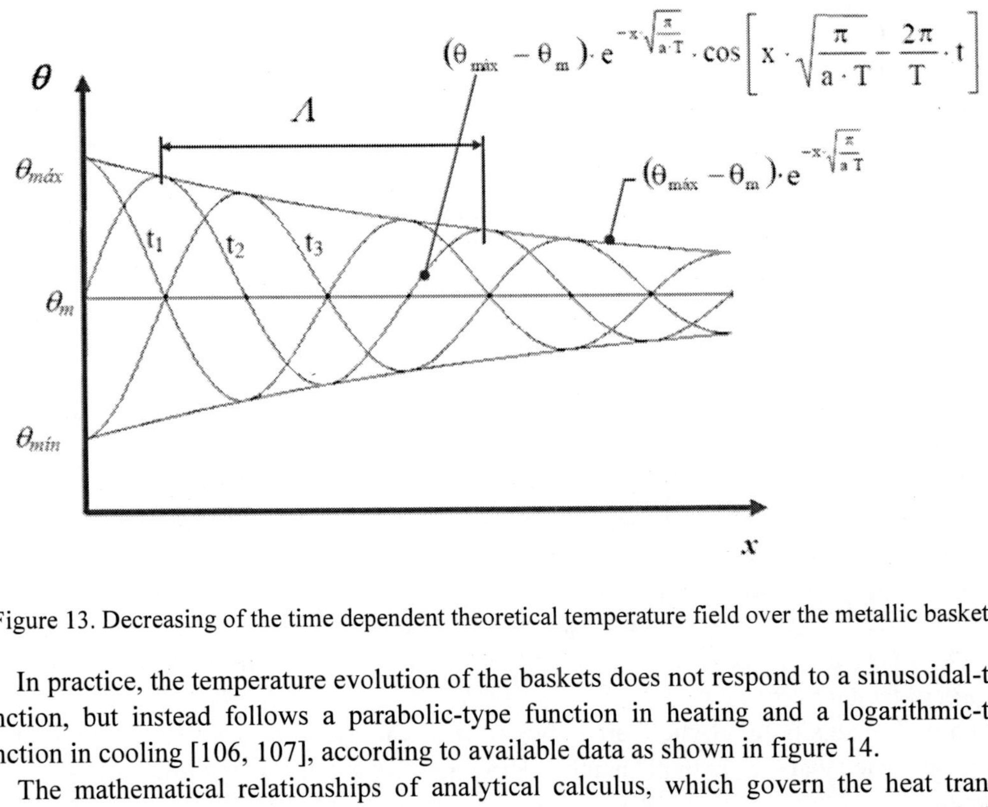

Figure 13. Decreasing of the time dependent theoretical temperature field over the metallic baskets.

In practice, the temperature evolution of the baskets does not respond to a sinusoidal-type function, but instead follows a parabolic-type function in heating and a logarithmic-type function in cooling [106, 107], according to available data as shown in figure 14.

The mathematical relationships of analytical calculus, which govern the heat transfer between the air and gas areas, are extremely complex in the real case of a *LAP*; nevertheless, there are approximate calculation methods. The most appropriate is the following expression [108].

$$\frac{q}{A} = \frac{\Delta\theta m / (\varphi_g - \varphi_a)}{\frac{1}{h_g\cdot\varphi_g} + \frac{1}{h_a\cdot\varphi_a} + \frac{1}{2{,}5\cdot Cp_s\cdot\rho_s\cdot r_B} + \frac{r_B}{k\cdot(\varphi_g + \varphi_a)}} \tag{61}$$

The third term in the denominator considers the accumulation of heat between the plates (thermal inertia) and the fourth term is a correction factor regarding conductivity. When these two latter terms are insignificant, with regard to the first two, this equation is reduced to that of a heat exchanger operating in steady state. Accordingly, the higher the coefficient for film *"h"*, the greater the heat transfers by convection [109, 110].

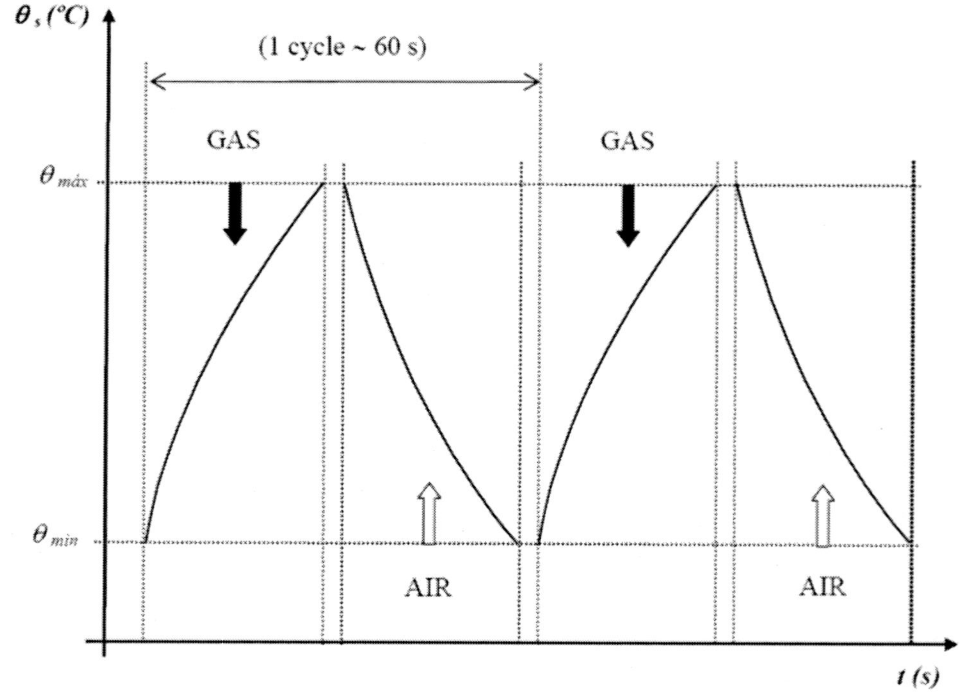

Figure 14. Time dependent empirical surface temperature field over the metallic baskets.

The determination of the film coefficients, in the gas and air areas respectively in the baskets, responds to the typical case of forced convection along flat surfaces [111] (The Reynolds number, in both the air and gas areas, clearly exceeds the value 5 10^5, which for gases constitutes the critical value for the transition from laminar to turbulent flow [112]). Nonetheless, it should be taken into account that there must be a certain area of laminar flow at the beginning of the metallic plates, before the critical Reynolds number is reached, with the onset of the turbulent flow.

With all of this in mind, the more appropriate calculation expression [101] could be as follows, provided that the "Prandtl" number does not differ greatly from unity:

$$N_{NU} = \frac{h \cdot L}{k} = (N_{PR})^{1/3} \cdot \left[(N_{RE})^{0,8} - 23.100 \right] \tag{62}$$

In order to facilitate the presentation to the reader, it will be accepted that the heat transferred from the heater to the combustion air has the same value as the heat transferred from the gases to the heater, which is tantamount to considering a lossless heater, according to the previous consideration in this way. In such a case, the mass and energy balance in the heater corresponds to the following expression [113, 114]:

$$\dot{Q} = \dot{m}_a \cdot c_{p(a)} \cdot \left(\theta_a^o - \theta_a^i \right) = \dot{m}_g \cdot c_{p(g)} \cdot \left(\theta_g^o - \theta_g^i \right) = U \cdot S \cdot \Delta\theta_m \tag{63}$$

Considering a nominal combustion air flow of 1.457 t/h and an output of reference for the thermoelectric power plant of 480 MW, it was noted that in the last cleaning operations that the temperature of the ashes held in the hoppers of mechanical precipitator *"A"* was significantly lower than the temperature of the ashes contained in the hoppers of precipitator *"B"*. In these working conditions (normal behaviour and clogging situation respectively), available data measured at inlet and outlet for both heaters in an ambient temperature of 26 °C, are shown in table 7.

Table 7. Comparative inlet and outlet temperatures of air and gases respectively for both air heaters

θ (°C)	Normal conditions		Clogging situation	
	LAP "A"	LAP "B"	LAP "A"	LAP "B"
Air inlet	34.80	35.90	37	37
Air outlet	271	235	286	233
Gas inlet	340	340	337	348
Gas outlet	150	150	91	168

This drop in temperature at the gas outlet and the respective increase in the air outlet temperature on *LAP "A"* for the so called "clogging situation", can only be due to a significant increase in the value of the film coefficients for the gases and air respectively.

Accordingly, the Prandtl numbers of the fluids, both from the air and gases in both heaters are also similar, given that they are located between common sources and heat sinks. Consequently, the only reasonable explanation for the increase in film coefficients is that the Reynolds numbers for *LAP "A"*, in both gas and air areas, have reached significantly higher values than those corresponding to *LAP "B"*, and this can only be due to a simultaneous increase in air and gas flows in the former.

In addition, it seems reasonable to consider that the significant drop in the flow velocity of both air and gases is due to a considerable drop in load [115] between the metal plates on the baskets on *LAP "A"*. This may be due to two circumstances whose effects are compounded:

1. On the one hand, to the clogging of the open spaces between plates due to the unexpected destruction of the boiler top (dome).
2. On the other (considering the long time elapsed since the last inspection of the heater baskets), to the partial destruction by acid corrosion [116, 117] of the metal plates on these baskets (formation of acid dew point, *ADP*), that could cause the displacement of the plates when cleaning with a hydrolaser, with the subsequent overlapping of some plates on top of each other, deforming the passages between them, which clearly magnified the formation of turbulent flow in those passages [118], with the ensuing increase in loss of power.

Subsequently, and on the basis of this study, the baskets on *LAP "A"* were carefully examined, which has indeed been showed that the expectations outlined here have been readily fulfilled, regarding clogging between metal plates, proceeding this way to repair them immediately.

3.4. Analysis of the Ash Accumulation Inside

When the exhaust gases leave the LAP, they are fed into the mechanical precipitators, where the ashes are separated out from the gases by the effect of the centrifugal forces they exert, and from here the "cleaned" gases are finally fed to the stack, as it was previously shown in figure 3a. We shall now proceed to analyse the accumulation of ash in the area between the gas outlet on LAP "A" and its corresponding mechanical precipitator, with the subsequent entrainment current and clogging of the precipitator hoppers, as it was previously shown in figure 3b. This is why it is deduced that there is a significant loss of power, together with a significant increase in furnace pressure and, therefore, also in the incoming gases on both heaters as it was previously shown in figure 12.

Whereas it can be considered in LAP "B" that the through section is free in practical terms and, therefore, the flow across the metal heat transfer plates may be considered slightly turbulent and inconsistent [119]. Nevertheless, in the case of LAP "A", the flow can be considered as turbulent, inconsistent and clearly chaotic, with the presence of "restrictions" and, consequently, the existence of a certain "lamination".

The laminate phenomenon is an irreversible process that is, in theory, carried out at constant enthalpy. Besides not producing any useful kind of energy, it is characterised by a significant increase in entropy and, consequently, by a sharp drop in "exergy" and, therefore, in the operating capacity of the circulating fluid. Yet, in addition, after the restriction there is turbulent flow in the sections close to said restriction [120, 121], with the formation of a large number of eddies or vortexes, and finally returning to turbulent flow, although slightly more consistent.

In practice, lamination with real gases cannot in any way be considered a process at constant enthalpy [122], which means taking the kinetic energy of the circulating fluid to be constant, before and after the restriction that gives rise to the lamination, and it is widely known that this is not the case. It is precisely the formation of the aforementioned turbulent flow after the restriction that renders the process utterly irreversible, and in practice the flow velocity in the section located immediately upstream is very high, and the corresponding kinetic energy will have been reached at the expense of a drop in enthalpy [123]. However, as the fluid flows through the widening of the outlet section, the velocity again drops whilst the original enthalpy is not re-established.

The confused, turbulent flow at the gas outlet on the *LAP "A"* means that there are a large number of particles in suspension, colliding against the "boundary layer" of the conduits where the flow velocity decreases significantly and, consequently, so does the velocity of the aforementioned particles in suspension. In 1904, Prandtl proposed that the study of the movement of a low viscosity fluid could be likened to that of a perfect fluid [124], except in a layer close to the boundary, of tiny thickness, which was the focus of friction phenomena, which has been called the "boundary layer". Thus, on the outside of this layer, the tangential stresses are insignificant, with the forces of inertia prevailing over those of viscosity, whereas inside the boundary layer, the proximity of the boundary means that the velocity gradient is very high and, therefore, the tangential stress is also significant; the friction forces in this situation can be considered of the same order of magnitude as the forces of inertia.

Thus, the force upholding a submerged solid in a fluid flow is given by Buckingham's "pi" theorem [125]:

$$F_s = S \cdot \rho \cdot V^2 \cdot C_s \tag{64}$$

Defining the lift coefficient as:

$$C_s = C \cdot \left(\frac{\mu}{L \cdot \rho \cdot V} \right)^b \cdot \left(\frac{E}{\rho \cdot V^2} \right)^c \tag{65}$$

when the Mach number is equal to or greater than unity, the lift coefficient may be considered solely a function of the aforementioned Mach number. Simplifying, and including the velocity height, the lift force of the suspended particles in the gases produced by combustion that flow through the gas conduits is:

$$F_s = C_s \cdot \left(\frac{\omega}{g} \right) \cdot A_p \cdot \left(\frac{V^2}{2} \right) \tag{66}$$

For a particle to remain suspended in a fluid, the condition of equilibrium in a vertical direction must be fulfilled, namely:

$$\begin{cases} \sum F_y = 0 \\ F_s - W = 0 \end{cases} \tag{67}$$

Obviously, when any particle strikes the flow's "boundary layer" its velocity is reduced significantly and therefore so is its lift force, and in the end it falls through the effect of gravity. In the case of laminar flow, analytical means can be used to obtain the velocity distribution in a straight section [126]. The velocity profile follows a law of parabolic variation, whereby the velocity can be considered zero on the conduit wall and maximum at the conduit axis, where its value could reach twice the average velocity in this section.

In the case of turbulent flows, it is not possible to obtain the velocity distribution by analytical means. Nonetheless, on the basis of experimental data provided by Nikuradse and other researchers, it is known that the velocity profile is more uniform, yet as with laminar flows, the minimum velocity is recorded close to the boundary layer and the maximum at the pipe axis. An experimental expression that provides the velocity profile for turbulent flows, on the basis of average velocity, is as follows [127]:

$$\frac{v - V}{V - \sqrt{f}} = 2 \cdot log\left(\frac{y}{r_0} \right) + 1'32 \tag{68}$$

This equation makes it clear that it would not be necessary for the particles to directly strike the boundary layer [128, 129], instead it would be enough for them to get sufficiently close to it, in order to be finally decanted.

It is reasonable to suppose that the turbulent outlet flow of the combustion gases from *LAP "A"* will lead to the renewal of suspended particles, either colliding directly with the boundary layer or getting sufficiently close to it, which will also result in a significant increase in the number of particles finally decanted.

At the end, the mass of particles decanted would be entrained by the gas flow towards the entrance to the mouth tools of the mechanical precipitators on *LAP "A"*. This entrainment effect explains why the mechanical precipitators of this heater, become clogged, even after short operating times.

3.5. Results and Discussion

The presence of high furnace pressure, with a load of 85% and even lower, denotes the existence of a serious problem in the gas conduits. Furthermore, the fact that the mechanical precipitator *"A"* registers major clogging after only short operating times confirms the previous hypothesis and, in addition, focuses the problem of this gas conduit.

Furthermore, the temperature differences in the air and gas conduits, before and after the heaters, constitute the main grounds for analysis in this study, which is based on the air-gas heat transfer in these heaters, according to the data previously shown in table 1.

The specifications of the air and gases are similar in both air heaters, as they both operate between common sources and heat sinks. However, the air leaving *LAP "A"* is hotter and the gases are significantly cooler. For this to happen, a significant increase in film *"h"* coefficients has to be considered, and the only reasonable explanation for this is that the Reynolds numbers in this heater, in both the air and gas areas, have reached values that are markedly higher than those corresponding to the other one and, in turn, this can only be due to a simultaneous increase in the velocity of air and gas flows in the former.

Accordingly, the fulfilment of the equation (63) necessarily requires a decrease in both mass flows, in air and gases alike.

The initial cause of the high loss of power in *LAP "A"* is undoubtedly the partial destruction of the boiler top (dome), with its component materials becoming detached and falling onto the heater, as this signals a turning-point in the trends for the differential pressures in both heaters. Likewise, the destruction, also partial, by acid corrosion of the metal plates on the heater baskets (formation of *ADP*), as well as the displacement of the same due to the subsequent cleaning with a hydro laser, with them overlapping each other, constitutes the ultimate cause of this pressure drop through the formation of a highly chaotic turbulent flow.

Finally, the increasing in pressure drop recorded by the gases through *LAP "A"* means that a greater number of suspended particles in the gases collide and get sufficiently close to the boundary layer of the gas conduit to lose lift force and become finally decanted. The gas flow then entrains the particle mass to the inlet mouth tools of the mechanical precipitators, clogging them up in short operating times, as we previously showed.

3.6. Conclusion

The following conclusions may be drawn from this paper:

1. The relative differences of temperature in outgoing air and combustion gases from *LAP "A"* are explained, accepting a significant increase in the flow velocity of both air and gases due to the deformation of the gas flow passages between the metal plates on the heater, with formation of inconsistent turbulent flow inside, and also allowing for laminations due to significant clogging of the spaces between the metal plates on the baskets in the aforementioned heater.
2. On the basis of this study, *LAP "A"* was finally dismantled and repaired, fully confirming the expectations held here.
3. Once repaired and under subsequent operation of the thermoelectric power station, it was verified that its load of power was once again within design limits. Naturally enough, the problem of the clogging of *MP "A"* with ash was resolved and the load limitation for high furnace pressure was corrected automatically, as can be seen at the end of figure 12.
4. This served as a prompt to examine and clean *LAP "B"*. At the time of writing, the furnace pressures at full load remain below 600 millimetres of water column and there is no limitation whatsoever on the load of the thermoelectric power plant, as shown by the readings taken for it.

4. Part C: Performance Efficiency of a Cogeneration Power Plant with Natural Gas Engines

4.1. Introduction

The performance of a cogeneration power plant is always a complex matter because such a plant supplies the amount of exhaust steam required by processes and simultaneously generates the electricity either being consumed in the factory or being exported to power supply when there is a production excess [130]. Here, the company supports a working regime consisting of four shifts and 7.680 hours / year which is suitable for conversion into a cogeneration scheme.

The factory is located at sea level, in a place where the mean temperature and relative humidity through the whole year are, respectively, around 14 °C and 82 %. Before the performance of this cogeneration project, the factory was connected to a 30 kV line with a contracted power of 350 kW. The electric energy consumed during last year was around 2.070 kWh and the yearly consumption of natural gas was about 3.200 Nm^3.

Prior to the designing process of the cogeneration power plant, a modelling of the energy load has been carried out, so a "typical day" (from the point of view of energy consumption) was established, as a starting point [130]. Starting from the fuel consumption and after assessing the outputs of the boilers, the steam load every two hours was determined.

By studying the fluctuations of those two-hour loads, we concluded that such a load profile could be adjusted with this single "typical day", although proportionally correcting each monthly load by means of its corresponding fuel consumption.

In order to suitably design a cogeneration plant, it is very important to assess accurately the heat load, because it will make the power (as well as the electricity) output conditional on that assessment [131].

4.2. Heat and Electricity Loads

Steam Loads

To determine the profile of the actual steam load, we started from the knowledge of the natural gas consumption every two hours, concerning all the boilers and applied to the most representative winter and summer months, such are February and July, respectively.

After a series of trials, we reached the conclusion that it was a random fuel consumption distribution. Nevertheless, representing the accumulated two-hour consumptions regarding each of the above-mentioned months, we can check that they are very similar. That is why we decided to choose only one "typical day" profile, representative of fuel consumptions in boilers and, therefore, the actual steam load in the factory.

Hot Water Load

To preheat the coal, a hot water flow is required with the following data:

- Outflow: 8 t/h
- Input temperature: 34 °C
- Output temperature: 75 °C
- Specific heat capacity: 0,586 kcal/kg °C

Therefore, the required thermal power turns to be about 216 kW.

4.3. Methodology

Before selecting the cogeneration system to be implemented, other possible alternatives were analyzed, taking into account the criteria as we show next. When selecting the most suitable cogeneration system, it is necessary to find out which the fittest one is and which the aim of such a project is. Sure enough, there are a lot of factors taking part, such are: property, user, electricity company, equipment and service suppliers, financial entities and, of course, the competent Public Administration.

The whole of these factors will have their prospects in view of the new project and no all of their objectives will have to match as it has been shown in [130]. Maybe, the most profitable plant either are unable to comply with the requirements established by the Administration or it turns out to be detrimental to the user due, for instance, to the fact of requiring excessive maintenance operations.

A period of 20 years is a normal lapse of life for a suitably working cogeneration plant. It is logical to view that along such a time, the energy and economy scenes will change. That is why it is essential to be sure, as far as possible, that new technologies concerning the energy sector will not affect the cogeneration plant.

Plant Capacity Criteria

Starting with the analysis of the steam electricity loads, defined in last section, this chapter goes on the study of possible cogeneration systems according to the criteria described below.

Aiming to select the most suitable cogeneration system, we started with a preliminary idea, this is the fact that the installation under study will be designed depending firstly on the thermal profile (in this case, the steam load). Therefore, electricity will be considered as a by-product, although viewing the possibility of selling the excess to power supply, so that the plant will be able to satisfy the appropriate technical and energy requirements.

Once established this first criterion, when designing the main equipments there will be taken into account that the greater number of working hours are and the greater amount of work under design conditions is, the more profitable the cogeneration plant will be.

In the current market, there are two types of thermal engines consuming natural gas as fuel: "gas turbines" and "internal combustion engines". In order to design the corresponding cycle, the following criteria have been taken:

- One of the boilers will be used as support when the requirements of steam cannot be matched by the corresponding production.
- The cogeneration system will work as a full load configuration for the whole year.

Economy and Working Criteria

The true objective of any cogeneration system is the increasing of industrial competitiveness. So, the savings and profits must be according to the economical environment of the property.

The selected alternatives offer in any case greater electrical power than the corresponding requirements, so then always become exporters. In this way, the cogeneration plant turns to be able to face up to emergency situations when supply failures occur, what makes much simpler its design.

Technological Criteria

The technological level of a cogeneration plant basically depends on the main engine. The selection of this element must be carried out according to thermal requirements and depends on the kind of fuel to be used. Concerning these aspects we know that the thermal load mainly consists in saturated steam at 7 bar pressure and that the fuel is natural gas.

4.4. Alternatives

Two alternatives have been analyzed in this paper: Internal Combustion Engines and Gas Turbines.

Table 8. Energy data of groups engine-alternator

Alternative	Number of engines	Electric power per unit *(kW)*	Exhaust gases themal power per unit *(kW)*	Water thermal power per unit *(kW)*	Thermal consumption per unit *(kW)*	Speed *(r.p.m.)*	η *(%)*
M1	2	2.100	1.758	1.200	5.250	1.000	40
M2	1	4.860	3.510	2.437	11.627	750	41,8
M3	2	1.696	1.374	997	4.484	1.500	37,8
M4	2	1.540	1.463	621	4.043	1.000	38,1

Alternative 1: Internal Combustion Engines

Electrical efficiencies of the *ICE* are greater than those of the *GT* and, besides, the heat from the cooling system can be used. Data are:

- Post cooler in supercharged engines: 40 ºC - 50 ºC
- Heat from lubricating oil system: 60 ºC - 70 ºC
- Heat from cylinders cooling: 80 ºC - 90 ºC
- Exhaust gases: 400 ºC - 500 ºC

In order to provide the cogeneration plant with a greater degree of flexibility and to increase its reliability, it is very easy to set out several engines in parallel, because such a plant can be adapted without difficulty to a hypothetical rise of the loads, just including new engines.

Gas engines, when belonging to the group of natural draft, require gas supply at low pressure (about 500 mm. w. c); when supercharged, they have to work under a relative pressure of 1.5 bar. This fact means an interesting advantage compared to the turbines, since these ones require a gas supply above 14 bar.

The ICE presents a greater maintenance degree than GT, mainly when referring to the cost of oil consumption. Nevertheless, ICE are devices better known than GT, so their maintenance can be accomplished by less specialized staff. Table 8 shows the main features of the engines we have studied focus on energy aspects.

Concerning recovery boilers, we have rejected those supporting post-combustion, since the oxygen contents in the exhaust gases of the engine, in any case, turns out to be less than 12 % in volume, what renders compulsory to mix the exhaust gases flow with atmospheric air. This increases the cost and the complexity of the boiler, so it is not advisable the inclusion of a recovery boiler with post combustion to take advantage of the thermal power of the exhaust gases of an ICE.

In table 9 the main characteristics of different boilers involved in alternative M1 (Gas mass flow: 8.28 kg/s; Input gas temperature: 410 ºC), are shown.

Table 9. Characteristics of boilers for engines involved in alternative *M1*

Alternative	Tipe of boyler	Exhaust gases temperature (º C)	Steam flow ($kg\ h^{-1}$)
CR-1	Water-tube	132	3.650
CR-2	Fire-tube	150	3.445
CR-3	Water-tube	140	3.590

Alternative 2: Gas Turbines

Waste heat in *GT* is always related to exhaust gases, whose temperature rises up to 500 ºC and energy flow means about 60 % of the fuel energy. Heat contents of those gases will be used to produce steam through a recovery boiler. Due to the fact that combustion process is pressurized, fuel must be injected into the combustion chamber at a pressure between 12 and

20 bar. Since the pressure of the gas supply is usually lower, a gas compressor is often required.

Environmental factors have a great influence on the efficiency and the output power of *GT*. Particularly, air temperature and head losses at admission and exhaust ducts are especially sensible parameters. The exhaust gases in a *GT* contain a high percentage of oxygen (15 %), what turn them to become reusable as a combustible in a later combustion process. Thus, the ratio heat/electricity can be increased. In this way, the steam load profile can be adjusted by modulating the power of burners. One of the alternatives studied considers post-combustion by means of an air-blast burner. Table 10 shows the main features of the *GT* analyzed here.

Table 10. General data for the *GT*

Alternative	Electric power per unit *(kW)*	Thermal consumption per unit *(kW)*	η *(%)*	Gas flow *(kg s^{-1})*	Exhaust gases temperature *(° C)*
T1	1.069	4.150	25,8	5,5	505
T2	1.069	4.150	25,8	5,5	505
T3	1.174	4.916	23,9	6,4	513

4.5. Definition of the Auxiliary Equipment

Apart from these main equipment above mentioned, there are other auxiliary ones such are:

- Water supply line.
- Steam line.
- Compressed air line.
- Natural gas layout.
- Electrical installation.
- Control and automation system.
- Civil works.

The "water supply line" includes a series of devices such as pumps, pipes, valves and so on from the actual system for water treatment up to the new recovery boiler.

The "steam line" covers the stretch from the output of the recovery boiler up to the collector of the actual boilers. It will suitably allow heat outflow and will be provided with adequate instruments and other auxiliary devices. The "compressed air line" starts from the actual air mains, being necessary the installation of pipes and suitable auxiliary devices in order to make the air flow towards all the final consumption points. The "Natural gas layout" covers the stretch from input section of the gas mains up to zone of supply to the engines [131]. This line consists of:

- Outer input section.
- *MCS*, able to comply with requirements of the whole factory or a stand-alone controller for the engines.

The "Electrical installation" must include:

- Generation equipment with its respective protection and connection systems that includes generator with their protection boxes and cells containing automatic connection switches and other auxiliary elements. Electrical generators must be synchronous at the generation voltage of 6.000 V, 50 Hz, with a power factor of 0.8.
- Power converter and suitable devices to adequate the actual installation. It covers from the power converter up to the connection to power supply. The power converter is used to raise generation voltage up to that of power supply. The voltage ratio is 30/6 kV.

Besides the instruments supplied together with the main equipment, we shall find additional instruments, by-passes and so on; special attention must be paid to the control system to be integrated in that one already existing in the plant. Regarding the engines, the manufacturer supplies a standard package including control logics, panel, field transmissions as well as the final elements. The "control and automation system", which forms also part of that supply, consists of:

- Start/stop system, synchronism, load control, etc.
- Supervision system.
- Protection and alarm system.

Concerning the boilers, is usual that the manufacturer supplies another standard package including the protection logic.

A centralized management system which would execute control functions when required has been foreseen, as well as the man-machine interface and the basic data regulation. It counts on a series of software packages aiming to control and optimize the overall efficiency of the system. At the same time, it will allow to control the efficiency of the main equipment by supplying information to quickly identify different problems as well as the equipment badly working, so making possible to accomplish a planning to fix the corresponding device and basically to maintain the installation.

Finally, the "civil works" to be fulfilled in this installation are:

- Adaptation of the building providing space for main equipments of the cogeneration plant, such is engines, alternators and recovery boiler.
- Piping networks for gas, steam, etc.
- Installation of the converter and the air cooling units.

4.6. Energy Analysis

Once the heat load modelling has been performed, the integration of the cogeneration process within the factory production has been achieved and the starting premises have been established, the energy flows obtained from the different proposed alternatives had to be simulated [132]. By using the thermal load profiles, the best possible modelling was carried out; as a result, the thermal balance could be set, including the assessment of fuel consumptions, the output of the engine/turbine according to the hourly temperature, the mass and energy balances [133] in the recovery boiler, etc.

Beginning with the energy balances every two hours referred to fuel consumption and electricity production, the accumulated both monthly and yearly values were obtained. The estimation of the *EEE* was performed for each alternative. In accordance with the Spanish laws, any alternative must match that value as a necessary condition to be considered feasible from a technical point of view. This requirement turns out to be one of the main determining factors restricting the size and working hours for the plant.

Energy Balance

The performance of the energy modelling has been carried out for each alternative, assessing this way the fuel consumptions and electricity productions. For this, the steam and electricity load profiles have been taken into account. In this way, the energy balances [134] have been established every two hours.

In accordance with the SSR, concerning "Production of electric power by hydraulic, cogeneration and other installations supplied with renewable energies", it is compulsory for such an installation to be registered under the special regime, to render an EEE greater than a certain value formerly established for the above-mentioned document.

In particular, when referring to gas natural engines, the condition to be fulfilled is that EEE > 0.55. The value of *EEE* is obtained through the Eq. (69):

$$EEE = \frac{E}{Q - \dfrac{V}{0,9}} \tag{69}$$

For alternative M1, using the warranted values as performance data for the engines, we get:

$$E = 31,783\ MWh,\ Q = 79,457\ MWh$$

and $V = 20,240\ MWh$, what yields: $EEE = 0,558\ (> 0,55)$.

There are also other indexes referring to energy saving degree which can be useful too. They allow a better assessment of the efficiency for each alternative. Such is, for instance, the case of the term ΔE, which for this alternative M1 is shown in Eq. (70):

$$\Delta E = \left(\frac{E}{0,32} + \frac{V}{0,9} \right) - Q = 42.353\ MWh \tag{70}$$

and the *IE* term is shown in Eq. (71)

$$IE = \left(\frac{\Delta E}{\frac{E}{0{,}32} + \frac{V}{0{,}9}} \right) \cdot 100 = 34{,}8\ \% \tag{71}$$

Finally, the term *R*, is shown in Eq. (72):

$$R = \frac{E + V}{Q} = 65{,}3\ \% \tag{72}$$

Summary of the Energy Assessment

The energy assessment of this cogeneration alternative compared with the reference situation (with no cogeneration scheme), is summarized in table 11.

Table 11. Summary of energy assesment for alternative *M1*

Item	Reference situation *(MWh)*	Cogeneration *(MWh)*
1. Fuel consumption.		
- Gas engines:	--	79.457
- Boyler:	22.206	784
2. Electric consumption	3.294	3.294
3. Self-generation:	--	31.783
4. Energy request.		
- Natural gas:	22.206	80.241
- Electricity:	3.294	3.060
5. Energy sales.		
- Electricity:	---	28.489
6. Gas consumption:	---	58.035

4.7. Economical Analysis

Regardless of the method used to assess the economical interest of an investment in cogeneration, it is always necessary to determine the expected saving [135]. This saving is a consequence of the change we intend to achieve, from the conventional installation to the cogeneration system. When assessing the new option we must be aware of the existence of a power supply (from which the plant can import the deficit of self-generated electricity and to which the system can also export the excess) and also the existence of auxiliary boilers ensuring the thermal supply when load peaks occur.

When analyzing an investment project [136] we have to foresee the behaviour of different parameters, such as energy prices, operation and maintenance costs, taxes, money cost, residual value of the investment, etc. In this way, four return indexes, usually handled, have been selected, such are:

- *SRT*, which measures the required period of time to recover the investment, without considering updating aspects.
- *URT*, which shows the lapse of time for the updated net flow of liquid assets, generated by the investment, to be able to refund the money initially spent.
- *NAV*, which indicates the updated excess in liquid assets flow with reference to the whole invested capital.
- *IRR*, which is the rate yielding a *NAV* equal to zero.

Initial Hypothesis

To deal with such an analysis, the following hypotheses have been considered:

- The capital investment is done during the first period, which corresponds to the so-called "1st year".
- The replacement cost period covers up six years and is considered to be lineal, what means, yearly speaking, at least the sixth of the difference between current and remaining values.
- The time unit is one year and a time horizon of ten years is viewed. At the end of the tenth year the remaining value becomes zero.
- To estimate *NAV*, different updating rates have been used. It is actually difficult to calculate accurately its value in this project, since it refers more to the opportunity cost rather than to the money cost.

Aiming to assess the effects on the value of these indexes due to the different suppositions bearing on the profitability of the project (so detecting the most critical ones), we start next section carrying out a special economy study.

Assessment of the Yearly Expected Saving

Saving is the result of a complex interaction between the heat and electricity loads, fuel and electric energy costs, types and performances of the main equipment, operation modes, etc. Our objective is the determination of that saving in the most accurate way as possible. This is the precise calculation of the cost difference between "before" and "after" the implementation of the cogeneration system.

Consumptions and Productions Estimation

Beginning with the load profiles corresponding to required heat and electricity and bearing in mind the characteristic curves of each one of the main equipment, the two-hour values as well as both the monthly and yearly accumulated values have calculated, concerning the following items:

- Fuel consumption in engines/turbines.
- Fuel consumption in auxiliary boiler.
- Fuel consumption in post combustor.
- Electricity generated in valley, plain and peak hours.
- Electricity exported to the power supply (sold) in those same periods.

Thermal and Electricity Bill

Gas natural has been charged at standard rate, established by the "Euskadi Gas Co.". The price considered has been got from the *SSR*. We assess the fuel cost required to satisfy the heat load in the reference situation (without cogeneration). When cogeneration exists, the thermal bill through both the energy and economy balances for each alternative was estimated.

With reference to this, electric energy has been charged at current rate and taken into account the *SSR* about self-generation. For each alternative, the contracted power has been kept and the guaranteed electric power available to export is also set. During the first year, electricity selling prices were:

- Power term: 10,734 € $kW^{-1}.month^{-1}$.
- Energy term: 0,048 € kWh^{-1}.

In the reference situation (without cogeneration) the cost of the electricity to be supplied by the company is also assessed. By means of the energy and economy balances, the electric bill can be obtained with the cogeneration system already established.

Recent normative introduce a coefficient to be applied to the excess of electric energy being sold. The value of this coefficient varies depending on the year (from the first to the fifth one). As a result of this, yearly savings are not the same for each one of the first five years of working life of the installation.

Working and Maintenance Costs

Working costs included in our study are only viewed for turbines and engines. Because of a lack of maintenance offers at this moment, a mean value of 0.0042 € kWh^{-1} in the case of *GT* and 0.0084 € kWh^{-1} for *ICE*, have been estimated.

Investment

The whole investment corresponding to the "complete and ready" project amounts to 3.000.000 € (three million Euros). This value covers the following items:

- Engines and recovery boiler.
- Electrical equipment.
- Civil works.
- Auxiliary mechanical equipments.
- Fuel system.
- Instruments and control system.
- Engineering and works charging and,
- Assembly works.

4.8. Results and Discussion

After having studied several alternatives, the so-called M1 has finally chosen because of the fact that it issues in a greater return on investment and a better energy use. Figure 15 shows a simplified Sankey diagram corresponding to this alternative.

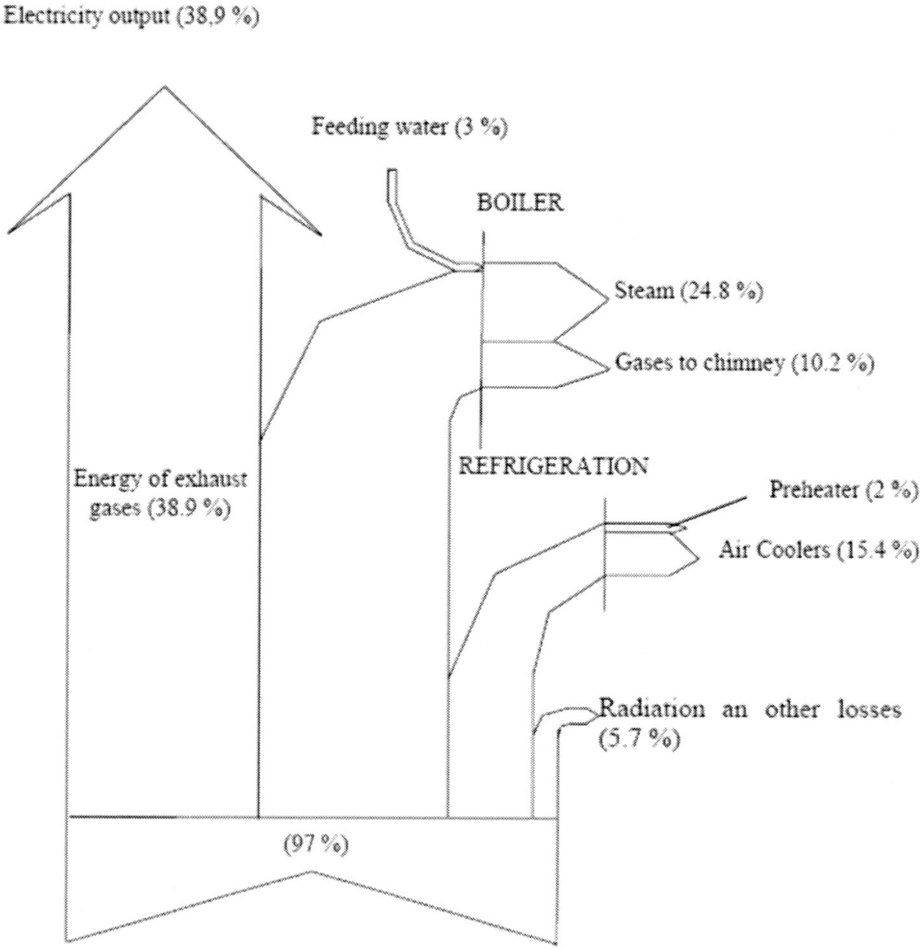

Figure 15. Sankey diagram of the alternative M1.

Besides, figure 16 represents a simple scheme of this installation with the most important items.

Finally, through the table 12, it is possible to establish a comparison regarding the investment-saving relationship.

Table 12. Comparative final results

Alternative	Investment *(MM€)*	1st year saving *(MM€)*	SRT *(Years)*	NAV *(MM€)*	IRR *(%)*
M1	3,24	1,14	2,86	1,48	21,2
M2	3,89	1,32	3,09	1,57	20,0
M3	2,78	0,96	3,00	1,23	20,9
M4	3,32	0,85	4,23	0,45	13,5
T1	1,82	0,31	6,33	--	6,8
T2	1,89	0,32	6,47	--	6,5

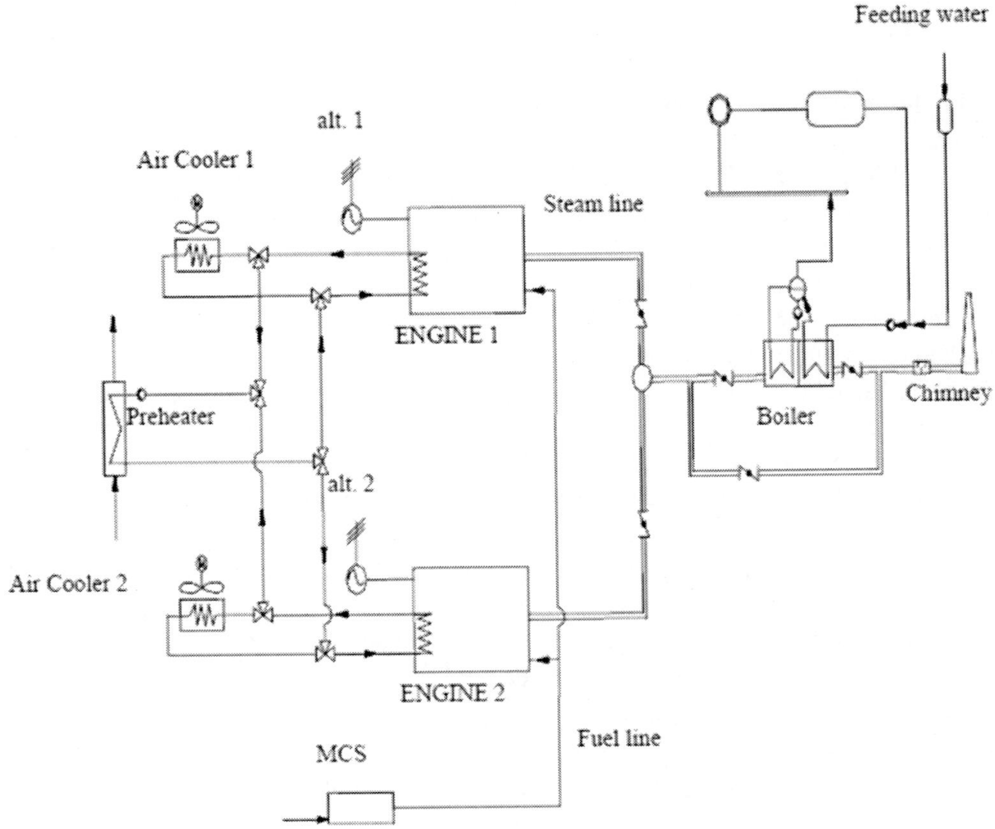

Figure 16. Installation scheme of the alternative M1.

4.9. Conclusion

1. The alternatives involving *GT*, despite their better *EEE*, issues in worse returns, rendering cost/power ratios of the order of 1,600 - 1,800 € kW^{-1}. Besides, steam requirements are low, so *GT* turning out to be too small.
2. On the contrary, when using ICE, although their EEE values are lower and even close to the minimum required, the corresponding returns become very attractive. Only alternative M4 offers worse results than the others due to the extra investment caused by its dual type engines.

5. Part D: Optimising the Process for Emptying the CO_2 Bottles in High-Power Generators

5.1. Introduction

The use of a hydrogen atmosphere (H_2) for cooling power producing generators is commonplace in large power plants, given that its cooling capacity is far better than air, whereby the size of these alternators can be reduced considerably for the same target power. The process for both filling and emptying the H_2 in the alternators is necessarily performed in an inert atmosphere. Operating experience has shown that carbon dioxide (CO_2) is the ideal fluid for this purpose, given that its molecular weight is considerably higher than that of hydrogen.

Each bottle delivers a load of around 30 kg of CO_2, which under normal conditions (273 K and 1 atm.) corresponds to a volume of about 16 Nm^3. It may be assumed that under these conditions approximately 90% of the contents are in liquid state. Thus, the process of discharging the bottle involves a continuous vaporisation of the liquid whereby the temperature inside the bottle will gradually fall unless a thermodynamic equilibrium is established around the bottle [137].

In accordance with the instructions of a major generator manufacturer as is "General Electric", use should only be made of CO_2 bottles that unload in the vapour phase. Bottles fitted with a siphon, and which therefore unload liquid and require a vaporiser, are not to be used with their generators [138 ,139]. During the operations for sweeping the generators with air or H_2, cooling the CO_2 bottles in the natural flow of ambient air, it has repeatedly been shown that thermodynamic equilibrium is not attained, leading to a rapid cooling of the bottles to the freezing point of CO_2 (-56.61 °C), with the formation of what is referred to as "dry ice", which obviously interrupts the flow. Thus, the unloading capacity is below 40% of the contents. In other words, the carbon dioxide bottles are rendered useless even though their useful load exceeds 60% of their capacity.

This involves a significant loss of resources. In particular, the operations for filling H_2 are usually undertaken during the power plant's start-up procedures, when staff has to attend to both the complex start-up procedure and the filling of H_2. This makes it essential to optimise the unloading time for CO_2 bottles. When this is compounded by the cost of the underused bottles, it seems reasonable to consider the introduction of the necessary improvements to stop them freezing during the operations for gas-sweeping the electrical generators.

Furthermore, a rapid application of CO_2 to the electrical generators may give rise to turbulent flow, and if this occurs, despite the huge difference in molecular mass between the fluids, the gas will become mixed [140] and then both sweeping times and the consumption of CO_2 bottles will far exceed the optimum.

This study describes the study and implementation of a solution that optimises the emptying of bottles by solving the problem of CO_2 freezing, thereby making a considerable saving through the better exploitation of the resources available that will also mean a reduced environmental impact [141,142].

5.2. Aims and Methodology

The purpose of this work is to analyse the operating procedures for traditional systems of applying CO_2 in gaseous state to the alternators in power plants, as we have just explained, with a detail of the appropriate actions proposed for minimising the cost regarding both the times involved in the operating sequences for filling and emptying and the financial resources associated with such operations. The study has involved two generator units (unit 1 and unit 2) in a power plant referred to as "reference thermal power plant", whereby a calculation will be made of the number of bottles required and a theoretical study will be conducted on the process for cooling their interior during the unloading process, and finally proposing a solution that will stop any freezing.

5.3. Description of the Generator Loading and Unloading Process

Both the loading and unloading processes are undertaken in 2 stages as can be seen in diagrammatic form in figure 17.

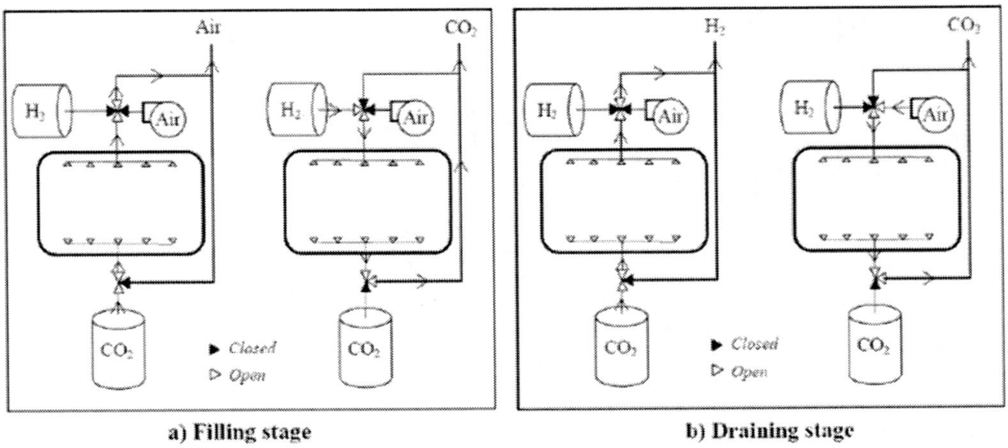

Figure 17. Process of filling and draining of the alternator.

Regarding the filling process, the first stage involves introducing CO_2 through the lower part of the generator, removing the air through its upper part, until the concentration of air in CO_2 is less than 5%. The second stage involves introducing H_2 through the upper part of the generator, whilst CO_2 is removed through its lower part, until the concentration of hydrogen in CO_2 exceeds 97%.

Regarding the emptying process, the first stage involves introducing CO_2 through the lower part of the generator, whilst simultaneously removing H_2 through its upper part, until the concentration of H_2 in CO_2 is less than 5%. The second stage involves introducing air through the upper part of the generator, whilst CO_2 is removed through its lower part, until the concentration of air in CO_2 exceeds 95%.

5.4. Cooling the Inside of the Bottles during the Unloading Process

In order to avoid the probable abrupt cooling of the coils in the electrical generator, which would pose a clear risk of detachment of their insulating varnish, and in the light of on-site experience, it was noted that the maximum flow of CO_2 application is somewhere between 3.5 and 3.7 m^3/min.

Under ambient conditions, the carbon dioxide inside the bottle is in a state of saturation. Thus, for example, and as we have also mentioned earlier, at a temperature of 21 °C, around 90% of the contents of the bottle is in a liquid state. If we proceed to unload the bottle under these conditions, the gas removed will be replaced by a continuous vaporisation of the liquid, which will lead to cooling during the process as a result of the gas flow and latent heat of vaporisation [143]. Table 13 shows the latent heat of vaporisation as well as the other properties of CO_2 for several temperatures.

Table 13. Vaporization latent heat of saturated carbon dioxide for diverse temperatures

T (K)	h_f (kJ/kg)	h_{fg} (kJ/kg)	h_g (kJ/kg)
216.54	0.0	351.87	351.87
220	7.0	346.12	353.12
225	17.52	337.21	354.73
230	28.07	328.02	356.09
235	38.52	318.67	357.19
240	48.90	309.10	358.00
245	59.32	299.16	358.48
250	69.85	288.74	358.59
255	80.57	277.72	358.29
260	91.49	266.04	357.53
265	102.66	253.57	356.23
270	114.16	240.13	354.29
275	126.12	225.45	351.57
280	138.72	209.16	347.88
285	152.21	190.69	342.90
290	166.95	169.14	336.09
295	183.67	142.61	326.28
300	205.31	104.89	309.95
304.21	257.31	0.00	257.31

Courtesy of Mechanical Engineering Department; Stanford University.

Thus, considering a temperature inside the bottle of 295 K (21.85 °C) with a corresponding pressure of 59.77 bar, the loss of heat for each kg of CO_2 that turns from liquid into gas will be:

$$h_{fg} = h_f - h_g = 142.61\ kJ/kg \tag{73}$$

This cooling of the carbon dioxide inside the bottle needs to be offset outside the bottle in order to maintain a thermodynamic equilibrium. Otherwise, as tends to be the norm in current practice, vaporisation will take place within the bottle at a temperature below the previous one and consequently, as it is in saturation, also at a lower pressure, and then the enthalpy gap will be greater and so therefore will be the cooling by mass unit [144, 145].

Therefore, considering a lower temperature, this loss of heat for each kg that evolves will be greater. If we continue dropping the temperature to 216.54 K (-56.61 °C), the latent heat of vaporisation records a value of 351.87 kJ/kg whereupon the CO_2 solidifies, forming what is referred to as "dry ice", and thereby rendering the bottle useless even though its useful load exceeds 60% of its capacity. This paper describes a system that stops such low temperatures being reached and compares the results between subjecting the CO_2 bottles to cooling by a natural air flow (traditional method) and the method proposed in this paper that involves cooling with a natural water flow at 20 °C using a spray system.

5.5. Assessing the Number of CO_2 Bottles Required

Table 14 shows the design flows for the alternator filling and emptying stages on two units in the reference power plant.

Table 14. Flows for the phases of filling and emptying of the two power units

(Nm^3)	U_1	U_2
V_g	81.41	86.37
V_{CO2}^F (*)	101.76	107.96
V_{CO2}^D (**)	162.82	172.74

(*) $V_{CO2}^F = 1.25 \cdot V_g$

(**) $V_{CO2}^D = 2 \cdot V_g$

Targeting a utilisation factor for the CO_2 bottles of 85%, and considering a load per bottle of 30 kg, gives us the number of bottles required for each unit.

Bottles for Unit 1

The necessary mass for the filling will be:

$$m_{U1}^F = 101.76\ Nm^3 \times 44.6189\ {}^{mol}\!/_{Nm^3} \times 44.01\ {}^{g}\!/_{mol} = 199{,}823.85\ g \approx 199.83\ kg \tag{74}$$

The number of bottles required for the filling is therefore:

$$N_{U1}^F = \frac{199.83}{30 \times 0.85} = 7{,}84 \approx 8\ bottles \tag{75}$$

The necessary mass for the emptying will be:

$$m_{U1}^D = 162.82\ Nm^3 \times 44.6189\ {}^{mol}\!/_{Nm^3} \times 44.01\ {}^{g}\!/_{mol} = 319{,}726.02\ g \Leftrightarrow 329.73\ kg \tag{76}$$

The number of bottles required for the emptying of this unit will be:

$$N_{U1}^D = \frac{329.73}{30 \times 0.85} = 12.93 \approx 13\ bottles \tag{77}$$

Bottles for Unit 2

Following the same procedure, we obtain the number of CO_2 bottles required for both the filling and emptying of unit 2. Table 15 provides a summary of the results for both units.

Table 15. Number of bottles necessary for the phases of filling and emptying of both units

(Units)	U_1	U_2
N^F	10	10
N^D	15	16

5.6. Study of the Arrangement for Introducing CO_2 into the Generator

If the CO_2 enters the electrical generator with a clearly turbulent regime, this will inevitably lead to a mixture of gases, despite the considerable difference between the molecular masses of carbon dioxide and air and particularly hydrogen [146], which should not be permitted.

The characterisation of the flow regime requires obtaining the non dimensional Reynolds number, which is calculated by applying Buckingham's Pi Theorem, which as is common knowledge represents the relationship between the forces of inertia and those of viscosity [147, 148]. Thus,

$$Re = \frac{\rho \cdot v \cdot d}{\mu} = \frac{v \cdot d}{\upsilon} \begin{cases} < 2{,}300 \Rightarrow Laminar\ flow \\ > 4{,}000 \Rightarrow Turbulent\ flow \end{cases} \tag{78}$$

Furthermore, the supply pressure of the CO_2 manifold should be between 0.14 and 0.35 kg/cm^2 higher than the generator pressure. The lower limit corresponds to the minimum pressure required for overcoming load losses in the supply pipe whereas the upper limit impedes the formation of gas turbulent flow at the generator inlet, which would lead to a mixture of gases and a sharp increase in the time required for the sweeping, thereby diminishing the operation's efficiency as noted earlier.

Thus, this study will apply a differential of 0.3 kg/cm^2 so that the generator pressure is around 0.5 kg/cm^2 and the pressure of the gas inlet pipe on the generator will therefore be 0.8 kg/cm^2. Accordingly, the density of CO_2 in the inlet conduit on the electrical generator can be obtained as [149-151]:

$$\rho_{0.8} = \frac{P}{R \cdot T} = \frac{(0.8 + 1.03) \cdot 10^4\ {}^{kg}\!/\!_{m^2}}{19.2\ {}^{m}\!/\!_{K} \times (20 + 273)\ K} = 3.253\ {}^{kg}\!/\!_{m^3} \tag{79}$$

For CO_2 under ambient conditions (at 20 °C and 1 atm.) the tables show that: $\upsilon = 0.846 \times 10^{-5}\ {}^{m^2}\!/\!_{s}$ and $\rho = 1.8359\ {}^{kg}\!/\!_{m^3}$. Dynamic viscosity can therefore be obtained as:

$$\mu = \frac{\rho}{g} \cdot \upsilon = \frac{1.8359}{9.81} \cdot 0.846 \cdot 10^{-5} = 0.1583 \cdot 10^{-5}\ kg\ {}^{s}\!/\!_{m^2} \tag{80}$$

Therefore, kinematic viscosity at a pressure of 0.8 kg/cm^2 in the supply pipeline is:

$$\upsilon_{0.8} = \mu \cdot \frac{g}{\rho_{0.8}} = 1.1583 \cdot 10^{-5} \cdot \frac{9.81}{3.253} = 1.07378 \cdot 10^{-5}\, {}^{m^2}\!/_{s} \tag{81}$$

The mass flow, taking into account that 16 CO_2 bottles can be used as shown in table 3, with a utilisation in principle of 73% and considering that the unloading time for obtaining a very low purity of H_2 in CO_2 (between 2% and 3%) is around 8 hours, is:

$$\dot{m} = \frac{30 \times 16 \times 0.73}{8 \times 3{,}600} = 0.012\, {}^{kg}\!/_{s} \tag{82}$$

Furthermore, the CO_2 velocity in the 2" Sh 40 inlet pipe on the electrical generator, considering the inner diameter to be d = 52.5 mm:

$$v = \frac{\dot{m}}{S \cdot \rho_{0.8}} = \frac{0.012}{\left(\pi \cdot \frac{(0.0525)^2}{4}\right) \cdot 3.253} = 1.70\, {}^{m}\!/_{s} \tag{83}$$

Substituting these values in equation (78) shows that the flow is clearly turbulent (Re = $8{,}000$). Thus, the maximum unloading velocity to stop the flow from being turbulent can be calculated according to expression (78) as:

$$4{,}000 = \frac{v_{máx} \cdot d}{\upsilon_{0.8}} \tag{84}$$

This gives us ($v_{máx}$ = 0.82 m/s). The recommended unloading velocity should be below this figure. This study will use a reference value of 0.8 m/s, which means significantly reducing the flow of CO_2 delivery and considerably increasing operating time. Table 16 shows the main parameters for the CO_2 supply installation with a comparison of current and recommended values.

Table 16. Characteristics of the equipment of carbon dioxide supply

CO_2 supply equipment	Specifications
Duct material	Steel 2" Sh 40
ΔP	0.3 kg/cm^2
d	52 mm
v	1.70 m/s
$\dot{m}$	0.012 kg/s
t	8 h
v_{max}	0.80 m/s
$\dot{m}_{max}$	0.0056 kg/s
t_{max}	29.8 h

5.7. Thermal Study of the Proposed Solution

The solution involves the heat exchange in the CO_2 bottles by convection with a film of water [152], which is sprayed on with a shower-type perforated ring fitted onto the neck of each bottle, whereby the curtain of water douses the entire perimeter of the cylinder that basically makes up the bottle. The CO_2 bottles are arranged vertically on a grid that is in turn placed over a drainage channel, so the water runs off to the nearest drainage point, as depicted schematically in figure 18.

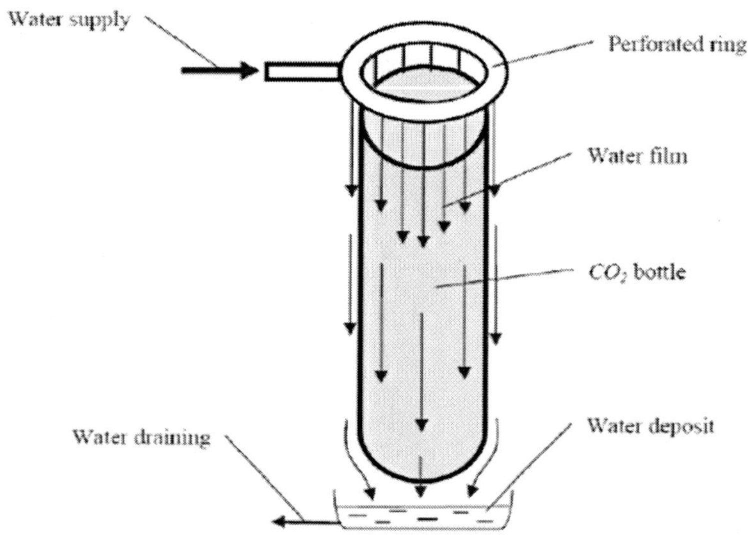

Figure 18. Simplified scheme of the refrigeration device proposed.

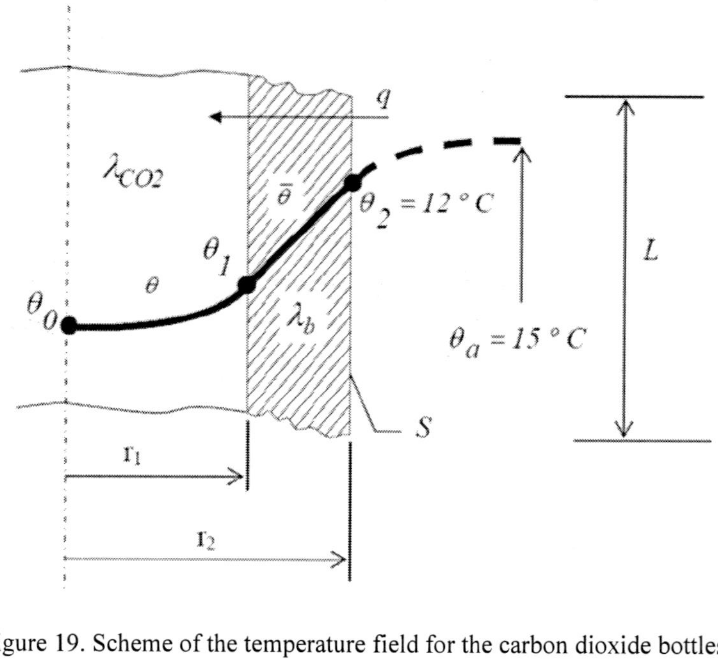

Figure 19. Scheme of the temperature field for the carbon dioxide bottles.

A mass and energy balance is considered, for which we shall accept the following assumptions, as verified by operating practice:

- The film of water that douses the bottle is sprayed on at a temperature of 20 °C and runs off at an average temperature of 15 °C.
- The temperature gradient between the cooling water and the bottle's outer metal surface is 3 °C.

A rigorous calculation requires considering the general equation for heat by conduction between the bottle's outer and inner surfaces. There follows a detail of the calculation methodology for the rigorous determination of the minimum temperature for CO_2 (which is reached on the bottle's geometric axis), according to the diagram shown in figure 19.

Given that we are dealing with a cylindrical container, the heat equation should be addressed in terms of cylindrical coordinates, with the corresponding boundary conditions, to provide the "temperature range" and based on this the negative heat generated (sink), which should be equal to the heat provided by the natural convection of a flow of water. Accordingly, we base our approach on the general equation for the transmission of heat by conduction [153]:

$$a \cdot \nabla^2 \theta + \frac{\pm \dot{q}_v}{\rho \cdot c_p} = \frac{\partial \theta}{\partial t} \tag{85}$$

Although in fact the process is initially undertaken in a transitory regime, the real aim is to achieve a constant CO_2 unloading flow, and as thermal equilibrium is also to be achieved, the regime will be stationary and can therefore be written as:

$$\frac{\partial \theta}{\partial t} = 0 \tag{86}$$

To calculate the volumetric thermal power ($\dot{q}_v$), we have based ourselves on the reference unit 2 of around 86 m^3 of gas volume (see table 2), capable of generating 540 MW. As we have already noted, we are going to use a value of around 0.8 m/s for CO_2 delivery velocity, which means a flow of 6.22 m^3/h under ambient conditions (20 °C and 1 atm.). The equivalence of this volume under normal conditions (0 °C and 1 atm.) is:

$$\frac{P_1 \cdot V_1}{T_1} = \frac{P_2 \cdot V_2}{T_2}; \quad \frac{1 \cdot 6.22}{293.15} = \frac{1 \cdot V_2}{273.15}; \quad V_2 = 5.8\, {}^{Nm^3}\!/_{h} \tag{87}$$

The required volume of CO_2 that is the most unfavourable condition is, according to table 2, 172.74 m^3, which means that the time required for H_2 sweeping is:

$$t = \frac{172.74}{5.8} = 29.8\ h \tag{88}$$

Consequently, the necessary heat per bottle of CO_2 to keep the temperature at a steady state of 285 K (for 85% utilisation) is:

$$q = m \cdot h_{fg} = 30\,{}^{kg}/_{bottle} \cdot 190.69\,{}^{kJ}/_{kg} \cdot 0.85 = 4{,}862.63\,{}^{kJ}/_{bottle} \tag{89}$$

Considering that for this temperature $\rho_{CO2} = 2.7197\,{}^{kg}/_{Nm^3}$, the volume of CO_2 in a bottle under normal conditions will be:

$$V_{CO2} = {}^{30}/_{2.7197} = 11.03\,{}^{Nm^3}/_{bottle} \tag{90}$$

Therefore, the volumetric thermal power ($\dot{q}_v$) is finally:

$$\dot{q}_v = {}^{q}/_{t \cdot V_{CO2}} = 4.11\,{}^{W}/_{Nm^3} \tag{91}$$

The approach for obtaining the "temperature ranges" will distinguish between, on the one hand, that corresponding to the inside of the cylinder, that is, the CO_2 and, on the other, the metal wall of the actual cylindrical container itself. Thus, for the liquid CO_2, it is considered a heat sink, so that we base ourselves on the following second-order general differential equation on cylindrical coordinates [154, 155]:

$$\frac{d^2\theta}{dr^2} + \frac{1}{r} \cdot \frac{d\theta}{dr} - \frac{\dot{q}_v}{\lambda_{CO2}} = 0 \tag{92}$$

when suitably integrated, this gives:

$$\theta = \frac{\dot{q}_v \cdot r^2}{4 \cdot \lambda_{CO2}} + C_1 \cdot \ln r + C_2 \tag{93}$$

In addition, for the metal wall of the cylindrical container with no heat generation, we use the following second-order general differential equation also on cylindrical coordinates:

$$\frac{d^2\overline{\theta}}{dr^2} + \frac{1}{r} \cdot \frac{d\overline{\theta}}{dr} = 0 \tag{94}$$

when suitably integrated, this gives:

$$\overline{\theta} = \overline{C}_1 \cdot \ln r + \overline{C}_2 \tag{95}$$

There follows a consideration of the boundary coordinates so that they can be used to obtain the value of the four constants of integration $C_1, C_2, \overline{C}_1, \overline{C}_2$. The following data will be used

[149-151]: $\lambda_{CO_2} = 0.020\ W / mK$ and $\lambda_b = 36\ W / mK$. The bottles have the following dimensions: $r_1 = 20\ cm$ and $r_2 = 21.5\ cm$. The boundary conditions are therefore:

$$r = 0 \Rightarrow \left.\frac{d\theta}{dr}\right|_{r=0} = 0 \tag{96}$$

$$r = r_1 \Rightarrow \theta_{r=r_1} = \overline{\theta}_{r=r_1} \tag{97}$$

$$r = r_1 \Rightarrow -\lambda_{CO2} \cdot \left.\frac{d\theta}{dr}\right|_{r=r_1} = -\lambda_b \cdot \left.\frac{d\overline{\theta}}{dr}\right|_{r=r_1} \tag{98}$$

$$r = r_2 \Rightarrow \overline{\theta}_{r=r_2} = 12\ ^{o}C \tag{99}$$

5.8. Results and Discussion

Once the preceding equations have been solved, the values obtained for the four integration constants are:

$$C_1 = 0 \quad C_2 = 9.944 \quad \overline{C}_1 = 0.00228 \ \ and \ \ \overline{C}_2 = 12.003 \tag{100}$$

So replacing these values in equations (93) and (95), respectively, finally gives us the temperature distributions:

$$\theta = 51.37 \cdot r^2 + 9.944 \tag{101}$$

$$\overline{\theta} = 0.00228 \cdot ln r + 12.003 \tag{102}$$

These equations provide us with the temperatures θ_0 and θ_1 that correspond respectively to the axis of the bottle in which verification is made of the minimum CO_2 temperature and the internal wall of the cylindrical container (see figure 3), whereby:

$$\theta_0 = 9.944\ ^{o}C = 282.94\ K \quad and \quad \theta_1 = 11.99\ ^{o}C = 284.99\ K \tag{103}$$

This proves that the CO_2 does not reach its freezing point, which means that the bottle's contents can indeed be better used. We shall now calculate the mass of cooling water required for maintaining these operating conditions.

Let us consider the case involved in emptying 90% of each bottle, maintaining a differential pressure between the CO_2 unloading manifold and the electrical generator of 0.3 kg/cm^2, keeping a laminar gas flow at the generator inlet, the quantity of water required per hour for cooling each bottle from 20 °C to 12 °C is provided by the following expression:

$$\dot{q} = \dot{m}_w \cdot c_{p_w} (\theta_{w2} - \theta_{w1}) \tag{104}$$

Firstly, for calculating the thermal power absorbed by the cooling water, we use equation (89) whereby:

$$\dot{q} = \frac{q}{t} = \frac{4{,}862.63\ {}^{kJ}\!/_{bottle}}{29.8\ h} = 163.17\ {}^{kJ}\!/_{h\ botlle} \tag{105}$$

By substituting equation (104), we finally obtain the flow of water for cooling each one of the bottles:

$$\dot{m}_w = 4.87\ {}^{kg}\!/_{h\ botlle} \tag{106}$$

So by taking the sum of 16 bottles, we obtain the total water flow. Table 17 presents the main design parameters for the proposed cooling installation and figure 20 shows the final arrangement of the CO_2 bottles with the cooling system proposed in this paper.

Figure 20. Experimental refrigeration device proposed.

Table 17. Characteristics of the refrigeration device proposed

Water refrigeration device	Specifications
Water supply ring material	PVC
Ring internal diameter	25 mm
Theoretical water flow	77.92 l/h
Experimental water flow	100 l/h
Water supply temperature	20 °C
Water outlet temperature	15 °C
External wall temperature	12 °C
Environment temperature	20 °C

5.9. Conclusion

It has repeatedly been shown in practice that the unloading of CO_2 exposed to natural air convection produces a rapid freezing of the fluid and therefore a substantial increase in both the turnover of bottles (as their effective utilisation is around 73%) and operating times (given the need to change bottles as well as the generation of internal turbulences).

The method proposed involving the use of spraying water at ambient temperature for exposing the bottles of CO_2 to natural convection has proven to be a suitable procedure for optimising resources. According to the calculations made in this study, around 78 l/h of water are required for tempering a unit's CO_2 bottles and so avoid them freezing, until CO_2 purity is achieved in the alternator.

It has been experimentally verified that with an ambient temperature of around 20 °C, the application of a flow of water in the form of a spray of 100 l/h, at a temperature also of 20 °C, whereby the curtain of water covers the entire outside of the bottles, the carbon dioxide can be unloaded with a differential pressure between the CO_2 unloading manifold and the electrical generator of 0.3 kg/cm^2. This takes place on a continuous basis, which means the maximum flow in laminar regime, keeping the temperature of the outer metal surface of the CO_2 bottles at around 12°C. As can be seen, the theoretical value of the flow of water calculated differs somewhat from the experimental value, as a higher flow than the theoretical one is generally required for effectively covering the outside of the bottles.

Regarding the time required for the process, it has been shown that this needs to be a lot better than the one undertaken in daily practice in order to reduce the turbulent flow at the CO_2 unloading outlet as much as possible, so as to avoid mixing with hydrogen. Accordingly, keeping the flow of CO_2 delivery to the electrical generators within a range that ensures the best performance requires the installation of a flow detector-indicator on the CO_2 unloading manifold.

Acknowledgements

The authors are deeply thankful to the Staff of the Power Plants of Reference, who were involved in the measuring and and building of both the physical models (the experimental model and the prototype one), for their help and access to testing and experimenting facilities and their assistance at all times in the data acquisition process.

References

[1] Nakicenovic, N., Grübler, A. and McDonald, A., Editors, *Global energy perspectives*, Cambridge University Press, Cambridge, UK, 1998.

[2] Colombo, U., *Development and the global environment.* In: J.M. Hollander, Editor, The energy–environment connection, Island Press, Washington, 1992, p. 3-14.

[3] Tester, J.W., Wood, D.O. and Ferrari, N.A., Energy and the environment in the 21st century, *MIT Press,* New York, 1991.

[4] Strehlow, R.A., Combustion fundamentals, McGraw-Hill, New York, 1985.

[5] Beér, J.M., Combustion technology developments in power generation in response to environmental challenges. *Prog. Energy Combust. Sci.* **26**, 2000, p. 301–27.

[6] Demirbas, A., Global energy sources, energy usage and future developments, *Energy Sources.* **26**, 2004, p. 191–204.

[7] European Union, Common Position (EC) No 28/2003 adopted by the Council on 17 March 2003 with a view to adopting Directive 2003/…/CE of the European Parliament and of the Council establishing a scheme for greenhouse gas emission allowance trading within the Community and Amending Council Directive 96/61/EC, *Official J. Eur. Union.* C **125 E**, 2003, p. 72–95.

[8] European Communities, Directive 2001/81/EC of the European Parliament and of the Council of 23 October 2001 on national emission ceilings for certain atmospheric pollutants, *Official J. European Communities. L* 309, 2001, p. 22–30.

[9] European Communities, Directive /80/EC of the European Parliament and of the Council of 23 October 2001 on the limitation of emissions from certain pollutants into the air from large combustion plants, *Official J. European Communities L.* **309**, 2001, p. 1–20.

[10] Warnatz, J., Maas, U. and Dibble, R.W., *Combustion.* (3rd ed.), Springer, Heidelberg, 2001.

[11] Bartok, W. and Sarofim, A.F., *Fossil fuel combustion; a source book,* Wiley, New York, 1991.

[12] Glassman, I., *Combustion* (3rd ed), Academic Press, San Diego, 1996.

[13] Wünning, J.A. and Wünning, J.G., Flameless oxidation to reduce thermal NO formation. *Prog. Energy Combust. Sci.* **23**, 1997, p. 81-97.

[14] Bolz, S. and Gupta, A.K. Effect of air preheated temperature and oxygen concentration on flame structure and emission. *Int. Joint Power Generat. Conf. ASME* **1**, 1998, p. 193–205.

[15] Ishiguro, T., Tsuge, S., Furuhata, T., Kitigawa, K. Arai, N., Hasegawa, T., Tanaka, R. and Gupta, A.K. Homogenization and stabilization during combustion of hydrocarbons with preheated air. *Proc. Combust. Inst.* **27**, 1998, p. 3205–13.

[16] Katsuki, M. and Hasegawa, T., The science and technology of combustion in highly preheated air. *Proc. Combust. Inst.* **27**, 1998, p. 3135–46.

[17] Weber, R., Verlaan, A.L., Orsino, S. and Lallemant, N. On emerging furnace design methodology that provides substantial energy savings and drastic reductions in CO_2, CO and NO_x emissions. *J. Inst. Energy.* **72** 492, 1999, p. 77–83.

[18] Hasegawa, T., Mochida, S. and Gupta, A.K., Development of advanced industrial furnace using highly preheated combustion air. *J. Propulsion Power.* **18** 2, 2002, p. 233-67.

[19] Xu, H., Smoot, L.D. and Hill, S.C., Computational modelling for NOx reduction by advanced reburning. *Energy Fuel.* **13** 2, 1999, p. 411–20.

[20] Akinyemi, O., Toqan, M.A., Beér, J.M., Syska, A., Thijsson, J., Benson, C. and Moreland, D. Development of a high air preheated low NO_x burner; experimental studies aided by computational modelling. *Proceedings of the Second International High Temperature Air Combustion Symposium*, Taiwan, 1999.

[21] Bychkov, V.V., Goldberg, S.M., Liberman, M.A., Kleev, A.I. and Eriksson, L.E., Numerical simulation of curved flames in cylindrical tubes, *Combust. Sci. Technol.* **129**, 1997, p. 217–42.

[22] Veynante, D. and Vervish, L., Turbulent combustion modeling, *Prog. Energy Combust. Sci.* **28**, 2002, p. 193–266.

[23] Tan, Y., Douglas, M.A. and Thambimuthu, K.V., CO_2 capture using oxygen enhanced combustion strategies for natural gas power plants, *Fuel.* **81**, 2002, p. 1007–16.

[24] Kazantis, N. and Kravaris, C., Nonlinear observer design for process monitoring, *Ind. Eng. Chem. Res.* **39**, 2000, p. 408–19.

[25] Perry, R.H. and Chilton, C.H., *Chemical engineer's handbook* (5th ed.), McGraw-Hill Book Company, New York, 1973.

[26] Demirbas, A., Determination of combustion heat of fuels by using non-calorimetric experimental data, *Energy Edu. Sci. Technol.* **1**, 1998, p. 7–12.

[27] Kuo, K.Y., Principles of Combustion. John Wiley and Sons, New York, 1986.

[28] Yan, Z. and Holmstedt, G., Numerical Heat Transfer Part B: *Fundamentals.* **31**, 1997, p. 61–71.

[29] http://www.me.berkeley.edu/gri_mech/version3.0/text30.html

[30] Kee, R.J., Rupley, F.M., Miller, J.A., Coltrin, M.E., Grcar, J.F., Meeks, E., Moffat, H.K.,Lutz, A.E., Dixon-Lewis, G., Smooke, M.D., Warnatz, J., Evans, G.H., Larson, R.S, Mitchell, R.E., Petzold, L.R., Reynolds, W.C. Caracotsios, M., Stewart, W.E., Glarborg, P., Wang, C., Adigun, O., Houf, W.G. Chou, C.P., Miller, S.F., Ho, P. and Young, D.J., Chemkin v. 4.0, Technical Report San Diego, CA, Reaction Design, Inc., 2004.

[31] Peters, N., *Turbulent combustion*, Cambridge University Press, Cambridge, 2000.

[32] Warnatz, J., Maas, U. and Dibble, R.W. Combustion. (3rd ed.), Springer, Heidelberg , 2001.

[33] Coelho, P.J. and Peters, N. Numerical simulation of a Mild Combustion burner. *Combust Flame.* **124**, 2001, p. 503–18.

[34] Kong, S.C. and Reitz, R.D. Application of detailed chemistry and CFD for predicting direct injection HCCI engine combustion and emissions. *Proc. Combust. Inst.* **29**, 2002, p. 663–69.

[35] Hinze, J.O., *Turbulence*, New York, McGraw-Hill Publishing Co., 1975.

[36] Kelkar, K.M. and Patankar, S.V., Development of Generalized Block Correction Procedures for the Solution of Discretized Navier-Stokes Equations, Computer Physics Communications, V. 53, 1989, p. 329- 36.

[37] Özdemir, I.B. and Peters, N. Characteristics of the reaction zone in a combustor operating at Mild Combustion. *Exp. Fluids.* **30** 6, 2001, p. 683–95.

[38] Launder, B.E., and Spalding, D.B., Lectures in Mathematical Models of Turbulence, London, England: Academic Press, 1972.

[39] Launder, B.E., and Spalding, D.B., The Numerical Computation of Turbulent Flows, Imperial College of Science and Technology, London, England, NTIS N74-12066, 1973.

[40] Peters, N., Laminar diffusion flamelet models in non-premixed turbulent combustion. *Prog. Energy Combust. Sci.* **10**, 1984, p. 319–39.

[41] Launder, B.E., Second-Moment Closure and Its Use in Modelling Turbulent Industrial Flows. *International Journal for Numerical Methods in Fluids*. **9**, 1989, p.963-85.

[42] Kays. W.M., Turbulent Prandtl Number - Where Are We?. J. Heat Transfer. **116**, 1994, 284-95.

[43] Launder, B.E., and Spalding, D.B., The Numerical Computation of Turbulent Flows. *Computer Methods in Applied Mechanics and Engineering.* **3**, 1974, p. 269-89.

[44] Kim, S.E. and Choudhury, D., A Near-Wall Treatment Using Wall Functions Sensitized to Pressure Gradient. In *ASME FED* Vol. 217, Separated and Complex Flows. *ASME.* 1995.

[45] Viegas, J.R., Rubesin, M.W. and Horstman, C.C., On the Use of Wall Functions as Boundary Conditions for Two-Dimensional Separated Compressible Flows, Technical Report AIAA-85-0180, *AIAA 23rd Aerospace Sciences Meeting*, Reno, Nevada, 1985.

[46] Jayatilleke, C., The Influence of Prandtl Number and Surface Roughness on the Resistance of the Laminar Sublayer to Momentum and Heat Transfer. *Prog. Heat Mass Transfer.* **1**, 1969, p. 193-321.

[47] Siegel, R. and Howell, J.R. *Thermal Radiation Heat Transfer*. Hemisphere Publishing Corporation, Washington D.C., 1992.

[48] Siegel, R. and Spuckler, C.M., Effect of Refractive Index and Diffuse or Specular Boundaries on a Radiating Isothermal Layer. J. Heat Transfer. **116**, 1994, p 787-90.

[49] Raithby, G.D. and Chui., E.H. A Finite-Volume Method for Predicting a Radiant Heat Transfer in Enclosures with Participating Media. *J. Heat Transfer.* **112**, 1990, p. 415-23.

[50] Peters, N., *Laminar Flamelet Concepts in Turbulent Combustion. In 21st Symp. (Int.) on Combustion,* The Combustion Institute. 1986, p. 1231-50.

[51] Bray, K.N. and Peters, N. Laminar Flamelets in Turbulent Flames. In Libby, P.A. and Williams, F.A. editors, *Turbulent Reacting Flows,* Academic Press, 1994, p. 63-114.

[52] Dixon-Lewis. G., Structure of Laminar Flames. In *23rd Symp. (Int.) on Combustion,* The Combustion Institute, 1990, p. 305-24.

[53] Binniger, B., Chan, M., Paczkko, G. and Herrmann, M. Numerical Simulation of Turbulent Partially Premixed Hydrogen Flames with the Flamelet Model. Technical report, *Advanced Combustion Gmbh,* Internal Report, 1998.

[54] Muller, C.M., Breitbach, H. and Peters. N., Partially Premixed Turbulent Flame Propagation in Jet Flames. Technical report, *25th Symposium (Int) on Combustion*, The Combustion Institute, 1994.

[55] Flower, W.L., Hanson, R.K., and Kruger, C.H. In *15th Symp. (Int.) on Combustion,* The Combustion Institute, 1975, p. 823.

[56] Monat, J.P. Hanson, R.K. and Kruger, C.H. In *17th Symp. (Int.) on Combustion*. The Combustion Institute, 1979, p. 543.

[57] Hanson, R.K. and Salimian, S., Survey of Rate Constants in H/N/O Systems. In Gardiner, W.C., editor, *Combustion Chemistry*. 1984, p. 361.

[58] Fennimore, C.P. Formation of Nitric Oxide in Premixed Hydrocarbon Flames. In *13th Symp. (Int.) on Combustion,* The Combustion Institute, 1971, p. 373.

[59] Barnes, F.J., Bromly, J.H., Edwards, T.J. and Madngezewsky, R., NOx Emissions from Radiant Gas Burners. *Journal of the Institute of Energy*. **155**, 1988, p.184-88.

[60] Schefer, R.W., Namazian, M. and Kelly., J. In *Combustion Research Facility News*, volume 3, number 4. Sandia, 1991.

[61] Backmier, F., Eberius, K.H. and Just., T. *Comb. Sci. Tech.,* **7**, 1973, p. 77.

[62] De Soete, G.G. Overall Reaction Rates of NO and N_2 Formation from Fuel Nitrogen. In 15th Symp. (Int.) on Combustion, The Combustion Institute, 1975, p. 1093-102.

[63] Miller, J.A. and Bowman, C.T., Mechanism and Modeling of Nitrogen Chemistry in Combustion. *Prog. in Energy and Comb. Sci.*, **15**:287-338, 1989.

[64] Miller, J.A. and Fisk, G.A., *Chemical and Engineering News*. **31**, 1987.

[65] Houser, T.J., Hull, M., Alway, R. and Biftu, T. *Int. Journal of Chem. Kinet.*, **12**, 1980, p. 579.

[66] Lockwood, F.C. and Romo-Millanes, C.A., Mathematical Modelling of Fuel - NO Emissions From PF Burners. *J. Int. Energy.* **65**, 1992, p. 144-52.

[67] Melte, P.C. and Pratt, D.T. Measurement of Atomic Oxygen and Nitrogen Oxides in Jet Stirred Combustion. In *15th Symposium (Int) on Combustion*, The Combustion Institute, 1974, p. 1061-70.

[68] Barlow, R.S., Fiechtner, G.J., Carter, C.D. and Chen, J.Y. Experiments on the Scalar Structure of Turbulent CO/H_2 /N_2 Jet Flames. *Combustion and Flame*. **120**, 2000, p. 549-69.

[69] Correa, S.M. A Review of NOx Formation Under Gas-Turbine Combustion Conditions. *Combustion Science and Technology.* **87**, 1992, p. 329-62.

[70] Glarborg, P., Johnsson, J.E. and Dam-Johansen, K. Kinetics of Homogeneous Nitrous Oxide Decomposition, *Combustion and Flame*. **99**, 1994, p.523-32.

[71] Steele, R.C. Malte, P.C., Nichol, D.G. and Kramlich., J.C. NOx and N_2 O in Lean-Premixed Jet-Stirred Flames. *Combustion and Flame*. **100**, 1995, p. 440-49.

[72] Fluent Inc., Fluent 6.0, Documentation User's Guide, 2001.

[73] Chou, C.J, Chen J.Y, Clement G., and Mark K.D., Numerical Modelling of NO Formation in Laminar Bunsen Flames-*A Flamelet Approach, Combustion and Flame.* **114**, 1998, p 420-35.

[74] Haynes, J.M., Beér, M., Toqan, M.A., Nicholson J.N., Duby, T.G. Narula R.N., Thornock, D.E., Borio, R.W., Sutton J.P., and LaFlesh R.C., Aerodynamic Modelling and Scale-Up of Low NOx Oil Burner, *Joint Power Generator Conference*, EC-Vol. 4/FACT Vol. 21, 1996.

[75] Pesenti, B., Evrard, P., and Lybaert, P., *NOx Production and Radiactive Heat Transfers from a Self-Regenerative Flameless Oxidation Burners*, http://stecwww.fpms.proj/noxflox.html, 2002.

[76] Stopford, P.J., Recent Applications of *CFD* Modelling in the Power Generation and Combustion Industries, *Applied Mathematical Modelling*, Vol. 26, 2002, p. 351-74.

[77] Bebar, L., Kermes, V., Stehlik, P., Canek J., and Oral, J., Low NOx Burners-Prediction of Emissions Concentration Based on Design, Measurements and Modelling, *Waste Management* **22**, 2002, p. 443-51.

[78] Willian, A., Woolley, R., Lawes, M., The formation of NOx Surface Burners, *Combustion and Flame* **89**, 1992, p. 157-66.

[79] Gregory P. Smith, David M. Golden, Michael Frenklach, Nigel W. Moriarty, Boris Eiteneer, Mikhail Goldenberg, C. Thomas Bowman, Ronald K. Hanson, Soonho Song, William C. Gardiner, Jr., Vitali V. Lissianski, and Zhiwei Qin, http://www.me.berkeley.edu/gri_mech/.

[80] Hanson, R.K., and Salimian, S., In Combustion Chemistry, W.C. Gardiner, Jr., Edit., Springer-Verlag, New York, 1984, p. 361.

[81] Davidson, D.F., and Hanson, R.K., *Int. J. Chem. Kinet.,* **22**, 1990, p. 843.

[82] Röhrig, M., Petersen, E.L., Davidson, D.F., and Hanson, R.K., *Int. J. Chem. Kin.,* 28, 1996, p. 599.

[83] Roht, P., and Just, T., 20th Symposium (international) on Combustion, 1984, p. 807-08.

[84] Michael,J.V., and Lim, K. P., *J. Chem. Phys.,* **97**, 1992, p. 3228-29.

[85] Westbrook, C. and Dryer, F., Simplified Reaction Mechanisms for Oxidation of Hydrocarbon Fuels in Flames. *Comb. Sci. Tech*, **27**, 1981, p. 31-43.

[86] Zimont, V., Gas Premixed Combustion at High Turbulence. Turbulent Flame Closure Model Combustion Model, *Experimental Thermal and Fluid Science*, **21**, 2000, p.179-86.

[87] Zimont, V.L., Biagioli, F. and Syed, K.J., Modelling Turbulent Premixed Combustion in the Intermediate Steady Propagation Regime. *Progress in Computational Fluid Dynamics.* **1**(1), 2001, p. 14-28.

[88] Shah, N.G., A New Method of Computation of Radiant Heat Transfer in Combustion Chambers. PhD. thesis, Imperial College of Science and Technology, London, England, 1979.

[89] Singer, J.G., *Combustion fossil power systems*, Combustion engineering Inc., 4th edition, 1991.

[90] Ahmad, T., Plee, S.L. and Myers, J.P. Computation of Nitric Oxide and Soot Emissions from Turbulent Diffusion Flames. *J. of Engineering for Gas Turbines and Power.* **107**,1985, p. 48-53.

[91] Drake, M.C., Correa, S.M. Pitz, R.W., Shyy, W. and Fenimore, C.P., Superequilibrium and Thermal Nitric Oxide Formation in Turbulent Diffusion Flames. *Combustion and Flame.* **69**, 1987, p. 347-65.

[92] Barnes F.J., Bromly, J.H. Edwards, T.J. and Madngezewsky, R., NOx Emissions from Radiant Gas Burners. *Journal of the Institute of Energy*. **155**, 1988, p.184-88.

[93] Perera, G., Antifora, A. Sala, M. and Vigevano, L. NOx Emissions in Combustion Systems of Coal Fired Furnaces with a Reducing Environment: Predictions and Measurements. In *Fourth International Conference on Technologies and Combustion for a Clean Environment*, Lisbon, Portugal, 1997.

[94] Boyd, R.K. and Kent, J.H. Three-dimensional furnace computer modeling. In *21st Symp. (Int.) on Combustion*, The Combustion Institute, 1986, p. 265-74.

[95] Glarborg, P., Johnsson, J.E. and Dam-Johansen, K. Kinetics of Homogeneous Nitrous Oxide Decomposition. *Combustion and Flame,* **99**, 1994, p. 523-32.

[96] Miller, J.A. and Bowman, C.T., Mechanism and Modeling of Nitrogen Chemistry in Combustion. *Prog. in Energy and Comb. Sci.,* **15**, 1989, p. 287-338.

[97] Miller, J.A., Branch, M.C. McLean, W.J. Chandler, D.W., Smooke, M.D., and Kee, R.J. In *20th Symp. (Int.) on Combustion*, The Combustion Institute, 1985, p. 673.

[98] A.F. Mills, *Heat Transfer* (second ed.), University of California, Los Angeles, California, 1999.

[99] R.M. Cotta, M.D. Mikhailov and M.N. Özisik, Transient conjugated forced convection in ducts with periodically varying inlet temperature, *Int. J. Heat Mass Transfer.* **30**, 10, 1986, p. 2073-82.

[100] F.W. Schmidt, A.J. Willmot, Thermal Energy Storage and Regeneration, ISBN 0-07-055346-7, McGraw-Hill, New York, 1981.

[101] P.M. Park, H.C. Cho, H.D. Shin, Unsteady thermal flow analysis in a heat regenerator with spherical particles, *Int. J. Energy Res.,* **27**, 2003, p. 161-72.

[102] M.T. Zarrinehkafsh, S.M. Sadrameli, Simulation of fixed bed regenerative heat exchangers for flue gas heat recovery, *Applied thermal Engineering.* **24**, 2004, p. 373-82.

[103] C.E. Baukal, *Heat transfer in industrial combustion*, ISBN 0-8493-1699-5, CRC Press LLC, USA, 2000.

[104] A.J. Chapman, *Heat transfer*, Ed. Interciencia, second ed., 1968.

[105] W. Li and K. Sadik, Unsteady thermal entrance heat transfer in laminar with a periodic variation inlet temperature, *Int. J. Heat Mass Transfer.* **34**, 10, 1991, p. 2581-92.

[106] H.Y. Zhang, M.A. Ebadian and A. Campo, Effects of heat generation and axial heat conduction in laminar flow inside a circular pipe with a step change in wall temperature, Trans. *ASME J. Heat Transfer.* **102**, 1980, p. 58-63.

[107] E.K. Zariffeh, H.M. Soliman and A.C. Trupp, The combined effect of wall and fluid axial conduction on laminar heat transfer in circular tubes, *Heat Transfer.* **4**, 1982, p. 131-35.

[108] K. Rummel, The Calculation of the Thermal Characteristics of Regenerators, *J. Institute of Fuel.* p. 160-74, 1931.

[109] N.E. Rafidi, *Thermodynamic aspects and heat transfer characteristics of HiTAC furnaces with regenerators*, Doctoral Dissertation, Royal Institute of Technology, School of Industrial Engineering and Management, Stockholm, ISBN 91-7178-223-0, 2005.

[110] W. Yang and W. Blasiak, combustion and heat flux evaluation in a furnace equipped with a high-cycle regenerative system, *Proc. of challenges in reheating furnaces,* London, 2002.

[111] W. Blasiak, W. Yang, and N. Rafidi, Physical properties of a LPG flame with high-temperature air on a regenerative burner, *Combustion and Flame* **136**, 2004, p. 567–69.

[112] H. Hausen, Heat Transfer in Counter flow, Parallel Flow and Cross Flow, ISBN 0-07-027215-8, McGraw-Hill, New York, 1983.

[113] D. R. Athey and P.E.Chew, Calculation of High Temperature Regenerative Heat Exchangers, *Proc. Of the International Centre for Heat and Mass Transfer*, edited by Sheindlin A. E., Hemisphere Publication Corp., USA, 1986, p. 73-113.

[114] P.J. Heggs Calculation of High Temperature Regenerative Heat Exchangers, *Proc. of the International Centre for Heat and Mass Transfer,* edited by Sheindlin A. E., Hemisphere Publication Corp., USA, 1986, p. 115-49.

[115] H. Klein, G. Eigenberger, Approximate Solution for metallic regenerative heat exchangers, *International Journal of Heat and Mass Transfer.* Vol. 44, 2001, p. 3553-63.

[116] V. Ganapathy, Cold-end corrosion: causes and cures, *Hydrocarbon Processing* **68**, 1989, p. 57-9.

[117] J.M. Blanco, F. Mendía and F. Peña, Comparative analysis of CO_2 and SO_2 emissions between combined and conventional cycles with natural gas and fuel oil consumption over the Spanish thermal power plants, *Fuel*, **85**, 2006, p.1280-85.

[118] Bejan, G. Tsatsaronis and M. Moran, Thermal design and optimization, John Wiley & Sons, Inc, New York, 1996.

[119] Vincent and M. Meneguzzi, The spatial structure and statistical properties of homogeneous turbulence, *J. Fluid Mech.* **225**, 1991, p. 1-20.

[120] F.P. Incropera and D.P. Dewitt, Fundamentals of Heat and Mass Transfer (fifth ed.), John Wiley and Sons, New York, 2002.

[121] T. M. Liou, J.J. Hwang and S.H. Chen, Turbulent transport phenomena in a channel with periodic rib turbulators, *J. Thermophys. Heat Transfer.* **6**, 1992, p. 513-21.

[122] E.A. Sewall and D.K. Tafti, Large Eddy Simulation of Flow and Heat Transfer in the 180° bend Region of a Stationary Ribbed Gas Turbine internal Cooling Duct. In: GT 2005-68518, ASME Turbo Expo, 2005, USA.

[123] Murata and S. Mochizuki, Comparison between laminar and turbulent heat transfer in stationary square ducts with transverse or angled rib turbulators, *Int. J. Heat Mass Transfer.* **44**, 2001, p. 1127-41.

[124] V. Ekkad and J.C. Han, Detailed heat transfer distributions in tow-pass square channels with rib turbulators, *Int. J. Heat Mass Transfer.* **40**, 1997, p. 2525-37.

[125] W.D. Curtis, J.D. Logan and W.A. Parker, Dimensional analysis and the pi theorem. *Lin. Alg. Appl.* **47**, 1982, p. 117-26.

[126] D. Naylor and J.O. Tarasuk, Natural convective heat transfer in a divided vertical channel: Part I-numerical study, *ASME J. Heat Transfer.* **115**, 1993, p. 377-87.

[127] S.W. Churchill and R. Usagi, A general expression for the correlation of rates of transfer and other phenomena, *AIChE J.*, **18**, 1972, p. 1121-28.

[128] R.A. Antonia and P.A. Krogstad, Turbulence structure in boundary layers over different types of surface roughness, *Fluid Dynamics Research.* **28**, 2001, p. 139-57.

[129] A.J. Grass, R.J. Stuart and M. Mansour-Tehrani, Common vortical structure of turbulent flows over smooth and rough boundaries, *AIAA Journal.* **331**, 1993, p. 437-47.

[130] J.M Sala, *Cogeneration: Thermodynamic, Technological and Economy Aspects*, UPV/EHU, Publishing Services, Bilbao, 1994.

[131] American Gas Association, *Guide to Natural Gas Cogeneration,* 2nd edition, 1991, p. 234-237.

[132] ASHRAE, Cogeneration, *Fuels and Combustion*, Technical Data Bulletin (ASHRAE Winter Meeting, San Francisco), 1996, p. 43-49.

[133] J.P. Holman, Heat transfer, seventh ed., McGraw-Hill, 1990, p. 340-348.

[134] A.J. Chapman, Heat Transfer, *The Macmillan Company.* 1977, 145-147.

[135] N. Berglin; T. Bertntsson, CHP in the pulp industry using black liquor gasification: thermodynamic analysis, *Applied Thermal Engineering.* **18**, 1998, p. 947-961.

[136] S. Consoni; E.D. Larson; T.G. Kreutz; N. Berglin, Black liquor gasifier/gas turbine cogeneration, *ASME Journal Engineering for Gas Turbines and Power.* **120**, 1998, p. 442-449.

[137] F.W. Schmidt, A.J. Willmot, Thermal Energy Storage and Regeneration, ISBN 0-07-055346-7, McGraw-Hill, New York, 1981.

[138] Hydrogen and Carbon Dioxide Gas Control System; Operating and maintenance reference instructions for the combined cycle turbo generator. *Ref. General Electric.* GEK-103763.

[139] Maintenance reference instructions for the turbo generators from the PR. *General Electric.* GEI-74441C, Gas Control, 2003.

[140] Bejan, G. Tsatsaronis and M. Moran, Thermal design and optimization, John Wiley & Sons, Inc, New York, 1996.

[141] N. Nakicenovic, A. Grübler and A. McDonald, Editors, *Global energy perspectives*, Cambridge University Press, Cambridge, UK, 1998.

[142] U. Colonbo, Development and the global environment. In: J.M. Hollander, Editor, *The energy–environment connection*, Island Press, Washington,1992, p. 3-14.

[143] F.P. Incropera and D.P. Dewitt, Fundamentals of Heat and Mass Transfer (fifth ed.), John Wiley and Sons, New York, 2002.

[144] A.F. Mills, Heat Transfer (second ed.), University of California, Los Angeles, California, 1999.

[145] W. Li and K. Sadik, Unsteady thermal entrance heat transfer in laminar state with a periodic variation inlet temperature, *Int. J. Heat Mass Transfer.* **34**, 10, 1991, p. 2581-92.

[146] Y. Tan, M.A. Douglas and K.V. Thambimuthu, CO_2 capture using oxygen enhanced combustion strategies for natural gas power plants, *Fuel.* **81**, 2002, p. 1007-1016.

[147] E.K. Zariffeh, H.M. Soliman and A.C. Trupp, The combined effect of wall and fluid axial conduction on laminar heat transfer in circular tubes, *Heat Transfer.* **4**, 1982, p. 131-35.

[148] Hinze, J.O., *Turbulence*, New York, McGraw-Hill Publishing Co., 1975.

[149] M.J. Moran and H.N. Shapiro, *Fundamentals of engineering thermodynamics*, Wiley, New York 1995.

[150] G.J. Van Wylen and R.E. Sonntag, *Fundamentals of classical thermodynamics* SI version (2nd ed), Wiley, New York, 1985, p. 464–74.

[151] C.L. Yaws, *Handbook of transport property data—viscosity, thermal conductivity, and diffusion coefficients of liquids and gases*, Gulf Publishing Company, Houston, 1995.

[152] D. Naylor and J.O. Tarasuk, Natural convective heat transfer in a divided vertical channel: Part I-numerical study, *ASME J. Heat Transfer.* **115**, 1993, p. 377-87.

[153] A.J. Chapman, *Heat transfer*, Ed. Interciencia, second ed., 1968.

[154] C.E. Baukal, *Heat transfer in industrial combustion*, ISBN 0-8493-1699-5, CRC Press LLC, USA, 2000.

[155] H.Y. Zhang, M.A. Ebadian and A. Campo, Effects of heat generation and axial heat conduction in laminar flow inside a circular pipe with a step change in wall temperature, Trans. *ASME J. Heat Transfer.* **102**, 1980, p. 58-63.

In: Thermal Engineering Research Developments
Editors: J. Evgova and O. Kostadinov, pp. 85-152
ISBN: 978-1-60741-497-1

Chapter 2

Study on Adsorption and Thermoelectric Cooling Systems Using Boltzmann Transport Equation Approach

***Bidyut Baran Saha*[1,a], *Anutosh Chakraborty*[2,b], *Kim Choon Ng*[1,c] and *Ibrahim I. El-Sharkawy*[3,d]**

[1] Department of Mechanical Engineering, National University of Singapore; 9 Engineering Drive 2, Singapore 117576.

[2] School of Mechanical and Aerospace Engineering, Nanyang Technological University; 50 Nanyang Avenue, Singapore 639798.

[3] Mechanical Power Engineering Department, Faculty of Engineering, Mansoura University, El-Mansoura 35516, Egypt

Abstract

In this chapter, the Boltzmann Transport Equations (BTE) is used to formulate the transport laws for equilibrium and irreversible thermodynamics and these BTE equations are suitable for analyzing system performance that are associated with systems ranging from macro to micro dimensions. In this regard, particular attention is paid to analyze the energetic processes in adsorption phenomena as well as in semiconductors from the view point of irreversible thermodynamics. The continuity equations for (i) gaseous flow at adsorption surface, and (ii) electrons, holes and phonons movements in the semiconductor structures are studied. The energy and entropy balances equations of (i) the adsorption system for macro cooling, and (ii) the thermoelectric device for micro cooling are derived that lead to expressions for entropy generation and system's bottlenecks. The BTE equation is applied to model the adsorption cooling processes for single-stage, multi-stage and multi-bed systems, and the simulated results are compared with experimental data. This chapter also presents a thermodynamic framework for the estimation of the minimum driving heat source temperature of an advanced adsorption cooling device from the rigor of Boltzmann distribution function. From this

[a] E-mail address: mpebbs@nus.edu.sg. Phone: +65-6516-4545; Fax: +65-6774-1459
[b] E-mail address: c.anutosh@gmail.com
[c] E-mail address: mpengkc@nus.edu.sg. Phone: +65-6516-2214; Fax: +65-6774-1459.
[d] E-mail address: ib_sharkawy@yahoo.com

thermodynamic analysis, an interesting and useful finding has been established that it is possible to develop an adsorption cooling device as a green and sustainable technology that operates with a driving heat source temperature of near ambient. Moreover, the Onsager relations are applied to model the thermoelectric transport equations and, after coupling with Gibbs law and BTE, the temperature-entropy flux derivations are further developed and presented the energetic performances of thermoelectric cooling systems.

List of Symbols

A	area	m^2
A_{te}	cross sectional area of a thermoelectric element	m^2
COP	the Coefficient of Performance	--
c	the mass fraction	$kg\ kg^{-1}$
c_p	specific heat capacity	$J\ kg^{-1}\ K^{1}$
$c_{p,sg}$	specific heat of silica gel	$J\ kg^{-1}\ K^{-1}$
$c_{p,te}$	specific heat capacity of thermoelectric material	$J\ kg^{-1}\ K^{-1}$
c_v	specific heat capacity at constant volume	$J\ kg^{-1}\ K^{-1}$
D_{so}	a kinetic constant for the silica gel water system	
d^3	3-D operator	
E	electric field	volt
E_a	activation energy of surface diffusion	$J\ kg^{-1}$
e	the total energy per unit mass	$J\ kg^{-1}$
F	the Faraday constant	$Coulomb\ mol^{-1}$
$\mathbf{F}$	the external force	N
f	the statistical distribution function	
h	enthalpy per unit mass	$J\ kg^{-1}\ K^{-1}$
h	convective heat transfer coefficient	$Wm^{-2}\ K^{-1}$
I	the current	amp
$\mathbf{J}$	heat flux	Wm^{-2}
$\mathbf{J}_s$	entropy flux	$W\ m^{-2}\ K^{-1}$
JE	electric field	$W\ m^{-3}$
$J(=I/A)$	the current density	$amp\ m^{-2}$
$\mathbf{J}_k$	diffusion flow of component k	
$\mathbf{J}_p$	the momentum flux	$N\ m^{-3}$
$\mathbf{J}_{s,pn}$	the entropy flux both the p and n leg	$W\ m^{-2}\ K^{-1}$
k	the thermal conductivity	$W\ m^{-1}\ K^{-1}$
K_o	pre-exponential coefficient	kPa^{-1}
k_{te}	thermal conductivity of thermoelectric element (equations 5.28 and 5.31)	$W\ m^{-1}\ K^{-1}$
L	length	m
L_{te}	characteristics length of thermoelectric element	m
M	current pulse magnitude	--
m^*	the mass of a single molecular particle	kg
m	mass	kg
$m_a{}^*$	mass of adsorbate under equilibrium condition	kg
n	carrier density	m^{-3}

n_m	number of thermoelectric modules	
$\mathbf{p}$	the momentum vector	
P	pressure	Pa
P_1	saturated pressure	Pa
$P_{in,ncolli}$	power input with non-collision effect	mW
$P_{in,colli}$	power input with collision effect	mW
Q	the total heat or energy	W or J
Q_o	the heat transport of electrons/ holes	J mol^{-1}
$Q_{H,ncolli}$	heating power at hot side with non-collision effect	mW
$Q_{H,colli}$	heating power at hot side with collision effect	mW
$Q_{L,ncolli}$	cooling power at cold side with non-collision effect	mW
$Q_{L,colli}$	cooling power at hot side with collision effect	mW
Q_{colli}	power dissipation due to collisions	mW
R_p	average radius of silica gel particle	m
R_g	Universal gas constant	J kg^{-1} K^{-1}
$\mathbf{r}$	the particle position vector	
$\mathbf{r}'$	the new position vector at an infinitesimally time dt	
s	total entropy	J kg^{-1} K^{-1}
S_{te}	the total Seebeck coefficient	V K^{-1}
s	entropy per unit mass	J kg^{-1} K^{-1}
T	temperature	K
t	time	s
t_1	the tóth constant	-
t_{cycle}	cycle time	s
$\mathbf{U}$	a peculiar velocity	m s^{-1}
U	total internal energy of the system	J.kg^{-1}
u	internal energy per unit mass	J kg^{-1} K^{-1}
V	volume of the system	m^3
V	Input voltage to the TE modules	volt
$\mathbf{v}''$	the particle velocity vector	m s^{-1}
$\mathbf{v}'$	the particle velocity vector at an infinitesimally time	dtm s^{-1}
$\mathbf{v}$	velocity	m s^{-1}
υ (= 1/ρ)	specific volume	m^3 kg^{-1}
w	energy flow (the combination of thermal and drift energy)	J
$\mathbf{X}$	any sufficiently smooth vector field	
$\mathbf{Y}$	energy dissipation during collision (= ρ e)	J m^{-3}
Z	thermoelectric Figure of merit	K^{-1}
z	the total electric charge	coulomb
ΔT	temperature difference between the hot and cold junctions	K
ΔH_{ads}	enthalpy of adsorption	kJ kg^{-1}

Φ	the source of momentum	
Π	tensor part of a pressure tensor	
π	Peltier coefficient	V
α	the Seebeck coefficient	VK^{-1}
χ	the substantive derive of any variable	
ρ	the effective density	kg m^{-3}
τ	average collision time	fs
Γ	the Thomson coefficient	VK^{-1}
λ	the thermal conductivity	$Wm^{-1}K^{-1}$
λ	the phonon wavelength	nm
ψ	potential energy per unit mass	J kg^{-1}
μ	the thermodynamic or chemical potential	
μ_l	liquid viscosity (equation 5.35)	kg/ms
σ	the entropy source strength	W m^{-3} K^{-1}
σ_s	Surface tension (equation 5.35)	N m^{-1}
σ'	the electrical conductivity	ohm^{-1} m^{-1}
$\nabla\phi$	the electric potential gradient	V m^{-1}

Subscripts

a	adsorbate
ads	adsorption
b	barrier
b,e	electrons in the barrier of superlattice
b,h	holes in the barrier of superlattice
b,g	phonons in the barrier of superlattice
colli	the collision term
colli e-h	collision between electrons and holes
colli e-g	collision between electrons and phonons
colli h-e	collision between holes and electrons
colli h-g	collision between holes and phonons
colli g-e	collision between phonons and electrons
colli g-h	collision between phonons and holes
cond	condenser
des	desorption
e	electron
evap	evaporator
f	liquid phase
g	phonons or gas
gas	gaseous phase
H or hj	hot junction
HX	heat exchanger
h	hole
in	inlet

irr	irreversible
L or *cj*	cold junction
l	liquid phase
max	maximum
min	minimum
np	non-pulsed period
o	outlet
p	the pulse period
pr	Prediction
q	Joulean heat flux
$\dot{\mathbf{r}}$	rate of change of position vector
rev	reversible
sys	system
TE	thermoelectric modules
te	thermoelectric device
th	theoretical
tu	tube
$\dot{\mathbf{v}}''$	rate of change of velocity vector
w,e	electrons in the well of superlattice
w,h	holes in the well of superlattice
w,g	phonons in the well of superlattice
w	well or water refrigerant
w,i	water inlet
w,o	water outlet

Chill	Chilled water
fin	Fin
Ref	Refrigerant
Cycle	Cycle time
F	Fluid
M	Maximum value
f-m	Fluid-metal
m-s	Metal-silica gel
fin-s	Fin-silica gel
eff	Effective value
s	Saturation

Superscripts

sg	Silica gel
evap	Evaporator
cond	Condenser
bed	Sorption bed
chill	Chill water
f	Liquid phase

g	Gaseous phase
in	Inlet
tube	Heat exchanger tube
a	Adsorbate
cool	Cooling fluid
heating	Heating fluid

1. Introduction

In the past few decades, significant progress has been made in development of adsorbents and semiconductor materials for cooling applications, for example, (i) the adsorption cooler employing silica gel, zeolites, activated carbon; and (ii) the thermoelectric coolers using the Bismuth-Telluride (Bi2Te3), Antimony-Telluride (Sb2Te3), Silicon-Germanium (SiGe) [1]. The adsorption chillers are also miniaturized with micro-technology based miniature adsorbers and desorbers for heat pump and cooling applications [2-4]. With greater needs for miniaturization of micro or mini-chillers, these semi-conductors are fabricated to within a few microns thick using the chemical-vapor deposition (CVD) techniques for making the thin-film materials in layers of the positive-negative or p-n junctions. The law of equilibrium thermodynamics as well as the conservation laws, which are well established for analyzing energy and mass transfers of macro control volumes or systems such as adsorption or bulk thermoelectric device, may not be sufficient to handle the systems of micro dimensions such as thin film superlattice thermoelectric element [5, 6], in particular, their inability to handle the dissipation phenomena associated with high density electrons, holes and phonons collisions or flows and these dissipations are known to have set real limits to the performance of micro systems.

In this chapter, the Boltzmann Transport Equation (BTE) [7, 8] is used to formulate the transport laws for equilibrium and irreversible thermodynamics. The BTE equations are deemed to be suitable for analyzing micro-scaled systems because they account for the collisions terms associated with the high density fluxes of electrons and phonons in the thin films. In figure 1, the relative merits of applying the classical Gauss theorem (control volume approach), BTE and the molecular dynamics model is elaborated with respect to the length and time scales. For example, at wavelengths ranging from a phonon (1~2 nm) to its mean free path (300 nm), the molecular dynamic model is usually employed [9]. In regions higher the mean free path of phonons, the Boltzmann Transport Equation (BTE) models could be utilized to capture the collision contributions of particles and the BTE model can be extended into the realm of the Gauss theorem or control volume approach.

In this regard, particular attention is paid to the energy, momentum and mass conservation properties of the collision operator in sub-micron semiconductor devices. It is noted that the dissipative effects from the collision terms are translated into unique expressions of entropy generation where one could analysis systems performance using simply the classical temperature-entropy (T-s) diagrams that capture the rate of energy input, the dissipation and the useful effects. Another advantage of using the BTE approach is that when the systems to be analyzed enters into a macro-scale domain, the collision terms could simply be omitted (as the effects are known to be additive) and the conservation laws revert back to that of the classical thermodynamics, a method similar to the direction taken by the works of de Groot and Mazur. [10]

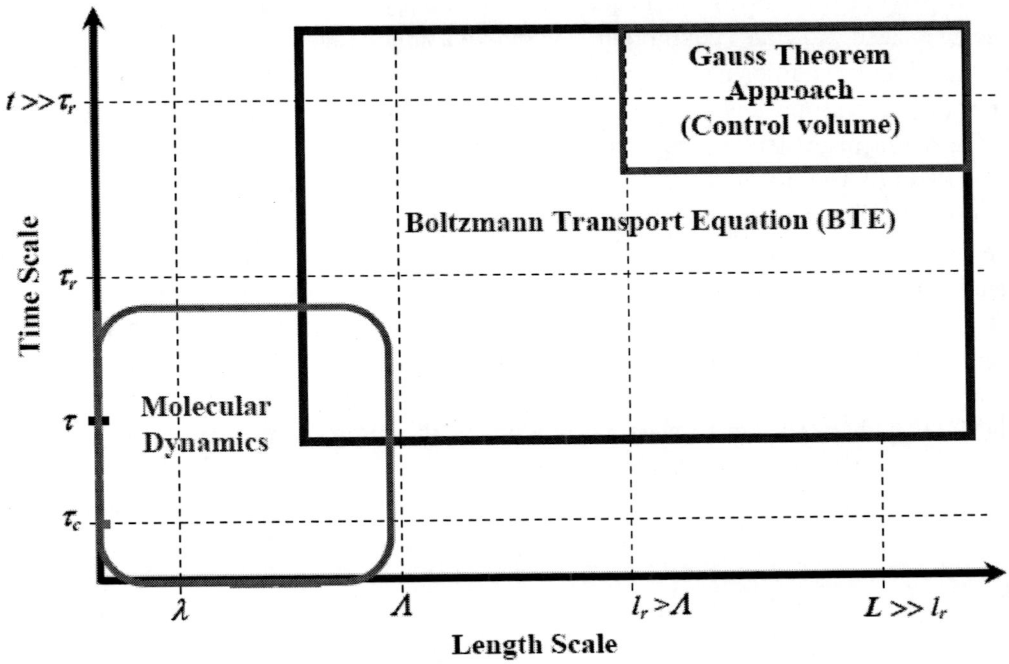

Figure 1. Schematic diagram of a thermal transport model showing the applicability of Boltzmann Transport Equation (BTE) in the thin films, where τc is the wave interaction time (100 fs), τ defines the average time between collisions, τr is the relaxation time, λ indicates the phonon wavelength (1~2 nm), Λ is the mean free path (~300 nm) and lr is the distance corresponding to relaxation time. (Lee, 2005). Hence when L ~ λ, wave phenomena exhibit, when L ~ Λ and t >> τ, τr, ballistic transport and there is no local thermal equilibrium, when L >> lr and t ~ τ, τr, statistical transport equations are used, when L >> lr and t >> τ, τr, local thermal equilibrium can be applied over space and time, leading to macroscopic transport laws.

Prior to writing down the BTE formulation below, some aspects of classical thermodynamic development in the recent years are first reviewed. One such topic that has been greatly discussed in the literature is the finite time thermodynamics (FTT) [11], which has been applied to problems of isothermal transfers to chiller cycles called the reversible cycle. In the FTT analysis, the processes of heat transfer (from and to the heat reservoir) are time dependent but all other processes (not involving heat transfers) are assumed reversible. Take for example, an endo-reversible chiller that has been commonly modelled [12]. Invariably, they assumed reversibility for the flow processes of its working fluid but irreversible processes are arbitrary applied and restricted to the heat interactions with the environment or heat reservoirs. Owing to the arbitrary assumption of reversibility for the flow processes of working fluid within the chiller cycle, the finite time thermodynamics (FTT) faces an “uncertain treatment” [13] and hence, it renders itself totally impractical in the real world.

On the other hand, there has been an excellent development in the general argument of equilibrium and irreversible that applies to macro systems. It is built upon an approach following the method of de Groot and Mazur [10]. They proposed the theory of fluctuations that describes non-equilibrium (irreversible) phenomena, based on the basic reciprocity relations [14], and relied on the microscopic as well as drawing phenomenon about macroscopic behavior. Most non-equilibrium thermodynamics assumes linear processes occurring close to an equilibrium thermodynamic state and assumes that the phenomenological coefficients are constant [15].

For the microscopic and sub-microscopic domains, the transport equations are developed from the rudiments of Boltzmann Transport Equation (BTE) but conforming to the First and Second Laws of Thermodynamics. In the sections to follow, the main objective is to discuss and formulate the basic conservation laws (mass, momentum and energy balances) for the thin films and based on the specific dissipative losses encountered in these layers, the entropy generation with respect to the electrons, holes and phonons fluxes will be formulated.

The organization of this chapter is as follows: Section 2 describes the general form of conservation equations. The equations to be discussed are the mass, momentum and energy conservation and the properties of the collision operator. In section 3, the mathematical rigor of the BTE with respect to irreversible production of entropy is presented. Following the Gibbs expression on entropy, the entropy balance equation is established in relation with the conservation equations, which comprise a source term or entropy source strength term. Section 3 tabulates a list of the common entropy generation mechanisms and demonstrates the manner the terms are deployed in the BTE formulation. For application examples of this demonstration presented in the previous sections, section 4 demonstrates the mathematical modeling and the performance analysis of adsorption cooling systems. Based on the conservation laws and the entropy balance equations, section 5 develops the temperature-entropy formulations of a solid state thermoelectric cooling device, a pulsed thermoelectric cooler and a micro-scale thermoelectric cooler. The thermodynamic performances of these solid state coolers are discussed graphically with T-s diagrams. In this section, the mathematical modelling of the thin film superlattice thermoelement is developed from the basic Boltzmann Transport Equation where the collision terms are included. This chapter finally ends with a conclusion.

2. General Form of Balance Equations

In this section, the transport processes involving the mass, momentum, and energy are described from the basic conservation laws, developed together with the transport of heat or energy by molecular fluxes such as electrons, holes and phonons within the micro layers in a system.

2.1. Derivation of the Thermodynamic Framework

Figure 2 shows a schematic of an ensemble or a group of molecular particles, such as an electrons, holes or phonons, represented initially at a time t where their positions and velocities are expressed in the range $(d^3\mathbf{r})$, $(d^3\mathbf{v}'')$ near $\mathbf{r}$ and $\mathbf{v}''$, respectively. The operator, d^3 indicates the three dimensional spaces of the variables $\mathbf{r}$ and $\mathbf{v}''$.

At an infinitesimally small time dt later, the molecular particles move, under the presence and influence of an external force $\mathbf{F}(\mathbf{r},\mathbf{v}'')$, to a new position $\mathbf{r}'=\mathbf{r}+\dot{\mathbf{r}}dt$ and attaining a velocity $\mathbf{v}'=\mathbf{v}''+\dot{\mathbf{v}}''\,dt$. Owing to collisions of molecular particles, the number of molecules in the range $d^3\mathbf{r}\,d^3\mathbf{v}''$ could be also changed where particles or molecules originally outside the range $(d^3\mathbf{r})(d^3\mathbf{v}'')$ can be scattered into the domain under consideration. Conversely, molecules originally inside the domain range could be scattered out. Following the framework of Reif, F. [16], the generic form of the conservation laws is additive and is given by,

$$\overbrace{f(\mathbf{r}',\mathbf{v}',t')\,d^3(\mathbf{r}')d^3(\mathbf{v}')}^{|Molecules\_at\_(t'),(r')\&v'|}=\overbrace{f(\mathbf{r}\;,\mathbf{v}'',t\;)\,d^3(\mathbf{r}\;)d^3(\mathbf{v}'')}^{|Molecules\_at\_(t\;),(r\;)\&v\;|}+\overbrace{f_{colli}(\mathbf{r}\;,\mathbf{v}'')d^3(\mathbf{r}')d^3(\mathbf{v}'')}^{|Molecules\_at\_(r')\&v''|} \quad (1)$$

where $t'=t+dt$, the range $(\mathbf{r}'=\mathbf{r}+\dot{\mathbf{r}}dt)$ and $(\mathbf{v}'=\mathbf{v}''+\dot{\mathbf{v}}''\,dt)$. $f(\mathbf{r},\mathbf{v},t)$ is the statistical distribution function of an ensemble particle, which varies with time t, particle position vector **r**, and velocity vector **v**.

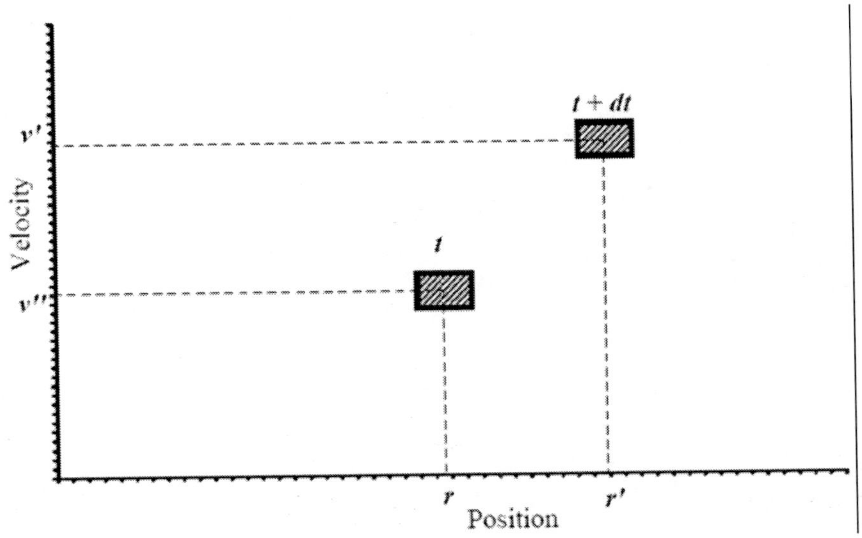

Figure 2. The motion of a particle specified by the particle position r and its velocity v''.

For convenience, we drop the implied 3-D operator, d^3, the equation reduces to a more readable form, i.e.,

$$f(\mathbf{r}+\dot{\mathbf{r}}dt,\mathbf{v}''+\dot{\mathbf{v}}''\,dt,t+dt)=f(\mathbf{r},\mathbf{v}'',t)+f_{colli}(\mathbf{r},\mathbf{v}'',t)\,dt$$

$$or\quad the\quad collision\quad term\quad is$$

$$f_{colli}(\mathbf{r},\mathbf{v}'',t)dt=\;f(\mathbf{r}+\dot{\mathbf{r}}dt,\mathbf{v}''+\dot{\mathbf{v}}''\,dt,t+dt)\;-f(\mathbf{r},\mathbf{v}'',t)$$

By invoking the partial derivatives, the basic Boltzmann transport equation is now expressed for a situation that handles the molecular flux over a thin layer of materials, i.e.,

$$\frac{\partial f_{colli}}{\partial t} = \left(\frac{\partial f}{\partial r}\frac{\partial r}{\partial t} + \frac{\partial f}{\partial v''}\frac{\partial v''}{\partial t}\right) + \frac{\partial f}{\partial t}$$

rearranging

$$\frac{\partial f}{\partial t} = -\left(\frac{\partial f}{\partial t}\right)_{drift} + \frac{\partial f_{colli}}{\partial t}$$

and the subscript "drift" refers to the flow of flux through space (**r**) and convective velocity (***v***) and the second term is the effect of collisions with time. It is noted that the term "drift" is a non-equilibrium transport mechanism that associates with the external force $\mathbf{F}\ (\mathbf{r}, \mathbf{v}'')$, i.e.,

$$\left(\frac{\partial f}{\partial t}\right)_{Drift} = \frac{\partial f}{\partial \mathbf{r}}\frac{d\mathbf{r}}{dt} + \frac{\partial f}{\partial \mathbf{v}''}\frac{d\mathbf{v}''}{dt}, \tag{2}$$

where $\mathbf{r}$ is the position vector, $\mathbf{v}''$ is the velocity vector, i.e., $\mathbf{v}'' = \frac{d\mathrm{r}}{dt}$, and the rate of change of $\mathbf{v}''$ is $\frac{d\mathbf{v}''}{dt} = \frac{\mathbf{F}}{m*}$, where m^* is the mass of a single molecular particle.

The physical meanings of velocity and electric field are described in the drift term of equation (2) is now fully elaborated as,

$$\left(\frac{\partial f}{\partial t}\right)_{Drift} = \mathbf{v}''\frac{\partial f}{\partial \mathbf{r}} - \frac{\mathbf{F}}{m*}\frac{\partial f}{\partial \mathbf{v}''} = \mathbf{v}''\frac{\partial f}{\partial \mathbf{r}} - \mathbf{F}\frac{\partial f}{\partial \mathbf{p}},$$

where the momentum $\mathbf{p} = \mathrm{m} * \mathbf{v}''$, is that associated with a molecular particle. Substituting back into equation (2), the general form of the transport equation with respect to **r**, **p** and *t* becomes

$$\frac{\partial f}{\partial t} = -\mathbf{v}''\frac{\partial f}{\partial \mathbf{r}} + \mathbf{F}\frac{\partial f}{\partial \mathbf{p}} + \frac{\partial f_{colli}}{\partial t}. \tag{3}$$

Having formulated the basic form of the Boltzmann transport equation or BTE in short, and two other functions will be used along the BTE and they are: Firstly, the carrier density of the molecular flux is now incorporated by defining the carrier density as the integral of the product between the molecular particles and the change in their momentum [7], i.e.,

$$n = \int f d\mathbf{p} . \tag{4}$$

Secondly, the substantive derive of any variable, χ, that varies both in time and in space can be expressed in vector notation as $\frac{d}{dt} = \frac{\partial}{\partial t} + \nabla\chi . v = \frac{\partial}{\partial t} + (v. \nabla)\ \chi$, or simply can be written as $\frac{\mathrm{d}}{\mathrm{dt}} = \frac{\partial}{\partial \mathrm{t}} + \mathbf{v} \cdot \nabla$. The following conservation equations can now be formulated with the BTE format and they are elaborated here below.

2.2. Mass Balance Equation

For a given space and time domain, the basic Boltzman transport equation (3) is now invoked to give

$$\frac{\partial f}{\partial t} = -\mathbf{v}''\frac{\partial f}{\partial \mathbf{r}} + \mathbf{F}\frac{\partial f}{\partial \mathbf{p}} + \frac{\partial f_{colli}}{\partial t}$$

where the partials of f to the momentum and space in 3-dimensions can be represented by applying the "Del" operator, i.e., $\nabla_p f = \frac{\partial f}{\partial \mathbf{p}}$, $\nabla f = \frac{\partial f}{\partial \mathbf{r}}$. Thus, integrating on both sides of equation (3) with respect to the momentum space, and noting the definition of the carrier density n (equation 4), yields:

$$\frac{\partial n}{\partial t} = -\nabla(\mathbf{v}''n) + \mathbf{F}\int \nabla_p f d\mathbf{p} + \frac{\partial n_{colli}}{\partial t} \tag{5}$$

Should $d\mathbf{p}$ be small in the space, the second term of right hand side of equation (5) can be omitted and equation (5) reduces to a familiar expression of the form;

$$\frac{\partial n}{\partial t} = -\nabla(\mathbf{v}''n) + \frac{\partial n_{colli}}{\partial t}, \tag{6}$$

where the collision effects from molecular particles are still retained. Defining the effective density as $\rho = m^* n$, where m^* is the mass of single molecular particle. For simplicity, assuming the particles move with an average velocity, then $\mathbf{v} = \mathbf{v}''$, and rearranging gives the mass conservation as

$$\frac{\partial \rho}{\partial t} = -\nabla(\mathbf{v}\rho) + \left.\frac{\partial \rho}{\partial t}\right|_{colli} . \tag{7}$$

The advantage of equation (7) is that it is perfectly valid for describing electrons or phonons flow within submicron semiconductor devices and the collision term takes them into account. In the case of a macro-scale domain, the mass balance omits the collision term, i.e.,

$$\frac{\partial \rho}{\partial t} = -\nabla.(\rho\, \mathbf{v}). \tag{8}$$

The form of equation (8) can also be derived using the Gauss theorem within a control volume method [17]. For example, the rectangular coordinates nomenclature for the component velocities for $\mathbf{v} = u\hat{\mathbf{i}} + v\hat{\mathbf{j}} + w\hat{\mathbf{k}}$ would yield the continuity equation as

$$\frac{\partial \rho}{\partial t} + \frac{\partial(\rho u)}{\partial x} + \frac{\partial(\rho v)}{\partial y} + \frac{\partial(\rho w)}{\partial z} = 0. \tag{9}$$

Only when steady state is considered, i.e., $\frac{\partial \rho}{\partial t} = 0$; and a case of constant-fluid density, the following simplified continuity equation reduces to

$$\frac{\partial u}{\partial x} + \frac{\partial v}{\partial y} + \frac{\partial w}{\partial z} = 0 \tag{10}$$

In vectorial notation, the mass conservation for a macro domain and constant density situation is simply expressed as

$$\begin{aligned} \frac{\partial \rho}{\partial t} &= -\nabla \cdot (\rho \mathbf{v}) \\ \frac{d\rho}{dt} &= -\rho \nabla \cdot \mathbf{v} \end{aligned} \tag{11}$$

2.3. Momentum Balance Equation

Starting from the basic transport equation,

$$\frac{\partial f}{\partial t} = -\mathbf{v}'' \frac{\partial f}{\partial \mathbf{r}} + \mathbf{F} \frac{\partial f}{\partial \mathbf{p}} + \frac{\partial f_{colli}}{\partial t},$$

and multiply the momentum, p, on both sides before integrating over the momentum space yields;

$$\frac{\partial}{\partial t} \int \mathbf{p} f\, d\mathbf{p} = -\int \mathbf{p}(\mathbf{v}''.\nabla f) d\mathbf{p} + \int \mathbf{p}(\mathbf{F}.\nabla_p f) d\mathbf{p} + \int \mathbf{p} \frac{\partial f_{colli}}{\partial t} d\mathbf{p} \tag{12}$$

Invoking the definition of the carrier density, n. and the left hand term of equation (12) can be expanded as,

$$\frac{\partial}{\partial t}\int \mathbf{p} f \, d\mathbf{p} = \frac{\partial(n\mathbf{p})}{\partial t} = \frac{\partial(nm^*\mathbf{v})}{\partial t} = \frac{\partial(\rho\mathbf{v})}{\partial t}$$

whilst the first term on the right hand side of (12) is reduced to

$$\int \mathbf{p}(\mathbf{v}''.\nabla f)d\mathbf{p} = \nabla.\int(\mathbf{v}''\mathbf{p}f)d\mathbf{p} = \nabla.\mathbf{J}_p,$$

where $\mathbf{J}_p$ is a tensor. This flux density of momentum can further be transformed to

$$\nabla.\mathbf{J}_p = \nabla.\left(\mathbf{v}''\mathbf{p}\int f\, d\mathbf{p}\right) = \nabla.(\rho\mathbf{v}''\mathbf{v}'') = \nabla.(\rho\mathbf{v}\mathbf{v} + \mathbf{P}),$$

where according to Reif, F. [17], the velocity vector $\mathbf{v}''(= \mathbf{v} + \mathbf{U})$ consists of a mean velocity v and a peculiar velocity U, and the pressure tensor is given by, $\mathbf{P} = \rho\mathbf{U}_\alpha\mathbf{U}_\gamma = \rho\mathbf{U}\mathbf{U}$.

The second term on the right hand side of equation (12) is given by,

$$\int \mathbf{p}\left(\mathbf{F}.\nabla_p f\right)d\mathbf{p} = \mathbf{F}\int \nabla_p.(\mathbf{p}f)\,d\mathbf{p} + \mathbf{F}\int f\,\nabla_p.\mathbf{p}\,d\mathbf{p} = \mathbf{\Phi},$$

known as the source of momentum, and the last term on the right hand side refers to as the collision integral,

$$\int \mathbf{p}\left(\frac{\partial f}{\partial t}\right)_{colli} d\mathbf{p} = \left.\frac{\partial(\rho\mathbf{v})}{\partial t}\right|_{colli}.$$

Combining these expressions from above, equation (7) is now summarized in terms of the physical variables ($\rho, \mathbf{v}, \mathbf{p}$) as

$$\frac{\partial(\rho\mathbf{v})}{\partial t} = -\nabla.(\mathbf{P} + \rho\mathbf{v}\mathbf{v}) + \mathbf{\Phi} + \left.\frac{\partial(\rho\mathbf{v})}{\partial t}\right|_{colli}. \tag{13}$$

The vector product vv, is also known as the Dyadic product which is given by a set of matrix functions with respect to the component velocities. Detailed manipulation of the Dyadic product can be found in many mathematics texts. In general, two kinds of external forces may act on a fluid: (i) the body forces which are proportional to the volume such as the gravitational, centrifugal, magnetic, and/or electric fields, and (ii) the surface forces which are proportional to area, for example, the static pressure drop due to viscous stresses. From a microscopic viewpoint, the pressure tensor, P, is a related to short-range interaction between the particles of system, whereas the body forces, Φ, pertains to the external forces in the system. Equation (13) describes a balance for the momentum density ($\rho\mathbf{v}$), the momentum flow, $(\mathbf{P}+\rho\,\mathbf{v}\,\mathbf{v})$ with a convective portion $(\rho\,\mathbf{v}\,\mathbf{v})$, and Φ, is a source of momentum. Some examples of the type of source terms employed in two different application scenarios are outlined in table 1.

For a macro scale flow situation, the collision term is omitted and equation (13) reduces to the conventional form as

$$\begin{aligned}\frac{\partial(\rho\,\mathbf{v})}{\partial t} &= -\nabla.(\mathbf{P}+\rho\,\mathbf{v}\,\mathbf{v})+\mathbf{\Phi} \\ \rho\frac{d\mathbf{v}}{dt} &= -\nabla\cdot\mathbf{P}+\mathbf{\Phi}\end{aligned} \tag{14}$$

It is noted that the total pressure tensor can be further divided into a scalar hydrostatic part p, and a tensor part, Π.

Table 1. Examples of the source term, Φ, for two common applications

Field	Φ
Fluid mechanics [18]	$-\nabla.\boldsymbol{\tau}+\rho g$ **, the first term indicates the viscous force per unit volume where τ is the stress tensor. The second term shows the gravitational force.**
Electric flow [10]	$\rho z\mathbf{E}$**, where z is the total electric charge per unit mass in the control volume and E is the electric field acting on the conductor.**

2.4. Energy Balance Equation

Associated with the flow of a carrier density, n, there is a corresponding energy flow, w, such as the thermal energy and the drift energy term. By multiplying both sides of the basic BTE, i.e., equation (3), by w and integrating over the momentum space (product of mass and velocity) yields

$$\frac{\partial}{\partial t}\int \mathbf{w} f\,d\mathrm{p} = -\int \mathbf{w}\left(\mathbf{v}''.\nabla f\right)d\mathrm{p} + \int \mathbf{w}\left(\mathbf{F}.\nabla_p f\right)d\mathrm{p} + \int \mathbf{w}\frac{\partial f_{colli}}{\partial t}\,d\mathrm{p} \tag{15}$$

Expanding the left hand side of equation (15),

$$\frac{\partial}{\partial t}\int \mathbf{w} f\,d\mathrm{p} = \mathbf{w}n = nm^{*}\mathbf{e} = \rho\mathbf{e},$$

where e is the total energy per unit mass and

$$e = \frac{1}{2}\mathbf{v}^2 + \psi + u.$$

On the right hand side of equation (2.15), the spatial gradient (i.e., the first term) can be placed outside the integral to give,

$$\int \mathbf{w}\left(\mathbf{v}''.\nabla f\right)d\mathrm{p} = \nabla.\int \mathbf{v}''\mathbf{w}f\ d\mathrm{p} = \nabla.\mathbf{J}_w,$$

where $\mathbf{J}_w$ represents the energy flux, defined here as

$$\mathbf{J}_w = \int \mathbf{v}''\mathbf{w}f\ d\mathrm{p} = \rho\,\mathrm{e}\,\mathbf{v} + \mathbf{P}.\mathbf{v} + \sum_{k=1}^{n}\psi_k\mathbf{J}_k + \mathbf{J}_q,$$

and $\rho\mathbf{e}\mathbf{v}$ is a convective term, ($\mathbf{P}.\mathbf{v}$) is the mechanical work in the system, $\sum_{k=1}^{n}\psi_k J_k$ equals to a potential energy flux due to diffusion, $\mathbf{J}_q$ is the heat flow. The second term on the right hand side of equation (15) can be expanded as

$$\int \mathbf{w}\left(F.\nabla_p f\right)d\mathrm{p} = \mathbf{F}.\int \mathbf{w}\nabla_p f\,d\mathbf{p} = \frac{\rho\,\mathbf{v}\mathbf{F}}{m^*} = \mathbf{\Phi}.\mathbf{v}$$

The collision term can also be expressed in terms of the energy dissipated Y, i.e.,

$$\int \mathbf{w}\frac{\partial f_{colli}}{\partial t}d\mathrm{p} = \left.\frac{\partial(n\mathbf{w})}{\partial t}\right|_{colli} = \left.\frac{\partial \mathbf{Y}}{\partial t}\right|_{colli}.$$

Combining the above expressions, the energy balance equation is now summarized that includes the collision terms as

$$\frac{\partial(\rho\mathbf{e})}{\partial t} = -\nabla.\mathbf{J}_w + \mathbf{\Phi}.\mathbf{v} + \left.\frac{\partial \mathbf{Y}}{\partial t}\right|_{colli} \tag{16}$$

To further understand the usage of the energy balance equation (i.e., equation 16), a minor digression is needed to demonstrate the presence of kinetic and potential energy terms. Multiplying the equation of motion [equation (13)] by v, a balance equation for the kinetic energy can be obtained, i.e.,

$$\frac{\partial(\rho\mathbf{v})}{\partial t}.\mathbf{v} = -\left[\nabla.(\rho\mathbf{v}\mathbf{v}+\mathbf{P})\right].\mathbf{v} + \Phi.\mathbf{v} \Rightarrow \frac{\partial\left(\frac{1}{2}\rho\mathbf{v}^2\right)}{\partial t} = -\nabla.\left(\frac{1}{2}\rho\mathbf{v}^2\mathbf{v} + \mathbf{P}.\mathbf{v}\right) + \mathbf{P}:\nabla\mathbf{v} + \Phi.\mathbf{v} \tag{17}$$

which is commonly known as the mechanical energy term. Note that the vectorial reduction could be performed on equation (14), as given by de Groot, [10]

$$[\nabla\cdot(\rho\mathbf{vv})]\cdot\mathbf{v}=\left(\sum_{i=1}^{3}\frac{\partial}{\partial x_i}\rho v_i v_j\right)v_j \quad i,j=1,\cdots,n \quad =\sum_{i=1}^{n}\frac{\partial}{\partial x_i}\frac{1}{2}\rho v_i v_j^2 \quad =\nabla\cdot\left(\frac{1}{2}\rho\mathbf{v}^2\mathbf{v}\right)$$

and

$$-\nabla\cdot(P\cdot\mathbf{v})+P:\nabla\mathbf{v}=-\sum_{i=1}^{n}\frac{\partial}{\partial x_i}\left(\sum_{j=1}^{n}P_{ij}v_j\right)+\sum_{i=1}^{n}\sum_{j=1}^{n}P_{ij}\frac{\partial}{\partial x_i}v_j \quad i,j=1,2,3,\cdots,n$$

$$=-\sum_{i=1}^{n}\sum_{j=1}^{n}\left(v_j\frac{\partial}{\partial x_i}P_{ij}+P_{ij}\frac{\partial}{\partial x_i}v_j\right)+\sum_{i=1}^{n}\sum_{j=1}^{n}P_{ij}\frac{\partial}{\partial x_i}v_j \quad =-\sum_{j=1}^{n}v_j\sum_{i=1}^{n}\left(\frac{\partial}{\partial x_i}P_{ij}\right) \quad =-(\nabla\cdot P)\cdot\mathbf{v}$$

Similar simplifications can also be made on the rate of change of potential energy density, as shown below;

$$\begin{aligned}\frac{\partial(\rho\psi)}{\partial t}&=-\nabla.\left(\rho\psi\,\mathbf{v}+\sum_{k=1}^{n}\psi_k J_k\right)-\sum_{k=1}^{n}J_kF_k+\sum_{k=1}^{n}\sum_{j=1}^{r}\psi_k\mathbf{v}_{kj}J_j\\ \frac{\partial(\rho\psi)}{\partial t}&=-\nabla.(\rho\psi\,\mathbf{v}+\psi\,J)-\mathbf{JF}\end{aligned} \tag{18}$$

The above results indicate that should

$$\sum_{k=1}^{n}\psi_k\mathbf{v}_{kj}=0 \quad (j=1,\cdots r),$$

this implies that the potential energy is conserved into the chemical reaction situation.

$$\sum_{k=1}^{n}J_kF_k=\mathbf{JF}$$

which is referred to the source term where the particles interact with a force field such as the gravitational field or the electric charge experienced within an electric field. Adding equations (17) and (18), the combined effects become

$$\frac{\partial\left(\frac{1}{2}\rho\mathbf{v}^2\right)}{\partial t}+\frac{\partial(\rho\psi)}{\partial t}=-\nabla.\left(\frac{1}{2}\rho\mathbf{v}^2\mathbf{v}+\mathbf{P.v}\right)+P:\nabla\mathbf{v}+\Phi.\mathbf{v}-\nabla.(\rho\psi\mathbf{v}+\psi J)-\mathbf{JF}$$

$$\frac{\partial\rho\left(\frac{1}{2}\mathbf{v}^2+\psi\right)}{\partial t}=-\nabla.\left(\frac{1}{2}\rho\mathbf{v}^2\mathbf{v}+\mathbf{P.v}+\rho\psi\mathbf{v}+\psi J\right)+\mathbf{P}:\nabla\mathbf{v}-\mathbf{JF}+\Phi.\mathbf{v}$$

From the energy balance equation (16) and substituting for these derived functions, one obtains working form of the energy equation which gives in terms of the internal energy, *u,* heat flux, J_q, and the associated source and collision terms;

$$\frac{\partial(\rho \mathbf{e})}{\partial t} = -\nabla.\mathbf{J}_w + \Phi.\mathbf{v} + \left.\frac{\partial(\rho \mathbf{e})}{\partial t}\right|_{colli}$$
$$\therefore \frac{\partial(\rho u)}{\partial t} = -\nabla.(\rho u\mathbf{v} + \mathbf{J}_q) - \mathbf{P}:\nabla\mathbf{v} + \mathbf{JF} + \left.\frac{\partial \mathbf{Y}}{\partial t}\right|_{colli} \tag{19}$$

It is noted that $\rho u\text{v}+J_q$ is the energy flow due to convection and thermal energy transfers which contributes to the internal energy change, while the vectors P:∇v + JF is deemed as the source term associated with the internal energy.

Further simplification of the energy balance Equation (19) leads to its concise form of having only 4 terms and it is applicable to micro or thin layer flow of electrons or phonons;

$$\frac{\partial}{\partial t}(\rho u) = -\nabla\cdot(\rho u\mathbf{v} + \mathbf{J}_q) - \mathbf{P}:\nabla\mathbf{v} + \mathbf{JF} + \left.\frac{\partial \mathbf{Y}}{\partial t}\right|_{colli}$$
$$\rho\frac{du}{dt} = -\nabla\cdot\mathbf{J}_q + \mathbf{\Psi} + \left.\frac{\partial \mathbf{Y}}{\partial t}\right|_{colli} \tag{20}$$

It is noted that the last term of the right hand side indicates the rate of change of energy as a result of collisions of the molecular particles such as electrons or phonons. Collision occurs by four processes in the micro-sublayer, i.e., (i) the increase in carrier kinetic energy due to interactions between electrons and holes or phonons, (ii) the heat dissipation arising from electrons or phonons interaction with the lattice vibrations, (iii) the heat dissipation by the electrons to the holes and vice versa, and (iv) the heat generation arising from the recombination and generation of the molecular particles.

Equation (20) could be extended to macro analysis by simply dropping the collision term which has its form known conventionally as the energy balance equation;

$$\rho\frac{du}{dt} = -\nabla\cdot\mathbf{J}_q + \mathbf{\Psi} \tag{21}$$

Another name for equation (20) is the equation of internal energy, as it is known in some texts [10, 18].

At this juncture, it is perhaps appropriate to elaborate these terms in terms of some common application examples. For example, the source term Ψ, in equilibrium and irreversible thermodynamics has two parts: (i) One aspect is the reversible rate of the internal energy attributed by effects of compression and (ii) the other is the irreversible rate of internal energy contributed by any dissipative effect. For most engineering applications involving real gasses, the internal energy is expressed in terms of the fluid temperature and the heat capacity. In such cases, the internal energy u may be considered as a function of υ and T, and assuming constant specific heats, i.e.,

$$du = \left(\frac{\partial u}{\partial \upsilon}\right)_T d\upsilon + \left(\frac{\partial u}{\partial T}\right)_\upsilon dT = \left[-p + T\left(\frac{\partial p}{\partial T}\right)_\upsilon\right] d\upsilon + c_v dT$$

and the substantial derivative term becomes

$$\rho\frac{du}{dt} = \left[-p + T\left(\frac{\partial p}{\partial T}\right)_\upsilon\right]\rho\frac{d\upsilon}{dt} + \rho c_v\frac{dT}{dt} = \left[-p + T\left(\frac{\partial p}{\partial T}\right)_\upsilon\right]\nabla.\mathbf{v} + \rho c_v\frac{dT}{dt}$$

Invoking the energy balance equation (21), it becomes

$$\rho c_v\frac{dT}{dt} = -\nabla\cdot\mathbf{J}_q - T\left(\frac{\partial p}{\partial T}\right)_{\hat{V}}(\nabla.\mathbf{v}) + \mathbf{JF} \tag{22}$$

Should the gas be assumed ideal, then $\left(\frac{\partial p}{\partial T}\right)_{\hat{V}} = \frac{p}{T}$ and the second term of right hand side equation (22) becomes zero, i.e.,

$$\rho c_v\frac{dT}{dt} = -\nabla\cdot\mathbf{J}_q + \mathbf{JF}. \tag{23}$$

In another example, the internal energy u is considered to be a function of p and T, as in a real gas. Hence,

$$h = u + p\upsilon \Rightarrow \frac{dh}{dt} = \frac{du}{dt} + p\frac{d\upsilon}{dt} + \upsilon\frac{dp}{dt}.$$

Invoking the constant pressure assumption, it can be shown that

$$\frac{dh}{dt} = \frac{du}{dt} + p\frac{d\upsilon}{dt} \Rightarrow \rho\frac{dh}{dt} = \rho c_p\frac{dT}{dt} \tag{24}$$

From equation (24) and substituting for the energy balance equation for $\rho\frac{du}{dt}$, it can be demonstrated that the substantial derivative for the enthalpy, $\rho c_p\frac{dT}{dt}$, comprise only two terms, i.e.,

$$\rho c_p \frac{dT}{dt} = \rho \frac{du}{dt} + \rho p \frac{d\upsilon}{dt}$$
$$\rho c_p \frac{dT}{dt} = -\nabla \cdot \mathbf{J}_q - p\nabla \cdot \mathbf{v} + \mathbf{JF} + \rho p \frac{d\upsilon}{dt} \tag{25}$$

From the continuity equation (11) we show that

$$\rho p \frac{d\upsilon}{dt} = p\nabla \cdot \mathbf{v}$$

Equation (25) is written as,

$$\rho c_p \frac{dT}{dt} = -\nabla \cdot \mathbf{J}_q + \mathbf{JF} \tag{26}$$

The right hand side of equation (26) comprises (i) the heat flux term, $\mathbf{J}_q$ and (ii) the energy source term JF. Table 2 tabulates three scenarios where the source energy could be elaborated. For example, in the study of adsorption, where many chapters of the thesis are devoted, the source energy is derived by considering adsorbate mass and the isosteric heat of adsorption, i.e., $\dot{q}\Delta H_{ads}$. Other examples in fluid mechanics and electrically powered semi-conductors are also explained.

It should be emphasized that equations (21-26) are valid for macro-scale devices simulation. However, for the analysis of sub-micron devices, the collision terms, which are provided in equation (20), must be included.

Table 2. Explanation of energy source in different fields

Field	Source of energy JF
Fluid mechanics	$\tau : \nabla v$; This term indicates the irreversible internal energy increase due to viscous dissipation.
Electrical	$\rho_{elect}J^2$; this is the joulean heat.
Chemical / Adsorption	$\dot{q}\Delta H_{ads}$, where $\dot{q}$ is the adsorption or desorption rate and ΔH_{ads} is the isosteric enthalpy of adsorption.

2.5. Summary of Section 2

In section 2, the conservation laws, i.e., mass, momentum and energy are derived from the transportation of molecules using the Boltzmann Transport equation. The thermodynamic framework for describing the conservation laws is applicable to both the macro-scale and micro-scale systems. This approach differs from the control volume approach of de Groot and Mazur [10] Bird, R. B. [18], etc., where the Gauss theorem was employed. The latter approach could not embrace the collision terms, as what has been shown here using the BTE approach. For thin films or micro-scale systems, the collisions of electrons with electrons, electrons with holes, holes with holes are deemed to have significant dissipative effects, a source of irreversibility as quantified in the next section.

3. Conservation of Entropy

The change of entropy of a system relates not only to the entropy (s) crossing the boundary between the system and its surroundings, but also to the entropy produced or generated by processes taking place within the system. Processes inside the system may be either reversible or irreversible [19]. The reversible process may occur during the transfer of entropy from one portion to another of the interior without entropy generation. On the other hand the irreversible process invariably leads to entropy generation inside the system.

The entropy per unit mass is a function of the internal energy u, the specific volume υ and the mass fractions c_k i.e. $s = s(u, \upsilon, c_k)$. In equilibrium the total differential of s is given by the Gibbs relation [10]

$$Tds = du + pd\upsilon - \sum_{k=1}^{n} \mu_k dc_k \,,$$

where p is the equilibrium pressure and μ_k is the thermodynamic or chemical potential of component k.

$$T\frac{ds}{dt} = \frac{du}{dt} + p\frac{d\upsilon}{dt} - \sum_{k=1}^{n} \mu_k \frac{dc_k}{dt} \tag{27}$$

From mass balance we get

$$\frac{\rho P}{T}\frac{d\upsilon}{dt} = \frac{P}{T}\nabla\cdot\mathbf{v} + \frac{P}{\rho T}\left.\frac{\partial\rho}{\partial t}\right|_{colli}$$

The mass balance equation, equation (7), is written as,

$$\frac{d\rho_k}{dt} = -\rho_k\nabla\cdot\mathbf{v} - \nabla\cdot\mathbf{J}_k + \left.\frac{\partial\rho_k}{\partial t}\right|_{colli}$$

where $\boldsymbol{J}_k$ $\left(= \rho_k\left(\mathbf{v}_k - \mathbf{v}\right)\right)$ is the diffusion flow of component k defined with respect to the 'barycentric motion'. Defining the component mass fractions c_k as

$$c_k = \frac{\rho_k}{\rho} \quad so\ that \sum_{k=1}^{n} c_k = 1,$$

and equation (7) can be simplified as

$$\frac{d\rho_k}{dt} = -\rho_k \nabla \cdot \mathbf{v} - \nabla \cdot \mathbf{J}_k + \left.\frac{\partial \rho_k}{\partial t}\right|_{colli}$$

$$\rho \frac{dc_k}{dt} = -\nabla \cdot \mathbf{J}_k$$

From equation (27) we have,

$$T\frac{ds}{dt} = \frac{du}{dt} + p\frac{d\upsilon}{dt} - \sum_{k=1}^{n} \mu_k \frac{dc_k}{dt}$$

$$\rho\frac{ds}{dt} = \frac{\rho}{T}\frac{du}{dt} + \frac{\rho p}{T}\frac{d\upsilon}{dt} - \sum_{k=1}^{n} \frac{\rho \mu_k}{T}\frac{dc_k}{dt}$$

$$= -\nabla \cdot \left(\frac{\mathbf{J}_q - \sum_{k=1}^{n} \mu_k \mathbf{J}_k}{T} \right) - \frac{1}{T^2}\mathbf{J}_q \cdot \nabla T - \frac{1}{T}\sum_{k=1}^{n} \mathbf{J}_k \cdot \left(T\nabla \frac{\mu_k}{T} \right) + \frac{1}{T}\mathbf{JF} + \frac{1}{T}\left.\frac{\partial \mathbf{Y}}{\partial t}\right|_{colli} + \frac{P}{\rho T}\left.\frac{\partial \rho}{\partial t}\right|_{colli} \tag{28}$$

where $J_k \left(= \rho_k (\mathbf{v}_k - \mathbf{v})\right)$ is the diffusion flow of component k defined with respect to the 'barycentric motion'.

From equation (28), one observes that the entropy flow is given by two terms:

$$\mathbf{J}_s = \frac{1}{T}\left(\mathbf{J}_q - \sum_{k=1}^{n} \mu_k \mathbf{J}_k \right), \tag{29}$$

and the entropy source strength by

$$\sigma = -\frac{1}{T^2}\mathbf{J}_q \cdot \nabla T - \frac{1}{T}\sum_{k=1}^{n} \mathbf{J}_k \cdot \left(T\nabla \frac{\mu_k}{T} \right) + \frac{1}{T}\mathbf{JF} + \frac{1}{T}\left.\frac{\partial \mathbf{Y}}{\partial t}\right|_{colli} + \frac{P}{\rho T}\left.\frac{\partial \rho}{\partial t}\right|_{colli} \tag{30}$$

$$\Rightarrow \sigma = \sigma_{irreversible} + \sigma_{collision}$$

The expression of entropy generation (equation 30) provides the key to irreversible thermodynamics. The first two terms indicate entropy generation as a result of irreversible transport processes whilst the third term represents entropy increase due to source applied to the system. The last two terms on the right hand side entropy increase due to (1) heat transfers from the other systems via collision processes, (2) free energy changes associated with recombination and generation of electron-hole pairs.

From (28) the general form of entropy balance equation is written as,

$$\frac{dS}{dt} = \nabla . J_s + \sigma . \tag{31}$$

It is noted here that equation (31) has a similar form to the expression for total entropy, i.e.,

$$\frac{dS}{dt} = \left(\frac{dS}{dt}\right)_{rev} + \left(\frac{dS}{dt}\right)_{irr}, \tag{32}$$

Comparing the first term of the right hand side of both equations, the reversible entropy flux is $\left[\left(\frac{dS}{dt}\right)_{rev} = \nabla.J_s\right]$. This is the net sum of quantities carried into the system by transfers of matter, plus the net sum of quantities of entropy carried in by the transfer of heat. The second term yields $\left[\left(\frac{dS}{dt}\right)_{irr} = \sigma\right]$, which is the entropy generation rate produced by irreversible processes taking place inside the system. Table 3 shows some common examples of the irreversible processes (hence the internal entropy generation) of common engineering processes.

Table 3. Example of processes leading to the irreversible production of entropy

Entropy generation $\sigma = \left(\frac{dS}{dt}\right)_{irr}$		Equations
Dissipation	Mechanical energy (due to friction or viscosity)	$\frac{1}{T}\left(\frac{dF}{dt}\right)$ (Richard et. al., 1948), where $\frac{dF}{dt}$ is the rate at which work is done against force or viscous force.
	Electrical energy (by the passage of electric current through a resistance)	$\frac{\mathbf{I}^2 R}{T}$, where $\mathbf{I}$ is the flow of current through a resistance R in conductor or semiconductor at temperature T.
Irreversible heat flow in a conducting medium		$\mathbf{q}.\nabla\left(\frac{1}{T}\right)$. Here $\mathbf{q}(= K\nabla T)$ is the vector which expresses the rate of heat flow. The final expression is $\left(\frac{dS}{dt}\right)_{irr} = \frac{k\nabla T\nabla T}{T^2}$.
Free expansion of gas without absorbing heat [19]		The approximate expression for the irreversible increase in entropy accompanying the transfer of any particle element of the gas , consisting of dN moles, from pressure P_1 to P_2 is $dS_{irr} = RdN\log\left(\frac{P_1}{P_2}\right)$

Table 3. Continued

Entropy generation $\sigma = \left(\frac{dS}{dt}\right)_{irr}$	Equations
Chemical reaction [19]	$\left(\frac{dS}{dt}\right)_{irr} = R\left(\log K_p - \log\frac{fc^c fD^d \ldots}{fA^a fB^b \ldots}\right)\frac{dx}{dt}$, where K_p is equilibrium constant at temperature T, R is the gas constant, *fAfB…fcfd*…etc., are the actual instantaneous values of the fugacities, a. b, …, c, d, … are the number of moles and $\frac{dx}{dt}$ is a factor.
Collisions (electron and hole transport in submicron semiconductor or thin film devices)	$\left(\frac{dS}{dt}\right)_{colli} = \frac{1}{T}\frac{\partial \mathbf{Y}}{\partial t}\Big\vert_{colli} + \frac{P}{\rho T}\frac{\partial \rho}{\partial t}\Big\vert_{colli}$, where the first term of right hand side indicates the collision due to change of energy between electrons-electrons, electrons-holes, holes-holes interactions, and the second term defines the collision because of change of carrier density.

Table 4. Summary of the conservation equations

	Gauss theorem approach (Appendix)	Boltzmann Transport Equation approach
	$\int_V \nabla.\mathbf{X}dV = \int_\Omega \mathbf{X}.\vec{n}d\Omega = \int_\Omega \mathbf{X}.d\Omega$	$\frac{\partial f}{\partial t} = -\left(\frac{\partial f}{\partial t}\right)_{drift} + \frac{\partial f_{colli}}{\partial t}$
Mass	$\frac{\partial \rho}{\partial t} = -\nabla.(\rho\,\mathbf{v})$	$\frac{\partial \rho}{\partial t} = -\nabla(\mathbf{v}\rho) + \frac{\partial \rho}{\partial t}\Big\vert_{colli}$
Momentum	$\frac{\partial \rho\mathbf{v}}{\partial t} = -\nabla\cdot(\rho\mathbf{v}\mathbf{v} + \mathbf{P}) + \Phi$	$\frac{\partial(\rho\mathbf{v})}{\partial t} = -\nabla.(\mathbf{P} + \rho\mathbf{v}\mathbf{v}) + \Phi + \frac{\partial(\rho\mathbf{v})}{\partial t}\Big\vert_{colli}$
Energy	$\rho\frac{du}{dt} = -\nabla\cdot\mathbf{J}_q + \mathbf{\Psi}$	$\rho\frac{du}{dt} = -\nabla\cdot\mathbf{J}_q + \mathbf{\Psi} + \frac{\partial \mathbf{Y}}{\partial t}\Big\vert_{colli}$
Entropy	$\rho\frac{ds}{dt} = -\nabla\cdot\left(\frac{\mathbf{J}_q - \sum_{k=1}^{n}\mu_k\mathbf{J}_k}{T}\right)$ $-\frac{1}{T^2}\mathbf{J}_q\cdot\nabla T - \frac{1}{T}\sum_{k=1}^{n}\mathbf{J}_k\cdot\left(T\nabla\frac{\mu_k}{T}\right)$ $+\frac{1}{T}\mathbf{JF}$	$\rho\frac{ds}{dt} = -\nabla\cdot\left(\frac{\mathbf{J}_q - \sum_{k=1}^{n}\mu_k\mathbf{J}_k}{T}\right) - \frac{1}{T^2}\mathbf{J}_q\cdot\nabla T$ $-\frac{1}{T}\sum_{k=1}^{n}\mathbf{J}_k\cdot\left(T\nabla\frac{\mu_k}{T}\right)$ $+\frac{1}{T}\mathbf{JF} + \frac{1}{T}\frac{\partial \mathbf{Y}}{\partial t}\Big\vert_{colli} + \frac{P}{\rho T}\frac{\partial \rho}{\partial t}\Big\vert_{colli}$

The details of the Boltzmann Transport equation for analyzing the mass, momentum, energy and entropy conservation equations in macro and micro scales have been discussed in this section. This section concludes with the summary of all conservation equations as shown in table 4 using Gauss and BTE approaches so that these two methods could be compared. This section is followed by the detailed thermodynamic modeling of adsorption cooling system (section 4), bulk thermoelectric module as examples of macro scale device, and thin film superlattice thermoelectric for micro scale system (section 5) to understand their thermal transport phenomena.

4. Adsorption Cooling

The physical adsorption process occurs mainly within the pores of adsorbent and at the external adsorbent surface, and is determined by its adsorption isotherms, thermodynamic property surfaces of energy and entropy, heats of adsorption and adsorption kinetics In a physisorption, the adsorbed phase is held near to the pores of the adsorbent by the existence of van der Waals forces as shown in figure 3.

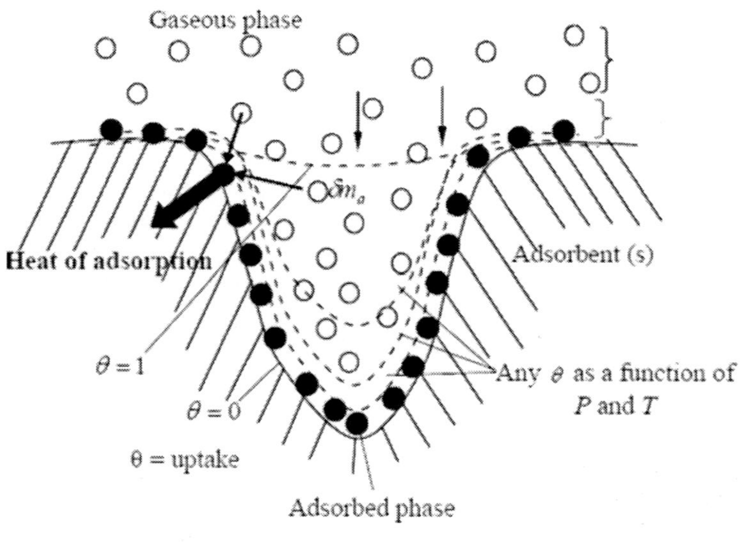

Figure 3. The schematic diagram of a single component adsorbent + adsorbate system for any uptake, x as a function of pressure, P and temperature, T. The effects of gaseous phase are also shown here.

The development of adsorption cooling system is based on the thermodynamic property surfaces of adsorbent-refrigerant system and the thermal compression of natural working refrigerants like water or alcohols and lies in its ability to operate with motive energy derived from fairly low temperature sources such as waste heat in process industries or sun light which indicates the adsorption process as an avenue for avoiding the use of ozone depleting substances [20]. The principal advantage of adsorption chillers is that it is amenable to regenerative use of adsorption heat. Adsorption system incorporates no mechanical moving parts and generates no noise or vibration. A certain number of adsorbent-adsorbate pairs have been tested theoretically and experimentally for evaluating the performances of adsorption chillers. These are silica gel-water, zeolite-water, silica gel-methanol, activated carbon-methanol, activated charcoal-NH_3, zeolite-CO_2, etc. Recently, a new family of composite sorbents called Selective Water Sorbents (SWSs) has been presented for sorption cooling and heat pumping [21]. It is based on a porous host matrix (silica, alumina, etc.) and an inorganic salt ($CaCl_2$, LiBr, $MgCl_2$, $MgSO_4$, $Ca(NO_3)_2$, etc.) impregnated inside pores [22, 23]. Building from the previous works, this article presents both the steady-state and dynamic behaviors of SWS-1L in a two-bed solid sorption cooling system using a transient distributed model. These results are compared with those of commercial adsorption cooler based on silica gel Fuji RD such that a device for new generation of cooling can be enlightened. Both the heat and mass transfer resistances of the adsorption heat exchanger as well as the temporal energetic behavior in the evaporator and condenser are also taken into account in the present model. In this section, we also elucidated the effects of the isosteric cooling and heating times as well as the total cycle time on the system performances, and demonstrate that the current cooler design tends at optimum conditions.

4.1. Description of Adsorption Cooling Model

The adsorption cooling system which utilizes the adsorbent-adsorbate characteristics and produces the useful cooling effects at the evaporator by the amalgamation of "adsorption-triggered-evaporation" and "desorption-resulted-condensation" was described elsewhere [24-26]. Figure 4 shows the schematic layout of the adsorption chiller.

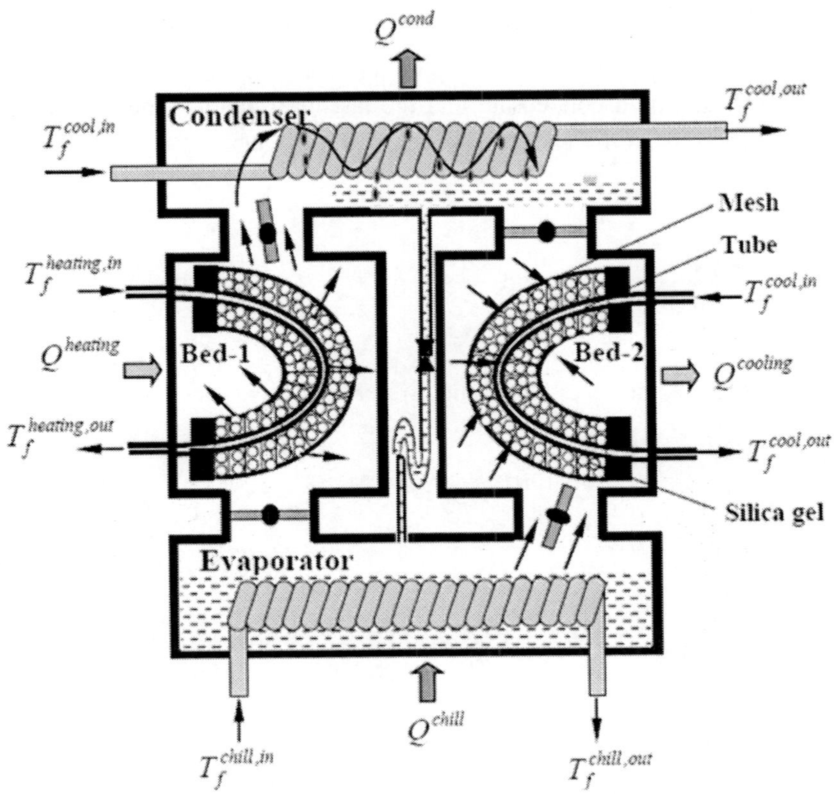

Figure 4. Schematic of a two-bed adsorption chiller.

It comprises the evaporator, the condenser and the reactors or adsorbent beds. For continuous cooling operation, firstly a low-pressure refrigerant (hence water) is evaporated at the evaporator due to external cooling load (or chill water) and is adsorbed into the solid adsorbent located in the adsorber. The process of adsorption results in the liberation of heat of adsorption at the adsorber providing a useful heat energy output and a cooling effect in the condenser/evaporator heat exchanger. Secondly, the adsorbed bed is heated by the external heat source and the refrigerant is desorbed from the adsorbent and goes to the condenser for condensation by pumping heat through the environment. The condensate (refrigerant) is refluxed back to the evaporator via a pressure reducing valve for maintaining the pressure difference between the condenser and the evaporator. Pool boiling is affected in the evaporator by the vapor uptake at the adsorber, and thus completing the refrigeration close loop or cycle. These phenomena are expressed mathematically using the mass and energy balances between major components of the adsorption chiller system.

Evaporator

The energy balance becomes

$$(\rho c_p)_{eff}^{evap}\frac{\partial T^{evap}}{\partial t}\left(\frac{d_{evap,o}^2-d_{evap,i}^2}{4}\right)=\lambda^{evap}\frac{\partial^2 T^{evap}}{\partial z^2}\left(\frac{d_{evap,o}^2-d_{evap,i}^2}{4}\right)-d_{evap,i}h_{evap}\left(T^{rvap}-T^{chill}\right) \quad (33)$$

where $(Mc_P)_{eff}^{evap}$ is the sum of all mass capacities of the evaporator. The first term on the right hand side defines the latent heat of evaporation that goes to adsorption bed, the second term is the enthalpy of liquid condensate and the last term denotes the cooling capacity of evaporator, which rises from the cooling of chilled water. Temperature boundary conditions are

$$\left.\frac{\partial T^{evap}}{\partial z}\right|_{z=0}=0 \text{ and } \left.\frac{\partial T^{evap}}{\partial z}\right|_{z=L^{cond}}=0\text{, respectively.}$$

The energy balance equation on the chilled water control volume is written as

$$\rho_f^{chill}c_{p,f}^{chill}\frac{\partial T^{chill}}{\partial t}=-u_f^{chill}\rho_f^{chill}c_{p,f}^{chill}\frac{\partial T^{chill}}{\partial z}+\lambda_f^{chill}\frac{\partial^2 T^{chill}}{\partial z^2}-\frac{(UA)_{chill}}{V_f^{chill}}\left(T^{chill}-T^{evap}\right) \quad (34)$$

Hence the terms,

$$\frac{dx_{ads}^{bed}}{dt} \text{ and } \frac{dx_{des}^{bed}}{dt}$$

indicate the adsorption rate and the desorption rate. The boundary conditions of the chilled water tube are

$$T^{chill}(z=0,t)=T^{chill,in} \text{ and } \frac{\partial T^{chill}}{\partial z}(z=L^{tube},t)=0.$$

Adsorption Isotherms and Kinetics

The adsorption/desorption rate is calculated from the knowledge of adsorption equilibrium and kinetics and is given by the conventional linear driving force (LDF) equation [27]

$$\frac{dx}{dt}=\frac{15D_{so}e^{-\frac{E_a}{RT}}}{R_p^2}\{x^*-x\}, \tag{35}$$

where D_{so} defines a pre-exponential factor of the efficient water diffusivity in the adsorbent, E_a represents the activation energy, R is the universal gas constant and R_p is the average radius of the adsorbent grains. Kinetic data were taken from [28, 29]. Hence the adsorption uptake at equilibrium condition is expressed as a function of pressure (P) and temperature (T). The authors have measured the isotherms of water adsorption on silica gel Fuji Davison type RD, type A [30] and on SWS-1L [31, 32]. These experimentally measured data are fitted using the Tóth's equation [33], i.e.,

$$x^*=\frac{x_m K_o\cdot exp\{\Delta H_{ads}/(R\cdot T)\}\cdot P}{\left[1+\{K_0\cdot exp(\Delta H_{ads}/(R\cdot T))\cdot P\}^{t_1}\right]^{1/t_1}}, \tag{36}$$

where x^* is the adsorbed adsorbate at equilibrium conditions, x_m denotes the monolayer capacity, ΔH_{ads} the isosteric enthalpies of adsorption, K_o the pre-exponential constant, and t_1 is the dimensionless Tóth's constant. These values are furnished in table 4.

Table 4. Adsorption isotherms of silica gel-water systems

Type	K_o (Pa^{-1})	ΔH_{ads} (kJ/kg)	x_m ($kg\ kg^{-1}$)	t_1
RD (Tóth isotherm)	7.3×10^{-13}	2800	0.4	8
SWS-1L (Tóth isotherm)	2×10^{-12}	2760	0.8	1.1
Type	a	b	$c\times10^9$	ΔF kJ/mol
SWS-1L (polynomial equation)	2.83378 5.80313	-3.0133×10^{-4} -0.00119	7.46702 69.9054	1-5 5-9

The most significant difference between these two adsorbents (silica type RD and SWS-1L) lies in their water vapor uptake characteristics. From table 4, one could observe that the water vapor uptake capacity of SWS-1L is higher than that of type RD. On the other hand, the adsorption uptake of water vapor on SWS-1L is expressed in terms of mole/mole by the polynomial equation, i.e., $x^* = exp\left(a + b\Delta F + c\Delta F^2\right)$ [28, 32], where ΔF indicates the Polanyi adsorption potential and is represented by $\Delta F = -RT\,ln\left(P/P_s\right)$. The values of the adjustable coefficients a, b and c that provide the best approximation of the two segments of the temperature-independent curve of sorption of water by SWS-1L composite [28, 32] are also furnished in table 4. It should also be noted here that the values of D_{so}, E_a and R_p are found same for both RD silica gel-water and SWS-1L-water systems, i.e., the adsorption/desorption rate for both systems are nearly same but the difference between the two systems are occurred due to their different uptake capacities.

Bed

As the evaporated refrigerant is associated onto the solid adsorbent by the flow of cooling fluid at ambient conditions during adsorption period, and the desorbed refrigerant is dissociated from the solid adsorbent by the flow of heating fluid during desorption period. The heat energy is exchanged between cooling/heating fluid and the adsorption bed. A schematic of the adsorption bed with heat exchanger is shown in figure 1. The energy balance for the heat transfer fluid is given by

$$\frac{\partial T_f^j}{\partial t} = -u_f^j \frac{\partial T_f^j}{\partial z} + \frac{\lambda_f^j}{\rho_f^j c_{p,f}^j} \frac{\partial^2 T_f^j}{\partial z^2} - \frac{h_{f-m}^j A_f^j}{\rho_f^j c_{p,f}^j V_f^j}\left(T_f^j - T_m\right), \tag{37}$$

where u defines the flow rate of cooling/heating fluid, j indicates heating or cooling for desorption or adsorption, h_{f-m}^j represents the heat transfer coefficient. The boundary condition for fluid flow becomes

during adsorption: $T_f^j\left(z=0,t\right) = T_f^{cool,in}$ and $\frac{\partial T_f^j}{\partial z}\left(z = L^{tube}, t\right) = 0$,

during desorption: $T_f^j\left(z=0,t\right) = T_f^{hot,in}$ and $\frac{\partial T_f^j}{\partial z}\left(z = L^{tube}, t\right) = 0$.

The energy balance of the metal tube that contains heat transfer fluid is written as

$$\frac{\partial T^{tube}}{\partial t} = \frac{\lambda^{tube}}{\rho^{tube} c_p^{tube}} \frac{\partial^2 T^{tube}}{\partial z^2} - \frac{h_{f-m}^j A^{tube}}{\rho^{tube} c_p^{tube} V^{tube}}\left(T^{tube} - T_f^j\right) - \frac{h_{m-s} A^{tube}}{\rho^{tube} c_p^{tube} V^{tube}}\left(T^{tube} - T^{sg}\right) - \varsigma \frac{A^{tube}}{\rho^{tube} c_p^{tube} V^{tube}} \lambda^{fin} \frac{\partial T^{fin}}{\partial r} \tag{38}$$

Here the value of ς is equal to 1 when any fin is attached with the tube, otherwise $\varsigma = 0$. The boundary conditions of the heat exchanger tube inside the adsorber are $\frac{\partial T^{tube}}{\partial z}\left(z = 0, t\right) = 0$ and $\frac{\partial T^{tube}}{\partial z}\left(z = L^{tube}, t\right) = 0$, respectively.

The fin thickness is very small and the heat transfer in the fin is assumed to be one dimensional in the radial direction. The energy balance equation of the fin is given by

$$\frac{\partial T^{fin}}{\partial t}=\frac{\lambda^{fin}}{\rho^{fin}c_p^{fin}}\frac{\partial\left(\frac{\partial T^{fin}}{\partial r}r\right)}{r\partial r}-\frac{h_{fin-s}A^{fin}}{\rho^{fin}c_p^{fin}V^{fin}}\left(T^{fin}-T^{sg}\right). \tag{39}$$

The boundary conditions are $T^{fin}(r=r_o)=T^{tube}$ and $\frac{\partial T^{fin}}{\partial r}(r=r^{fin})=0$.

The energy balance of the adsorbent control volume can be written as (heat flow is considered in both r and z directions)

$$\begin{aligned}\left(\rho^{sg}c_p^{sg}+\rho^{g}xc_p^{a}\right)\frac{\partial T^{sg}}{\partial t}&=\nabla.\left(\lambda^{eff}\nabla T^{sg}\right)+\rho^{sg}\Delta H_{ads}\frac{dx}{dt}-\frac{h_{fin-s}A^{fin}}{V^{fin}}\left(T^{fin}-T^{sg}\right)\\&-\frac{h_{m-s}A^{tube}}{V^{tube}}\left(T^{tube}-T^{sg}\right)\end{aligned} \tag{40}$$

where the specific heat capacity of the adsorbed phase is given by [34]

$$c_p^a=c_p^l+\Delta H_{ads}\left\{\frac{1}{T^{sg}}-\frac{1}{v^g}\frac{\partial v^g}{\partial T^{sg}}\right\}-\frac{\partial(\Delta H_{ads})}{\partial T^{sg}}.$$

The first term in the right hand side indicates the specific heat capacity at liquid phase, and the other terms occur due to the non ideality of gaseous phase, which incorporate two additional inputs from the properties of adsorbent + adsorbate system, namely the ΔH_{ads} and the isotherms (*P-T-x*) data [35]. On the other hand, the effective thermal conductivity of the adsorbed phase is

$$\lambda^{eff}=\lambda^{g}\Big/\left(\phi+2/3\frac{\lambda^{g}}{\lambda^{sg}}\right)\text{[36]},$$

where ϕ indicates the porosity of bed. Here λ^{eff} is defined as the total thermal conductivity of adsorbent particles stacked together in the adsorber. The boundary condition at radial direction becomes

$$-\lambda^{eff}\left.\frac{\partial T^{sg}}{\partial r}\right|_{r=r_o}=h_{m-s}\left(T^{tube}-T^{sg}\right)\text{ and }\left.\frac{\partial T^{sg}}{\partial r}\right|_{r=r^{fin}}=0.$$

Condenser

After desorption, the desorbed refrigerant is delivered to the condenser as latent heat and this amount of heat is pumped to the environment by the flow of external cooling fluid. In the modelling, we assume that the condenser tube bank surface is able to hold a certain maximum amount of condensate. Beyond this the condensate would flow into the evaporator via a U-tube. This ensures that the condenser and the desorber are always maintained at the saturated pressure of the refrigerant. The energy balance of the condenser is expressed as

$$(\rho c_p)_{eff}^{cond} \frac{\partial T^{cond}}{\partial t}\left(\frac{d_{con,o}^2 - d_{con,i}^2}{4}\right) = \lambda^{cond} \frac{\partial^2 T^{cond}}{\partial z^2}\left(\frac{d_{con,o}^2 - d_{con,i}^2}{4}\right) + d_{con,i} h_{con,i}\left(T^{cond} - T^{cool}\right) \quad (41)$$

where $(Mc_P)_{eff}^{cond}$ is the sum of all mass capacities of the condenser. The first term on the right hand side defines the latent heat of condensation, the second term is the enthalpy of liquid condensate and the last term denotes the sensible cooling of the condenser.

Temperature boundary conditions are

$$\left.\frac{\partial T^{cond}}{\partial z}\right|_{z=0} = 0 \text{ and } \left.\frac{\partial T^{cond}}{\partial z}\right|_{z=L^{cond}} = 0 .$$

The energy balance equation of the cooling water control volume is written as

$$\rho_f^{cool} c_{p,f}^{cool} \frac{\partial T_f^{cool}}{\partial t} = -u_f^{cool} \rho_f^{cool} c_{p,f}^{cool} \frac{\partial T^{cool}}{\partial z} + \lambda_f^{cool} \frac{\partial^2 T^{cool}}{\partial z^2} - \frac{(UA)^{cool}}{V_f^{cool}}\left(T^{cool} - T^{cond}\right) \quad (42)$$

The boundary conditions of the cooled water tube are $T^{cool}(z=0,t) = T^{cool,in}$ and $\frac{\partial T^{cool}}{\partial z}(z = L^{tube}, t) = 0$.

Mass Balance

The mass balance of refrigerant in the adsorption chiller is expressed by

$$\frac{dM_{ref}}{dt} = -M^{sg}\left\{\frac{dx_{des}^{bed}}{dt} + \frac{dx_{ads}^{bed}}{dt}\right\}, \quad (43)$$

where M^{sg} is the mass of adsorbents packed in each of the two adsorbent beds, and M_{ref} is the mass of refrigerant in liquid phase.

The roles of the beds (containing the adsorbent) are refreshed by switching which is performed by reversing the direction of the cooling and the heating fluids to the designated sorption beds and similarly, the evaporator and condenser are also switched to the respective adsorber and desorber. It is noted that during switching interval, no mass transfers occur between the hot bed and the condenser or the cold bed and the evaporator.

The cycle average cooling capacity Q^{chill}, heating capacity $Q^{heating}$ and COP are, respectively calculated as

$$Q^{chill} = \rho_f^{chill} u_f^{chill} A_f^{tube,chill} c_{p,f}^{chill} \int_0^{t_{cycle}} \frac{T_f^{chill,in} - T_f^{chill,out}}{t_{cycle}} dt , \tag{44}$$

$$Q^{heating} = \rho_f^{heating} u_f^{heating} A_f^{tube,bed} c_{p,f}^{heating} \int_0^{t_{cycle}} \frac{T_f^{heating,in} - T_f^{heating,out}}{t_{cycle}} dt , \tag{45}$$

and

$$COP = \frac{Q^{chill}}{Q^{heating}} . \tag{46}$$

4.2. Discussion

The adsorption bed design incorporates a circular finned tube heat exchanger. The values for the parameters used in the present model are furnished in table 5. Figure. 5 features the temperature histories at the outlets of the type RD (red lines in figure 3) and SWS-1L based chiller system (thick black lines in figure 5) and these are compared with experimental data (thick blue lines and circles) of RD silica gel and water based adsorption chiller. Due to the positioning of the temperature sensors, the experimentally measured outlet temperatures are affected by the time constant of downstream mixing valves in the pipeline. It is evident that our present simulation results exhibit a sufficiently good agreement with the experimental data stemming from a distributed parameter model. Figure 5 also shows the temperature histories at the outlets of the condenser and chilled water. It should be noted here that the delivered chilled water temperature is slightly lower for adsorption chiller employing SWS-1L as can be seen in figure 5.

Table 5. Values adopted for adsorption chiller simulation used in the present model [37]

D_{so}	2.54×10^{-4} m^2/s for SWS-1L and RD silica gel
E_a	4.2×10^4 J/mol for SWS-1L and RD silica gel
R_p	1.7×10^{-4} m for type RD and 1.74×10^{-4} m for SWS-1L
c_p^{sg}	924 J(/kg K)
h_{m-s}	36 W/(m^2 K) [16]
h_{fin-s}	36 W/(m^2 K) [16]
$(UA)^{chill}$	(2557 W/(m^2 K) × 1.37 m^2)
$(UA)^{cond}$	(4115 W/(m^2 K) × 3.71 m^2)
$(Mc_p)_{eff}^{evap}$	(8.9 kg × 386 J/(kg K) + 40 kg× $c_{p,water}$ J/(kg K))

Table 5. Continued

$\left(Mc_p\right)_{eff}^{cond}$	(24 kg × 386 J/(kg K) + 5 kg× $c_{p,water}$ J/(kg K))
r_i	7.94 mm
$A^{tube} = 2\pi r_o L^{tube}$	r_o = 8.64 mm, L^{tube} = 1 m
$V_f^{chill} = \pi r_o^2 L^{evap}$	L^{evap} = 2.1 m
$V_f^{cool} = \pi r_o^2 L^{cond}$	L^{cond} = 2.34 m
$V^{fin} = \pi N h\left(r_{fin}^2 - r_o^2\right)$	N = 100, h = 0.1 mm, r_{fin} = 22 mm
u_f^{chill}	0.18 m/s
u_f^{cool}	0.198 m/s
$u_f^{heating}$	0.14 m/s
M^{sg}	20 kg
$T_f^{cool,in}$	31 °C
$T_f^{hot,in}$	85 °C
$T_f^{chill,in}$	14.8 °C

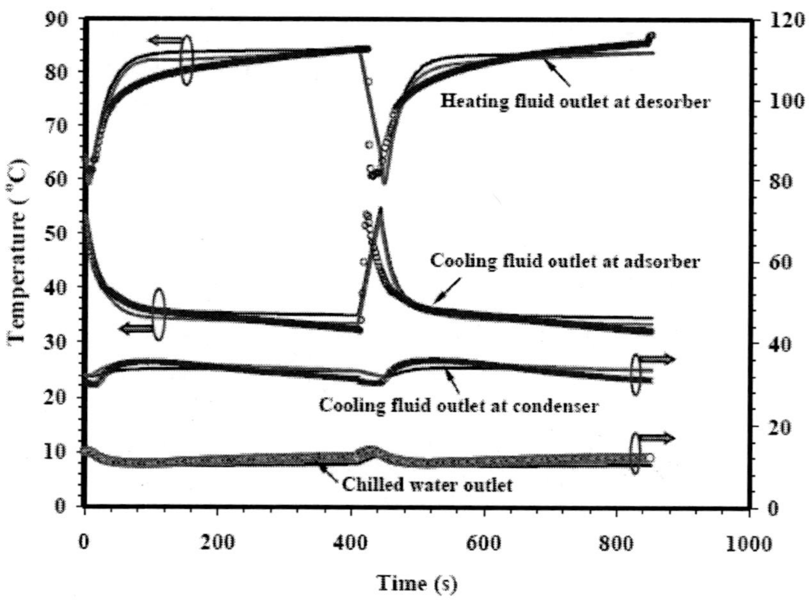

Figure 5. Simulated (black for SWS-1L-water system; red line for type RD silica gel-water system) and experimental (blue circles for type RD silica gel-water based adsorption system) heating fluid and cooling fluid outlet temperatures from the adsorption chiller system. Simulated and experimentally (for type RD silica gel only) measured fluid outlet temperatures from the condenser and evaporator are also shown here.

Figure 6 (a) presents the simulated Dühring diagram of the cyclic steady state condition of an entire bed comprising SWS-1L, from which one observes that during cold-to-hot thermal swing of the bed, momentary adsorption takes place although heating source has already been applied to heat up the bed in pre-heating mode. The entire bed is observed to be essentially following an isosteric path (constant x) during switching. In contrast, the local spatial points in the bed are not evolving in an isosteric manner, which is confirmed by the present analysis. This shows that while some parts of the bed may continue to adsorb, other parts desorb, resulting in the entire bed following an isosteric path. At the end of hot-to-cold thermal swing, there is a pressure drop in the bed. This causes the adsorbate in the cool bed to desorb momentarily and condense into the evaporator. The P-T-x diagram for adsorption bed employing RD type silica gel during steady state is also imposed in figure 6(b) for comparison.

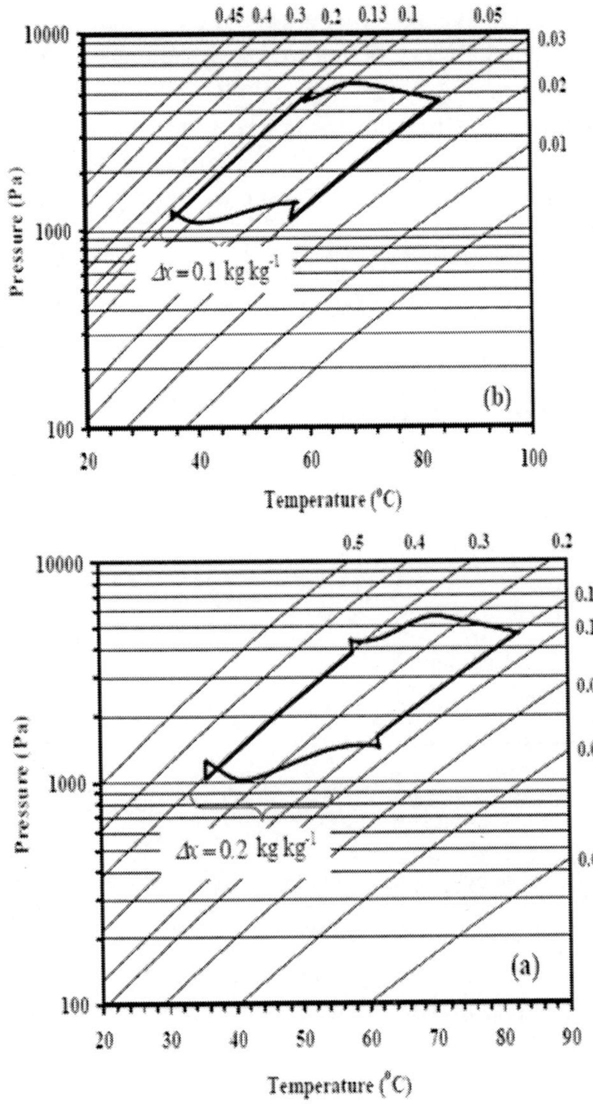

Figure 6. Dühring diagram of the whole bed under cyclic steady state condition for (a) SWS-1L and water based adsorption chiller, (b) type RD silica gel and water based adsorption cooling cycle.

Figure 7 presents the effects of cycle time on COP and cycle average chiller cooling capacity for type RD and SWS-1L based adsorption chiller systems. It is clearly seen that the COP increases monotonically with the cycle time. The reason is that with a longer cycle time, the relative time frame occupied bed switching which involves a significant sensible heat exchange is reduced vis-à-vis that of a shorter cycle time. This will lead to a favorable effect on the COP. The variation of cooling capacity is not monotonic. For SWS-1L based adsorption chiller, the cooling power increases steeply up to 500 s, and it begins to decrease with a similar slope at the cycle time of over 500 s. Lower cooling capacity under a relative shorter cycle time is caused by a reduced extent of adsorption, which is also related to a reduced extent of desorption due to the insufficient heating of the desorber. At a certain cycle time, the maximum adsorption/ desorption capacity is achieved at the prevailing heating and cooling source temperatures. Extending the cycle time further brings forth unfavorable effect on useful cooling as the cycle average cooling capacity decreases. For the SWS-water based pair, cycle times longer than 300 s can be realized with the higher cooling power and COP as compared to the adsorption cooling system based on RD silica gel-water pair.

The system performance at different driving heat source temperatures in case of optimum conditions [(i) cycle time 420 s, switching time 30 s for type RD silica gel-water system, and (ii) cycle time 630 s, switching time 30 s for SWS-1L – water system] is shown in figure 8 for the same heat sink temperature of 31 °C.

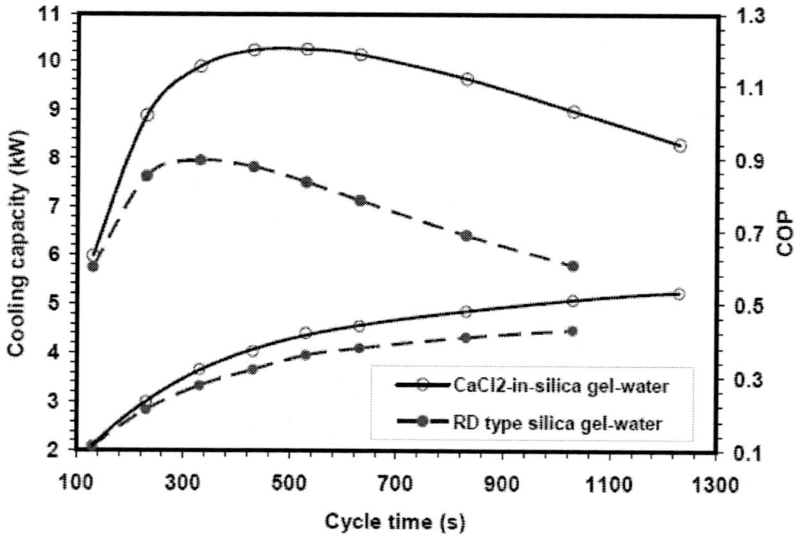

Figure 7. Effect of cycle time on COP and cycle average cooling capacity.

For type RD silica gel-water system the COP reaches the maximum value of 0.35 at T = 75-80°C. For SWS-1L - water system at this temperature range the COP is larger (0.42-0.45) and remains almost constant at higher T. The COP of SWS-1L - water system is higher than that of RD-water system because of high cooling and less driving heat generation powers, which may occur due to larger Δx for the same heat source and heat sink temperatures. This much larger COP shows significant advantage of the new working pair as compared with the conventional unit. Thus, from the present simulation results, it is found that the newly SWS-1L based adsorption chiller provides a promising unit for cooling applications. This theoretical conclusion should be confirmed practical testing of the new adsorbent in the optimized ADC configuration.

4.3. Summary of Section 4

We have successfully modeled and predicted the performances of SWS-1L and water based adsorption chiller using a simplified distributed approach such that both the transient and steady state behaviors of ADC can be captured. It is found that the performances of ADC incorporating SWS-1L as adsorbents, in terms of cooling capacity, coefficient of performance and peak chilled water temperature, are better than those of the commercial available silica gel-water based adsorption chiller. An optimum switching time of 30 s is obtained for both cycles and the cycle performance improves with increasing hot water inlet temperature.

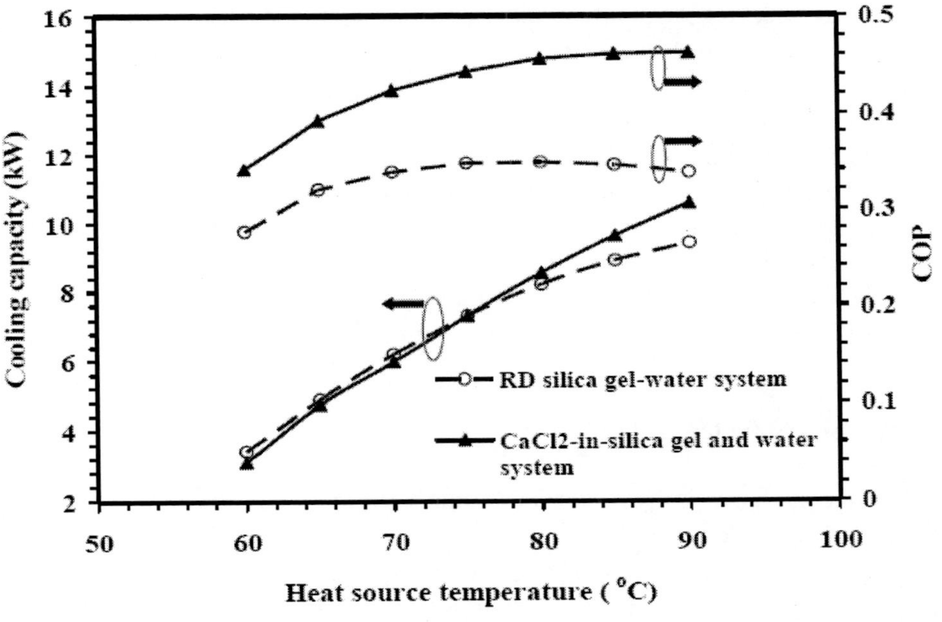

Figure 8. Influence of driving heat source temperature on cyclic average cooling capacity and COP.

5. Macro and Micro Thermoelectric Coolers

The reversible thermoelectric effects are the Seebeck, Peltier and Thomson effects [38]. These are related to the transformation of thermal into electrical energy, and vice versa. The physical nature of the thermoelectric effects is not well explained in the literature, as well as the dependence of Seebeck coefficient on temperature and materials. From the electron theory, the absorption of Thomson heat in the interior of a thermally non-uniform conductor has been reported to be the additive superposition of two effects: Firstly, a part of Thomson effect is the internal Peltier effect which is caused by the non-equilibrium electron distribution functions in a thermally non-uniform conductor. Secondly, the other part is heat absorbed due to current flow against the drift potential difference. These effects are assumed to be reversible and hence, the sign of the Thomson heat changes with the reversal of current direction. The thermoelectric effects in semi-conductors are stronger than those found in metals. This is attributed to the fact that electron gas in semi-conductors is far from being degenerated and obeys the classical Boltzmann statistics. The non-equilibrium changes in the distribution function are more noticeable than in a degenerated gas and this affects the magnitude of the thermoelectric effects. The magnitude of non-equilibrium change in the distribution function (and consequently the magnitudes of the thermoelectric effects) depends on the relative importance of the factors (such as electric field, thermal non-uniformity) that are responsible for the departure from equilibrium, as well as the mechanism for re-establishing the equilibrium. When an electric field is applied to a thermoelectric device, the following irreversible processes are deemed to occur in each elemental volume:

i. Electric conduction (electric current due to an electric potential gradient).
ii. Heat conduction (heat flow due to a temperature gradient).
iii. A cross effect (electric current due to a temperature gradient).
iv. The appropriate reciprocal effect (heat flow due to an electric potential gradient).

The last two belongs to a new class of irreversible process [14].

The organization of this section is as follows:

Section 5.1 describes the thermodynamic modelling of the bulk thermoelectric cooler and in addition, the entropy analysis of a transient thermoelectric cooler is also discussed. The thermodynamic behavior of a pulse thermoelectric cooler is investigated in section 5.2. The governing equations employed in this section are derived from the Boltzmann Transport Equation (BTE) that is reported in the previous section.

5.1. Thermoelectric Cooling

The refrigeration capability of a semiconductor material depends on parameters such as the Peltier effect, Joule heat, Thomson heat and material's physical properties over the operational temperatures between the hot and cold ends. A thermoelectric device is composed of p-type and n-type thermoelements such as Bi_2Te_3 and they are connected electrically in series and thermally in parallel. These thermoelectric elements are sandwiched between two ceramic substrates. As shown in figure 9, thermoelectric cooling is generated by passing a direct current through one or more pairs of n and p-type thermoelements, the temperature of cold reservoir decreases because the electrons and holes pass from the low energy level in p-type material through the interconnecting conductor to the higher energy level in the n-type material. Similarly, the arrival of electrons and holes in the opposite end results in an increase in the local temperature forming the hot junction.

When a temperature differential is established between the hot and cold junctions, a Seebeck voltage is generated and the voltage is directly proportional to the temperature differential. The amount of heat absorbed at the cold-end and dissipated at the hot end depends on the product of Peltier coefficient (π), Fourier effect (λ_{te}) and the current (I) flowing through the semiconductor material, whilst the heat generation due to the Thomson effect occurs along the thermoelectric element. Thermoelectric effects are caused by coupling between charge transport and heat transport and from which the basic mass, energy and entropy equations can be formulated to form a thermodynamic framework.

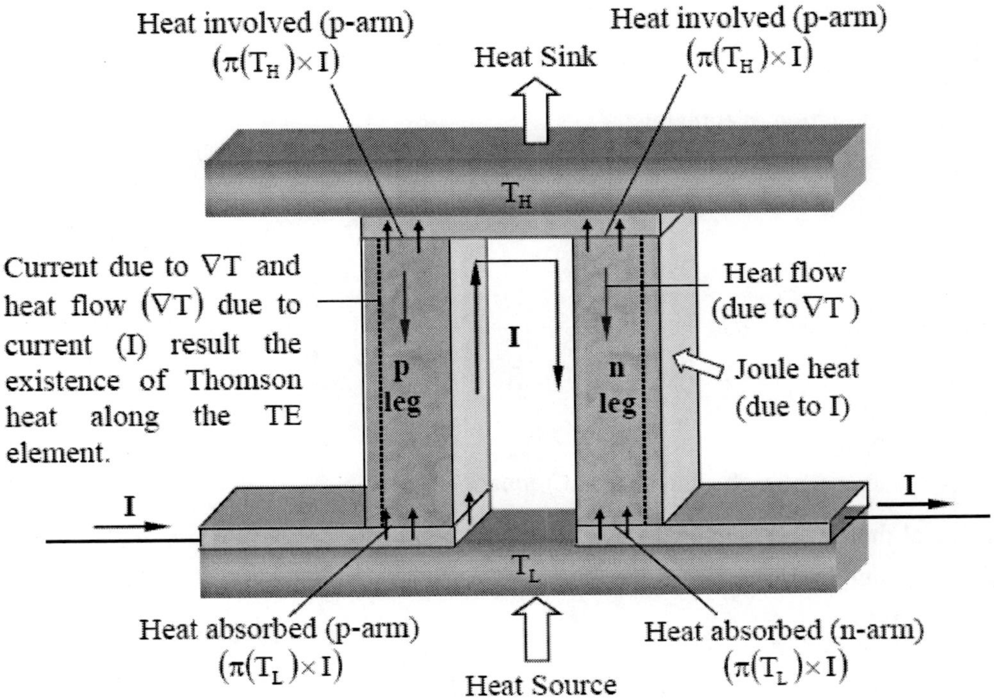

Figure 9. The schematic view of a thermoelectric cooler.

5.1.1. Energy Balance Analysis

The Peltier and the Thomson effects enhance the cooling effect in thermoelectric materials. The Thomson effect is a function of the gradients from the Seebeck coefficient and the temperature in an operating thermoelectric pallet or element. The energy balance equation of the thermoelectric cooler, as shown in figure 9, follows the methodology of the Boltzmann transport equation which has been discussed in the previous section, namely:

$$\frac{\partial(\rho u)}{\partial t} = -\nabla.(\rho u\mathbf{v} + \mathbf{J}_q) - \mathbf{P} : \nabla\mathbf{v} + \mathbf{J}E + \left.\frac{\partial(\rho u)}{\partial t}\right|_{colli}, \tag{47}$$

where u is the internal energy per unit mass, ρ is the density, **v** defines velocity, $\mathbf{J_q}$ indicates the heat flow, **P** represents the pressure and **JE** is the energy field term.

For the bulk thermoelectric material analysis, the contributions from the collision terms are deemed negligible and can be neglected;

$$\frac{\partial(\rho_{te} u)}{\partial t} = -\nabla.(\rho_{te} u\mathbf{v} + \mathbf{J}_q) - \mathbf{P} : \nabla\mathbf{v} + \mathbf{J}E\,. \tag{48}$$

Hence the subscript $_{te}$ indicates the thermoelectric cooler. As there is no velocity and pressure fields in a thermoelectric device,

$$\rho_{te} \frac{\partial u}{\partial t} = -\nabla.\mathbf{J}_q + \mathbf{J}E \tag{49}$$

Invoking the definition of enthalpy;

$$h = u + pV \quad \Rightarrow \frac{\partial h}{\partial t} = \frac{\partial u}{\partial t} + \frac{\partial (pV)}{\partial t} = \frac{\partial u}{\partial t}$$

$$h = h(T, p) \Rightarrow \frac{\partial h}{\partial t} = \left(\frac{\partial h}{\partial T_{te}}\right)_p \frac{\partial T_{te}}{\partial t} + \left(\frac{\partial h}{\partial p}\right)_{T_{te}} \frac{\partial p}{\partial t} = c_{p,te} \frac{\partial T_{te}}{\partial t}$$

where the pressure gradient term is zero.

The first term of right hand side of equation (3.1) is the heat current density, $\mathbf{J}_q$ (in W/m^2), and the second term is the electric current field, $\mathbf{J}E(= J\nabla\phi)$ (in W/m^3). Using the thermodynamic-phenomenological treatment [39], the heat current density can be written as

$$\mathbf{J}_q = -\frac{Q_o}{F} J - \lambda_{te} \nabla T, \tag{50}$$

where J (in amp/m^2) denotes the electric current density, Q_o indicates the heat transport of electrons/ holes, λ_{te} defines the thermal conductivity of the bulk thermoelectric element and F is the Faraday constant and the electric current density can be expanded as

$$J = \sigma'_{te} \nabla\phi + \frac{\sigma'_{te} Q_o}{F T_{te}} \nabla T_{te}, \tag{51}$$

where σ'_{te} defines the electrical conductance and ϕ is the electrical potential.

Equation (51) is re-arranged as,

$$\nabla\phi = \frac{J}{\sigma'_{te}} - \frac{Q_o}{F T_{te}} \nabla T_{te} \tag{52}$$

Hence,

$$-\nabla.\mathbf{J}_q + \mathbf{J}E = \nabla.(\lambda_{te} \nabla T_{te}) + \frac{J^2}{\sigma'_{te}} + \frac{J}{F} T_{te} \nabla\left(\frac{Q}{T_{te}}\right)$$

where $\left(\frac{Q}{T_{te}}\right)$ is equivalent to the entropy term.

Now, substitute all these relations into equation (49), the expression becomes [40]

$$\rho_{te} c_{p,te} \frac{\partial T_{te}}{\partial t} = \nabla.(\lambda_{te} \nabla T_{te}) + \frac{J^2}{\sigma'_{te}} + \frac{J T_{te}}{F} \left(\frac{\partial s_{te}}{\partial T_{te}} \right)_p \nabla T_{te}, \tag{53}$$

where s_{te} introduces the transported entropy of the bulk thermoelectric arms. Defining the Thomson coefficient (Γ) as [41],

$$-\Gamma \equiv \frac{T_{te}}{F} \left(\frac{\partial s_{te}}{\partial T_{te}} \right)_p,$$

the governing thermal transport in the bulk semiconductor arms is given by

$$\rho_{te} c_{p,te} \frac{\partial T_{te}}{\partial t} = \nabla.(\lambda_{te} \nabla T_{te}) + \frac{J^2}{\sigma'_{te}} - J\Gamma \nabla T_{te} \tag{54}$$

Using the Kelvin relations [17],

$\Gamma = -\frac{\pi}{T_{te}} - \left(\frac{\partial \pi}{\partial T_{te}} \right) =$, where $\pi (= \alpha T)$ is the Peltier coefficient.

Equation (54) can now be written as (derivations are shown in Appendix),

$$\rho_{te} c_{p,te} \frac{\partial T_{te}}{\partial t} = \lambda_{te} \nabla^2 T_{te} + \frac{J^2}{\sigma'_{te}} + J T \nabla \alpha|_{T_{te}} - J T \left(\frac{\partial \alpha}{\partial T} \right) \nabla T_{te} \tag{55}$$

5.1.2. Entropy Balance Analysis

The basic entropy balance equation has been introduced in the previous chapter, as shown below:

$$\begin{aligned} \rho \frac{ds}{dt} &= -\nabla \cdot \mathbf{J}_s + \sigma \\ \frac{\partial (\rho s)}{\partial t} &= -\nabla.(\mathbf{J}_s + \rho s v) + \sigma \end{aligned} \tag{56}$$

Introducing the phenomenological relations:

$$\mathbf{J}_{s,te} + \rho_{te} s_{te} v = -\frac{\lambda_{te}}{T_{te}} \nabla T_{te} + \frac{\pi}{T_{te}} J, \tag{57}$$

and invoking the Kelvin laws, one could observe

$$T\alpha = \pi\,. \tag{58}$$

Assuming only the bulk thermoelectric materials, the collision terms are usually neglected and the entropy source strength is given by (as discussed in the previous sections);

$$\sigma = -\frac{1}{T^2}\mathbf{J}_q \cdot \nabla T - \frac{1}{T}\sum_{k=1}^{n}\mathbf{J}_k \cdot \left(T\nabla\frac{\mu_k}{T}\right) + \frac{1}{T}J \cdot E$$

Hence, the entropy generation of thermoelectric arm can now be modified into a form that is inclusive of the Peltier coefficient and the local temperature as [17]

$$\sigma = -\frac{1}{T_{te}}\left(-\frac{\lambda}{T_{te}}\nabla T_{te} + \frac{\pi}{T_{te}}J\right)\cdot \nabla T_{te} + \frac{1}{T_{te}}J\cdot\left(\alpha\nabla T_{te} + J/\sigma'_{te}\right)$$

Finally, the entropy balance equation (56) becomes;

$$\frac{\partial(\rho_{te}s_{te})}{\partial t} = \frac{\nabla\cdot(\lambda_{te}\nabla T_{te})}{T_{te}} - \nabla\cdot\frac{\pi J}{T_{te}} + \frac{J^2}{T_{te}\sigma'_{te}} \tag{59}$$

The first term on the right-hand side of equation (59) represents the change of entropy due to heat conduction; the second term indicates the entropy flux due to Seebeck effect and the last term is the entropy generation due to Joule heat.

A more elegant method of writing the entropy fluxes of thermoelectric arm is to express them in terms of (i) the heat and carrier fluxes, (ii) the internal dissipation, and (iii) the entropy generation that comes from conduction and Joulean heat transfer:

$$\frac{\partial(\rho_{te}s_{te})}{\partial t} = \overbrace{\nabla\cdot\left(\frac{\lambda_{te}\nabla T_{te}}{T_{te}} - \frac{\pi J}{T_{te}}\right)}^{entropy\ \ flux} + \left\{\overbrace{-\lambda_{te}\nabla T_{te}.\nabla\left(\frac{1}{T_{te}}\right)}^{internal\ \ dissipation} + \overbrace{\frac{J^2}{T_{te}\sigma'_{te}}}^{\substack{entropy\ source\ strength\ due\ to\\ Joulean\ heat}}\right\} \tag{60}$$

where the units of entropy flux is in W/m^2 K and it is given by,

$$S_{flux} = \frac{\lambda_{te}\nabla T_{te}}{T_{te}} - \frac{\pi J}{T_{te}}, \tag{61}$$

and entropy generation per unit volume is in W/m^3 K which is expressed as:

$$S_{gen} = -\lambda_{te}\nabla T_{te}.\nabla\left(\frac{1}{T_{te}}\right) + \frac{J^2}{T_{te}\sigma'_{te}}. \tag{62}$$

Equation (62) is always a positive term.

5.1.3. Temperature-Entropy Plots of Bulk Thermoelectric Cooling Device

From the work of Carnot and Clasius, Kelvin deduced that the reversible heat flow discovered by Peltier must have entropy associated with it. It has been shown that the coefficient discovered by Seebeck was a measure of entropy associated with electric current [41]. A temperature-entropy (T-s) has been proven to be a pedagogical tool in analyzing the performance of thermoelectrics undergoing both reversible and irreversible processes. The parameters used in the computation are listed in table 6. For an ideal thermoelectric device, the Seebeck (α) coefficient is constant and the heat flux at the hot and cold junctions are given by αJT_i where the subscript i refers either to the hot or cold junction. Thus, the temperature Seebeck coefficient (T-α) plot is simply a simple rectangle. However, the temperature effect on the Seebeck (α) coefficient induces the dissipative losses along the p and n legs of thermoelectrics and an indication of these losses are shown by the shaded area on the T-α plot, as shown in figure 10, in which the physical properties of the Bismuth Telluride thermoelectric pairs are used. The two isotherms are the hot and cold junctions whilst the inclined lines indicate the effect of temperature on the α values when the current (I) travels along the p- and n-legs.

For thermodynamic consistency, the temperature-entropy (or temperature-entropy flux) diagram is. In an optimum current operation of the thermoelectric cooler, the presence of entropy loss (due to conduction heat transfer and Joule heat) along the p-n legs is given by $K\nabla T/T_i$ and they are manifested by enclosed areas "k-i-u-r-d-c-k" and "j-i-n-q-a-b-j" of figure 11.

Table 6. Physical parameters of a single thermoelectric couple

Property	value
Hot junction temperature, T_H	300 K
Cold junction temperature, T_L	270 K
Thermoelectric element length, L	1.15×10^{-3} m *
Cross-sectional area, A	1.96×10^{-6} m^2 *
Electrical conductivity, σ'	97087.38 ohm^{-1}.m^{-1} *
Seebeck coefficient α (V/K) $\alpha(T) = \alpha_o + \alpha_1 \ln(T/T_o)$	$\alpha_o = 210 \times 10^{-6}$ V/K, $\alpha_1 = 120 \times 10^{-6}$ V/K $T_o = 300$ K (Seifert et al., 2002)
Thermal conductivity, λ	1.70 W/m K *

* Remarks Melcor thermoelectric catalogue, MELCOR CORPORATION, 1040 Spruce Street, Trenton, NJ 08648, USA. Web site: www.melcor.com.

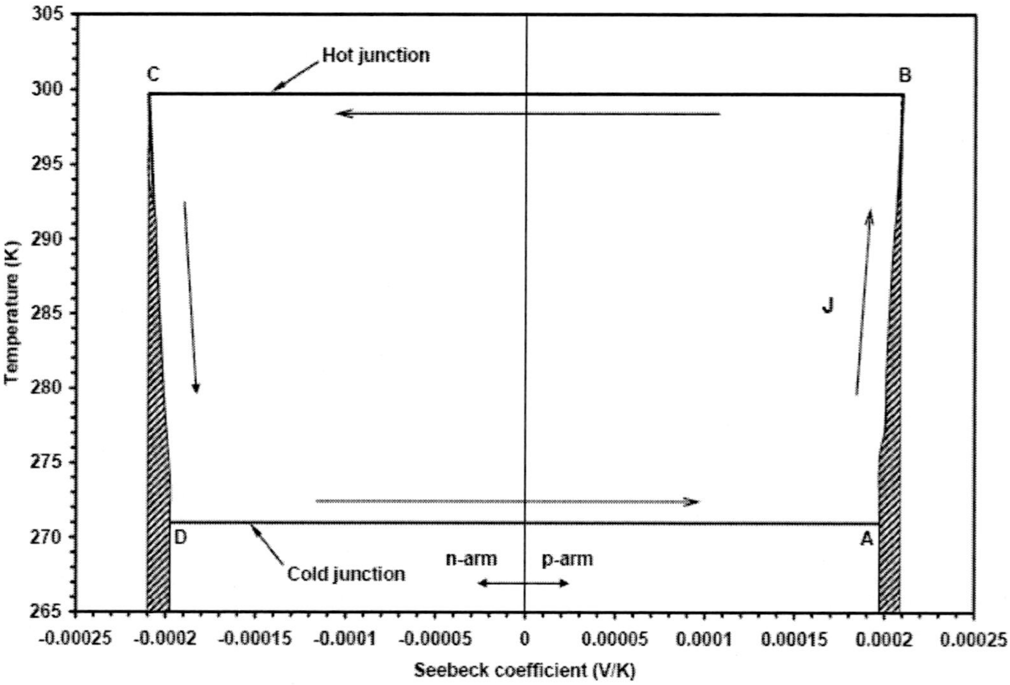

Figure 10. T-α diagram of a thermoelectric pair.

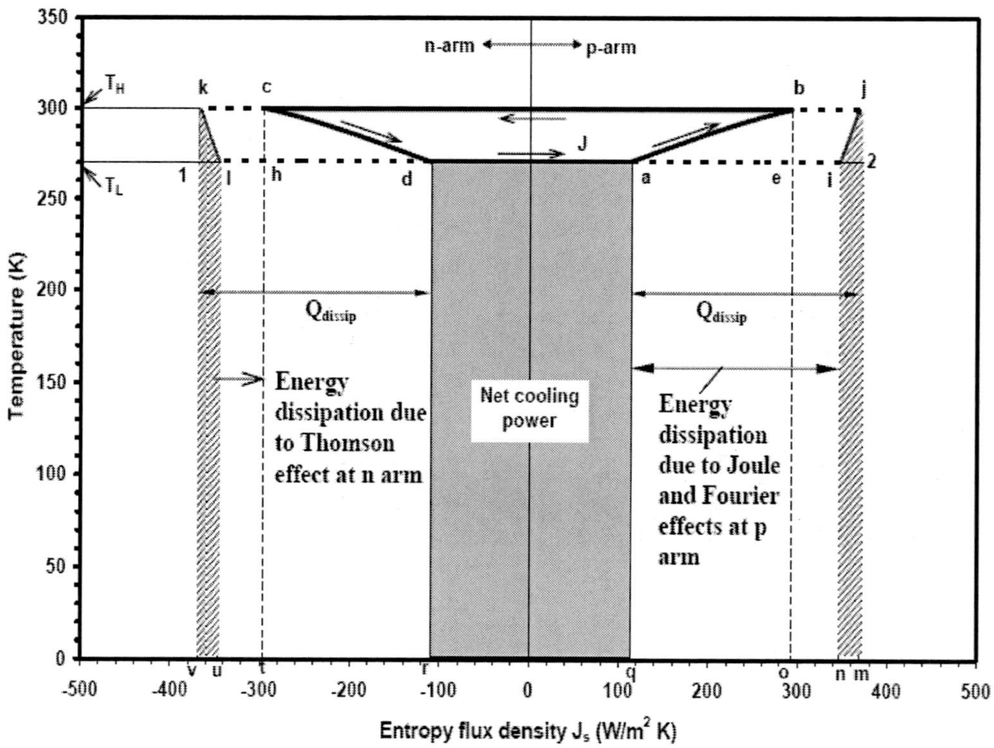

Figure 11. Temperature-entropy flux diagram for the thermoelectric cooler at the maximum coefficient of performance identifies the principal energy flows.

Both mentioned dissipative losses have the consequence of reducing the cooling effect, represented by area “a-d-r-q”. Within the internal framework (adiabatic chiller), the cycle “a-b-c-d” operates in an anti-clockwise manner with process with two isotherms b-c and d-a, as well as two non-adiabats a-b and c-d. Details of these processes are described in table 7. Thus, the T-s diagram provides an effective means of analyzing how the real chiller cycle of thermoelectrics is inferior to the ideal Carnot cycle by capturing the key losses and useful effects in the cycle.

Table 7. Description of close loop a-b-c-d-a (figure 11)

Process	Description
a--b	This process is developed along P-arm of the thermoelectric pair. Points a and b are the states at the ends of the thermoelement of which the temperatures are T_L and T_H respectively (as current J flows from P-arm to N-arm). The process a-b can not be adiabatic because it absorbs heat due to the Thomson effect which depends on the Seebeck coefficient of the thermoelectric element. The areas beneath the a-b process represent dissipation due to heat conduction, and heat absorbed by the Thomson effect. Area 2-i-j represents the energy dissipation due to Seebeck coefficient, which varies along the thermoelectric p-arm.
b--c	This process represents the heat exchange between system and medium (environment) which occurs at the union between P-arm and N-arm at temperature T_H.
c--d	The process c-d is developed along N-arm, where points c and d are at the ends of the thermo-element whose temperature are T_H and T_L. The areas beneath c-d process indicate heat losses due to dissipation. If this process is adiabatic, the Thomson effect would be null. Area 1-l-k indicates the energy losses due to Seebeck effect along the n-arm.
d--a	This process represents the heat exchange between the system and the heat load at isothermal condition; however it occurs in the physic union between thermoelectric elements at temperature T_L.

From temperature-entropy flux diagram, the thermodynamic performance variables can be easily tallied with the enclosed rectangles. The heat fluxes at the cold (process d-a) and hot (process b-c) reservoirs are as follows (for details please see Appendix)

$$Q_L = \boldsymbol{Area}(adrq) = T_L \left| J_{S,pn} \right| = T_L \left(\alpha \boldsymbol{I} - \frac{\lambda A \nabla T|_{cj}}{T_L} \right) = \alpha \boldsymbol{I} T_L - \lambda A \nabla T|_{cj} ,$$

$$= \overbrace{\alpha I T_L}^{Peltier\_cooling} - \overbrace{\frac{1}{2} I^2 R}^{Joulean\_heat} - \overbrace{K \Delta T}^{conduction\_heat} - \overbrace{\frac{1}{2} I \tau \Delta T}^{Thomson\_heat} \quad (63)$$

$$Q_H = \boldsymbol{Area}(bcto) = T_H \left| J_{S,pn} \right|_{hj} = T_H \left(\alpha \boldsymbol{I} - \frac{\lambda A \nabla T|_{hj}}{T_H} \right) = \alpha I T_H - \lambda A \nabla T|_{hj} ,$$
$$= \overbrace{\alpha I T_H}^{Peltier\_heat} + \overbrace{\frac{1}{2} I^2 R}^{Joulea\_heat} - \overbrace{K \Delta T}^{Conduction\_loss} + \overbrace{\frac{1}{2} I \tau \Delta T}^{Thomson\_heat} \tag{64}$$

where ΔT is the temperature difference across the hot and cold junctions. The first term of right hand side of both equations (63) and (64) indicates Peltier cooling and heating, the second term is the Joulean heat, the third term represents heat conduction between hot and cold junctions and the heat absorbed by the Thomson effect. The Thomson heat is transmitted by thermal conduction across the thermo-element and this effect is not reversible. From the first law of thermodynamics, the required electrical power input is given by,

$$P = \boldsymbol{Area}(bcto) - \boldsymbol{Area}(adrq) = T_H \left| J_{S,pn} \right|_{hot} - T_L \left| J_{S,pn} \right|_{cold}$$
$$= \overbrace{\alpha I T_H + \frac{1}{2} I^2 R - K\Delta T + \frac{1}{2} I \tau \Delta T}^{Q_H} - \overbrace{\alpha I T_L + \frac{1}{2} I^2 R + K\Delta T + \frac{1}{2} I \tau \Delta T}^{Q_L} \tag{65}$$
$$P = \alpha I (T_H - T_L) + I^2 R + I \tau \Delta T$$

The equivalent cycle at the same T_H and T_L is denoted by a rectangle 1-2-j-k (see figure 11), for which

$$COP_{carnot} = Area(1vm2) / [Area(kvmj) - Area(1vm2)]$$

or

$$COP_{carnot} = Area(1vm2) / Area(12jk) = T_L / \Delta T .$$

The irreversibility is the heat dissipation, which is shown in figure 11, occurs due to the Joule heat and to the heat conduction along the thermoelectric arm.

5.2. Transient Behavior of Thermoelectric Cooler

The performance of a Peltier or thermoelectric cooler can be enhanced by utilizing the transient response of a current pulse [42-45]. At fast transient, a high but short-period (4 seconds) current pulse is injected so that additional Petlier cooling is developed instantly at the cold junction: The joule heating developed in the thermoelectric element is a slow phenomenon as heat travels slowly over the materials and could not reached the cold junction in the short transient. Hence, the cold junction temperature is at the lowest possible value and heat flows only from the source to the cold junction. An example of such an application of the short transient cooling is an infrared detector operating in a 'winking mode, which needs to be super cold only during the wink. Figure 12 shows the schematic diagram of a current "pulse train" thermoelectric cooler with the p- and n- doped semiconductor elements are electrically connected in series and are thermally in parallel. The hot side of the device is mounted directly onto a substrate to reject heat to the environment and the thermal contact between the cold junction of thermoelectric device and the cold substrate (load) is made periodically. There are two periods of operation: Firstly, an "open contact" period with t_{no} and current flow J_{np} to maintain the lowest temperature T_{min} at the cold junction but the cold substrate (load) is not connected with the cold junction. Secondly, a large current pulse J_p (where J_p=MJ_{np}, $M > 1$) is applied for a duration t_p and concomitantly, the cold substrate makes contact with the cold junction. Intense Peltier cooling is achieved during t_p, which reduces the cold junction temperature below T_{min}. However, Joule and Thomson heating are developed in the bulk thermoelectric module and diffuse towards the hot and cold ends of the device.

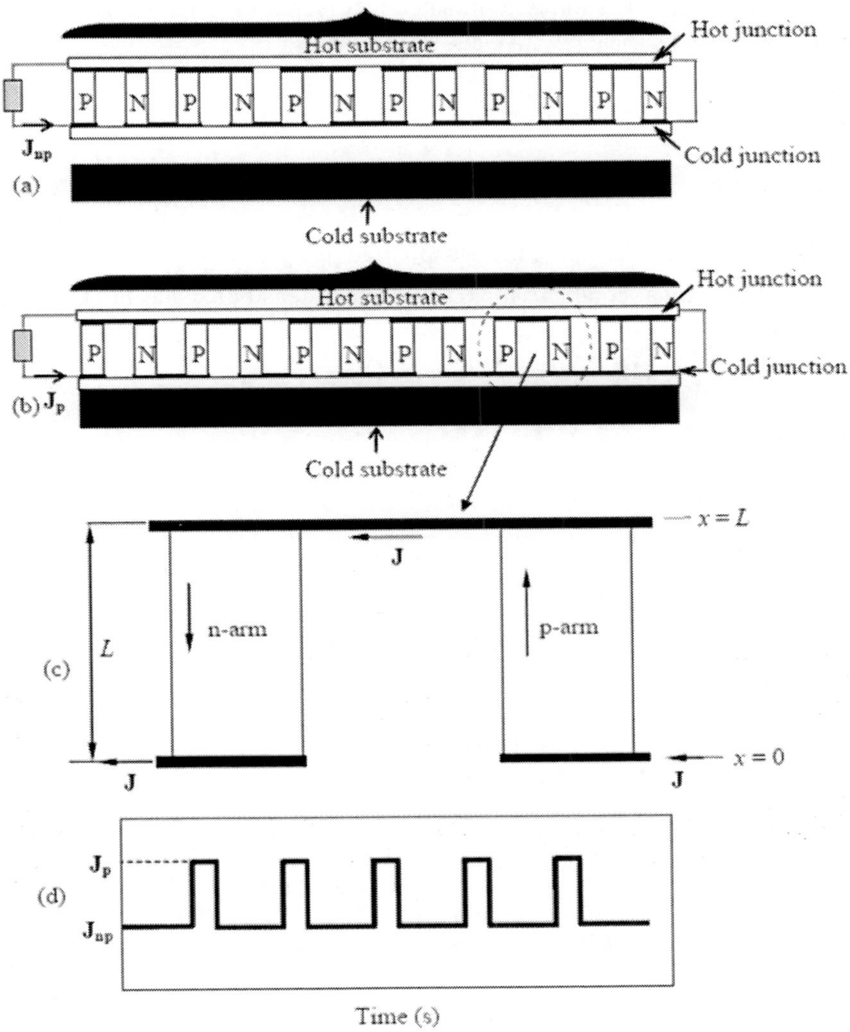

Figure 12. Schematic diagram of the first transient thermoelectric cooler. (a) indicates the normal operation and cold substrate is out of contact, (b) defines the thermoelectric pulse operation and cold substrate is in contact with the cold junction, (c) shows the thermoelectric couple consisting of p and n arms, (d) indicates that the pulse current is applied to the device.

Thermal contact of the cold substrate with the cold end of the thermoelectric module is maintained over the short current pulse period but these surface break contact before the cooling could be degraded by the conduction heat. In the open contact manner, the cold substrate temperature could reach below that of the cold junction temperature under the normal thermoelectric operation. The total operation consists of two periods; one is the normal thermoelectric operation and the other is the pulse or effective operation. The cycle of (normal operation + pulse operation) is repeated such that the device is continuously produced cooling.

This section formulates a non-equilibrium thermodynamic model to demonstrate the energy and entropy balances of a transient thermoelectric cooler. In analyzing the fast transient cycle in the classical *T-s* plane, one distinguishes the entropy fluxes of the working fluid (which is the current flux *J*) from the external entropy fluxes in the hot and cold reservoirs and at the interface with the reservoirs. The T-s relation is formulated that identifies the key heat and work flows, the heat conduction and electrical resistive dissipative losses.

5.2.1. Derivation of the T-S Relation

The methodology of energy and entropy balances of the commercial available thermoelectric device has been discussed briefly in the previous sections. Starting from the general entropy balance equation, it can be expressed in terms of the key parameters such as Peltier effect (π), electrical conductivity (σ'), current density (*J*) and Fourier effect (λ), as shown below [46]:

$$\frac{\partial(\rho_p s_p)}{\partial t} = \overbrace{\nabla\cdot\left(\frac{\lambda_p \nabla T_p}{T_p} - \frac{\pi J_p}{T_{p,te}}\right)}^{entropy\ \ flux} + \overbrace{\left\{-\lambda_p \nabla T_p . \nabla\left(\frac{1}{T_p}\right) + \frac{J_p^2}{T_p \sigma'_p}\right\}}^{entropy\ \ generation}, \tag{66}$$

hence the subscript p defines the pulsed thermoelectric cooler, the first term of the right hand side indicates the entropy flux in a given control volume and the second terms represent the total entropy generation. The total entropy flux density (in W/m^2 K) is expressed by

$$\mathbf{J}_{s,tot}\Big|_p = \frac{\lambda_p \nabla T_p}{T_p} - \alpha J_p, \tag{67}$$

and the entropy generation (in W/m^3 K) is written as

$$\sigma_{p,tot} = -\lambda_p \nabla T_p . \nabla\left(\frac{1}{T_p}\right) + \frac{J_p^2}{T_p \sigma'_p}. \tag{68}$$

The energy conservation equation is as follows

$$\rho_p c_p \frac{\partial T_p}{\partial t} = \lambda_p \nabla^2 T_p + \frac{J_p^2}{\sigma'_p} + J_p T_{p,te} \nabla\alpha\big|_{T_p} - J_p T_p \left(\frac{\partial \alpha}{\partial T_p}\right)\nabla T_p. \tag{69}$$

To solve the problem, two boundary conditions are defined: Firstly, during the normal (open-contact) operation, there is only Peltier cooling at the cold junction

$$\left.\frac{\partial T_p}{\partial x}\right|_{at\,cj} = \frac{\alpha J T_{cj}}{\lambda_p}, \tag{70}$$

and at hot junction, a hot side heat sink is represented by, $T_p\big|_{at\,hj} = T_{hj}$. But during the pulse (close-contact) operation, at the cold junction the boundary condition becomes

$$\left.\frac{\partial T_p}{\partial x}\right|_{at\,cj} = \frac{\alpha J T_{cj}}{\lambda_p} + \frac{q^{\cdot}_{subs}}{\lambda_p}, \tag{71}$$

where $q^{\cdot}_{subs}$ is the cooling load of the cold substrate.

5.2.2. Discussion

Figure 13 shows the transient behavior of the pulsed thermoelectric cooler where the hot junction is assumed to be constant at T_{hj} = 308 K. Based on the physical properties of a thermoelectric cooler, as tabulated in table 1, the simulations with an initial current density of J_{np} = 0.675 A/mm^2 yield an exponential decrease of the cold junction temperature to $T_{c,A'}$ = 240 K (denoted by point A′), and beyond which a pulsed current density of J_p = 2.025 A/mm^2 for a duration of 4 s is applied for the pulsed or contact mode. The high pulsed current in the thermoelectrics produces instantaneous Peltier cooling, as shown by the decrease in the cold junction temperature to point B. Concomitantly, the pulsed current generates the Joule and Thomson heating within the p-n legs and this is reflected by the rise of cold junction temperature from point B to C. As no cooling could be achieved by the cold substrate, the cold junction breaks contact with it at point C. However, the residual heat transfer from the mentioned effects is generally a slower phenomenon as compared to Peltier cooling, the temperature of the cold junction continues to rise until the cooling by J_{np} overcomes the residual heat within the legs, culminating in a maxima for the temperature of the cold junction, as depicted by point B'.

For the non-contact duration, the energy flows in the thermoelectric device can best be followed by tracking the temperature versus entropy fluxes along the p-leg in a anti-clockwise direction, i.e., points "a" to "g" of figure 14: Point "a" denotes the cold junction temperature and the entropy flux here is taken to be zero or the datum. As holes migrates along the p-leg, both entropy flux and temperature increase with the spatial length until point "g" is reached, due primarily to the property changes with the local temperature of Bi_2Te_3 materials such as the electrical resistivity (ohm-cm), Seebeck coefficient (V/K) and thermal conductivity (W/cm K). For the given configuration and current density, the maximum entropy flux is about 300 W/m^2.K. At the hot junction which is assumed to remain constant at T_{hj}, the sign of the entropy flux changes from positive to negative due to negative temperature gradient along the n-leg, resulting in the isotherm "g" to "h" line on the T-s diagram. Following the path in an anti-clockwise direction, the local entropy flux increases from point "h" to point "a", completing the cycle (denoted by A′) on the T-s diagram.

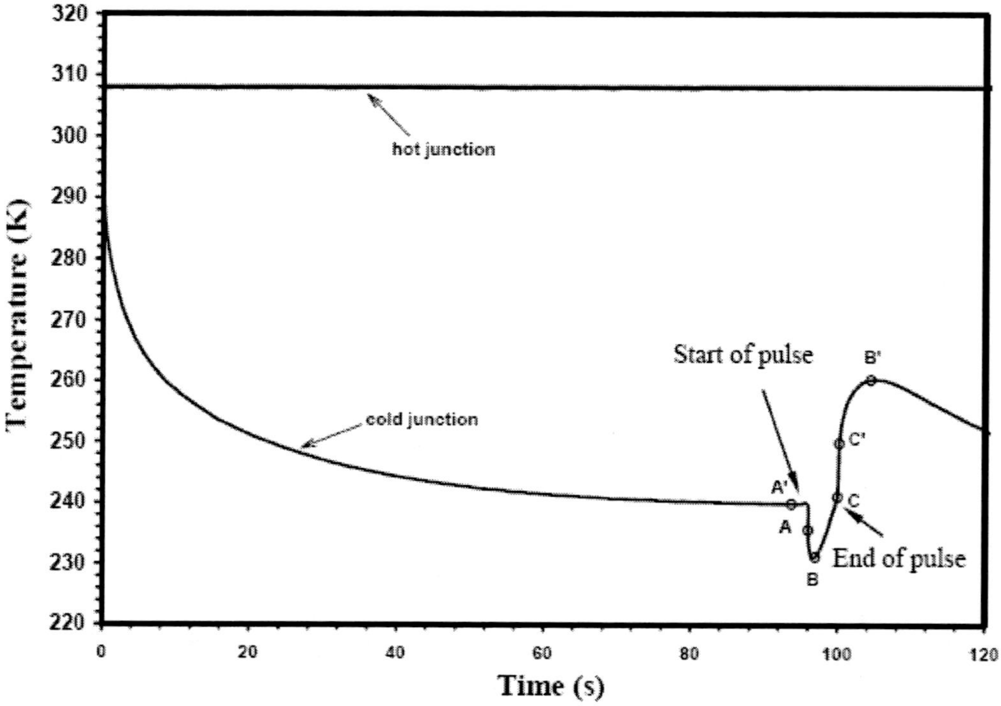

Figure 13. The transient temperature-time trace of cold reservoir temperature of a pulsed thermoelectric cooler. Point "A" indicates the beginning of pulse period, "B" denotes the lowest temperature reached by the cooler and "C" represents the end of pulse period, "C' " shows the beginning of non-contact period, "B' " indicates maximum temperature rise due to energy balance between the residual heat from the pulsed period and the steady state cooling.

Two other process cycles could also be observed for the non-contact duration, namely cycles "B" and "C": Both Cycles are subjected to J_{np} and they show the presence of residual Joulean and Thomson heat within the p-n legs of the thermoelectrics. The balance between the cooling rate generated by the J_{np} and the amount of residual Joule and Thomson heat from the previous pulsed injection yield a local thermal wave front in each of the legs (at distances closer to the hot junction) where the maximum local temperature is higher than that of the T_{hj}.

During closed contact operation, as shown in figure 15, point A indicates the commencement of contact interval where the imposed current pulse (J_p) generates a substantial cooling, depressing the cold junction and the cold template temperature from "f" to "a" (see the smaller insert). As opposed to the non-contact duration, the cold junction on the T-s plot has an isotherm path, and the area enclosed below the isotherm a-a' indicates the amount of cooling produced at point A of the contact operation. Similar isotherms c-c' and d-d' correspond to the points B and C during the transients. Although the enclosed area for point C is slightly larger than that of point B, but the cold junction temperature of point C is found to be higher, caused by the diffusion of Joule heat from the legs of the thermoelectrics.

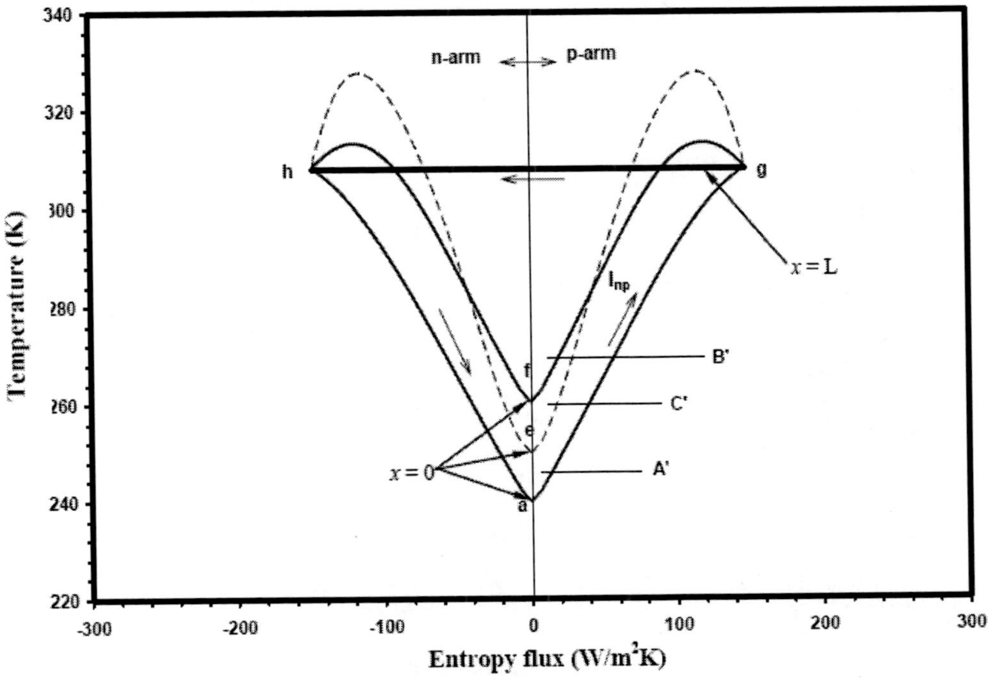

Figure 14. Temperature-entropy flux (T-s) diagram showing the anti-clockwise loop of a pulsed thermoelectric cooler for the non-pulse duration for three operation points [46].

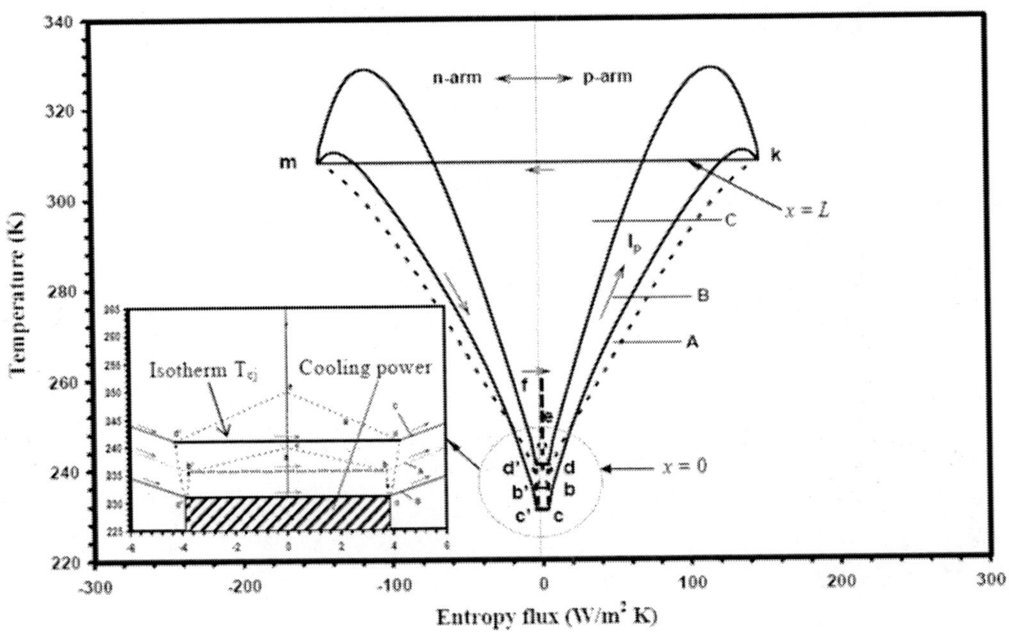

Figure 15. Temperature-entropy flux (T-s) diagram showing the energy flow details of a pulsed period of thermoelectric cooler. Cycles A to C refer to the beginning of cooling, the lowest temperature reached and the maximum cooling power by the cooler. The details of the processes within the cold junction of thermoelectrics are shown by the small insert, denoted correspondingly by "a" to "c". The enclosed area below T_c isotherms in the T-s diagram indicates the cooling power.

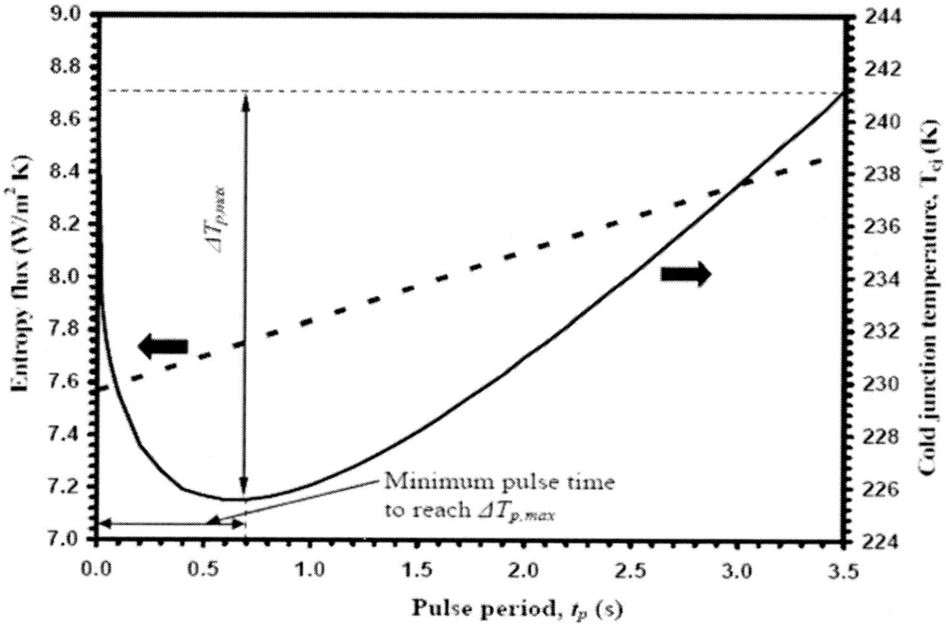

Figure 16. The effects of pulse periods on entropy flux and the cold junction temperature subjected to a step current pulse of magnitude M = 4.

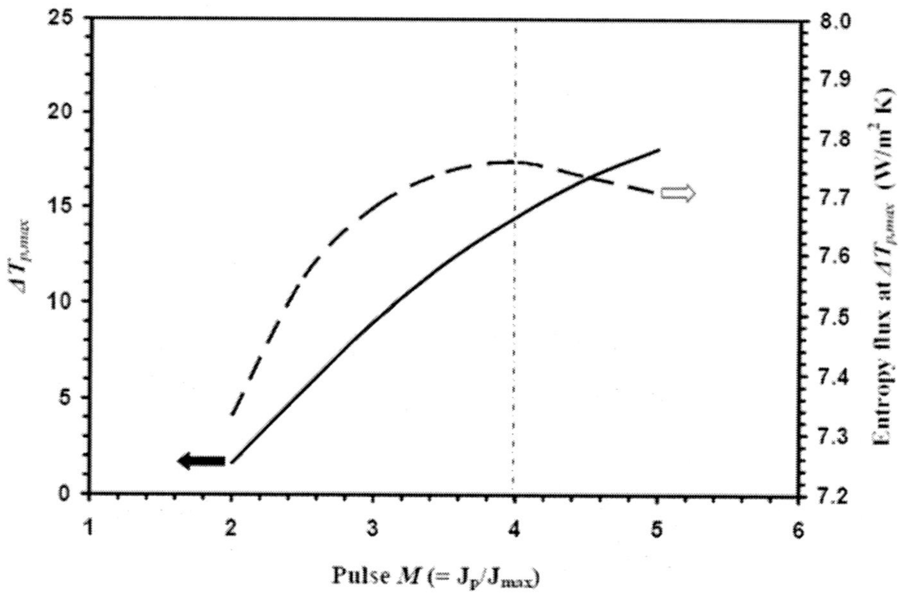

Figure 17. Maximum pulse cooling difference in temperatures and entropy fluxes as a function of pulse factor, exhibits the optimization.

No cooling power is developed during non-contact period but the cooling power is generated during pulse period. Figure 16 shows the time evolution of entropy flow and temperature at the cold junction and calculates the minimum time to reach the maximum temperature difference between the super-cooling and the steady state or $\Delta T_{p,\max}$. The characteristics time, entropy flux and cold end temperature as shown in figure 16 may be used to characterize the pulse cooler as functions of length of pulse t_p, pulse factor M, and pulse current density, J_p. This graph is helpful to design a cooler where the user would easily calculate the amount of increasing cooling, cooling behaviors, the needed pulse current and the time between pulses.

The effects of pulse factors on entropy flux at cold end and the maximum temperature drop or $\Delta T_{p,\max}$ are shown in figure 17 and the optimization is obtained at pulse factor 4 [46].

5.2.3. Summary of Section 5.2

We have successfully plotted the T-s diagrams for the transient features of a pulsed thermoelectric cooler. The thermodynamic formulation (based on the Gibbs law) provides the necessary expression for the entropy flux that is expressed in terms of the basic variables describing the device, namely, the current density, Seebeck coefficient, the local temperature and temperature gradient. Using the T-s diagram, the process paths of the pulsed and non-pulsed operations of thermoelectric cooler are accurately mapped and it has immense pedagogical value for cycle evaluation, depicting the useful energy flows and "bottlenecks" of cycle operation.

5.3. Microscopic Analysis: Super-Lattice Type Devices

The development of semi-conductor materials for cooling and heating applications has made steady progress with the thermoelectric (TE) coolers being made of the Bismuth-Telluride (Bi_2Te_3), Antimony-Telluride (Sb_2Te_3), and Silicon-Germanium (SiGe) materials [47]. Two factors, however, limit the wide spread application of such coolers, namely (i) the low efficiency or coefficient of performance (COP) at higher temperature differential across the cold and hot junctions, and (ii) the cost of the semi-conductor devices per unit cooling capacity. The cooling efficiency of TE elements, comprising the p-n layers of semi-conductors, is characterized by the dimensionless ZT parameter: T is the absolute operational temperature and the thermoelectric figure of merit Z is given by $\alpha^2/\rho\lambda$ where α is the Seebeck coefficient, ρ is the density and λ is the thermal conductivity. The best thermoelectric materials available, hitherto, have a $ZT \approx 0.9$, but for commercial viability as cooling devices at ambient conditions, the ZT value should exceed a value 3. Recently, there were reports of semi-conductor materials with multi-quantum well materials [48], PbSeTe-based quantum dot superlattice structures [49] and thin-film Bi_2Te_3/Sb_2Te_3-based superlattice thermoelectric element for both refrigeration and power generation applications [50, 51]. Thin-film thermoelectric coolers using silicon materials such as Si / $Si_{0.8}Ge_{0.2}$, have been fabricated by alternating barrier and conducting layers which generate the quantum wells and increase the electron mobility. The procedures of making thermoelectric modules with p and n-type (Si/ $Si_{0.8}Ge_{0.2}$) superlattices have been reported [52].

The size of thermoelectric device characterizes the transport of electrons, holes and phonons. In a very thin film with a thickness less than mean free path, electrons go through the film ballistically and experience little or no collision [53]. Scattering or collision dominants the transport process of electrons, holes and phonons in thick films or bulk materials. In case of bulk thermoelectric material such as Bi_2Te_3, SiGe, SbTe, the limiting speed of thermoelectric element responses from milliseconds to a few second, so the energy dissipation due to collisions of electrons, holes and phonons is not significant and heat transport is purely diffusive [54]. The reduction of device sizes to the sub-micrometer range not only increases the device switching speed from nano-second to pico-second but also increases the local rate of heat generation due to electron-phonon interactions or hole-phonon interactions [54, 55].

The motivation of this section is to model the thin-film thermo-electric cooling devices [50] as shown in figure 18 using the BTE where both macro and micro phonological dissipations are rigorously captured at the micro-scaled dimensions. The present model treats the non-equilibrium transport of electrons-phonons (in case of n-type such as $Bi_2Te_3/Bi_2Te_{2.83}Se_{0.17}$ thin film thermoelectric element) and holes-phonons (for p-type such as Bi_2Te_3/Sb_2Te_3 thin film thermoelectric element), and their non-equilibrium coupling or the collision contributions at different electric fields using the practical temperature-entropy (T-s) diagrams. This T-s diagram quantifies the net useful cooling effects at the cold junction and the COP. The proposed model of the thin film thermoelectric cooler that accounts for the collision contributions is validated with a recently published experimental data [50].

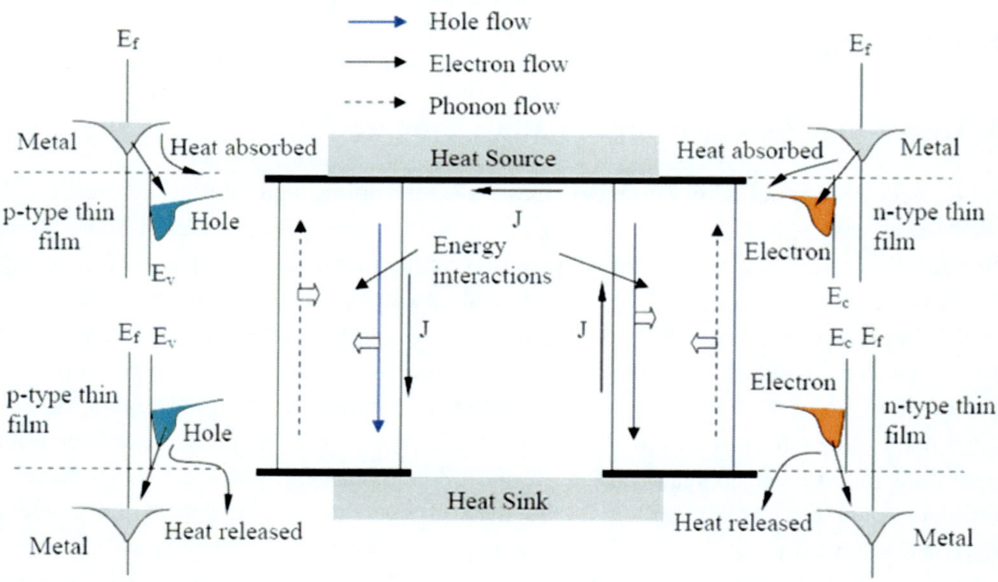

Figure 18. Schematic diagram of a thin film thermoelectric couple.

5.3.1. Thermodynamic Modeling for Thin-Film Thermoelectrics

The interactions of electrical and thermal phenomena are important to simulate the superlattice thermoelectric element. The most important thermal effect is the thermal runaway, in which the electrical dissipation of energy causes an increase in temperature which in turn causes a greater dissipation of energy. Without repeating the basic derivation (as derived in section 2), a complete model of electron, hole and phonon conservation equations for a thin film thermoelectric element is summarized in table 7.

Table 7. The balance equations

Particles		Equations
Electrons	mass	$\frac{\partial n_e}{\partial t} = -\nabla(v_e n_e) + \left.\frac{\partial n_e}{\partial t}\right\|_{colli}$ (72)
	energy	$\frac{\partial u_e}{\partial t} = -\nabla.Q_e + \mathbf{JF}\|_e + \left.\frac{\partial u_e}{\partial t}\right\|_{colli}$ (73)
	entropy	$\frac{\partial s_e}{\partial t} = \nabla.J_{se} + \sigma_{irr,e} + \sigma_{colli,e-h} + \sigma_{colli,e-g}$ (74)
Holes	mass	$\frac{\partial n_h}{\partial t} = -\nabla(v_e n_h) + \left.\frac{\partial n_h}{\partial t}\right\|_{colli}$ (75)
	energy	$\frac{\partial u_h}{\partial t} = -\nabla.Q_h + \mathbf{JF}\|_h + \left.\frac{\partial u_h}{\partial t}\right\|_{colli}$ (76)
	entropy	$\frac{\partial s_h}{\partial t} = \nabla.J_{sh} + \sigma_{irr,h} + \sigma_{colli,h-e} + \sigma_{colli,h-g}$ (77)
Phonons	energy	$\frac{\partial u_g}{\partial t} = -\nabla.Q_g + \left.\frac{\partial u_h}{\partial t}\right\|_{colli}$ (78)
	entropy	$\frac{\partial s_g}{\partial t} = \nabla.J_{s,g} + \sigma_{irr,g} + \sigma_{colli,g-e} + \sigma_{colli,g-h}$ (79)
Combined (Electron + Hole + Phonon)	energy	$\frac{\partial u}{\partial t} = -\nabla.Q + \mathbf{JF} + \left.\frac{\partial u}{\partial t}\right\|_{colli}$ (80)
	entropy	$\frac{ds}{dt} = \nabla.J_s + \sigma$ (81)

As seen in table 7, the Boltzmann Transport Equation is employed together with the phenomenological relationship [14]. The energy balance equations for electrons, holes and phonon are computed using Gibbs law. Based on these formulation, the entropy flux and entropy generation for electrons, holes and phonons in the well and barrier of the superlattice thermoelement are developed and they are summarized in table 8.

Table 8. The energy, entropy flux and entropy generation equations of the well and the barrier [4]

well			
Particles	**Energy balance**	**Entropy flux W/m² K**	**Entropy generation W/m³K**
Electron	$c_w \frac{\partial T_{w,e}}{\partial t} = \nabla.\left(\lambda_{w,e} \nabla T_{w,e}\right) + \frac{J_{w,e}^2}{\sigma'_{w,e}} + c_w \left.\frac{\partial T_{w,e}}{\partial t}\right\|_{colli}$	$-\frac{\lambda_{w,e} \nabla T_{w,e}}{T_{w,e}} + \frac{\pi_e J_{w,e}}{T_{w,e}}$	$-\lambda_{we} \nabla T_{we} \nabla\left(\frac{1}{T_{we}}\right) + \frac{J_{we}^2}{T_{we}\sigma'_{we}} + \frac{c_w}{T_{we}} \left.\frac{\partial T_{we}}{\partial t}\right\|_{colli}$
Hole	$c_w \frac{\partial T_{w,h}}{\partial t} = \nabla.\left(\lambda_{w,h} \nabla T_{w,h}\right) + \frac{J_{w,h}^2}{\sigma'_{w,h}} + c_w \left.\frac{\partial T_{w,h}}{\partial t}\right\|_{colli}$	$-\frac{\lambda_{w,h} \nabla T_{w,h}}{T_{w,h}} + \frac{\pi_h J_{w,h}}{T_{w,h}}$	$-\lambda_{wh} \nabla T_{wh} . \nabla\left(\frac{1}{T_{wh}}\right) + \frac{J_{wh}^2}{T_{wh}\sigma'_{wh}} + \frac{c_w}{T_{wh}} \left.\frac{\partial T_{wh}}{\partial t}\right\|_{colli}$
Phonon	$c_w \frac{\partial T_{w,g}}{\partial t} = \nabla.\left(\lambda_{w,g} \nabla T_{w,g}\right) + c_w \left.\frac{\partial T_{w,g}}{\partial t}\right\|_{colli}$	$-\frac{\lambda_{w,g} \nabla T_{w,g}}{T_{w,g}}$	$-\lambda_{wg} \nabla T_{wg} . \nabla\left(\frac{1}{T_{wg}}\right) + \frac{c_w}{T_{wg}} \left.\frac{\partial T_{wg}}{\partial t}\right\|_{colli}$
Combined	$c_w \frac{\partial T_w}{\partial t} = \nabla.\left(\lambda_w \nabla T_w\right) + \frac{J_w^2}{\sigma'_w} + c_w \left.\frac{\partial T_w}{\partial t}\right\|_{colli}$	$-\frac{\lambda_w \nabla T_w}{T_w} + \frac{\pi J_w}{T_w}$	$-\lambda_w \nabla T_w . \nabla\left(\frac{1}{T_w}\right) + \frac{J_w^2}{T_w\sigma'_w} + \frac{c_w}{T_w} \left.\frac{\partial T_w}{\partial t}\right\|_{colli}$

Table 8. Continued

Barrier			
	Energy balance	**Entropy flux**	**Entropy generation**
Electron	$c_b \frac{\partial T_{b,e}}{\partial t} = \nabla.\left(\lambda_{b,e} \nabla T_{b,e}\right) + c_b \frac{\partial T_{b,e}}{\partial t}\Big\vert_{colli}$	$-\frac{\lambda_{b,e} \nabla T_{b,e}}{T_{b,e}}$	$-\lambda_{be} \nabla T_{be}.\nabla\left(\frac{1}{T_{be}}\right) + \frac{c_b}{T_{be}} \frac{\partial T_{be}}{\partial t}\Big\vert_{colli}$
Hole	$c_b \frac{\partial T_{b,h}}{\partial t} = \nabla.\left(\lambda_{b,h} \nabla T_{b,h}\right) + c_b \frac{\partial T_{b,h}}{\partial t}\Big\vert_{colli}$	$-\frac{\lambda_{b,h} \nabla T_{b,h}}{T_{b,h}}$	$-\lambda_{bh} \nabla T_{bh}.\nabla\left(\frac{1}{T_{bh}}\right) + \frac{c_b}{T_{bh}} \frac{\partial T_{bh}}{\partial t}\Big\vert_{colli}$
Phonon	$c_b \frac{\partial T_{b,g}}{\partial t} = \nabla.\left(\lambda_{b,g} \nabla T_{b,g}\right) + c_b \frac{\partial T_{b,g}}{\partial t}\Big\vert_{colli}$	$-\frac{\lambda_{b,g} \nabla T_{b,g}}{T_{b,g}}$	$-\lambda_{bg} \nabla T_{bg}.\nabla\left(\frac{1}{T_{bg}}\right) + \frac{c_b}{T_{bg}} \frac{\partial T_{bg}}{\partial t}\Big\vert_{colli}$
Combined	$c_b \frac{\partial T_b}{\partial t} = \nabla.\left(\lambda_b \nabla T_b\right) + c_b \frac{\partial T_b}{\partial t}\Big\vert_{colli}$	$-\frac{\lambda_b \nabla T_b}{T_b}$	$-\lambda_b \nabla T_b.\nabla\left(\frac{1}{T_b}\right) + \frac{c_b}{T_b} \frac{\partial T_b}{\partial t}\Big\vert_{colli}$

5.3.2. Discussion

The simulation results presented herein are for the transient heat transport in the direction of hot and cold junctions of Bi_2Te_3/Sb_2Te_3 thin-film thermoelectric element. Venkatasubramanian et al. [50] reported the properties and dimensions of a prototype super-lattice, and the values are tabulated in table 9. In the presence of high electric fields within the thin film structure, lesser equilibrium between electrons-phonons or holes-phonons is expected. Thus, the collision effect from electrons and phonons is increased.

Table 9. The thermo-physical properties of thin film thermoelectric element

Sample	Electrical resistivity Ω-m	Seebeck coefficient μV/°C	Phonon Thermal conductivity W/m K	Electron Thermal conductivity W/m K
Bi_2Te_3/Sb_2Te_3 (P-arm) [50]	1.2×10^{-5}	238	0.22	0.37
$Bi_2Te_3/Bi_2Te_{2.83}Se_{0.17}$ (n-arm) [50]	1.23×10^{-5}	-238	0.58	0.37

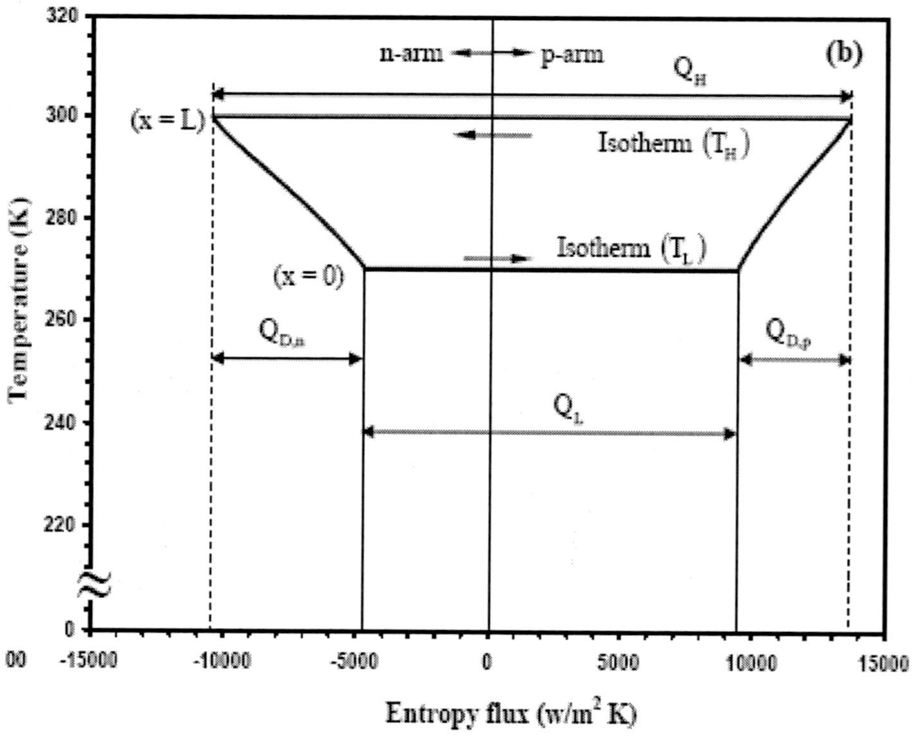

Figure 19. The temperature-entropy diagram of a thin film thermoelectric couple (both the p and n-type thermoelectric element are shown here) at optimum cooling capacity or maximum COP.

Following the classical T-s diagram methodology, which demarcates accurately how energy input to a device is consumed in overcoming the dissipative and finite-rate heat transfer losses, the figures 19 (a) and 19 (b) show the temperature and entropy fluxes of electrons, holes and phonons for a set of cold and hot temperatures of 270K and 300K, respectively. As can be seen, the electrons flow in n-type element (cold to hot junctions or negative Seebeck coefficient) which is opposite to the current flow, whilst the holes flow through the p-type element (cold to hot junctions or positive Seebeck coefficient) in the same direction of current flow. As the electrons of n-type thin film and holes of p-type thin film are known to have lesser dissipative losses, the based area of cold reservoir is larger, i.e. $T_L \times AD$ [in figure 19 (a)], reflecting a larger potential for cooling. However, the accompanying phonons, which have larger internal and dissipative losses, reduce the cooling capacity, i.e., $(T_L \times OE)$ at the p-arm and $(T_L \times OH)$ at the n-arm, as shown in figure 19 (a). Taking the phonon losses into account, the combined diagram gives the flow of current density, J, from p to n arms, as given by the close loop "a-b-c-d-a" which establishes the T-s diagram (denoted by the full lines) and this is shown in figure 19 (b). Thus, the cooling capacity is given by the area, T_L x (ad), and the heating power is given by T_H x (bc), and power input is the area bounded by their differences, i.e., $T_H \times (bc) - T_L \times (ad)$. The net coefficient of performance (COP) is the ratio of useful effect to energy input which is simply given by $T_L \times (ad)/ [T_H \times (bc) - T_L \times (ad)]$.

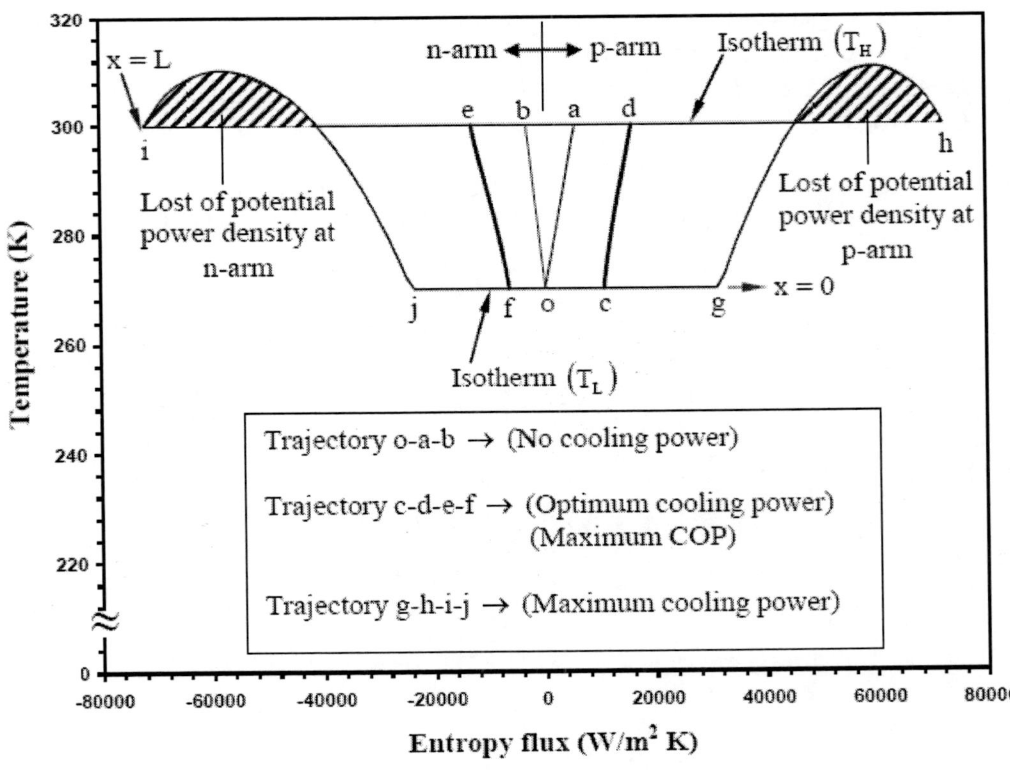

Figure 20. Temperature-entropy flux (T-J_s) diagrams for the thin film thermoelectric cooler at three operating points.

Figure 20 constitutes the temperature-entropy flux plots for different electric fields of the thin film thermoelectric cooler, where the close loop o-a-b-a defines zero cooling capacity and zero COP at very low electric field (0.05 A). The close loop c-d-e-f-c represents the maximum COP or the optimum cooling capacity at high electric field (6 A). The close loop g-h-i-j-g indicates the T-s diagram for the maximum cooling capacity.

The effect of collisions arising from electrons and phonons (n-arm) and holes and phonons (p-arm) is compared with the same cycle without the collision terms for different currents and these simulation results are tabulated in table 10. For a given dimensions of thin film thermoelectric element [50] and a current range of 1 - 10 A [50], the contribution of collision dissipation is found to be less than 7.4% of the total energy input. It is noted that the COP increases with the current density increases in the thin-films, reaching an optima at 2 A when the energy lost to collision effect is 11.8%. However, at higher input current densities, the cooling capacity of the thin-films decreases to near zero due to three key effects, namely (i) the Joulean heat loss, (ii) the Fourier loss and (iii) the collision loss. The contribution from collisions becomes increasing important at these high current densities, reaching as high as 32% of the input power.

As depicted on a T-s diagram, the comparative results are shown in figure 21. Cycle A-B-C-D-A is without collision terms and the cycle is superimposed onto the cycle A′-B′-C′-D′- A′ that includes collision effect. At the optimal conditions, the net heat dissipation at the hot junction with and without collisions are computed to be 208 mW and 200 mW, respectively, whilst the corresponding cooling capacities are 112 mW and 120 mW, respectively. The effect of collision reduces the COP of the thin film thermoelectrics by 20%. This increase is attributed to two effects, namely, (i) the reduction of cooling power as indicated by the shaded area, $T_L \times (AA'+DD')$, and secondly, (ii) the energy input is increased by shaded area, $T_H \times (CC'+BB')$.

Table 10. Performance analysis of the thin film thermoelectric cooler (excluding collisions effect and including collisions effect) [56]

I kA/cm^2	With out collision effects			With collision effects			Input energy loss due to collisions (%)
	$J_{s,hj}$ W/cm^{2}K	$J_{s,cj}$ W/cm^{2}K	COP	$J_{s,hj}$ W/cm^{2}K	$J_{s,cj}$ W/cm^{2}K	COP	
40	0.55	0.281	0.85	0.57	0.28	0.78	7%
60	1.62	0.96	1.14	1.70	0.95	1.0	12%
72	2.30	1.47	1.35	2.41	1.42	1.13	16%
80	2.75	1.80	1.40	2.89	1.72	1.15	18%
100	3.91	2.56	1.43	4.13	2.43	1.12	22%
120	5.16	3.40	1.45	5.44	3.08	1.04	27%
160	7.85	5.10	1.40	8.20	4.19	0.85	35%
240	13.4	7.60	1.04	14.41	5.56	0.53	45%

For the case of optimum operation 2 A, the area under the T-s diagram are (figure 6)			
$Q_{H, ncolli}$	200 mW	$Q_{H, colli}$	208 mW
$Q_{L, ncolli}$	120 mW	$Q_{L,colli}$	112 mW
$P_{input, ncolli} = Q_H - Q_L$	80 mW	$P_{input, colli} = Q_H - Q_L$	96 mW
Loss due to collisions Q_{colli} = 16 mW			

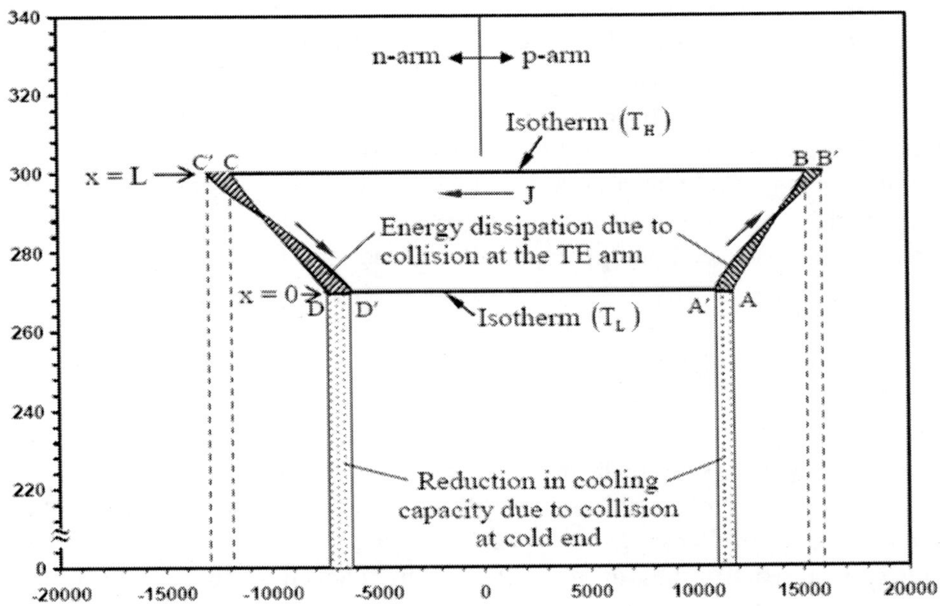

Figure 21. The temperature-entropy diagrams of p and n-type thin film thermoelectric elements. The close loop A′-B′-C′-D′ shows the T-s diagram when the collision terms are taken into account and the close loop A-B-C-D indicates the T-s diagram when collisions are not included. Energy dissipation due to collisions is shown here.

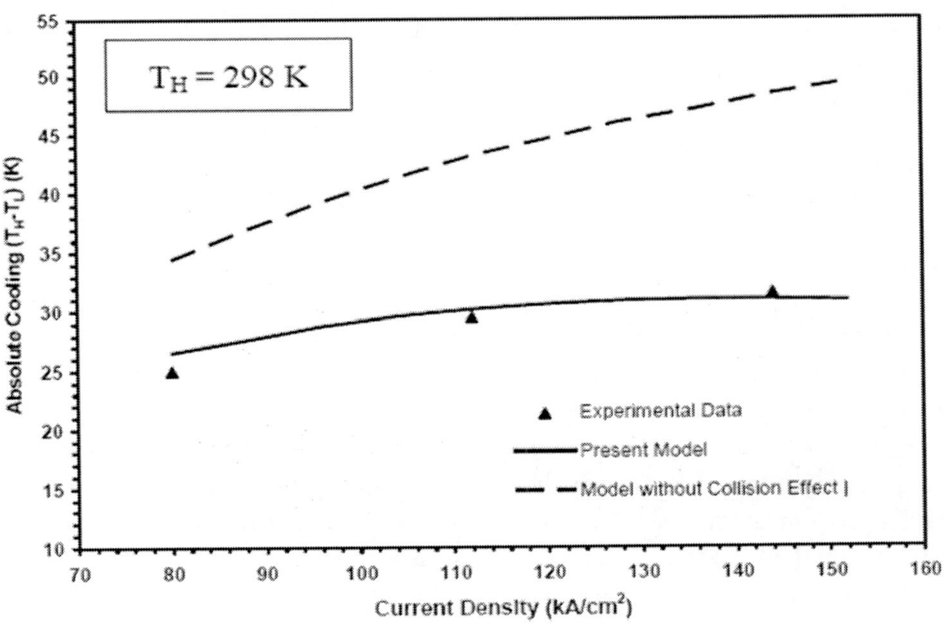

Figure 22. Absolute cooling (T_H-T_L) as a function of current density in a p-type thin film superlattice. Hence (▲) defines the experimental data for p-type superlattice device [6], (**—**) represents the simulation of present modelling for p-type superlattice, where the energy transports between holes and phonons are taken into account and (—) depicts the simulation of p-type thin film superlattice without energy transfer between phonons and holes.

All results presented herein are the predictions of the performances of next generation solid state micro-coolers. As it is beneficial to compare the predictions with experiments, the measured ΔT across the hot and cold junctions of a *p-type* thin film superlattice [50] is depicted against the imposed current density, as shown in figure 22. For this comparison, the thermophysical properties of the thin film thermoelectric element are extracted from table 9. As seen from comparison, the proposed BTE model with the collision effect shows good agreement with the experimental data whilst the conventional thermoelectric model [1, 21] over predicts the absolute cooling power and could not capture the true behaviors of thin-films.

5.3.3. Summary of Section 5.3

This section presents the uniquely combined conservations laws where the classical irreversible (Joule heat and conduction) and electrons/holes-phonons collision losses are captured. A finite difference code for solving the BTE for finite time collisions of the particles within the thin-film super-lattice thermo-electric cooler has been demonstrated. For a broad range of current densities, the dissipative contributions from both the electron-phonons and holes-phonon collisions within the thin-film, have been derived theoretically with BTE approach and the actual data from literature were incorporated here for analysis. Given a finite dimension of the thermo-element, the COP exhibits an optimal behaviour at medium current density level where the collision dissipation is about 10% of the power input, and increasing the current densities beyond the optimum point would accelerate the increase in the dissipative losses from collision and mitigates the COP values to naught. The practical and pedagogical values of a novel Temperature-entropy flux diagram for thin film thermoelectrics have also been demonstrated. To predict the performance of the thin film thermoelectric device, it is important to understand the thermal behavior, which occurs in nanosecond range. This is also useful to design a micro-structure and understand the mechanisms of phonon transport.

6. Conclusions

In this chapter, a universal thermodynamic framework that is suitable for both the macro and micro scale modeling of thermoelectric cooler has been derived. It is based on the Boltzmann Transport Equation (BTE) approach and the classical Gibbs law, where their amalgamation yields the form of the temperature-entropy flux (T-s) formulations. This modelling is applied to demonstrate the energy flows of adsorption cooling system and calculate its performances. The modeling is also applied to the energy flows and performance investigations of the solid state cooling devices such as the thermoelectric coolers, the pulsed-mode thermoelectric coolers and the thin-film superlattice thermoelectric device. It demonstrates the regimes of entropy generation that are associated to irreversible transport processes arising from the spatial thermal gradients, as well as the contributions from the collisions of electrons, phonons and holes in a micro- scaled type solid state cooler. The pedagogical value of the general temperature-entropy flux diagram has been successfully demonstrated. The area under the process paths of a thermoelectric element indicates how energy input has been consumed by the presence of both the reversible (Peltier, Seebeck and Thomson) and the irreversible (Joule and Fourier) effects.

The present chapter also demonstrates successfully the transient operation of a pulsed thermoelectric cooler using the same thermodynamic approach and it provides the necessary expression for the entropy flux, expressed in terms of the basic measurable variables. Using the “time-frozen” T-s diagrams, the cyclic processes of the pulsed and non-pulsed operations of thermoelectric cooler show how the input energy has been distributed - depicting the useful energy flows and “bottlenecks” of cycle operation. The thesis presents also the T-s diagrams of a thin film thermoelectric, where the effects of electrons flow in n-leg and holes flow in p-leg generate the total cooling effects at the cold junction. However, the phonons flow in both legs of the semi-conductors contributes to dissipative losses at the cold junction and the energy dissipation due to collisions is about 5-10%.

Appendix

Gauss Theorem Approach

V is the volume of an element (solids, liquids, viscoelastic materials and rigid bodies) which is bounded by a closed surface **Ω**. The differential element of the volume and surface area is written as d**V** and d**Ω** = n dΩ respectively, where n is the unit normal to Ω and is defined as pointing out of **V**. with this notation the Gauss' Theorem is written as,

$$\int_V \nabla.\mathbf{X} dV = \int_\Omega \mathbf{X}.\vec{n} d\Omega = \int_\Omega \mathbf{X}.d\Omega \tag{A1}$$

where **X** is any sufficiently smooth vector field and is defined on the volume V and the boundary surface **Ω** as shown below.

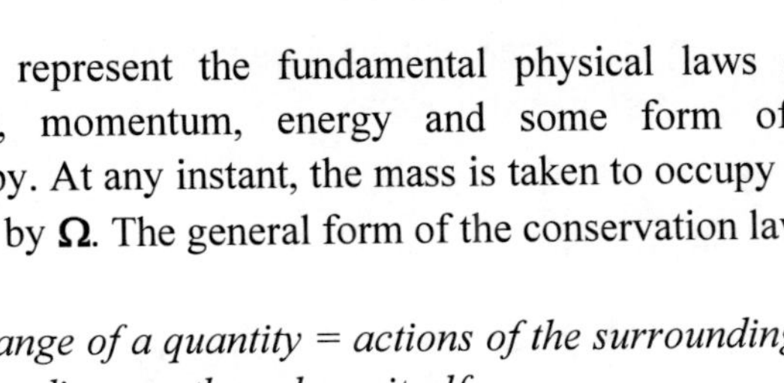

Balance equations represent the fundamental physical laws and these include the conservation of mass, momentum, energy and some form of the second law of thermodynamics, entropy. At any instant, the mass is taken to occupy the volume V and has a bounding surface given by **Ω**. The general form of the conservation laws is given by,

The time rate of change of a quantity = actions of the surroundings on the surface of V + the actions of the surroundings on the volume itself.

The word "actions" means the change of quantity. Logically, the general form of the general form of the right hand side of the conservation principle always include all possible ways to influence the time rate of change of the quantity of interest.

The Thomson Effect in Equation (55)

The electrical and thermal currents are coupled in a thermoelectric device. The Peltier coefficient of the junction is a property depends on both the materials and is the ratio of power evolved at the junction to the current flowing through it. When current is applied to a thermoelectric element, thermal energy is generated or absorbed at the junction due to Peltier effect. The Seebeck coefficient depends on temperature and this is different at different places along the TE material. So the thermoelectric element is thought of as a series of many small Peltier junctions and each of which is generating or absorbing heat. This is called Thomson power evolved per unit volume $(J\Gamma\nabla T)$.

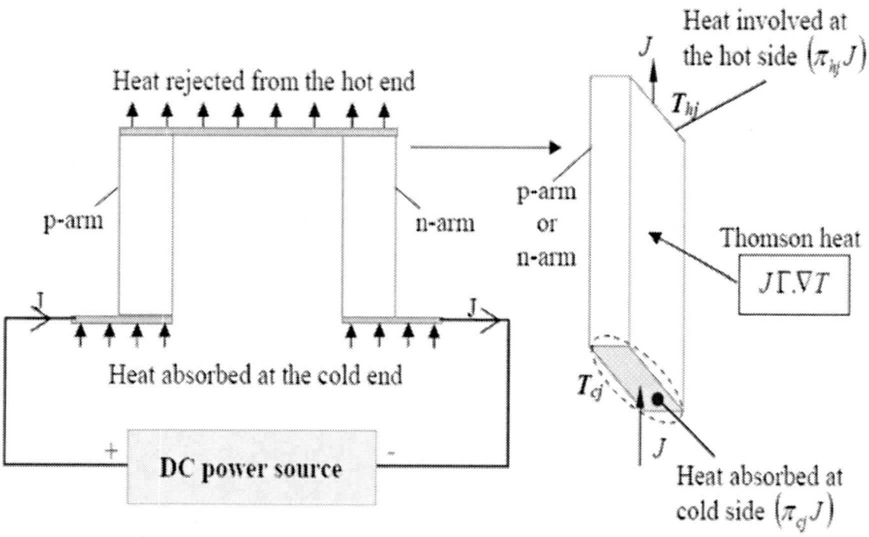

Figure B1. A thermoelectric element showing the transportation of heat and current.

In a Thomson effect, heat is absorbed or evolved when current is flow in a thermoelectric element with a temperature gradient. The heat is proportional to both the electric current and the temperature gradient. The proportionality constant is known as Thomson coefficient.

If the heat current density of a thermoelectric element is only a result of temperature gradient and the electric current density occurs due to electric potential gradient. Then, $\mathbf{J}_q = -\lambda.\nabla T$ and $J = \sigma\nabla\phi$. Using equation (3.1) the energy balance equation of a thermoelectric element is obtained.

$$\rho c_p \frac{\partial T}{\partial t} = \nabla.(\lambda\nabla T) + \frac{J^2}{\sigma'} \tag{B1}$$

The first term of right hand side of equation (B1) is Fourier conduction term and the second term indicates the Joule heat. But in a thermoelectric cooling device, the junctions are maintained at different temperatures and an electromotive force will appear in the circuit. This flow of heat has a tendency to carry the electricity along the thermoelectric arm.

When two or more irreversible processes take place in a thermodynamic system, they may interfere with each other. A completely consistent theory of thermoelectricity has been developed using the Onsager symmetry relationship (Onsager, 1931). When a current is applied to a thermoelectric device, the heat current of TE arm is developed as a result of electric current and temperature gradient (cross flow) and the electric current is generated due to electric potential and the temperature gradient.

Using the thermodynamic-phenomenological treatment, the heat current density is

$\mathbf{J}_q = -\frac{Q_o}{F} J - \lambda \nabla T$, and the electric current density in amp per m^2 becomes

$$J = \sigma' \nabla \phi + \frac{\sigma' Q_o}{FT} \nabla T$$

$$\nabla \phi = \frac{J}{\sigma'} - \frac{Q_o}{FT} . \nabla T$$

Equation (55) now becomes

$$\rho c_p \frac{\partial T}{\partial t} = -\nabla . \left(-\lambda \nabla T - \frac{Q_o}{F} J \right) + J \left(\frac{J}{\sigma'} - \frac{Q_o}{FT} . \nabla T \right)$$

$$= \nabla . (\lambda \nabla T) + \frac{J}{F} \nabla Q_o + \frac{J^2}{\sigma'} - \frac{J Q_o}{FT} . \nabla T$$

$$= \nabla . (\lambda \nabla T) + \frac{J^2}{\sigma'} + \frac{JT}{F} \nabla \left(\frac{Q_o}{T} \right)$$

$$\rho c_p \frac{\partial T}{\partial t} = \nabla . (\lambda \nabla T) + \frac{J^2}{\sigma'} + \frac{JT}{F} \nabla \left(\frac{Q_o}{T} \right) \qquad \text{(B2)}$$

Hence the third term $\frac{JT}{F} \nabla \left(\frac{Q_o}{T} \right)$ occurs due to current flow for temperature gradient (cross flow) and heat flow for the electric potential gradient (reciprocal effect). This term indicates the least dissipation of energy [14]

$\frac{JT}{F} \nabla \left(\frac{Q_o}{T} \right) = \frac{JT}{F} \nabla (s)$, where $s \left(= \frac{Q_o}{T} \right)$ is the entropy (in J/mol.K) along the thermoelectric arm.

Hence

$$\frac{JT}{F}\nabla\left(\frac{Q_o}{T}\right)=\frac{JT}{F}\nabla(s)$$
$$=\frac{JT}{F}\frac{\partial s}{\partial T}.\nabla T$$
$$=J\left(\frac{T}{F}\frac{\partial s}{\partial T}\right).\nabla T$$

The Thomson heat is defined as

$$Q_T=J\left(\frac{T}{F}\frac{\partial s}{\partial T}\right).\nabla T$$
$$Q_T=J(-\Gamma)\nabla T$$

where Thomson coefficient $-\Gamma=\frac{T}{F}\frac{\partial s}{\partial T}$.

The temperature gradient is bound to cause a certain degradation of energy by conduction of heat. The electric potential gradient must cause an additional degradation of energy, making the total rate of increase of entropy along the thermoelectric arm.

So the electric current due to temperature gradient (cross flow) and the heat current due to electric potential gradient (reciprocal effect) indicate the Thomson effect, which will be discussed in the following section. Without Thomson effect, the physics of thermoelectricity is not completed.

Derivation of Equation (55)

The energy balance of the thermoelectric arm is

$$\rho c_p\frac{\partial T}{\partial t}=\nabla.(\lambda\nabla T)+\frac{J^2}{\sigma'}-J\Gamma\nabla T \tag{B3}$$

Expression of the third term

Using Kelvin relations $\left[\Gamma=\frac{\pi}{T}-\frac{\partial\pi}{\partial T}, and \quad \pi=\alpha T\right]$

$$J\Gamma\nabla T=J\left(\frac{\pi}{T}-\frac{\partial\pi}{\partial T}\right).\nabla T \tag{B4}$$

$$\nabla\pi=\nabla\pi\big|_T+\frac{\partial\pi}{\partial T}.\nabla T$$ (de Groot and Mazur, 1962)

From equation (B4) we have

$$\begin{aligned}
J\Gamma\nabla T &= J\frac{\pi}{T}.\nabla T - J\frac{\partial \pi}{\partial T}.\nabla T \\
&= J\frac{\pi}{T}.\nabla T - J\left(\nabla\pi - \nabla\pi|_T\right) \\
&= J\frac{\pi}{T}.\nabla T - J\nabla\pi + J\nabla\pi|_T \\
&= J\alpha\nabla T - J\nabla(\alpha T) + JT\nabla\alpha|_T \\
&= J\alpha\nabla T - J\alpha\nabla T - JT\frac{\partial\alpha}{\partial T}.\nabla T + JT\nabla\alpha|_T \\
&= -JT\frac{\partial\alpha}{\partial T}.\nabla T + JT\nabla\alpha|_T
\end{aligned}$$

Equation (B3) is now written as

$$\rho c_p \frac{\partial T}{\partial t} = \nabla.(\lambda\nabla T) + \frac{J^2}{\sigma'} - JT\frac{\partial\alpha}{\partial T}.\nabla T + JT\nabla\alpha|_T \qquad \text{(B5)}$$

Equation (B5) Is Same as Equation (55)

References

[1] Rowe, D. M. (ed). *CRC handbook of thermoelectrics*. Boca Raton FL: CRC Press. 1995.

[2] Ng, K. C., Sai, M. A., Chakraborty, A., Saha, B. B., Koyama, S., The electro-adsorption chiller: Performance rating of a novel miniaturized cooling cycle for electronics cooling, *Journal of Heat Transfer*. Vol. 128, No. (9), pp. 889-896, 2006.

[3] Saha, B. B., Chakraborty, A., Koyama, S., Ng, K. C., and Sai, M. A., Performance modelling of an electro-adsorption chiller, *Philosophical Magazine*. Vol. 86, No. 23, pp.3613-3632, 2006.

[4] Chakraborty, A., Thermoelectric cooling devices: thermodynamic modelling and their applications in adsorption cooling cycles, Ph. D. Thesis, National University of Singapore, November 2005.

[5] Chen, G. Ballistic-Diffusive Heat-Conduction Equations, *Physical Review Letters*. Vol. 86, No. 11, pp. 2297-2300. 2001.

[6] Chen, G. Ballistic-Diffusive Equations for Transient Heat Conduction From Nano to Macroscales, *Journal of Heat Transfer*. Vol. 124, pp. 320-328. 2002.

[7] Tomizawa, K. *Numerical simulation of Submicron semiconductor Devices*. pp. 171-209, Boston: Artech House. 1993.

[8] Parrott, J.E, The interpretation of the stationary and transient behaviour of refrigerating thermocouples, *Solid state Electronics.*. Vol. 1, pp. 135-143, 1960.

[9] Lee, J.S. Micro/Nano scale Energy Transport and conversion A State-of-Art Review. *The 6th KSME-JSME Thermal and Fluids Engineering Conference*, March 20-23, Jeju, Korea, 2005.

[10] de Groot, S.R. and P. Mazur. *Non-equilibrium Thermodynamics*. pp. 11-42, Amsterdam: North-Holland Pub co. 1962.

[11] Novikov, I. I. The Efficiency of Nuclear Power Stations, *Journal of Nuclear Energy II.* (U.S.S.R.), Vol 7, pp.125-128, 1958.

[12] Mey, G.D. and A. D. Vos. On the optimum efficiency of endoreversible thermodynamic processes, *Journal of Physics D: Applied Physics.* Vol. 27, No. 4, pp. 736-739, 1994.

[13] Moron, J. S. On second-law analysis and the failed promise of finite-time thermodynamics. *Energy.* Vol. 23, No. 5, pp. 517-519, 1998.

[14] Onsager, B. L. Reciprocal relations in irreversible Processes. I. *Physical Review.* Vol. 37, pp.405-426, 1931.

[15] Denton, J. S. Thermal cycles in classical thermodynamics and non-equilibrium thermodynamics in contrast with finite time thermodynamics, *Energy Conversion and Management.* Vol. 43, pp.1583-1617. 2002.

[16] Reif, F. Fundamentals of Statistical and Thermal Physics. New York: McGraw-Hill. 1965.

[17] de Groot, S.R., On the Thermodynamics of irreversible heat and mass transfer, *International Journal of Heat and Mass Transfer.* Vol. 4, pp.63-70, 1961.

[18] Bird, R. B. Transport Phenomena. New York: Wiley. 1960.

[19] Richard, C.T and P. C. Fine. On the irreversible production of entropy. *Reviews of Modern Physics.* Vol. 20, No. 1, pp. 51-77, 1948.

[20] Meunier, F., Solid sorption: An alternative to CFCs, Heat recovery systems & CHP Vol. 13, pp. 289-295. 1993.

[21] Aristov, Y. I., Restuccia, G., Cacciola, G., and Parmon, V.N., A family of new working materials for solid sorption air conditioning systems, *Applied Thermal Engineering.* Vol. 22, No. 2, pp. 191-204, 2002.

[22] Aristov, Y. I., New composite adsorbents for conversion and storage of low temperature heat: Activity in the Boreskov Institute of Catalysis, *J. Heat Transfer Society of Japan.* Vol. 45, No. 192, pp. 12-19, 2006.

[23] Aristov, Y. I., New family of materials for adsorption cooling: material scientist approach, *Journal Engineering Thermophysics.* Vol. 16, No. 2, pp. 63-72, 2007.

[24] Saha, B. B., Chakraborty, A., Koyama, S., Ng, K. C., Sai, M. A., Performance modelling of an electro-adsorption chiller, *Philosophical Magazine.* Vol. 86, No. 23. pp. 3613-3632, 2006.

[25] Chua, H. T., Ng, K. C., Malek, A., Kashiwagi, T., Akisawa, A. and Saha, B. B., Multi-bed regenerative adsorption chiller - improving the utilization of waste heat and reducing the chilled water outlet temperature fluctuation, *International Journal of Refrigeration.* Vol. 24, No. 2, pp. 124-136, 2001.

[26] Chua, H. T., Ng, K. C., Wang, W., Yap, C., and Wang, X. L., Transient modeling of a two-bed silica gel-water adsorption chiller, *International Journal of Heat and Mass Transfer.* Vol. 47, No. 4, pp. 659-669, 2004.

[27] Glueckauf, E., Formulae for diffusion into spheres and their application to chromatography, *Trans. Faraday Soc.,* Vol. 51, pp. 1540-1551, 1955.

[28] Aristov, Y. I., Tokarev, M. M., Freni, A., and Restuccia, G., Comparative study of water adsorption on SWS-1L and microporous silica: equilibrium and kinetics, *Proc. Sorption Heat Pumps Conference, Denver, U.S.A.* (2006) (CD-ROM).

[29] Aristov, Y. I., Glaznev, I. S., Freni, A. and Restuccia, G., Kinetics of water sorption on SWS-1L (Calcium chloride confined to mesoporous silica gel): Influence of grain size and temperature, *Chemical Engineering Science.* Vol. 61, pp. 1453-1458, 2006.

[30] Chua, H.T., K. C. Ng, A. Chakraborty, N. M. Oo and M A. Othman, Adsorption Characteristics of Silica Gel + Water System, *Journal of Chemical and Engineering Data.* Vol. 47, pp. 1177-1181. 2002.

[31] Aristov, Y. I., Tokarev, M. M., Cacciola, G., and Restuccia, G., Selective water sorbents for multiple applications: 1. $CaCl_2$ confined in mesopores of the silica gel: sorption properties, *React. Kinet. Cat. Lett.* Vol. 59, No. 2, pp. 325-334, 1996.

[32] Tokarev, M. M., Okunev, B. N., Safonov, M. S., Kheifets, L. I., and Aristov, Y. I., Approximation equations for describing the sorption equilibrium between water vapor and a SWS-1Lcomposite sorbent, *Russian Journal of Physical Chemistry.* Vol. 79, No. 9, pp. 1490-1493, 2005.

[33] Tóth, J., State equations of the solid-gas interface layers, *Acta Chem. Acad. Sci. Hung.* Vol. 69, pp. 311-338, 1971.

[34] Chakraborty, A., Saha, B. B., Koyama, S., and Ng, K. C., The specific heat capacity of a single component adsorbent-adsorbate system, *Applied Physics Letters.* Vol. 90, art no. 171902, 2007.

[35] Chakraborty, A., Saha, B. B., Koyama, S., and Ng, K. C., On the thermodynamic modeling of isosteric heat of adsorption and comparison with experiments, *Applied Physics Letters.* Vol. 89, art no. 171901, 2006.

[36] Ruthven, D. M., Principles of Adsorption and Adsorption Processes, London: J. Wiley & Sons, (1984).

[37] Saha, B. B., Chakraborty, A., Koyama, S., and Aristov, Y. I.., A new generation cooling device employing CaCl2-in-silica gel-water system, *International Journal of Heat and Mass Transfer.* Article in Press, 2008.

[38] Burshteyn, A. I. Semiconductor Thermoelectric Device. pp. 1-20, London: Temple Press. 1964.

[39] Hasse, R., Thermodynamics of irreversible processes, New York: Dover Publications Inc. 1990.

[40] Chakraborty, A., Saha, B.B., Koyama, S., and Ng, K.C., Thermodynamic modelling of a solid state thermoelectric cooling device: Temperature-entropy analysis, *International Journal of Heat and Mass Transfer.* Vol. 49, No. 19-20, pp. 3547-3554, 2006.

[41] Nolas, G. S., J. Sharp. and H. J. Goldsmid. *Thermoelectrics Basic Principles and New Materials Developments.* pp. 1-15, New York: Springer. 2001.

[42] Snyder, G.J., J.P. Fleurial, T. Caillat, R. Yang and G. Chen. Super-cooling of Peltier cooler using a current pulse, *Journal of Applied Physics.* Vol. 92, No. 3, pp. 1564-1569. 2002.

[43] Miner, A., A. Majumdar and U. Ghosal. Thermoelectromechanical refrigeration based on transient thermoelectric effects, *Applied Physics letters.* Vol. 75, No. 8, pp.1176-1178. 1999.

[44] Landecker, K., and A. W. Findlay, Study of the fast transient behaviour of Peltier junctions, *Solid state electronics.* Vol. 3, pp. 239-260, 1961.

[45] Field, R. L. and H. A. Blum, Fast transient behavior of thermoelectric coolers with high current pulse and finite cold junction, *Energy Conversion.* Vol. 19, No. 3, pp. 159-165, 1979.

[46] Chakraborty, A., and Ng, K.C., Thermodynamic formulation of temperature-entropy diagram for the transient operation of a pulsed thermoelectric cooler, *International Journal of Heat and Mass Transfer.* Vol. 49, No. 11-12, pp. 1845-1850, 2006.

[47] Goldsmid, H.J., and Douglas R.W., The use of semiconductors in thermoelectric refrigeration, *British J. Appl. Phys.* Vol. 5, pp. 386-390, 1954.

[48] Disalvo, F. J., Thermoelectric cooling and power generation, *Science.* 1999, Vol. 285, pp. 703-706, 1999.

[49] Harman, T. C., Taylor, P. J., Walsh, M. P., and LaForge, B. E. Quantum Dot Superlattice Thermoelectric materials and devices, *Science.* Vol. 297, pp. 2229-2232, 2002.

[50] Venkatasubramanian, R., Silvola, E., Colpitts, T., and Q'Quinn, B. Thin film thermoelectric devices with high room temperature figures of merit, *Nature.* Vol. 413, pp. 597-602, 2001.

[51] Venkatasubramanian, R., Thin film thermoelectric device and fabrication method of same, US Patent no. 6300150, 2001.

[52] Elsner, N. B., and Ghamaty, S. *Superlattice quantum well material,* US Patent no. 5550387, 1996.

[53] Swe, S. M. Physics of semiconductor device, 1981, Second edition, Wiley, New York.

[54] Joshi, A. A., and Majumdar, A. Transient ballistic and diffusive phonon heat transport in thin films*, J. Appl. Phys.,* Vol. 74, pp. 31-39, 1993.

[55] Zeng, T., and Chen, G., Nonequilibrium electron and phonon transport and energy conversion in heterostructures, *Microelectronics Journal.* Vol. 34, pp. 201-206, 2003.

[56] Chakraborty, A., Saha, B.B., Koyama, S., and Ng, K.C., Thin-film thermoelectric cooler: Thermodynamic modelling and its temperature-entropy flux formulation, Proceedings of the Institution of Mechanical Engineers, Part E: *Journal of Process Mechanical Engineering.* Vol. 221, No. 1, pp. 33-46, 2007.

In: Thermal Engineering Research Developments
Editors: J. Evgova and O. Kostadinov, pp. 153-204

ISBN: 978-1-60741-497-1

Chapter 3

THE HEAT TRANSFER, THERMAL STRESS AND THERMAL STRAIN IN HIGH-POWER SOLID-STATE LASERS

Zhigang Li[1,2,a] ***and Xiulan Huai***[1,b]

[1] Institute of Engineering Thermophysics, Chinese Academy of Sciences, Beijing 100190, China.

[2] Graduate School of Chinese Academy of Sciences, Beijing 100190, China.

Abstract

Thermal effects are among the major barriers obstructing the technological development in high-power solid-state laser. In this chapter, the problems of heat exchange in laser medium and pump DLAs, as well as the associated thermal stress and thermal strain are studied to seek the methods for enhancing the heat transfer, lowering the device temperature, improving the temperature uniformity and reducing the thermal stress, in order to promote the laser output and the beam quality.

For a side-pumped, side-cooled laser rod, mathematical models of the fluid flow, the heat transfer and the thermoelastic problems were developed. Both analytical and numerical methods were used to solve the 2-D axisymmetric fields of temperature, thermal stress and thermal strain. The various influencing factors are easily observed from the analytical expressions. The numerical results agree well with the analytical solution, while better exhibiting the conjugate heat transfer, the entrance effects and the end effects. Therefore the two methods should complement each other in engineering design.

For an orthotropic laser medium, mathematical models are presented for the heat conduction and the thermoelastic problems under different pumping schemes. The 3-D temperature, 2-D thermal stress and thermal strain fields are analytically derived. Full 3-D numerical simulations are performed on the coupled thermal-mechanical problem. Comparisons of temperature, thermal stress and thermal strain components are made among different pumping schemes. The maximum temperature and thermal stress under a Tophat pump beam are lower than under a Gaussian pump beam, whereas the thermal effects under

[a] E-mail address: hxl@mail.etp.ac.cn
[b] E-mail address: jager_li@sina.com

Uniform pumping are the least significant of all; in the absence of external forces, the orthotropic thermal properties has more significant effects on the thermal strain than on the thermal stress.

For a pump DLA, three-dimensional conjugate numerical simulations using the inlet, average and variable thermal properties, respectively, were performed for the laminar water flow and heat transfer in rectangular microchannels with D_h of 0.333mm at *Re* of 101-1775. The calculated local and average characteristics of flow and heat transfer were compared among different methods, and also with the experiments, correlations and simplified theoretical solution data from published literatures. The results indicate that the "Variable property" method is superior in engineering application.

For a typical pump DLA, a liquid-flow microchannel heat sink was designed and fabricated by adopting dense, thin and short channels instead of the originally designed sparse, wide and long passages in order to increase the cooling capacity, improve the temperature uniformity, while achieving a small flow resistance. Under the maximum flow rate of 70ml/s in this experiment, the pressure drop across the channels is only 10.3kPa, the cooling capacity of the heat sink reaches 146.9W/cm^2.

Keywords: Solid-state laser, Thermal effects, Temperature, Thermal stress, Thermal strain.

1. Introduction

1.1. Background and Significance of Research

High-power solid-state lasers play an important role in industrial, agricultural, defense, biomedical, scientific research and many other fields due to their short wavelength, high material absorption, small volume and optic-fiber transportability. The laser technology is constantly developing, forming multiple applied areas by combining with other disciplines, promoting the development of both traditional and burgeoning industries. Meanwhile, the output energy and power are continuously increasing.

Only a small portion of the total energy input into the solid-state laser is converted into laser output, and the most part are converted into waste heat. The total electro-optic efficiency is only 1~3% for a typical flash-lamp pumped solid-state laser, and only 10~20% for a typical diode pumped solid-state laser (DPSSL or DPL), causing serious thermal effects to the working medium, seriously degrading the laser performance and stability, even leading to medium break and component destruction. With the constant increase of laser power and the continuous miniaturization of device size, the thermal effects become a major obstacle that restricts further improvement of laser power and performance, and a major bottle-neck to the development of laser technology. Therefore, how to effectively eliminate the waste heat from an extremely limited space, becomes a crucial problem for the further development of high-power solid-state laser, and become the focus of international research on laser technology in recent years [1, 2].

The solid-state laser medium is simultaneously subject to both the irradiation / heating by pump beam and the cooling by surrounding fluid during its operation. This results in non-uniform temperature distribution and thermal stress inside the laser medium, and gives rise to variation of optical properties of the medium, Consequently leads to birefringence, thermal lensing effect and depolarization effect, and even medium break, which seriously degrades the stability of the resonant cavity, laser output and beam quality, electro-optic efficiency and life expectancy. The thermal effects of laser medium affect almost all aspect of the laser performance. The influence of thermal effects becomes even more outstanding with the increase of pump power. Therefore, it is an urgent need to investigate the distributions of temperature and thermal stress in the laser medium and its influence on the laser performance, so as to find effective ways to eliminate or reduce the thermal effects, in order to obtain high-power laser with high beam quality, which is also a hot area in the research of solid-state laser.

Diode Laser Array (DLA) has the advantages of high power, low threshold, high conversion efficiency, small volume, low weight, easy modulation, high reliability, long life, etc., is used as the pump source to provide pump energy for solid-state laser , and is also widely applied in laser machining, medical treatment, information processing and military weapons and other field.

The output laser power of the 2-D array of Quasi-continuous-wave (QCW) laser diodes reaches an order of kilowatts, and the Continuous-wave (CW) laser diodes has an output of dozens of Watts. The Electro-optic conversion efficiency is up to 50% ~ 60%, and the peak value of the heat flux on the interface between the DLA and its carrier reaches an order of 1kW/cm^2 due to the compact structure, so the thermal loading becomes one of the main limiting factors to the operation of the high-power laser diodes. Currently, the cooling of high-power DLA has become a hotspot issue in the research of laser technology [3].

The thermal properties are among the most important properties of the diode lasers. The threshold of a diode laser increases exponentially with the change of junction temperature. With the increase of junction temperature, the output power decreases parabolically and the slope efficiency decreases negative-exponentially. The life expectancy of a diode laser will typically decrease by about 50% when the temperature increases by 10°C. The performance of diode lasers will deteriorate and quickly fail when the junction temperature exceeds 50°C. The output wavelength of a diode laser usually changes with the junction temperature, which will cause the output wavelength deviate from the absorption spectrum center of the laser medium. The wavelength shift is typically 0.3nm/°C. The non-uniformity of temperature distribution in the active layer of a DLA also leads to the widening of the output spectrum and the thermal stress, which will degrade the life expectancy of the DLA. Therefore, it is demanded that not only the cooling capacity be high, but also the temperature distribution be as uniform as possible in the design of a DLA cooler.

In this chapter, the analytical, numerical and experimental methods are adopted to study the laminar and turbulent convective heat transfer in the cooling channels, the heat conduction and thermal stress problem of the laser medium and the conjugate heat transfer in the microchannel heat sink for DLAs. The influence of various factors on the distributions of temperature and thermal stress in the laser medium are analyzed, in order to seek effective measures to lower the device temperature, to improve the temperature non-uniformity and to reduce the thermal stress, so as to ensure the beam quality and the laser output.

This research is to provide important guidance for the optimized design, control and operation of cooling system in high-power solid-state lasers. The results can also be widely applied in various other electronic components with high heat flux.

1.2. Simplified Analysis of the Thermal Effects in the Laser Medium

In as early as 1967, Belostotskii et al. [4] derived a simplified analytical solution for the temperature and thermal stress in a cylindrical laser medium, which has been widely cited. It is assumed that the laser medium are uniformly pumped, i.e. the internal heat source is uniformly distributed, the side surface are uniformly cooled, the axial temperature variation is ignored, the thermal properties are constant, then the 1-D temperature profile in the laser medium can be obtained from the 1-D heat conduction equation as

$$t = t_f + \frac{q_V r_i}{2h_i} + \frac{q_V}{4k}\left(r_i^2 - r^2\right) \tag{1-1}$$

for cylindrical laser medium, and

$$t(y) = t_f + \frac{q_V \delta}{h} + \frac{q_V}{2k}\left(\delta^2 - y^2\right) \tag{1-2}$$

for slab laser medium.

The thermal stress profile is obtained as

$$\sigma_{rr} = \frac{1}{16}\frac{E\alpha_T}{1-\nu}\frac{q_V}{k_s}\left(r^2 - r_i^2\right) \tag{1-3a}$$

$$\sigma_{\theta\theta} = \frac{1}{16}\frac{E\alpha_T}{1-\nu}\frac{q_V}{k_s}\left(3r^2 - r_i^2\right) \tag{1-3b}$$

$$\sigma_{zz} = \frac{1}{8}\frac{E\alpha_T}{1-\nu}\frac{q_V}{k_s}\left(2r^2 - r_i^2\right) \tag{1-3c}$$

for cylindrical laser medium, and

$$\sigma_{xx}(y) = \sigma_{zz}(y) = \frac{E\alpha_T}{1-\nu}\frac{q_V \delta^2}{2k}\left[\left(\frac{y}{\delta}\right)^2 - \frac{1}{3}\right] \tag{1-4a}$$

$$\sigma_{yy} = 0 \tag{1-4b}$$

for slab laser medium.

1.3. Influential Factors of the Thermal Effects

The thermal effects of solid-state laser are closely associated with the type of pumping source, the pumping scheme, and the shape and material of the laser medium. The influence of thermal effects on the laser performance depends on a variety of factors, including the thermo-mechanical properties, the thermo-optical properties of the laser medium, the structure of the resonant cavity, the intensity profile of the pump beam and the cooling condition of the laser medium.

1.4. Distinctive Features of this Chapter

1. In addition to the medium geometry, the pumping and cooling scheme, the multi-dimensionality, the following factors are also taken into account in the analytical and numerical model of the temperature and thermal stress problems of the laser medium:

- the interaction between the heat conduction inside the laser medium and the convective heat transfer in the cooling channel;
- the influence of non-uniform boundary conditions, i.e. the influence of longitudinal temperature rise of the coolant fluid (non-uniform fluid temperature) and the entrance effects (non-uniform convective heat transfer coefficient) on the temperature and thermal stress distributions inside the laser medium;
- the influence of orthotropy (the orthotropic heat conductivity and thermal expansion coefficient) on the temperature and thermal stress distributions in the laser medium.

2. In the investigation of the microchannel heat sink for the DLA pump source,

- the variations of the thermal properties (viscosity, heat conductivity, etc.) with temperature and their influence on the flow and heat transfer characteristics in the microchannel heat sink are considered and analyzed;
- conjugate numerical simulation is adopted to study the conjugate heat transfer between the liquid and the solid in the DLA microchannel heat sink, more accurate solutions are obtained, overcoming the disadvantages of the 1st, the 2nd and the 3rd Type B.C.s;
- through the optimized design of the microchannel configuration and the subsequent experimental verification, the measures to enhance the heat transfer, to improve the temperature uniformity and to reduce the flow resistance are proposed.

In general, the research of this chapter is able to truly reflect the real working condition of the concerned components, to reduce the solution errors and to provide clear guidance and theoretical basis for the optimization, control and operation of the high-power solid-state lasers.

2. The 2-D Heat Transfer and Thermoelastic Problems of a Cylindrical Laser Rod

The theoretical solutions of the heat conduction or thermal stress in the cylindrical and slab laser media based on the simplified one-dimensional model were derived by Belostotskii et al. [4] in 1967. This model results in relatively large error, since the coolant temperature-rise along the flow direction has been ignored. Multi-dimensional analytical solutions have been derived by Belostotskii and Kudryashev [5], Farrukh, et al. [6], Hello, et al. [7], Usievich, et al. [8], Chénais, et al. [9]. However, the longitudinal variation of coolant temperature has not been taken into account in these analytical models. The axial temperature gradient within the laser medium in their solutions was caused by some other reasons, including the end-pumping and the end-cooling effects. Multi-dimensional numerical solutions for the temperature and thermal stress distribution can be found in the papers by Parfenov, et al. [10], Xie, et al. [11], Peng, et al. [12], Chen, et al. [13], Ostermeyer, et al. [14], in which complicated geometries, variable thermophysical properties, realistic heat source distributions and boundary conditions were incorporated, but the influence of the spatial variations of the fluid temperature due to heat absorption on the longitudinal temperature gradient, the thermal stress and strain inside the laser medium were not paid attention to.

Analytical methods possess several advantages over numerical methods, e.g., clear physical concepts, logical reasoning, and rigorous mathematical base in the derivation process. In addition, the analytical solutions are usually expressed in an explicit function form that can indicate the influence of various factors. So they are still widely adopted to solve heat conduction and thermal stress problems up to date. However, usually some approximations have to be made to treat the finite solution conditions, e.g., physical properties are regarded as constant values, and surface heat transfer coefficient is considered as an average value. Errors are also inevitable because only a limited number of terms can be taken when calculating the solution in a series form. Although in principle, numerical simulations provide only approximate solutions, they could accommodate complex geometries, complicated boundary conditions and variable physical parameters, thereby have been broadly applied, and are especially applicable to the prediction of some real world problems.

In this section, the conjugate problem of fluid flow and heat transfer was divided into two problems. The first is the convective heat transfer of the fluid in an annular passage under given heat fluxes. The second is the 2D (i.e., radial and axial) heat conduction of the solid with uniform internal heat source under non-uniform convective boundary condition. To solve the first problem using analytical methods or experimental correlations, we could obtain the flow resistance parameters, the coolant temperature distribution and the surface convective heat transfer coefficient. To solve the second problem, the convective heat transfer coefficient and the spatially varying fluid temperature acquired from the first problem are taken as the convective boundary condition. Then the temperature profile in the laser rod can be analytically solved. A mathematical model of the thermal stress and thermal strain in the laser medium will be developed. Subsequently an analytical solution will be derived, so that the influence of the longitudinal temperature rise of the coolant fluid on the thermal stress and thermal strain in the laser rod can by analyzed.

Afterwards, a conjugate numerical simulation that couples up the fluid convection and the solid conduction is performed to better reflect the entrance effect of flow and heat transfer. In addition, for the purpose of verification and comparison, numerical simulation of the thermal stress and thermal strain will be conducted by using the temperature distribution obtained from previous numerical results of the conjugate heat transfer as the thermal load.

2.1. Problem Description and Mathematical Model

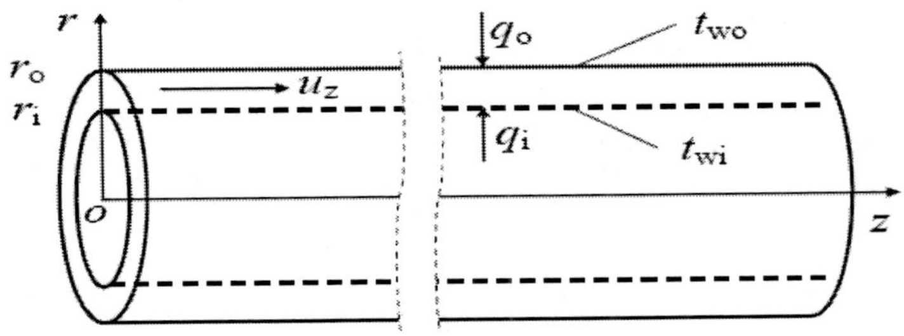

Figure 2.1. Schematic of the laser rod and the cooling channel.

As shown in figure 2.1, the intensity of the pump radiation is uniformly distributed in the axial direction. A quartz glass tube is concentrically arranged surrounding the laser rod to guide the coolant. The coolant flows in the annular channel between the glass tube and the laser rod, carrying away the heat generated by the laser medium and the glass tube. Assuming the physical properties of the laser rod are constant, and consider the internal heat source as uniform. Then, for the problem of steady state axisymmetric heat conduction in the laser rod under convective boundary condition, the mathematical description is given as

$$\frac{\partial^2 t_s}{\partial r^2}+\frac{1}{r}\frac{\partial t_s}{\partial r}+\frac{\partial^2 t_s}{\partial z^2}+\frac{q_V}{k_s}=0 \tag{2-1a}$$

$$r=0: \qquad t_s=\text{finite value} \tag{2-1b}$$

$$r=r_i: \qquad -k\,\partial t_s/\partial r=h_i\left[t_s-t_b(z)\right] \tag{2-1c}$$

$$z=0: \qquad \partial t_s/\partial z=0 \tag{2-1d}$$

$$z=L: \qquad \partial t_s/\partial z=0 \tag{2-1e}$$

where r_i is the radius of the laser rod, L is the rod length, $t_s(r,z)$ is the local temperature in the laser medium, k_s is the thermal conductivity, h_i is the convective heat transfer coefficient on the cylindrical surface of the laser rod, and $t_b(z)$ is the bulk fluid temperature, q_V is the heat generation rate (W/m^3).

The stress-strain relations in the laser medium are given as

$$\sigma_{rr} = \frac{E}{1+\nu}\left[\frac{1-\nu}{1-2\nu}\varepsilon_{rr} + \frac{\nu}{1-2\nu}\left(\varepsilon_{\phi\phi} + \varepsilon_{zz}\right)\right] - \frac{E\alpha_T\theta_s}{1-2\nu} \tag{2-2a}$$

$$\sigma_{\phi\phi} = \frac{E}{1+\nu}\left[\frac{1-\nu}{1-2\nu}\varepsilon_{\phi\phi} + \frac{\nu}{1-2\nu}\left(\varepsilon_{rr} + \varepsilon_{zz}\right)\right] - \frac{E\alpha_T\theta_s}{1-2\nu} \tag{2-2b}$$

$$\sigma_{zz} = \frac{E}{1+\nu}\left[\frac{1-\nu}{1-2\nu}\varepsilon_{zz} + \frac{\nu}{1-2\nu}\left(\varepsilon_{rr} + \varepsilon_{\phi\phi}\right)\right] - \frac{E\alpha_T\theta_s}{1-2\nu} \tag{2-2c}$$

$$\sigma_{rz} = \frac{E}{1+\nu}\varepsilon_{rz} \tag{2-2d}$$

The thermal strain components are defined as

$$\varepsilon_{rr} = \frac{\partial u_r}{\partial r}, \quad \varepsilon_{zz} = \frac{\partial u_z}{\partial z}, \quad \varepsilon_{\phi\phi} = \frac{u_r}{r} \tag{2-3}$$

The equilibrium equation is written as

$$\frac{\partial \sigma_{rr}}{\partial r} + \frac{\partial \sigma_{zr}}{\partial z} + \frac{\sigma_{rr} - \sigma_{\phi\phi}}{r} = 0 \tag{2-4a}$$

$$\frac{\partial \sigma_{zr}}{\partial r} + \frac{\partial \sigma_{zz}}{\partial z} + \frac{\sigma_{zr}}{r} = 0 \tag{2-4b}$$

In order to reduce the thermal stress in the laser medium, it is common to subject the laser rod to a retention force close to zero, i.e. the laser rod is under the traction free boundary condition given as

$$\left(\sigma_{rr}\right)_{r=r_i} = \left(\sigma_{rz}\right)_{r=r_i} = 0 \tag{2-5}$$

2.2. Forced Convective Heat Transfer in an Annular Passage

For the hydrodynamically developed region of axisymmetric laminar flow (*Re*<2300), when the tube is horizontally positioned, the momentum conservation equation can be simplified as

$$\partial p/\partial r = -\rho g \sin\phi \tag{2-6a}$$

$$\mu \frac{1}{r}\frac{d}{dr}\left(r\frac{du_z}{dr}\right) = \frac{\partial p}{\partial z} \tag{2-6b}$$

If the tube is vertically positioned, $\partial p/\partial r = 0$, and the $\partial p/\partial z$ in equation (2-6b) should be replaced by $\partial p/\partial z - \rho g$. The velocity distribution can be solved as

$$u_z(r) = -\frac{1}{4\mu}\frac{\partial p}{\partial z}\left[r_o^2 - r^2 + \left(r_o^2 - r_i^2\right)\frac{\ln(r/r_o)}{\ln(r_o/r_i)}\right] \tag{2-7}$$

Then the flow resistance in terms of Fanning friction coefficient is obtained as

$$C_f Re = 16\left(1-r^*\right)^2 / \left[1 + r^{*2} + \left(1-r^{*2}\right)/\ln r^*\right] \tag{2-8}$$

where $C_f = 2\tau_{ave,w}/\left(\rho u_m^2\right)$ is the Fanning friction coefficient, $Re = 2\rho u_m\left(r_o - r_i\right)/\mu$ is the Reynolds number, where $\tau_{ave,w} = \left(r_i\tau_{wi} + r_o\tau_{wo}\right)/\left(r_i + r_o\right)$ is the average shear stress, $\tau_{wi} = \mu\left(du_z/dr\right)_{r=r_i}$ and $\tau_{wo} = -\mu\left(du_z/dr\right)_{r=r_o}$ are the shear stress on the inner and outer wall of the flow channel, and $r^* = r_i/r_o$ is the normalized *r* coordinate.

For the low velocity steady fluid flow with constant physical properties in hydrodynamically developed region under given heat fluxes, we have $\partial t_f/\partial z = dt_{wi}/dz = dt_{wo}/dz = dt_b/dz = \text{constant}$. So the boundary layer energy equation can be simplified as

$$\frac{1}{r}\frac{\partial}{\partial r}\left(r\frac{\partial t_f}{\partial r}\right) = \frac{u_z}{a}\frac{dt_b}{dz} \tag{2-9a}$$

the boundary conditions are

$$r = r_i:\ V = 0;\ \ -k_f\,\partial t_f/\partial r = q_i \tag{2-9b}$$

$$r = r_o:\ V = 0;\ \ k_f\,\partial t_f/\partial r = q_o \tag{2-9c}$$

where q_i and q_o are the heat fluxes passing through these two surfaces, and *a* is the heat dissipation rate of the fluid.

The convective heat transfer coefficients on the inner and outer surfaces of the annular channel are defined as $h_i = q_i / (t_{wi} - t_b)$, and $h_o = q_o / (t_{wo} - t_b)$, and the corresponding Nusselt numbers are defined as $Nu_i = h_i D_h / k_f$ and $Nu_o = h_o D_h / k_f$. If the convective heat transfer coefficients on the inner and outer surfaces for heating only from the inner side and only from the outer side are defined as $h_{ii} = q_i / (t_{wi,A} - t_{b,A})$ and $h_{oo} = q_o / (t_{wo,B} - t_{b,B})$, respectively, and the corresponding Nusselt numbers are defined as $Nu_{ii} = h_{ii} D_h / k_f$ and $Nu_{oo} = h_{oo} D_h / k_f$, then the Nu_i and Nu_o can be rearranged as

$$Nu_i = \frac{Nu_{ii}}{1 - \theta_i^* q_o / q_i}, \qquad Nu_o = \frac{Nu_{oo}}{1 - \theta_o^* q_i / q_o} \tag{2-10}$$

where θ_i^* and θ_o^* are respectively called inner and outer surface correction coefficients. Detailed deduction procedures and some results can be found in the works of Lundberg, et al. [15].

For the hydrodynamic entrance region of the laminar flow, Shah and London [16] offered the friction coefficients. The results are rearranged into the form $C_{f,app} Re = \Phi\left[r^*, (z / D_h)/Re\right]$ by Kays and Crawford [17], where $C_{f,app}$ is the Apparent Average Friction Coefficient, and $r^* = r_i / r_o$. Then the pressure drop in the passage is calculated by

$$\Delta p = 4C_{f,app} \frac{L}{D_h} \frac{\rho u_m^2}{2} = C_{f,app} \frac{L}{r_o - r_i} \rho u_m^2 \tag{2-11}$$

For the thermal entrance region of the laminar flow in an annular passage, if the heat fluxes from inner and outer surfaces have been given, the Nusselt number can also be expressed in the form of formula (2-10), where Nu_{ii}, Nu_{oo}, θ_i^* and θ_o^* have already been derived by Kays and Crawford [17], and Lundberg, et al. [18]. All of them are functions of $r^* = r_i / r_o$ and $z^+ = 2(z/D_h)/(RePr)$. The length of thermal entrance region can be determined by $L/D_h \approx 0.05 RePr$.

The friction coefficient of hydrodynamically developed turbulent flow in an annular passage can be approximated by adding 10% to the corresponding correlation for a circular tube as

$$C_f = 0.085 Re^{-0.25} \tag{2-12}$$

as given by Rohsenow, et al. [19]. The applicable range is 6000<Re<300,000. The hydrodynamic entrance region in the channel is typically less than 10~15 times of hydraulic diameter. The effect of this region is not much, as described by Rohsenow, et al. [19].

As a good approximation, the Dittus-Boelter formula can be used to calculate Nu as suggested by Yang and Tao [20]:

$$Nu = 0.023Re^{0.8}Pr^{n} \tag{2-13}$$

where Pr is the Prandtl Number. For the fluid heated by the laser medium, $n = 0.4$. The characteristic temperature is the arithmetic mean value of the average temperature on the inlet and outlet cross-sections. The characteristic length is $D_{\mathrm{h}} = 2(r_{\mathrm{o}} - r_{\mathrm{i}})$. The application range is $10^4 < Re < 1.2\times10^5$, $0.7 < Pr < 120$, $L/D_{\mathrm{h}} \geq 60$. The length of the turbulent thermal entrance region is determined by $L/D_{\mathrm{h}} \approx 60$. And for the thermal entrance region, the following correction coefficient can be applied, as recommended by Yang and Tao [20].

$$c_{\mathrm{L}} = 1 + \left(D_{\mathrm{h}}/L\right)^{0.7} \tag{2-14}$$

In other words, the average Nusselt number for the entire tube with a total length of l including the thermal entrance region becomes $Nu_{\mathrm{ave}} = c_{\mathrm{L}}Nu$.

To sum up, the flow resistance coefficient C_{f} or $C_{\mathrm{f,app}}$, the pressure drop Δp for laminar flow or for turbulent flow in an annular passage, the Nusselt number Nu_{i} and the corresponding convective heat transfer coefficient h_{i} on the cylindrical surface of the laser rod can be accurately obtained from the analytical solutions or available experimental correlations. In addition, the bulk fluid temperature $t_{\mathrm{b}}(z)$ can be obtained from the heat balance equation.

2.3. The Analytical Solution of the Temperature in the Laser Rod

From the energy balance equation

$$\frac{dt_{\mathrm{b}}}{dz} = \frac{q_i \cdot 2\pi r_i}{\dot{m}c_{\mathrm{p}}} \tag{2-15}$$

where $\dot{m}$ is the coolant mass flow rate, it can be derived that $t_{\mathrm{b}}(z)$ is a linear function under the boundary condition of constant heat flux, i.e.,

$$t_{\mathrm{b}}(z) = t_{\mathrm{b,in}} + \Delta t_{\mathrm{b}}z/L \tag{2-16}$$

where $t_{b,in}$ is the inlet coolant temperature, and $\Delta t_b = \dot{Q}/(\dot{m}c_p)$ is the fluid temperature-rise from the inlet to the outlet, where $\dot{Q}$ is the heat generation rate.

Substituting h_i and $t_b(z)$ obtained above into the boundary conditions in equation (2-1), the temperature field can be solved by the method of variable substitution and the method of separation of variable as

$$\begin{aligned} t_s(r,z) &= -\frac{q_V}{4k_s}r^2 + C_1 I_0(\beta_1 r)\cos(\beta_1 z) + \sum_{m=2}^{\infty} C_m I_0(\beta_m r)\cos(\beta_m z) \\ &= \left(-\frac{q_V}{4k_s}r^2 + C_1\right) + \sum_{m=2}^{\infty} C_m I_0\left(\xi_m \frac{r}{L}\right)\cos\left(\xi_m \frac{z}{L}\right) \end{aligned} \tag{2-17a}$$

where

$$C_1 = t_{ave,b} + \left(1 + \frac{2}{Bi_r}\right)\frac{q_l}{4\pi k_s} \tag{2-17b}$$

$$C_m = \frac{2\Delta t_b Bi_r\left[(-1)^{m-1} - 1\right]}{\xi_m^3 \frac{r_i}{L} I_1\left(\xi_m \frac{r_i}{L}\right) + Bi_r \xi_m^2 I_0\left(\xi_m \frac{r_i}{L}\right)} \qquad m = 2,3\cdots \tag{2-17c}$$

where $H_i = h_i/k_s$, $Bi_r = h_i r_i/k_s$ is the radial Biot number, q_l (W/m) is the heat generation rate per unit length, $\beta_m = (m-1)\pi/L$, $m = 1,2,3\cdots$ are the eigen values, $\xi_m = \beta_m L = (m-1)\pi$, and $t_{ave,b} = \frac{1}{L}\int_0^L t_b(z)dz = t_{b,in} + \frac{\Delta t_b}{2}$ is the average fluid temperature.

In equation (2-17), all the odd terms of C_m $(m = 2,3,4\cdots)$ are 0. The absolute values of the even terms have positive relations with the fluid temperature rise Δt_b, the axial Biot number Bi_r and the length-to-radius ratio L/r_i of the laser rod (note that this conclusion is drawn under the condition that the two end surfaces of the laser rod are adiabatic, without considering the hydrodynamic and thermal entrance effects). With the increase of the three parameters Δt_b, Bi_r and L/r_i, the axial temperature gradient in the laser rod increases. Consequently, there is a larger difference between the 2-D axisymmetric model and the 1-D model.

It can be concluded from the energy equation $\Delta t_b = \dot{Q}/(\dot{m}c_p)$ that when the total heat generation rate in the laser rod is unchanged, the axial temperature gradient can be reduced by raising the coolant mass flow rate and by adopting a coolant of high specific heat capacity.

From the definition $Bi_r = h_i r_i / k_s$, it can be concluded that a primary measure for reducing the axial temperature gradient in the laser rod is to enhance the convective heat transfer, while from the formula (2-13) one will find the various factors that influence the convective heat transfer coefficient h_i: $h_i \propto u_m^{0.8} D_h^{1.8} \rho^{0.8} \mu^{0.8} c_p^{0.4} k_f^{0.6}$.

Apparently, when Δt_b, Bi_r and L/r_i are small enough, taking only the first term of the series (2-17) will be satisfactory, then expression (2-17) is simplified as

$$t_s = -\frac{q_V}{4k_s} r^2 + C_1 = t_{ave,b} + \frac{q_V r_i}{2h_i} + \frac{q_V}{4k_s}\left(r_i^2 - r^2\right) \tag{2-18}$$

This result is identical with the temperature field obtained from the radial 1-D heat conduction model under the convective boundary condition.

As shown in formulas (2-17) and (2-18), the overall temperature level in the laser rod is determined by the values of q_V, h_i, k_s, r_i and $t_{ave,b}$.

2.4. The Analytical Solution of the Thermoelastic Problem

The temperature rise in the laser medium can be expressed as

$$\theta_s(r,z) = t_s - t_{ave,b} = \theta_s^{(1)} + \sum_{m=2}^{\infty} \theta_s^{(m)} \tag{2-19}$$

where

$$\theta_s^{(1)} = \frac{q_V r_i}{2h_i} + \frac{q_V}{4k_s}\left(r_i^2 - r^2\right),\ \theta_s^{(m)} = C_m I_0(\beta_m r)\cos(\beta_m z),\ m = 2,3,4,\cdots.$$

In the absence of external forces, the thermal stress components caused by the temperature rise $\theta_s(r,z)$ can be obtained by summing up the thermal stress components $\sigma_{rr}^{(m)}$, $\sigma_{rz}^{(m)}$, $\sigma_{\theta\theta}^{(m)}$, and $\sigma_{zz}^{(m)}$ caused by each term $\theta_s^{(m)}$, $m = 1,2,3,\cdots$ of the temperature series (2-19), i.e.

$$\sigma(r,z) = \sigma^{(1)} + \sum_{m=2}^{\infty} \sigma^{(m)} \tag{2-20}$$

Since $\theta_s^{(1)}$ is only a function of r, the thermal stress caused by $\theta_s^{(1)}$ under the traction free condition can simply be obtained by the displacement method or the thermal stress function method as

$$\sigma_{rr}^{(1)} = \left\{E\alpha_T q_V / \left[16(1-\nu)k_s\right]\right\}\left(r^2 - r_i^2\right) \tag{2-21a}$$

$$\sigma_{\theta\theta}^{(1)} = \left\{E\alpha_T q_V / \left[16(1-\nu)k_s\right]\right\}\left(3r^2 - r_i^2\right) \tag{2-21b}$$

$$\sigma_{zz}^{(1)} = \left\{E\alpha_T q_V / \left[8(1-\nu)k_s\right]\right\}\left(2r^2 - r_i^2\right) \tag{2-21c}$$

Since $\theta_s^{(m)}(r,z)$, $m=1,2,3,\cdots$ is axisymmetric and varies in the longitudinal direction, the resulting thermal stress under the traction free condition can be determined by the Thermoelastic displacement potential method by supposing a Love displacement function [21].

A differential equation in terms of the thermoelastic displacement potential Φ can be written as

$$\nabla^2\Phi = \left[(1+\nu)/(1-\nu)\right]\alpha_T\theta_s \tag{2-22}$$

The particular solution σ' obtained from the equation (2-22) needs to be superimposed to the solution σ'' of an isothermal elasticity problem, and allow the final solution $\sigma = \sigma' + \sigma''$ to satisfy the boundary conditions.

The Love displacement function $L(r,z)$ will satisfy the biharmonic equation

$$\nabla^2\nabla^2 L = \left[\partial^2/\partial r^2 + (1/r)\partial/\partial r + \partial^2/\partial z^2\right]^2 L = 0 \tag{2-23}$$

Therefore the final stress components become

$$\sigma_{rr} = \frac{E}{1+\nu}\left[\frac{\partial^2\Phi}{\partial r^2} - \nabla^2\Phi + \frac{\partial^2}{\partial r^2}\left(\frac{\partial L}{\partial z}\right) - \nu\nabla^2\left(\frac{\partial L}{\partial z}\right)\right] \tag{2-24a}$$

$$\sigma_{rz} = \frac{E}{1+\nu}\left[\frac{\partial^2\Phi}{\partial r\partial z} + \frac{\partial^2}{\partial r\partial z}\left(\frac{\partial L}{\partial z}\right) - (1-\nu)\frac{\partial}{\partial r}\left(\nabla^2 L\right)\right] \tag{2-24b}$$

$$\sigma_{\phi\phi} = \frac{E}{1+\nu}\left[\frac{1}{r}\frac{\partial\Phi}{\partial r} - \nabla^2\Phi + \frac{1}{r}\frac{\partial}{\partial r}\left(\frac{\partial L}{\partial z}\right) - \nu\nabla^2\left(\frac{\partial L}{\partial z}\right)\right] \tag{2-24c}$$

$$\sigma_{zz} = \frac{E}{1+\nu}\left[\frac{\partial^2\Phi}{\partial z^2} - \nabla^2\Phi + \frac{\partial^2}{\partial z^2}\left(\frac{\partial L}{\partial z}\right) - (2-\nu)\nabla^2\left(\frac{\partial L}{\partial z}\right)\right] \tag{2-24d}$$

Substituting the expression of $\theta_s^{(m)}$, $m=2,3,4,\cdots$ into the equation (2-22), we will obtain a particular solution of Φ_m, $m=2,3,4,\cdots$ as

$$\Phi_m(r,z)=\frac{1+\nu}{1-\nu}\alpha_T\frac{C_m}{2\beta_m}rI_1(\beta_m r)\cos(\beta_m z) \tag{2-25}$$

According to the characteristics of the expression of $\theta_s^{(m)}$, $m=2,3,4,\cdots$, the form of the Love displacement function L_m is assumed as

$$L_m(r,z)=\frac{1+\nu}{1-\nu}\frac{\alpha_T}{2\beta_m^{\ 3}}\left[B_{1m}I_0(\beta_m r)+B_{2m}\beta_m rI_1(\beta_m r)\right]\sin(\beta_m z) \tag{2-26}$$

Substituting the expressions of Φ_m and L_m into the formulas (2-24), the thermal stress components caused by $\theta_s^{(m)}$, $m=2,3,4,\cdots$ are obtained as

$$\begin{aligned}\sigma_{rr}^{(m)}=&\frac{E\alpha_T}{2(1-\nu)}\{\left[-C_m+B_{1m}+(1-2\nu)B_{2m}\right]I_0(\beta_m r)\\&+\left[C_m\beta_m r-B_{1m}/(\beta_m r)+B_{2m}\beta_m r\right]I_1(\beta_m r)\}\cos(\beta_m z)\end{aligned} \tag{2-27a}$$

$$\begin{aligned}\sigma_{rz}^{(m)}=&-\frac{E\alpha_T}{2(1-\nu)}\{(C_m+B_{2m})\beta_m rI_0(\beta_m r)\\&+\left[B_{1m}+2(1-\nu)B_{2m}\right]I_1(\beta_m r)\}\sin(\beta_m z)\end{aligned} \tag{2-27b}$$

$$\begin{aligned}\sigma_{\phi\phi}^{(m)}=&-\frac{E\alpha_T}{2(1-\nu)}\{\left[C_m-(1-2\nu)B_{2m}\right]I_0(\beta_m r)\\&-\left[B_{1m}/(\beta_m r)\right]I_1(\beta_m r)\}\cos(\beta_m z)\end{aligned} \tag{2-27c}$$

$$\begin{aligned}\sigma_{zz}^{(m)}=&-\frac{E\alpha_T}{2(1-\nu)}\{\left[2C_m+B_{1m}+2(2-\nu)B_{2m}\right]I_0(\beta_m r)\\&+(C_m+B_{2m})\beta_m rI_1(\beta_m r)\}\cos(\beta_m z)\end{aligned} \tag{2-27d}$$

Substituting the formulas (2-27) into the following traction free condition,

$$\left[\sigma_{rr}^{(m)}\right]_{r=r_i}=\left[\sigma_{rz}^{(m)}\right]_{r=r_i}=0 \tag{2-28}$$

The coefficients in formulas (2-26) are determined as

$$\begin{aligned} B_{1m} &= \left[D_m P_m + G_m C_m \beta_m r_i I_0(\beta_m r_i)\right] / \left[R_m P_m - G_m I_1(\beta_m r_i)\right] \\ B_{2m} &= \left[R_m C_m \beta_m r_i I_0(\beta_m r_i) + D_m I_1(\beta_m r_i)\right] / \left[G_m I_1(\beta_m r_i) - R_m P_m\right] \end{aligned} \quad (2\text{-}29a)$$

where

$$\begin{aligned} &D_m = C_m\left[I_0(\beta_m r_i) - \beta_m r_i I_1(\beta_m r_i)\right], \quad G_m = (1-2\nu) I_0(\beta_m r_i) + \beta_m r_i I_1(\beta_m r_i), \\ &P_m = \beta_m r_i I_0(\beta_m r_i) + 2(1-\nu) I_1(\beta_m r_i), \quad R_m = I_0(\beta_m r_i) - I_1(\beta_m r_i)/\beta_m r_i \end{aligned} \quad (2\text{-}29b)$$

So far we have obtained all the thermal stress components. By substituting the formulas (2-29) into the formulas (2-27), then substituting the formulas (2-21) and (2-27) into the equation (2-20), the total thermal stress field in the laser rod is obtained.

If we define the temperature rise as

$$\Delta t_s = t_s - t_{in,b} \quad (2\text{-}30)$$

Then the total thermal strain is obtained as

$$\varepsilon_{rr} = \left[\sigma_{rr} - \nu\left(\sigma_{\phi\phi} + \sigma_{zz}\right)\right]/E + \alpha_T \Delta t_s \quad (2\text{-}31a)$$

$$\varepsilon_{\phi\phi} = \left[\sigma_{\phi\phi} - \nu\left(\sigma_{rr} + \sigma_{zz}\right)\right]/E + \alpha_T \Delta t_s \quad (2\text{-}31b)$$

$$\varepsilon_{zz} = \left[\sigma_{zz} - \nu\left(\sigma_{rr} + \sigma_{\phi\phi}\right)\right]/E + \alpha_T \Delta t_s \quad (2\text{-}31c)$$

Assuming the temperature increase inside the laser rod is approximately linear in the longitudinal direction, just as the coolant temperature pattern does, i.e.

$$\frac{\Delta t_b}{2} + \sum_{m=2}^{\infty} C_m I_0(\beta_m r)\cos(\beta_m z) \approx \Delta t_b \frac{z}{L} \quad (2\text{-}33)$$

Then the temperature field in the laser rod is approximately written as

$$\Delta t_s \approx \frac{q_V r_i}{2h_i} + \frac{q_V}{4k_s}\left(r_i^2 - r^2\right) + \Delta t_b \frac{z}{L} \quad (2\text{-}34)$$

Therefore the total thermal strain components become

$$\varepsilon_{rr} \approx \frac{1}{16}\frac{\alpha_T}{1-\nu}\frac{q_V}{k_s}\left[(3-\nu)r_i^2 - 3(1+\nu)r^2\right] + \alpha_T\left[\frac{q_V r_i}{2h_i} + \Delta t_b \frac{z}{L}\right] \quad (2\text{-}35a)$$

$$\varepsilon_{\phi\phi} \approx \frac{1}{16}\frac{\alpha_T}{1-\nu}\frac{q_V}{k_s}\left[(3-\nu)r_i^2-(1+\nu)r^2\right]+\alpha_T\left[\frac{q_V r_i}{2h_i}+\Delta t_b\frac{z}{L}\right] \tag{2-35b}$$

$$\varepsilon_{zz} \approx \frac{1}{8}\alpha_T\frac{q_V}{k_s}r_i^2+\alpha_T\left[\frac{q_V r_i}{2h_i}+\Delta t_b\frac{z}{L}\right] \tag{2-35c}$$

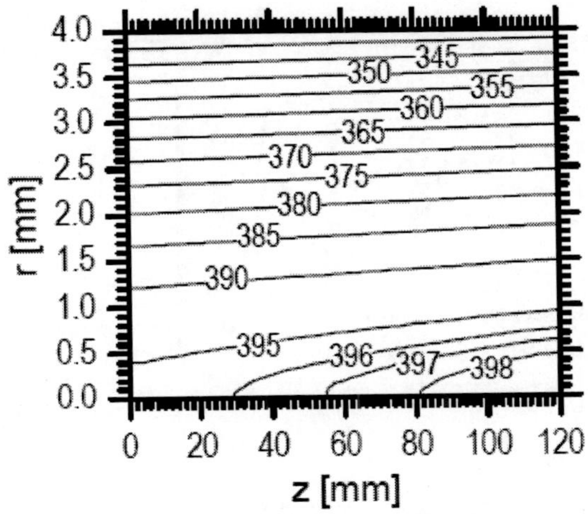

Figure 2.2. Temperature contours in the longitudinal section of the laser rod plotted from analytical solution [K].

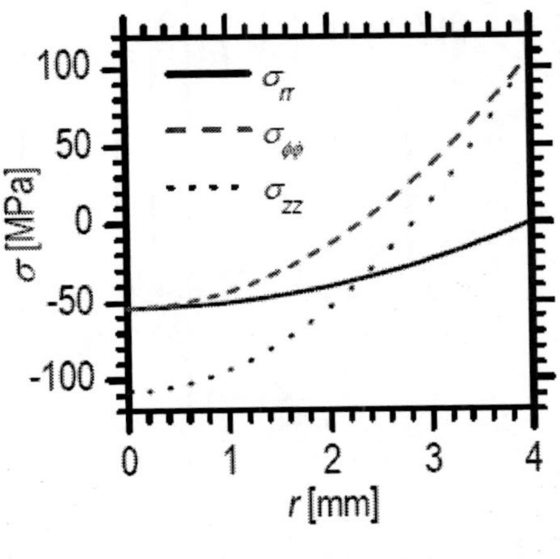

Figure 2.3. Thermal stress profile in the laser rod plotted from analytical solution [MPa].

Take the Nd:YAG laser rod as an example. The rod radius is $r_i = 4\text{mm}$, and its length is $L = 120\text{mm}$. The inner radius of quartz glass tube is $r_o = 6\text{mm}$. The total heat generation rate of the laser rod is $\dot{Q} = 1200W$. The mass flow rate of cooling water is $\dot{m} = 0.10\text{kg/s}$. The inlet coolant temperature is $t_{\text{b,in}} = 298\text{K}$. The Young's modulus of Nd:YAG medium is $E = 3\times10^{11}$Pa , the Poisson's ratio is $\nu = 0.3$, and the thermal expansion coefficient is $\alpha_T = 8.2\times10^{-6}$Pa .

The temperature contours in the longitudinal section of the laser rod are plotted in figure 2.2 according to the analytical expression (2-17). The temperature increase in the longitudinal direction is in a range of 2.73~2.78K at different radius.

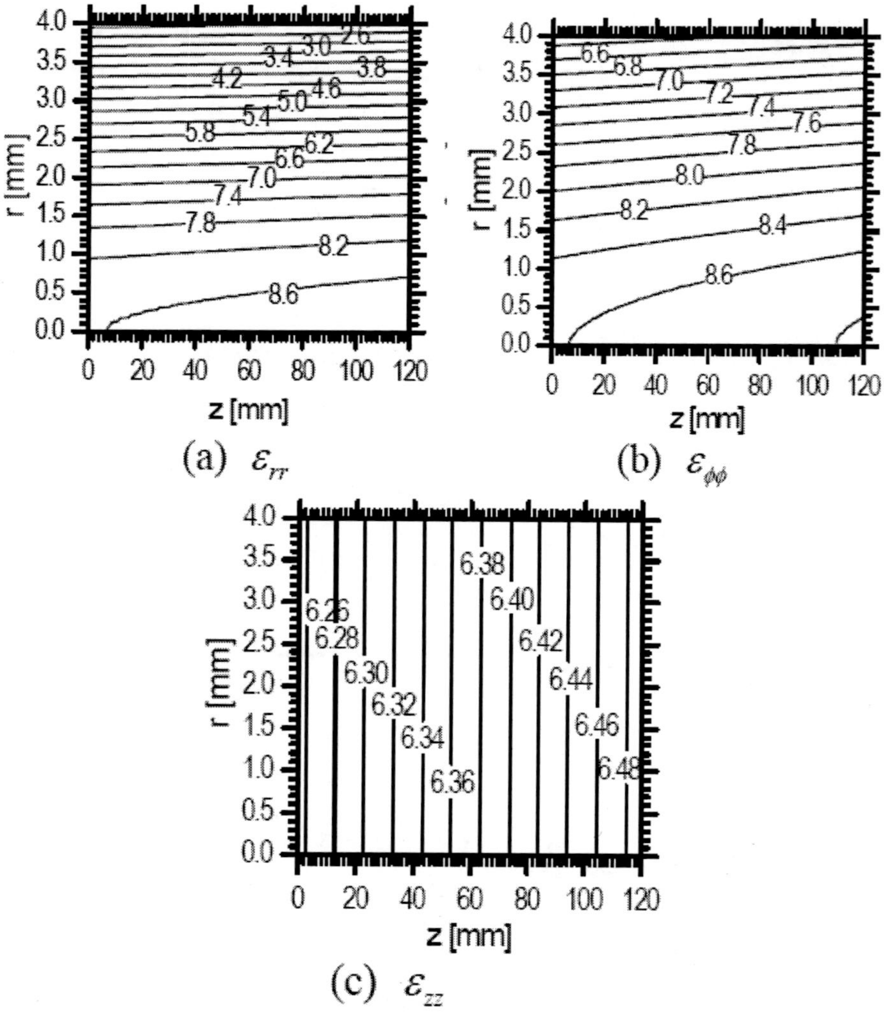

Figure 2.4. Thermal strain distribution in the laser rod (Unit: 10^{-4}).

Figure 2.3 shows the thermal stress components as functions of r drawn from the analytical solutions. Since the laser rod is under the traction free boundary condition without any external force, and the difference in longitudinal temperature variation inside the laser rod is negligible among different radius locations, giving rise to negligible longitudinal variation of the thermal stress profile. However, the thermal strain field in the laser rod caused by the temperature distribution and the traction free boundary condition has an important influence on the light transmission characteristics and the beam quality of the laser medium. Figure 2.4

shows the thermal strain components in the laser rod plotted from the analytical solution. As shown in the figure, significant variation of thermal strain exists in the longitudinal direction.

2.5. Comparison with the Conjugate Numerical Simulation

In view that the numerical method can deal with more complicated boundary conditions, a conjugate numerical simulation was also performed to verify and compare with the above analysis. The "axi (axial), segregated, ske (standard k-ε model)" solver of the commercial CFD code FLUENT is selected, which is based on finite volume method. Coupled wall boundary condition is defined on the surface of convective heat transfer, i.e.,

$$r = r_i: \quad V_{\mathrm{f}} = 0,\ T_{\mathrm{f}} = T_{\mathrm{s}},\ k_{\mathrm{f}} \frac{\partial t_{\mathrm{f}}}{\partial r} = k_{\mathrm{s}} \frac{\partial t_{\mathrm{s}}}{\partial r} \tag{2-36}$$

Figure 2.5 shows the temperature contours in the longitudinal section of both the laser rod and the fluid obtained from the FLUENT simulation. The numerical results agree very well with the analytical solution, but are able to better exhibit the thermal entrance effect. The temperature near the fluid inlet is significantly lower. This can be explained in connection with figure 2.6, which shows the variation of the local convective heat transfer coefficient in the flow direction obtained from the FLUENT simulation. These curves well exhibits the entrance effects of fluid flow and heat transfer due to the thinner boundary layer at the entrance region.

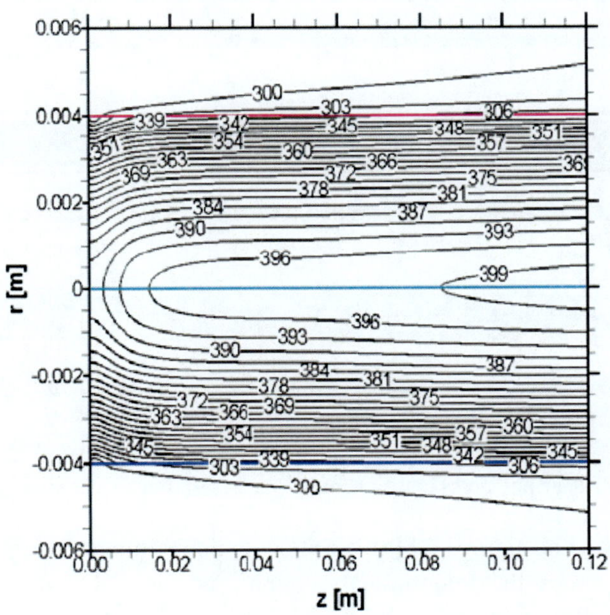

Figure 2.5. Temperature contours in the longitudinal section of both the laser rod and the fluid plotted from the conjugate numerical simulation [K].

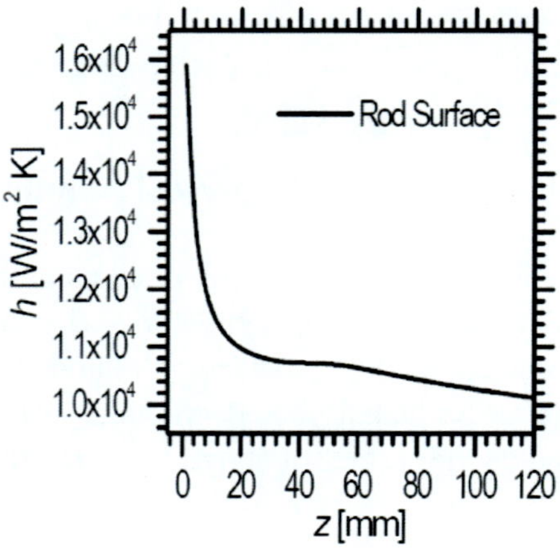

Figure 2.6. The local convective heat transfer coefficient along the channel length.

Subsequently, a commercial FEA code ANSYS was used to calculate the thermal stress and thermal strain fields. A Coupled-field element type "PLANE13" and the option "Axisymmetric" were selected to treat the coupled thermal-mechanical problem. The temperature distribution obtained from the above numerical results of FLUENT is applied to the laser rod as the thermal load. Traction free boundary conditions are applied to both the cylindrical outer surface and the two end surfaces of the laser rod.

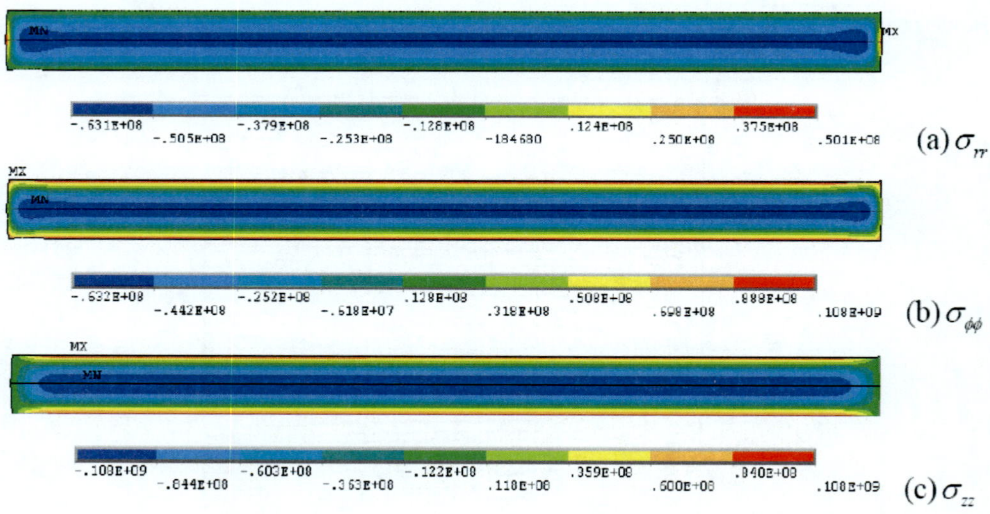

Figure 2.7. Thermal stress components plotted from numerical simulation [Pa].

Figures 2.7(a)~(c) and figures 2.8(a)~(c) show the contours of thermal stress and thermal strain components across the longitudinal section. As can be seen from the figures, the numerical results of all thermal stress and thermal strain components agree very well with the analytical results (see figure 2.3 and figure 2.4) in the major section (working section) of laser rod. However, the numerical results better exhibit the entrance effects of fluid flow and heat transfer due to its adoption of the real temperature distribution obtained from the FLUENT simulation as the thermal load, and also better reflect the end effects since it adopts two more traction free boundary conditions than the analytical method at the two rod ends, i.e.,

$$\left(\sigma_{zz}\right)_{z=0} = \left(\sigma_{zr}\right)_{z=0} = 0; \ \left(\sigma_{zz}\right)_{z=L} = \left(\sigma_{zr}\right)_{z=L} = 0 \tag{2-37}$$

These two boundary conditions are neglected in the analytical solution for simplification due to the elongate shape ($L/r_i = 30$) of the laser rod.

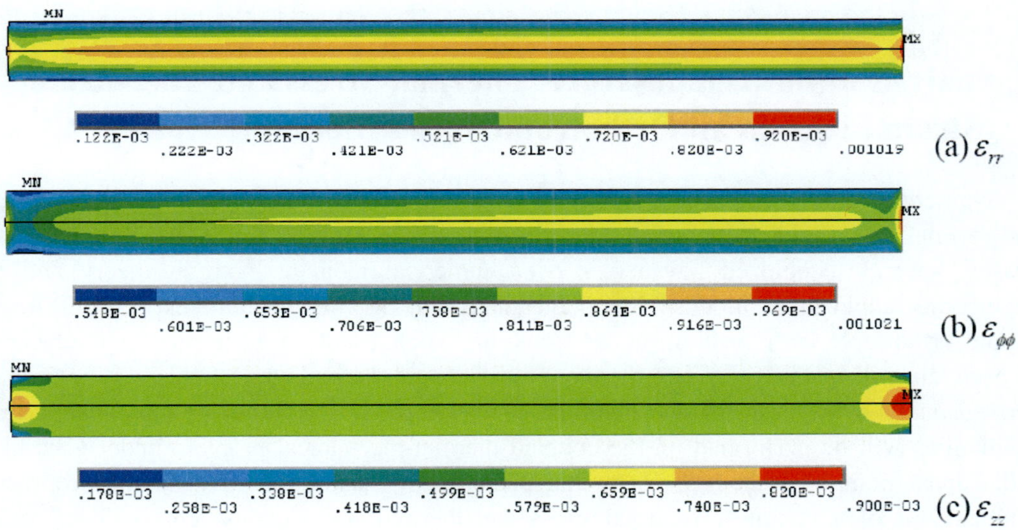

Figure 2.8. Thermal strain components plotted from numerical simulation.

2.6. Summary

In this section, two-dimensional axisymmetric analytical solutions are derived for the temperature, thermal stress and thermal strain in an elongate laser rod, with significant longitudinal rise of coolant temperature taken into account. Conjugate numerical simulations were also performed and compared with the analytical solution. The major conclusions are:

1. If the two ends of the laser rod are considered adiabatic, and the thermal entrance effects are ignored, the axial temperature gradient in the laser rod has positive relations with the fluid temperature-rise Δt_{b}, the axial Biot number Bi_r and the length-to-radius ratio L/r_i of the laser rod.
2. The values of q_l (or q_{V}), h_i (or Bi_r), k_{s}, r_i and $t_{\mathrm{ave,b}}$ determine the overall temperature level of the laser medium.
3. When the total heat generation rate is constant, the axial temperature gradient can be reduced by raising the coolant mass flow rate, selecting a coolant of high specific heat capacity, and improving the convective heat transfer coefficient.

4. Under the traction free boundary conditions, although there is significant longitudinal temperature increase inside the laser rod, the thermal stress variation in in the longitudinal direction is negligible. However, the thermal strain distribution has an evident variation in the longitudinal direction, which will affect the light transmission characteristics and the beam quality of the laser medium to a certain extent.
5. The numerical method agree well with the analytical solution in the main section of the laser rod, but better exhibits the entrance effects of fluid flow and heat transfer and the end effects at the rod ends. Therefore the two methods should complement each other in the design of laser cooling system.

3. Analysis of the Temperature, Thermal Stress and Thermal Strain Fields in an Orthotropic Laser Medium

Numerical method has been adopted by some researchers to investigate the temperature and thermal stress fields in orthotropic laser media [12, 13, 22, 23]. Since analytical methods possess several recognized advantages over numerical methods both in theory and in engineering application, in this section the analytical 3-D temperature expressions for a cuboid orthotropic laser medium under Gaussian, Tophat and Uniform pumping schemes, respectively will be derived. Then the thermal stress and thermal strain fields will be further derived analytically based on the plain-strain assumption. In addition, full 3-D numerical simulations will be performed on the coupled thermal-mechanical problem under Gaussian and Tophat pumping schemes, where the plane-strain assumption may be inappropriate. Comparisons of temperature, thermal stress and thermal strain components will be made among different pumping schemes, and also between orthotropic and isotropic media, therefore the influence of orthotropic thermal properties on the thermal effects could be analyzed.

3.1. Analytical Solutions of the Temperature Fields

As shown in figure 3.1, the dimension of the laser medium is $2a \times 2b \times L$. A pump beam is incident onto the central area of one end surface, and passes through the laser medium. The four side surfaces are subject to forced convective cooling by coolant liquid. The two end surfaces are both adiabatic. Due to the symmetric characteristics of the problem itself, only 1/4 model is selected for analysis. The mathematical description for the heat conduction in the orthotropic laser medium is written as [24]

$$\frac{\partial^2\theta}{\partial x^2}+\frac{1}{\varepsilon^2}\frac{\partial^2\theta}{\partial y^2}+\frac{1}{\eta^2}\frac{\partial^2\theta}{\partial z^2}+\frac{q_V\left(x,y,z\right)}{k_x}=0 \tag{3-1a}$$

$$\partial\theta/\partial x=0, \qquad x=0; \tag{3-1b}$$

$$\partial\theta/\partial x+H_x\theta=0, \qquad x=a; \tag{3-1c}$$

$$\partial\theta/\partial y = 0, \quad y = 0; \tag{3-1d}$$

$$\partial\theta/\partial y + H_y\theta = 0, \quad y = b; \tag{3-1e}$$

$$\partial\theta/\partial z = 0, \quad z = 0; \tag{3-1f}$$

$$\partial\theta/\partial z = 0, \quad z = L. \tag{3-1g}$$

where $\theta = T - T_f$ is the temperature rise relative to the fluid temperature T_f; $\varepsilon^2 = k_x/k_y$, $\eta^2 = k_x/k_z$, $H_x = h_x/k_x$, $H_y = h_y/k_y$, $H_z = h_z/k_z$, where k_x, k_y, k_z are the thermal conductivity components in three orthotropic directions, respectively, and h_x and h_y are the convective heat transfer coefficients in x and y directions, respectively; and $q_V(x,y,z)$ represents the heat loading profile in the laser medium.

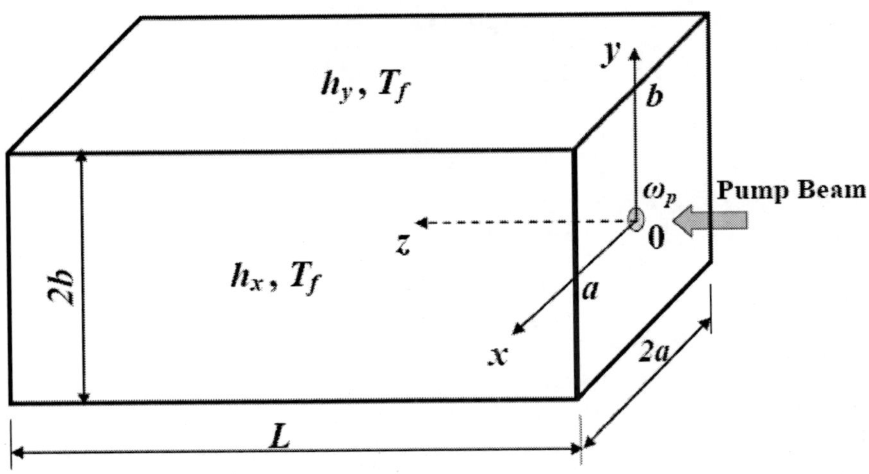

Figure 3.1. Schematic of an end-pumped YVO_4 laser medium.

(I) When the pump beam is a circularly symmetric Gaussian type, which is corresponding to the case of optic fiber coupled output, the heat loading profile is expressed as [25]

$$q_V(x,y,z) = \frac{2Q\alpha}{\pi\omega_p^{\,2}\left(1-e^{-\alpha L}\right)}\exp\left(-2\frac{x^2+y^2}{\omega_p^{\,2}} - \alpha z\right) \tag{3-2}$$

where $Q = \xi_p P_{in}$ is the total heat energy absorbed by the laser medium, ξ_p is the fractional thermal loading, P_{in} is the absorbed pump power, ω_p is the $1/e^2$ intensity radius of the Gaussian pump beam, α is the absorption coefficient at the pump wavelength.

The equation system (3-1) can be solved by an integral transform method. By applying two steps of integral transform (for the variables x and y sequentially) to the equation system (3-1) and then two steps of inverse transform, we can obtain the temperature expression as

$$\theta(x,y,z)=\sum_{m=1}^{\infty}\sum_{n=1}^{\infty}\frac{\cos(\beta_m x)\cos(\gamma_n y)}{N(\beta_m)N(\gamma_n)}\left(A_{mn}e^{\lambda_{mn}z}+B_{mn}e^{-\lambda_{mn}z}+C_{mn}e^{-\alpha z}\right) \quad \text{(3-3a)}$$

where β_m and γ_n are eigenvalues determined by the transcendental equations $\beta_m\tan(\beta_m a)=H_x$ and $\gamma_n\tan(\gamma_n b)=H_y$; $\lambda_{mn}{}^2=\beta_m{}^2\eta^2+\gamma_n{}^2\eta^2/\varepsilon^2$; $N(\beta_m)$ and $N(\gamma_n)$ are the norms of the eigenfunctions defined by

$$N(\beta_m)=\frac{a(\beta_m{}^2+H_x{}^2)+H_x}{2(\beta_m{}^2+H_x{}^2)},\ N(\gamma_n)=\frac{b(\gamma_n{}^2+H_y{}^2)+H_y}{2(\gamma_n{}^2+H_y{}^2)} \quad \text{(3-3b)}$$

$$A_{mn}=\frac{M_{mn}\alpha\left(e^{-\lambda_{mn}L}-e^{-\alpha L}\right)}{2\lambda_{mn}\left(\lambda_{mn}{}^2-\alpha^2\right)\sinh(\lambda_{mn}L)} \quad \text{(3-3c)}$$

$$B_{mn}=\frac{M_{mn}\alpha\left(e^{\lambda_{mn}L}-e^{-\alpha L}\right)}{2\lambda_{mn}\left(\lambda_{mn}{}^2-\alpha^2\right)\sinh(\lambda_{mn}L)} \quad \text{(3-3d)}$$

$$C_{mn}=-M_{mn}/\left(\lambda_{mn}{}^2-\alpha^2\right). \quad \text{(3-3e)}$$

where

$$M_{mn}=-\frac{\eta^2 Q\alpha}{4k_x\left(1-e^{-\alpha L}\right)}\exp\left(-\frac{\beta_m{}^2+\gamma_n{}^2}{8}\omega_p{}^2\right) \quad \text{(3-3f)}$$

(II) When the pump beam is a top-hat type, which is corresponding to the case of direct pumping from diode lasers or dye lasers, the heat loading profile is given by [26]

$$q_V(x,y,z)=\begin{cases}\dfrac{Q(\alpha L)e^{-\alpha z}}{\pi r_0{}^2L\left(1-e^{-\alpha L}\right)}, & x^2+y^2\le r_0{}^2\\ 0, & x^2+y^2>r_0{}^2\end{cases} \quad \text{(3-4)}$$

where r_0 is the pump beam radius.

The solution under the top-hat scheme is obtained by the same method and has the same form with formula (3-3) if only we replace the parameter M_{mn} contained in the coefficients A_{mn}, B_{mn} and C_{mn} with the parameter P_{mn} defined by

$$P_{mn} = \frac{-\eta^2 Q\alpha}{k_x \pi r_0^2 \left(1-e^{-\alpha L}\right)} \frac{1}{\gamma_n} \int_0^{r_0} \cos\left(\beta_m x\right) \sin\left(\gamma_n \sqrt{r_0^2 - x^2}\right) dx \tag{3-5}$$

(III) For Uniform pumping, which is corresponding to the case of Uniform side pumping, and when both the parameters $2a\alpha$ and $2b\alpha$ are relatively small [27], the heat loading profile becomes

$$q_V\left(x, y, z\right) = Q/\left(2a \cdot 2b \cdot L\right) \tag{3-6}$$

With the same method, we will obtain the temperature expression as

$$\theta\left(x, y\right) = \sum_{m=1}^{\infty}\sum_{n=1}^{\infty} \frac{\cos\left(\beta_m x\right)\cos\left(\gamma_n y\right)}{N\left(\beta_m\right)N\left(\gamma_n\right)} \frac{\eta^2}{\lambda_{mn}^2} \frac{Q}{4abLk_x} \frac{\sin\left(\beta_m a\right)}{\beta_m} \frac{\sin\left(\gamma_n b\right)}{\gamma_n} \tag{3-7}$$

To sum up, the temperature expressions under different pumping schemes can be generalized into the form of

$$\theta\left(x, y, z\right) = \sum_{m=1}^{\infty}\sum_{n=1}^{\infty} K_{mn}\left(z\right)\cos\left(\beta_m x\right)\cos\left(\gamma_n y\right) \tag{3-8a}$$

where for the Gaussian and Top-hat schemes,

$$K_{mn}\left(z\right) = \frac{A_{mn} e^{\lambda_{mn} z} + B_{mn} e^{-\lambda_{mn} z} + C_{mn} e^{-\alpha z}}{N\left(\beta_m\right)N\left(\gamma_n\right)} \tag{3-8b}$$

while for the Uniform pumping scheme,

$$K_{mn}\left(z\right) = \frac{1}{N\left(\beta_m\right)N\left(\gamma_n\right)} \frac{\eta^2}{\lambda_{mn}^2} \frac{Q}{4abLk_x} \frac{\sin\left(\beta_m a\right)}{\beta_m} \frac{\sin\left(\gamma_n b\right)}{\gamma_n} \tag{3-8c}$$

It is noted that the temperature solutions for an isotropic medium can be deduced from the above temperature expressions for an orthotropic medium under different pumping schemes, if only we let the parameters $k_x = k_y = k_z = k$.

Take Nd:YVO_4, one of the most commonly used orthotropic laser medium as an example. The related parameters are $k_x = 5.23\text{W}/\left(\text{m}\cdot\text{K}\right)$, $k_y = 5.10\text{W}/\left(\text{m}\cdot\text{K}\right)$, $k_z = 5.10\text{W}/\left(\text{m}\cdot\text{K}\right)$,

$\alpha = 7.0\ \text{cm}^{-1}$ @808nm, $2a \times 2b \times L = 3 \times 3 \times 6\ \text{mm}^3$, $\omega_p = 0.4$ mm, $h_x = h_y = 12\ \text{KW}/\left(\text{m}^2 \cdot \text{K}\right)$, and $Q = 5$ W.

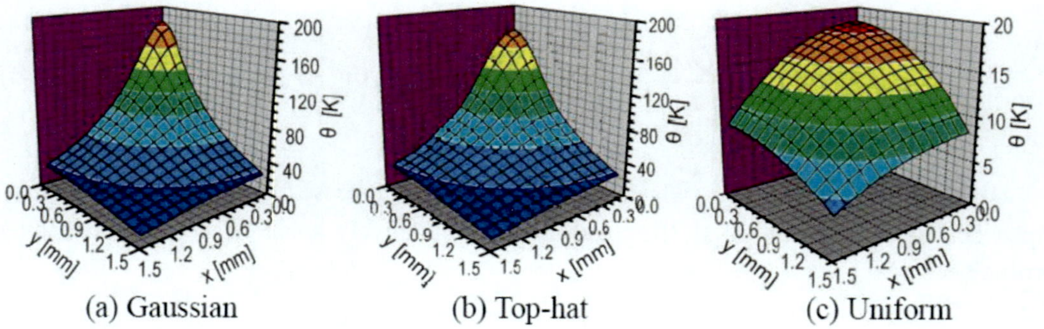

Figure 3.2. Temperature profiles in the pumping end surface ($z = 0$) under different pumping schemes.

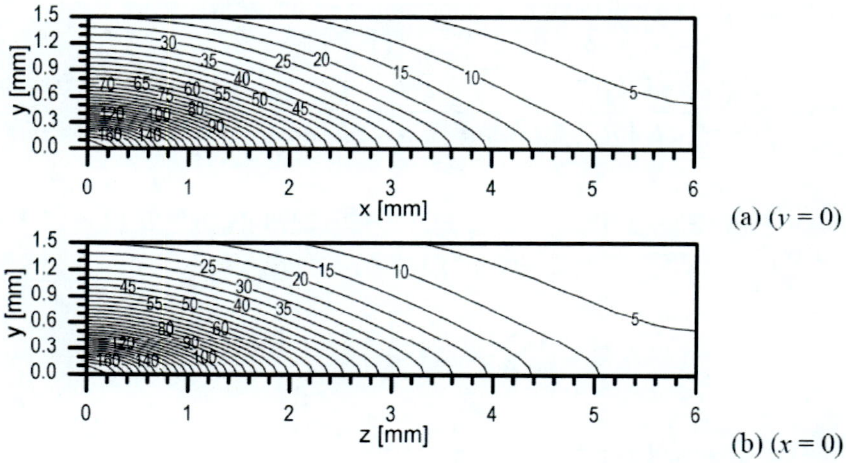

Figure 3.3. Temperature contours in the longitudinal central planes [K].

The temperature profiles in the $z = 0$ plane (corresponding to the pumping end surface in the end-pumping scenarios, and an arbitrary cross-section in the Uniform side-pumping scheme) under the Gaussian, top-hat and Uniform pumping schemes, respectively, are plotted in figures 3.2(a)~(c), by assuming that $r_0 = \omega_p$, and all the other conditions are the same. The maximum temperature rise $\theta_{\max}$ occurs at the center point $\left(x = y = 0\right)$ of the cross section in all cases, which is 180.72K for the Gaussian scheme, 168.96 K for the top-hat scheme, and only 18.29 K for the Uniform scheme. This indicates that, given an equal total heat loading Q, the Gaussian beam has the highest peak intensity, thus the medium has the highest heat generation rate $q_V\left(x, y, z\right)$ (W/m^3) at the point of $x = y = 0$, and consequently leading to the highest maximum temperature; the top-hat scheme has a Uniformly spread heat loading in a central area, resulting in a relatively lower $q_V\left(0,0,0\right)$ value relative to the Gaussian scheme, and causing a relatively moderate maximum temperature; while the Uniform scheme has the same total heat loading Q Uniformly dispersed in the overall medium, there exists no spot of high $q_V(x,y,z)$ within the medium, therefore bring about a significantly lower maximum temperature than the previous two schemes, giving rise to insignificant thermal effects.

Figure 3.3 shows how the 3-D temperature field varies in the longitudinal direction, under the Gaussian pumping scheme as an example. Even if the thermal conductivity of Nd:YVO_4 medium has a weak orthotropy ($\varepsilon = 1.0255$), the temperature contours in the x direction is slightly sparser than in the y direction. This is caused by the relatively higher k_x than k_y, thus the temperature gradient component $\partial T/\partial x$ is relatively smaller than $\partial T/\partial y$.

3.2. Determination of the Thermal Stress and Thermal Strain Fields

3.2.1. Two-Dimensional Analytical Solutions

For the ease of mathematical treatment, the plane-strain hypothesis is assumed in modeling the thermoelastic problem, which is applicable when the medium length L is quite large relative to the cross-sectional dimensions $2a$ and $2b$ and when there is no variation of thermal or mechanical loadings in the longitudinal direction. This is true for the Uniform pumping scheme, but may bring about relatively large error for the Gaussian and Tophat pumping schemes when the longitudinal temperature variation can not be ignored.

Under the plane-strain assumption, the strain-stress relations become

$$\varepsilon_{xx} = \frac{1-\nu^2}{E}\left(\sigma_{xx} - \frac{\nu}{1-\nu}\sigma_{yy}\right) + (1+\nu)\alpha_x\theta \tag{3-9a}$$

$$\varepsilon_{yy} = \frac{1-\nu^2}{E}\left(\sigma_{yy} - \frac{\nu}{1-\nu}\sigma_{xx}\right) + (1+\nu)\alpha_y\theta \tag{3-9b}$$

$$\varepsilon_{xy} = \frac{1+\nu}{E}\sigma_{xy} \tag{3-9c}$$

where α_x, α_y and α_z are the three components of thermal expansion coefficient of the orthotropic laser medium, k_x, k_y and k_z are the thermal conductivity components in the three orthotropic directions, E is the Young's Modulus, and ν is the Poisson's ratio.

If an Airy's thermal stress function ϕ is defined such that

$$\sigma_{xx} = \frac{\partial^2\phi}{\partial y^2}, \sigma_{yy} = \frac{\partial^2\phi}{\partial x^2}, \sigma_{xy} = -\frac{\partial^2\phi}{\partial x\partial y} \tag{3-10}$$

then we will obtain the governing partial differential equation in terms of ϕ as

$$\nabla^4\phi = -\frac{E}{1-\nu}\left(\alpha_y\frac{\partial^2\theta}{\partial x^2}+\alpha_x\frac{\partial^2\theta}{\partial y^2}\right) \tag{3-11a}$$

where $\nabla^4 = \partial^4/\partial x^4 + 2\partial^4/\partial x^2\partial y^2 + \partial^4/\partial y^4$.

In the absence of externally applied forces, the mechanical boundary conditions expressed in terms of ϕ become [21]

$$\phi = \partial\phi/\partial x = 0, \quad x = a; \tag{3-11b}$$

$$\phi = \partial\phi/\partial y = 0, \quad y = b \tag{3-11c}$$

Considering the symmetry of the original problem and the characteristics of the temperature expressions (3-8), the thermal stress function ϕ can be written as

$$\begin{aligned}\phi &= \left[b_1\cosh(\beta_1 y)+b_4 y\sinh(\beta_1 y)\right]\cos(\beta_1 x)\\ &+\left[c_1\cosh(\gamma_1 x)+c_4 x\sinh(\gamma_1 x)\right]\cos(\gamma_1 y)\\ &+\sum_{m=1}^{\infty}\sum_{n=1}^{\infty}\frac{E}{1-\nu}K_{mn}(z)\frac{\left(\alpha_y\beta_m{}^2+\alpha_x\gamma_n{}^2\right)}{\left(\beta_m{}^2+\gamma_n{}^2\right)^2}\cos(\beta_m x)\cos(\gamma_n y)\end{aligned} \tag{3-12}$$

where b_1, b_4, c_1 and c_4 are unknown constants to be determined by boundary conditions.

Substituting the function (3-12) into the boundary conditions (3-11b) and (3-11c), utilizing the orthogonality of the eigenfunctions, yields a system of linear algebraic equations, from which the unknown constants b_1, b_4, c_1, c_4 are determined. Finally, the thermal stress components are analytically derived by substituting Eq. (3-12) into Eq. (3-10) as

$$\begin{aligned}\sigma_{yy} &= -\left[b_1\cosh(\beta_1 y)+b_4 y\sinh(\beta_1 y)\right]\beta_1^2\cos(\beta_1 x)\\ &+\left\{c_1\gamma_1^2\cosh(\gamma_1 x)+c_4\left[2\gamma_1\cosh(\gamma_1 x)+\gamma_1{}^2 x\sinh(\gamma_1 x)\right]\right\}\cos(\gamma_1 y)\\ &-\sum_{m=1}^{\infty}\sum_{n=1}^{\infty}\beta_m{}^2 R_{mn}(z)\cos(\beta_m x)\cos(\gamma_n y)\end{aligned} \tag{3-13a}$$

$$\begin{aligned}\sigma_{xx} &= \left\{b_1\beta_1^2\cosh(\beta_1 y)+b_4\left[2\beta_1\cosh(\beta_1 y)+\beta_1^2 y\sinh(\beta_1 y)\right]\right\}\cos(\beta_1 x)\\ &-\left[c_1\cosh(\gamma_1 x)+c_4 x\sinh(\gamma_1 x)\right]\gamma_1^2\cos(\gamma_1 y)\\ &-\sum_{m=1}^{\infty}\sum_{n=1}^{\infty}\gamma_n{}^2 R_{mn}(z)\cos(\beta_m x)\cos(\gamma_n y)\end{aligned} \tag{3-13b}$$

$$\sigma_{xy} = \left\{b_1\beta_1 \sinh(\beta_1 y) + b_4\left[\sinh(\beta_1 y) + \beta_1 y \cosh(\beta_1 y)\right]\right\}\beta_1 \sin(\beta_1 x)$$
$$+\left\{c_1\gamma_1 \sinh(\upsilon_j x) + c_4\left[\sinh(\gamma_1 x) + \gamma_1 x \cosh(\gamma_1 x)\right]\right\}\gamma_1 \sin(\gamma_1 y) \quad \text{(3-13c)}$$
$$-\sum_{m=1}^{\infty}\sum_{n=1}^{\infty}\beta_m\gamma_n R_{mn}(z)\sin(\beta_m x)\sin(\gamma_n y)$$

The analytical expressions of the thermal strain components can be obtained accordingly by substituting the formulas (3-13) into the formulas (3-9).

It is noted that the above formulas for orthotropic medium are also applicable to the isotropic medium, if only we let the parameters $k_x = k_y$, $\alpha_x = \alpha_y$ in the above formulas.

3.2.2. Full 3-D Coupled Numerical Simulation

As previously mentioned, the plane-strain assumption may cause large error for the Gaussian and Tophat pumping schemes when the thermal loadings significantly vary in the longitudinal direction. Full 3-D numerical simulations are necessary in these cases. In this section, a commercial FEA code ANSYS is used to determine the thermal stress and thermal strain fields for the Gaussian and Tophat pumping schemes. An element type of "SOLID 5" is selected to treat the coupled thermal-mechanical problem. Orthotropic thermal properties are set in the material model. Only 1/4 geometry is selected as the computational domain due to the symmetry of the laser medium. A mesh of 50×50×80 elements is used, with increasing node intervals in the x, y and z directions, respectively to appropriately reflect the large temperature gradient adjacent to the incidence point of the pumping beam. "Symmetry" conditions are applied to the $x=0$ and $y=0$ central planes, "Convection" boundary conditions are applied to the $x=a$ and $y=b$ planes. The functions of heat generation rates corresponding to expressions (3-2) and (3-4) are applied as body loads for the Gaussian and Tophat pumping schemes, respectively. Adiabatic boundary conditions are applied to the end surfaces, and traction free conditions are applied to all medium surfaces. Then Static simulations are performed.

3.2.3. Typical Examples

Still take the Nd:YVO_4 laser medium as an example. The thermo-mechanical properties are $\alpha_x = 11.37\times10^{-6}\text{K}^{-1}$, $\alpha_y = 4.43\times10^{-6}\text{K}^{-1}$, $E = 1.33\times10^{11}\text{Pa}$, and $\nu = 0.33$, and the other conditions are assumed the same as the examples in the section 3.1.

Figures 3.4~3.6 show the distributions of thermal stress components under different pumping schemes. As shown in the figures, in all these pumping schemes, the inner regions around the medium center are subject to compressive stress, whereas the outer regions adjacent to the medium surfaces bear tensile stress. This is caused by the direction of temperature gradients, which are always pointing inwards from the coolant liquid to the medium center. The different pumping schemes have similar patterns of thermal stress distribution, whereas the primary difference lies in their specific values. In addition, it can be seen that the absolute values of shear stress in all the schemes are far smaller than the corresponding normal stress.

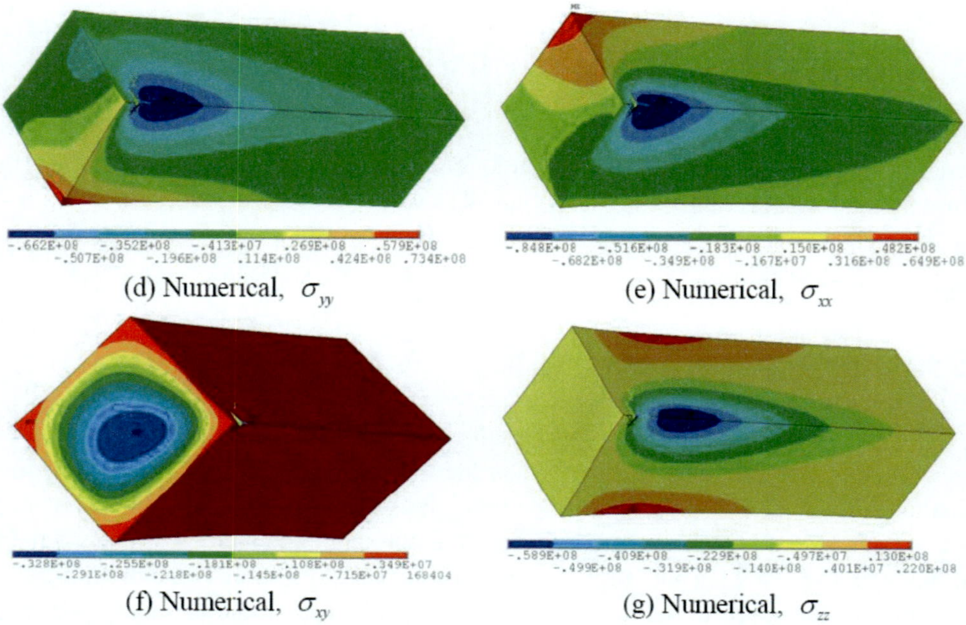

Figure 3.4. Contours of thermal stress components under Gaussian pump beam (z = 0) [Pa].

Figure 3.4 shows the numerical results of the thermal stress components at the pumping end (z = 0) under the Gaussian pump beam. The maximum compressive stress occurs at the point of $x = y = z = 0$, where σ_{yy}=-66.2MPa and σ_{xx}=-84.8MPa, in which the negative signs denote compressive stress. The maximum tensile stress occur at the boundary surfaces, where σ_{yy}=73.4MPa ($x = a,\ y = 0$) and σ_{xx}=64.9MPa ($x = 0,\ y = b$). Obviously, the difference between σ_{yy} and σ_{xx} is caused by the orthotropic properties of the laser medium since all the other conditions have been assumed the same in both directions. However, in the absence of external forces and constraints, the orthotropic thermal expansion coefficient will firstly influence the thermal strain components, while the laser medium is allowed to expand freely in both directions to some extent, so the difference in the thermal stress components becomes less significant.

Figure 3.5 shows the numerical results of the thermal stress components in the plane of z = 0 under the top-hat pump beam, provided that the top-hat beam radius equals the Gaussian beam waist, i.e. $r_0 = \omega_p$, and the other conditions are assumed the same with the Gaussian pump beam. The analytical results are not shown because of its large error. The maximum compressive stress at the center point ($x = y = z = 0$) is σ_{yy}=-60.4MPa and σ_{xx}=-76.0MPa, the maximum tensile stress at the boundary surfaces is σ_{yy}=73.0MPa ($x = a,\ y = 0$) and σ_{xx}=64.5MPa ($x = 0,\ y = b$), all of which are lower than the Gaussian scheme. The reasoning can be made based on the characteristics of the heat loading function $q_V(x, y, z)$ in a similar manner with the explanation concerning the temperature distribution in the section 3.1. In a similar way to the Gaussian scheme, the difference between σ_{yy} and σ_{xx} is also caused by the orthotropic properties of the laser medium.

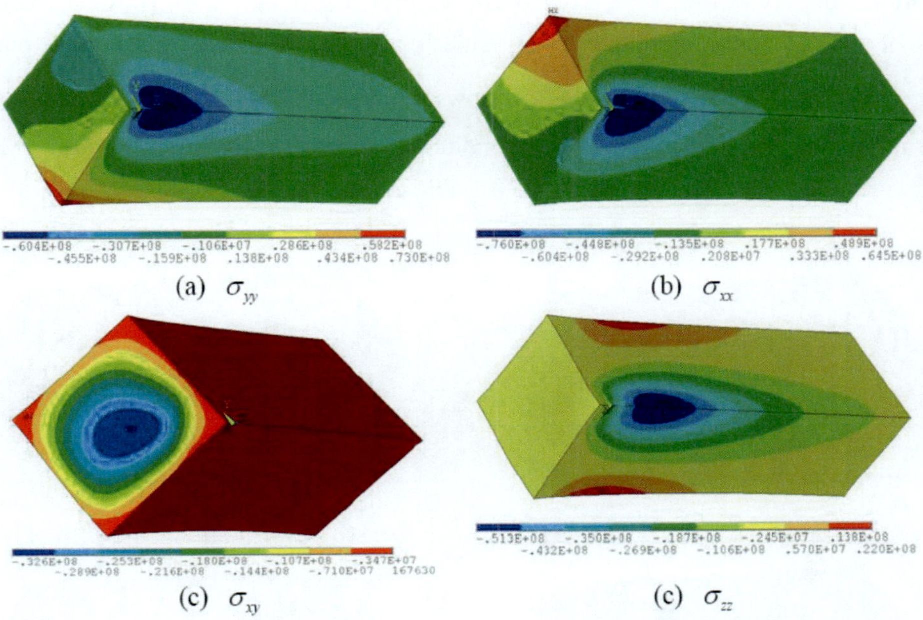

Figure 3.5. Contours of thermal stress components under top-hat pump beam (z = 0) [Pa].

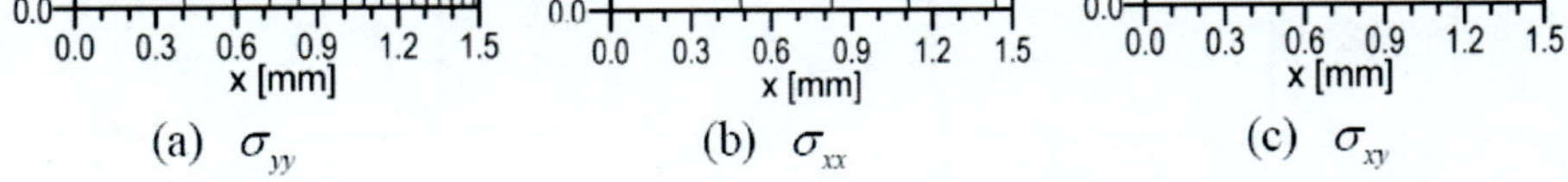

Figure 3.6. Contours of thermal stress components under Uniform pumping [MPa].

Figure 3.6 shows the analytical results of thermal stress components in an arbitrary cross section under Uniform pumping with the same total heat loading. All the other conditions are assumed the same with the Gaussian and the Top-hat schemes. The maximum compressive stress at the medium central axis ($x = y = 0$) is σ_{yy}=-4.85MPa and σ_{xx}=-4.60MPa, the maximum tensile stress at the boundary surfaces are σ_{yy}=10.53MPa ($x = a,\ y = 0$) and σ_{xx}=11.33MPa ($x = 0,\ y = b$). These results are significantly lower than the Gaussian and the top-hat schemes, leading to insignificant thermal effects. This is because the same total heat loading Q is scattered uniformly in the whole medium, there exists no spot of high heat generation rate $q_V(x, y, z)$. The difference between σ_{yy} and σ_{xx} can also be explained by the orthotropic properties of the laser medium.

The thermal strain distribution in a solid-state laser medium is an important factor affecting the quality of the output laser beam. Figures 3.7(a) and 3.7(b) show the numerical results of thermal strain components ε_{yy} and ε_{xx} under Gaussian and Tophat pumping schemes, respectively, while figure 3.7(c) shows the analytical results of ε_{yy} and ε_{xx} under Uniform pumping scheme with the same conditions as the cases of figures 3.7(a) and 3.7(b). As shown, ε_{xx} is significantly higher than ε_{yy} due to the higher α_x than α_y.

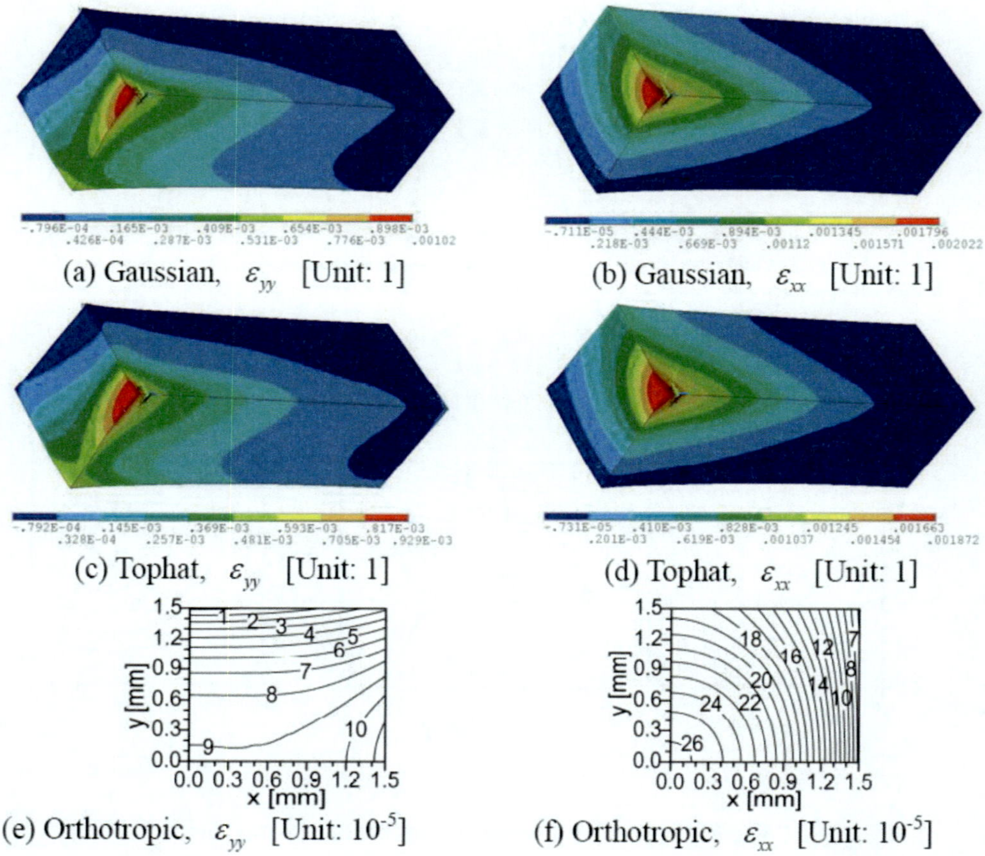

Figure 3.7. Contours of thermal strain components in an orthotropic laser medium.

3.3. Summary

Analytical and numerical solutions are obtained for the temperature, thermal stress and thermal strain fields in an orthotropic laser medium under different pumping schemes. The main conclusions are drawn as follows.

1. The different pumping schemes result in similar thermal stress patterns with different values. The inner regions of the laser medium are subject to compressive stress, whereas the outer regions bear tensile stress. The absolute values of shear stress in all these schemes are far smaller than the corresponding normal stress.
2. Supposing a same total heat loading, and provided that the radius of the top-hat pump beam equals to the e^{-2} intensity radius of the Gaussian pump beam, then the maximum temperature and thermal stress under the top-hat pump beam is lower than that of the Gaussian pump beam, whereas the thermal effects under Uniform pumping is the least serious of all.
3. In the absence of external forces and constraints, the orthotropic thermal expansion coefficient has significant influence on the thermal strain components, but has less significant influence on the thermal stress components.
4. The analytical method and the numerical method should complement each other in the design process due to their respective advantages and disadvantages.

4. 3-D Conjugate Numerical Simulation of DLA Microchannel Heat Sinks Considering the Varying Thermal Properties

In a high-power Diode Pumped Solid Laser, the stringent temperature constraint of Laser Diode array is critical for ensuring beam quality and laser output. In addition, the trend of ever-increasing miniaturization and circuit integration in electronic devices leads to higher power densities and higher heat generation rates. The microchannel heat sinks with single-phase liquid flow provide a promising cooling approach that is miniaturized, reliable and highly efficient, with a heat removal capacity usually in the order of hundreds of W/cm^2. The microchannel heat sinks combine the merits of high material conductivity and liquid forced convection, together with the advantages of huge heat exchange surface per unit volume, so as to efficiently utilize the limited space available for deploying cooling devices, therefore the heat transfer performance is significantly enhanced. To profoundly understand and accurately describe the fluid flow and heat transfer characteristics in microchannels will provide valuable reference for engineering application.

In this section, three-dimensional conjugate numerical simulations are performed for the water flow and heat transfer in rectangular microchannels including the entrance region. Three different methods in treating the thermal properties, namely the "Inlet property" method, the "Average property" method and the "Variable property" method are adopted. The three methods have completely the same boundary conditions except for the different ways in treating thermal properties. Local and average flow and heat transfer characteristics are obtained and compared among different methods, and also compared with the recently reported experimental data, the frequently cited correlations and the simplified analytical solution data from the published literatures. Discussions are made on the influence of the above different methods on the results of flow and heat transfer characteristics under the conditions of low Reynolds number and high heat flux, which combined to result in a high longitudinal temperature rise in microchannels.

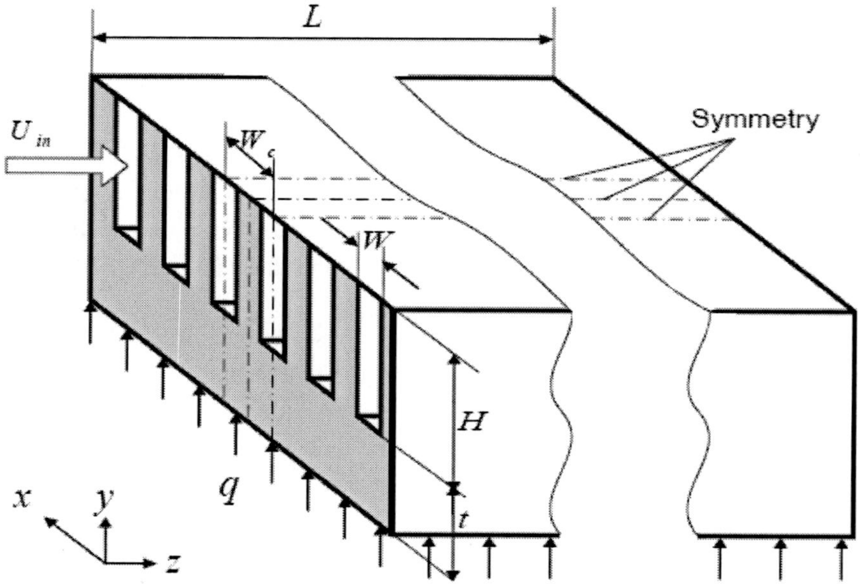

Figure 4.1. Schematic of microchannel heat sink.

4.1. Problem Description and Mathematical Model

A schematic of a microchannel heat sink is shown in figure 4.1, with 2 sets of dimensions shown in table 4.1. The bottom surface of the microchannel heat sink is subject to a uniform heat flux, and the cooling water flows in the z direction. The governing equations for the steady state 3D viscous flow and heat transfer can be expressed in the vector form as [28]

$$\nabla \cdot \left(\rho \vec{V}\right) = 0 \quad (4\text{-}1a)$$

$$\vec{V} \cdot \nabla(\rho \vec{V}) = \rho \boldsymbol{F} - \nabla p + \nabla \cdot \left(\mu \nabla \vec{V}\right) \quad (4\text{-}1b)$$

$$\vec{V} \cdot \nabla(\rho c_{\mathrm{p}} T_{\mathrm{f}}) = (\vec{V} \cdot \nabla) p + 2\mu[\varepsilon]^2 + \nabla \cdot \left(k_{\mathrm{f}} \nabla T_{\mathrm{f}}\right) + \rho Q \quad (4\text{-}1c)$$

$$\nabla \cdot \left(k_{\mathrm{s}} \nabla T_{\mathrm{s}}\right) = 0 \text{ (for solid)} \quad (4\text{-}1d)$$

Table 4.1. Model description

Model No.	**Dimensions (mm)**					***q***
	W	***W*$_c$**	***H***	***L***	***t***	**(W/cm^2)**
A	0.2	0.4	1	20	0.8	150
B	0.2	0.4	1	12	0.5	250

The computational domain includes both the solid and the liquid regions. Both the center planes of the channel and the solid fin are set as "Symmetry" boundary conditions ($\partial/\partial n = 0$), as shown in figure 4.1. A uniform inlet velocity u_{in} (m/s) is given at the channel inlet. The inlet temperatures are arbitrarily all set to a constant value of 293K in all cases. A uniform heat flux q is applied to the bottom surface of the heat sink. Coupled wall boundary conditions ($T_s = T_f$, $-k_s\left(\frac{\partial T_s}{\partial n}\right)_w = -k_f\left(\frac{\partial T_f}{\partial n}\right)_w$) are assigned to the fluid-solid interface to couple up the fluid convection and solid conduction since the physical parameters on the interface are unknown beforehand. No-slip and no-penetration condition is also assigned to this interface ($\vec{V} = 0$). All the other walls were treated as adiabatic ($\left.(\partial T/\partial n)\right|_w = 0$). Constant and variable thermal properties are adopted for the fluid, respectively. The grid sizes (for half domain) are $x\times y\times z$=50×150×200 for model A, and $x\times y\times z$=50×150×150 for model B. The longitudinal node intervals are allowed a successive ratio of 1.01, in order to capture the abrupt change of flow and temperature fields in the entrance region, while avoiding too large cell aspect ratios near the outlet. The node intervals in the x and y directions are uniform. A series of gird-dependence tests indicates that the variation of the calculated average temperature of the heating surface after convergence is less than 0.1% if further refining the mesh. A commercial CFD code FLUENT is used for the numerical simulation, in which the implicit method is adopted for discretizing the governing equations, the SUS (Second Order Upwind Scheme) is used for discretizing the convective terms in the momentum and energy conservation equations. The SIMPLEC algorism is selected for the pressure-velocity coupling owing to its excellent converging efficiency. The axial heat conduction and the conjugate heat transfer are automatically taken into account in the 3D conjugate simulations.

4.2. Analysis on the Simulation Results

The apparent flow friction coefficient, defined as Eq. (4-2) incorporates both the pressure drops caused by the higher wall shear stress due to the higher velocity gradient normal to the wall, and by the momentum flux variation due to the change of velocity field from a uniform profile at the inlet to a specific profile somewhere in the channel.

$$f_{app}(z) \equiv \frac{1}{2}\frac{D_h}{z}\frac{p_{in} - p_z}{\rho u_{ave}^2} \tag{4-2}$$

where the hydraulic diameter is defined as $D_h = 2WH/(W+H)$ for a channel with a rectangular cross section; u_{ave} is the mean longitudinal velocity; p_z can be taken as a function of the longitudinal position z in the data post-processing stage, because the transverse pressure gradients $\partial p/\partial x$ and $\partial p/\partial y$ can be neglected as we do a dimensional analysis on the momentum equation (Eq. 4-1(b)) for the flow field in such a long thin channel.

The "Variable property" method are conducted first at a specified inlet velocity of 0.300 m/s, in which all thermal properties are set as piecewise-linear functions of water temperature. Then in the "Inlet property" method, all thermal properties are determined at the inlet water temperature of 293K, while the inlet velocity are set to the averaged velocity over the entire water space obtained from the calculation results of "Variable property" method. In the "Average property" method, all thermal properties and the inlet velocity are set to the averaged values over the entire water space acquired from the calculation results of the "Variable property" method. Thus the three methods have a substantially equal mass flow rate of 3.0×10^{5}kg/s, while the "Average property" method and the "Variable property" method have substantially equal average properties. So the three cases have completely the same boundary conditions except for the different ways in treating the thermal properties.

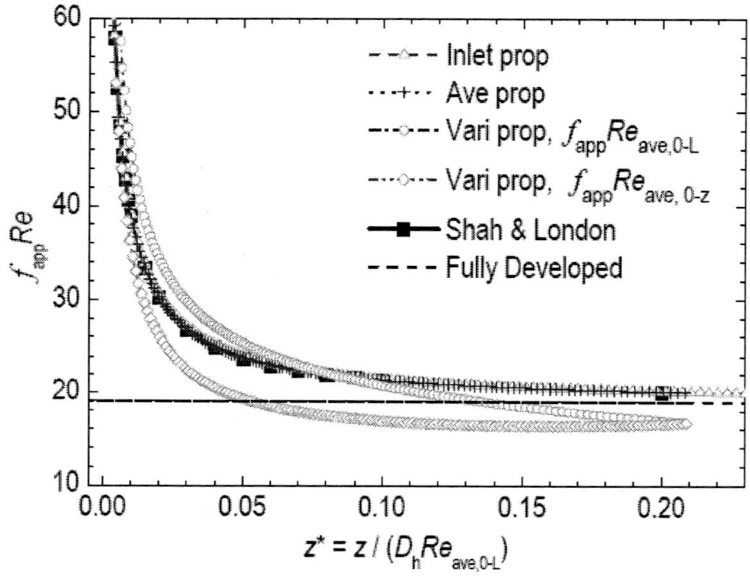

Figure 4.2. Comparison of local $f_{app}Re$ among different methods and with the Semi-analytical solution (for model B, u_{ave}= 0.30376m/s).

Figure 4.2 shows the calculation results of $f_{app}Re$ using the above-mentioned three methods, respectively, in comparison with the well-known simplified analytical solution data presented by Shah and London [16] and Curr et al. [29]. When the "Variable property" method is used, there are two options to calculate the $f_{app}Re$ in the data post-processing stage: one is to use the Reynolds number $Re_{ave,0\text{-}L}$ that is averaged over the entire channel length 0~L, the other is to use the Reynolds number $Re_{ave,0\text{-}z}$ that is averaged over the channel length of 0~z, i.e. from the inlet to a local longitudinal position. The latter option is more compatible with the definition of f_{app}, since they both describe the flow characteristics over the channel length of 0~z. For reference purpose, the theoretical value of $f_{app}Re$=19.07 for fully developed laminar flow in a rectangular channel with an aspect ratio of 0.2 presented by the Shah and London [16] is also shown in figure 4.2. The following features can be observed from this figure:

1. Both the "Inlet property" and the "Average property" methods agree very well with the data of Shah and London [16] and Curr et al. [29]. Compared with the "Inlet property" method, the "Average property" method has a lower friction coefficient,

but a higher Reynolds number due to the lower average viscosity, while the product of friction coefficient and the Reynolds number keeps unchanged. This indicates that these $f_{\text{app}}Re$ data are applicable in all constant property situations whether they are constant inlet properties or constant average properties.

2. The $f_{\text{app}}Re_{\text{ave,0-L}}$ values of the "Variable property" method are higher than that of the constant property methods at the beginning section of the channel, but drops below that of the constant property methods at the later section of the channel. This is because it has the same $Re_{\text{ave,0-L}}$ level with the "Average property" method, but its friction coefficient is decreasing with a greater slope along the channel due to the viscosity drop caused by the temperature rise.
3. The $f_{\text{app}}Re_{\text{ave,0-z}}$ value is always lower than $f_{\text{app}}Re_{\text{ave,0-L}}$, but their difference is decreasing along the channel length until they become equal at the outlet, where $Re_{\text{ave,0-z}}=Re_{\text{ave,0-L}}$.
4. The $f_{\text{app}}Re$ value of the constant property methods will approach an asymptote, i.e. the theoretically fully developed value of 19.07 with the development of the flow field, but the $f_{\text{app}}Re$ value of the variable property method may drop below this theoretical value.

From the above analysis, we can find that the "Variable property" method using $Re_{\text{ave,0-z}}$ is the most reasonable approach in predicting the local $f_{\text{app}}Re$ values, though it may have evident deviations relative to known methods and the "Variable property, $Re_{\text{ave,0-L}}$" approach.

The local convective heat transfer coefficient and the local Nusselt number are defined as

$$h_{\text{xyz}}(x,y,z) \equiv \frac{q_{\text{w}}(x,y,z)}{T_{\text{w}}(x,y,z)-T_{\text{b}}(z)} \tag{4-3}$$

$$Nu_{\text{xyz}}(x,y,z) \equiv \frac{h_{\text{xyz}}(x,y,z)D_{\text{h}}}{k_{\text{z}}(z)} \tag{4-4}$$

where

$$T_{\text{b}}(z) \equiv \frac{\int_{A_{\text{c}}} \rho u c_{\text{p}} T_{\text{f}} dA}{\int_{A_{\text{c}}} \rho u c_{\text{p}} dA} \tag{4-5}$$

is the mixed mean temperature.

The longitudinal local convective heat transfer coefficient h_{z} is defined as

$$h_{\text{z}} \equiv \int_{\Gamma} h_{\text{xyz}}(x,y,z) d\Gamma' \tag{4-6}$$

where Γ is the perimeter of the channel cross section at a local longitudinal position.

The longitudinal local Nusselt number is defined as

$$Nu_{\text{z}} \equiv \frac{h_{\text{z}} D_{\text{h}}}{k_{\text{z}}} \tag{4-7}$$

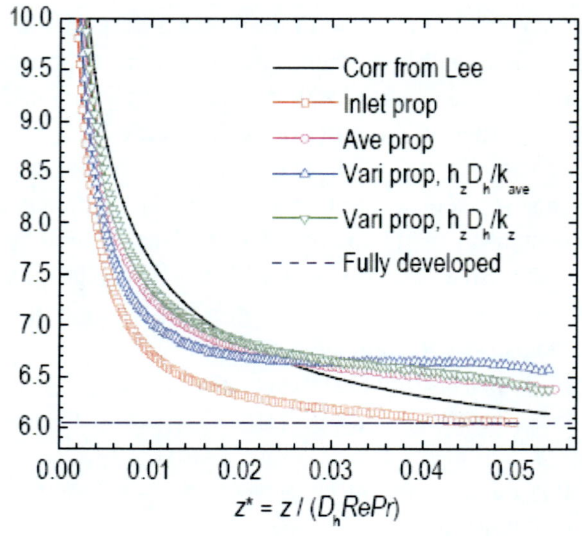

Figure 4.3. Comparison of the longitudinal local Nusselt number among different methods and with available correlations (for model B, u_{ave}= 0.30376m/s).

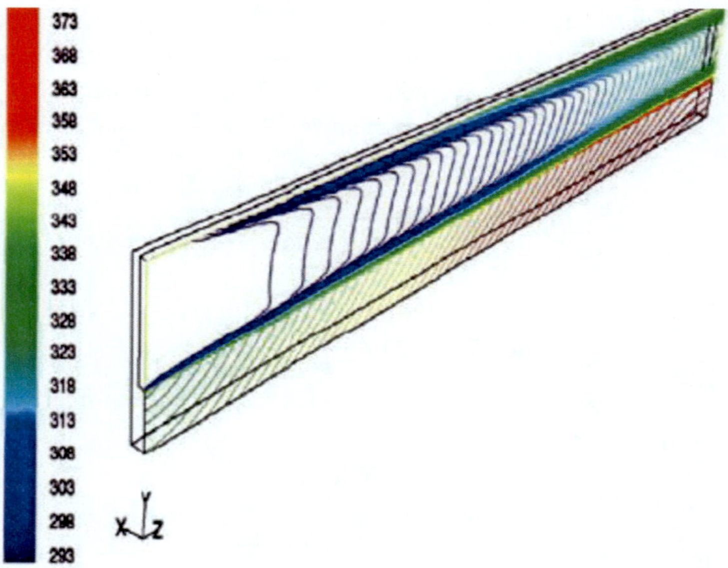

Figure 4.4. Temperature contours on the symmetry plane through the channel center (unit: K; for model B, u_{ave}= 0.30376m/s).

Figure 4.3 shows the calculation results of Nusselt number using different methods in treating thermal properties, in comparison with the correlation from Lee [30], which agree well with the experimental data of Lee [31].

As for the "Inlet property" method, k_z=k_{in}. For the "Average property" method, k_z=k_{ave}. For the "Variable property" method, two options are available regarding the treatment of thermal conductivity for calculating Nu_z during the data post-processing: one is to use the k_{ave} that is averaged over the entire water space inside the channel, hereinafter referred to as the "Variable property, h_zD_h/k_{ave}" method; the other is to use the longitudinal local value k_z, hereinafter referred to as the "Variable property, h_zD_h/k_z" method. The latter option better matches the definition of h_z, since they both describe the local heat transfer characteristics. As shown in figure 4.3, the following trends can be observed:

1. Both the "Average property" and the "Variable property" methods have notably higher Nusselt numbers than the "Inlet property" method, due to the higher h_z brought by lower viscosity and higher thermal conductivity levels. A lower viscosity contributes to a higher water velocity and a smaller boundary layer thickness. A higher thermal conductivity directly enhances the heat transfer. Therefore both factors are favorable for intensifying the convective heat transfer.
2. Compared to the "Average property" method, the "Variable property, h_zD_h/k_{ave}" method has a lower Nu_z at the beginning stage of the channel, but flattens out earlier and exceed the "Average property" method at the later stage. This is because the two methods share substantially equal levels of k_{ave} and average h_z, while the h_z values of the "Variable property" method are lower than that of the "Average property" method at the beginning section of the channel, but flatten out earlier and exceed that of the "Average property" method. This can be explained by the continuous decrease of viscosity and increase of thermal conductivity due to the continuous temperature rise along the channel, though the two methods share a substantially equal level of average thermal properties.
3. The "Variable property, h_zD_h/k_z" method has a Nu_z pattern very close to that of the "Average property" method. Both the h_z and k_z values in the "Variable property, h_zD_h/k_z" method are lower than that of the "Average property" method in the beginning section, but are higher than that of the "Average property" method in the later section of the channel, while their quotient keeps similar to that of the "Average property" method. Consequently, the Nu_z difference between the two methods is quite small according to the definition of Eq. (4-7).
4. Relative to the "Variable property, h_zD_h/k_{ave}" method, the "Variable property, h_zD_h/k_z" method produces a higher Nu_z at the beginning, and a lower Nu_z at the later stage due to the continuous increase of k_z along the channel.
5. The correlation from Lee [30] is closer to the "Variable property, h_zD_h/k_z" method and the "Average property" method, but deviations still exist, which might be attributed to the differences in the computational domain, the hydraulic and thermal boundary conditions and the treatment of thermal properties. In contrast to the present work, Lee [30] use only the fluid region as the computational domain, hydrodynamically fully developed inlet velocity, H1 thermal boundary condition, and constant thermal properties. Due to the conjugate effects, the three-dimensional heat conduction in the solid part has a large z component of temperature gradient, as can be seen from figure 4.4. Maranzana [32] and Morini [33] suggest respectively, that the axial conduction in the walls can not be neglected when the criteria

$$M = \frac{k_s A_s/L}{\rho c_p A_f V} = \frac{k_s\left[(H+t)W_c - HW\right]}{\rho c_p HWVL} > 0.01$$

or

$$N = \frac{k_s}{k_f}\frac{D_h}{L}\left(\frac{W_c(H+t)}{HW}-1\right)\frac{1}{RePr} > 0.02$$

hold. In this case, these two parameters, which reasonably estimate the importance of axial conduction relative to the convection heat transfer, are $N = 0.05$ and $M = 0.05$ respectively. In fact, as a result of low *Re* number and high axial temperature rise, the solid part with high conductivity conducts a portion of the heat from the rear part back to the front part of the channel, leading to the redistribution of local wall heat flux, local wall temperature, local bulk temperature and the associated water thermal properties. Therefore, the final Nu_z values calculated according the definitions (4-7) may have difference with the correlation from Lee [30].

6. For reference purpose, the fully developed Nusselt number under the H1 thermal boundary condition [34], is also shown in the figure. As can be seen from the figure, the Nu_z values of the different methods in this work and the correlation of Lee [30] lie above and approach the fully developed Nusselt number with the development of the hydraulic and thermal boundary layers, but at distinctly different paces and with different thermal entrance lengths, for the reasons described above.

It can be concluded from the above analysis that the "Variable property, h_zD_h/k_z" approach is the most reasonable and advantageous approach in predicting the local Nusselt numbers, though it may have a little deviation relative to known methods and the "Variable property, h_zD_h/k_{ave}" approach.

The averaged Nusselt number over the entire channel length is defined as

$$Nu_{ave} \equiv \frac{1}{L}\int_0^L Nu_z(z)dz \tag{4-8}$$

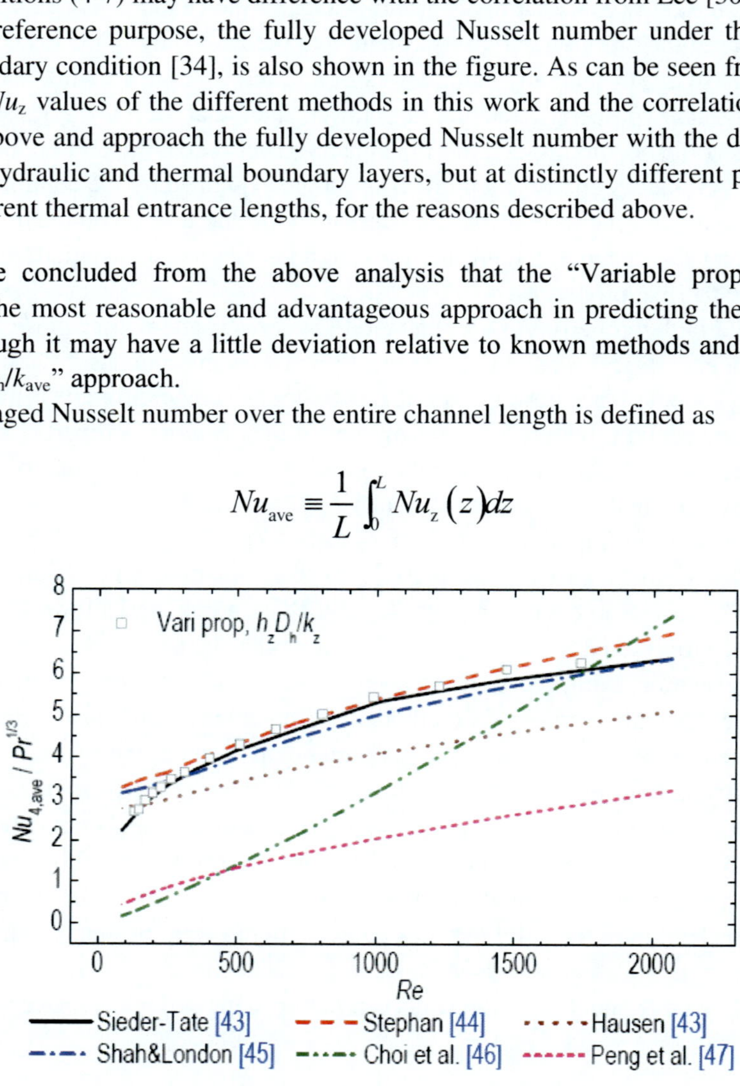

Figure 4.5. Comparison of averaged $Nu_{4,ave}/Pr^{1/3}$ versus Re between the variable property simulation and the frequently cited correlations.

Figure 4.5 shows the calculated results of $Nu_{4,ave}/Pr^{1/3}$ using the "Variable property, h_zD_h/k_z" method in comparison with the well-known correlations listed in table 2 from different sources. Among them, the Eq.s (4-9)~(4-12) were developed for conventional size channels, while the Eq.s (4-13) and (4-14) were developed specially for microscale channels.

Table 5. Frequently cited correlations for laminar convective heat transfer inside channels

Reference	Correlation		Conditions
Incropera & DeWitt [35]	$Nu = 1.86\left(RePr\,D/L\right)^{1/3}\left(\mu_f/\mu_w\right)^{0.14}$	(4-9)	Simultaneously developing, $Re<2200$
Stephan & Preuβer [36]	$Nu = 4.364 + \dfrac{0.086\left(RePr\,D/L\right)^{1.33}}{1+0.1Pr\left(Re\,D/L\right)^{0.83}}$	(4-10)	Simultaneously developing, constant wall heat flux, $0.7<Pr<7$, or $RePrD/L<33$ for $Pr>7$
Incropera & DeWitt [35]	$Nu = 3.66 + \dfrac{0.19\left(RePr\,D/L\right)^{0.8}}{1+0.117\left(RePr\,D/L\right)^{0.467}}$	(4-11)	Thermally developing laminar, constant wall temperature, $Re<2200$
Shah & London [34]	$Nu = \begin{cases} 1.953\left(RePr\dfrac{D}{L}\right)^{1/3}, & RePr\dfrac{D}{L} \geq 33.3 \\ 4.364 + 0.0722RePr\dfrac{D}{L} & RePr\dfrac{D}{L} \geq 33.3 \end{cases}$	(4-12)	Thermally developing laminar, constant wall heat flux
Choi et al. [37]	$Nu = 0.000972Re^{1.17}Pr^{1/3}$	(4-13)	For microscale channels, $Re<2000$
Peng et al. [38]	$Nu = 0.1165\left(D_h/W_c\right)^{0.81}\left(H/W\right)^{-0.79}Re^{0.62}Pr^{1/3}$	(4-14)	For microscale channels laminar flow

Relatively large deviations among them can be observed from the figure, while the calculated $Nu_{4,ave}/Pr^{1/3}$ results of the present work is closer to the Sieder-Tate correlation (Eq. (4-9)), possibly because it more efficiently accounts the influence of viscosity variation through the form of "$(\mu_f / \mu_w)^{0.14}$". As can be seen from the figure, the Stephan correlation (Eq. (4-10)) and the Shah and London correlation (Eq. (4-12)) also provide predictions generally consistent with the results of the present 3D conjugate simulation. Meanwhile, the disparities between the present simulation and the other three correlations are remarkably large. Since the influencing factors on the laminar convective heat transfer inside tubes are very complicated, including the detailed flow and thermal boundary conditions, the cross-sectional geometries and the thermal property variations, etc., there are great difficulties in fully and appropriately incorporating all the complex factors into a single correlation. However, due to the convenience in engineering application, better correlations are still in demand and are in pursuit of by many researchers.

4.3. Summary

"Inlet property", "Average property", and "Variable property" methods are respectively used in the 3D conjugate numerical simulations on the laminar water flow and convective heat transfer in rectangular microchannels. The following conclusions can be drawn from the results:

1. The "Average property" and the "Variable property" methods have significantly lower f_{app} than the "Inlet property" method. Compared with the "Average property" method, the "Variable property" method has higher f_{app} at the beginning, but lower f_{app} at a later section of the channel, owing to viscosity variations along the channel.
2. Both the "Average property" and the "Variable property" methods have notably higher h_z and Nu_z than the "Inlet property" method, due to the lower μ and the higher k levels. Compared with the "Average property" method, the "Variable property" method has a lower h_z at the beginning, but a higher h_z at the later section of the channel, due to the continuous decrease of μ_z and increase of k_z along the channel, though they share substantially equal μ_{ave} and k_{ave} levels.
3. The calculated $Nu_{4,ave}/Pr^{1/3}$ results of the present work is closer to the Sieder-Tate correlation (Eq. (4-9)). The conventional correlations of Stephan (Eq. (4-10)) and Shah and London (Eq. (4-12)) are generally consistent with the present work. The $Nu_{4,ave}$~Re patter of this work agree well with the recently published experimental data from Lee [31].
4. The above results indicate that the present 3D conjugate simulation based on conventional macroscale theory is capable of properly predicting the flow and heat transfer characteristics in microchannels for the dimension and Re ranges of the present work. Further, the "Variable property" method is superior in engineering application since it more accurately and reasonably characterizes the actual phenomenon by nature, while causing insignificant computational cost when the thermal properties are solved in an iterative way.

5. Experimental Study of a DLA Microchannel Heat Sink

In this section, a liquid-flow microchannel heat sink is introduced, which is designed and fabricated in order to overcome the problems of an existing DLA heat sink, to promote the cooling capacity, to improve the temperature uniformity of the heating surface, and simultaneously to reduce the flow resistance.

Figure 5.1 is a schematic of an existing DLA pump source and its liquid-flow heat sink. 12 DLA chips are arranged and soldered on the copper-based heat sink. 4 water passages with channel width of 1mm, fin width of 1mm and length of approx. 80mm are machined in the heat sink. The heat generation rate of the DLA in operation reaches 500W, corresponding to an interfacial heat flux of 100.6W/cm^2. The existing heat sink is unable to meet this heat removal requirement. Moreover, the long water passage may cause relatively large longitudinal temperature rise, resulting in a series of other problems.

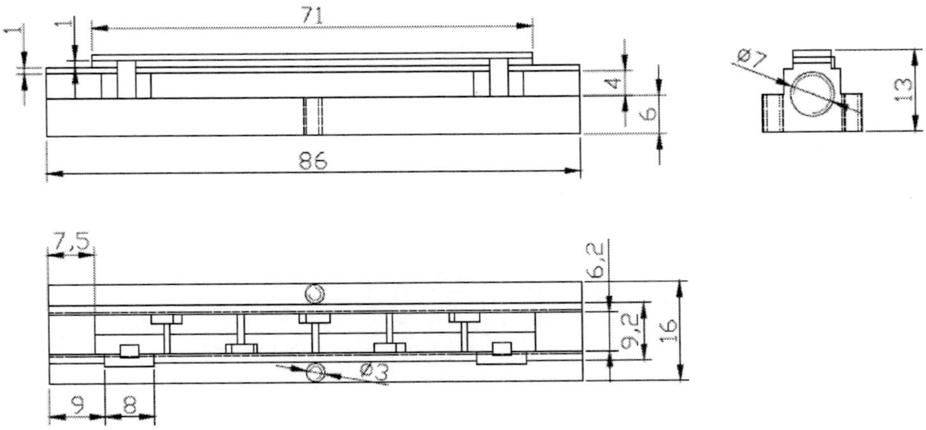

Figure 5.1. Schematic of an existing DLA pump source and its liquid-flow heat sink.

The main idea is to adopt densely arranged short microchannels to replace the originally designed long wide passages, so as to promote the cooling capacity by significantly enlarging the convective heat transfer area in the limited space and by utilizing the thermal entrance effect. On the other hand, the short channels allow lower flow resistance, higher flow rate and lower fluid temperature-rise, so as to improve the temperature uniformity of the heating surface.

5.1. Experimental Setup

Figure 5.2 is a schematic of the newly designed microchannel heat sink. It consists of a microchannel core made from copper with high thermal conductivity [$k = 398\text{W}/(\text{m}\cdot\text{K})$], an inlet manifold and an outlet manifold. The outline dimension of the core is 8mm×11mm×91mm. 188 rectangular microchannels with channel width of W=200μm, fin width of W_w = 200μm and length of L = 8mm are machined by wire cutting process. In order to provide a heat exchange area as large as possible in the limited space, we adopt a high aspect ratio of $\alpha = 10$, corresponding to a channel height of H = 2mm.

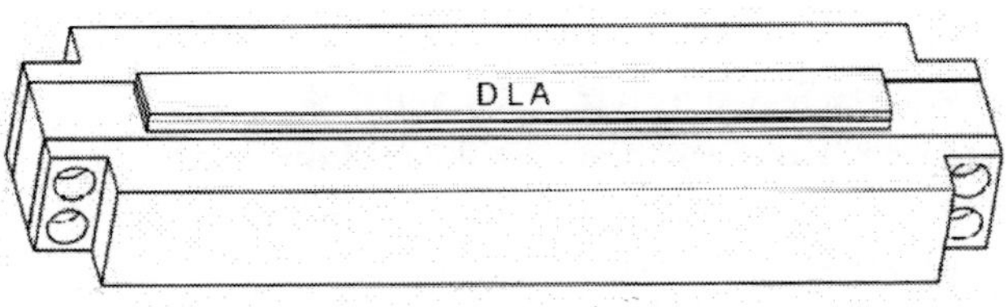

(a) Isometric view

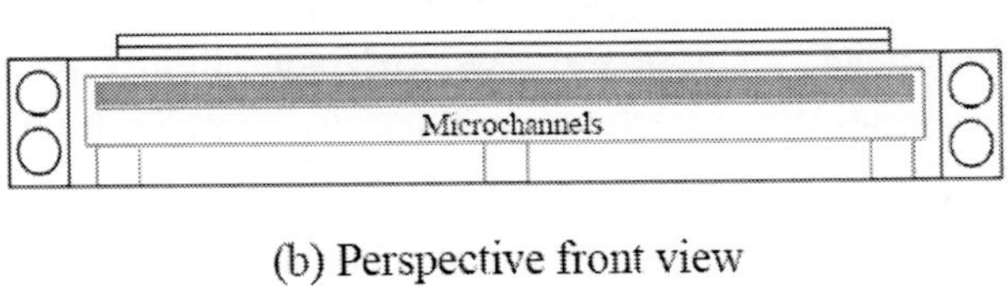

(b) Perspective front view

Figure 5.2. Schematic of the microchannel heat sink.

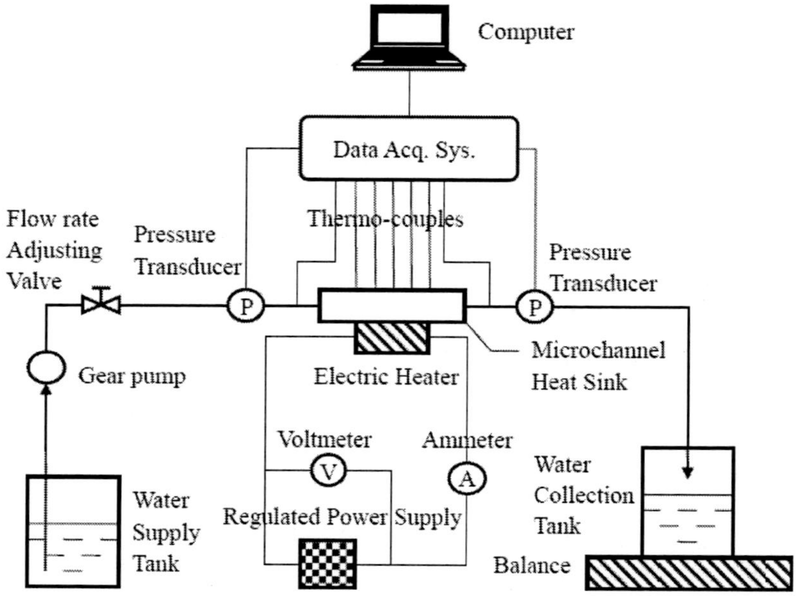

Figure 5.3. Schematic of the Experimental Installation.

Figure 5.3 is a schematic of the experimental installation, comprising a microchannel heat sink, an electric heater, a regulated power supply, a liquid supply system, a sensoring-measuring system and a data acquisition-processing system.

10 thermocouples are used to measure the temperature of the heating surface, and 3 other thermocouples are used to measure the inlet and outlet water temperature. A data acquisition system NR-1000 from KEYENCE is used to dynamically record the temperature information. a computer and a special software are used for the data processing. The flow rate is accurately measured by the traditional chronograph and balance-weighting method.

5.2. Analysis on the Experimental Results

The experiment is carried out under room temperature. The inlet water temperature is between 18℃ and 33℃, and the volume flow rate is between 22 and 70 ml/s, corresponding to Reynolds numbers of 105.7~336.5. If $z^+ = 0.05$ and $z^* = 0.05$ are respectively regarded as the criteria for judging whether the laminar flow convective heat transfer has hydrodynamically and thermally fully developed, then the working conditions in the this experiment all fall within the hydrodynamic and thermal entrance region. The heating power is 70~730W. The variation of water temperature is less than 8℃.

This experiment measures the steady state characteristics of the heat sink. It needs 10~40 minutes to reach a new steady state once the working condition is varied, depending on the specific change of condition. When the maximum temperature variation of the thermocouples is less than 0.2℃, it is considered that a steady state is reached.

Since short channels are adopted in this experiment, while the pressure gauges are installed at the inlet and outlet pipes adjacent to the heat sink, the pressure loss due to the abrupt contraction and abrupt expansion of the flow channel shall be subtracted from the measured values of pressure drop in the data reduction.

Since it is difficult to directly measure the wall temperature of the microchannels, the wall temperature values are extrapolated from the measurement values of the thermocouples according to the Fourier heat conduction law.

5.2.1. Flow Resistance Characteristics

Figure 5.4 shows the experimental results of the pressure drop in the microchannels versus the flow rate of the cooling water, which has an approximately linear relationship. The pressure drop of the microchannel portion is only 10.3kPa under the maximum flow rate of 70ml/s, exhibiting the advantages of the short channels. Figure 5.5 shows the data reduction results of the apparent flow friction coefficient f_{app} versus the Reynolds number. The result of the traditional Shah & London correlation (5-1), which is applicable to the laminar flow in ordinary scale channels whether it is fully developed or not at the outlet, is also shown in the figure for comparison. It agrees quite well with the present experimental data.

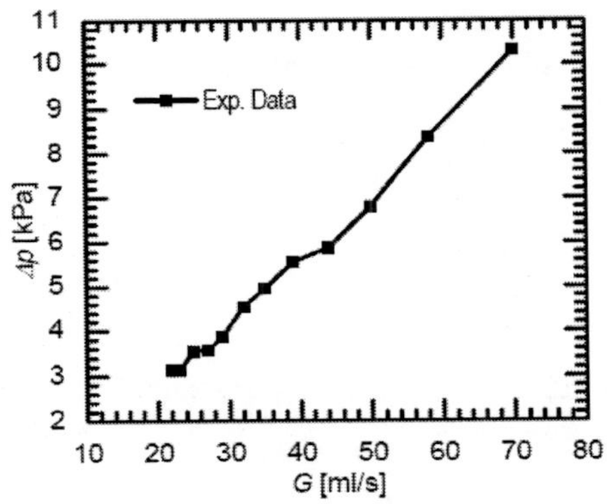

Figure 5.4. Pressure drop of the microchannel vs. the flow rate.

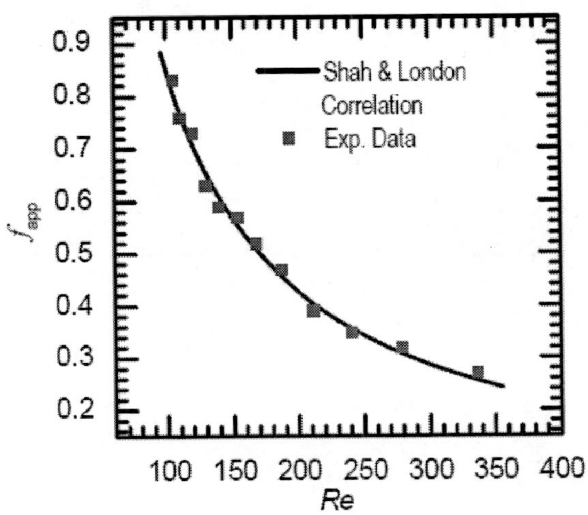

Figure 5.5. Apparent friction coefficient vs. the Reynolds number.

$$f_{\text{app}}(L)Re_{\text{ave,0-L}} = \sqrt{\left\{\frac{3.2}{\left[L/(D_{\text{h}}Re)\right]^{0.57}}\right\}^2 + (fRe)_{\text{fd}}^2} \quad \text{(5-1a)}$$

where

$$(fRe)_{\text{fd}} = 96(1 - 1.3553\alpha + 1.9467\alpha^2 - 1.7012\alpha^3 + 0.9564\alpha^4 - 0.2537\alpha^5) \quad \text{(5-1b)}$$

is applicable to the fully developed laminar flow, where

$$f \equiv 2\frac{D_{\text{h}}}{L}\frac{\Delta p}{\rho u_{\text{ave}}^2} \quad \text{(5-1c)}$$

is the Darcy flow friction coefficient.

5.2.2. Heat Transfer Characteristics

The cooling capacity of the heat sink is calculated by

$$\Phi = \rho c_p G(T_o - T_i) \quad \text{(5-2)}$$

where G is the water flow rate (m^3/s), Φ is the total heat removed by the water (W).

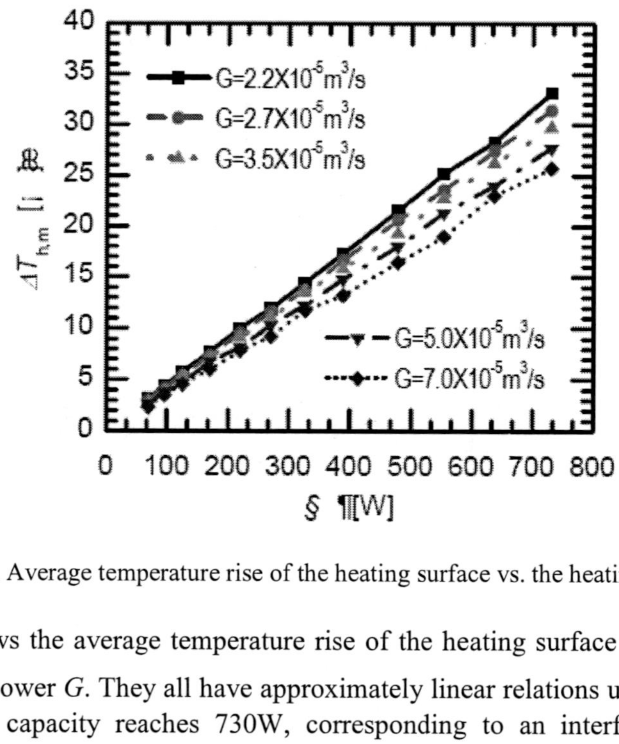

Figure 5.6. Average temperature rise of the heating surface vs. the heating power.

Figure 5.6 shows the average temperature rise of the heating surface $\Delta T_{h,m} = T_{h,m} - T_b$ versus the heating power G. They all have approximately linear relations under different flow rates. The cooling capacity reaches 730W, corresponding to an interfacial heat flux of

146.9W/cm^2 when the average temperature rise is 25.7℃ under the maximum flow rate, which is sufficient to meet the heat removal requirement.

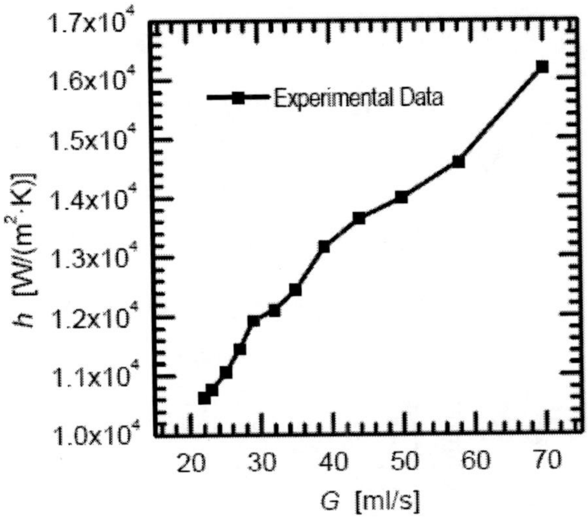

Figure 5.7. Variation of the heat transfer coefficient with the flow rate.

The convective heat transfer coefficient is calculated by

$$h = \frac{\Phi}{nS\left(T_w - T_b\right)} \tag{5-3}$$

where n is the total channel number, S is the heat exchange area of each channel, T_w is the average wall temperature of the microchannel. Figure 5.7 shows the variation of the average convective heat transfer coefficient with the flow rate of the cooling water. The relation is also approximately linear.

Accordingly, the Nusselt number are calculated

$$Nu = \frac{hD_h}{k_f} \tag{5-4}$$

where k_f is the thermal conductivity under the average water temperature T_b. Figure 5.8 shows the variation of the Nusselt number with the Reynolds number. The calculated *Nu-Re* results of the Sieder-Tate correlation (4-9) and the Shah & London correlation (4-12) are also shown in the figure for comparison, which agree well quite well with the present experimental data.

The thermal resistance of the heat sink is defined by

$$R_T = \frac{T_{h,m} - T_b}{\Phi} \tag{5-5}$$

Accordingly, the thermal resistance per unit area is $R_t = R_T S_2$. As shown in figure 5.9, the thermal resistance has an approximately linear relationship with the flow rate of the cooling water. Under the maximum flow rate of G = 70ml/s, the total thermal resistance is 0.035K/W or 0.199K·cm^2/W.

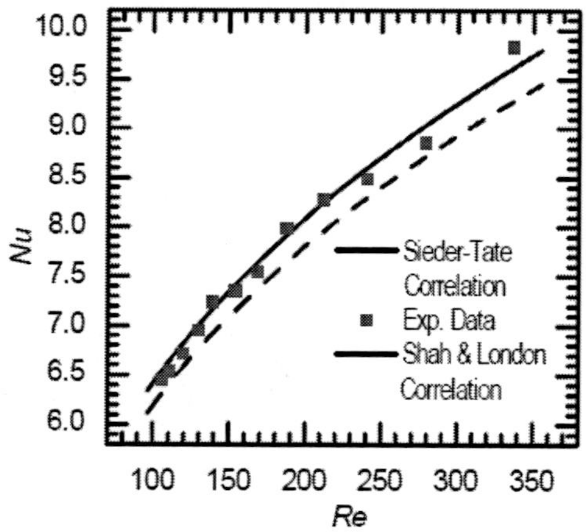

Figure 5.8. *Nu-Re* relations.

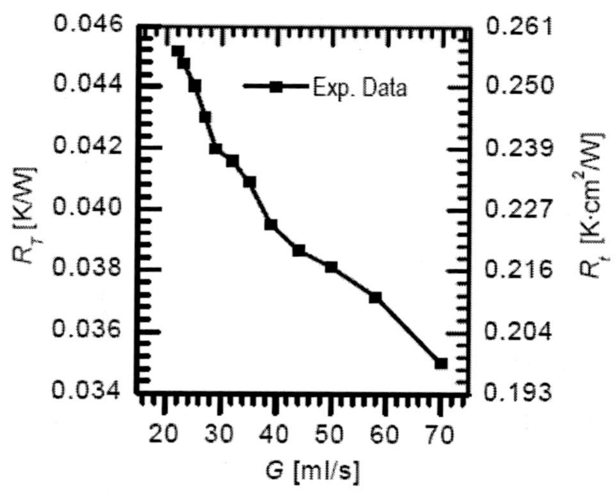

Figure 5.9. Variation of the thermal resistance of the heat sink with the flow rate of the cooling water.

5.2.3. Simplified Analysis of the Thermal Resistance Composition

In order to further enhance the heat transfer, the composition of the thermal resistance of the microchannel heat sink is analyzed as follow, using the Thermal Resistance Network method and the Fin model.

The total thermal resistance is

$$R_T = R_{cond} + R_{conv} \tag{5-6}$$

where $R_{cond} = \dfrac{\delta}{nk_{Cu}W_cL}$ is the conductive thermal resistance between the heating surface and the flow channel. $R_{conv} = \dfrac{1}{nhA_{eff}}$ is the thermal resistance of the convective heat transfer between the channel wall and the cooling water, where A_{eff} is the effective heat transfer area of each microchannel, i.e.

$$A_{eff} = \mathrm{A}_{base} + \eta A_{fin} \tag{5-7}$$

where A_{base} is the bottom area of each microchannel, i.e. the area that directly exchanges heat with the liquid without the help of the fins, A_{fin} is the fin surface area of each microchannel, and η is the fin efficiency, which is calculated by

$$\eta = \frac{\tanh(mH)}{mH} \tag{5-8}$$

where

$$m = \sqrt{\frac{2h}{k_{Cu}W_w}} \tag{5-9}$$

where $h = NuD_h/k_f$, and Nu can be calculated by the Sieder-Tate correlation (4-9), therefore R_{conv} is obtained.

Both the analysis and the experimental data indicate that the main thermal resistance of the present experiment lies in the convective thermal resistance. R_{cond} is only 0.00835K/W, whereas R_{conv} is 0.035K/W~0.045 K/W in the flow rate range of this experiment. Therefore in order to further increase the cooling capacity, higher pump power should be used to reach higher Reynolds numbers.

5.3. Uncertainty Analysis of the Experimental Data

The uncertainty of the directly measured quantities are δG±1.1%, δL±0.5%, δH±1.0%, δW±1.5%, δp±0.5%, Copper wall temperature δT_B±1.7%, Fluid temperature T_i±1.2%, T_o±1.2%。

The relative error of the indirectly measured quantities are calculated as Re±6.8%~12.6%, f_{app}±7.4%~14.9%, Nu±9.3%~16.7%。The major contribution to the f_{app} uncertainty comes from the size error of the microchannels, while the major contribution to the Nu uncertainty comes from the measurement error of the copper wall temperature and the fluid temperature at the inlet and the outlet. Especially under high flow rate conditions, the difference of water temperature between the inlet and the outlet is very small, causing relatively large error to the calculation of the total heat Φ, while the measurement error of the metal temperature significantly affects the calculation of the heat transfer coefficient h.

5.4. Summary

A liquid-flow microchannel heat sink is designed and fabricated for a pump DLA by adopting dense, narrow and short channels instead of the originally designed sparse, wide and long passages, in order to increase the cooling capacity, improve the temperature uniformity over the heating surface, while achieving a very low flow resistance, thus overcoming the problems involved in the old heat sink design. Under the maximum flow rate of 70ml/s in this experiment, the pressure drop across the channels is only 10.3kPa; when the average temperature rise of the heating surface is 25.7℃, the cooling capacity of the heat sink reaches 730W, corresponding to an interfacial heat flux of 146.9W/cm^2, while the overall thermal resistance is as low as 0.035K/W, or 0.199 K·cm^2/W. The cooling capacity is expected to be further increased if the Reynolds number is increased by using a pump with higher power. It is also found that the Shah and London correlation of apparent flow friction coefficient and the Sieder-Tate correlation of Nusselt Number agree well with the present experimental data.

Acknowledgments

This research is supported by National Natural Science Foundation of China (Grant No.50776087), the Chinese National Key Foundation Research Subject (Grant No.2006AA05Z203), and the Institute of Engineering Thermophysics, Chinese Academy of Sciences. This chapter is a derivative works of a PHD dissertation [39] of the author. Portions of this work are excerpts from originally published papers of the authors [40-46], so grateful acknowledgement is made to the respective publishers Springer, AIP, Elsevier, OSA, and the China Science Press.

References

[1] Hügel H. New solid-state lasers and their application potentials. *Opt. Lasers Eng.,* 34 (2000) 213-229

[2] Payne S A, Beach R J, Bibeau C, et al. Diode arrays, crystals, and thermal management for solid-state lasers. *IEEE J. Sel. Top. Quant.,* 1997, 3 (1): 71-81

[3] Chung Te Yuan. Thermal management, beam control, and packaging designs for high power diode laser arrays and pump cavity designs for diode laser array pumped rod shaped lasers. PhD Dissertation, University of Central Florida, 2004

[4] Belostotskii B R, Ogorodnikov N N. Temperature-induced stresses in the active isotropic medium of a laser. *Zhurnal Prikladnoi Spektroskopii,* 1967, 7 (3): 423-426

[5] Kudryashev L I, Belostotskii B R, Kudryashova N L. Temperature distribution in a triangular laser rod. *Zhurnal Prikladnoi Spektroskopii,* 1967, 6 (5): 614-619

[6] Farrukh U O, Buoncristiani A M, Byvik C E. An analysis of the temperature distribution in finite solid-state laser rods. *IEEE J. Quantum Elect.*, 1988, 24 (11): 2253-2263

[7] Hello P, Durand E, Fritschelt P K, et al. Thermal effects in Nd:YAG slabs, 3D modelling and comparison with experiments. *J. Mod. Opt.*, 1994, 41 (7): 1371-1390

[8] Usievich B A., Sychugov V A, Pigeon F, et al. Analytical treatment of the thermal problem in axially pumped solid-state lasers. *IEEE J. Quantum Elect.*, 2001, 37 (9): 1210-1214

[9] Chénais S, Balembois F, Druon F, et al. Thermal lensing in diode-pumped ytterbium lasers - part I: theoretical analysis and wavefront measurements. *IEEE J. Quantum Elect.*, 2004, 40 (9): 1217-1234

[10] Parfenov V, Shashkin V, Stepanov E. Numerical investigation of thermally induced birefringence in optical elements of solid-state lasers. *Appl. Opt.*, 1993, 32 (27): 5243-5255

[11] Xie Wen Jie, Kwon Y, Hu Wen Tao, et al. Thermal modeling of solid state lasers with super-Gaussian pumping profiles. *Opt. Eng.*, 2003, 42 (6): 1787-1794

[12] Peng Xiaoyuan, Xu Lei, Anand Asundi. Thermal lensing effects for diode-end-pumped Nd:YVO_4 and Nd:YAG lasers. *Opt. Eng.*, 2004, 43 (10) 2454-2461

[13] Chen Y, Chen B, BASS M, et al. Calculation of three-dimensional thermal-gradient-induced stress birefringence in slab lasers. *Appl. Phys. B* - Lasers O., 2005, 81 (1): 75-82

[14] Ostermeyer M, Mudge D, Veitch P J, et al. Thermally induced birefringence in Nd:YAG slab lasers. *Appl. Opt.*, 2006, 45 (21): 5368-5376

[15] Lundberg E, McCuen P A, Reynolds W C. Heat transfer in annular passages: hydrodynamically developed laminar flow with arbitrarily prescribed wall temperatures or heat fluxes. *Int. J. Heat Mass Transfer.*, 1963, 6 (6): 495-529

[16] Shah R K, London A L. Laminar flow forced convection in ducts, *Advances in heat transfer*. New York: Academic Press, 1978

[17] Kays W M, Crawford M E. *Convective heat and mass transfer*. 2nd Ed. New York: McGraw Hill, 1980

[18] Lundberg R E, Reynolds W C, Kays W M. *Heat transfer with laminar flow in concentric annuli with constant and variable wall temperature with heat flux*. Washington: NASA technical note; NASA TND-1972. 1963

[19] Rohsenow W M, Hartnett J P, Cho Y I. *Handbook of heat transfer*. New York: McGraw-Hill, 1998

[20] Yang, S. and Tao, W., 1998, *Heat Transfer*, Beijing: Higher Education Press, 162-172

[21] Takeuchi, Y., 1977, *Thermal Stress*, Science Press, Beijing.

[22] Xiong Z, Zhigang G Li, Moore N, et al. Detailed investigation of thermal effects in longitudinally diode-pumped Nd:YVO_4 lasers. *IEEE J. Quantum Elect.*, 2003, 39 (8): 979-986

[23] Lindberg H, Strassner M, Gerster E, et al. Thermal management of optically pumped long-wavelength InP-based semiconductor disk lasers. *IEEE J. Sel. Top. Quant.*, 2005, 11 (5): 1126-1134

[24] Özışik M N. *Heat conduction*. New York: John Wiley and Sons, 1980. 611-650

[25] Frauchiger J, Albers P, Weber H P. Modeling of thermal lensing and higher order ring mode oscillation in end-pumped CW Nd lasers. *IEEE J. Quantum Elect.*, 1992, 28 (4): 1046-1056

[26] Cousins A K. Temperature and thermal stress scaling in finite-length end-pumped laser rods. *IEEE J. Quantum Elect.,* 1992, 28 (4): 1057-1069

[27] J. Wang. *Thermal analysis and thermal control of high-power pumped laser medium.* PhD Dissertation (Tsinghua Univ., Beijing, 2005)

[28] F.M. White, *Viscous Fluid Flow.* 2nd ed., McGraw-Hill Publishers, 1991.

[29] Curr R M, Sharma D, Tatchell D G. Numerical predictions of some three dimensional boundary layers in ducts. *Comput. Meth. Appl. Mech. Eng.,* 1972, 1: 143-158.

[30] Lee P S, Garimella S V. Thermally developing flow and heat transfer in rectangular microchannels of different aspect ratios. *Int. J. Heat Mass Transfer*, 2006, 49 (17-18): 3060-3067

[31] Lee P S, Garimella S V, Liu D. Investigation of heat transfer in rectangular microchannels. *J. Int. Heat Mass Transfer*, 2005, 48 (9): 1688-1704

[32] Maranzana G, Perry I, Maillet D. Mini- and micro-channels: influence of axial conduction in the walls. *Int. J. Heat and Mass Transfer,* 2004, 47 (17-18): 3993-4004

[33] Morini G L. Scaling effects for liquid flows in micro-channels. *Heat Transfer Eng.*, 2006, 27 (4): 64-73

[34] Shah R K, London A L. Laminar flow forced convection in ducts, *Adv. Heat Transfer, Suppl.* I. New York: Academic Press, 1978

[35] Incropera F P, DeWitt D P. *Fundamentals of heat and mass transfer*. New York: John Wiley and Sons, 1996

[36] Stephan K, Preuβer P. Wärmeübergang und maximale wärmestromdichte beim behältersieden binärer und ternä-rer flüssigkeitsgemische. *Chem. Ing. Tech.,* 1979, 51: 37

[37] Choi S B, Barron R F, Warrington R O. Fluid flow and heat transfer in microtubes. In: Proc. Micromech. *Sensors Actuat. Syst. ASME DSC* .32, 1991. 123-134

[38] Peng X F. Peterson G P. Convective heat transfer and flow friction for water flow in microchannel structures. *Int. J. Heat Mass Transfer,* 1996, 39 (12): 2599-2608

[39] Li Z G. Study on the heat transfer and the thermal stress in high-power solid-state lasers. PhD dissertation. Graduate School of Chinese Academy of Sciences. 2008

[40] Li Z G, Huai X L, Tao Y J. A study of the heat transfer characteristics of a laser rod in a concentric circular tube. *Appl. Phys. B - Lasers O.*, 2007, 87 (2): 301-309

[41] Huai X L, Li Z G. Thermal Stress Analysis of a Nd:YVO4 Laser Medium End-pumped by a Gaussian Beam. *Appl. Phys. Lett.,* 2008, 92 (4): 041121

[42] Li Z G, Huai X L, Tao Y J, Chen H Z. Effects of thermal property variations on the liquid flow and heat transfer in microchannel heat sinks. A*ppl. Therm. Eng.,* 2007, 27 (17-18): 2803-2814

[43] Li Z G, Huai X L, Wang L, Tao Y J. Influence of longitudinal rise of coolant temperature on the thermal strain in a cylindrical laser rod. *Opt. Lett.*, 2009, 34 (2): 187-189

[44] Li Z G, Huai X L, Tao Y J, Guo Z Y. Analysis of thermal effects in an orthotropic laser medium. *Appl. Opt.*, 2009, 48 (3): 598-608

[45] Li Z G, Huai X L, Tao Y J. A study of the axisymmetric thermal strain in a laser rod with longitudinal temperature rise. *Appl. Therm. Eng.*, 2009, 29 (14-15): 2927-2934

[46] Li Z G, Huai X L, Tao Y J, Wang L. An experimental study on the microchannel cooling of diode laser array. *J. Eng. Thermophysics*, 2009, 30 (5): 817-820

In: Thermal Engineering Research Developments
Editors: J. Evgova and O. Kostadinov, pp. 205-243
ISBN: 978-1-60741-497-1

Chapter 4

THERMAL AND FLUID DYNAMICS IN JET FLOW

Fabio Gori* and Ivano Petracci
Department of Mechanical Engineering,
University of Rome "Tor Vergata"
Via del Politecnico 1, 00133 Rome, Italy

Abstract

The fluid dynamics of a free submerged jet is reviewed. Different conclusions are found in the literature for what concerns the lengths of the potential core region and the line source of the jet. The submerged slot jet of air is the research subject of the present authors to investigate its evolution with the distance from the slot exit. Hot anemometry is used to measure velocity and turbulence distributions on the exit as well as at several distances from the slot exit. The shadowgraph visualization technique is also employed to confirm the hot anemometry measurements.

The existence of an undisturbed region of flow, where the fluid dynamics characteristics are equal to those on the exit, was measured and visualized in a preliminary experiment with a small wind tunnel, named first, at the turbulent Reynolds number, based on the hydraulic diameter, equal to about Re=10,600-12,100. With a slot of height H=4.25 mm the dimensionless length of the undisturbed region, L_u, is equal to about $L_u/H=1.6$-2. The length of the undisturbed region decreases with the increase of the Reynolds number from the laminar to the turbulent regime. Further measurements of velocity and turbulence, carried on in the Reynolds number range from Re=13,500 to Re=22,100, confirmed that the dimensionless length of the undisturbed region span in the range $L_u/H=1.6$–2. The length of the undisturbed region of flow was visualized with the shadowgraph technique in the Reynolds range from Re=1,700 to Re=22,100. The length of the undisturbed region is $L_u/H>3$ for the laminar Reynolds number of Re=1,700 and is $L_u/H=1$ for the turbulent Reynolds number of Re=22,100.

The velocity and turbulence measurements were performed also in a different small wind tunnel, named second, with a slot height of 10 mm, for the range from Re=7,600 to Re=37,300. At Re=7,600 the undisturbed length is $L_u/H=3$–4, at Re=23,300 is $L_u/H=1$–2 and is zero at Re=37,300. The flow visualizations with the shadowgraph technique were performed in the same wind tunnel and slot, for the range from Re=7,100 to Re=33,400, confirming the dimensionless length of the undisturbed region of flow is $L_u/H=3.8$ for Re=7,100, and $L_u/H=0$ for Re =33,400. The flow visualizations were performed also in the

* E-mail address: fammannati@yahoo.com. (Corresponding author)

first wind tunnel, with a slot of height H=17 mm. The length of the undisturbed region is $L_u/H=2.35$ for Re=8,400, $L_u/H=1$ for Re=31,200 and $L_u/H=0$ for Re=52,900.

The introduction of a metallic grid inside the wind tunnel, between the converging duct and the slot, and on the slot exit, produces a shorter length of the undisturbed region, due to the higher turbulence. Hot anemometry was employed to measure velocity and turbulence in a free jet up to an impinged cylinder, set at two distances from the slot exit, equal to L/H=4 and 6, for the Reynolds number of Re=23,300. The undisturbed region is equal to $L_u/H=1$ when the cylinder is set at L/H=4 and $L_u/H=2$ when the cylinder is set at L/H=6.

The experimental results found by the present authors for the length of the undisturbed region of flow, obtained in turbulent as well as in laminar regime with hot anemometer and shadowgraph visualizations, are collected together with previous results of the literature, obtained with shadowgraph, Schlieren and PIV visualization in laminar regime, with a fairly good agreement.

The experimental heat transfer, or Nusselt number, measured by the present authors are reviewed in the present paper at the light of the evolution of a submerged jet, i.e. in order to study the influence of the length of undisturbed and potential core regions.

Preliminary experiments of heat transfer with the first small wind tunnel without the three metallic screens, i.e. at relatively high level of turbulence, were carried out to cool a cylinder of diameter D=10 mm with a slot of height H=5 mm. The experiments showed a qualitative increase of the average and local Nusselt number at Re=17,000. Heat transfer experiments were performed with the same wind tunnel, slot and circular cylinder for Reynolds number ranging from Re=4,000 to Re=20,000. The influence of the distance between the exit of the slot and the cylinder, X, was studied in the range of X/H=2–10. The local Nusselt number is a function of X/H, the angle from the impinging point and the Reynolds number.

Local heat transfer was investigated with the second wind tunnel at reduced turbulence and slot jets of different heights. The slot jet with the highest ratio between cylinder diameter and slot height, i.e., D/H, realizes the highest mean Nusselt number at the same Reynolds number. The distance from the slot exit where is present the highest mean Nusselt number is X/H=8 for D/H=4. The similar distance observed in a slot jet with D/H=2 is X/H=6.

Local heat transfer along the cylinder perimeter was measured in an upgraded version of the second wind tunnel, with lower turbulence level, for a large range of Reynolds numbers, i.e. from Re=7,400 to Re=37,300. Local Nusselt number, measured at different distances from the slot exit along the perimeter of the cylinder, presented interesting features from the point of view of the presence of the undisturbed region.

A. Fluid Dynamics of a Submerged Jet

A.1. Introduction

The evolution of a two-dimensional submerged free jet can be summarized by figure 1, according to [1] and the current literature. In the potential core region the velocity of the fluid has the same value as on the slot exit. In the mixing fluid region the jet interacts with the stagnant fluid and the velocity decreases transversally to the value of the stagnant fluid, which is zero in a submerged jet.

The centerline velocity of an air jet from a two-dimensional slot, consisting of a rectangular aperture without a converging duct, was measured versus the distance from the slot exit [2]. The jet was reported to interact with the stagnant fluid immediately after the slot exit. The centerline velocity, along the flow development, remained at the value equal to the velocity on the slot exit, independently of the velocity and the Reynolds number, for a dimensionless distance equal to X/H=5.2, where X is the distance from the slot exit and H the slot height, as in figure 1. Consequently, this distance was considered as the length of the potential core, L_P, i.e., the region where the fluid maintains the same maximum velocity on the centerline of the jet. The wideness of the jet, where the velocity is the same as on the slot exit, decreases with the distance X because of the shear interaction with the stagnant fluid.

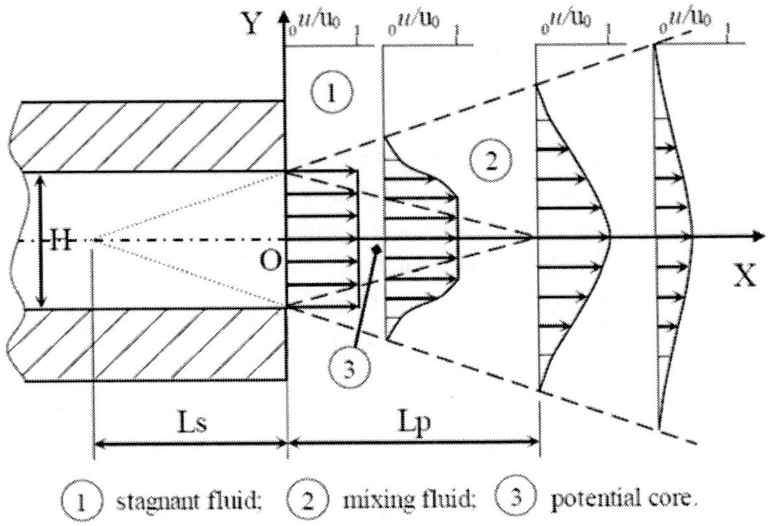

Figure 1. Evolution of a submerged free jet from the literature [1].

Measurements of the velocity distribution in a plane turbulent jet of air were carried on in [3] at Re=13,300. From the ratio of the velocities between the centerline and the slot exit, the concept of the hypothetical origin of the jet was introduced, and the distance of the origin upstream the slot exit is indicated by L_S in figure 1. The distance L_S is positive if the origin source is downstream the slot exit and negative if is upstream, as in figure 1. The experiments were done with two types of nozzle, with a form similar to that used in [2]. In the slot of height H=5 mm the origin of the jet was found upstream the slot exit at a distance $L_S/H=-0.6$, while in the slot of height H=10 mm, $L_S/H=-1.7$. Besides the fact that the Reynolds number, based on the hydraulic diameter of the slot, was the same, Re=13,300, the different distances of the jet origin in the two slots were not explained, with the conclusion that a larger jet has a larger spread than a smaller one. The end of the potential core was reported at different distances for the two slots of the experiments. Namely, it was $L_P/H=7$ for the smaller slot and $L_P/H=4.33$ for the larger one, according to the definition of the end of the potential core, as the position where the ratio of the maximum (U_M) velocity on the centerline to the exit velocity (U_E), was equal to $U_M/U_E = 0.9$.

Hot wire measurements in a plane incompressible jet from a converging duct were done in [4] to investigate the structure of an air jet. The jet was investigated with the downstream distance up to X/H=120. The jet was found to spread linearly with the distance from the slot exit. The line source of the jet was located upstream the slot exit, at $L_S/H=-2$, for Re=30,000, but the end of the potential core was not stated explicitly.

The development and the structure of a jet, obtained from a rectangular slot followed by a long rectangular channel (40 mm of length and 3 mm of height), was studied experimentally in [5] for Re=12,000. The velocity measurements showed a length of the potential core equal to $L_P/H=3\text{-}4$, the two dimensional jet-type region extends up to L/H=60 and the asymmetric region is beyond L/H=60. Schlieren visualizations of the jet were carried on but no conclusions were done about the structure of the jet.

The deformation of a three-dimensional rectangular jet was studied in [6] where it was shown that the short side of the jet increases with the distance from the exit while the larger side shrinks, changing the respective dimensions at some distances downstream the jet. The origin of the jet was upstream the slot exit at $L_S/H=-4.5$. The end of the potential core was investigated for nozzle and rectangular hole type jets. Conclusions were quantitatively different for nozzle and rectangular hole jets, with the trend of the nozzle to provide smaller distances than the rectangular hole jet. Different values for the end of the potential core were found for different nozzle and hole (orifice), but no number was reported.

The review of turbulent mixing and diffusion of air or water jets, [7], mentioned that in a nozzle jet the end of the potential core was found at $L_P/H=12$.

The disagreement in the lengths of potential core and origin of rectangular jets, reported in the literature, promoted the research of the present authors, where velocity and turbulence measurements with hot anemometry, and visualizations with shadowgraph technique were carried on in a jet flow of air emerging from a rectangular channel after a wind tunnel.

Preliminary velocity and turbulence measurements with hot wire anemometry and shadowgraph visualization were done in an air jet following a small wind tunnel, called Wind Tunnel 1 (WT1), with a rectangular slot of height H=4.25 mm [8]. The main conclusion was the discovery that a turbulent jet at low turbulence has an undisturbed region of flow, just outside the slot exit, where velocity and turbulence remain equal to those measured on the exit. The experiments, carried out at the Reynolds number, evaluated with the hydraulic diameter of the slot, equal to Re=12,100 in the hot wire anemometry and Re=10,600 in the shadowgraph visualization, showed a length of the undisturbed region equal to $L_U/H=1.6$-2.

Measurements of velocity and turbulence intensity were carried out also with a slightly different small wind tunnel, called Wind Tunnel 2 (WT2), with a rectangular slot of height H=10 mm, [9]. The experiments confirmed the presence of the undisturbed region of flow and its dependence on the Reynolds number. For Re=7,600 the length of the undisturbed region is in the range $L_U/H=3$–4, for Re=23,300 decreases to $L_u/H=1$–2 and disappears for Re=37,300. The measurements allowed to determine also the length of the potential core region of the jet, L_P/H, which includes also the undisturbed one. The length of the potential core region is $L_P/H=8$-9 for Re=7,600, $L_P/H=5$-6 for Re=23,300 and $L_P/H=4$-5 for Re=37,300. In conclusion, the difference between the length of the potential core and the undisturbed region is around L/H=4-5.

Further fluid dynamics measurements, carried on with the wind tunnel 1, WT1, and the rectangular slot of height H=4.25 mm, [10], showed an undisturbed length equal to $L_U/H=2$ for Re=13,500 and $L_U/H=1.6$ for Re=22,100.

Flow visualizations of a submerged free slot of air were carried on with the shadowgraph technique using the two wind tunnels WT1 and WT2, [10-11]. The flow visualizations with a rectangular slot of height H=4.25 mm and the wind tunnel WT1 showed that the length of the undisturbed region decreases with the increase of the Reynolds number from $L_u/H>3$ for Re=1,700, to $L_u/H=3.3$ for Re=3,900, to $L_u/H=2.2$ for Re=6,300, to $L_u/H=1.3$ for Re=8,400, to $L_u/H=1.2$ for Re=13,500, to $L_u/H=1$ for Re=17,300 and Re=22,100 [10]. Flow visualizations were carried out with a rectangular slot of height H=17 mm in the wind tunnel WT1 for the range Re=8,400-52,900 and with a rectangular slot of height H=10 mm in the wind tunnel WT2 for the range Re=7,100-33,400 [11]. The length of the undisturbed region decreases from L_U/H=2.35 for Re=8,400, to L_U/H=2.1 for Re=15,000, to L_U/H=1.3 for Re=20,000, to L_U/H=1 for Re=31,200, to L_U/H=0 for Re=52,900 in the H=17 mm slot and from L_U/H=3.8 for Re=7,100, to L_U/H=3.4 for Re=8,600, to L_U/H=2.2 for Re=12,200, to L_U/H=1.6 for Re=20,600, to L_U/H=0 for Re=33,400 in the H=10 mm slot [11].

The presence of the undisturbed region of flow has never been mentioned in the literature but the flow visualizations reported in [12] can be observed at the light of the new discovery. The experiment with a D=¼ inch round jet of carbon dioxide, flowing at 127 ft/s into air, Re=30,000, showed a flow region of length L_U/D=1, just outside the exit, which can be considered similar to the undisturbed region of flow. In the flow outside a round jet, of diameter D at Re=13,000, is visible a similar region of length L_U/D=0.375. In the experiment of a water jet of diameter D, impinging a plate at Re=4,000, the undisturbed region has a length of L_U/D=1. The experiment with a subsonic air jet, flowing into ambient air at 12 m/s from a nozzle of diameter D=5 cm, presented a very small undisturbed region of length L_U/D=0.16. The experiment of a supersonic axis-symmetric jet of helium into air, from a converging nozzle of diameter D=0.26 cm, displayed an undisturbed region of length L_U/D=1. The experiments reported in [12] are mentioned as a source of information on the state of art of the literature, but they are not comparable each to the other, neither to those performed by the authors of the present paper, because no detail is given on the turbulence levels of the jet flow.

The visualizations reported in [13] were performed with Schlieren and shadowgraph techniques. The experiments are relative to helium jet entering into air at low Reynolds numbers, i.e. in laminar flow. The lengths of the undisturbed region, measured by the present authors from the graphs of [13], are equal to L_U/D=11.7 for Re=80, L_U/D=9 for Re=200, L_U/D=3 for Re=470 and L_U/D=2.3 for Re=630 in the shadowgraph pictures and equal to L_U/D=15.7 for Re=80, L_U/D=11 for Re=200, L_U/D=4 for Re=470 and L_U/D=2.7 for Re=630 in the Schlieren pictures.

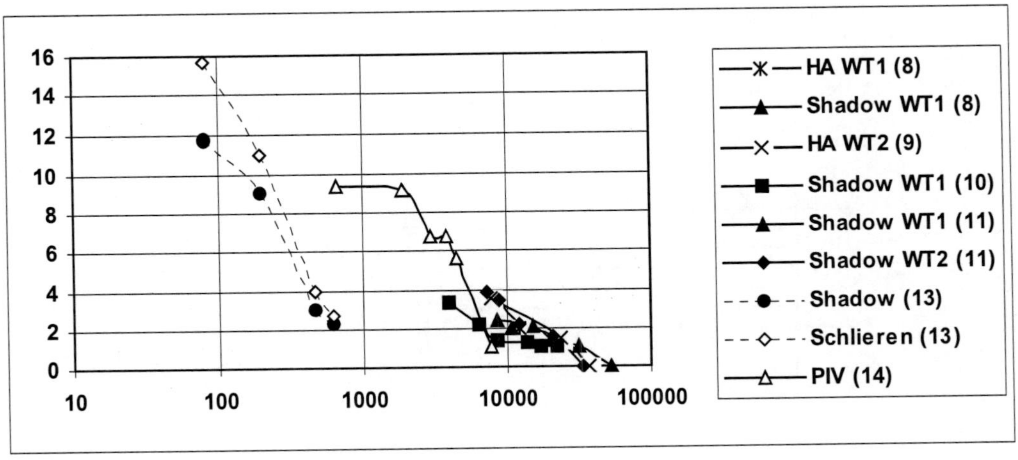

Figure 2. Length of the undisturbed region, L_U/H, versus Reynolds number, Re [11].

Laminar jets were investigated with the Particle Image Velocimetry (PIV) technique [14]. The PIV visualizations of the wall jet were compared to the free jet, both covering the region from the slot exit to about X/H=16 for Reynolds numbers, based on H, Re(H) between 300 and 3,800. The PIV visualizations of the free jet show the undisturbed region exists in the laminar flow and its length decreases with the increase of the Reynolds number. The undisturbed region disappears once the Reynolds number increases above Re(H)=3,800.

The results of the mentioned experiments of Hot Anemometry, Shadow, Schlieren and PIV visualizations are reported in figure 2 where the undisturbed length, L_U/H, is presented versus the Reynolds number, calculated with the hydraulic diameter or 2H in case of the experiments [14]. The experiments of [12] are not presented in figure 3 for clarity reasons but they are in agreement with the data presented. The important conclusion of figure 2 is the qualitative agreement of the PIV visualizations of [14] and the experiments/shadowgraphs visualizations of [8-11]. It is also possible to see a reasonable agreement between the PIV visualizations of [14] and the shadowgraphs/Schlieren visualizations of [13] at very low Reynolds numbers.

The conclusion of the present introductory review on the state of art of the literature gives a different picture on the evolution of a submerged free jet which can be now represented by figure 3, where the symbols are those used previously.

The potential core region, L_P/H, includes also the undisturbed region, L_U/H, as the region where the velocity on the centreline remains equal to the value on the slot exit. The region of flow where the jet starts to interact with the stagnant fluid, called mixing flow region, has a length which is the difference between the potential core and the undisturbed region, $L_M=L_P-L_U$.

Hot anemometry was employed to measure velocity and turbulence in a rectangular jet flow up to the impinged cylinder with the wind tunnel WT2, [15]. The cylinder was set at two distances from the slot exit, respectively equal to L/H=4 and 6, and the Reynolds number was equal to 23,300. The conclusion was that the presence of the impinged cylinder reduces the length of the undisturbed region. If the cylinder is set at a distance equal to L/H=6 the length of the undisturbed region is $L_U/H=2$, while the length of the undisturbed region is $L_U/H=1$ if the cylinder is set at L/H=4. The length of the potential core region is $L_P/H=5$ when the cylinder is set at L/H=6 and $L_P/H=3$ when the cylinder is set at L/H=4.

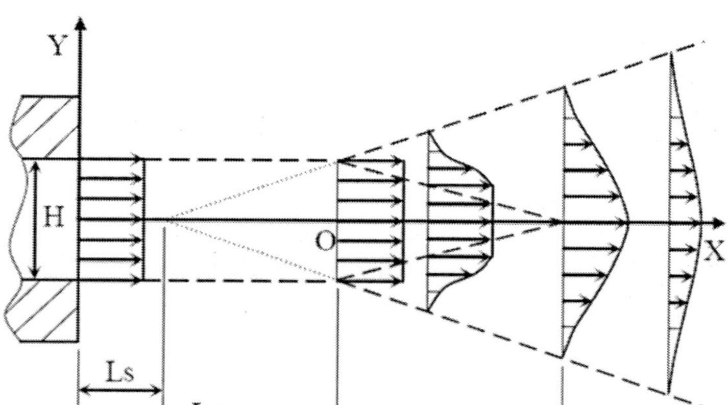

Figure 3. Evolution of a submerged jet in presence of the undisturbed region [11].

A.2. Experimental Apparata

A.2.1. Wind Tunnel 1

The first experimental set-up generating the jet flow was used in [8, 10-11] and is referred as wind tunnel 1, WT1,figure 4.

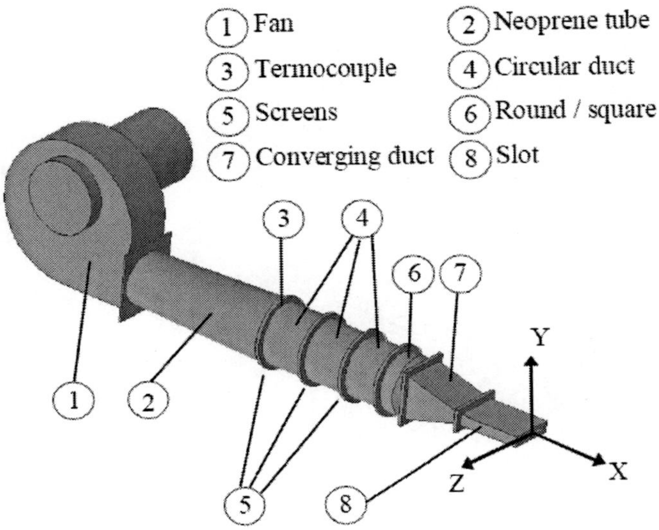

Figure 4. Wind tunnel 1 (WT1) [11].

The fan, with a power of 750 W, can move a maximum mass flow rate of 0.45 kg/s. The inlet duct of the fan has a circular cross section with diameter 160 mm and the exit duct has a square section with side 135 mm. The fan is set on a different base than the rest of the wind tunnel in order to decrease vibrations. After the fan is present a settling chamber made of a non-metallic (neoprene) tube, in order to reduce vibrations, where the cross section is changed from rectangular (at the outlet of the fan) to circular, with diameter 160 mm. After the settling chamber, an ensemble of honeycomb is located before three metallic screens. The honeycomb ensemble consists of several small-size ducts, with circular section that fill the entire cross sectional area of the duct, to minimize the transversal flow, while the screens reduce the large-scale turbulence generated by the fan, decreasing the eddies size to values comparable to the square mesh. After the last screen, the cross section of the wind tunnel is changed into a square one with side 106 mm. A duct converts the square cross section to a rectangular one, with sides 106 mm by 17 mm. Rectangular slots with height H=17 mm or 4.25 mm and length 300 mm follow the converging nozzle.

A.2.2. Wind Tunnel 2

The second experimental set-up generating the jet flow, used in [9, 11, 15] and referred as wind tunnel 2, WT2, is presented in figure 5.

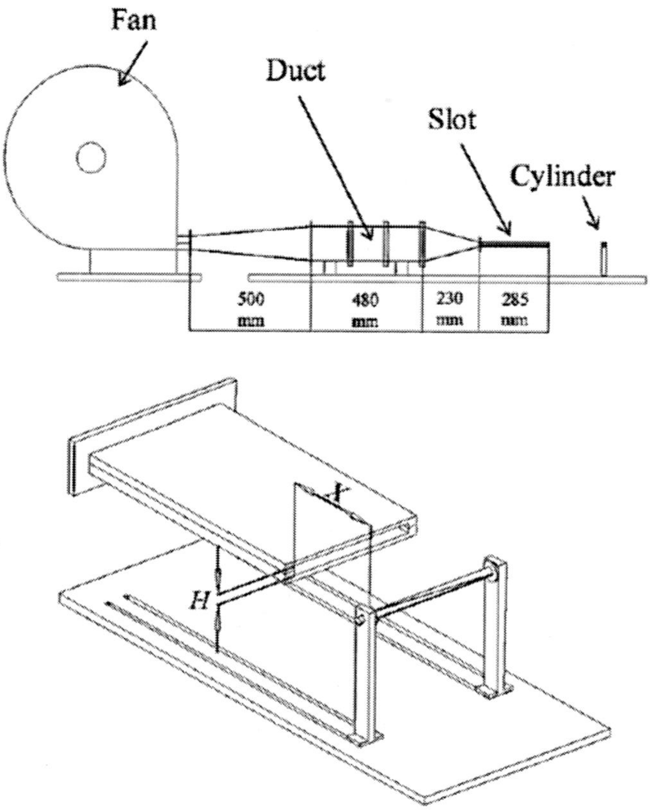

Figure 5. Top: Wind tunnel 2 (WT2). Bottom: detail of slot and cylinder [15].

The radial fan has an electric power of 1.45 kW and a maximum volume flow rate 0.15 m^3/s. A smooth diverging duct is located after the fan and before the settling chamber which is made with a rectangular channel, long 480 mm, and has a square section of 135 by 135 mm. After the settling chamber an ensemble of honeycomb is located before the three metallic screens. The converging duct following the last screen has a contraction ratio of 13.5:1, an angle of 30° and a rectangular final section of 135 mm by 10 mm. The slot jet is made with a horizontal channel, 135 mm wide, 285 mm long and H = 10 mm high.

A.2.3. Hot Anemometry

The hot film anemometer used is a CTA, constant temperature anemometer, made by TSI, mod. 1201, with a single platinum wire, 152.4 mm thick and 2.03 mm long. The anemometer is mounted, with an appropriate angular support, on a separate base, in order to minimize the vibrations. The base of the anemometer is on a micrometric sledge controlled by the personal computer to move the anemometer of small distances. The anemometer can be moved in X and Y direction, with steps of about 0.5 mm, in order to measure velocity and turbulence along stream-wise (X) and transverse directions (Y). The anemometer voltage signal output is collected by a high speed HP voltmeter. The output signal is filtered at 14 kHz. The mass flow rate of air is measured, through the pressure loss, across the converging duct. The data acquisition system is made with a card DAC 3852 of HP, installed on the personal computer. The uncertainty estimation has been carried out following the procedure suggested in [16]. The optimum sampling rate, the number of independent samples, and the total sampling time are chosen in order to obtain an accuracy for the measured mean velocity of 1%, with a 98% confidence level (probability).

A.3. Velocity and Turbulence Measurements in a Free Jet

A.3.1. Wind Tunnel 1

The structure and the evolution of the jet were investigated with the hot anemometer to measure velocity and turbulence in each point of the flow [8]. Measurements were carried out at Reynolds number 12,100 with a rectangular slot of height H=4.25 mm. The velocity measurements are reported in figure 6 versus the transverse coordinate (Y), at several distances from the slot exit, X/H, reported on the legend of figure 6.

The velocity measurements are taken stepwise at small X/H from the slot exit. The velocity is almost flat in the main core of the jet and zero outside for about 2 times the slot height or, in other words, the jet penetrates into the stagnant air without interacting with it for X/H=2. From X/H=2, the jet starts to interact with the stagnant air and the velocity distribution is less sharp near the zero value. Inside the jet the velocity distribution is less flat. For higher values of X/H the velocity distribution has a maximum on the centerline and decreases with the increase of the distance X/H

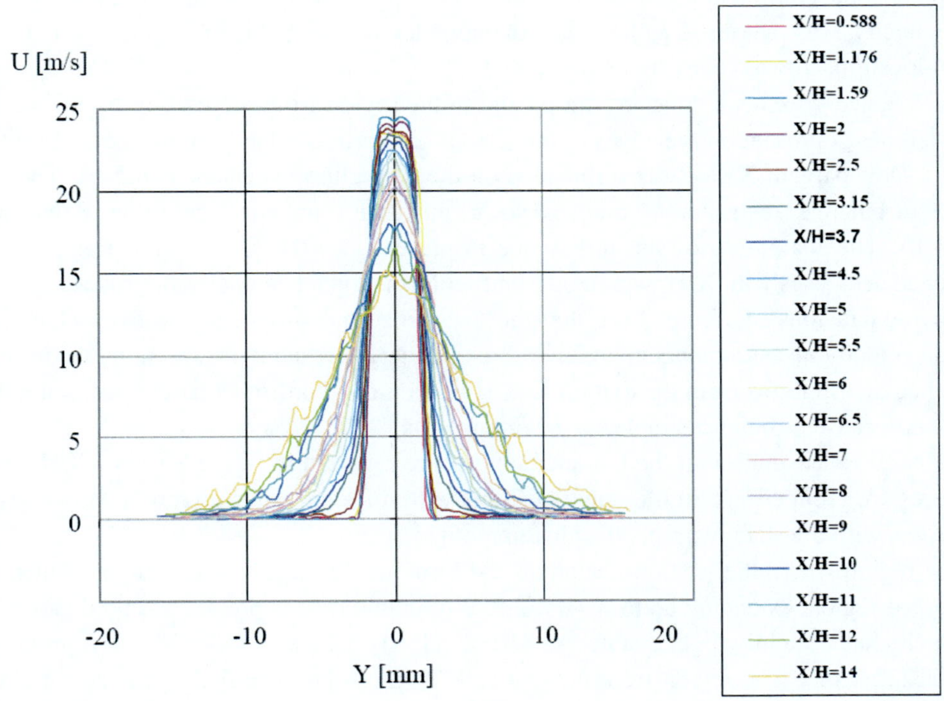

Figure 6. Velocity across the jet flow for Re=12,100 [8].

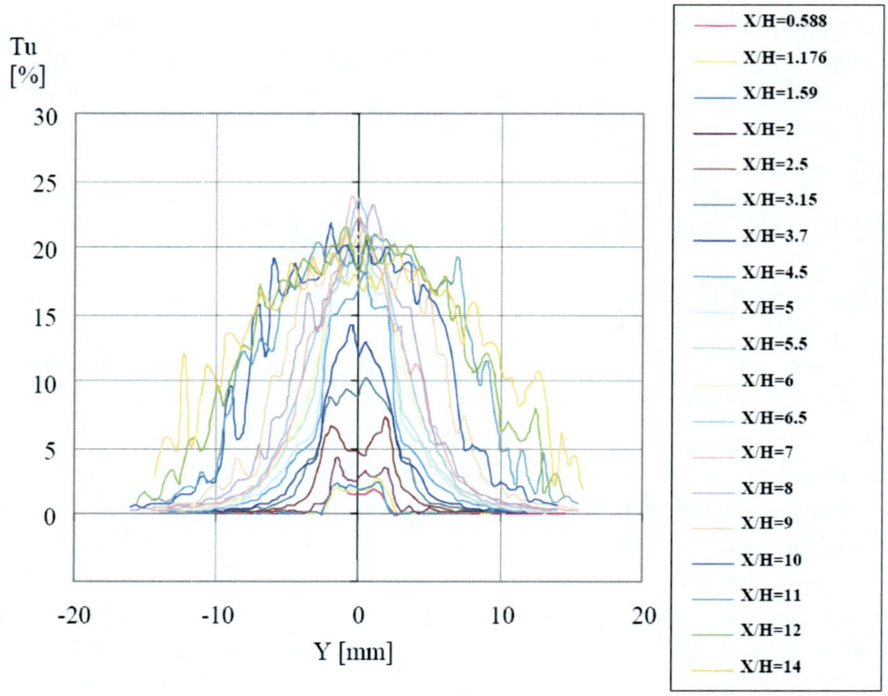

Figure 7. Absolute turbulence across the jet flow for Re=12,100 [8].

The absolute turbulence, $T_U=u'/U_E$, where u′ is the velocity fluctuation and U_E the exit mean velocity, was measured with the hot anemometer and is reported in figure 7 versus Y, at several locations downstream the jet flow, X/H.

Turbulence presents a constant flat profile in the region from the exit up to X/H=2 with two small peaks present on the edges of the jet. The centerline value is almost equal to that on the exit. Downstream X/H=2 the turbulence on the centerline increases with X/H. The form of the turbulence distribution changes to a maximum on the centerline downstream X/H=3.15. The maximum of the turbulence is present at X/H=7. For X/H greater than 7 turbulence decreases with X/H, with a distribution form which becomes more round.

The conclusions of figure 7 are that the turbulence measurements confirm what found with the velocity measurements up to X/H=2, i.e. the jet flow maintains the same velocity and turbulence distributions from the exit up to X/H=2. Figure 7 confirms that the position X/H=7 is critical because turbulence reaches a maximum of about 23-24%.

Velocities measured with the hot anemometer are reported in figure 8 versus X/H and Y. Velocities are grouped within the range of 2 m/s with different color. Figure 8 shows several similarities with the shadowgraph visualization of [8].

The region where the jet flow maintains the same wideness can be roughly considered as undisturbed region, extending up to X/H=2-2.5. Within this zone is present a central part where the velocity has a slightly higher value. At X/H=8 velocity decreases to 18-20 m/s from the exit value of 22-24 m/s. The region from the exit to X/H=8 can be identified as the potential core region. The evaluation of the line-source of the jet flow can be done extending the lines of minimum velocity of flow on the left. The line-source of the jet is before the slot exit, i.e. at X/H=-1. It is smaller than that measured in previous investigations, confirming that the present experiments have a zone of undisturbed flow which decreases the length of the line source.

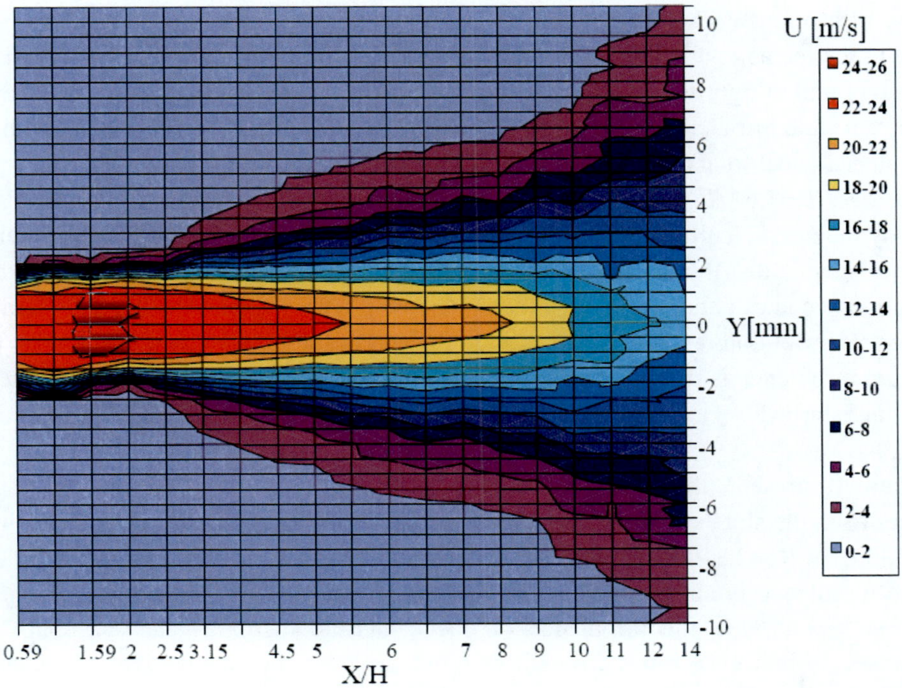

Figure 8. Visualization of the velocity measurements for Re=12,100 [8].

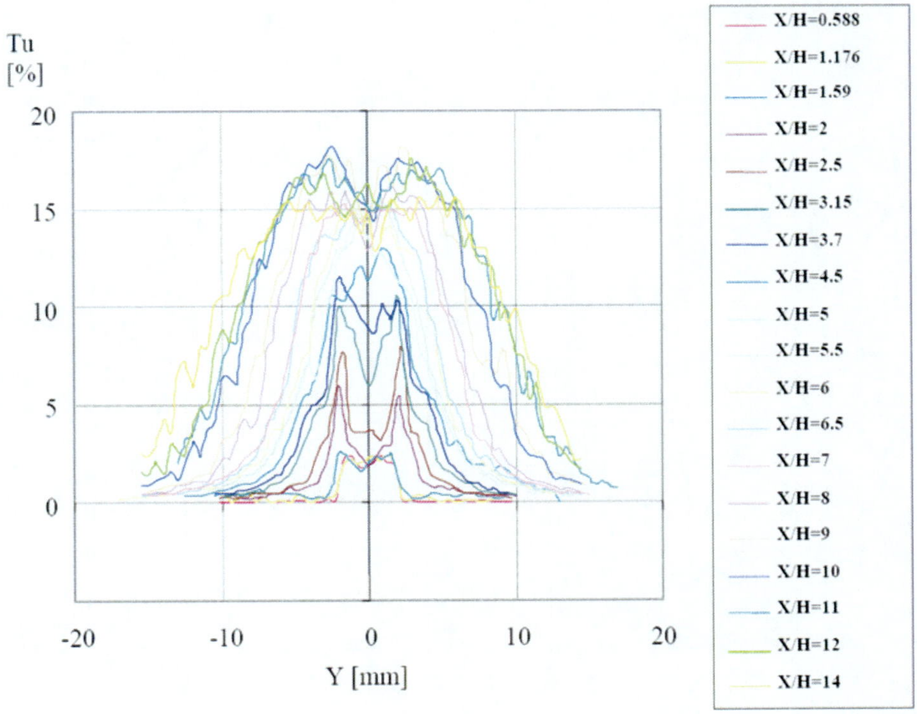

Figure 9. Absolute turbulence across the jet flow for Re=17,200 [10].

The measurements with the hot wire anemometer of the absolute turbulence, T_U, were done with the rectangular slot of height H=4.25 mm in the wind tunnel WT1, for Re=17,200, and are reported in figure 9 at several distances from the slot exit X/H [10],.

The absolute turbulence close to the slot exit is almost constant across the jet with a level of turbulence equal to about 2%. The turbulence on the centerline remains constant up to the distance of X/H=1.6. An increase of turbulence is present at X/H=2 with two lateral spikes, almost symmetrical. Turbulence increases on the centerline from X/H=2.5. A different form of the turbulence distribution is observable at X/H=3.15 where the shape of the turbulence distribution becomes parabolic-like with the maximum on the axis of the jet. This form of the turbulence distribution remains constant up to the end of the measurements, i.e. X/H=14. The maximum turbulence is measured at X/H=7. In conclusion, it is possible to define the length of the undisturbed region of flow, from the turbulent measurements, as equal to about $L_U/H=1.6$.

Figure 10 presents the absolute turbulence measurements, across the jet flow at several distances from the slot exit, X/H, with the wind tunnel WT1 and the rectangular slot of height H=4.25 mm for Re=22,100 [10].

The turbulence is almost constant within the jet up to X/H=1.6 with a value equal to about 2%. For X/H=2 the turbulence on the centerline increases and the shape of the turbulence distribution becomes different with two peaks on the lateral edges of the jet where the interaction with the stagnant air is stronger. The shape of the turbulence distribution does not change with the increasing X/H up to the maximum distance X/H of the experiments, i.e. X/H=14.

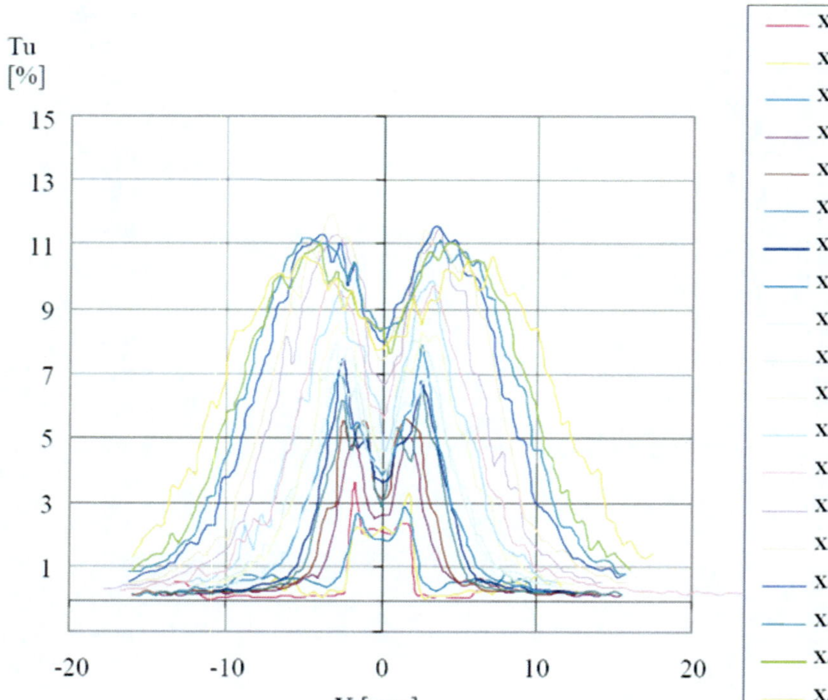

Figure 10. Absolute turbulence across the jet flow for Re=22,100 [10].

A.3.2. Wind Tunnel 2

Velocity and turbulence were measured in the wind tunnel WT2 and with the rectangular slot of height H=10 mm for several Reynolds numbers [9].

Figure 11 reports the velocity measurements across the jet versus the distance from the centerline of the slot, Y, at several distances from the jet exit, X/H, and for Re=7,600. The first measurement is taken at 3 mm, i.e. X/H=0.3, close to the slot exit. A stepwise variation of the velocity is present outside the slot. The velocity on the centerline remains constant up to about X/H=8. At higher X/H decreases continuously up to the last measured point, X/H=16. Increasing the distance from the inlet, i.e. X/H, the velocity outside the slot is higher because the jet flow penetrates more and more into the stagnant fluid.

The absolute turbulence T_U is reported in figure 12 versus the distance from the centerline, Y, at several locations downstream the jet flow. At X/H=0.3, turbulence is lower than 1% on the centerline and lower than 1.5% on the two lateral edges of the jet. The form of the turbulence distribution is unchanged up to X/H=3, with a minimum on the centerline and two little peaks, equal to 1.5-2%, on the two lateral edges. At X/H=4 the form of the turbulence distribution in the jet is almost flat with values of the turbulence in the range 2-4%. At X/H=5 the turbulence level increases up to about 6.5%. At X/H=8 turbulence presents a maximum on the centerline. At higher values of X/H turbulence has a minimum on the centerline and higher values on the two edges of the jet. From figure 12 it can be concluded that the undisturbed region of flow is equal to L_U/H=3-4 for Re=7,600.

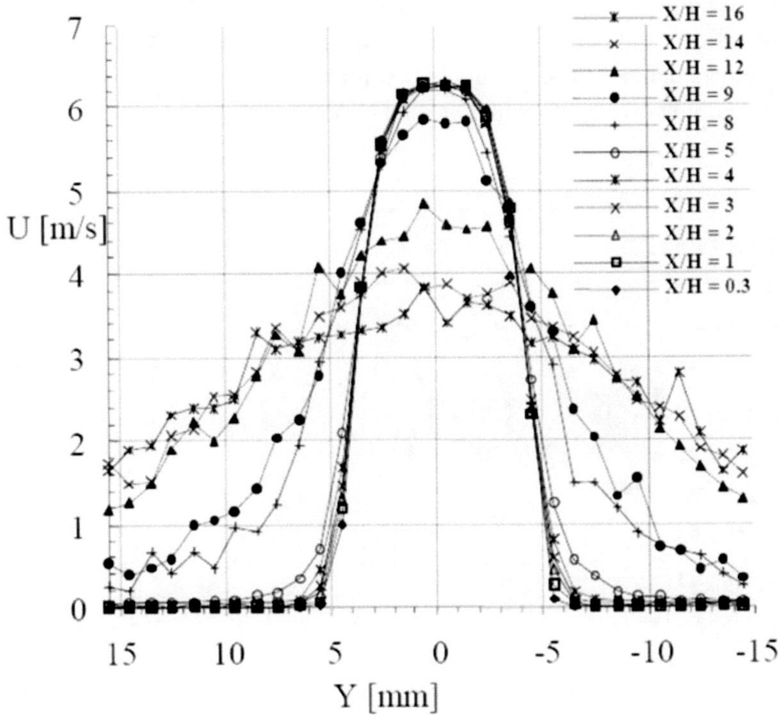

Figure 11. Velocity across the jet flow for Re=7,600 [9].

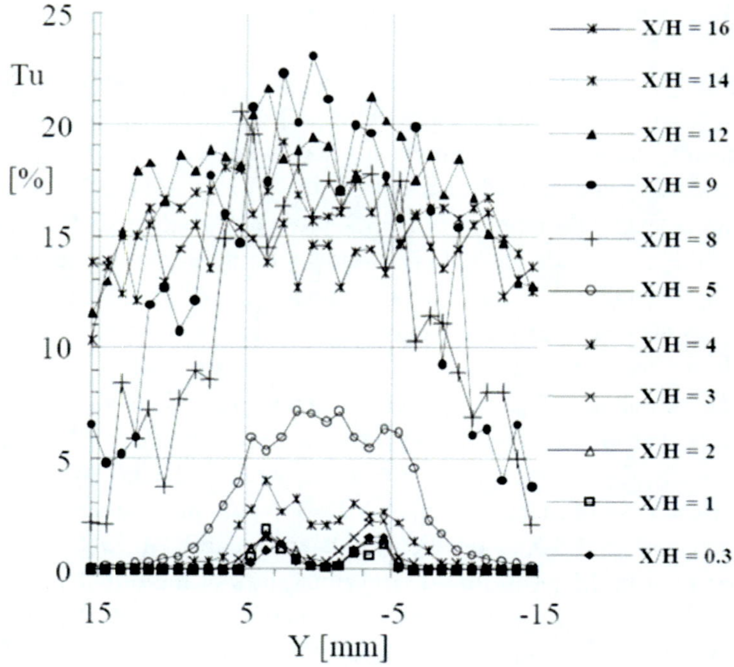

Figure 12. Absolute turbulence across the jet flow for Re=7,600 [9].

The velocity measurements of figure 12 are presented in figure 13 to visualize the jet evolution. Velocities are grouped within ranges reported in the legend of the same figure 13. Figure 13 visualizes the length of the undisturbed region which is equal to L_U/H=3-4 and the length of the potential core region which is X/H=8-9.

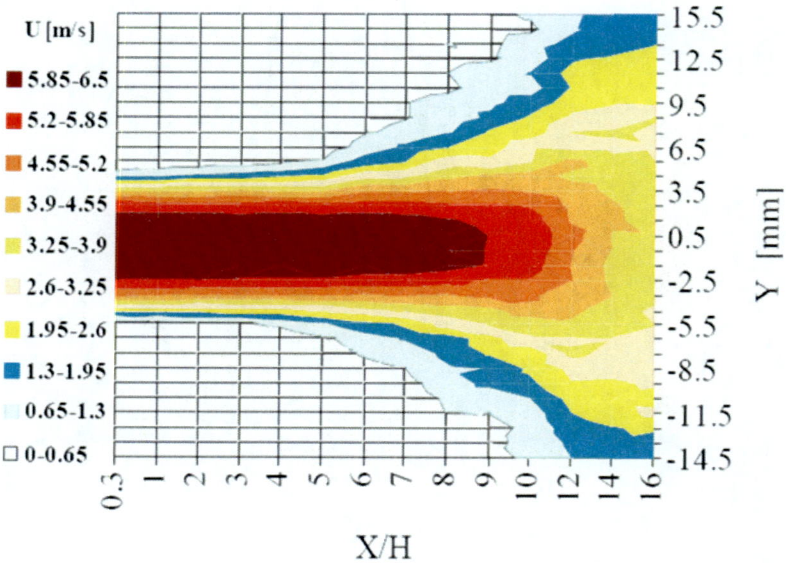

Figure 13. Visualization of the velocity measurements for Re=7,600 [9].

Figures 14-15 present the measurements of velocity and turbulence on the centerline for Re=7,600. Figure 14 shows that the velocity on the centerline maintains a value above 6 m/s up to X/H=8, which characterizes the end of the potential core region, $L_P/H=8$. After X/H=8 the velocity decreases to the minimum, measured at X/H=16. The measurements of the absolute turbulence, T_U, along the centerline are reported in figure 15 versus X/H. The absolute turbulence remains constant up to X/H=3, increases up to X/H=9, where has a maximum, and then decreases.

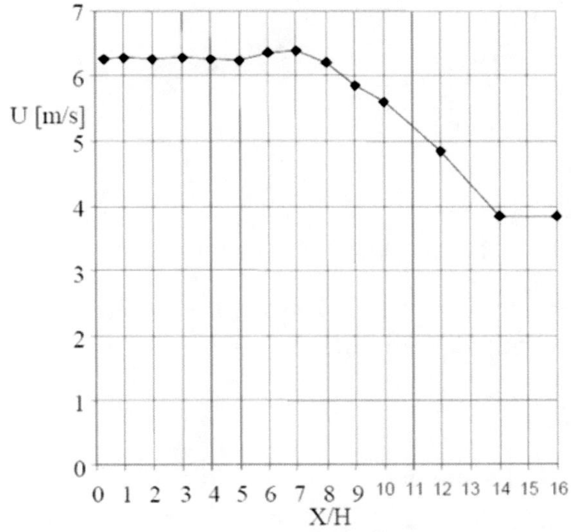

Figure 14. Velocity along the centerline for Re=7,600 [9].

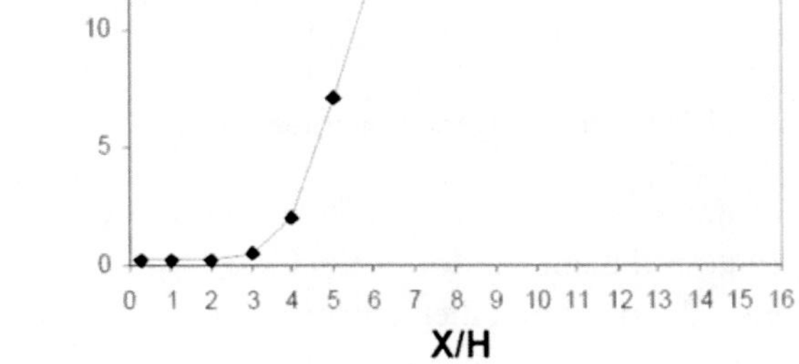

Figure 15. Absolute turbulence along the centerline for Re=7,600 [9].

At Re=23,300, the velocity measurements are visualized in figure 16 similarly to figure 13. The visualization of the velocity measurements shows that the undisturbed region is about $L_U/H=1\text{-}2$ and the potential core region about $L_P/H = 5\text{-}6$.

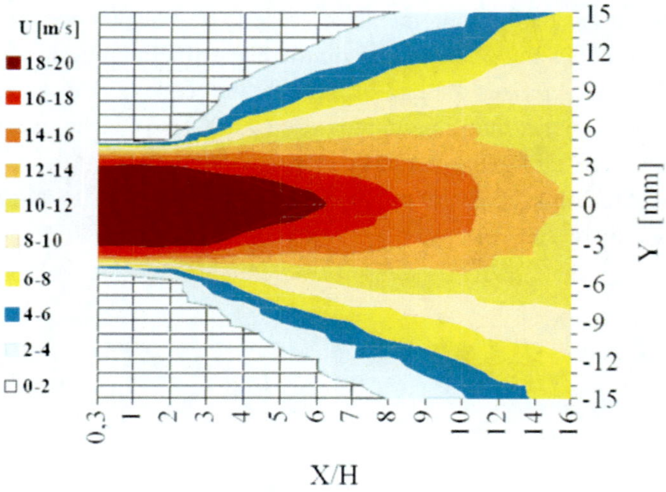

Figure 16. Visualization of the velocity measurements for Re=23,300 [9].

Figure 17 shows the evolution of velocity on the centerline for Re=23,300. Velocity maintains the value of the exit for about X/H=5.

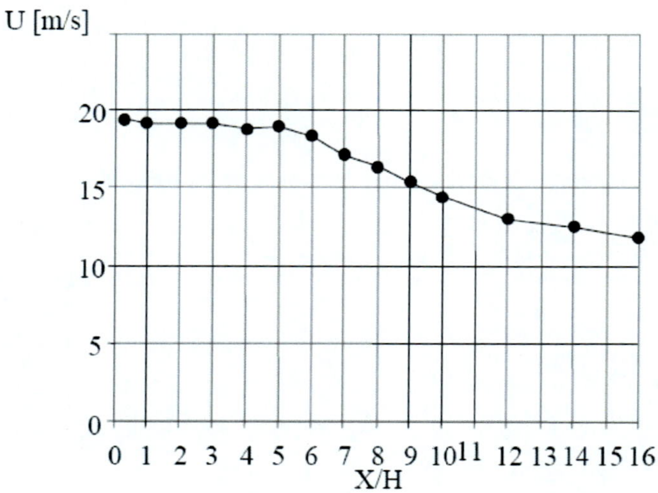

Figure 17. Velocity along the centerline for Re=23,300 [9].

The measurements of the absolute turbulence along the centerline are reported in figure 18 versus X/H. The absolute turbulence is lower than 2% up to X/H=1, increases with X/H reaching the maximum around X/H=9 and then decreasing to the minimum at X/H=16.

Velocity and turbulence measurements at Re=37,300 are reported in figures 19-21. Figure 19 presents the velocity measurements across the jet, i.e. versus the distance from the centerline, Y, at several distances from the slot exit. The velocity distribution is steeply variable on the two edges of the jet on the slot exit, X/H=0.3. After X/H=0.3 the jet starts to interact with the stagnant fluid and the velocity distribution is more round edged. The velocity on the centerline remains equal to the exit value for a distance of about X/H=3-4, then it starts to decrease to the minimum, which is at X/H=16.

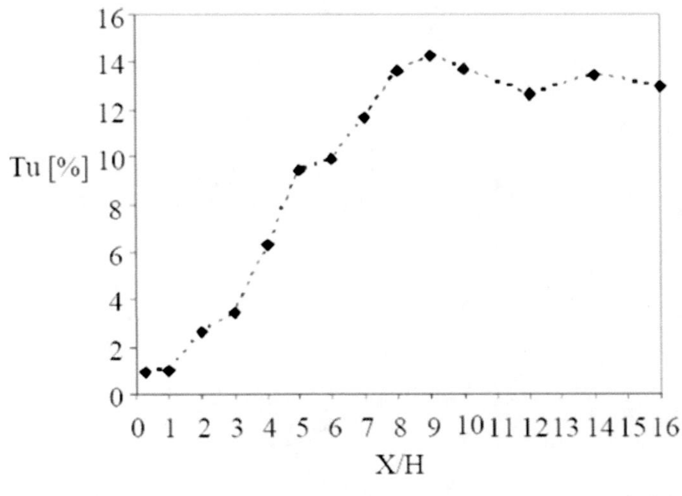

Figure 18. Absolute turbulence along the centerline for Re=23,300 [9].

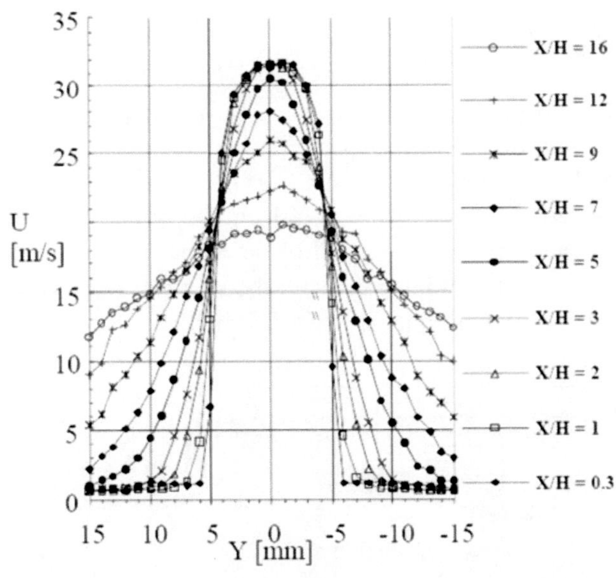

Figure 19. Velocity distribution across the jet flow for Re=37,300 [9].

Absolute turbulence measurements are reported in figure 20 at several distances from the slot exit. On the exit, i.e. X/H= 0.3, absolute turbulence has a minimum on the centerline of about 2 %, but on the edges of the slot the turbulence has two peaks, almost symmetrical, which are equal to 12-13%. The turbulence is then at a high level immediately after the exit from the slot with the consequence to interact with the stagnant fluid from the beginning of the jet evolution. Figure 20 shows the turbulence increases downstream the jet up to the maximum level of about 16-17 % at X/H=8. The absolute turbulence has never a shape with a maximum on the centerline.

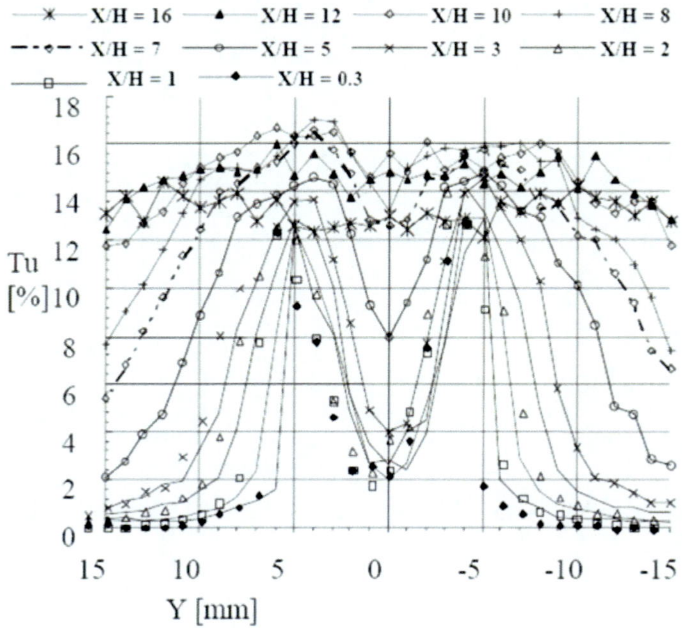

Figure 20. Absolute turbulence across the jet flow Re=37,300 [9].

Visualization of the velocities for Re=37,300 is shown in figure 21 where the measurements are reported versus the distance Y from the centerline. Velocities are grouped within values which are reported in the legend of the same figure. Figure 21 shows that the undisturbed region is absent, the jet interacts with the stagnant fluid immediately outside the exit and the potential core region is equal to about L_P/H=4-5.

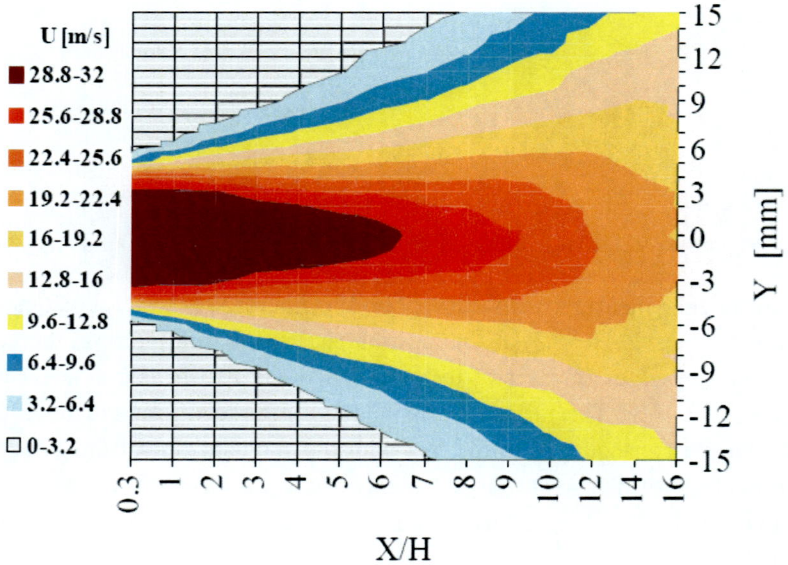

Figure 21. Visualization of the velocity measurements for Re=37,300 [9].

The measurements of velocity and turbulence on the centerline are reported respectively in figures 22-23 for Re=37,300.

Figure 22 shows the velocity distributions on the centerline with the conclusion that the velocity maintains the exit value up to about X/H=4 and a value of the velocity higher than 30 m/s up to X/H=5. In other words the potential core region is equal to L_P/H=4-5.

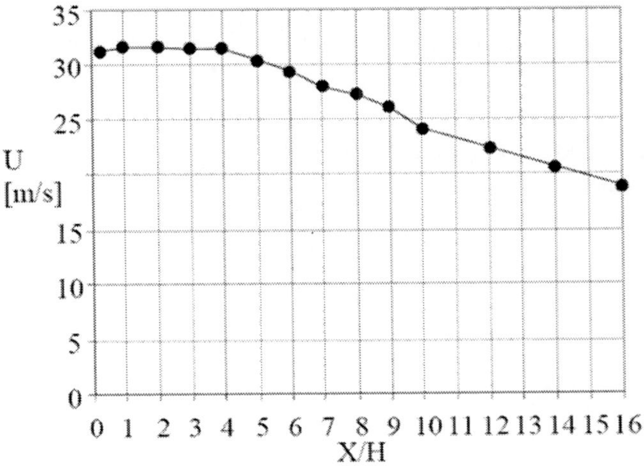

Figure 22. Velocity along the centerline for Re=37,300 [9].

Figure 23 presents the absolute turbulence along the centerline versus X/H. The absolute turbulence is below 5% up to X/H=4, presents a maximum at X/H=10 and a subsequent slight decrease.

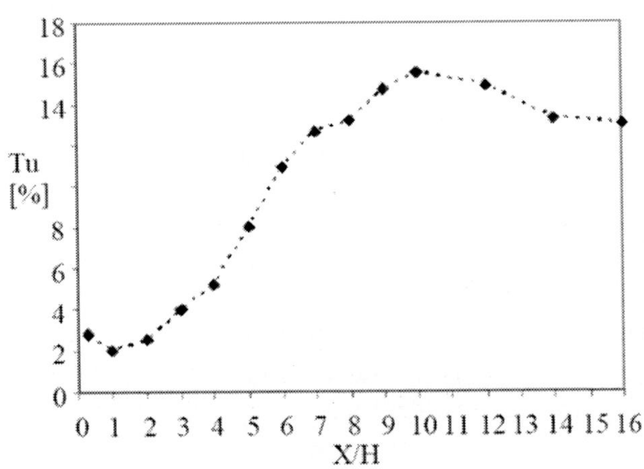

Figure 23. Absolute turbulence along the centerline for Re=37,300 [9].

A.4. Velocity and Turbulence in a Jet Impinging a Cylinder

Velocity and turbulence measurements in presence of a cylinder were carried out in [15] at a Reynolds number equal to Re=23,300 with the wind tunnel WT2. The diameter D of cylinder, D=10 mm, is equal to the slot height, H=10 mm. The impinged cylinder is set at two distances from the slot exit, namely X/H=4 and 6.

Figure 24 presents velocity and relative turbulence, Tu, when the cylinder is set at H/S=4. The relative turbulence is defined as the velocity fluctuation, u', divided the time-average velocity at the same point, U_X, i.e. Tu=u'/U_X. Figure 24 shows that the velocity is almost constant up to the impingement of the cylinder, showing that the potential core region is still present in the jet flow with the cylinder set at X/H=4. Behind the cylinder the velocity increases with the distance up to X/H=7, then a slight decrease is present and a successive increase up to X/H=18, close to the final measurement, X/H=20. The relative turbulence on the slot exit is lower than 1% but the turbulence increases to 7% close to the impingement on the cylinder, while, behind the cylinder, the turbulence increases up to 48%. Then is present a decrease, in correspondence to the increase of velocity and a subsequent increase up to the maximum value of 80% around X/H=9. After X/H=9 the turbulence decreases to about 20% at X/H=20.

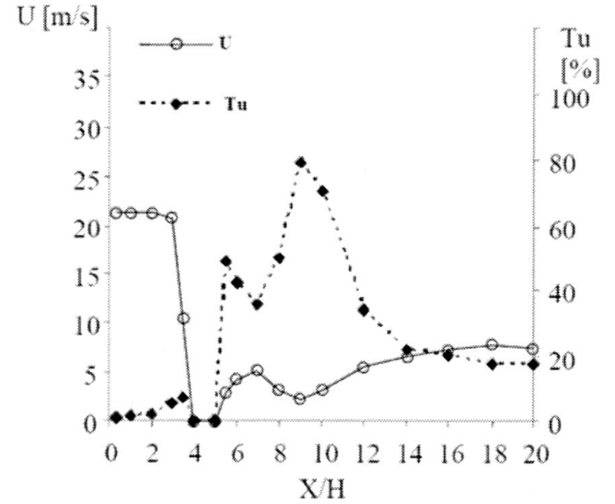

Figure 24. Velocity and relative turbulence along the centerline with the cylinder set at X/H=4, for Re=23,300 [15].

Velocity versus the angle from the impinging point is presented in figure 25 at several distances from the cylinder wall. At 60° the velocity at 1mm from the wall is equal to the undisturbed value on the centerline, 22 m/s. Increasing the distance from the wall the velocity decreases, due to the shear interaction with the stagnant air. The measurement at 1 mm from the wall of the cylinder indicates that the transition in the flow field is present at about 90° from the impinging point, i.e. is almost similar to the uniform flow case. This conclusion is in agreement with the presence of a potential core region longer than X/H=4.

The measurements of the relative turbulence, at some distances from the cylinder wall, are presented in figure 26 versus the angle from the impinging point. Turbulence at 1 mm from the wall is around 10% on the impinging point, than it decreases slightly along the wall up to about 90° when a large increase is measured. The high value of turbulence, 140%, around 90° must be compared to the time-average velocity on the same point, which is equal to about 1 m/s. Then the turbulence decreases with the angle and reaches the value of about

50% on the rear point of the cylinder. The measurements at 1 mm confirm the conclusion already stated that a transition in the flow field is present at 90°, in agreement to an uniform flow. At higher distances from the wall the turbulence presents lower values and a sudden increase happens at higher angles, e.g. at 100° at 2 mm.

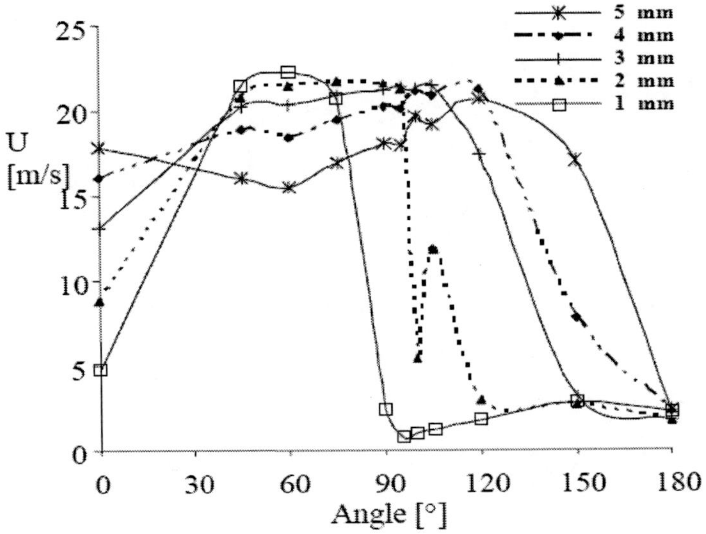

Figure 25. Velocity at distances from the wall of a cylinder set at X/H=4 for Re=23,300 [15].

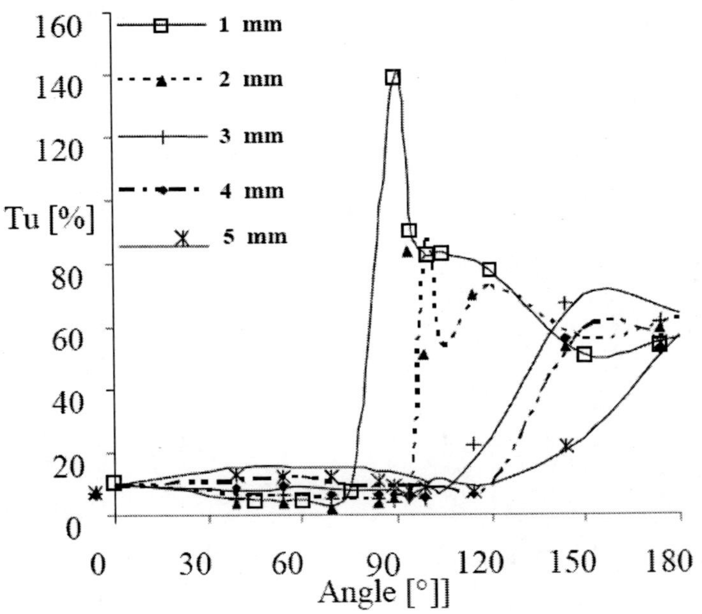

Figure 26. Relative turbulence versus the angle from impingement at several distances from the wall of a cylinder set at X/H=6 for Re=23,300 [15].

To summarize velocity and relative turbulence measurements at 1 mm from the wall, figure 27 shows the occurrence of the flow field transition near 90° from the impinging point.

It can be noticed that the minimum in the velocity corresponds to a maximum in the turbulence.

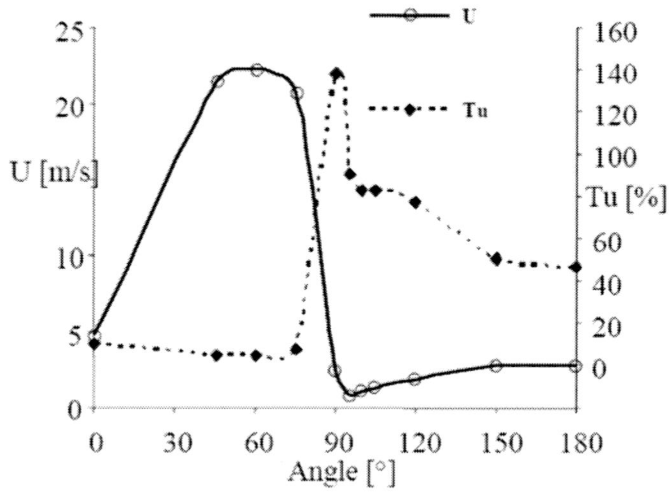

Figure 27. Velocity and relative turbulence versus the angle from impingement at 1 mm from the cylinder set at X/H=4 for Re=23,300 [15].

Figures 28-32 report the velocity and the relative turbulence measurements with the cylinder set at X/H=6. Figure 28 reports velocity and turbulence on the centerline of the jet.

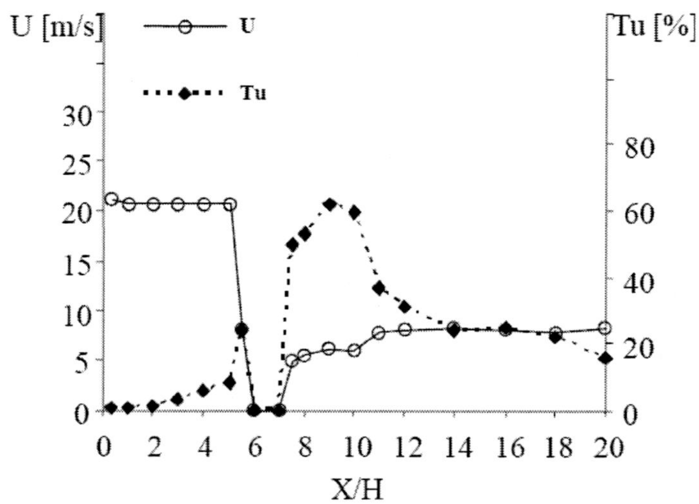

Figure 28. Velocity and relative turbulence along the centerline with the cylinder set at X/H=6 for Re=23,300 [15].

Velocity is almost constant at 20.6 m/s up to X/H=4.5, when starts to decrease, showing that the potential core region is still present in the jet flow with the cylinder set at X/H=6. Behind the cylinder the velocity increases up to the last measuring point, X/H=20. The relative turbulent on the slot exit is lower than 1%. Close to the impingement of the cylinder the turbulence is equal to about 25%. Behind the cylinder the turbulence increases up to 62%

at about X/H=9. Behind X/H=9 the turbulence decreases down to values below 20% at X/H=20.

Figure 29 presents the velocity measurements at some distances from the wall versus the angle from the impinging point. At 1mm from the wall velocity increases with the angle from the impinging point, where is around 6 m/s, to the maximum around 70°, to decrease further reaching the minimum at 110°. Then a slight increase is present. In conclusion, with the cylinder set at X/H=6, the flow field transition happens at a higher angle than with the cylinder set at X/H=4. This conclusion shows the difference with the uniform flow case. At an angle of 70°, with the increase of the distance from the wall, velocity decreases because of the shear interaction of the jet with the stagnant air.

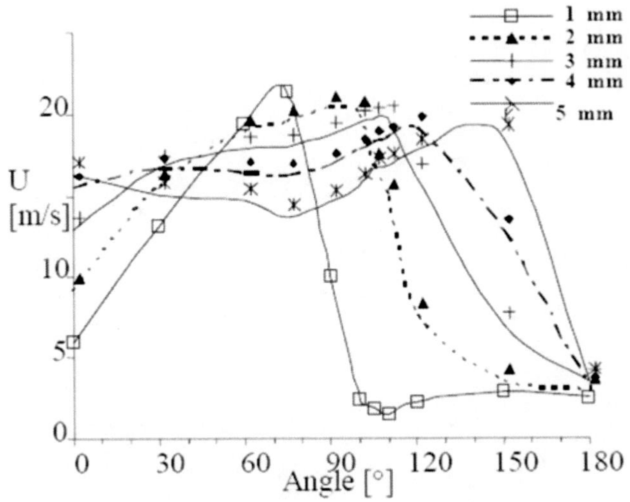

Figure 29. Velocities versus the angle from the impingement at several distances from the wall of the cylinder set at X/H=6 for Re=23,300 [15].

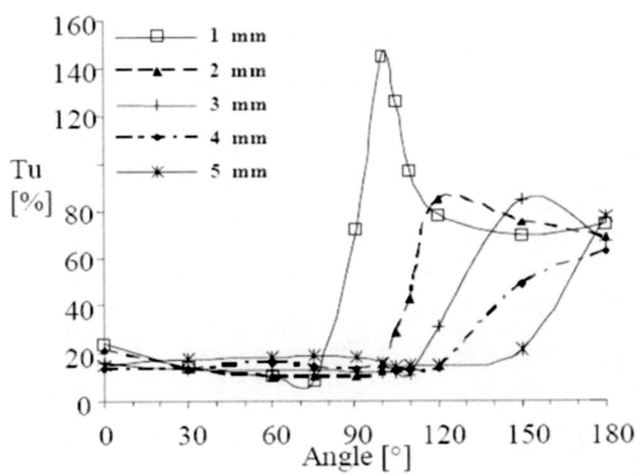

Figure 30. Relative turbulence versus the angle from impingement at several distances from the wall of the cylinder set at X/H=6 for Re=23,300 [15].

Relative turbulence measurements, at several distances from the wall, are reported in figures 30. Relative turbulence on the impinging point is about 25% at 1 mm. A slight decrease is observable with the increasing angle, reaching a minimum at about 75° and a strong increase at about 100°, followed by a decrease to about 80% on the rear point. The high value of turbulence, 150%, around 100° must be compared to the small time-average velocity on the same point, which is equal to about 1 m/s. At higher distances from the wall the turbulence on the impinging point remains practically constant up to the angle where the flow transition is present.

Velocity and relative turbulence measurements at 1 mm from the wall are reported in figure 31, where the angle of the flow transition is clearly evidenced by the joint decrease of velocity and the increase of relative turbulence. For X/H=6 this angle is equal to about 110°.

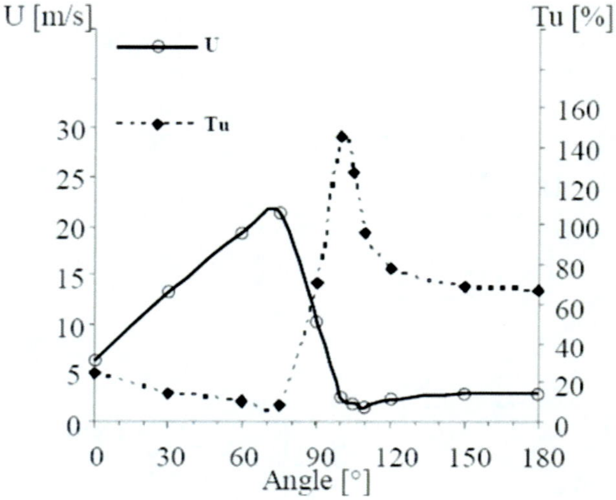

Figure 31. Velocity and relative turbulence versus the angle from impingement at 1 mm from the cylinder set at X/H=6 for Re=23,300 [15].

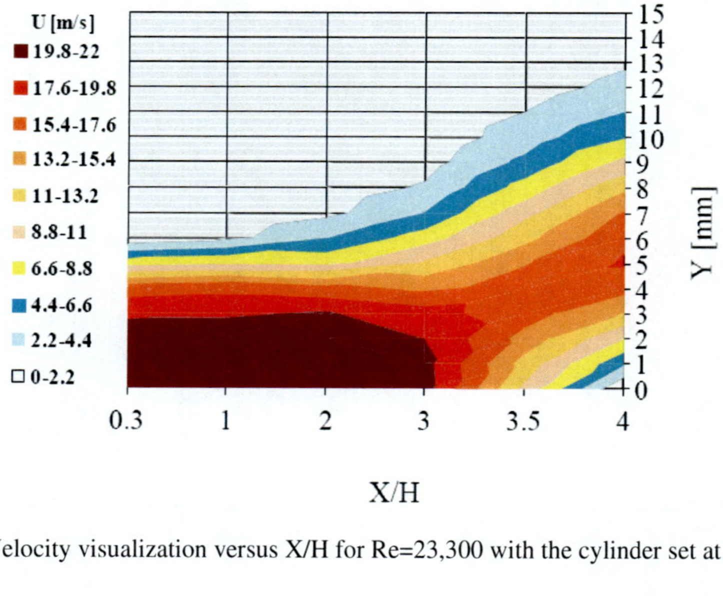

Figure 32. Velocity visualization versus X/H for Re=23,300 with the cylinder set at X/H=4 [15].

Velocity measurements are visualized in figures 32-33, respectively with the cylinder set at X/H=4 and X/H=6, reporting the values grouped within the ranges reported in the legend of the figures, and represented with different colors,.

Figure 32 presents the velocity visualization with the cylinder set at X/H=4. Figure 32 visualizes the length of the undisturbed region, which is equal to about $L_U/H = 1$ and the length of the potential core region, defined as the position where U_M/U_E is equal to 0.9, as equal to about $L_P/H = 3$.

Figure 33 presents the velocity visualization with the cylinder set at X/H=6. The length of the undisturbed region is equal to $L_U/H=2$ and the length of the potential core region is equal to $L_P/H = 5$.

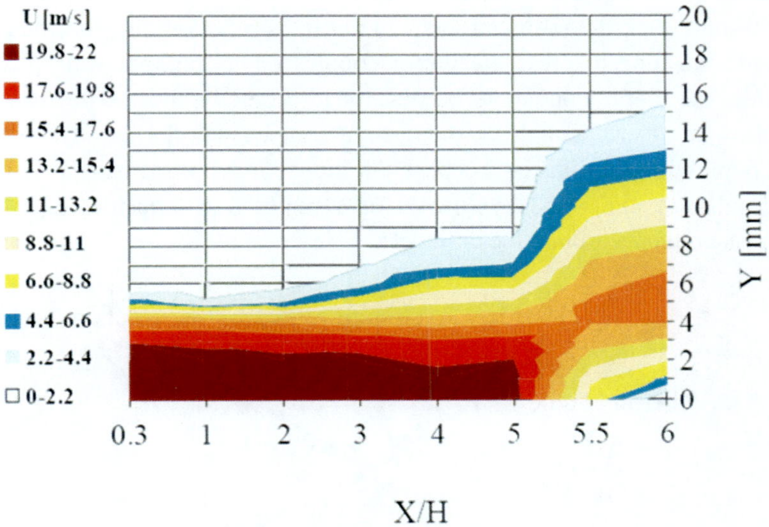

Figure 33. Velocity visualization for Re=23,300 with the cylinder set at X/H=6 [15].

B. Heat Transfer on a Cylinder Cooled by a Submerged Jet

B.1. Introduction

Wind tunnel WT1 was employed in the preliminary setup, i.e. without the three metallic screens, to study the influence of an increase of the level of turbulence in the jet flow. Preliminary results in the cooling of a cylinder with diameter D=10 mm and a slot of height H=5 mm showed an increase of the average and local Nusselt number due to the introduction of the metallic screen for Re=17,000 [17].

The experiments with the preliminary set-up and without the metallic screen on the exit were carried out in an extended Reynolds number range, i.e. from Re=4,000 to Re=20,000, and for X/H=2-10 [18].

Experiments with slot of different heights were done according to the flow transition along the circumference, [19]. The slot jet with the highest ratio of cylinder diameter to slot height realizes the highest heat transfer coefficients. The distance from the slot exit with the highest mean Nusselt number is equal to X/H=8 for D/H=4 and to X/H=6 for D/H=2. The mean Nusselt number measured in a slot jet with D/H=4 at X/H=8 is comparable to the heat transfer realized by a uniform flow with a very high turbulence intensity.

Measurements of the local heat transfer coefficient were carried out with the wind tunnel WT1, modified to reduce turbulence [20], i.e. in presence of the three metallic screens. Comparisons were made with the results of the literature with a uniform flow.

Numerical simulations with a commercial code have been carried out to predict the experimental mean heat transfer, [21].

B.2. Experimental Apparatus

The cylinder to be cooled is made of an AISI 304L stainless steel tube of external diameter D=10 mm, thick 0.2 mm, and wide 135 mm. The Teflon bar, sketched in figure 34, is located inside the tube to press the thermocouples on its internal wall. Grooves on the Teflon bar allow the thermocouples to be positioned correctly and the decreasing depths of the grooves allow the thermocouples to be in tight contact with the internal side of the wall. Five thermocouples are introduced because of the cylinder dimensions. This number of thermocouples allows, under the hypothesis of symmetry, to consider eight positions of temperature measurements around the cylinder.

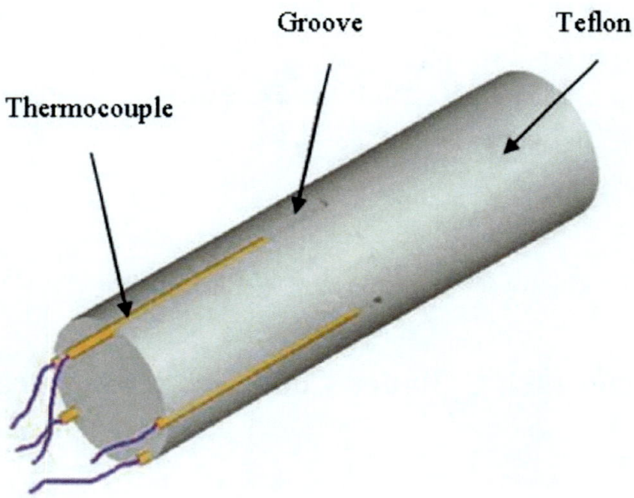

Figure 34. Teflon bar with the five thermocouples in the grooves [20].

Figure 35 shows the positions of the five thermocouples along the circumference; at the angles of 0°, 90°, 180°, 225° and 315° from the impinging point. The steel tube is heated by electrical current with a dissipated power variable in the range 5-11 W. The mass flow rate of air is measured through the pressure drop across the converging duct. The data acquisition system is made with an A/D converter and a multiplexer for the thermocouples readings. The electric current was measured through the voltage drop on the 0.1 ohm reference shunt of the feeder.

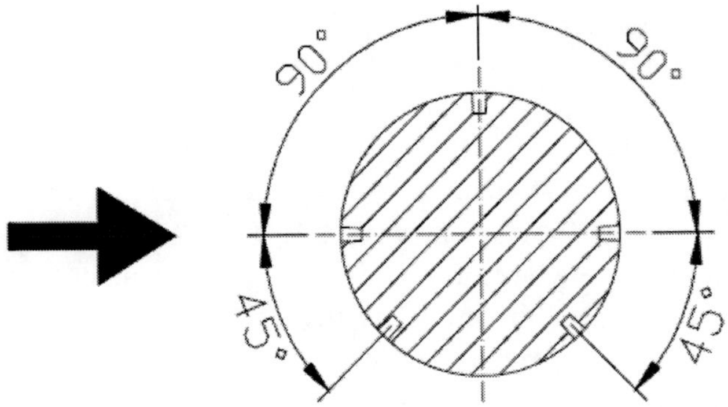

Figure 35. Positions of the five thermocouples [20].

B.2.1. Experimental Analysis

In each experiment, temperatures, mass flow rate, electrical resistance of the cylinder, and dissipated power were measured. The power dissipated locally on each thermocouple was assumed constant, along the circumference of the tube, because the temperature differences among the five thermocouples were enough small to neglect the variation of the electric resistance of the stainless steel tube.

The power dissipated per unit area of the cylinder, *P/A*, is given by

$$P/A = R I^2 / (p L) \tag{1}$$

where *R* is the electrical resistance of the steel tube, *I* the electrical current, $p = \pi D$, the perimeter of the cylinder, *D* the diameter and *L* the length.

The local convective heat transfer coefficient is given by

$$h_i = P / [A (T_i - T_a)] = R I^2 / [\pi D L (T_i - T_a)] \tag{2}$$

where T_a is the air jet temperature and T_i the average local temperature of the thermocouple on the cylinder.

The local Nusselt number is given by

$$Nu_{loc} = h_i D / k = R I^2 / [\pi k L (T_i - T_a)] \tag{3}$$

where *k* is the thermal conductivity of air.

Reynolds number is defined as

$$Re = w D / \nu \tag{4}$$

where *w* is the average velocity on the slot exit, and ν the kinematics viscosity.

B.2.2. Experimental Uncertainties

The copper-constantan thermocouples have wires of 0.25 mm diameter. Their calibration uncertainty was lower than 0.05°C. The local temperature, registered by each thermocouple, was the ten minutes average, where the maximum variations are ± 0.15°C. In conclusion, the uncertainty of the temperature measurement was ± 0.15°C.

The stainless steel tubes were electrically heated by passing direct current throughout the thin thickness of the tube by a stabilized (10^{-4} relative value) power supplier KEPCO, model JQE. The power dissipated allowed to evaluate the temperature gradient in the radial direction. The radial temperature gradient was evaluated for the maximum power dissipated. The maximum temperature drop of 0.035 K has been evaluated in the thickness of the steel cylinder, i.e. lower than the uncertainty of the temperature measurement.

The experimental temperature difference along the circumference of the tube was function of the angle from the impinging point. The difference between the thermocouple at 0° (on the impinging point) and at 180° (on the rear point) is variable between 3°C and 5°C for the different powers applied and the different air velocities. The corresponding variation of the electric resistance along the outer wall of the tube was lower than 0.5%. The circumferential heat flux along the wall was evaluated by the measurement of the temperatures on the impinging point and the rear point. The heat flux was evaluated as 1.2% of the total power dissipated. Considering that the difference between the temperature of the impinging air and the wall was variable between 16 °C and 24 °C, the contribution of the circumferential heat flux was considered negligible.

The measurements of the electric power dissipated in the tube gave a maximum uncertainty of 10^{-2} Watts. The evaluation of the power dissipated in each experiment was carefully controlled taking into account the influence of natural convection, thermal radiation, heat conduction at the edge of the cylinder, and boundary condition of the casing. The uncertainty in the evaluation of the Nusselt number was considered as 6%.

B.2.3. Experimental Results

Local heat transfer was measured along the circumference of the cylinder, at several distances from the slot exit and for different Reynolds numbers, with the wind tunnel WT1 at low turbulence [20].

Re=37,300 - Local Nusselt number on the impinging point is reported versus X/H in figure 36 for Re=37,300. For a free jet at Re=37,300 the undisturbed region is zero, $L_U/H=0$, and the potential core region is about $L_P/H=4$.

Figure 36 shows the local Nusselt number is almost constant from the jet exit up to X/H=4.25, which is the end of the potential core region. Nusselt number increases from X/H=4.25 to X/H=8, which is the position of the maximum Nusselt number, and then decreases with X/H. The variation of the local Nusselt number with X/H, figure 36, can be compared with the evolutions of the centerline velocity and turbulence (figures 22 and 23). The centerline velocity is constant from the exit until X/H=4 (end of the potential core region) while the turbulence is lower than 5%. The turbulence increases from 5% at X/H=4 up to the maximum of 15% at X/H=10 while the centerline velocity at X/H=10 decreases to about 24 m/s from the maximum of about 32 m/s.

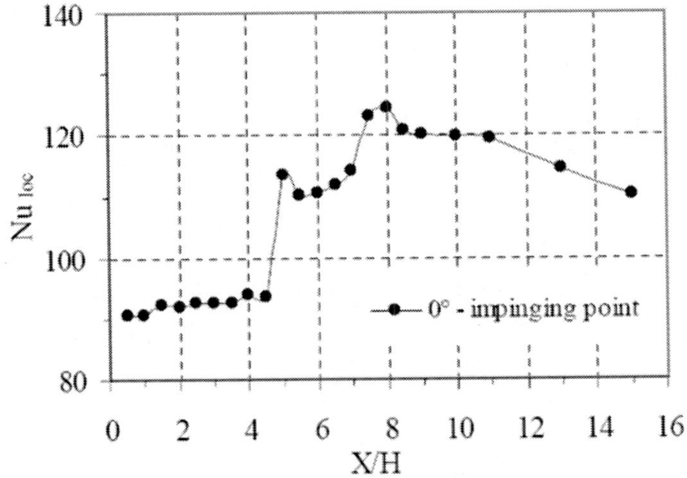

Figure 36. Local Nusselt number on the impinging point for Re=37,300 [20].

The conclusions that can be done from figure 36 are:

- In the potential core region, i.e. X/H=0-4, local Nusselt number on the impinging point is constant because velocity on the centerline is constant and turbulence remains lower than 5%.
- In the mixing region, which follows the potential core region, local Nusselt number depends mainly on turbulence which increases from 5% at X/H=4 up to 15% at X/H=10.
- Maximum of the local Nusselt number is at X/H=8, i.e. before the maximum of the turbulence at X/H=10, because the decrease of velocity is compensating the turbulence increase, which remains above 13%.

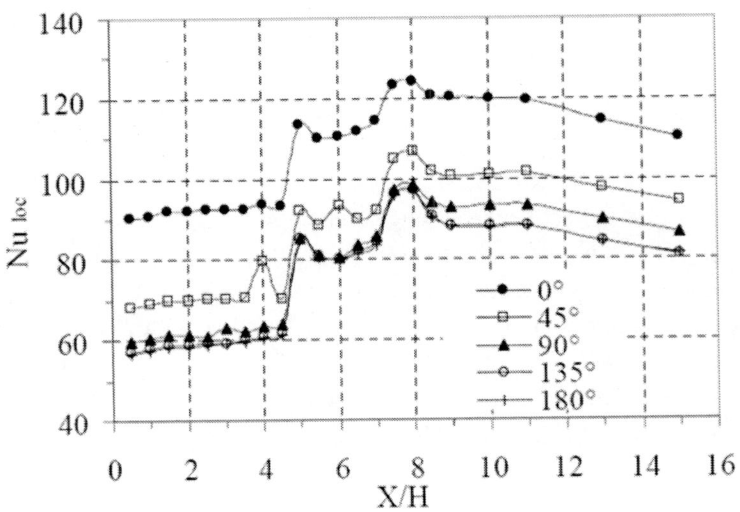

Figure 37. Local Nusselt number versus X/H at different angles from the impingement for Re=37,300 [20].

Local Nusselt numbers are reported at different angles from the impinging point in figure 37 where the sequence of angles is respectively: 0°, 45°, 90°, 135° and 180°. Nusselt number decreases with the increase of the angle reaching the minimum at 180° from the impingement. Values of Nusselt numbers at 135° and 180° are almost coincident. Variation of the Nusselt number with X/H at different angles is very similar to that on the impinging point, except for 45°.

Re=29,300 - Local Nusselt number on the impinging point is reported versus X/H in figure 38 for Re=29,300. For a free jet at Re=29,300 the undisturbed region is about L_U/H=0-1, and the potential core region is in the range of L_P/H=4-5. Local Nusselt is almost constant from the jet exit up to X/H=1, end of the undisturbed region. Local Nusselt number increases at very low rate up to X/H=4, end of the potential core region. The rate of increase of Nusselt number is higher from X/H=4 to X/H=6 and lower from X/H=6 to X/H=8, which is the position of the maximum Nusselt number. Nusselt number decreases after X/H=8. The position of the maximum Nusselt number is equal for Re=29,300, i.e. H/S=8, than for Re=37,300. The conclusions on the variation of the Nusselt number with X/H for Re=29,300 are similar to those for Re=37,300.

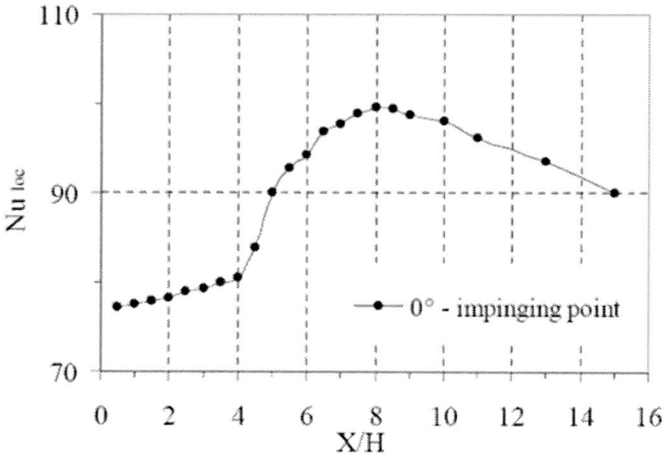

Figure 38. Local Nusselt number on the impinging point for Re=29,300 [20].

Local Nusselt numbers for Re=29,300 are reported at different angles from the impinging point in figure 39 where the sequence of angles is respectively: 0°, 45°, 90°, 135° and 180°. Nusselt number decreases with the increase of the angle, registering a strong decrease from the impinging point to 45° and a similar decrease at 90°. The results at 90°, 135° and 180° are very close each to the other. Variation of the Nusselt number with X/H at different angles is similar to that on the impinging point.

Re=23,300 - Local Nusselt number on the impinging point is reported versus X/H in figure 40 for Re=23,300. For a free jet at Re=23,300 the undisturbed region is around L_U/H=1-2, and the potential core region is in the range of L_P/H=5-6. The trend of the local Nusselt number can be remarked on the basis of the fluid dynamics evolution presented in figures 16-18 for a free jet and in figures 24 and 28 for a jet impinging a cylinder at X/H=4 and X/H=6.

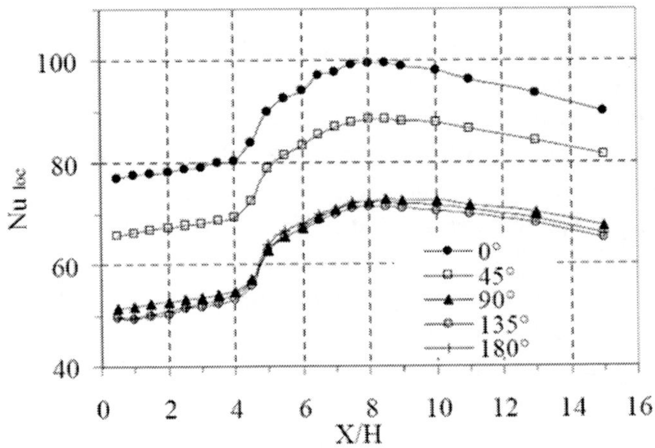

Figure 39. Local Nusselt number versus X/H at different angles from the impingement for Re=29,300 [20].

Local Nusselt is almost constant from the jet exit up to X/H=1-1.5, end of the undisturbed region. Local Nusselt number increases at low rate from X/H=1.5 to X/H=4, at high rate from X/H=4 to X/H=6, and at low rate from X/H=6 to X/H=7, which is the position of the maximum Nusselt number. Then the Nusselt number decreases with X/H. In a free jet the centerline velocity (figure 17) is constant from the exit to X/H=5-6 while the turbulence (figure 18) is equal to 1% from the exit up to X/H=1 and lower than 5% up to X/H=3.5. The rate of increase of turbulence from X/H=3.5 to X/H=5 is high, while is smaller above X/H=6, up to X/H=9, where is present the maximum turbulence of 14%.

The conclusions are that the local Nusselt number is constant in the undisturbed region because the velocity is constant and the turbulence is very low. Nusselt number increases slowly up to X/H=4 because velocity is constant and turbulence is low, but the rate of increase is higher from X/H=4 to 6 where the turbulence increases up to about 10%. The maximum of the Nusselt number is located around X/H=7 because the centerline velocity decreases despite turbulence increases up to the maximum at X/H=9.

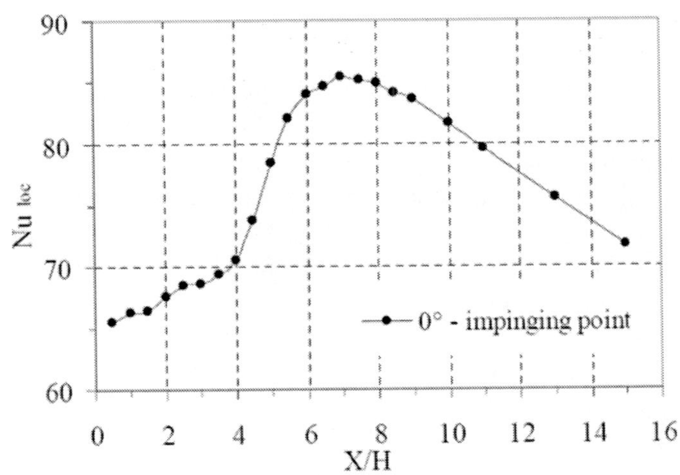

Figure 40. Local Nusselt number versus X/H on the impinging point for Re=23,300 [20].

Fluid dynamics measurements in presence of a cylinder can explain the reason of a greater increase of the Nusselt number when the cylinder is set at X/H=6 rather then when is set at X/H=4. When the cylinder is set at X/H=4 the centerline velocity is constant up to X/H=3 while the turbulence is lower than 7%. When the cylinder is set at X/H=6 the centerline velocity is constant up to X/H=5 while the turbulence is around 10% at X/H=5 but increases above 20% at X/H=5.5, i.e. just before the cylinder. The increase of local Nusselt number is then higher if the cylinder is set at X/H>4 up to X/H=6 because the turbulence is higher.

Local Nusselt numbers are reported at different angles from the impinging point in figure 41 where the sequence of angles is respectively: 0°, 45°, 90°, 135° and 180°. Nusselt number decreases with the increase of the angle reaching the minimum at 180° from the impingement. Variation of the Nusselt number with X/H at different angles is very similar to that on the impinging point.

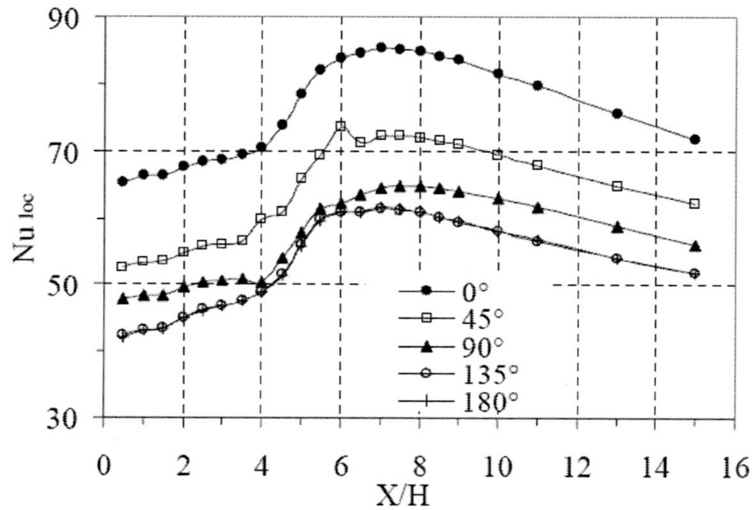

Figure 41. Local Nusselt number versus X/H at different angles from the impingement for Re=23,300 [20].

Re=17,400 - Local Nusselt number on the impinging point is reported versus X/H in figure 42 for Re=17,400. For a free jet at Re=17,400 the undisturbed region is L_U/H=1-2, and the potential core region is L_P/H=6-7. Local Nusselt number is averagely constant from the jet exit up to X/H=1.5, end of the undisturbed region. Nusselt number increases at low rate from X/H=2 to X/H=4.25, at high rate from X/H=4.25 to X/H=6, and at low rate from X/H=6 to X/H=8, which is the position of the maximum Nusselt number. Then, Nusselt number decreases with X/H. The conclusions for this Reynolds number are similar to those for Re=23,300.

Local Nusselt numbers are reported at different angles from the impinging point in figure 43, where the sequence of angles is respectively: 0°, 45°, 90°, 135° and 180°. Nusselt number decreases with the increase of the angle reaching the minimum at 180° from the impingement. Variation of the Nusselt number with X/H at different angles is very similar to that on the impinging point.

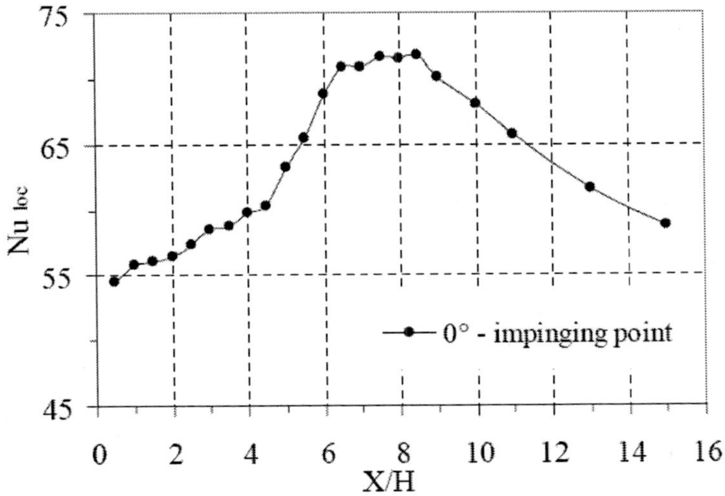

Figure 42. Local Nusselt number versus X/H on the impinging point for Re=17,400 [20].

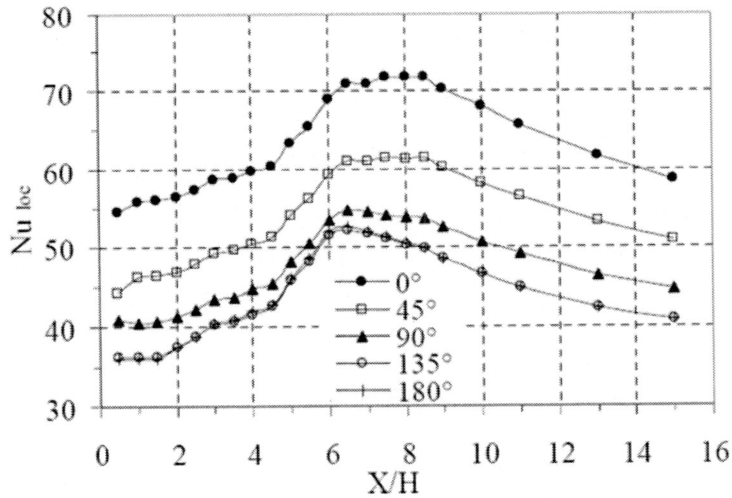

Figure 43. Local Nusselt number versus X/H at different angles from the impingement for Re=17,400 [20].

Re=13,500 - Local Nusselt number on the impinging point versus X/H is reported in figure 44 for Re=13,500. For a free jet at Re=13,500 the undisturbed region is $L_U/H=1\text{-}2$, and the potential core region is $L_P/H=6\text{-}7$.

Local Nusselt number is constant from the jet exit to about X/H=1.5, end of the undisturbed region. The rate of increase is lower from X/H=2 to X/H=4, higher from X/H=4.25 to X/H=6, and lower from X/H=6 to X/H=9, which is the position of the maximum Nusselt number. Then, Nusselt number decreases with X/H. It can be observed that the position of the maximum Nusselt number is higher for Re=13,500, i.e. X/H=9, than for Re=17,400, i.e. X/H=8.5. The conclusions at this Reynolds number are similar to those for Re=17,400.

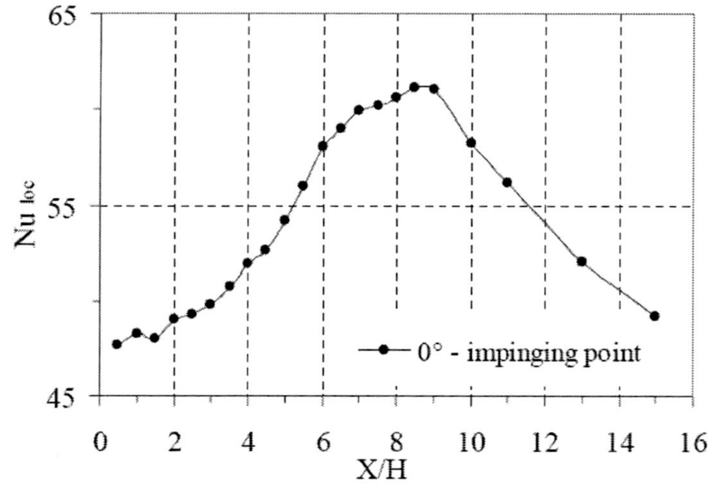

Figure 44. Local Nusselt number versus X/H on the impinging point for Re=13,500 [20].

Local Nusselt numbers are reported at different angles from the impinging point in figure 45 where the sequence of angles is respectively: 0°, 45°, 90°, 135° and 180°. Nusselt number decreases with the increase of the angle reaching the minimum at 180° from the impingement. The trend of the Nusselt number with X/H at different angles is very similar to that on the impinging point.

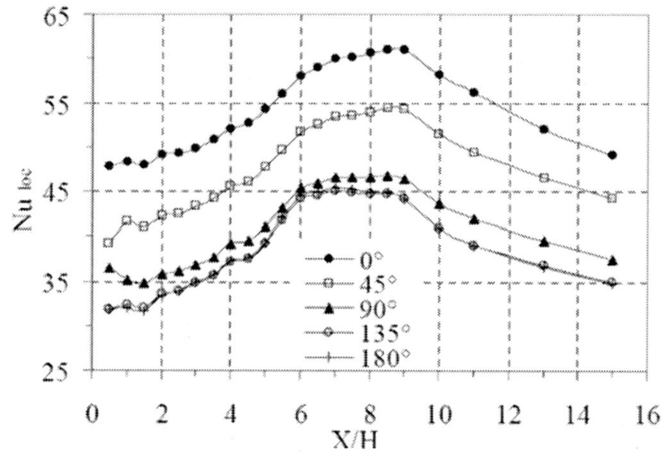

Figure 45. Local Nusselt number versus X/H at different angles from the impinging point for Re=13,500 [20].

Re=9,400 - Local Nusselt number on the impinging point versus X/H is reported in figure 46 for Re=9,400. For a free jet at Re=9,400 the undisturbed region is about $L_U/H=3$. Local Nusselt is averagely constant from the jet exit up to X/H=3, end of the undisturbed region. The rate of increase is higher from X/H=3.5 to X/H=6, also with some oscillations, but is lower from X/H=6 to X/H=11, which is the position of the maximum Nusselt number. Then, Nusselt number decreases with X/H. It can be observed that the position of the maximum Nusselt number is greater for Re=9,400, i.e. X/H=11, than for Re=13,500, i.e. X/H=9.

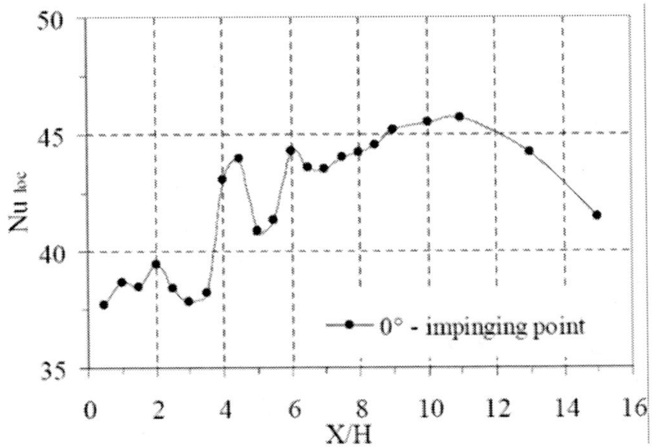

Figure 46. Local Nusselt number versus X/H on the impinging point for Re=9,400 [20].

Local Nusselt numbers are reported at different angles from the impinging point in figure 47, where the sequence of angles is respectively: 0°, 45°, 90°, 135° and 180°. Nusselt number decreases with the increase of the angle. The results at 135° and 180° are very similar with a slight trend of being smaller at 135°, especially at greater X/H. The trend of the Nusselt number with X/H at different angles is very similar to that on the impinging point.

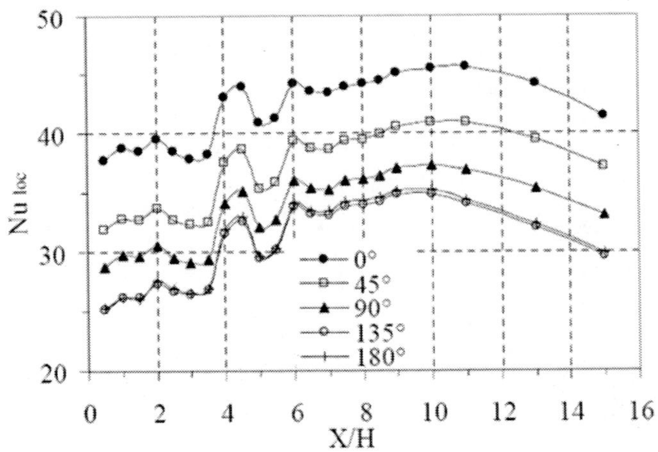

Figure 47. Local Nusselt number versus X/H at different angles from the impinging point for Re=9,400 [20].

Re=7,400 - Figure 48 reports the local Nusselt number on the impinging point versus X/H for Re=7,400. For a free jet at Re=7,400 the undisturbed region is L_U/H=3-4 and the potential core region is L_P/H=8-9. It is possible to make the following remarks on the local heat transfer, despite some oscillations, due to the uncertainty of the measurements. Local Nusselt number is averagely constant from the jet exit up to X/H=3.5, end of the undisturbed region. The rate of increase of Nusselt number with X/H is higher from X/H=4 to X/H=6, but is lower from X/H=6 to X/H=13, which is the position of the maximum Nusselt number. Then, Nusselt number decreases with X/H.

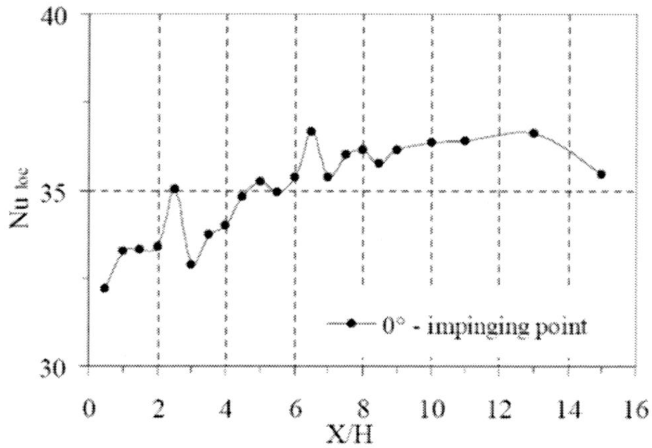

Figure 48. Local Nusselt number versus X/H on the impinging point for Re=7,400 [20].

The following remarks on the fluid dynamics evolutions (i.e. velocity and turbulence) for Re=7,400, as a function of X/H on the centerline, can be reminded:

- Velocity on the centerline (figure 14) is constant up to H/S=8 (end of the potential core).
- Turbulence on the centerline (figure 15) is below 2% in the undisturbed region (i.e. up to X/H=3-4). From X/H=4 to X/H=6 the rate of increase of turbulence is higher up to the maximum at X/H=9. From X/H=9 to X/H=16 turbulence decreases with X/H with a constant region from X/H=10 to X/H=11.

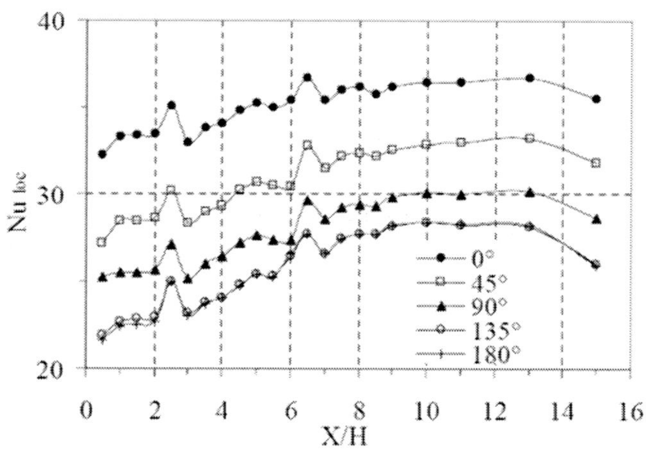

Figure 49. Local Nusselt number versus X/H at different angles for Re=7,400 [20].

The conclusion on figure 48 is the main dependence of local Nusselt number on turbulence up to X/H=8, because velocity on the centerline is constant from the jet exit up to X/H=8, while turbulence increases from the very low value on the inlet up to about 15% at X/H=8 and 23% at X/H=9. From X/H=9 the decrease of velocity is not so strong, as compared to the decrease of turbulence, which makes the local Nusselt number to increase slightly up to X/H=13, with a following slight decrease up to X/H=15.

Local Nusselt numbers are reported at different angles from the impinging point in figure 49 where the sequence of angles is respectively: 0°, 45°, 90°, 135° and 180°. Nusselt number decreases with the increase of the angle reaching the minimum at 180° from the impingement. The trend of the Nusselt number with X/H at different angles is very similar to that on the impinging point.

Conclusions

The experiments on the fluid dynamics of a submerged turbulent free jet of air at low turbulence show that velocity and turbulence on the slot exit maintain constant values for a certain length in the flow development of the jet before to interact with the stagnant fluid. This region of constant fluid dynamics characteristics, observed for the first time in the turbulent regime, is called "undisturbed region of flow" and its length increases with the decrease of the Reynolds number. The undisturbed region of flow disappears for Reynolds number greater than about Re=30,000–40,000. In the undisturbed region of flow velocity on the centerline remains equal to the exit value as in the potential core region. The presence of the undisturbed region of flow can justify some differences reported in the literature about the length of the potential core region and the length of the jet origin. The length of the undisturbed region of flow, measured by the present authors at several turbulent Reynolds numbers, is in qualitative agreement to the length measured previously in the literature for laminar submerged free jet.

Local heat transfer on a cylinder, impinged by a submerged free jet, was measured with the cylinder set at different distances from the slot exit. Local Nusselt number is strongly related to the fluid dynamics of the free jet and, specifically, to the presence of the undisturbed region of flow and the potential core region. In conclusion, local Nusselt number is almost constant versus the distance from the slot exit if the cylinder is located inside the undisturbed region of flow, while increases with the distance if the cylinder is located inside the potential core region, where turbulence on the centerline increases with the distance from the slot exit. Local Nusselt number increases at higher rate with the distance from the slot exit if the cylinder is set after the end of the potential core region. The maximum value of the local Nusselt number is obtained if the cylinder is set close to the maximum turbulence, after the end of the potential core region.

Acknowledgment

The present work has been co-funded by MIUR under PRIN 2006, n. 200693719.

References

[1] S. Corrsin, Investigation of flow in an axially symmetrical heated jet of air, *NACA Wartime Reports*, Series WR-94, 1943.

[2] M. L. Albertson, Y.B. Dai, R. A. Jensen and H. Rouse, Diffusion of submerged jets, *ASCE Papers*, vol. 74, pp. 1571-1596, 1948.

[3] B. G. Van der Hegge Zijnen, Measurements of the velocity distribution in a plane turbulent jet of air, *Applied Science Research*, Sec. A, vol. 7, pp. 257-276, 1958.

[4] E. Gutmark and I. Wygnanski, The planar turbulent jet, *Journal of Fluid Mechanics*, vol. 73, part 3, pp. 465-495, 1976.

[5] Krothapalli, D. Baganoff and K. Karamcheti, Development and structure of a rectangular jet in a multiple jet configuration, *AIAA Journal*, vol. 18, n. 8, pp. 945-950, Aug., 1980.

[6] G. N. Abramovich, On the deformation of the rectangular turbulent jet cross-section, *International Journal of Heat and Mass Transfer*, vol. 25, n. 12, pp. 1885-1894, 1982.

[7] N. Rajaratnan, Turbulent mixing and diffusion of jets, *Encyclopedia of Fluid Mechanics, Dynamics of Single-Fluid Flows and Mixing*, Gulf Publishing Company, Houston, Texas, 1986.

[8] F. Gori, F. De Nigris and E. Nino, Fluid Dynamics Measurements and Optical Visualization of the Evolution of a Submerged Slot Jet of Air, *Proceedings of the 12th International Heat Transfer*, Grenoble (France), vol. 2, pp. 303-308, August 2002.

[9] F. Gori and I. Petracci, Fluid Dynamics Measurements In A Submerged Free Slot Jet Of Air, *Proceedings of the 14th International Symposium on Transport Phenomena*, Bali, Indonesia, pp. 491-496, 6-10 July, 2003.

[10] F. Gori and E. Nino, Fluid Dynamics Measurements And Flow Visualizations Of A Free Slot Jet Of Air, ASME, FED, vol. 259, pp. 187-192, *Proceedings of IMECE '03*, 2003 ASME International Mechanical Engineering Congress & Exposition, Washington, D.C., November 16-21, 2003.

[11] F. Gori, E. Nino and I. Petracci, Shadowgraph Visualizations of a Submerged Free Slot Jet of Air, *International Journal of Heat and Technology*, 2007, vol. 25 (n. 1), pp.157-164.

[12] M. van Dyke, *An Album of Fluid Motion*, The Parabolic Press Stanford California, 1982.

[13] R. J. Goldstein, *Fluid Mechanics Measurements*, Taylor & Francis, 1996.

[14] S. Gogineni, C. Shih. Experimental investigation of the unsteady structure of a transitional plane wall jet. *Experiments in Fluids,* vol. 23 (1997), pp. 121-129.

[15] F. Gori and I. Petracci, Fluid Dynamics Measurements And Numerical Simulations Around A Circular Cylinder Impinged By A Submerged Slot Jet Of Air, ASME, FED, vol. 259, pp. 179-185, *Proceedings of IMECE '03*, 2003 ASME International Mechanical Engineering Congress & Exposition, Washington, D.C., November 16-21, 2003.

[16] H. H. Bruun, *Hot-Wire Anemometry*, Oxford University Press, Oxford, 1995.

[17] F. Gori and L. Bossi, Influence of the turbulence level in the jet flow on the heat transfer upon cylinders: preliminary results, *Proceedings of 17th UIT National Heat Transfer Conference*, vol. I, pp. 63-74, 1999.

[18] F. Gori and L. Bossi, On the Cooling Effect of an Air Jet Along the Surface of a Cylinder; *International Communications in Heat and Mass Transfer*, vol. 27, n. 5, pp. 667-676, 2000.

[19] F. Gori and L. Bossi; Optimal Slot Height in the Jet Cooling of a Circular Cylinder; *Applied Thermal Engineering,* vol. 23 (n. 7) pp. 859-870, May 2003.

[20] F. Gori and I. Petracci, Local Heat Transfer on a Circular Cylinder Impinged by a Submerged Slot Jet of Air, *Eurotherm Seminar 74*, March 23-26, pp. 63-69, 2003.

[21] F. Gori and I. Petracci; Heat Transfer Measurements And Numerical Simulations In The Cooling Of A Circular Cylinder By A Slot Jet Of Air; ASME, HTD, vol. 374 (n. 1), pp. 3-9. *Proceedings of IMECE '03*, 2003 ASME International Mechanical Engineering Congress & Exposition, Washington, D.C., November 16-21, 2003.

In: Thermal Engineering Research Developments
Editors: J. Evgova and O. Kostadinov, pp. 245-282 ISBN: 978-1-60741-497-1

Chapter 5

MASS AND ENERGY-CAPITAL CONSERVATION EQUATIONS FOR THE TIME EVOLUTION OF NON-RENEWABLE ENERGY RESOURCES

Fabio Gori*

Department of Mechanical Engineering, University of Rome "Tor Vergata"
Via del Politecnico 1, 00133 Rome, Italy

Dedication

The present work is dedicated to my mother,

Cesarina AMMANNATI in GORI,

who, during the preparation of this work, passed away at the age of 93, close to her birth place near Florence, Italy, on August 30th, 2008.

Cesarina was descendent of the AMMANNATIs, a family established in Tuscany since Middle Age and Renaissance periods. Amongst the well-known AMMANNATIs are Bartolomeo, architect and sculpturer, and Giulia, mother of Galileo GALILEI. AMMANNATIs are known also for having established a bank in Pistoia, Italy, in the late XIII century.

Cesarina, together with my late father Cav. Uff. Emilio, actively encouraged my research work in the University from the beginning of my career until the very last day of her life when she asked me news about this work few hours before ...

"Grazie MAMMA".

* E-mail address: fammannati@yahoo.com. Fax +39-06-7259-7129.

Abstract

Mass conservation equation is employed to study the time evolution of the amount of non-renewable energy resources according to the extraction policy, while the energy conservation equation is transformed into an energy-capital conservation equation to study the time evolution of the price of non-extracted resources. The Hotelling rule for non-extracted resources, i.e. an exponential increase of the price of non-renewable resources at the rate of the current interest and time, is shown to be a special case of the general energy-capital conservation equation when the mass flow rate of extracted resources is unity. Mass and energy-capital conservation equations are coupled and solved jointly to investigated the time evolution of the extracted resources. The parameter PIFE, "Price Increase Factor of Extracted resources", is introduced as the difference between the interest rate of capital, typically the inflation rate, and the mass flow rate of extraction of non-renewable resources. The price of extracted resources increases if PIFE is greater than zero, i.e. if the mass flow rate of extraction is smaller than the inflation rate. The price is constant if PIFE is zero, i.e. if the mass flow rate of extraction is equal to the inflation rate. The price is decreasing with time if PIFE is smaller than zero, i.e. if the mass flow rate of extraction is greater than the inflation rate. The price evolution with time of the resources sold to the market is investigated in case of no-accumulation and no-depletion of the resources; i.e. when the resources are extracted and sold to the market at the same mass flow rate. The price of sold resources varies with time according to the relation between the parameters PIFE and PIFS, "Price Increase Factor of Sold resources", which is the difference between the extraction rate and the interest rates of sold resources, typically prime rate. The price of sold resources increases with time if the initial price is greater than CIPS, "Critical Initial Price of Sold resources", which depends on the initial price of extracted resources, the interest rate of non-extracted resources, and the difference between PIFS and PIFE or greater than CIPES, "Critical Initial Price Extreme of Sold resources", which depends on the initial price of extracted resources, the interest rate of non-extracted resources, and PIFS. The price of sold resources increases temporarily with time if the interest rates of non-extracted and extracted resources are equal, i.e. PIFE is equal to PIFS, and the initial price is greater than CIPES, "Critical Initial Price Extreme of Sold resources". Examples for oil and gas are discussed extensively in the paper.

Nomenclature

Latin

$G_k = G\,p\,r$	capital flow rate
G	mass flow rate
$K=G\,p$	capital flow
M	mass
p	price
$p^*_{S0}=r_N\,p_{E0}/(r_E-r_N)$	critical initial price of sold resources, CIPS
$p^{**}_{S0}=r_N\,p_{E0}/\beta'$	critical initial extreme price of sold resources, CIPES
r	interest rate (annum^{-1})
t	time

Greek

$\alpha = \frac{1}{G}\frac{dG}{dt}$	extraction rate
$\beta = r_N - \alpha$	price-increase factor of extracted resources, PIFE
$\beta' = r_E - \alpha$	price-increase factor of sold resources, PIFS

Subscript

0	initial
E	extracted resources
f	final
k	capital
m	maximum or minimum
max	maximum
min	minimum
N	non-extracted resources
S	sold resources

Introduction

The price evolution with time of non-renewable resources, also known as Hotelling rule, [1-2], states that the price of a resource increases exponentially with the product of time and the interest rate of capital.

The Hotelling rule was generalized in [3] with the conclusion that the price of an extracted resource increases exponentially with the product of time and the parameter PIFE, "Price Increase Factor of Extracted resources", i.e. the difference between the interest rate of capital (usually the inflation rate) and the extraction rate of the resource. A further generalization was done in [4] with the introduction of the price of a sold resource, which depends on PIFE and the new parameter PIFS, "Price Increase Factor of Sold resources", i.e. the difference between the interest rate of capital (usually the prime rate) and the extraction rate of the resource. The energy supply curve, i.e. price versus consumption, of a non-renewable energy resource was constructed in [5] where new parameters were introduced to forecast the price evolution of non-renewable resources. Other new parameters have been introduced to forecast the time evolution of non-renewable resources as oil [6] and gas [7].

The approach of using mass and energy-capital conservation equations [3-7] is reviewed in the present work which presents several numerical applications to recent time evolutions of non-renewable energy resources.

Mass Conservation Equation for Non-Extracted Resources [3]

Let assume the non-extracted resources are in the reservoir, C, reported in figure 1, with mass, M. The mass flow rate of extraction from the reservoir is, G, with dimensions (mass/annum).

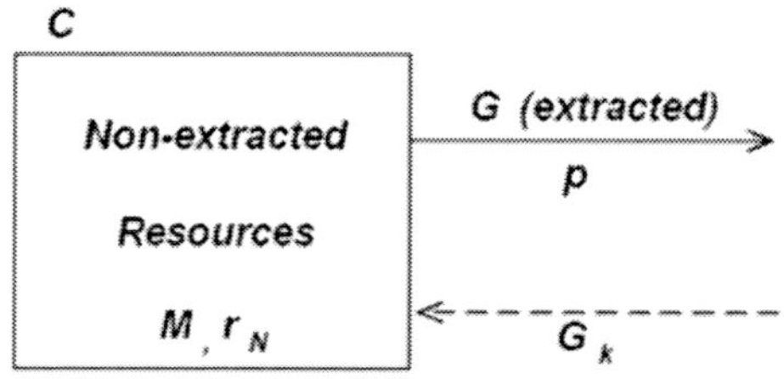

Figure 1. Non-extracted resources in the reservoir C, with mass, M and mass flow rate of extraction, G [3].

The mass conservation equation of the non-extracted resources is

$$\frac{dM}{dt} = -G \tag{1}$$

Integration of Eq. 1 gives

$$M = M_o - \int_0^t G\, dt \tag{2}$$

where M_0 is the mass of resources in the reservoir at the initial time $t = 0$.

Assuming the extraction rate, α, at the time t, given by

$$\frac{1}{G}\frac{dG}{dt} = \alpha \tag{3}$$

the mass flow rate of extraction, obtained from the integration of Eq. 3, with constant α in the time interval $0 - t$, is

$$G = G_o \exp(\alpha t) \tag{4}$$

The relative evolution of the mass flow rate of extraction is presented in figure 2. The evolution of the mass flow rate of extraction depends on the value of α: G is constant with time if $\alpha = 0$, G increases with time if $\alpha > 0$; G decreases with time if $\alpha < 0$.

The mass of non-extracted resources in the reservoir C, Eq. 2, becomes, using Eq. 4,

$$M=M_0+ G_0\,[1-exp(\alpha\, t)]/\alpha \tag{5}$$

and its evolution with time is presented in figure 3 , as M/G_0.

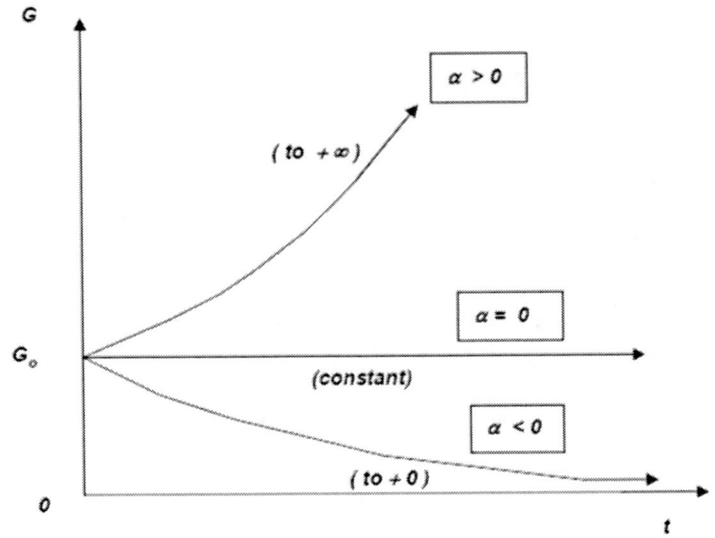

Figure 2. Evolution with time of mass flow rate of extraction, G, at the extraction rate α [3].

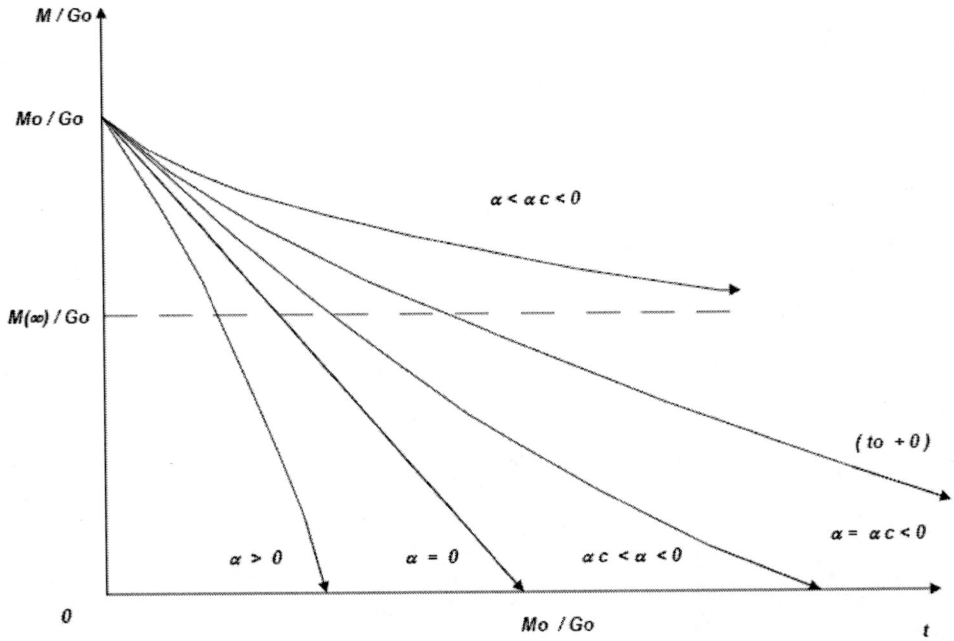

Figure 3. Evolution with time of M/G_0 (years) for non-extracted resources at the extraction rate α [3].

In case of a constant extraction rate, $G = G_0$, i.e. $\alpha = 0$, the integration of Eq. 1 gives

$$M = M_o - G_o\, t \tag{6}$$

i.e. the mass of resources in the reservoir C decreases linearly with time. Equation 6 gives the time $t_f = M_0 / G_0$ when M=0, i.e. the exhaustion time of the non-extracted resources. If, for example, the ratio M_0/G_0 is equal to 50 years, the time to exhaust the resources is t_f = 50 years.

If $\alpha > 0$, the mass of resources in the reservoir C is exhausted faster than for $\alpha = 0$, i.e. before 50 years. The time t_f, which gives $M = 0$, is given by Eq. 5 as:

$$t_f = \frac{1}{\alpha}\ln\left[1+\alpha\frac{M_o}{G_o}\right] \tag{7}$$

Assuming $Mo/Go=50$ years, the number of years for the exhaustion of the resources depends on the value of α. From Eq. 7 it can be found:

- t_f = 40.55 years for α = 0.01 (1/annum);
- t_f = 17.92 years for α = 0.1 (1/annum).

If $\alpha < 0$, Eq. 5 can be written as

$$M=M_0 - G_0\ [1-exp(|\alpha|\ t)]/|\alpha| \tag{8}$$

which, for an infinite time $t \rightarrow \infty$, gives

$$M(\infty) = M_o - \frac{G_o}{|\alpha|} \tag{9}$$

The extraction policy to exhaust the non-renewable resources in the infinite time $t \rightarrow \infty$, is to have

$$M(\infty) = M_o - \frac{G_o}{|\alpha|} = 0, \tag{10}$$

which gives the critical extraction rate, α_c,

$$|\alpha_c| = \frac{G_o}{M_o} \tag{11}$$

For $M_0/G_0 = 50$ years the critical extraction rate is α_c = - 0.02 (1/annum).

If $\alpha_c < \alpha < 0$, the time to exhaust the resources is given by

$$t_f = -\frac{1}{|\alpha|}\ln\left[1-|\alpha|\frac{M_o}{G_o}\right]. \tag{12}$$

which depends on α. For $M_0/G_0 = 50$ years the time t_f is:

- t_f = 69.31 years for α = - 0.01 (1/annum);
- t_f = 51.29 years for α = - 0.001 (1/annum).

The owner of the non-extracted resources does not find convenient to have

$$M(\infty)=(M_0 - G_0/|\alpha|)>0 \tag{13}$$

for an infinite time, $t \to \infty$, because the resources could remain non-extracted.

If $\alpha < \alpha_c < 0$ the non-renewable resources will never be extracted and a certain amount of resources will remain in the reservoir as non-extracted. For example, if α=-0.04 (1/annum) the amount of resources which will remain non-extracted is $M(\infty) / G_0$=25 years.

Examples for Oil in the World [6]

The real evolution of the mass flow rate of oil extraction in the world [8] is presented in figure 4 for the last 41 years, i.e. for the period 1965-2006. From 1965 until 1973 it was present an increase of G at high rate. From 1973 to 1975 the mass flow rate decreased while a subsequent increase was present from 1975 until 1979 with a trend similar to that of the period 1965-1973. From 1979 until 1983 the mass flow rate decreased while it was constant from 1983 until 1985. From 1985 until 2006 the mass flow rate has continuously increased at a lower rate than during the period 1965-1973, with the exception of 1999, 2001 and 2002.

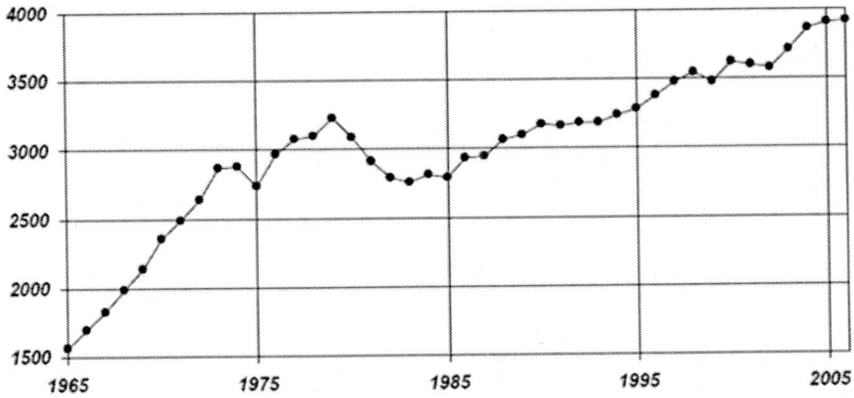

Figure 4. Mass flow rate of oil extraction, G (MT/year), versus year (1965-2006) [6].

The real extraction rate, α, can be evaluated from Eq. 4 using the data of two successive years. The extraction rate from 1966 until 2006 is presented in figure 5. The extraction rate oscillated around 8 (1/annum) from 1966 to 1976 with a minimum below -4 (1/annum) in 1975. From 1977 until 2006 the extraction rate oscillated between 5 (1/annum) and -5 (1/annum) with the trend to go to zero during the last two years 2005-2006.

The evolution of the proved reserves of oil in the world [8] is reported in figure 6 in the period 1980-2006. A consistent increase of the oil reserves in the world can be observed with almost a double value in 2006 as compared to 1980.

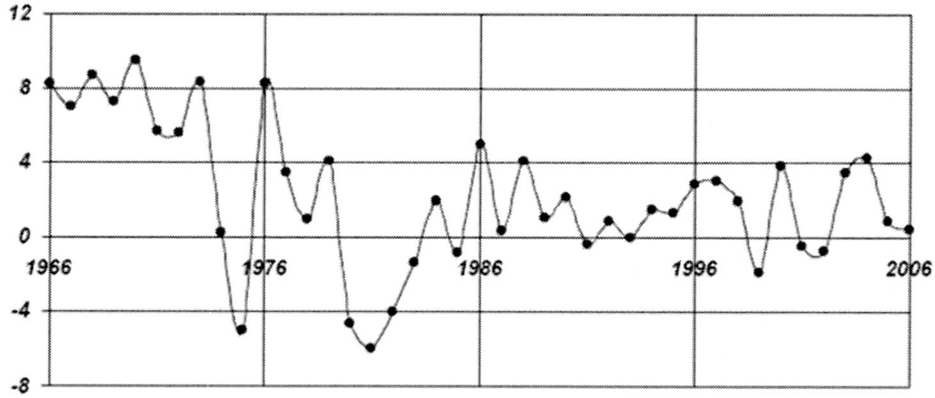

Figure 5. Extraction rate of oil, α (1/annum), versus year (1966-2006) [6].

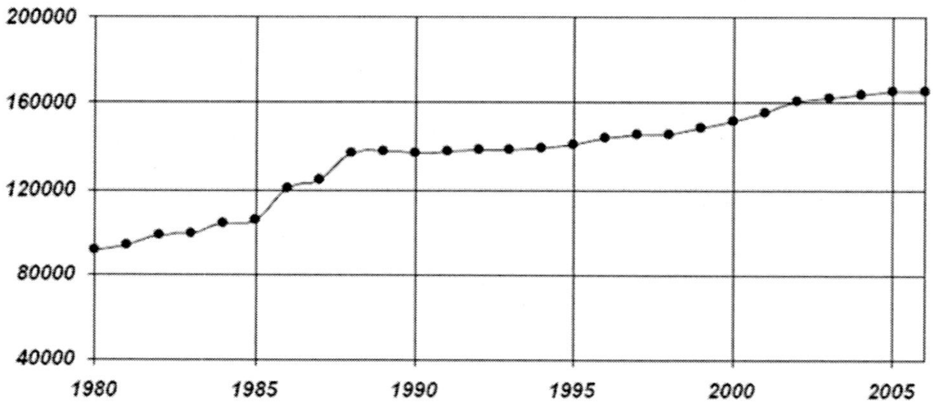

Figure 6. Evolution of the proved reserves of oil, *M* (M T), versus year (1980-2006) [6].

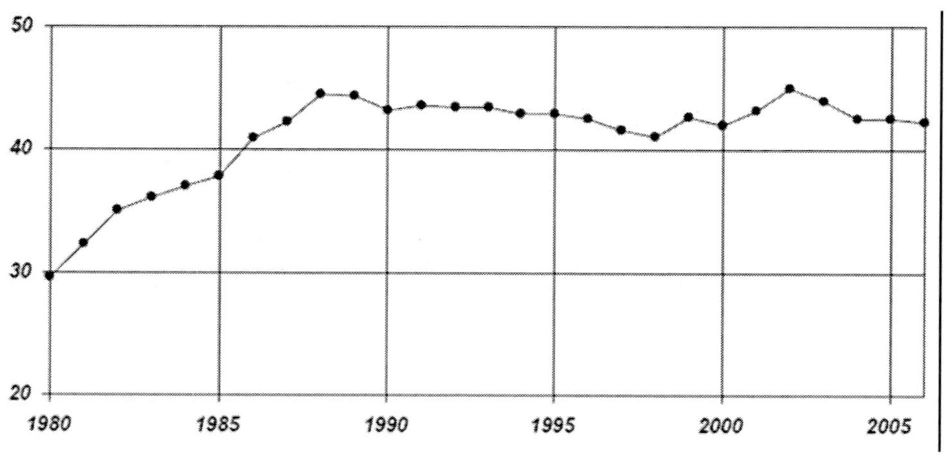

Figure 7. Evolution of *M/G* (years) of oil versus year (1980-2006) [6].

The evolution of the ratio *M/G* from 1980 until 2006, evaluated from figures 6 and 4, is reported in figure 7. The ratio *M/G* increased from 1980 until 1988 but from 1989 it is oscillating around the value of about 42-43 years.

It is possible to evaluate the critical extraction rate, α_c, from Eq. 11, which is presented in figure 8 together with the annual average extraction rate, α. The conclusion of figure 8 is that the extraction rate decreased from the high values of the 60's to values around zero, i.e. closer to the critical extraction rate.

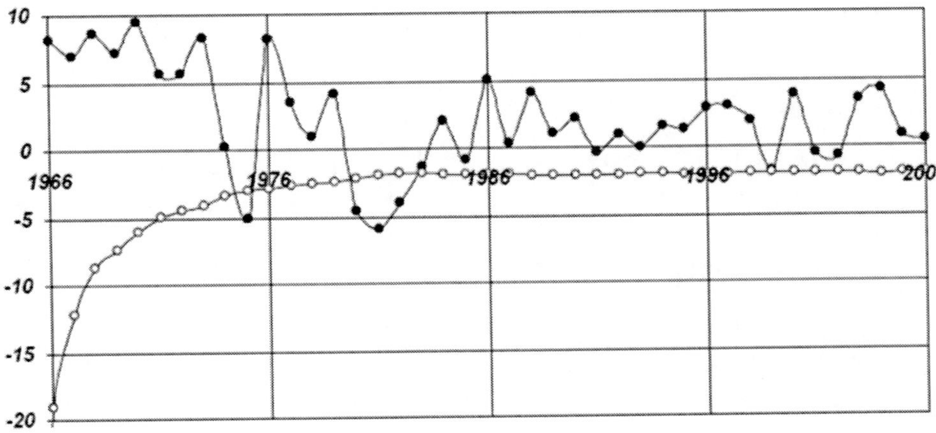

Figure 8. Evolution of extraction rate α (1/annum) and critical extraction rate α_c (1/annum) of oil in the world versus year (1966-2006) [6].

Examples for Gas in the World [7]

The real evolution of the mass flow rate of gas extraction in the world [8] is presented in figure 9 from 1970 until 2006. From 1970 to 1980 it was present a continuous increase of G. From 1981 to 1983 the mass flow rate was constant but from 1984 until 2006 a continuous increase is present, with an average trend similar to the period 1970-1980.

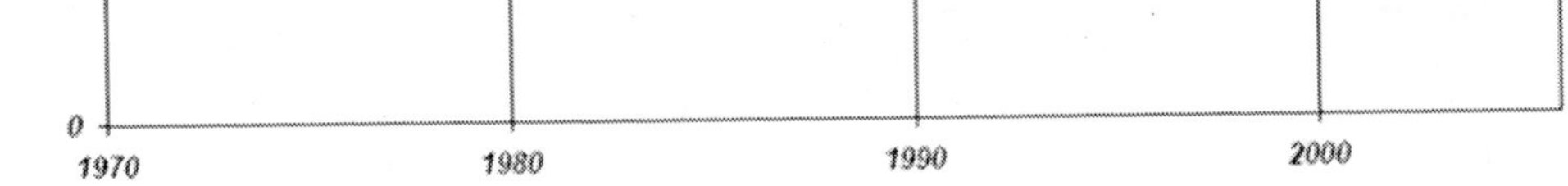

Figure 9. Mass flow rate of extraction of the gas in the world, G (M m^3/year), versus year (1970-2006) [7].

The extraction rate, α, of the gas in the world is reported in figure 10. The general conclusion of figure 10 is that the extraction rate oscillated from 0 to 8 (1/annum) from 1974 until 1984 while the oscillations are within the range 0-5 (1/annum) from 1985 until 2006.

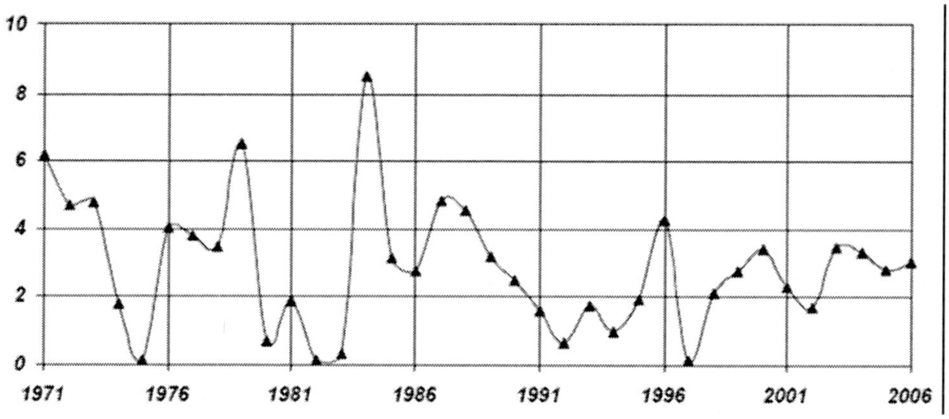

Figure 10. Extraction rate of the gas in the world, α (1/annum), versus year (1971-2006) [7].

The evolution of the proved reserves of gas in the world [8] is reported in figure 11 during the period 1980-2006. It can be observed a consistent increase of the proved reserves of the gas in the world which have more than doubled from 1980 until 2006.

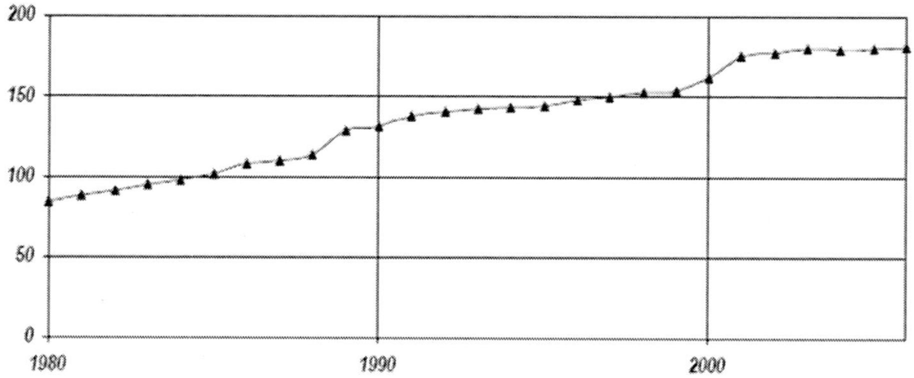

Figure 11. Evolution of the proved reserves of the gas in the world, M (Trillion m^3) versus year (1980-2006) [7].

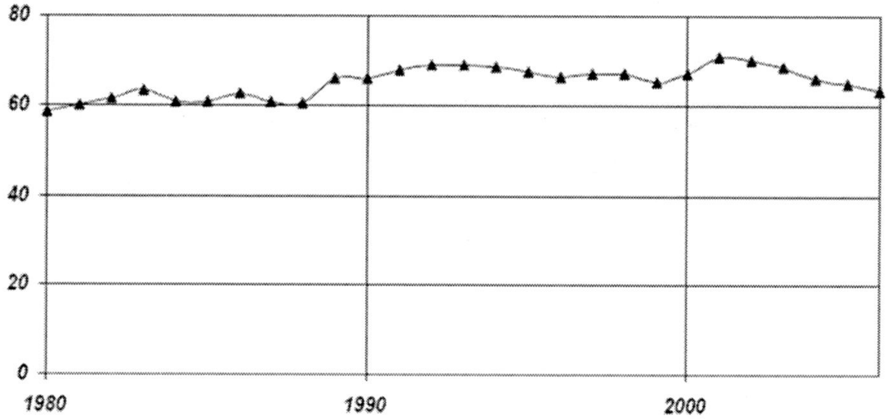

Figure 12. Evolution of M/G (years) of the gas in the world versus year (1980-2006) [7].

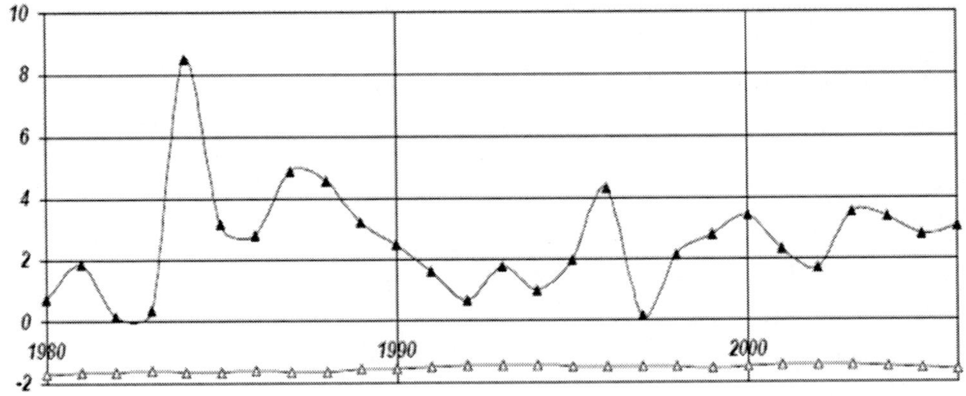

Figure 13. Evolution of extraction rate, α (1/annum), and critical extraction rate, α_c (1/annum), of the gas in the world versus year (1980-2006) [7].

The evolution of the ratio M/G of the gas in the world from 1980 until 2006,evaluated from figures 11 and 9, is reported in figure 12. The ratio is oscillating between 60 and 70 years.

The critical extraction rate, α_c, can be evaluated from Eq. 11 and is presented in figure 13, together with the annual extraction rate, α. Figure 13 shows that the extraction rate has oscillated, between 0 and 5 (1/annum) with the exception of the spike of 8 (1/annum) in 1984. In conclusion, the extraction rate of gas in the world is not showing any trend towards the critical extraction rate, as in the case of oil.

Examples for Gas in Europe at #25 [7]

The evolution of mass flow rate of gas extraction in Europe at #25 [8] is presented in figure 14 from 1970 until 2006. From 1970 to 1976 it was present a continuous increase of G. From 1977 until 1990 the mass flow rate of extraction slightly decreased with some oscillations. A continuous increase is present from 1990 until 1996 with a high rate. From 1997 until 2004 the mass flow rate has been practically constant while G has decreased from 2004 until 2006.

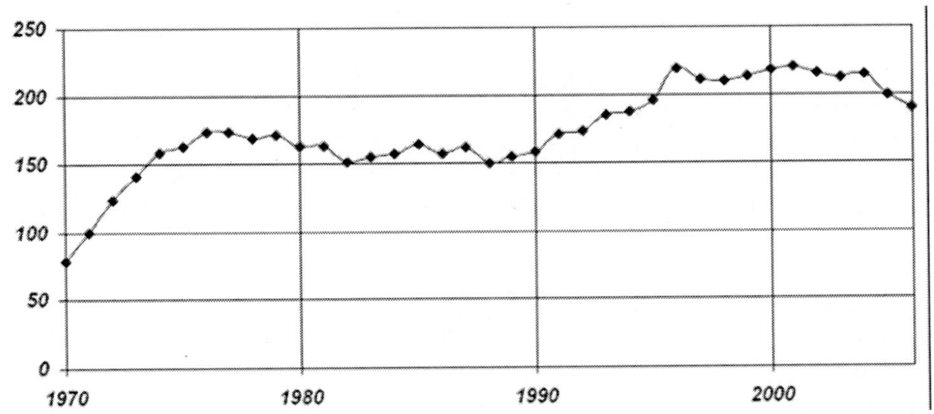

Figure 14. Mass flow rate of gas extraction in Europe at #25, G (G m^3/year), versus year (1970-2006) [7].

The extraction rate, α, of gas in Europe at #25 is reported in figure 15. The extraction rate was above 20 (1/annum) during 1971-72 and above 10 (1/annum) during 1973-74. From 1975 until 2006 the oscillations are in the range from -10 (1/annum) to +10 (1/annum) with the exception of 1996.

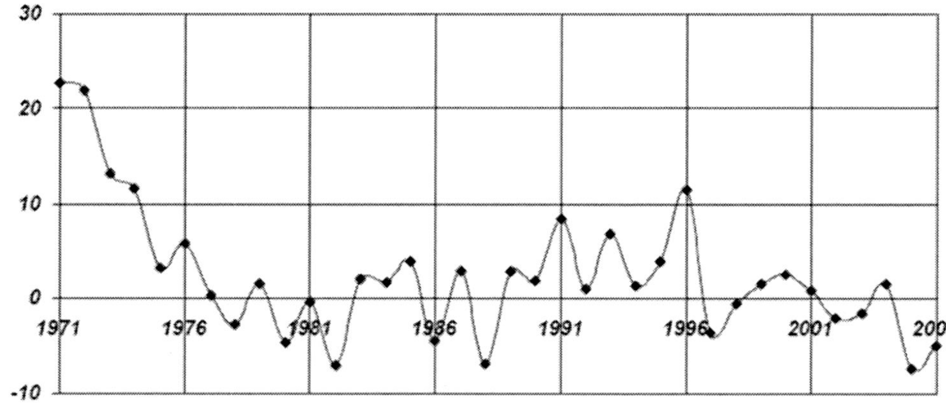

Figure 15. Extraction rate of gas in Europe at #25, α (1/annum), versus year (1971-2006) [7].

The evolution of the proved reserves of gas in Europe at #25 [8] is reported in figure 16 during the period 1980-2006. Differently from what observed for oil and gas in the world the proved reserves of gas in Europe at #25 were constant from 1980 until 1997 while they decreased from 1998 until 2006.

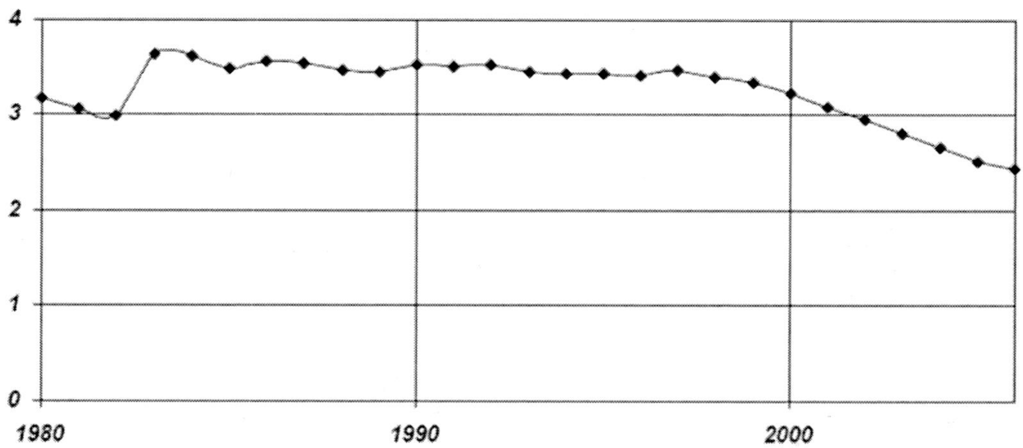

Figure 16. Evolution of the proved reserves, M (Trillion m^3), of gas in Europe at #25 versus year (1980-2006) [7].

The evolution of the ratio M/G of gas in Europe at #25, evaluated from figures 16 and 14, is reported in figure 17 from 1980 until 2006. The ratio M/G was oscillating around 20 from 1980 until 1992 but is continuously decreasing from 1993 and in 2006 is almost half the value it was in the middle of the 80's.

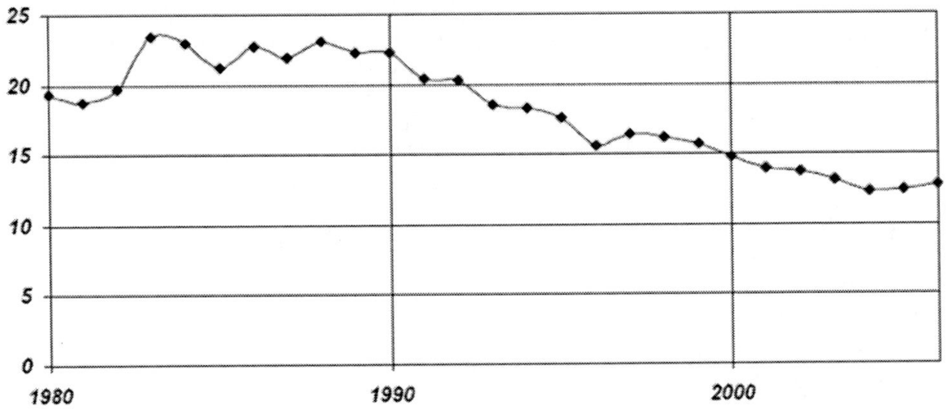

Figure 17. Evolution of M/G (years) for gas in Europe at #25 versus year (1980-2006) [7].

The critical extraction rate, α_c, can be evaluated from Eq. 11 and is presented in figure 18, together with the annual extraction rate, α. From figure 18 it can be concluded that the extraction rate was oscillating around the critical extraction rate in the 80's, it was higher than α_c from 1990 until 2004, and only in the last two years has become closer to α_c .

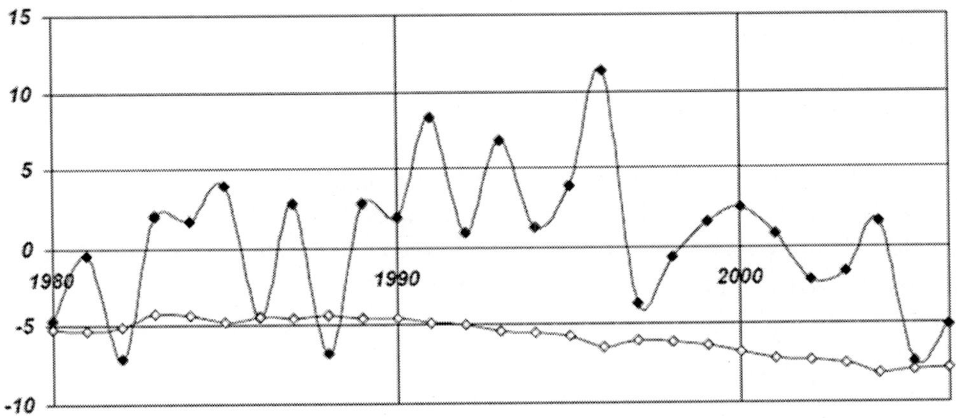

Figure 18. Evolution of extraction rate, α (1/annum), and critical extraction rate, α_c (1/annum), for gas in Europe at #25 versus year (1980-2006) [7].

Generalization of the Hotelling Rule [3]

The so-called Hotelling rule has been generalized in [3] with an approach which employs mass and energy-capital conservation equations normally used in thermal engineering. Let assume the non-renewable resource can be extracted without cost. Denote p_o the initial price per unit of resource at the initial time $t = 0$ and suppose the interest rate, r, is constant in the time interval from $t = 0$ to t. The owner of the resources unit has two options. To keep the unit of resources in the reservoir for the time interval t with the real expectation that the price at t will be higher, let say $p(t)$, or to sell the unit of resources and put the money, obtained from the sale, in a bank for the same time interval t, to get an interest return equal to $p(t)\ t\ r$.

Under competitive conditions the owner is indifferent between these two options, then the price of the unit at t, $p(t)$, must be equal to the price p_0 at the time t_0, increased of the interest return $p(t)\ t\ r$. In other words

$$p(t) = p_0 + p(t)\ t\ r \tag{14}$$

or, taking the limit as $t \rightarrow 0$, it can be obtained

$$\frac{dp}{dt} = pr\ , \tag{15}$$

After integration, Eq. (15) becomes

$$p(t) = p_0\ exp(r\ t) \tag{16}$$

also called Hotelling rule, [1], in honour of H. Hotelling [2], where, r, the interest rate of capital is assumed constant in the time period $0 - t$.

Equation (16) states that the price of non-renewable resources can only increase exponentially with time if the interest rate r is positive. The price evolution is independent of any other variable present in the energy game of non-renewable resources.

Interest Rates [6-7]

The interest rates considered in the present theoretical approach are the inflation, the discount and the prime rates, [9-10], and are reported in figures 19 and 20, respectively as average annual and average monthly. It can be observed that during some periods the average inflation rate has been higher than the discount and even the prime rate.

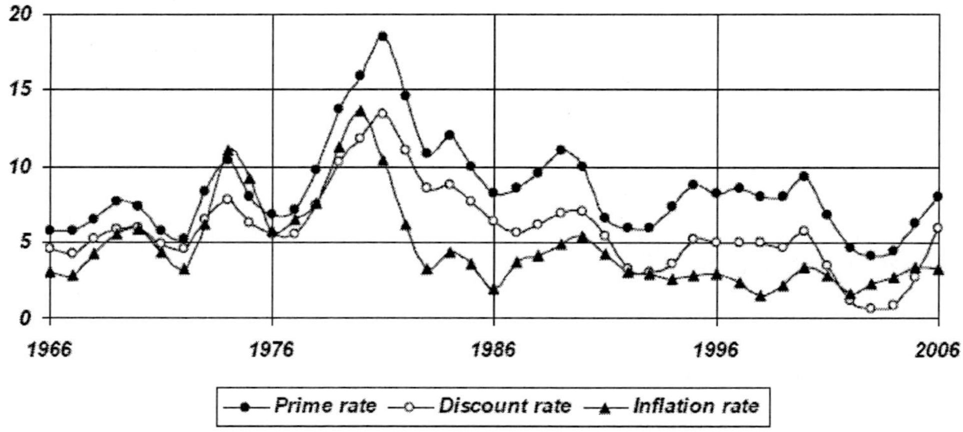

Figure 19. Evolution of the average annual interest rates (1/annum) versus year (1966-2006) [6-7].

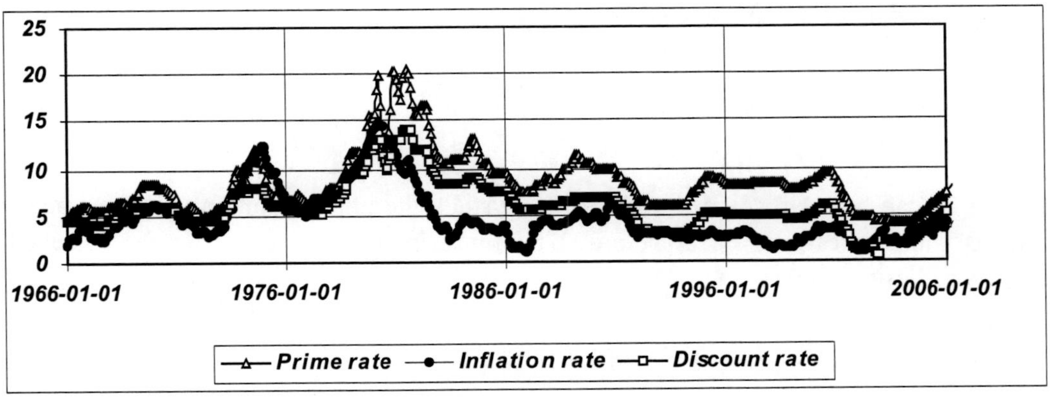

Figure 20. Evolution of the average monthly interest rates (1/annum) versus month (1966-2006) [6-7].

Energy-Capital Conservation Equation of Non-extracted Resources [3]

The energy-capital conservation equation can be written in unsteady state, similarly to the energy conservation equation used in the thermal engineering analysis, for the non-extracted resources of the reservoir C, figure 1, with interest rate r_N (1/annum), as

$$\frac{dK}{dt} = G_k \tag{17}$$

The left term expresses the time variation of the capital K of the resources in the reservoir C, while G_k is the capital flow rate entering into the reservoir C, which derives from the mass flow rate of extraction, G.

The capital K is written as

$$K = G\,p \tag{18}$$

where p is the current price of a unity of extracted resources and G the mass flow rate of extraction.

The right term of Eq. 17, i.e. the capital flow rate entering into the reservoir, is written as

$$G_k = K\,r_N = G\,p\,r_N \tag{19}$$

where r_N is the interest rate of non-extracted resources.

Equation 17 then becomes

$$d(G\,p)/(dt)=G\,p\,r_N \tag{20}$$

which, for a mass flow rate of extraction equal to 1, i.e. $G=1$, gives Eq. 15, i.e. the Hotelling rule before integration. In conclusion, the Hotelling rule can be obtained from the energy-capital conservation equation with a mass flow rate of extraction equal to $G = 1$.

Differentiation of the capital conservation equation, Eq. 20, and division by the capital K gives

$$\frac{1}{p}\frac{dp}{dt}+\frac{1}{G}\frac{dG}{dt}=r_N \tag{21}$$

Equation 21 can be used, coupled with Eq. 3, to study the price evolution of the extracted resources with interest rate r_N. Equation 21 becomes, indicating p as p_E for the extracted resources and using Eq. 3,

$$\frac{1}{p_E}\frac{dp_E}{dt}=r_N-\alpha \tag{22}$$

which, after integration, gives

$$p_E = p_{E0}\, exp\ (\,\beta t\,) \tag{23}$$

where,

$$\beta = r_N - \alpha, \tag{24}$$

i.e. the difference between the interest rate of non-extracted resources, r_N , and the extraction rate, α, is the new parameter "Price Increase Factor of Extracted resources", PIFE. The price evolution with time of the extracted resources, given by Eq. 23, is shown in figure 21.

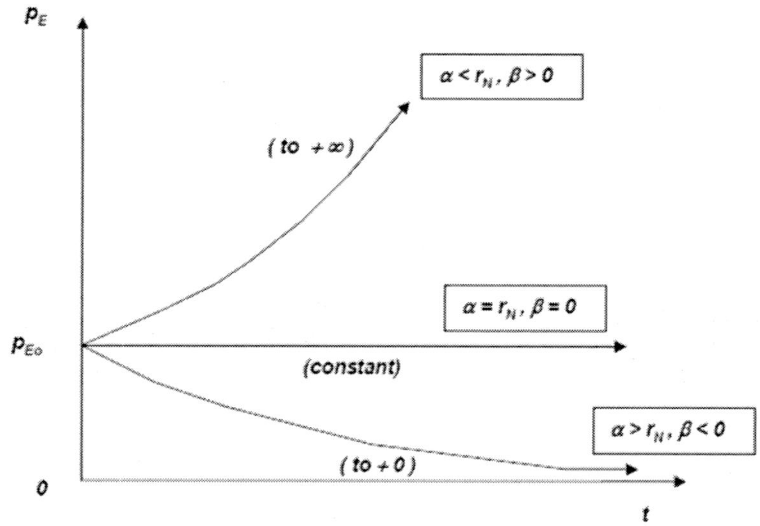

Figure 21. Price evolution with time of extracted resources according to PIFE [3].

The price of extracted resources evolves with time according to the value of PIFE, β. If the extraction rate α is equal to the interest rate of non-extracted resources, r_N , the price-increase factor, PIFE is equal to zero, $\beta = 0$, and the price p_E remains constant with time. If the extraction rate α is higher than r_N , the price-increase factor, PIFE is negative, $\beta < 0$, and the price p_E decreases with time. If the extraction rate α, is lower than r_N, the price-increase factor, PIFE is positive, $\beta > 0$, and the price p_E increases with time. In case of a conservative policy of extraction, i.e. $\alpha < 0$, the price-increase factor is greater than zero, $\beta > 0$ and the price p_E increases with time.

Examples for Oil [6]

The evolution of the oil price WTI (West Texas Intermediate) [10] is reported in figures 22 and 23 from 1966 until 2006, respectively as annual and monthly average. The oil price has the dimensions of US$ of the day per Giga Joule (US$/GJ). The oil price evolution is well know and is not necessary to make comments. It can be reminded that the oil price increase of the end of 2003 was forecasted in 2001 [11].

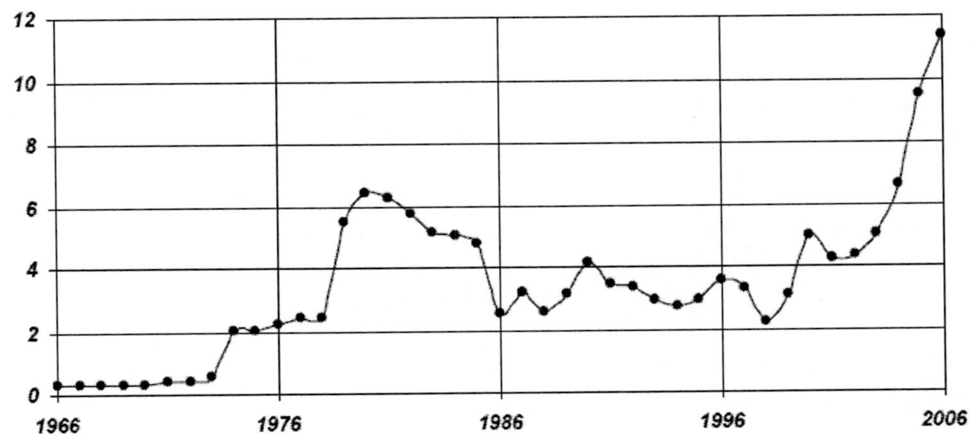

Figure 22. Evolution of the average annual price of oil (US$/GJ) versus year (1966-2006) [6].

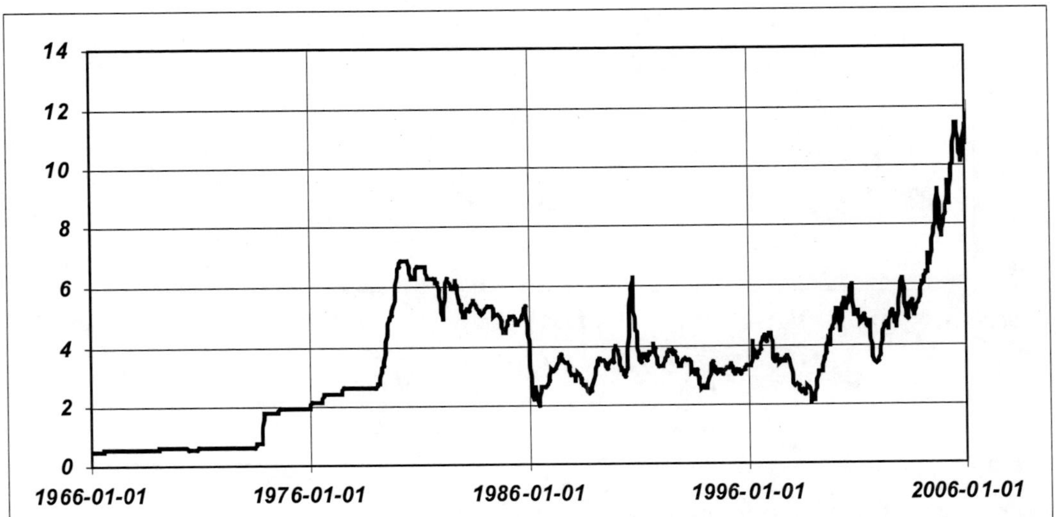

Figure 23. Evolution of the average monthly price of oil (US$/GJ) versus month (1966-2006) [6].

The evolution of the oil price is compared to the inflation rate in figures 24 and 25. From 1966 until 1973 there is no relation between oil price and inflation rate because the oil price remains constant despite a large oscillation of the inflation rate. The sharp increase of inflation during 1973 and 1974 is related to the oil price increase of 1974. The same happens with the increase of inflation in the late 70's which is related to the increase of oil price during 1979. From 1982 until 2003 the relation between oil price and inflation rate has been stronger while the relation has become weaker from 2004.

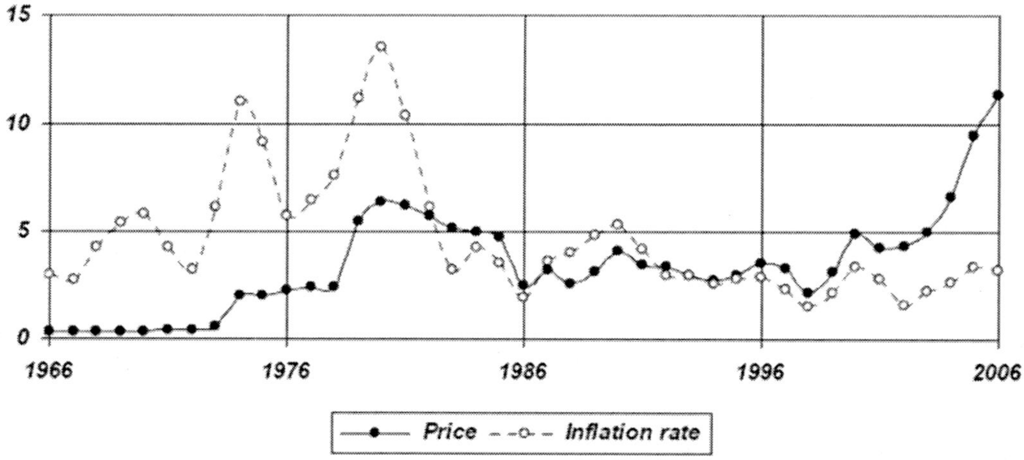

Figure 24. Evolution of average annual inflation rate (1/annum) and oil price (US$/GJ) versus year (1966-2006) [6].

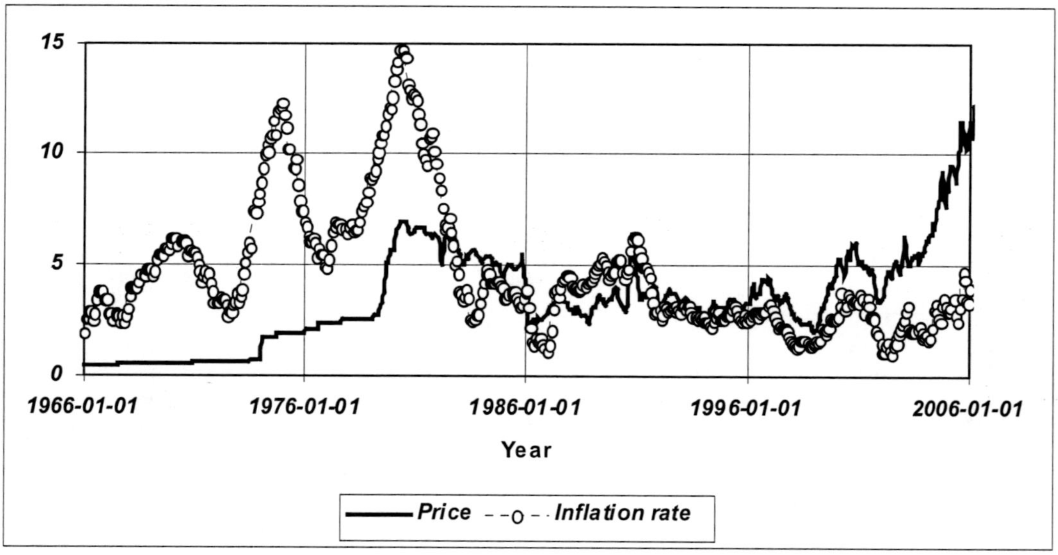

Figure 25. Evolution of average monthly inflation rate (1/annum) and oil price (US$/GJ) versus month (1966-2006) [6].

The evolutions of oil price and PIFE from 1966 until 2006 are reported in figures 26 and 27. The use of PIFE in the analysis of the oil price evolution from 1966 until 1973 is very interesting because a negative PIFE could have decreased the oil price, according to Eq. 23, which, on the other hand, was constant because of the monopolistic condition in the oil market. The sharp increase of PIFE in 1974 is well related to the increase of oil price in the same year. A similar increase of PIFE in 1980 is one year later than the price increase in 1979. From 1983 until 2006 the relation between oil price and PIFE is stronger than with the use of the inflation rate.

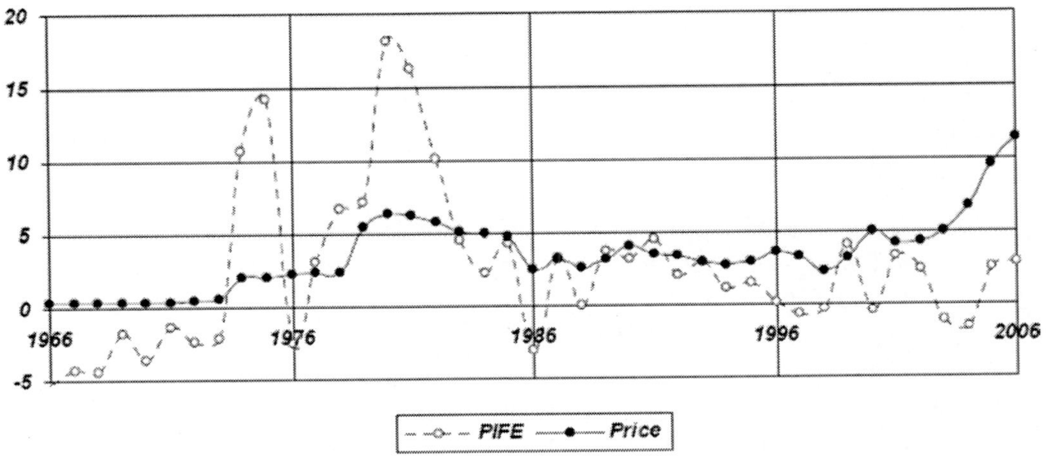

Figure 26. Evolution of PIFE and average annual oil price (US$/GJ) versus year (1966-2006) [6].

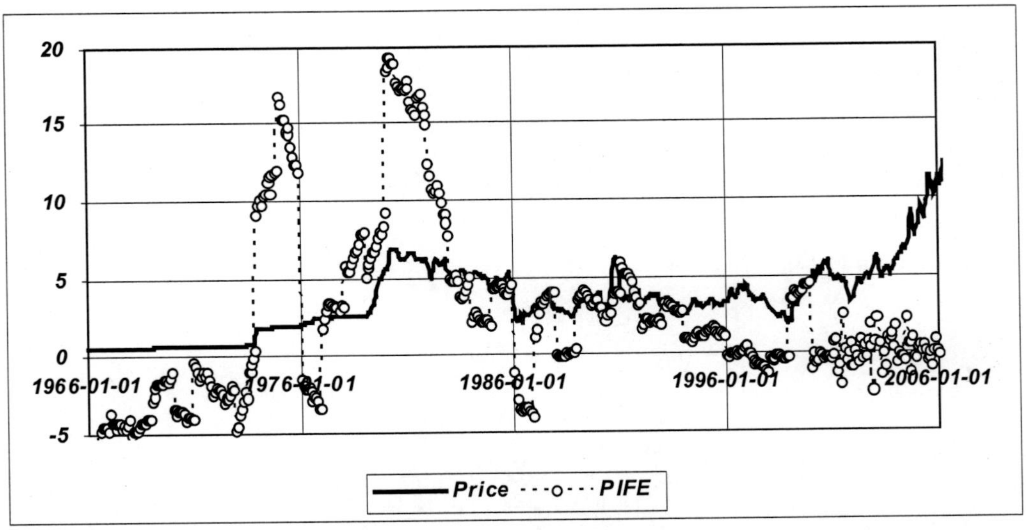

Figure 27. Evolution of PIFE and average monthly oil price (US$/GJ) versus month (1966-2006) [6].

Mass Conservation Equation of Extracted Resources and Energy-Capital Conservation Equation of Sold Resources [4]

Let assume non-extracted and extracted resources are in the two reservoirs, C_N and C_E, as reported in figure 28.

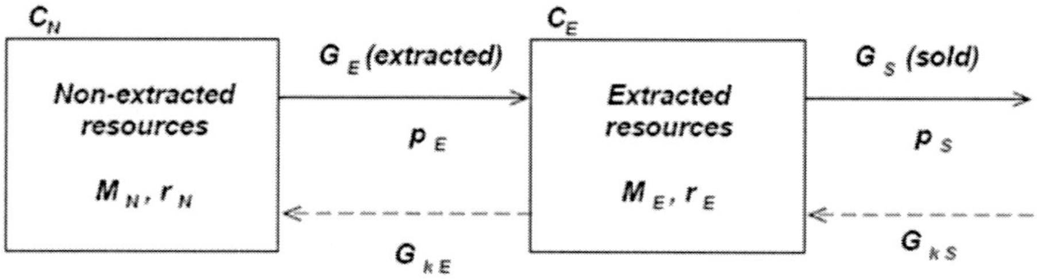

Figure 28. Mass, M, mass flow rate, G, and capital flow rate, G_k, of non-extracted, extracted and sold resources [4].

The non-extracted resources are in the reservoir C_N with mass M_N and interest rate r_N. The mass flow rate of extracted resources from the reservoir C_N is G_E while the capital flow rate, relative to the mass flow rate G_E, is G_{KE}. The resources of the reservoir C_N are extracted at the price of extracted resources, p_E, and allocated in the reservoir C_E where M_E is the mass of extracted resources. The resources of the reservoir C_E are sold at the price of sold resources, p_S. The mass flow rate of sold resources from the reservoir C_E is G_S while the capital flow rate, relative to the mass flow rate G_S, is G_{KS}.

The mass conservation equation of the extracted resources in the reservoir C_E is

$$\frac{dM_E}{dt} = G_E - G_S \tag{25}$$

while the relative energy-capital conservation equation is

$$\frac{d}{dt}(K_E) = G_{KS} - G_{KE} \tag{26}$$

which, according to Eqs. 18-19 , becomes

$$\frac{d}{dt}(G_S \cdot p_S) = G_S \cdot p_S \cdot r_E - G_E \cdot p_E \cdot r_N \tag{27}$$

and finally

$$\frac{1}{p_S} \cdot \frac{dp_S}{dt} + \frac{1}{G_S} \cdot \frac{dG_S}{dt} = r_E - \frac{G_E \cdot p_E}{G_S \cdot p_S} \cdot r_N \tag{28}$$

In case of no-accumulation and no-depletion of the extracted resources, $G_E = G_S$, the right term of Eq. 25 is zero, i.e. the mass M_E of extracted resources in the reservoir C_E is constant with time. The mass flow rates of extracted and sold resources have the same extraction rate α from the reservoirs C_N and C_E, i.e.

$$\frac{1}{G_E}\frac{dG_E}{dt} = \frac{1}{G_S}\frac{dG_S}{dt} = \alpha \tag{29}$$

which gives, after integration

$$G_E = G_S = G_{Eo} \cdot \exp(\alpha t) = G_{So} \cdot \exp(\alpha t) \tag{30}$$

and the time evolution is similar to that of figure 2.

Price Evolution of Sold Resources [4]

Equation (28) becomes, with Eq. (30),

$$\frac{1}{p_S}\frac{dp_S}{dt} + \alpha = r_E - \frac{p_E}{p_S} r_N \tag{31}$$

or

$$\frac{dp_S}{dt} = p_S(r_E - \alpha) - p_E \cdot r_N \tag{32}$$

The substitution of Eq. 23 into Eq. 32 and the solution of the relative differential equation, [12], gives the time evolution of the price of sold resources

$$p_S = p_{S0}\, exp\,(\beta' t) + p^*_{S0}\,[exp\,(\beta' t) - exp\,(\beta t)] =$$

$$p^*_{S0}\, exp\,(\beta t) + [p_{S0} - p^*_{S0}]\, exp\,(\beta' t) \tag{33}$$

where

$$\beta' = r_E - \alpha \tag{34}$$

is the "Price Increase Factor of Sold resources", PIFS, and

$$p^*_{S0} = r_N p_{E0} / (r_E - r_N) = r_N p_{E0} / (\beta' - \beta) \tag{35}$$

is the "Critical Initial Price of Sold resources", CIPS.

The price of sold resources, Eq. 33, has an extreme (maximum or minimum) for the time t_m given by

$$t_m = 1/(\beta' - \beta) ln\,(p^*_{S0}\,\beta/\beta'/(p^*_{S0} - p_{S0})) = 1/(\beta - \beta') ln\,(\beta'\,(p^*_{S0} - p_{S0})/\,\beta\,/\,p^*_{S0}) \tag{36}$$

The extreme is a maximum if

$$\beta'\,(\beta' - \beta)\,(p_{S0} - p^*_{S0}) < 0 \tag{37}$$

otherwise is a minimum.

The time t_m of the extreme is zero if p_{S0} is equal to

$$p^{**}_{S0} = p^{*}_{S0} [(\beta' - \beta) / \beta'] = r_N p_{E0} / \beta' \tag{38}$$

which is the "Critical Initial Price Extreme of Sold resources", CIPES.

The time t_m of the maximum is greater than zero if:

$$p_{S0} > p^{**}_{S0} \text{ for } \beta' > \beta \text{ (i.e. } r_E > r_N), \tag{39}$$

or

$$p_{S0} < p^{**}_{S0} \text{ for } \beta' < \beta \text{ (i.e. } r_E < r_N), \tag{40}$$

For $p_{S0} = p^{*}_{S0}$ the time of the maximum is :

$$t_m = + \infty \text{ for } \beta' > \beta \tag{41}$$

$$t_m = - \infty \text{ for } \beta' < \beta \tag{42}$$

Examples for Oil [6]

The evolutions of oil price and PIFS from 1966 until 2006 are reported in figures 29 and 30, where PIFS is evaluated as difference between prime rate and extraction rate. The use of PIFS in the analysis of the oil price evolution from 1966 until 1973 gives similar conclusions to those already remarked by the use of PIFE, i.e. a negative PIFE could have decreased the oil price according to Eq. 23, which, on the other hand, was constant because of the monopolistic condition in the oil market. The sharp increase of PIFS in 1974 is well related to the increase of the oil price in the same year. A similar increase of PIFS in 1980 is related to the price increase of the years 1979-1980 with one year of delay. From 1983 until 2006 the relation between oil price and PIFS is closer than with PIFE. It can remarked the good relation during the last three years.

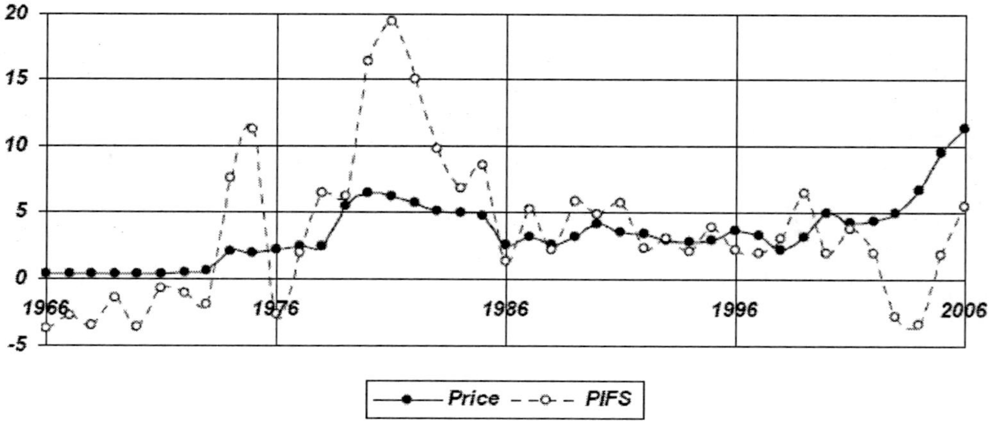

Figure 29. Evolution of PIFS and average annual oil price (US$/GJ) versus year (1966-2006) [6].

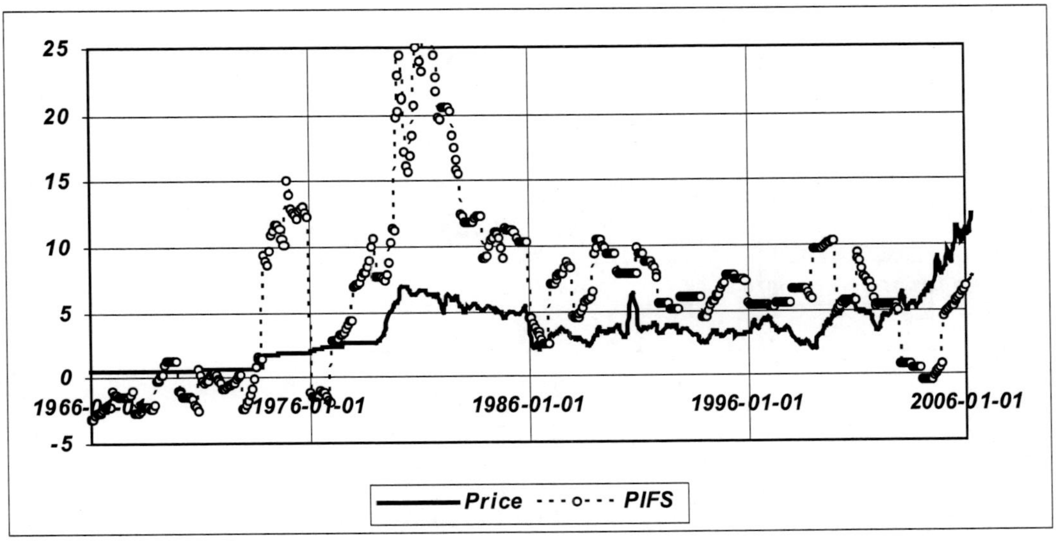

Figure 30. Evolution of PIFS and average monthly oil price (US$/GJ) versus month (1966-2006) [6].

The evolutions of oil price from 1966 until 2006 are compared in figures 31 and 32 to CIPS, p^*_{S0}, which is evaluated with the prime rate equal to r_E, the inflation rate equal to r_N and $p_{S0}=p_{E0}$. The use of CIPS, p^*_{S0}, in the analysis of the oil price evolution from 1966 until 1973 is very interesting because p^*_{S0} is almost constant in that period. During the years from 1974 to 1980 p^*_{S0} becomes negative and positive with large oscillations. From 1981 until 2006 the relation between oil price and p^*_{S0} is good, also after 2000.

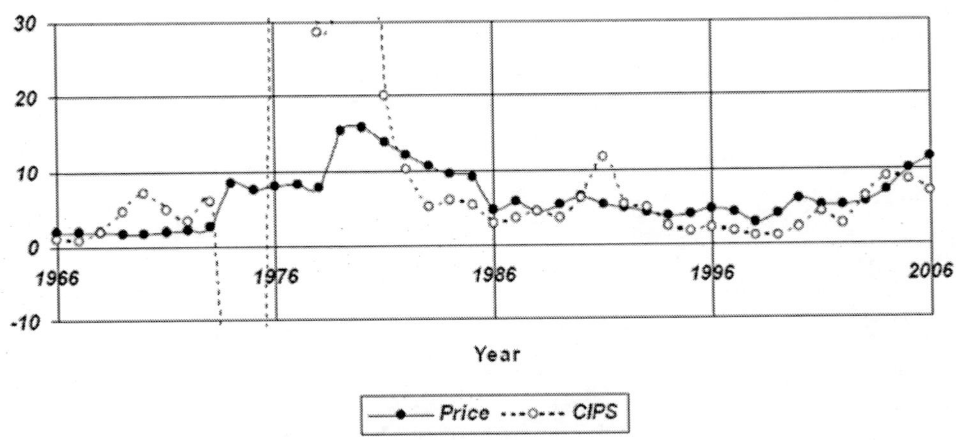

Figure 31. Evolution of average annual oil price (US$/GJ) and CIPS (US$/GJ), versus year (1966-2006) [6].

The evolution of oil price from 1966 until 2006 is compared in figures 33 and 34 to CIPES, p^{**}_{S0}, which is evaluated with the prime rate equal to r_E, the inflation rate equal to r_N and $p_{S0}=p_{E0}$. During the period from 1966 until 1968 CIPES remains constant while from 1969 until 1977 CIPES becomes negative and positive with large oscillations. From 1978 until 2002 the relation between CIPES and oil price is good while from 2003 to 2004 there are oscillations with high positive values.

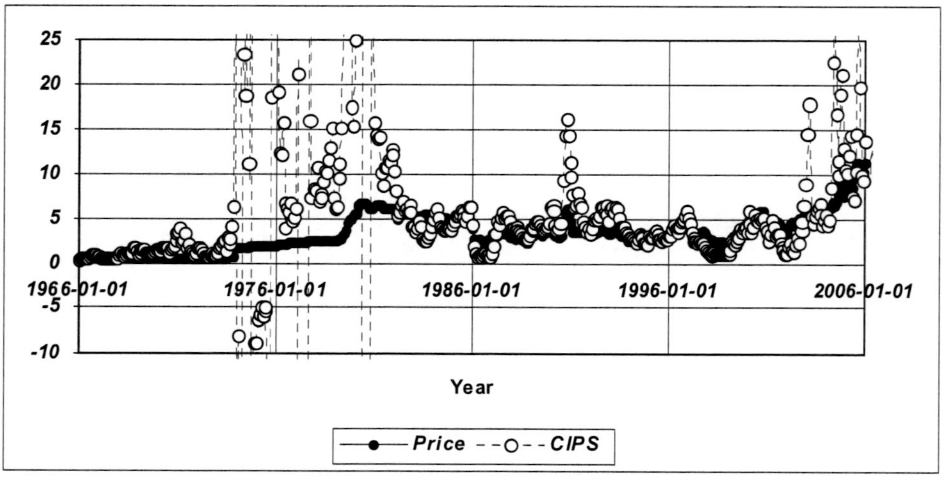

Figure 32. Evolution of average monthly oil price (US$/GJ) and CIPS (US$/GJ), versus month (1966-2006) [6].

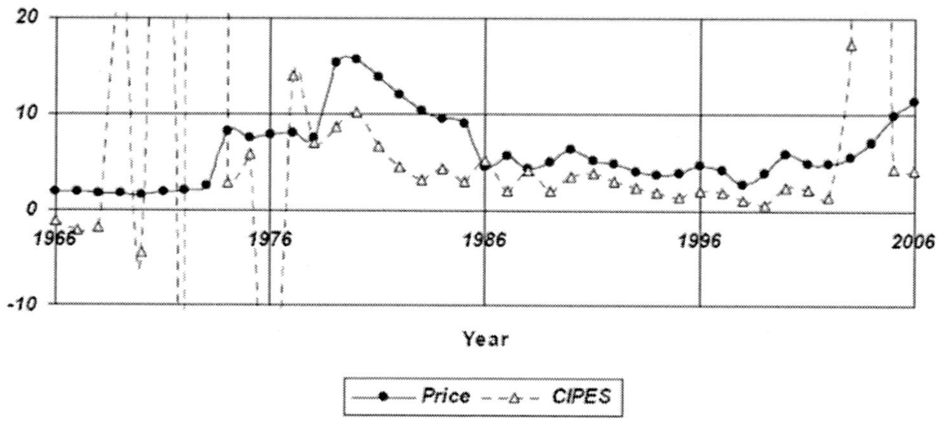

Figure 33. Evolution of average annual oil price (US$/GJ) and CIPES (US$/GJ,) versus year (1966-2006) [6].

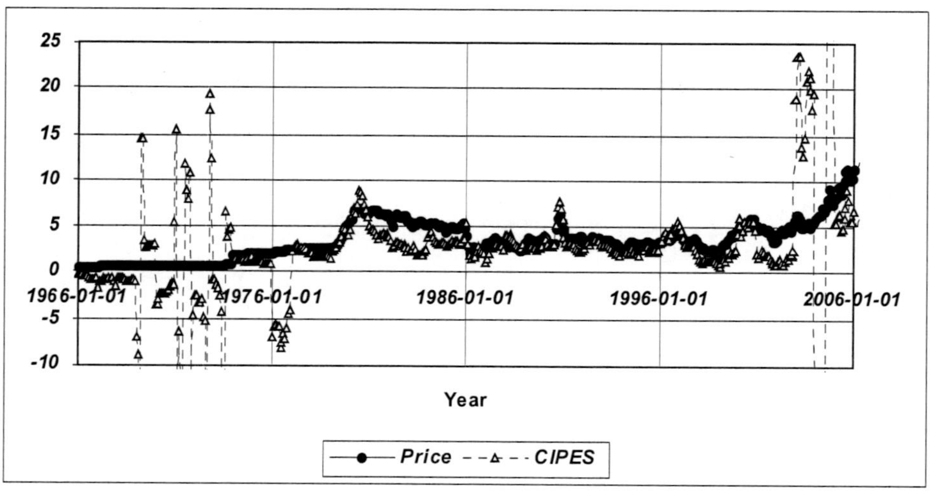

Figure 34. Evolution of average monthly oil price (US$/GJ) and CIPES (US$/GJ), versus month (1966-2006) [6].

The evolution of oil price from 1966 until 2006 is compared to CIPS and CIPES in figures 35 and 36. During the period from 1966 until 1968 both CIPS and CIPES remain constant with CIPS greater than CIPES while from 1969 until 1980 CIPES becomes greater or smaller than CIPS with large positive and negative oscillations. From 1981 until 2002 both CIPS and CIPES are well related to the oil price with CIPS generally greater than CIPES. From 2002 until 2004 CIPES has positive oscillations with higher values than CIPS.

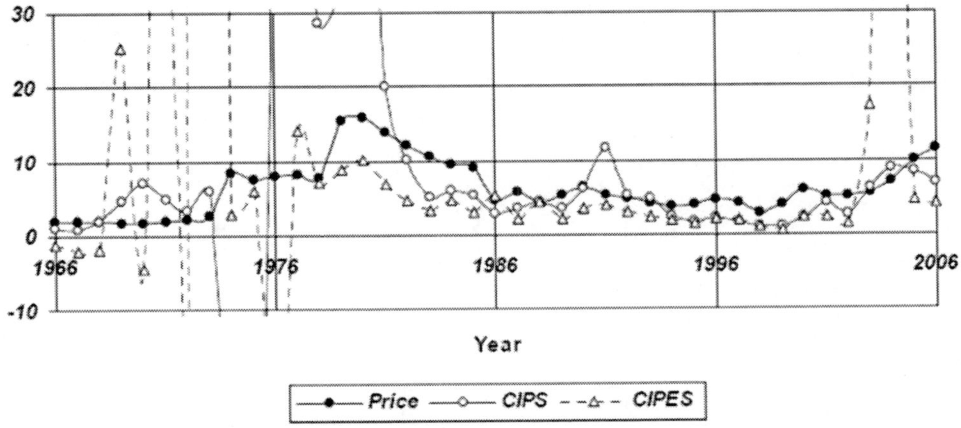

Figure 35. Evolution of average annual oil price (US$/GJ), CIPS and CIPES versus year (1966-2006) [6].

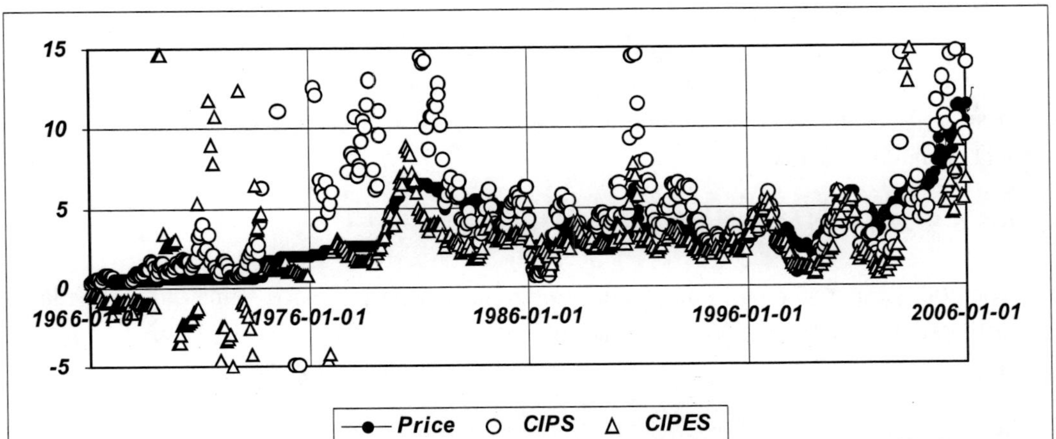

Figure 36. Evolution of average monthly oil price (US$/GJ), CIPS and CIPES versus month (1966-2006) [6].

Price Evolution of Sold Resources [4]

The price evolution of sold resources is presented in the following according to the value of the interest rate r_N and the relations between α, r_N and r_E.

Price Evolution for $r_N = 0$ [4]

If the interest rate of non-extracted resources is $r_N = 0$, (i.e. CIPS = CIPES = *0*), the price evolution of sold resources is given by:

$$p_S = p_{S0}\, exp(\beta'\, t) \tag{43}$$

where β' is the PIFS of sold resources.

Figure 37 presents the price evolution of sold resources versus month (fraction of a year) according to the relation between extraction rate, α, and interest rate of extracted resources r_E, i.e. according to PIFS.

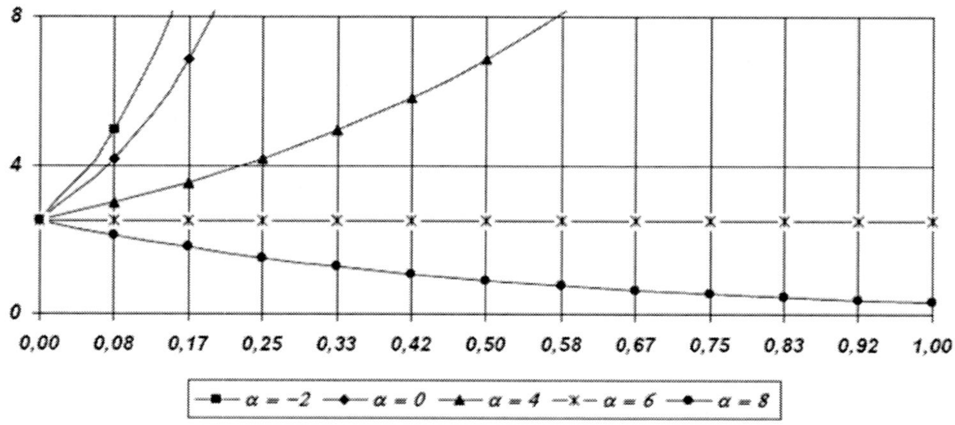

Figure 37. Price evolution of sold resources, p_S (US$/GJ), for r_N=0, r_E=6 according to the extraction rate, α (1/annum) versus fraction of a year.

a. $\alpha < r_E$, ***(i.e. PIFS =*** $\beta' > 0$***).***

The price p_S increases with time at different rates, according to the value of α. The rate of increase is higher for $\alpha < 0$, ($\beta > 0$), intermediate for $\alpha = 0$, ($\beta = 0$) and lower for $\alpha < r_E$, ($\beta < 0$).

Example for Oil

During 1986 the average discount rate was r_E=*6.33,* the average extraction rate α=*5.03* and the average inflation rate, r_N=*1.91*, which is assumed as r_N=*0* in the present example. The average world oil price increased from *2.52* US$/GJ in 1986 to *3.21* US$/GJ in 1987. The application of Eq. 43 gives the trend to increase because β'=*1.3*.

b. $\alpha = r_E$, ***(i.e. PIFS =*** $\beta' = 0$, **PIFE =** $\beta < 0$***).***

The price p_S is constant with time at p_{S0}.

c. $\alpha > r_E$, ***(i.e. PIFS =*** $\beta' < 0$, **PIFE =** $\beta < 0$***).***

The price p_S decreases with time towards $+ 0$.

In conclusion, for $r_N = 0$ the criteria to evaluate the price evolution of sold resources, p_S, is the PIFS $= \beta'$, of sold resources. If $\beta' > 0$ the price p_S increases, if $\beta' = 0$ the price p_S remains constant and if $\beta' < 0$ the price p_S decreases towards $+ 0$.

Price Evolution for $\alpha < r_N$ (PIFE = $\beta > 0$) [4]

The price evolution of sold resources, p_S, is investigated for an extraction rate, $\alpha < r_N$, (i.e. PIFE $= \beta > 0$). The price evolutions of p_S are reported in figures 38-39 for given values of α, r_N and r_E, i.e. PIFE and PIFS. The evolution depends on the relation between p_{S0}, CIPS and CIPES.

a. $r_N < r_E$ **(i.e. PIFS =** $\beta' >$ **PIFE =** $\beta > 0$ **and CIPS =** $p^*_{S0} >$ **CIPES =** $p^{**}_{S0} > 0$**).**

Three trends are possible according to the initial value p_{S0}, as shown in figure 38.

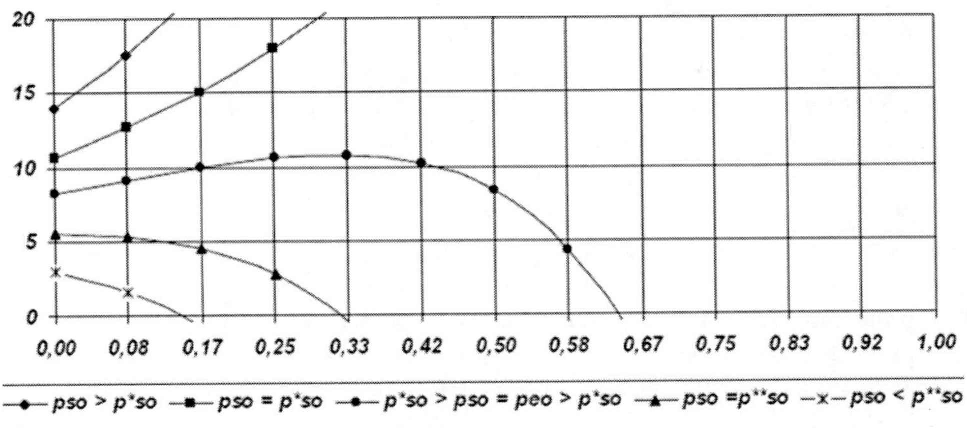

Figure 38. Price evolution of sold resources, p_S (US\$/GJ), versus fraction of a year, for $\alpha=0.88<r_N=2.97<r_E=5.25$, according to the relation between p_{S0}, $CIPS=p^*_{S0}=10.64$ and CIPES $= p^{**}_{S0}=5.55$.

a.1) If $p_{S0} => p^*_{S0}$, the price of sold resources, p_S, increases with time.

The price of sold resources increases with time for $p_{S0}= 14> p^*_{S0}=10.64$, or for $p_{S0}=p^*_{S0} = 10.64$, as reported in figure 38.

Examples for Oil

At the beginning of 1999 the average data of January were the followings: inflation rate $r_N=1.67$, prime rate $r_E=7.75$, and the annual average extraction rate was $\alpha=- 1.86$. The average monthly oil price increased from 2.174 US\$/GJ, in January 1999, to 4.546 US\$/GJ in December 1999, with some oscillations. The application of Eqs. 35 and 38 with the data of

January 1999, i.e. α=- *1.86*, r_N=*1.67* and r_E=*7.75*, gives p^*_{S0} =*0.68* and p^{**}_{S0}=*0.38* under the assumption p_{S0}=p_{E0}. Assuming the initial price of sold resource equal to p_{S0}=*2.174* US$/GJ it is predicted an increase because p_{S0}=*2.174* >p^*_{S0}=*0.68*> p^{**}_{S0}=*0.38*.

At the beginning of 2002 the average data of January were the followings: inflation rate r_N=*1.14*, prime rate r_E=*4.75*, and the annual average extraction rate was α=*-0.75*. The average monthly oil price increased from *3.429* US$/GJ, in January 2002, to *5.129* US$/GJ in December 2002, with some oscillations. The application of Eqs. 35 and 38 with the data of January 2002, i.e. α=*-0.75*, r_N=*1.14* and r_E=*4.75*, gives p^*_{S0} =*1.08* and p^{**}_{S0}=*0.71* under the assumption p_{S0}=p_{E0}. Assuming the initial price of sold resource equal to p_{S0}=*3.429* US$/GJ it is predicted an increase because p_{S0}=*3.429*>p^*_{S0}=*1.08*> p^{**}_{S0}=*0.71*.

a.2) If $p^{**}_{S0}<p_{S0}<p^*_{S0}$, the price p_S increases up to a maximum value and then decreases.

The price of sold resources increases with time up to a maximum and then decreases for p^*_{S0}=*10.64* > p_{S0} =*8.165* > p^{**}_{S0}=*5.55*, as reported in figure 38.

Examples for Oil

At the beginning of 2005 the average data in January were the followings: inflation rate r_N=*2.97*, prime rate r_E=*5.25*, and the annual average extraction rate was α=*0.88*. The average monthly price increased from *8.165* US$/GJ, in January 2005, to the maximum of *11.431* US$/GJ in September 2005, and then decreased in the rest of the 2005 with some oscillations. The application of Eqs. 35 and 38 with the data of January 2005, i.e. α=*0.88*, r_N=*2.97* and r_E=*5.25*, gives p^*_{S0} =*10.64* and p^{**}_{S0}=*5.55* under the assumption p_{S0}=p_{E0}. Assuming the initial price of sold resource equal to p_{S0}=*8.165* US$/GJ it is predicted an increase up to a maximum and then a decrease, because p^*_{S0}=*10.64*>p_{S0}=*8.165*>p^{**}_{S0}=*5.55*.

At the beginning of 2006 the average data in January were the followings: inflation rate r_N=*3.99*, prime rate r_E=*7.26*, and the annual average extraction rate was α=*0.44*. The average monthly price increased from *11.42* US$/GJ, in January 2006, to a maximum of *12.972* US$/GJ, in July 2006, and then decreased in the rest of the 2006 with some oscillations. The application of Eq. 35 and 38 with the data of January 2006, i.e. α=*0.44*, r_N=*3.99* and r_E=*7.26*, gives p^*_{S0}=*13.93* and p^{**}_{S0}=*6.68* under the assumption p_{S0}=p_{E0}. Assuming the initial price of sold resource equal to p_{S0}=*11.42* US$/GJ it is predicted an increase up to a maximum and then a decrease, because p^*_{S0}=*13.93*> p_{S0}=*11.42*> p^{**}_{S0}=*6.68*.

a.3) If p_{S0} = < p^{**}_{S0} the price p_S decreases. If p_{S0} = p^{**}_{S0} the maximum is at *t* =*0*.

The price of sold resources decreases with time for p_{S0}=p^{**}_{S0}=*5.55* or p_{S0}=*3*<p^{**}_{S0}=*5.55* as reported in figure 38.

Example for Oil

At the beginning of 1990 the average data in January were the followings: inflation rate r_N=*5.2*, discount rate r_E=*7* while the annual average extraction rate was α=*2.16*. The average monthly price decreased from *3.947* US$/GJ, in January 1990, to *2.941* US$/GJ in June 1990. The application of Eq. 35 and 38 with the data of January 1990, i.e. α=*2.16*, r_N=*5.2* and r_E=*7*, gives $p^*_{S0}=11.4$ and $p^{**}_{S0}=4.24$ under the assumption $p_{S0}=p_{E0}$. Assuming the initial price of sold resource equal to $p_{S0}=3.947$ US$/GJ it is predicted a decrease because $p^*_{S0}=11.4>p^{**}_{S0}=4.24>p_{S0}=3.947$.

b. $r_N > r_E > \alpha$, **(i.e.** $0 <$ **PIFS** $= \beta' <$ **PIFE** $= \beta$ **and** $p^*_{S0} =$ **CIPS** $< 0 <$ **CIPES** $=p^{**}_{S0}$**).**

Two trends are possible according to the initial value p_{S0}, as shown in figure 39.

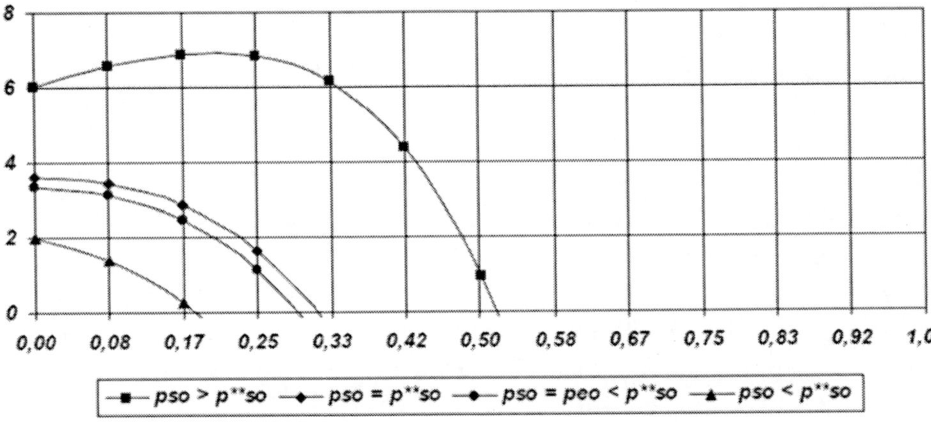

Figure 39 – Price evolution of sold resources, p_S (US$/GJ), versus fraction of a year, for $\alpha = -0.04 < r_E = 3.0 < r_N = 3.26$, according to the relation between p_{S0} and CIPS $= p^*_{S0} = -41.69$, CIPES $= p^{**}_{S0} = 3.57$.

b.1) If $p_{S0} > p^{**}_{S0}$ the price increases up to a maximum and then decreases.

The price of sold resources increases with time up to a maximum and then decreases for $p_{S0} = 6 > p^{**}_{S0}=3.57$ as reported in figure 39.

Example for Oil

At the beginning of 1980 the average data of January were the followings: inflation rate $r_N=13.91$, discount rate $r_E=12$, and the annual average extraction rate was $\alpha=-4.59$. The average monthly price increased from 5.666 US$/GJ in January 1980 to a maximum of 6.886 US$/GJ in the period from April to July 1980, and then decreased in the rest of the 1980 with some oscillations. The application of Eqs. 35 and 38 with the data of January 1980, i.e. $\alpha=-4.59$, $r_N=13.91$ and $r_E=12$, gives $p^*_{S0}=-41.26$ and $p^{**}_{S0}=4.75$ under the assumption $p_{S0}=p_{E0}$. Assuming the initial price of sold resource equal to $p_{S0}=5.666$ US$/GJ it is predicted an increase up to a maximum and then a decrease because $p_{S0}=p_{E0}=5.666> p^{**}_{S0}=4.75$.

b.2) If $p_{S0} = < p^{**}_{S0}$ the price p_S decreases. If $p_{S0} = p^{**}_{S0}$ the maximum is at $t = 0$.

The price of sold resources decreases with time for $p_{S0} = p^{**}_{S0}=3.57$ and $p_{S0} = 3.325 < p^{**}_{S0}=3.57$ or $p_{S0} = 2 < p^{**}_{S0}=3.57$ as reported in figure 39.

Example for Oil

At the beginning of 1993 the average data of January were the followings: inflation rate $r_N=3.26$, discount rate $r_E=3$, and the annual average extraction rate was $\alpha=-0.04$. The average monthly price decreased from 3.325 US$/GJ, in January1993, to 2.529 US$/GJ, in December 1993, with some oscillations. The application of Eqs. 35 and 38 with the data of January 1993, i.e. $\alpha=-0.04$, $r_N=3.26$ and $r_E=3$, gives $p^*_{S0}=-41.69$ and $p^{**}_{S0}=3.57$ under the assumption $p_{S0}=p_{E0}$. Assuming the initial price of sold resource equal to $p_{S0}=3.325$ US$/GJ it is predicted a decrease because $p_{S0}=p_{E0}< p^{**}_{S0}$.

c. $\alpha= r_E$, (i.e. PIFS = $\beta' = 0$, CIPS = $p^*_{S0} < 0 <$ CIPES = $p^{**}_{S0} = \infty$).

The price p_S decreases for every price p_{S0}.

In conclusion, for PIFE of extracted resources greater than zero, $\beta > 0$, the criteria to evaluate the price evolution of p_S is the relation between PIFS = β' and PIFE = β, i.e. r_N and r_E , and the initial price, p_{S0}. For $\beta' > \beta$ the price p_S increases if $p_{S0} >= p^*_{S0}$. The price p_S increases up to a maximum and then decreases if $p^{**}_{S0} < p_{S0} < p^*_{S0}$ and $\beta' > \beta$, or if $p_{S0} > p^{**}_{S0}$ and $\beta' < \beta$. For $p_{S0} =< p^{**}_{S0}$ and $\beta' = 0$ the price p_S decreases for every initial price p_{S0}.

Price Evolution for $0 < \alpha = r_N$ (PIFE = β = 0) [4]

If the extraction rate α is equal to the interest rate of non-extracted resources, r_N, i.e. $\beta=0$,

$$p^*_{S0} = r_N \, p_{E0} \,/\, \beta' = p^{**}_{S0} \,. \tag{44}$$

and Eq. 33 becomes

$$p_S = exp\,(\beta' \, t)\,[p_{S0} - p^*_{S0}\,] + p^*_{S0} \tag{45}$$

The price evolutions of sold resources, p_S ,are reported in figures 40 for given values of $\alpha=r_N$ and r_E, i.e. PIFE and PIFS. The evolution depends on the relation between p_{S0}, p^*_{S0} and p^{**}_{S0}.

a. $r_N < r_E$, (i.e. PIFS = β' > PIFE = $\beta = 0$, CIPS = p^*_{S0} = CIPES = $p^{**}_{S0} > 0$).

Three trends are possible according to the initial value p_{S0}, as shown in figure 40.

a.1) If $p_{S0} > p^*_{S0}$, the price of sold resources, p_S, increases with time.

The price of sold resources increases with time for $p_{S0}=8>p^*_{S0}=p^{**}_{S0}=6.21$, as reported in figure 40.

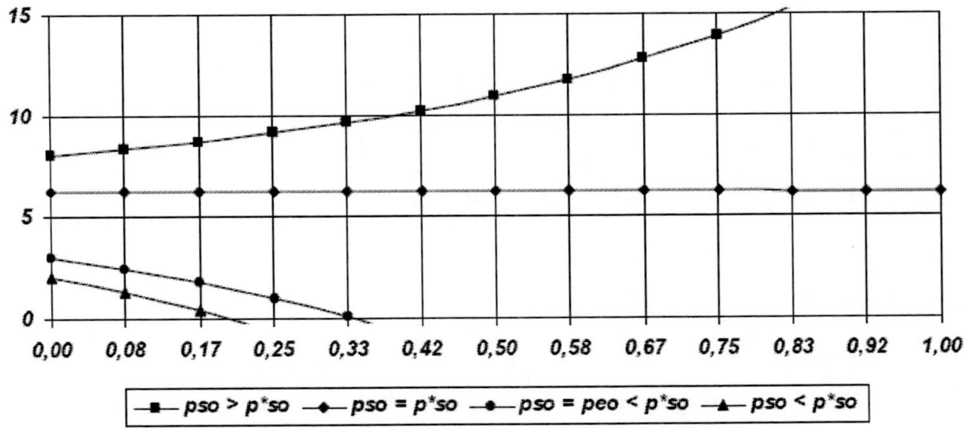

Figure 40. Price evolution of sold resources, p_S (US$/GJ), versus fraction of a year, for $\alpha=r_N=4.05<r_E=6.0$, according to the relation between p_{S0} and PIFE = p^*_{S0} = PIFS = p^{**}_{S0} = 6.21.

Example for Oil

At the beginning of 1996 the average data of January were the followings: inflation rate $r_N=2.73$, prime rate $r_E=8.5$, and the annual average extraction rate was $\alpha=2.87$. The average monthly price increased from *3.291* US$/GJ, in January 1996, to *4.426* US$/GJ in December 1996, with some oscillations. The application of Eqs. 35 and 38 with the data of January 1996, i.e. $\alpha=2.87$, $r_N=2.73$ and $r_E=8.5$, gives $p^*_{S0}=1.56$ and $p^{**}_{S0}=1.6$ under the assumption $p_{S0}=p_{E0}$. Assuming the initial price of sold resource equal to $p_{S0}=3.291$ US$/GJ it is predicted an increase because $p_{S0}=3.291>p^{**}_{S0}=1.6\approx p^*_{S0}=1.56$.

a.2) If $p_{S0} = p^*_{S0}$, p_S remains constant at p^*_{S0}.

The price of sold resources remains constant with time for $p_{S0} = p^*_{S0} = p^{**}_{S0}=6.21$, as reported in figure 40.

a.3) If $0 < p_{S0} < p^*_{S0}$, p_S decreases with time.
The price of sold resources decreases with time for $p_{S0} = 2.991 < p^*_{S0} = p^{**}_{S0}=6.21$ or for $p_{S0} = 2 < p^*_{S0} = p^{**}_{S0}=6.21$ as reported in figure 40.

Example for Oil

At the beginning of 1988 the inflation was $r_N=4.05$, the discount rate $r_E=6.0$ and the extraction rate $\alpha=4.05$. The oil price decreased from *2.991* US$/GJ in January 1988 to *2.836* US$/GJ in December 1988 with some oscillations. The application of Eqs. 35 and 38 with the data of January 1988, i.e. $\alpha=r_N=4.05$ and $r_E=6$, gives $p^*_{S0}=p^{**}_{S0}=6.21$ under the assumption $p_{S0}=p_{E0}$. Assuming the initial price of sold resource equal to $p_{S0}=2.291$ US$/GJ it is predicted a decrease because $p_{S0}=p_{E0}<p^{**}_{S0}$.

b. $r_N > r_E$, **(i.e. PIFS =** $\beta' <$ **PIFE =** $\beta=0$, **CIPS =** p^*_{S0} **= CIPES =** $p^{**}_{S0} < 0$**).**

The price of sold resources, p_S, decreases towards $p^*_{S0} < 0$ for every initial value p_{S0} .

In conclusion, for $\beta = 0$, the criteria to evaluate the price evolution of p_S is the PIFS of sold resources, β', the PIFE of extracted resources, β, and the initial price p_{S0}. For $\beta' > \beta$ the price p_S increases if $p_{S0} > p^*_{S0}$, p_S remains constant if $p_{S0} = p^*_{S0}$ and p_S decreases if $p_{S0} < p^*_{S0}$. For $\beta' < \beta$ the price decreases towards $p^*_{S0} < 0$.

Price Evolution for $\alpha > r_N > 0$, PIFE = $\beta < 0$ [4]

The extraction rate α is greater than the interest rate of non-extracted resources, r_N, i.e. $\beta < 0$. The price evolutions of sold resources, p_S ,are reported in figures 41-42 for given values of α, r_N and r_E, i.e. PIFE and PIFS. The evolution depends on the relation between p_{S0}, CIPS $= p^*_{S0}$ and CIPES $= p^{**}_{S0}$.

a. $r_E > \alpha > r_N$ **, (i.e. PIFS =** $\beta' > 0 > $ **PIFE =** β**,** $0 <$ **CIPS =** $p^*_{S0} <$ **CIPES =** p^{**}_{S0} **).**

Three trends are possible according to the initial value p_{S0}, as shown in figure 41.

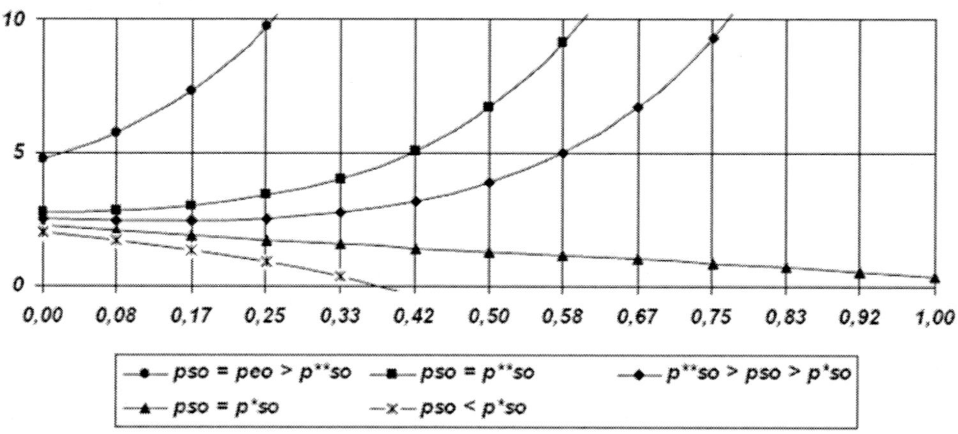

Figure 41. Price evolution of sold resources, p_S (US\$/GJ), versus the fraction of a year, for $r_E=8.5>\alpha=3.81>r_N=2.74$, according to the relation between p_{S0} and CIPS = $p^*_{S0}=2.25$, CIPES = $p^{**}_{S0}=2.77$.

a.1) If $p_{S0} >= p^{**}_{S0} > p^*_{S0}$ the price p_S increases.

The price of sold resources increases with time for $p_{S0}=4.738> p^{**}_{S0}=2.77>p^*_{S0}=2.25$ or for $p_{S0} = p^{**}_{S0}=2.77 > p^*_{S0}=2.25$ as reported in figure 41.

Example for Oil

At the beginning of 2000, January, the inflation rate was $r_N=2.74$, the prime rate $r_E=8.5$ and the annual average extraction rate $\alpha=3.81$. The oil price increased from *4.738* US\$/GJ in January 2000 to *5.997* US\$/GJ in November 2000 with some oscillations. The application of

Eqs. 35 and 38 with the data of January 2000, i.e. $r_E=8.5>\alpha=3.81>r_N=2.74$, gives $p^*_{S0}=2.25$ and $p^{**}_{S0}=2.77$ under the assumption $p_{S0}=p_{E0}$. Assuming the initial price of sold resource equal to $p_{S0}=4.738$ US$/GJ it is predicted an increase because $p_{S0}=p_{E0}>p^{**}_{S0}$.

a.2) If $p^{**}_{S0} > p_{S0} > p^*_{S0}$ the price p_S decreases to a minimum and then increases.

The price of sold resources decreases to a minimum and then increases with time for $p^{**}_{S0}=2.77>p_{S0} = 2.5 > > p^*_{S0}=2.25$ as reported in figure 41.

Example for Oil

At the beginning of 1986, February, the inflation rate was $r_N=3.89$, the discount rate $r_E=7.5$ and the annual average extraction rate $\alpha=5.02$. The oil price decreased from *2.692* US$/GJ, in February 1986, to the minimum of *2.018* US$/GJ, in July 1986, and then increased to *2.802* US$/GJ, in December 1986, with some oscillations. The application of Eqs. 35 and 38 with the data of February 1986, i.e. $r_E=7.5>\alpha=5.02>r_N=3.89$, gives $p^*_{S0}=1.91$ and $p^{**}_{S0}=3.38$ under the assumption $p_{S0}=p_{E0}$. Assuming the initial price of sold resource equal to $p_{S0}=2.692$ US$/GJ it is predicted a decrease to a minimum and then an increase because $p^{**}_{S0}=3.38>p_{S0}=p_{E0}=2.692>p^*_{S0}=1.91$.

a.3) If $p_{S0} = < p^*_{S0}$ the price p_S decreases.

The price of sold resources decreases with time for $p_{S0}=p^*_{S0}=2.25$ or $p_{S0}=2<p^*_{S0}=2.25$ as reported in figure 41.

Examples for Oil

At the beginning of 1997, March, the inflation rate was $r_N=2.76$, the discount rate $r_E=5$ and the annual average extraction rate $\alpha=3.03$. The oil price decreased from *3.659* US$/GJ, in March 1997, to *3.194* US$/GJ, in December 1997, with some oscillations. The application of Eqs. 35 and 38 with the data of March 1997, i.e. $r_E=5>\alpha=3.03>r_N=2.76$, gives $p^*_{S0}=4.51$ and $p^{**}_{S0}=5.13$ under the assumption $p_{S0}=p_{E0}$. Assuming the initial price of sold resource equal to $p_{S0}=3.659$ US$/GJ it is predicted a decrease because $p^{**}_{S0} = 5.13 > p^*_{S0} = 4.51 > p_{S0} = p_{E0} = 3.659$.

At the beginning of 2003, January, the inflation rate was $r_N=2.6$, the prime rate $r_E=4.25$ and the annual average extraction rate $\alpha=3.46$. The oil price decreased from *5.742* US$/GJ, in January 2003, to *5.605* US$/GJ, in December 2003, with some oscillations. The application of Eqs. 35 and 38 with the data of January 2000, i.e. $r_E = \alpha = 4 > r_N = 2$, gives $p^*_{S0} =9.05$ and $p^{**}_{S0}=18.9$ under the assumption $p_{S0}=p_{E0}$. Assuming the initial price of sold resource equal to $p_{S0}=5.742$ US$/GJ it is predicted a decrease because $p_{S0}=p_{E0} < p^*_{S0}$.

b. $\boldsymbol{\alpha=r_E>r_N}$**, (i.e. PIFS =** $\boldsymbol{\beta' = 0 >}$ **PIFE =** $\boldsymbol{\beta}$**, CIPES =** $\boldsymbol{p^{**}_{S0} = \infty >}$ **CIPS =** $\boldsymbol{p^*_{S0} > 0}$**).**

Three trends are possible according to the initial value p_{S0}, as shown in figure 42.

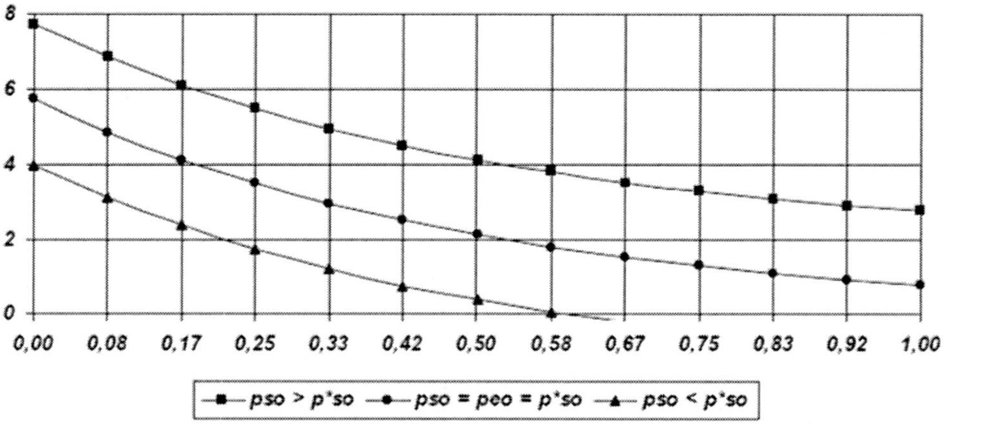

Figure 42. Price evolution of sold resources, p_S (US$/GJ), versus fraction of a year, for $r_E = \alpha = 4 > r_N = 2$, according to the relation between p_{S0} and CIPS = p^*_{S0}=5.74, CIPES = $p^{**}_{S0}=\infty$.

b.1) If $p_{S0} > p^*_{S0}$ the price p_S decreases towards $(p_{S0} - p^*_{S0})$.

The price of sold resources decreases with time to $(p_{S0} - p^*_{S0})$ for $p_{S0} = 7.74 > p^*_{S0} = 5.74$ as reported in figure 42.

b.2) If $p_{S0} = p^*_{S0}$ the price p_S decreases towards $(p_{S0} - p^*_{S0}) = 0$.

The price of sold resources decreases with time to $(p_{S0} - p^*_{S0}) = 0$ for $p_{S0} = p^*_{S0} = 5.74$ as reported in figure 42.

b.3) If $p_{S0} < p^*_{S0}$ the price p_S decreases towards $(p_{S0} - p^*_{S0}) < 0$.

The price of sold resources decreases with time to $(p_{S0} - p^*_{S0}) < 0$ for $p_{S0} = 4 < p^*_{S0} = 5.74$ as reported in figure 42.

c. $\alpha > r_N > r_E$, **(i.e. PIFS =** $\beta' <$ **PIFE =** $\beta < 0$, $0 >$ **CIPES =** $p^{**}_{S0} >$ **CIPS =** p^*_{S0} **).**

The price p_S decreases to a minimum and then increases for every initial price p_{S0}.

d. $\alpha > r_E > r_N$, **(i.e.** $0 >$ **PIFS =** $\beta' >$ **PIFE =** β**, CIPS =** $p^*_{S0} > 0 >$ **CIPES =** p^{**}_{S0} **).**

The price p_S decreases to a minimum and then increases for every initial price p_{S0}.

In conclusion, for $\beta < 0$, the criteria to evaluate the price evolution p_S is the PIFS of sold resources, β', and the initial price p_{S0}. If $\beta' > 0$ the price p_S increases for $p_{S0} >= p^*_{S0}$, increases after a minimum for $p^*_{S0} < p_{S0} < p^{**}_{S0}$, decreases towards $+0$ for $p_{S0} = p^*_{S0}$ and decreases indefinitely for $0 < p_{S0} < p^*_{S0}$. If $\beta' = 0$ the price p_S decreases towards $(p_{S0} - p^*_{S0})$ for $p_{S0} >=< p^*_{S0}$. If $\beta' < 0$ the price p_S decreases to a minimum for every initial price p_{S0}.

Price Evolution for $r_N = r_E$, PIFS = β' = PIFE = β [4]

Equation 33 is not valid for $r_N = r_E$, i.e. $\beta' = \beta$, when Eq. 32 becomes

$$\frac{dp_S}{dt} = p_S(r_E - \alpha) - p_{E0} \cdot r_E \exp\ (\beta\ t) = p_S\ \beta\ -\ p_{E0}\ r_E \exp(\beta\ t) \tag{46}$$

and p_S is finally given by

$$p_S = (\,p_{S0} - r_E p_{E0}\ t\,)\ exp(\,\beta t\,) = (p_{S0} - p^{**}_{S0}\ \beta t\,)\ exp\ (\beta t) \tag{47}$$

where p_{S0} is the initial price at $t = 0$.

CIPS, p^*_{S0}, is not defined while CIPES, p^{**}_{S0}, is defined according to the second part of Eq. 38.

The price of sold resources has an extreme for the time t_m given by

$$t_m = (\,p_{S0}\ \beta' - r_E\ p_{E0}\,) \,/\, (\,r_E\ p_{E0}\ \beta'\,) = (\,p_{S0} - p^{**}_{S0}\,) \,/\, p^{**}_{S0} \,/\, \beta' \tag{48}$$

The extreme is a maximum if $p^{**}_{S0}>0$ and a minimum if $p^{**}_{S0}<0$. The time t_m of the maximum is zero if the initial price p_{S0} is equal to

$$p_{S0} = p^{**}_{S0} = r_N\ p_{E0} \,/\, \beta' \tag{49}$$

The time t_m of the maximum is greater than zero if:

$$p_{S0} > p^{**}_{S0} \text{ and } p^{**}_{S0}\ \beta' > 0 \tag{50}$$

or

$$p_{S0} < p^{**}_{S0} \text{ and } p^{**}_{S0}\ \beta' < 0 \tag{51}$$

The time t_0 for $p_S = 0$ is

$$t_0 = p_{S0} \,/\, p^{**}_{S0} \,/\, \beta' = p_{S0} \,/\, (p_{E0}\ r_E) \tag{52}$$

The time evolution of sold resources, p_S, reported in figure 43, depends on the values of $r_N = r_E$ and α, and the relation between p_{S0}, and CIPS = p^*_{S0}, CIPES = p^{**}_{S0}.

a. $\alpha < r_N$, (i.e. PIFS = β' = PIFE = $\beta > 0$, CIPES = $p^{**}_{S0} > 0$).

Three trends are possible according to the initial value p_{S0}, as shown in figure 43.

a.1) If the initial price, $p_{S0} > p^{**}_{S0}$ the price p_S increases to a maximum and then decreases.

The price of sold resources increases with time to a maximum and then decreases for $p_{S0} = 5 > p^{**}_{S0} = 2.11$ and for $p_{S0} = 3.43 > p^{**}_{S0} = 2.11$ as reported in figure 43.

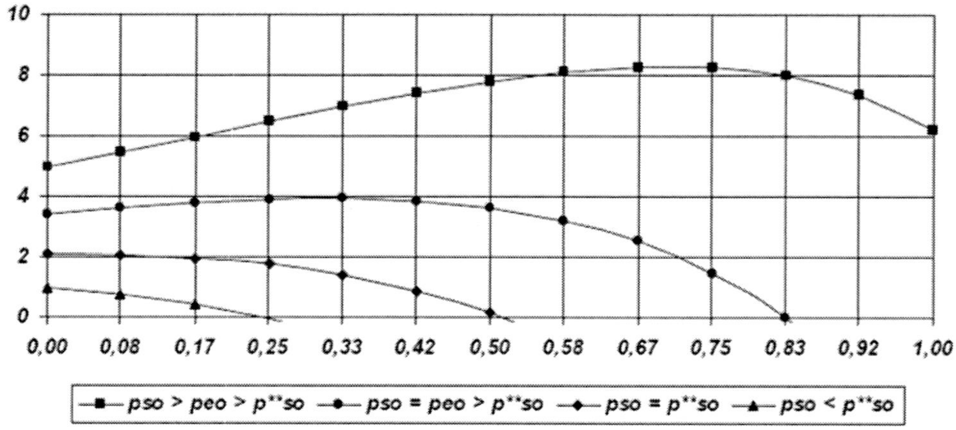

Figure 43. Price evolution of sold resources, p_S (US$/GJ), versus fraction of a year for $r_N=r_E=1.2>\alpha=-0.75$ according to the relation between p_{S0} and CIPS = $p^*_{S0}=\infty$, CIPES = $p^{**}_{S0}=2.11$.

Example for Oil

At the beginning of 2002, January, the inflation rate was $r_N=1.14$, the discount rate $r_E=1.25$ and the annual average extraction rate $\alpha=-0.75$. The oil price increased from 3.429 US$/GJ, in January 2002, to the maximum of 5.172 US$/GJ, in September 2002, and then decreased in the rest of the year, with some oscillations. The application of Eq. 38 with the data of January 2002, i.e. $r_N=r_E=1.2>\alpha=-0.75$, gives $p^*_{S0}=35.54$ and $p^{**}_{S0}=1.95$ under the assumption $p_{S0}=p_{E0}$. Assuming the initial price of sold resource equal to $p_{S0}=3.429$ US$/GJ it is predicted an increase to a maximum and then a decrease because $p_{S0}=p_{E0}>p^*_{S0}$.

a.2) If the initial price $p_{S0}=p^{**}_{S0}$ the price p_S decreases with the maximum at $t=0$.

The price of sold resources decreases with time for $p_{S0}=p^{**}_{S0}=2.11$ as reported in figure 43.

a.3) If the initial price $p_{S0}<p^{**}_{S0}$ the price p_S decreases.

The price of sold resources decreases with time for $p_{S0}=1<p^{**}_{S0}=2.11$ as reported in figure 43.

b. $\alpha>r_N$, **(i.e. PIFS =** β' **= PIFE =** $\beta<0$**, CIPES =** $p^{**}_{S0}<0$**).**

The price p_S decreases to a minimum and then increases.

c. $\alpha=r_E=r_N$, **(i.e. PIFS =** β' **= PIFE =** $\beta=0$**, CIPES =** $p^{**}_{S0}=\infty$**).**

The price p_S decreases linearly.

In conclusion, for $\beta'=\beta$, the criteria to evaluate the price evolution is PIFS, β', or PIFE, β, and the initial price p_{S0}. If $\beta'>0$ the price p_S increases up to a maximum for $p_{S0}>p^{**}_{S0}$

and then decreases, while for $p_{S0} <= p^{**}_{S0}$ the price p_S decreases. If $\beta' < 0$ the price p_S decreases to a minimum and then increases. If $\beta' = 0$ the price p_S decreases linearly.

Conclusions

The amount of reserves of a non-renewable resource depends on the extraction rate. The use of the mass conservation equation allows to define the critical extraction rate, which is the extraction rate which exhausts the reserves of a non-renewable resource in an infinite time. The critical extraction rate can be used in the analysis of the consumption rate of a non-renewable resource to verify the trend of the extraction rate towards the critical extraction rate.

The price evolution of non-renewable resources, in the form of extracted resources, depends on extraction rate, interest rate of non-extracted resources, typically the inflation rate, and initial price. If the extraction rate is negative or smaller than the inflation rate, i.e. the Price Increase Factor of Extracted resources, PIFE, is greater than zero, the price increases with time. If the extraction rate is equal to the inflation rate, i.e. PIFE is equal to zero, the price remains constant with time. If the extraction rate is greater than the inflation rate, i.e. PIFE is smaller than zero, the price decreases with time.

The price evolution of non-renewable resources, in the form of sold resources, depends on extraction rate, interest rate of non-extracted resources, typically the inflation rate, interest rate of extracted resources, typically the prime or the discount rate, and the initial price of the resources.

A special case is non-extracted resources with zero interest rate. The price of sold resources increases with time when the extraction rate is zero or smaller than the interest rate of the extracted resources, i.e. the Price Increase Factor of Sold resources, PIFS, is greater than zero. The price remains constant if the extraction rate is equal to the interest rate of extracted resources, i.e. PIFS is zero. The price decreases if the extracted rate is higher than the interest rate of the extracted resources, i.e. PIFS is smaller than zero.

When the extraction rate is smaller than the interest rate of non-extracted resources the price evolution of sold resources depends on the relation between PIFS and PIFE and the relation between initial price, CIPS and CIPES. If PIFS is greater than PIFE the price of sold resources increases with time if the initial price is greater or equal to CIPS, the Critical Initial Price of Sold resources. The price of sold resources increases up to a maximum and then decreases with time if the initial price is smaller than CIPS but greater than CIPES, the Critical Initial Price Extreme of Sold resources. If PIFS is greater than zero but smaller than PIFE, the price of sold resources increases to a maximum and then decreases with time if the initial price is greater than CIPES. In all the other cases the price of sold resources decreases with time.

When the extraction rate is equal to the interest rate of non-extracted resources, i.e. PIFE is equal to zero, the price evolution of sold resources depends on PIFS and the initial price. If PIFS is greater than PIFE, the price of sold resources increases with time if the initial price is greater than CIPS. The price of sold resources remains constant if the initial price is equal to CIPS. In all the other cases the price of sold resources decreases with time.

When the extraction rate is greater than the interest rate of non-extracted resources, i.e. PIFE is smaller than zero, the price evolution of sold resources depends on the relation between PIFS and PIFE and between initial price, CIPS and CIPES. If PIFS is greater than zero, the price of sold resources increases with time if the initial price is greater or equal than CIPES. The price of sold resources decreases to a minimum and then increases with time if the initial price is smaller than CIPES and greater than CIPS. In all the other cases the price of sold resources decreases with time.

A special case is present when the interest rates of non-extracted and extracted resources are equal, i.e. PIFE is equal to PIFS. If the extraction rate is smaller than the interest rate, i.e. PIFE and PIFS are greater than zero, the price can increase to a maximum before to decrease only if the initial price is greater than CIPES. In all other cases the price of sold resources decreases with time.

The general conclusion is that the price of a non-renewable resource sold to the market can increase with time if the initial price is greater than the critical initial price of sold resources, CIPS, or the critical initial price extreme of sold resources, CIPES. The increase is only temporary if PIFE is equal to PIFS and the initial price is greater than CIPES, the critical initial price extreme of sold resources.

References

[1] H. Hotelling, The Economics of Exhaustible Resources, *Journal of Political Economy,* (1931) 39, 137-175.

[2] P. S. Dasgupta and G. M. Heal, Economic Theory and Exhaustible Resources, Cambridge University Press, 1979.

[3] F. Gori, Mass and Energy-Capital Conservation Equations to Study the Price Evolution of non-Renewable Energy Resources. Part I-Generalization of the Hotelling Rule, *Applied Thermal Engineering*, **26** (2006) 1746-1750.

[4] F. Gori, Mass and Energy-Capital Conservation Equations to Study the Price Evolution of non-Renewable Energy Resources. Part II –Extension to resources sold to the market, *Applied Thermal Engineering*, **26** (2006) 1751-1770.

[5] F. Gori, Mass and Energy-Capital Conservation Equations to Study the Price Evolution of Non-Renewable Energy Resources. Part III–Energy supply curve, *Applied Thermal Engineering,* **29** (2009) 2172-2186.

[6] F. Gori, New Parameters to Forecast the Time Evolution of the Non-Renewable Energy Resource Oil, *Internal Report and Submitted for Publication*, 2008.

[7] F. Gori, New Parameters to Forecast the Time Evolution of the Non-Renewable Energy Resource Gas, *Internal Report and Submitted for Publication*, 2008.

[8] www.bp.com/statisticalreview.

[9] minneapolisfed.org/Research/data/us/calc/hist1913.cfm

[10] www.economagic.com/em-cgi/data.exe

[11] F. Gori, D. Ludovisi and P. F. Cerritelli, Forecast of oil price and consumption in the short term under three scenarios: Parabolic, linear and chaotic behavior, *Energy,* (2007) 32 (7), 1291-1296.

[12] M. Tenenbaum and H. Pollard; *Ordinary Differential Equations*; Dover Pub. Inc., New York, 1963.

In: Thermal Engineering Research Developments
Editors: J. Evgova and O. Kostadinov, pp. 283-316
ISBN: 978-1-60741-497-1

Chapter 6

ADVANCES IN THERMAL MODELING OF LASER CHEMICAL VAPOR DEPOSITION AND INFILTRATION

Yuwen Zhang
Department of Mechanical and Aerospace Engineering
University of Missouri; Columbia, MO 65211, USA

Abstract

Laser chemical vapor deposition and infiltration (LCVD and LCVI) are based on the concept of building functional structures by deposition of solid materials from gas precursors by a laser beam in an environmentally controlled chamber. The chemical vapor deposition during these two processes can be based on reactions initiated pyrolytically, photolytically or a combination of both. The most important issue that the thermodynamic analysis can address is whether a given chemical reaction is feasible. For chemical reactions that are feasible, both composition plot and deposition maps can be obtained. The distinction between the conventional Chemical Vapor Deposition (CVD) and LCVD is that a laser beam heats a very small spot on the substrate and vapor deposition occurs only in the heated spot in the LCVD process. A series of thermal models of LCVD of Titanium Nitride on a finite slab under irradiation by a moving laser beam have been developed. Heat transfer in the substrate and gases, the chemical reaction on the substrate's top surface, and the mass transfer of gases in the chamber were taken into account. The effect of natural convection induced by the temperature and concentration gradients during LCVD was instigated. A parametric study on the LCVD by moving laser beam was carried out and an empirical correlation was proposed. To study the transport phenomena in LCVI, heat transfer in a gas-saturated powder layer subject to a short-pulsed volumetric heat source is modeled using a two-temperature model. The results show that the non-equilibrium is significant when the heat source pulse width is less than 1 microsecond and, therefore, heat transfer during nanosecond laser- powder interaction should be modeled using a non-equilibrium model. For the case with long pulse width, on the other hand, the equilibrium model is sufficient.

Nomenclature

a_{ij}	stoichiometric coefficient for the ith component in the jth chemical reaction
C	concentration (kg/m^3)
c_p	specific heat (J/kgK)
D	mass diffussivity (m^2/s)
D	strain tensor
d_p	diameter of the powder particle (m)
E	activation energy (kJ/mol)
G	coupling coefficient (W/m^3-K)
h	thickness of the substrate (m)
h_0	heat transfer coefficient at the powder bed surface (W/m^2-K)
h_p	heat transfer coefficient at the powder particle surface (W/m^2-K)
I	unit tensor
J	mass flux vector due to ordinary diffusion (kg/m^2-s)
J	heat source fluence (J/m^2)
k	thermal conductivity (W/mK)
K_0'	Arrhenius constant
L	thickness of the powder layer
M	molecular weight (g/mol)
Nu	Nusselt number
$\dot{m}''$	mass flux (kg/m^2)
n_r	number of reactants
n_p	number of gaseous products
P	laser power (W)
p	pressure (Pa)
q''	heat flux (W/m^2)
q_s'''	intensity of the internal heat source (W/m^3)
r_o	radius of the laser beam (m)
R_u	universal gas constant (= 8.314 kJ/kmol)
S	source term in the energy equation
S_c	source term in the mass transfer equation
t	time (s)
t_p	HWHM (Half Width at Half Maximum) pulse width (s)
T	temperature (K)
u	velocity component in x direction (m/s)
v	velocity component in y direction (m/s)
V	velocity vector (m/s)
w	velocity component in z direction (m/s)
x	coordinate in length direction (m)
y	coordinate in width direction (m)
z	coordinate in height direction (m)

Greek Symbols

α	diffusivity (m^2/s)
α_a	absorptivity
β	coefficients of thermal expansion (1/K)
β_c	concentration expansion coefficients (m^3/kg)
γ	sticking coefficient
φ	porosity
δ	thickness of the deposited film (m), or optical penetration depth of the powder layer (m)
ΔH_R	heat of chemical reaction (J/kg)
ε	emissivity
ρ	density (kg/m^3)
σ	collision diameter (Å)
$\boldsymbol{\tau}'$	stress tensor (N/m^2)

Subscripts

g	gas
i	initial value
s	substrate, or solid
∞	infinite

1. Introduction

Gas-phase Solid Freeform Fabrication (SFF), which includes Laser Chemical Vapor Deposition (LCVD) and Laser Chemical Vapor Infiltration (LCVI), is an emerging manufacturing technology in which three-dimensional parts can be built from CAD data. The gas-phase SFF process consists of two steps (Marcus et al., 1993): (1) design a three-dimensional part with CAD and the CAD data is then transferred to a STL computer file, which is sliced to generate two-dimensional sections of the *virtual product,* (2) build the *physical product* by layering two-dimensional sections of finite thickness.

An extremely wide array of SFF technologies has been proposed for purposes of providing physical three-dimensional renderings of CAD data. However, only a few have been extended to build *structurally-sound* parts of near full density. Almost always, the SFF technologies that fit into the latter group are powered by *thermal fabrication* of three-dimensional objects from powders or gases. The heat source is nearly always a laser operating in infrared regions of the electromagnetic spectrum. To produce fully functional structural components, gas-phase SFF (such as LCVD and LCVI), seems to be very promising. The distinctive advantage of gas-phase SFF over other SFF techniques is that the local chemical compositions of the final product can be tailored by the control of the composition of precursors, deposition temperature and wavelength of the laser beam.

LCVD/LCVI are based on the concept of building functional structures by deposition of solid materials from gas precursors by a laser beam in an environmentally controlled chamber. The chemical vapor deposition during these two processes can be based on reactions initiated pyrolytically, photolytically or a combination of both (Marcus et al., 1993). The LCVD technique uses precursors to directly create a free-standing part or to join together simple shapes to create parts with higher complexity [see figure 1(a)]. On the other hand, LCVI uses gas precursors *and* powder particles to build three-dimensional parts [see figure 1(b)], which is similar to other powder-based SFF techniques, such as Selective Laser Sintering (SLS). The advantages of LCVI over LCVD include (a) uninfiltrated powder provides support for producing overhangs, (b) confining the deposition to thin powder layers provide dimensional control in the direction of growth, and (c) it is possible to tailor local chemistry and microstructures (Jakubenas et al., 1997).

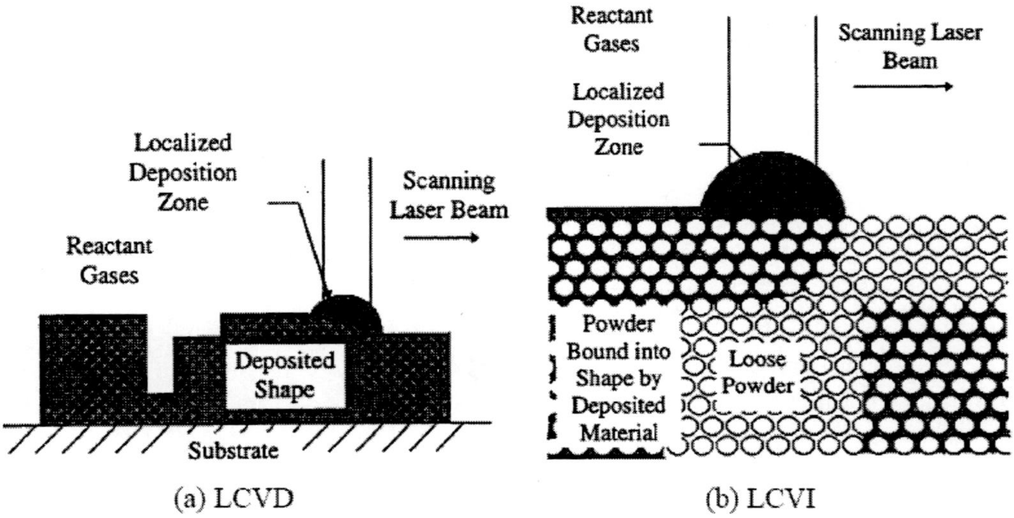

Figure 1. Schematic of SALD and SALDVI (Jakubenas et al., 1997).

The chemical reactions in LCVD and LCVI are the same as those occurring in the conventional Chemical Vapor Deposition (CVD; Mahajan, 1996), which is widely used to fabricate semiconductor devices. It depends on availability of a volatile gaseous chemical that can be converted to solid film through some thermally activated chemical reaction. The CVD can be used to produce a large variety of thin films with different precursors. It is very crucial that the chemical reaction takes place on the substrate surface only, so that a thin film can be deposited onto the substrate. If undesired chemical reactions occur in the gas phase, the solid particles can be formed which may fall onto the substrate or coat the chamber walls. To avoid the undesired chemical reaction, the substrate surface temperature, deposition time, pressure, and surface specificity should be carefully selected. The chemical reaction during a CVD process is usually accomplished in several steps and the path of chemical reactions can be altered by changing the substrate temperature. For example, when titanium tetrabromide ($TiBr_4$) is used as a precursor to deposit a titanium film, the chemical reaction is accomplished in the following steps (Mazumder and Kar, 1995):

$$TiBr_4(g) \rightarrow TiBr_2(s)+Br_2(g)$$

$$3TiBr_2(s) \rightarrow 2TiBr(s)+TiBr_4(g)$$

$$4TiBr(g) \rightarrow 3Ti(s)+TiBr_4(g)$$

The mechanisms of chemical reactions for many CVD processes are not clear, so the chemical reactions occurring in a CVD process are often represented by a single overall chemical reaction equation. Table 1 summarizes some examples of the overall chemical reactions occurring in CVD processes (including LCVD and LCVI).

Table 1. Overall chemical reaction of CVD processes

Thin films	Overall reaction	Temperature of reaction	References
Al_2O_3	$Al(l) + H_2O(g) = AlO(g) + H_2(g)$ $AlO(g) + H_2O(g) = Al_2O_3(s) + H_2(g)$	1230–1255°C	Powell et al. (1966)
C	$C_xH_y(g)=xC(s)+(y/2)H_2(g)$	700–1450°C	Taylor et al. (2004)
GaAs	GaCI(g) + (1/4)As4(g) = GaAs(s) + HCI(g)		Sivaram (1995)
GaAs	GaAs(g) + HCI(g) =GaCI(g) + 1/4(As_4(g)) + 1/2(H_2(g))	700–850°C	Sivaram (1995)
	$Ga(CH_3)_3+AsH_3 = GaAs+3CH_4$ $Al(CH_3)_3+AsH_3 = AlAs+3CH$	500–800°C	Ueda (1996)
GaN	$Ga(g) + NH_3 = GaN(s) + (3/2)H_2(g)$	650°C	Elyukhin et al. (2002)
Ge(s)	$GeH_4 = Ge(s) + 2H_2$		Herring (1990)
Si	$SiH_4(g)=Si(s)+2H_2(g)$	>600°C (polysilicon) >850-900°C (single crystal)	Herring (1990)
SiC	$Si(CH_3)_4(g)=SiC(s)+3CH_4(g)$	700–1450 °C	Sun et al. (1998b)
SiO_2	$SiH_4+ O_2 = SiO_2 + 2H_2$		Sivaram (1995)
	$SiH_4 + 2N_2O = SiO_2 + 2H_2O + 2N_2$	800°C	Sivaram (1995)
	$SiH_2Cl_2 + 2N_2O = SiO_2 + 2HCI + 2N_2$	> 900°C	Sivaram (1995)
TiO_2	$TiCl_4(g)+O_2(g)=TiO_2(S)+2Cl_2(g)$		Jakubenas et al. (1997)
TiN	$TiCl_4(g)+2H_2(g)+(1/2)N_2(g)=TiN(s)+4HCl(g)$	900°C	Mazumder and Kar (1995)

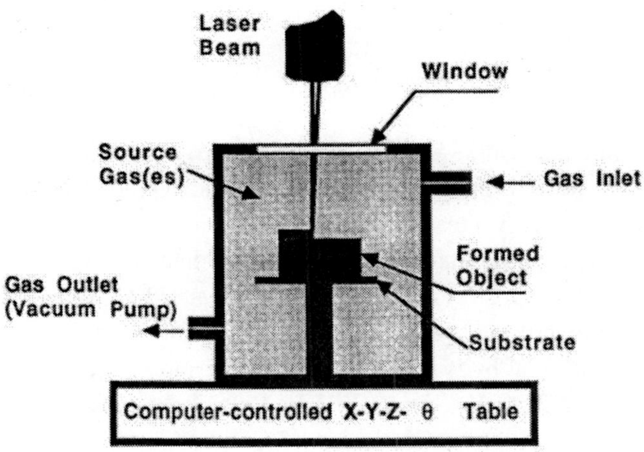

Figure 2. LCVD system (Marcus et al., 1993).

Figure 2 shows a reaction chamber for a LCVD process (Marcus et al., 1993). In contrast to conventional CVD, in which the entire susceptor is heated, only a very small spot on the substrate is heated by a directed laser beam. Scanning of the substrate surface is accomplished by a movable table. After the first layer of the solid is deposited, consecutive layers can be deposited to build the three-dimensional part based on the CAD design. The pressure inside the chamber is usually under 1 atm and the temperature of the spot under laser irradiation can range from 700 to 1500 °C. Successful deposition of various ceramic and metallic materials using various gaseous precursors has been reported.

Figure 3 shows a schematic diagram of a LCVI system. The mechanism to bind powder particles together in a LCVI process is deposition of solid material on the powder particle surface by decomposition of gas precursors. During the LCVI process, a thin (100–250 μm thick) powder layer is laser-scanned to form the two-dimensional slice to an underlying solid piece, which consists of a series of stacked two-dimensional slices. After laser scanning, a fresh powder layer is spread and the scanning process is repeated. Loose powder is removed after the part is extracted from its bin. The finished part has a composite structure consisting of starting powder bonded into a matrix of the deposited materials (Birmingham and Marcus, 1994).

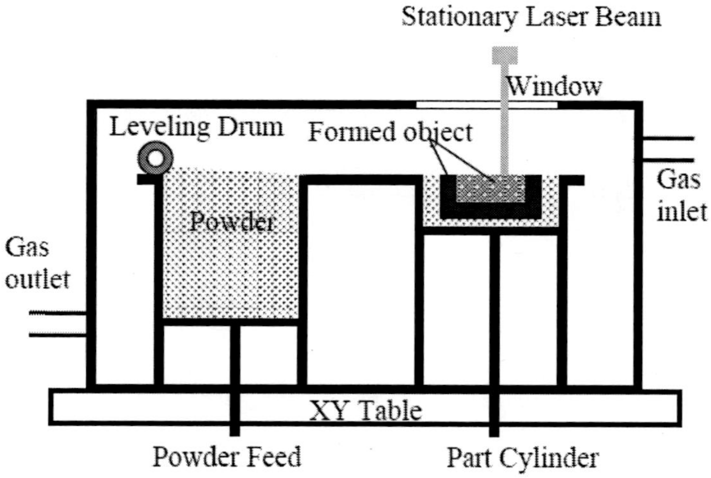

Figure 3. LCVI system (Zhang, 2006).

Extensive experimental works on LCVD/LCVI have been carried out and various applications of these new techniques have been discovered. The inherent nature of LCVD/LCVI makes them very ideal to fabricate ceramics and metal parts that are either impossible or very difficult to be manufactured by any other methods. Another advantage of LCVD/LCVI is that the composite of the parts can be varied by variation of several easily controlled process parameters, such as composition of gas precursors and the local deposition temperature. Sun et al. (1998a, b, c) fabricated *in situ* thermocouples into macro-components using combined LCVD and LCVI techniques. The bulk shape was built by using LCVI and the SiC/C thermocouples were deposited using LCVD technique. Two layers of Si_3N_4 were deposited below and above the SiC/C thermocouples so that the thermocouple legs are electronically insulated from the bulk shape into which the thermocouples is embedded. Crocker et al. (1998) embedded multiple SiC/C thermocouples in the SiC bulk shape using

LCVD technique. Crocker et al. (2000; 2001; 2002) presented LCVI of SiC into metal and ceramic powders to fabricate composite parts.

Jean et al. (1999) designed a LCVD system with a fixed 100 W CO_2 laser focused on a movable substrate. Temperatures and height were monitored to provide feedback for controlling the process. Jean et al. (2000) studied precision carbon deposition using the LCVD system described in Jean et al. (1999) and found that the carbon deposit grew uncontrollably if constant laser power was used. LCVD can also find its application in micro manufacturing. Deposition of rods, fibers, coils, and blocks as well as other complicated structures (such as microcoils) with different metal and ceramic materials have been achieved (Maxwell et al., 1997). Maxwell et al. (1998a) successfully fabricated high field density microsolenoids from 100 to 500 μm in diameter using 3-D LCVD technique. Two different ways in which 3D-LCVD can be used to generate a microcoil are investigated: (a) direct–writing of a conductive line on an insulated fiber, or (b) freeform growth (figure 4). Maxwell et al. (1999) and William et al. (1999) deposited carbon in the form of helical spring and the fiber diameter of the spring was as small as 5 μm and the spring inner diameter was as small as 60 μm. Maxwell et al. (1998b) deposited high-aspect-ratio Titanium Nitride needles up to 12 mm long using $Ti(N(CH_3)_2)_4$ as precursor.

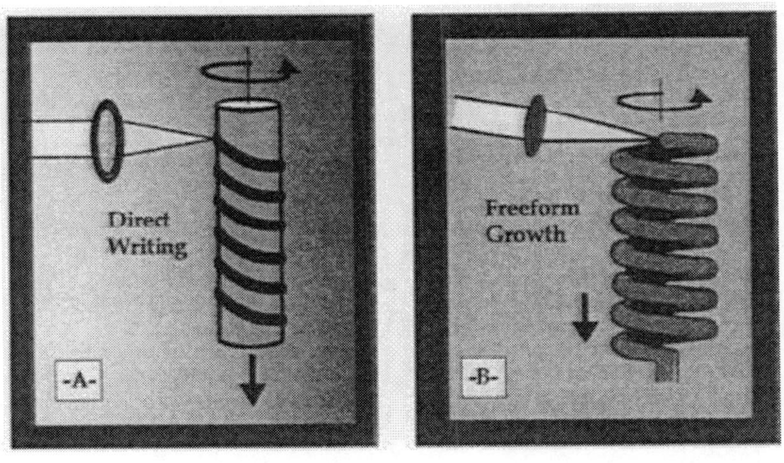

Figure 4. Techniques for growing coils using 3D-LCVD (Maxwell et al., 1998a).

2. Advances in LCVD/LCVI Research

2.1. Thermodynamics and Kinetics of Chemical Reactions

Accurate prediction and control LCVD/LCVI require a model of the processes that incorporate heat/mass transfer, chemical reaction kinetics, and chemical thermodynamics. Kinetic data that are important to modeling LCVD/LCVI are difficult to obtain and are currently compiled for only a limited number of chemical systems. Realizing these difficulties, Jakubenas et al. (1997, 1998) presented thermodynamic modeling of LCVD/LCVI for various precursors using a computer code CET89 based on minimization of system free energy (Gordon and McBride, 1976). Sun et al. (1998a) performed thermodynamic analysis using the same code for *in-situ* thermocouples in macro-components fabricated using LCVD and LCVI techniques. The gaseous precursors considered include tetramethylsilane (TMS) and methyltrichlorosilane (MTS) for deposition of SiC, and methane, ethylene, and acetylene for deposition of carbon. The most important issue that the thermodynamic analysis can address is whether a given chemical reaction is feasible. For chemical reactions that are feasible, both *composition plot* and *deposition maps* can be obtained (Jakubenas et al., 1998). The composition plot shows the equilibrium composition vs. a process parameter, such as temperature (see figure 5). Deposition maps, on the other hand, show only the equilibrium solid phase exists as fields in the plane defined by temperature and the initial reactant mixture (see figure 6). Thermodynamic modeling is very useful for precursor selection; however, it *cannot* predict the deposition rate that is very critical for design of LCVD/LCVI systems.

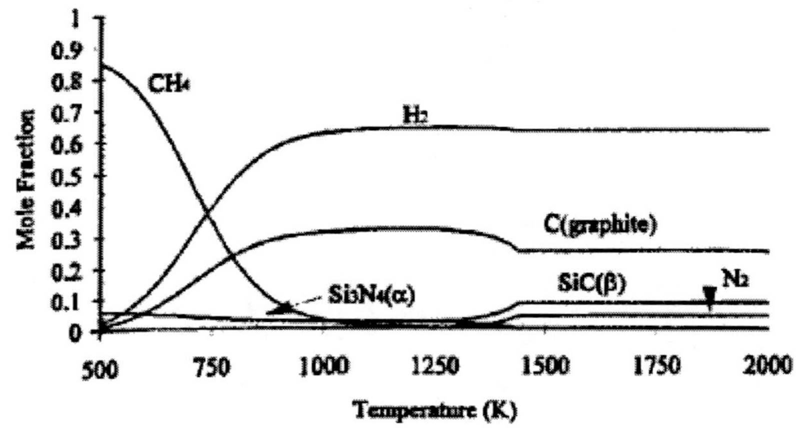

Figure 5. Composition vs. temperature for a 1:1 mixture of TMS and NH_3 at 0.5 atm and at thermodynamic equilibrium (Jakubenas et al., 1998).

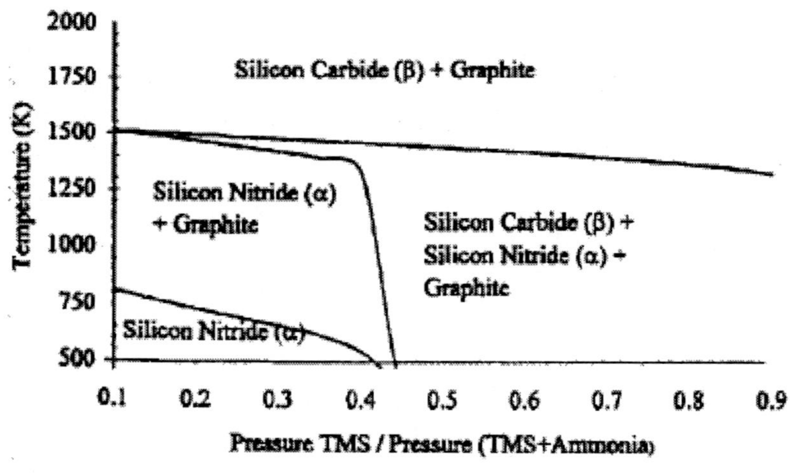

Figure 6. Deposition map for mixture of TMS and NH3 at 0.5 atm and at thermodynamic equilibrium (Jakubenas et al., 1998).

In order to accurately predict the reaction rate and the composition in the final product, the knowledge of chemical reaction and heat and mass transfer is needed. The kinetic equation, which express deposit rate as function of surface temperature and concentration, serves as constitutive relation in modeling of LCVD/LCVI process. For the deposition of TiN film using $TiCl_4$, N_2 and H_2 as precursors (Conde et al., 1992a; Zhang and Faghri, 2000), the deposit rate $d\delta/dt$ was obtained by:

$$\frac{d\delta}{dt} = \frac{K_0}{\rho_{TIN}} \exp\left(-\frac{E}{R_u T_s}\right) C_s \tag{1}$$

where $K_0 = (C_{H_2})_i (C_{N_2})_i^{1/2} K_0^{'}$ and K_0' is Arrhenius constant, ρ_{TiN} is the density TiN, E is the activation energy, R_u is the universal gas constant, T_s is the local surface temperature, and C_s represents the concentration of $TiCl_4$ at the surface of the substrate. For deposition of SiC with $Si(CH_3)_4$ as precursor gas, a similar kinetic equation was employed by Lee et al. (1995). In order to incorporate the kinetic equation into the model of LCVD/LCVI, both Arrhenius constant K_0' and activation energy E must be determined. For conventional chemical vapor deposition of SiC using TMS as precursor and N_2 as carrier gas, Figueras et al. (1991a,b; 1992) obtained the apparent activation energy of 90 kJ/mol. Conde et al. (1992b) developed an analytical model and determined the values of these two constants that best matched the analytical and experimental results. For deposition of SiC with $Si(CH_3)_4$ as precursor gas, Lee et al. (1995) determined Arrhenius constant by fitting experimental data. In order to develop global models for LCVD/LCVI, Arrhenius constant and the activation energy must be obtained.

2.2. Thermal Modeling of LCVD Processes

The distinction between the conventional Chemical Vapor Deposition and LCVD is that laser beam heats a very small spot on the substrate and vapor deposition occurs only in the heated spot in the LCVD process. Kar and Mazumder (1989) presented a 3-D transient thermal analysis for LCVD on a uniformly moving slab, in which a 3-D, non-linear, transient conduction problem was solved analytically. Kar et al. (1991) investigated laser chemical vapor deposition of Titanium on stationary finite slabs by solving 3-D, transient mass diffusion equations. Conde et al. (1992a) presented a model of laser chemical vapor deposition of Titanium Nitride (TiN) dots produced by the following overall chemical reaction: $TiCl_4(g) + 2H_2(g) + \frac{1}{2}N_2(g) \rightarrow TiN(s) + 4HCl(g)$. The 3-D transient diffusive mass transfer equation was solved using Fourier's transformation method. The volcano-like profile of the deposited film is found in certain conditions, which was in good agreement with the experimental data (Conde et al., 1992b). Conde et al. (1992a), however, did not consider the heat transfer of the gases in the chamber.

Marcus et al. (1993) studied the residual stresses in laser processed SFF that included LCVD. Heat conduction in the substrate was modeled as a pure conduction problem with a moving heat source. Jacquot et al. (1995) proposed a thermal model of the LCVD process using acetylene (C_2H_2) as precursor. Various phenomena, which included heat conduction in the substrate, chemical reaction during carbon deposition and mass diffusion of acetylene in the chamber, were taken into account. The temperature of the gases is assumed to be uniform and therefore the heat transfer in the gas phase was neglected. Shaarawi et al., (2000) presented a 2-D axisymmetric model of LCVD of SiC rods from tetramethylsilane (TMS), which included heat and mass transfer and chemical reaction in the model. Zhang and Faghri (2000) developed a very detailed model of LCVD process, which included the submodels of heat transfer, chemical reaction and mass transfer, and the model was employed to simulate the laser chemical vapor deposition of TiN film on a finite slab with stationary and moving laser beams. The above efforts assumed that the deposited film was very thin and heat transfer in the thin film was not taken into account, which was valid only for the first layer of deposition. In order to build 3-D part, multiple layer deposition will be needed and heat conduction in the continuously growing part needs to be accounted for in the model. This is especially the case when depositing layer and the substrate has significantly different thermal conductivity (Han and Jensen, 1994). To obtain the surface temperature that is critical to predict the growth rate [see eq. (1)], the heat transfer model must be able to account for the effect of part growth on the temperature distribution.

2.3. Thermal Modeling of LCVI

The physical and chemical phenomena occurred in LCVI are similar to that in LCVD. The only difference is that vapor deposition occurs on the surface of the powder particles during LCVI, instead of on the top of non-porous substrate in LCVD. Laser processing in LCVI is complicated by the fact that the irradiated material responds differently than in the case for a simple, fully dense material. Successful modeling of the LCVI process also requires knowledge about kinetics of chemical reaction, fluid flow, and heat and mass transfer in the precursors and in the porous powder bed. In addition, radiation heat transfer in powder bed also plays an important role in the heat transfer analysis due to the high temperature involved. Convective heat transfer in porous media has been extensively investigated in the past and very detailed reviews are available in the literature (Nield and Bejan, 1999; Kaviany, 1995). The distinctive feature of heat transfer and fluid flow in the LCVI process is that the porosity is not constant. The chemical reaction that occurs on the surface of the particle results in deposition on the surface of the powder particles and eventually joins the powder particles together. During the entire process, the porosity may change from a large value, up to 0.6, to nearly zero.

Dai et al. (2000) performed numerical simulation of LCVI using a finite element commercial code ANSYS. The dependence of the powder bed properties on the porosity was taken into account. The powder bed before laser densification was assumed to be 50% of its theoretical density. The densities of the powder bed were set to 80% and 100% of its theoretical value when the temperature of the powder reached to 1200K and 1273K, respectively. This treatment directly correlated the density of the powder bed to the temperature and avoided the chemical reaction model. Dai et al. (2001) presented a finite element modeling of SALDVI with the same method and an experiment was also performed. In order to simulate the closed loop control used in the experiment, the laser power was modified from one time step to the next time step to ensure the powder bed temperature was a constant. The models of Dai et al. (2000, 2001) were improved later by incorporating a densification model by vapor infiltration based on experimental growth rate (Dai et al., 2003). The effects of convection and mass transfer in the powder bed and precursors were not taking into account in any of the existing works. Compared to LCVD, the efforts on simulation of LCVI are extremely limited at this time.

3. Physical Models and Transport Properties

3.1. Governing Equations

Since the velocity of the precursors is generally very low and the characteristic length is also very small, the corresponding Reynolds number is under 100 and the Grashof number governing natural convection is under 10^6. Therefore, the transport phenomena in the LCVD process are laminar in nature. The temperature in a reactor varies significantly (typically from 300 to 900K), so the Boussinesq approximation is no longer appropriate. It is necessary to use the compressible model for transport phenomena in LCVD processes. The following assumptions can be made to obtain the governing equations:

1. The only body force is the gravitational force, which is the same for all components in the precursors.
2. Dilute approximation is valid because the partial pressure of the reactant is much lower than that of the carrier gas.
3. The deposited film is very thin (from nanometers to microns) and its effect on the flow field can be neglected.

The continuity equation is

$$\frac{D\rho}{Dt} + \rho\nabla \cdot \mathbf{V} = 0 \tag{2}$$

where the precursor gases are treated as a compressible fluid mixture.

The momentum equation is

$$\rho\frac{D\mathbf{V}}{Dt} = \nabla \cdot \boldsymbol{\tau}' + \rho\mathbf{g} \tag{3}$$

where the stress tensor is

$$\boldsymbol{\tau}' = -p\mathbf{I} + 2\mu\mathbf{D} - \frac{2}{3}\mu(\nabla \cdot \mathbf{V})\mathbf{I} \tag{4}$$

The energy equation is

$$\rho c_p \frac{DT}{Dt} = \nabla \cdot (k\nabla T) + T\beta\frac{Dp}{Dt} \tag{5}$$

where the effect of viscous dissipation and the Dufour effect have been neglected.

For a system containing n_r reactants and n_p gaseous products, the chemical equation can be expressed as

$$\sum_{i=1}^{n_r} a_{ri} A_{ri}(g) = a_p A_p(s) + \sum_{i=1}^{n_p} a_{pi} A_{pi}(g) \tag{6}$$

where A_{ri} $(i = 1, 2 \cdots n_r)$ are chemical symbols for the gaseous reactants, A_p and A_{pi} $(i = 1, 2 \cdots n_p)$ are chemical symbols for the deposit (solid) and other gaseous products. The coefficients preceding the chemical symbols in eq. (6) are referred to as stoichiometric coefficients and describe the proportion of the mole numbers of reactants disappearing and mole number of products appearing as a result of the reaction process. The number of components in the precursors is $n = n_r + n_p$.

The conservation of species mass in terms of the mass fraction is

$$\rho \frac{D\omega_i}{Dt} = -\nabla \cdot \mathbf{J}_i + \dot{m}_i''', \quad i = 1, 2, \ldots N-1 \tag{7}$$

where ω_i is the mass fraction of the ith component in the gaseous precursor. The mass flux $\mathbf{J}_i$ includes mass fluxes due to ordinary diffusion driven by the concentration gradient, and thermal (Soret) diffusion (Faghri and Zhang, 2006). The production rate of the ith species, $\dot{m}_i'''$, can be obtained by analyzing the chemical reaction. If the number of chemical reactions taking place in the system is N_g, the mass production rate is (Mahajan, 1996)

$$\dot{m}_i''' = \sum_{j=1}^{N_g} a_{ij}^g M_i \Re_j^g \tag{8}$$

where a_{ij}^g is the stoichiometric coefficient for the ith component in the jth chemical reaction in the gas phase, and $\Re_j^g$ is the net reaction rate of the jth chemical reaction in the gas phase (Faghri and Zhang, 2006).

The density of the gas is related to the pressure and temperature by the ideal gas law:

$$\rho = \frac{p}{R_g T} \tag{9}$$

The boundary conditions for the governing equations of a CVD process depend on the geometric configuration of the reactor. It is generally assumed that the nonslip condition is applicable to all solid walls. The normal velocity on a solid wall is zero (no penetration) for all walls except the susceptor where chemical reaction takes place. The total net mass flux of all species can lead to a normal velocity component on the susceptor as (Mahajan, 1996)

$$v_n = \frac{1}{\rho}\sum_{i=1}^{N}\sum_{j=1}^{N_s} a_{ij}^s M_i \Re_j^s \tag{10}$$

where a_{ij}^s is the stoichiometric coefficient for the ith component in the jth chemical reaction on the susceptor surface, and $\Re_j^g$ is the net reaction rate of the jth chemical reaction on the susceptor surface.

The net surface reaction rate $\Re_j^g$ is a product of the sticking coefficient γ_j (fraction of product that can be stuck on the substrate) and the effusive flux of the jth species (Mahajan, 1996), i.e.,

$$\Re_j^s = \gamma_j \frac{x_j p_j}{\sqrt{2\pi M_j RT}} \tag{11}$$

The growth rate of the deposit on the susceptor is

$$\frac{d\delta}{dt} = M_f \sum_{i=1}^{N}\sum_{j=1}^{N_s} a_{ij}^s \Re_j^s \alpha_{i,f} \tag{12}$$

where M_f is the molecular mass of the deposited film, and $\alpha_{i,f}$ is the number of film atoms in the ith species.

3.2. Transport Properties

The thermophysical properties of various gaseous precursors are necessary to utilize the transport models outlined above. Since the temperature varies significantly throughout a CVD system, the transport phenomena must be modeled using variable thermal physical properties. This requires knowledge of the dependence of the thermophysical properties on temperature. The thermophysical properties of the commonly used gas(es) in CVD are tabulated in table 2.

The viscosity and thermal conductivity of some precursors that are not readily available can be estimated using the method recommended by Bird et al. (2002). The viscosity is

$$\mu = 2.6693 \times 10^{-6} \frac{\sqrt{MT}}{\sigma^2 \Omega_\mu} \tag{13}$$

where M is the molecular mass and σ is the collision diameter (Å= 10^{-10} m) of the molecule that can be estimated by

$$\sigma = 0.841 \overline{v}_c^{1/3} \tag{14}$$

or

$$\sigma = 1.166 \overline{v}_{b,liq}^{1/3} \tag{15}$$

where $\bar{v}_c$ and $\bar{v}_{b,liq}$ are the specific volumes (cm^3/mol) of the precursor at critical point, and of the saturated liquid at normal boiling point, respectively.

Table 2. Transport properties of the common gas(es) for CVD (Mahajan, 1996)

Properties	Gas(es)	c_0	c_1	c_2
μ^a ($N\text{-}s/m^2$)	TMGa	-1.15×10^{-6}	3.35×10^{-8}	-6.68×10^{-12}
	AsH_3	-4.32×10^{-7}	5.94×10^{-8}	-1.46×10^{-11}
	H_2	2.63×10^{-6}	2.22×10^{-8}	-5.19×10^{-12}
	N_2	4.93×10^{-6}	4.55×10^{-8}	-1.08×10^{-11}
	SiH_4	1.47×10^{-6}	3.66×10^{-8}	-6.81×10^{-12}
k^a (W/m-K)	TMGa	-3.52×10^{-3}	3.85×10^{-5}	-3.84×10^{-8}
	AsH_3	-7.16×10^{-3}	6.53×10^{-5}	-3.47×10^{-9}
	H_2	5.77×10^{-2}	4.43×10^{-4}	-7.54×10^{-8}
	N_2	8.15×10^{-3}	6.24×10^{-5}	-4.48×10^{-9}
	SiH_4	-2.12×10^{-2}	1.45×10^{-4}	-1.31×10^{-8}
c_p^a (kJ/Kg-K)	TMGa	5.40×10^{2}	1.60	0
	AsH_3	2.45×10^{2}	1.08×10^{0}	-4.24×10^{-4}
	H_2	1.44×10^{4}	-2.61×10^{-1}	8.67×10^{-4}
	N_2	1.03×10^{3}	4.58×10^{-3}	1.34×10^{-4}
	SiH_4	4.74×10^{2}	3.26	-1.08×10^{-3}
D_{12}^b (m^2/s)	SiH_4, N_2	-9.64×10^{-1}	6.25×10^{-3}	8.50×10^{-6}
	SiH_4, H_2	-2.90	2.06×10^{-2}	2.81×10^{-5}
	N_2, H_2	-3.20	2.44×10^{-2}	3.37×10^{-5}
	TMGa, H_2	-1.87	1.64×10^{-2}	3.13×10^{-5}
	TMGa, N_2	-4.17×10^{-1}	2.89×10^{-3}	4.93×10^{-6}
	AsH_3, H_2	-2.26	1.73×10^{-2}	2.80×10^{-5}
	AsH_3, N_2	-6.15×10^{-1}	4.57×10^{-3}	7.49×10^{-6}
	TMGa, AsH_3	-2.26×10^{-1}	1.27×10^{-3}	3.18×10^{-6}
k_{12}^T c	H_2,SiH_4	-2.74×10^{-1}	-1.70	-6.35×10^{-3}
	N_2,SiH_4	-5.15×10^{-2}	-1.69	-4.94×10^{-3}
	H_2,N_2	-2.71×10^{-1}	-1.61	-9.15×10^{-3}
	TMGa, H_2	1.32	-1.54	-3.57×10^{-3}
	TMGa, N_2	6.36×10^{-1}	-1.58	-3.36×10^{-3}
	AsH_3,H2	8.86×10^{-1}	-1.57	-4.35×10^{-3}
	AsH_3, N2	3.09×10^{-1}	-1.55	-4.06×10^{-3}
	TMGa, AsH_3	1.94×10^{-1}	-1.79	-1.91×10^{-3}

[a] $\mu, k, c_p = c_0 + c_1 T + c_2 T^2$

[b] $D_{12} = D_{21} = \left(c_0 + c_1 T + c_2 T^2\right) / p$

[c] $k_{12}^T = -k_{12}^T = c_0 x_1 x_2 \left[1 + c_1 \exp(c_2 T)\right]$, for $x_1 \to 0$.

T is absolute temperature (K).

Table 3. Dependence of collision integral Ω_μ on dimensional temperature k_bT/ε (Bird et al., 2002)

k_bT/ε	Ω_μ	$\Omega_{D,12}$	k_bT/ε	Ω_μ	$\Omega_{D,12}$	k_bT/ε	Ω_μ	$\Omega_{D,12}$
0.30	2.840	2.649	1.70	1.249	1.141	4.2	0.9598	0.8748
0.35	2.676	2.468	1.75	1.235	1.128	4.3	0.9551	0.8703
0.40	2.531	2.314	1.80	1.222	1.117	4.4	0.9506	0.8659
0.45	2.401	2.182	1.85	1.209	1.105	4.5	0.9462	0.8617
0.50	2.284	2.066	1.90	1.198	1.095	4.6	0.9420	0.8576
0.55	2.178.	1.965	1.95	1.186	1.085	4.7	0.9380	0.8537
0.60	2.084	1.877	2.00	1.176	1.075	4.8	0.9341	0.8499
0.65	1.999	1.799	2.10	1.156	1.058	4.9	0.9304	0.8463
0.70	1.922	1.729	220	1.138	1.042	5.0	0.9268	0.8428
0.75	1.853	1.667	2.30	1.122	1.027	6.0	0.8962	0.8129
0.80	1.790	1.612	2.40	1.107	1.013	7.0	0.8727	0.7898
0.85	1.734	1.562	2.50	1.0933	1.0006	8.0	0.8538	0.7711
0.90	1.682	1.517	2.60	1.0807	0.9890	9.0	0.8380	0.7555
0.95	1.636	1.477	2.7	1.0691	0.9782	10.0	0.8244	0.7422
1.00	1.593	1.440	2.8	1.0583	0.9682	12.0	0.8018	0.7202
1.05	1.554	1.406	2.9	1.0482	0.9588	14.0	0.7836	0.7025
1.10	1.518	1.375	3.0	1.0388	0.9500	16.0	0.7683	0.6878
1.15	1.485	1.347	3.1	1.0300	0.9418	18.0	0.7552	0.6751
1.20	1.455	1.320	3.2	1.0217	0.9340	20.0	0.7436	0.6640
1.25	1.427	1.296	3.3	1.0139	0.9267	25.0	0.7198	0.6414
1.30	1.401	1.274	3.4	1.0066	0.9197	30.0	0.7010	0.6235
1.35	1.377	1.253	3.5	0.9996	0.9131	35.0	0.6854	0.6088
1.40	1.355	1.234	3.6	0.9931	0.9068	40.0	0.6723	0.5964
1.45	1.334	1.216	3.7	0.9868	0.9008	50.0	0.6510	0.5763
1.50	1.315	1.199	3.8	0.9809	0.8952	75.0	0.6140	0.5415
1.55	1.297	1.183	3.9	0.9753	0.8897	100.0	0.5887	0.5180
1.60	1.280	1.168	4.0	0.9699	0.8845			
1.65	1.264	1.154	4.1	0.9647	0.8796			

The collision integral Ω_μ in eq. (13) is a slowly-varying function of dimensionless temperature, k_bT/ε, and is tabulated in table 3. k_b is the Boltzmann constant and ε is a characteristic energy of interaction between molecules which can be estimated by

$$\frac{\varepsilon}{k_b} = 0.77T_c \tag{16}$$

or

$$\frac{\varepsilon}{k_b} = 1.15T_b \tag{17}$$

where T_c and T_b are critical temperature and normal boiling point, respectively.

The thermal conductivity of the polyatomic gas is related to its viscosity by

$$k = \left(c_p + \frac{5}{4} R_g \right) \mu \tag{18}$$

where R_g is the gas constant.

For a mixture of different gases, as is usually the case in CVD processes, the viscosity and the thermal conductivity of the mixture are related to those of the individual components by

$$\mu = \sum_{i=1}^{N} \frac{x_i \mu_i}{\sum_{j=1}^{N} x_i \phi_{ij}} \tag{19}$$

$$k = \sum_{i=1}^{N} \frac{x_i k_i}{\sum_{j=1}^{N} x_i \phi_{ij}} \tag{20}$$

where

$$\phi_{ij} = \frac{1}{\sqrt{8}} \left(1 + \frac{M_i}{M_j} \right)^{-1/2} \left[1 + \left(\frac{\mu_i}{\mu_j} \right)^{1/2} \left(\frac{M_j}{M_i} \right)^{1/4} \right]^2 \tag{21}$$

The specific heat of the gaseous mixture is related to those of the individual components by

$$c_p = \sum_{i=1}^{N} x_i c_{p,i} \tag{22}$$

For applications that involve unknown mass diffusivity, it can be estimated by

$$D_{12} = 1.8583 \times 10^{-7} \frac{\sqrt{T^3 \left(M_1^{-1} + M_2^{-1} \right)}}{p \sigma_{12}^2 \Omega_{D,12}} \tag{23}$$

where the unit for pressure is atm and

$$\sigma_{12} = \frac{1}{2} \left(\sigma_1 + \sigma_2 \right) \tag{24}$$

$\Omega_{D,12}$ is a function of $k_b T / \varepsilon_{12}$ that can be obtained from table 3 using

$$\varepsilon_{12} = \sqrt{\varepsilon_1 \varepsilon_2} \tag{25}$$

The concentration of the reactant is usually much lower than that of the carrier gas(es). When the reactant is a single gas diluted by the carrier gas, the diffusivity of the reactant to the carrier gaseous mixture is of interest. If the reactant is defined as component 1 in the precursor, and the carrier gases are components 2 through N, the diffusivity of the reactant – 1 – to the carrier gas mixture – m – can be obtained by

$$\frac{1-x_1}{D_{1m}}=\sum_{j=2}^{N}\frac{x_j}{D_{1j}} \tag{26}$$

4. Thermal Modeling of LCVD

4.1. Effects of Natural Convection

During LCVD, the spot on the substrate under laser irradiation is at a very high temperature (1200 K or higher). Temperature gradients in the source gases will cause natural convection in the chamber. The concentration of the gas mixture near the hot spot on the substrate is affected by the chemical reaction taking place on the substrate. Concentration differences in the chamber become another force driving natural convection in the chamber. For the case of LCVD by a stationary laser beam, Lee et al. (1995) concluded that the effect of natural convection on the thin film deposition rate was negligible and that the heat and mass transfer in the gases were dominated by diffusion. In the LCVD process, a laser beam scans the substrate and induces chemical reaction; the resulting product forms a line on the substrate. These lines, formed by multiple laser scans, are subsequently interwoven to form a part layer. To thoroughly understand the effects of various physical phenomena – including natural convection – on the LCVD process, natural convection during LCVD with a moving laser beam was investigated by Zhang (2003).

The physical model of LCVD under consideration is illustrated in figure 7. A substrate made of Incoloy 800 with a thickness of h is located in the bottom of a chamber. Before the vapor deposition starts, the chamber is evacuated and then filled with a mixture of H_2, N_2, and $TiCl_4$. A laser beam moves along the surface of the substrate with a constant velocity, u_b. The initial temperature of the substrate, T_i, is below the chemical reaction temperature. Vapor deposition starts when the surface temperature reaches the chemical reaction temperature. The chemical reaction that occurs on the top substrate surface absorbs part of the laser energy and consumes the $TiCl_4$. A concentration difference is thereby established and becomes the driving force for mass transfer. The physical model of the LCVD process includes: natural convection, heat transfer in the substrate and gases, and chemical reaction, as well as mass transfer in the gases.

The laser beam travels with a constant velocity u_b along the surface of the substrate, constituting a typical moving heat source problem. If the substrate is sufficiently large in comparison to the diameter of the laser beam, which has an order of magnitude of 10^{-3} m, a quasi-steady state occurs. The system appears to be in steady-state from the standpoint of an observer located in and traveling with the laser beam. By simulating LCVD with a moving laser beam in the moving coordinate system, the computational time will be substantially shortened, thereby enabling numerical simulation for a significant number of cases.

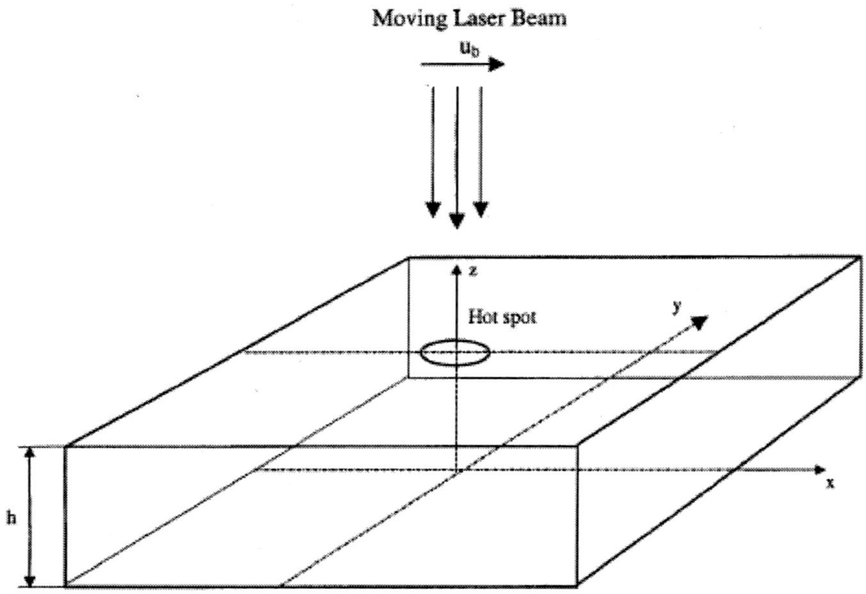

Figure 7. Physical model of Laser Chemical Vapor Deposition (Zhang, 2003).

Heat transfer in the substrate and gases is modeled as one problem with different thermal properties in each region. In the substrate region, the velocity is set to zero in the numerical solution. The advantage of modeling the heat and mass transfer in the substrate and the gases as one problem is that the temperature distribution in the substrate and gases can be obtained by solving one equation. This eliminates the iteration procedure needed to match the boundary condition at the substrate-gas interface. Since the model geometry is symmetric about the *xz* plane, only half of the problem needs to be investigated. For a coordinate system moving with the laser beam, as shown in figure 7, the laser beam is stationary but the substrate and the chamber move with a velocity $-u_b$. The heat and mass transfer in the substrate and gases is governed by eqs. (2) – (5) and (7), with buoyancy forces due to temperature and concentration gradients accounted for, but the Soret effect neglected (Zhang, 2003). For the substrate region, the thermal properties are those of Incoloy 800, the substrate material. For the gaseous region, the thermal properties are determined by the individual thermal properties of H_2, N_2, and $TiCl_4$ as well as their molar fractions [see eq. (19) – (20)]. The mass diffusivity of $TiCl_4$ in the gas mixture is determined by the Stefan-Maxwell equation, using the binary diffusivity of $TiCl_4$ with respect to all other species, which is calculated using the hard sphere model.

The heat flux at the substrate surface due to laser beam irradiation and chemical reaction is expressed as

$$q'' = \frac{2P\alpha_a}{\pi r_0^2}\exp\left[-\frac{2(x^2+y^2)}{r_0^2}\right] - \varepsilon\sigma(T^4 - T_\infty^4) + \rho_{TiN}\Delta H_R u_b \frac{d\delta}{dx}, \quad z = h \tag{27}$$

where ΔH_R is chemical reaction heat, and $d\delta/dt$ is the deposition rate. For a chemical reaction in the order of unity, the deposition rate is expressed as

$$\frac{d\delta}{dx} = -\frac{\gamma_{TiN} K_0}{u_b \rho_{TIN}} \exp\left(-\frac{E}{R_u T_s}\right)\omega_s \tag{28}$$

where ω_s represents the concentration of $TiCl_4$ at the surface of the substrate. The constant K_0 in eq. (28) is defined as $K_0 = (\omega_{H_2})_i (\omega_{N_2})_i^{1/2} K_0'$.

The coefficient γ_{TiN} in eq. (28) is a sticking coefficient defined as

$$\gamma_{TiN} = \begin{cases} 1 & T < T_m \\ 1 + (T_m - T_s)/(T_M - T) & T_m \le T \le T_M \\ 0 & T > T_M \end{cases} \tag{29}$$

where T_s is the surface temperature of the substrate, T_m is a threshold temperature below which the product of the chemical reaction can fully stick to the substrate, and T_M is another threshold temperature above which no product of chemical reaction can be stuck on the substrate. If the surface temperature is between T_m and T_M, the product of chemical reaction can only be partially stuck on the substrate. The values of T_m and T_M are chosen as 1473 K and 1640 K, respectively (Conde et al., 1992a).

The boundary conditions of the velocities are

$$u = -u_b, \quad v = w = 0, \quad |x| \to \infty \tag{30}$$

$$v = \frac{\partial u}{\partial y} = \frac{\partial w}{\partial y} = 0, \quad y = 0 \tag{31}$$

$$u = -u_b, \quad v = w = 0, \quad y \to \infty \tag{32}$$

$$u = -u_b, \quad v = w = 0, \quad z = 0, \infty \tag{33}$$

The governing equations are discretized using the finite volume method (Patankar 1980); the SIMPLEC algorithm (Van Doormaal and Raithby, 1984) was employed to handle the linkage between velocity and pressure.

In order to verify the validity of the code, the calculation is initially made with conduction only in the substrate with a moving laser beam. The conduction problem in the substrate is achieved by setting the thermal conductivity of the gas to zero so that there is no heat conduction in the vapor phase. The steady state surface temperature obtained by numerical solution is compared with the temperature distribution caused by a moving point heat source. The temperature expression of a semi-infinite body with a moving point heat source is (Carslaw and Jaeger, 1959)

$$T_s - T_i = \frac{P\alpha_a}{2\pi kr}\exp\left[-\frac{u(r+x)}{2\alpha}\right] \tag{34}$$

In order to simulate conduction in the substrate subject to a moving point heat source, the heat flux at the substrate surface due to laser beam irradiation is expressed as

$$q'' = \begin{cases} \dfrac{P\alpha_a}{\pi r_0^2} & r \le r_0 \\ 0 & r > r_0 \end{cases}, \quad z = h \tag{35}$$

where the moving heat source is assumed to be top-hat (uniform distribution) instead of a Gaussian distribution because the latter results in the energy spread in a large spot. Effects of surface radiation and chemical reaction on the surface heat flux are also neglected in eq. (35) in order to simulate pure conduction in the substrate.

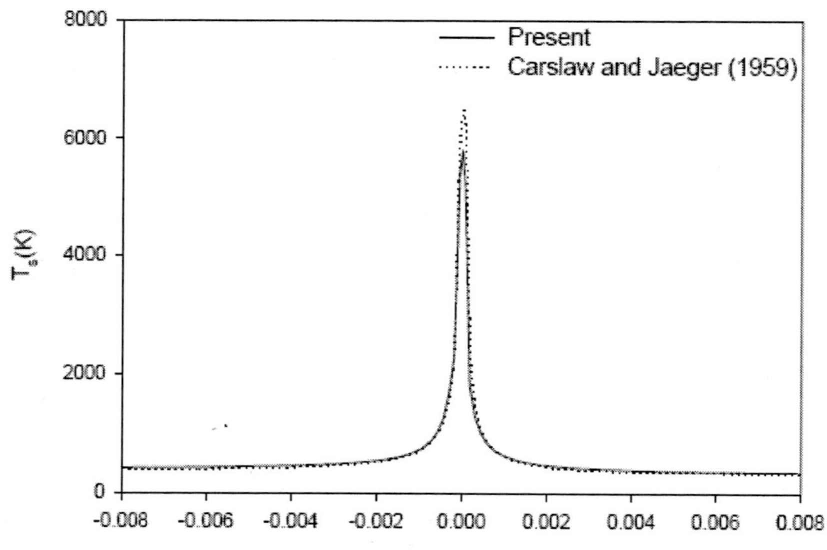

Figure 8. Comparison between analytical and numerical solutions for pure conduction.

The surface temperatures obtained by both methods are plotted in figure 8. The radius of the moving laser beam is r_0=1.41×10^{-4} mm, which is very small in order to simulate conduction in the substrate subject to a moving point heat source. It is seen that the overall agreement between the two solutions is very good except at the locations near x= 0. This discrepancy of the two results is due to the nature of the heat source modeled in the analytical and numerical solutions. The analytical result is for the temperature distribution caused by an infinitesimal heat source at x= 0, while the numerical temperature distribution result is for a finite size heat source. Considering the different models, the agreement between the numerical and analytical solutions is very good.

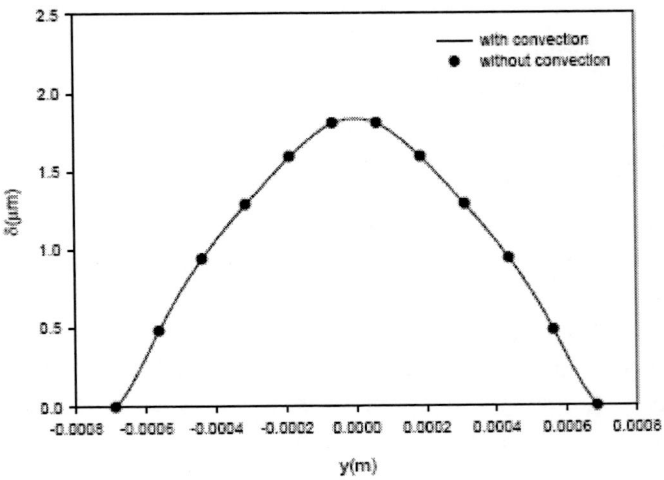

(a) P=300W, u_b=1.2mm/s

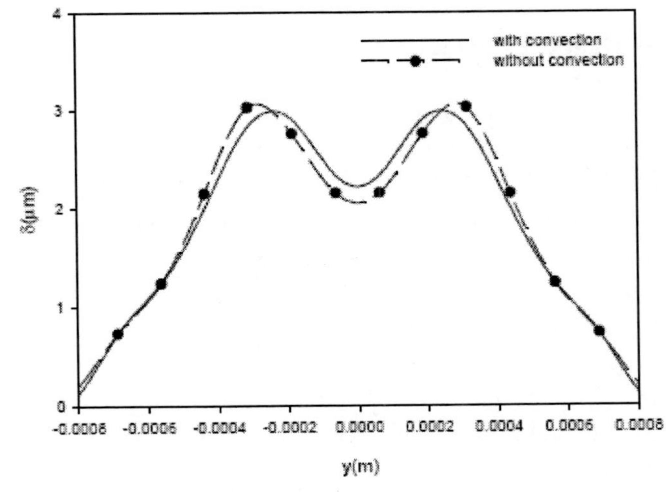

(b) P=360W, u_b=1.2mm/s

Figure 9. Comparison of cross-sections (Zhang, 2003).

Numerical simulations of LCVD process are then performed with parameters similar to those in Conde et al. (1992a) and Zhang and Faghri (2000). The radius of the laser beam is very small compared with that of the chamber and therefore, the LCVD problem can be modeled as LCVD on a substrate with infinite length and width. The validity of this assumption, of course, depends on the actual size of the substrate. On the other hand, it allows obtaining quasi-steady state solutions and conducting a parametric study. The computational domain (the size of the chamber) used in this paper is 0.5×0.5×0.125 m³ ($x \times y \times z$) to ensure that the effect of the computational domain on the temperature distribution and deposited film can be eliminated. The thickness of the substrate is 5 mm and the bottom of the substrate is assumed to be adiabatic. The radius of the laser beam, which is defined as the radius where the laser intensity is $1/e^2$ of the intensity at the center of the laser beam, is 1.0×10^{-3} m. The absorptivity of the laser beam at the substrate surface is taken to be 0.23 (Conde et al., 1992a; Mazumder and Kar, 1995). The activation energy of the chemical reaction is taken to be E = 51.02 kJ/mol. The total pressure in the chamber is 207 torr and the partial pressure of titanium tetrachloride is 7 torr. The partial pressures of N_2 and H_2 are the same. The initial temperature of the substrate and gas is 338 K. The concentrations of different species can be obtained by using ideal gas law. With the conditions specified above, the constant K_0 in eq. (28) is 8.4m/s. Chemical reaction heat, ΔH_R, as determined by using JANAF thermochemical tables (Chase, 1986), is 5.379×10^6 J/kg.

Figure 9(a) and (b) shows the shape of the deposited TiN film obtained by using a scanning velocity of 1.2 mm/s and laser power of 300 W and 360W, respectively. The results show that the effect of natural convection on the shape of deposited film is negligible for the laser power of P = 300 W [see figure 9(a)]. When the laser power is increased to 360W, the effect of natural convection on the shape of the cross-section becomes important, although the cross sectional area is almost unchanged [see figure 9(b)]. A groove is observed on the top of the deposited film for P = 360 W due to a low sticking coefficient.

4.2. A Simulation-Based Correlation

It was demonstrated from the above results that the effect of natural convection on the shape of deposited film is negligible when the laser power is less than 360W. The shape of the cross-section of the film is slightly affected by natural convection when the laser power is higher. However, the cross-sectional areas obtained by the models with and without natural convection are almost identical even for higher laser power. The numerical practices indicated that inclusion of natural convection would significantly increase the CPU time to obtain the converged solution, which would prevent a thorough parametric study. For the purpose of parametric study, the effect of natural convection can be neglected because it only has little effect on the shape of the deposited film and has no effect on the area of the cross-sectional area of the deposited film. Zhang (2004) carried out a thorough parametric study for LCVD with moving laser beams and a proposed a simulation-based correlation

The numerical simulations of the LCVD process are performed in a broader range. The total pressure in the chamber is 207 torr and the partial pressure of Titanium Chloride is maintained at 7 torr. The partial pressures of N_2 and H_2 are both maintained at 100 torr. The thickness of the substrate is 5 mm. The radius of the laser beam, which is defined as the radius where the laser intensity is $1/e^2$ of the intensity at the center of the laser beam, is 0.8 ~ 1.0 mm. The laser power varies from 180 to 360W and the scanning velocity ranged between 0.4 to 2 mm/s.

The effects of laser scanning velocity, laser power, and radius of the laser beam on the shapes of the deposited film were investigated. The results showed that a groove could be observed on the top of the film in conjunction with higher laser power and lower scanning velocity. The cross-sectional area, calculated by

$$A_c = \frac{2}{r_0^2} \int_0^\infty \delta dy \tag{36}$$

at different processing parameters, is shown in figure 10. It decreases with increasing scanning velocity. It also increases with increasing laser power and decreasing laser beam radius. In order to obtain a correlation for the dimensionless cross-sectional area, the results in figure 10 are replotted in figure 11. The dimensionless cross-sectional area is plotted as function of Biot number and Peclet number, which are defined by

$$\mathrm{Bi} = \frac{\alpha_a P}{\pi r_0 k_s (T_c - T_i)} \tag{37}$$

$$\mathrm{Pe} = \frac{u_b r_0}{\alpha_s} \tag{38}$$

where α_a is the absorptivity of the laser beam on the substrate surface, P is the laser power, r_0 is the radius of the laser beam, T_c is the chemical reaction temperature, T_i is the initial temperature of the gases, and u_b is the laser scanning velocity. The thermal conductivity k_s and the thermal diffusivity α_s in eqs. (37) – (38) are those of the substrate (Incoloy) at chemical reaction temperature.

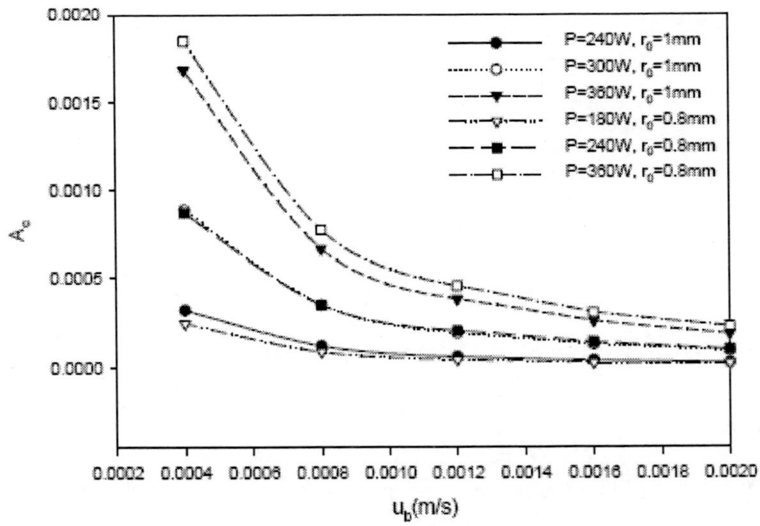

Figure 10. Dimensionless cross-sectional area vs. scanning velocity (Zhang, 2004).

It can be seen that the dimensionless cross-sectional areas are linear functions of $(\mathrm{Bi/Pe})^{1.35}$ for all cases. The dimensionless cross-sectional areas shown in figure 11 can be correlated into the following expression.

$$A_c = a_0 + a_1 \left(\frac{Bi}{Pe} \right)^{1.35} \tag{39}$$

where

$$a_0 = \begin{cases} -2.136\times10^{-5} + 6.126\times10^{-6}\text{Bi} & r_0 = 1.0\ \text{mm} \\ -9.206\times10^{-5} + 9.352\times10^{-5}\text{Bi} & r_0 = 0.8\ \text{mm} \end{cases} \quad (40)$$

$$a_1 = \begin{cases} 8.343\times10^{-7} + 1.423\times10^{-5}\text{Bi} & r_0 = 1.0\ \text{mm} \\ -2.587\times10^{-5} + 4.372\times10^{-5}\text{Bi} & r_0 = 0.8\ \text{mm} \end{cases} \quad (41)$$

The difference between the cross-sectional area obtained by numerical simulation and eq. (10) is less than 10%.

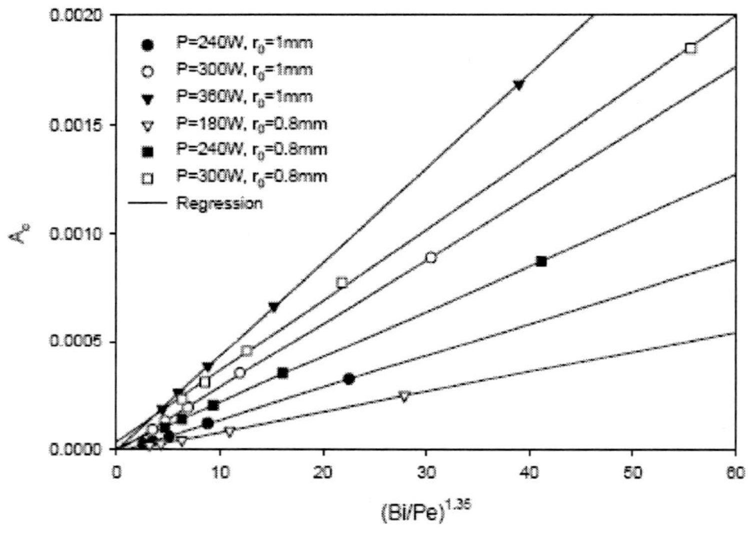

Figure 11. Dimensionless cross-sectional area vs. $(\text{Bi/Pe})^{1.35}$ (Zhang, 2004).

5. Non-Equilibrium Heat Transfer in Pulsed LCVI

5.1. Introduction

Heat transfer in fluid-saturated porous media has been investigated extensively and excellent reviews are available in the literatures (Kaviany, 1995; Nield and Bejan, 1999). Nithiarasu et al. (1996) studied double-diffusive natural convection in an enclosure filled with fluid-saturated porous media using a generalized Non-Darcy approach and the non-Darcy effects on flow, heat, and mass transfer were investigated. Murphy et al. (1996) studied the effect of surface undulations on the natural convection heat transfer from an isothermal surface in a Darcian fluid-saturated porous enclosure using a finite element method. Khanafer and Chamkha (1998) numerically studied unsteady, laminar, 2-D hydromagnetic natural convection in an inclined square enclosure filled with fluid-saturated porous medium in presence of transverse magnetic field. The flow in the porous region is modeled using Brinkman-extended Darcy's law, which enabled non-slip condition at the walls. Effects of thermal stratification on double-diffusive natural convection in vertical porous square and wavy enclosures were studied by Kumar et al. (2002) and Kumar and Shalini (2003), respectively. While local thermal equilibrium between the porous matrix and the fluid is assumed in the above studies, it is well recognized that local equilibrium between different phases in a multiphase system cannot be achieved when thermal properties for different phases differ widely (Jiang et al., 1999; Mohamad, 2000; Al-Amiri, 2002) or during rapid heating or cooling (Minkowycz et al., 1999).

Since laser beam interacts with powder particle only and the precursors are assumed to be transparent during a LCVI process, heat transfer occurs in two steps: (1) absorption of laser beam by the powder particles, and (2) heat transfer from the powder particle to the precursors. If the laser-powder particle interaction time is much longer than the time that it takes for the powder particles and the gas to reach to thermal equilibrium, which is the case for continuous wave (CW) or long-pulsed laser, the local thermal equilibrium assumption between the powder and the gas is valid. Although CW laser is exclusively used in the existing LCVI experimental studies, the short-pulsed laser posses some advantages that are not available with continuous laser. One advantage is that the porosity of the final product can be controlled by controlling the degree of LCVD on the powder particle surface via adjustment of pulse width and repetition rate. Other advantages include achieving LCVD at a much lower average laser power, more accurate control of the dimension of the final product, etc. Since the thermal physical property of the precursors and the particles differ significantly and LCVI with short-pulsed laser is a rapid process, the powder particles and precursors may not be able to reach local equilibrium and consequently, a non-equilibrium model for transport phenomena in the precursors and particles must be developed. In order to explore LCVI with short-pulsed laser, non-equilibrium phenomenon in a powder layer subject to short-pulsed volumetric heat source was investigated in by Zhang (2006).

5.2. Nonequilibrium Heat Transfer Model

The physical model of the problem under consideration is shown in figure 12. A powder layer with a thickness of L and an initial temperature of T_i is subjected to a temporal Gaussian heat source with a FWHM (Full Width at Half Maximum) pulse width of t_p and fluence of J (J/m^2). Since the powder layer is porous, the heat source can penetrate the powder layer and the thermal energy is absorbed within the powder layer, not only at the surface of the powder layer. It is assumed that heat transfer is 1-D along the thickness of the powder layer only. The effect of chemical reaction heat on the heat transfer in the powder layer is neglected because its effect on the deposition rate was negligible as shown by Zhang and Faghri (2000). The porosity of the powder layer during irradiation of one single pulse is assumed to be constant because the amount of deposition in the duration of one short pulse is very small. The effect of convection in the gas phase is also neglected because the laser pulse is very short. Therefore, the problem under consideration becomes heat conduction in a gas-saturated powder layer with internal heat source.

Since the precursors and the particles are not in thermal equilibrium, a two-energy equation model will be employed to describe heat transfer in precursors and particles. The energy equations of the powder particles (s) and the precursors (g) can be respectively expressed as

$$(1-\varphi)(\rho c_p)_s \frac{\partial T_s}{\partial t} = \frac{\partial}{\partial x}\left[k_{s,eff}\frac{\partial T_s}{\partial x}\right] + \dot{q}_s''' - G(T_s - T_g) \quad (42)$$

$$\varphi(\rho c_p)_g \frac{\partial T_g}{\partial t} = \frac{\partial}{\partial x}\left[k_{g,eff}\frac{\partial T_g}{\partial x}\right] + G(T_s - T_g) \quad (43)$$

where $k_{s,eff}$ and $k_{g,eff}$ are the effective thermal conductivities of the solid and gas phases, respectively. During the LCVI process, the powder particles are partially joined together and therefore, it can be assumed that heat conductions occurred in both phases are in parallel. The effective thermal conductivity for solid and gas phases are $k_{s,eff} = (1-\varphi)k_s$ and $k_{g,eff} = \varphi k_g$, respectively. Similar assumptions can be found in existing non-equilibrium models for fluid-saturated porous media (e.g., Al-Amiri, 2002).

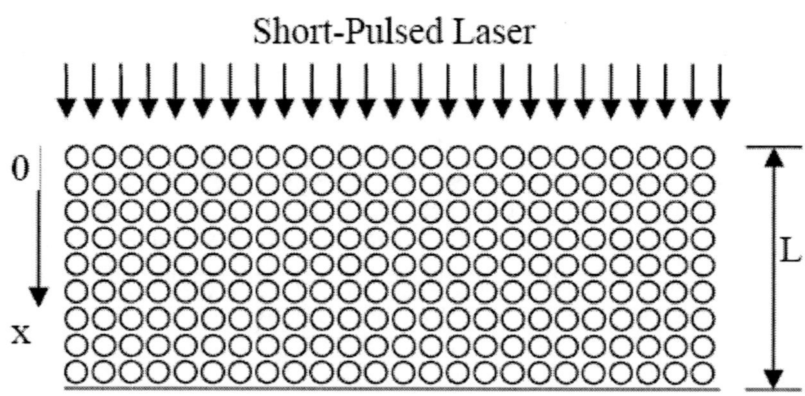

Figure 12. Nonequilibrium heat conduction in powder layer (Zhang, 2006).

The internal heat source term in eq. (42) is

$$q_s''' = 0.94\frac{1-R}{t_p\delta} J \exp\left[-\frac{x}{\delta} - 2.77\left(\frac{t}{t_p}\right)^2\right] \quad (44)$$

where R is reflectivity, and δ is the optical penetration depth. In arrival of eq. (44), the time t =0 is defined as the time when the peak of the heat source occurs. For spherical particle, the coupling coefficient between powder particle and the gas is determined by

$$G = \frac{6(1-\varphi)h_p}{d_p} \quad (45)$$

where d_p is the diameter of the powder particle, h_p is the heat transfer coefficient at the particle surface. Mohamad (2000) suggested that heat transfer between the powder particles and the fluid can be estimated by using the empirical correlation for natural convection from an isolated sphere (Churchill, 1983):

$$\mathrm{Nu} = \frac{h_p d_p}{k_g} = 2 + 0.589 \frac{\mathrm{Ra}_d^{1/4}}{[1 + (0.496/\mathrm{Pr})^{9/16}]^{4/9}} \tag{46}$$

In absence of natural convection in the gas phase, the Nusselt number based on the particle diameter is

$$\mathrm{Nu} = \frac{h_p d_p}{k_g} = 2 \tag{47}$$

The initial conditions of the problem are

$$T_s(x, -\infty) = T_g(x, -\infty) = T_i \tag{48}$$

The surface of the powder layer subjected to heating is also subject to convective cooling, i.e.,

$$-k_{s,eff} \left.\frac{\partial T_s}{\partial x}\right|_{x=0} = h_0 (T_s - T_\infty), \quad -k_{g,eff} \left.\frac{\partial T_g}{\partial x}\right|_{x=0} = h_0 (T_g - T_\infty) \tag{49}$$

while the bottom of the powder layer is adiabatic

$$\left.\frac{\partial T_g}{\partial x}\right|_{x=L} = \left.\frac{\partial T_s}{\partial x}\right|_{x=L} = 0 \tag{50}$$

The heat transfer during laser-powder layer interaction can be solved by using finite different method (Patankar, 1980). Equations (42) and (43) are discretized using an implicit scheme and the resulting algebraic equations are solved using Tri-Diagonal Matrix Algorithm (TDMA). Heat transfer in powder and gas are coupled and therefore iteration is needed. After the grid number independent test, the grid number for all computations is 502. Numerical solution is started from $t = -5t_p$ with a time step of $\Delta t = t_p / 100$ until $t = 5t_p$. The time step is changed to $\Delta t = t_p$ for $t > 5t_p$ and again to $\Delta t = 100t_p$ after $t = 100t_p$. The numerical solution for each time step begins with solution of the powder temperature, T_s, from eq. (42) based on an assumed gas temperature distribution, T_g. The powder temperature is then used to obtain the updated gas temperature T_g^{new} from eq. (43), which is then compared with the assumed gas temperature, T_g. If the maximum difference between T_g^{new} and T_g is less than a small tolerance value, 10^{-6} K, end the iteration for the current time step and go to the next time step. Otherwise, update T_g and repeat the iteration until the convergence criterion is met. Underrelaxation is not necessary during the iteration.

5.3. Effects of Pulse Duration on Nonequilibrium Heat Transfer

Numerical simulations are performed for a system with silicon carbide powder and tetramethylsilane (TMS) gas (Dai et al. 2003). The thermal physical properties of the silicon carbide are $\rho_s = 3.21\times10^3\,\text{kg/m}^3$, $c_{ps} = 660\,\text{J/kg-K}$, $k_s = 58.86\,\text{W/m-K}$. The porosity of the powder layer is set as 0.42, which is within the range of randomly packed powder bed (Kaviany, 1995). The specific heat of the TMS as obtained from JANAF Thermochemical tables (Chase, 1986) is $c_{pg} = 3438\,\text{J/kg-K}$. The thermal conductivity of TMS is $k_g = 0.536\,\text{W/m-K}$, which is derived by using the method in Bird et al. (2002) at 1273K. The density of the TMS is $\rho_g = 0.045\,\text{kg/m}^3$, which is obtained using ideal gas law with a pressure of 10 torr (Dai et al. 2003). The reflectivity of the laser beam is taken as $R = 0.6$. The initial temperature of the powder layer for all cases discussed below is $T_i = 300$ K. The optical penetration depth is related to the powder particle size, porosity, and the absorptivity of the powder material. Alexandre et al. (2004) measured the optical penetration depth of the silicon carbide powder under irradiation of Nd:YAG laser and found that the penetration depth is about 1.5 to 2 times of the particle diameter. The optical penetration depth in eq. (44) is therefore, taken as $\delta = 2d_p$ in this paper. The particle sizes to be studied in this paper are d_p = 15, 20 and 25 μm, which are consistent with the particle size used in the LCVI experiments (Crocker et al., 2002; Dai et al. 2003).

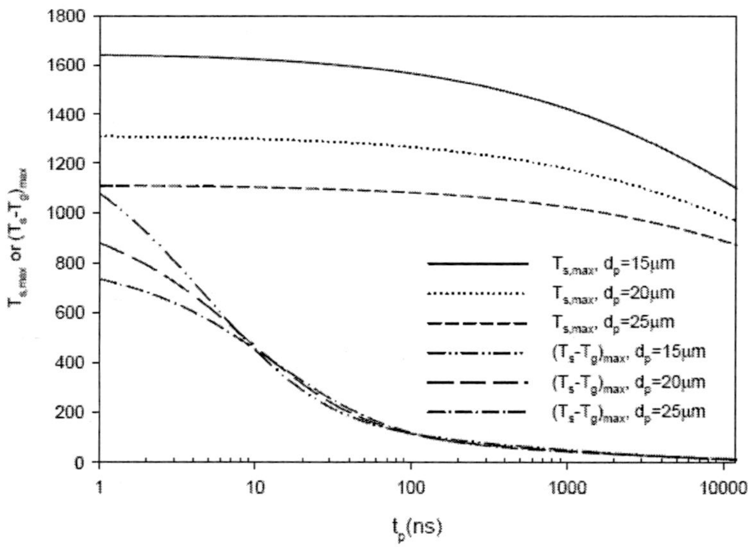

Figure 13. Maximum powder temperature and maximum temperature difference vs. pulse width ($J = 1.25\times10^5\,\text{J/m}^2$; Zhang, 2006).

The effect of pulse width on the maximum powder temperature and the maximum temperature differences is shown in figure 13. The amount of laser energy delivered to the powder layer is the same for all cases because the laser fluence is kept at $J = 1.25\times10^5\,\text{J/m}^2$. It can be seen that the powder temperature at the heating surface decreases with increasing powder particle diameter because the optical penetration depth increases with increasing powder particle diameter. The powder temperature at the heating surface decreases with increasing pulse width because the same laser energy spread over a longer period of time for larger pulse width. When the laser pulse width is shorter than 10 ns, the maximum temperature difference between the powder particle and gas decreases with increasing powder particle size. The effect of particle diameter on the maximum temperature difference becomes weaker with increase pulse width. After the laser pulse length is longer than 10 ns, the maximum temperature difference decreases with decreasing powder particle but the effect of particle size is very insignificant. When the laser pulse width is longer than 100ns, the effect of particle diameter is almost vanished. For all particle sizes studied in figure 13, the maximum temperature difference decreases with increase pulse width. After the laser pulse width is longer than $1\,\mu s$, the non-equilibrium phenomenon becomes very insignificant and the laser-powder interaction can be modeled using equilibrium model.

6. Conclusions

Advances in thermal modeling of laser chemical vapor deposition and infiltration are reviewed in this chapter. In contrast to conventional CVD, in which the entire susceptor is heated, only a very small spot on the substrate is heated by a directed laser beam during LCVD. The mechanism to bind powder particles together in a LCVI process is vapor deposition of solid material on the powder particle surface by decomposition of gas precursors. Extensive experimental works about LCVD/LCVI have been carried out and various applications of these new techniques have been discovered. The inherent nature of LCVD/LCVI makes them very ideal to fabricate ceramics and metal parts that are either impossible or very difficult to manufacture by any other methods. Accurate prediction and control LCVD/LCVI require a model of the processes that incorporate heat/mass transfer, chemical reaction kinetics, and chemical thermodynamics. Kinetic data that are important to modeling LCVD/LCVI are difficult to obtain and are currently compiled for only a limited number of chemical systems. Thermodynamic modeling is very useful for precursor selection; however, it cannot predict the deposition rate that is very critical for design of LCVD/LCVI systems. The generalized governing equations for transport phenomena and chemical reaction during LCVD as well as transport properties of the gaseous precursors are presented.

Natural convection in the LCVD of TiN films on a substrate with a moving laser beam was investigated numerically. The results show that the effect of natural convection on the shape of deposited films is negligible for a laser power of 300 W. When the laser power is increased to 360 W, a groove is observed on top of the thin film and the effect of natural convection on the shape of the cross-sectional area becomes important. A parametric study on the LCVD of TiN film on a substrate with moving laser beam is presented. The results showed that a groove could be observed on the top of the film for higher laser power and lower scanning velocity. The cross-sectional area decreases with increasing scanning velocity. It also increases with increasing laser power and decreasing laser beam radius. A simulation-based correlation on the dimensionless cross-sectional area is obtained.

Heat transfer in a thin powder layer subject to a short-pulsed volumetric heat source is modeled using a non-equilibrium model. The degree of non-equilibrium in the process, as measured using the maximum temperature difference between powder and gas, decreases with increase laser pulse width and it becomes insignificant for the laser pulse longer than 1 μs. The peak temperature and the maximum temperature difference between powder and gas increase with increasing powder particle diameter. The times at which peak temperatures and maximum temperature difference occur are not affected by increasing laser fluence, but the maximum temperatures and the maximum temperature difference between the powder and gas increase significantly with increasing laser fluence.

Acknowledgment

The work presented in this chapter is supported in part by the US National Science Foundation under grant number CBET -0730143.

References

Al-Amiri, A.M., 2002, "Natural Convection in Porous Enclosures: The Application of the Two-Energy Equation Model," *Numerical Heat Transfer*, Part A., Vol. 41, pp. 817-834.

Alexandre, T., Giovanola, J., Vaucher, S., Beffort, O., Vogt, U., 2004, "Layered Manufacturing of Porous Ceramic Parts from SiC Powders and Preceramic Polymers," *Proceedings of the 4th Laser Assisted Net Shape Engineering (LANE)*, Erlangen, Germany, pp. 497-505.

Bird, R.B., Stewart, W.E., and Lightfoot, E.N., 2002, *Transport Phenomena*, 2nd ed., John Wiley and Sons, New York.

Birmingham, B.R. and Marcus, H.L., 1994, "Silicon Carbide Shaped by Selective Area Laser Deposition Vapor Infiltration," *Proc. of the 5th Solid Freeform Fabrication Symposium*, Austin, TX, pp. 348-355.

Carslaw, H.S., and Jaeger, J.C., 1959, *Conduction of Heat in Solids*, Clarendon, Oxford.

Chase, W.M., 1986, JANAF Thermochemical Tables, 3rd Edition, *J. Phys. Chem. Ref. Data*, Vol. 14, Suppl. 1.

Churchill, S.W., 1983, "Free Convection Around Immersed Bodies," *Heat Transfer Design Handbook*, Schlunder, E.U. Ed., Hemisphere, New York, NY.

Conde, O., Kar, A., and Mazumder, J., 1992a, "Laser Chemical Vapor Deposition of TiN Dot: A Comparison of Theoretical and Experimental Results," *J. Applied Physics,* Vol. 72, pp. 754-761.

Conde, O., Ferreira, M.L.G., and Hochholdinger, P., and Silvestre, A.J., 1992b, "CO_2 laser induced CVD of TiN," *Applied Surface Science*, Vol. 54, pp. 130-134.

Crocker, J.E., Sun, L, Shaw, L.L., and Marcus, H.L., 1998, "Preparation and Properties of In-Situ Device Using the SALD and SALDVI techniques," *Proceedings of the 9th Solid Freeform Fabrication Symposium*, pp. 543-547.

Crocker, J.E., Shaw, L.L., and Marcus, H.L., 2000, "Gas-Phase Solid Freeform Fabrication of SALDVI of SiC Cermets," *Mat. Res. Soc. Proc., Vol. 625, pp. 211-215.*

Crocker, J.E., Wei, H., Shaw, L.L., and Marcus, H.L., 2001, "SALDVI of SiC into Metal and Ceramic Powders," *Proceedings of the 12th Solid Freeform Fabrication Symposium*, pp. 163-169.

Crocker, J.E., Shaw, L.L., and Marcus, H.L., 2002, "Powder Effects in SiC Matrix Layered Structures Fabricated Using Selective Area Laser Deposition Vapor Infiltration," *J. Materials Science*, Vol. 37, pp. 3149-3154.

Dai, K., Crocker, J., Shaw, L., and Marcus, H., 2000, "Finite Element Analysis of the SALDVI Process," *Proceedings of the 11th Solid Freeform Fabrication Symposium*, pp. 393-398.

Dai, K., Crocker, J., Shaw, L., and Marcus, H., 2001, "Modeling of Selective Area Laser Deposition Vapor Infiltration (SALDVI) of Silicon Carbide," *Proceeding of the 13th Solid Freeform Fabrication Symposium*, pp. 392-399.

Dai, K., Crocker, J., Shaw, L., and Marcus, H., 2003, "Thermal Modeling of Selective Area Laser Deposition (SALD) and SALD Vapor Infiltration of Silicon Carbide," *Rapid Prototyping Journal,* Vol. 9, pp. 231-239.

Elyukhin, V.A., Garcia-Salgado, G., and Pena-Sierra, R., 2002, "Thermodynamic Model for Low Temperature Metalorganic Chemical Vapor Deposition of GaN," *Journal of Applied Physics*, Vol. 91, pp. 9091-9094.

Faghri, A., and Zhang, Y., 2006, *Transport Phenomena in Multiphase Systems*, Elsevier, Burlington, MA.

Figueras, A., Garelik, S., and Rodroguez-Clememe, R. 1991a, "A morphological and Structural Study *of* SiC layers obtained by LPCVD using Tetramethylsilane," *Journal of Crystal Growth*, Vol. 100, pp. 528.

Figueras, A., Garelik, S., Santiso, J., and Rodrogucz-Clemente, R., Annas, B., Combescure, C. Berjoan, R., Saurel, J. M., and Caplain, R., 1991b, "Growth and Propcnics of CVD-SiC layers obtained by LPCVD using Tetramcthylsilane," *Journal of Materials Science*. Vol. 23, pp. 1632.

Figueras, A., Garelik, S., Santiso, J., and Rodrogucz- Clemente, R., 1992, "Growth and Properties of CVD-SiC layers using Tetramethylsilanc," *Material Science and Engineering*, B11, pp. 83.

Gordon, S., and McBride, B.J., 1976, Computer Program for Calculation of Complex Chemical Equilibrium Compositions, Rocket Performance, Incident and Reflected Shocks and Chapman-Jouguet Detonations, NASA Lewis Research Center, NASA SP-273, March, 1976.

Han, J., and Jensen, K.F., 1994, "Combined Experimental and Modeling Studies of Laser-assisted Chemical Vapordeposition of Copper from Copper(I)-hexafluoroacetylacetonate Trimethylvinylsilane," *J. Applied Physics*, Vol. 75, pp. 2240-2250.

Herring, R.B., 1990, "Silicon Epitaxy," In: *Handbook of Semiconductor Silicon Technology*, edited by O'Mara, W.C., Herring, R.B., and Hunt, L.P., Noyes Publications, Park Ridge, NJ, pp. 258-336.

Jakubenas, K.J., Birmingham, B., Harrison, S., Crocker, J., Shaarawi, M.S., Tompkins, J.V., Sanchez, J., and Marcus, H.L., 1997, "Recent Development in SALD and SALDVI," *Proceedings of 7th International Conference on Rapid Prototyping*, San Francisco, CA.

Jakubenas, K.J., Sanchez, J.M., and Marcus, H.L., 1998, "Multiple Material Solid Free-form Fabrication by Selective Area Laser Deposition," *Materials and Design*, Vol. 19, pp. 11-18.

Jacquot, Y., Zong, G.S., and Marcus, H.L., 1995, "Modeling of Selective Laser Deposition for Solid Freeform Fabrication," *Proceedings of Solid Freeform Fabrication Symposium 1995*, pp. 74-82.

Jean, D.L., Duty, C.E., Fuhrman, B.T., and Lackley, W.J., 1999, "Precision LCVD System Design with Real Time Process Control," *Proceedings of 10th Solid Freeform Fabrication Symposium*, pp. 59-65.

Jean, D.L., Duty, C.E., Fuhrman, B.T., and Lackley, W.J., 2000, "Precision Carbon Deposition Using Pyrolytic Laser Chemical Vapor Deposition," *Proceedings of 11th Solid Freeform Fabrication Symposium*, pp. 73-79.

Jiang, P.X., Ren, Z. P., and Wang, B.X., 1999, "Numerical Simulation of Forced Convection Heat Transfer in Porous Plate Channels Using Thermal Equilibrium and Nonthermal Equilibrium Models," *Numerical Heat Transfer,* Part A, Vol. 35, pp. 99-113.

Kar, A., and Mazumder, J., 1989, "Three-Dimensional Transient Thermal Analysis for Laser Chemical Vapor Deposition on Uniformly Moving Finite Slabs," *J. of Applied Physics*, Vol. 65, pp. 2923-2934.

Kar, A., Azer, M.N., and Mazumder, J., 1991, "Three- Dimensional Transient Mass Transfer Model for Laser Chemical Vapor Deposition of Titanium on Stationary Finite Slabs," *J. of Applied Physics* Vol. 69, pp. 757-766.

Kaviany, M., 1995, *Principles of Heat Transfer in Porous Media*, 2nd Edition, SpringerVerlag, New York.

Khanafer, K.M, and Chamkha, A. J., 1998, "Hydromagnetic Natural Convection from an Inclined Porous Square Enclosure with Heat Generation," *Numerical Heat Transfer,* Part A, Vol. 33, pp. 891-910.

Kumar, B.V.R., Singh, P., and Bansod, V.J., 2002, "Effect of Thermal Stratification on Double-Diffusive Natural Convection in a Vertical Porous Enclosure," *Numerical Heat Transfer,* Part A, Vol. 41, pp. 421-447.

Kumar, B.V.R., and Shalini, 2003, "Natural Convection in a Thermally Stratified Wavy Vertical Porous Enclosure," *Numerical Heat Transfer,* Part A, Vol. 43, pp. 753-776.

Lee, Y.L., Tompkins, J.V., Sanchez, J.M. and Marcus, H.L., 1995, "Deposition Rate of Silicon Carbide by Selected Area Laser Deposition," *Proceedings of Solid Freeform Fabrication Symposium 1995*, pp. 433-439.

Mahajan, R.L., 1996, "Transport Phenomena in Chemical Vapor-Deposition Systems," *Advances in Heat Transfer*, Academic Press, San Diego, 1996.

Marcus, H.L, Zong, G., and Subramanian, P.K., 1993, "Residual Stresses in Laser Processed Solid Freeform Fabrication, Residual Stresses in Composites," *Measurement, Modeling and Effect on Thermomechanical Properties*, Barrera, E.V. and Dutta, I., eds., TMS, pp. 257-271.

Maxwell, J., Krishnan, R., and Haridas, S., 1997, "High Pressure, Convectively-Ehhanced Laser Chemical Vapor Deposition of Titanium," *Proceedings of Solid Freeform Fabrication Symposium 1997*, pp. 497-504.

Maxwell, J., Larsson, K., Boman, M., Hooge, P., Willaims, K., and Coane, P., 1998a, "Rapid Prototyping of Functional Three-Dimensional Microsolenoids and Electromagnets by High-Pressure Laser Chemical Vapor Deposition," *Proceedings of Solid Freeform Fabrication Symposium 1998*, pp. 529-536.

Maxwell, J., Shah, J., Webster, T., and Mock, J., 1998b, "Rapid Prototyping of Titanium Nitride Using Three-Dimensional Laser Chemical Vapor Deposition," *Proceedings of Solid Freeform Fabrication Symposium 1998*, pp. 575-580.

Maxwell, J, Boman, M., Williams, K., Larsson, K. , Jaikumar, N., and Saiprasanna, G., 1999 "High-Speed Laser Chemical Vapor Deposition of. Amorphous Carbon Fibers, Stacked

Conductive Coils, and Folded Helical Springs," *Part of the SPIE Conference on Micromachining and Microfabrication Process Technology V, SPIE* vol. 3874, Santa Clara, CA.

Mazumder, J., and Kar, A., 1995, Th*eory and Application of Laser Chemical Vapor Deposition*, Plenum Publishing Co., New York, NY.

Minkowycz, W. J., Haji-Sheikh, A., and Vafai, K., 1999, "On Departure from Local Thermal Equilibrium in Porous Media due to a Rapidly Changing Heat Source: the Sparrow Number," *International Journal of Heat and Mass Transfer*, Vol. 42, pp. 3373-3385.

Mohamad, A.A., 2000, "Nonequilibrium Natural Convection in a Differentially Heated Cavity Filled with a Saturated Porous Matrix," ASME *J. Heat Transfer*, Vol. 122, pp. 380-384.

Murphy, P.V.S.N., Kumar, B.V.R., and Singh, P., 1996, "Natural Convection Heat Transfer from a Horizontal Wavy Surface in a Porous Enclosure," *Numerical Heat Transfer,* Part A, Vol. 31, pp. 207-221.

Nield, D.A., and Bejan, A., 1999, *Convection in Porous Media*, 2nd ed., Springger-Verlag, New York.

Nithiarasu, P., Seetharamu, K. N., and Sundararaian, L, 1996, "Double-Diffusive Natural Convection in an Enclosure Filled with Fluid-Saturated Porous Medium: A Generalized Non-Darcy Approach," *Numerical Heat Transfer,* Part A, Vol. 30, pp. 413-426.

Patankar, S.V., 1980, *Numerical Heat Transfer and Fluid Flow*, Hemisphere, Washington, DC.

Powell, C., Blocher, M., and Oxley, J., 1966, *Vapor Deposition*, John Wiley and Sons, New York.

Shaarawi, M.S., Sanchez, J.M., Kan, H., and Manthiram, A., 2000, "Modeling of Laser-Induced Chemical Vapor Deposition of Silicon Carbide Rids from Tetramethylsilane," J. Am. Ceram. Soc., Vol. 83, pp. 1947-1952.

Sivaram, S., 1995, *Chemical Vapor Deposition:* Th*ermal and Plasma Deposition of Electronic Materials*, Kluwer Academic Publishers, Bordrecht, Netherlands.

Sun, L., Jakubenas, K.J., Crocker, J.E., Harrison, S., Shaw, L.L., and Marcus, H.L. 1998a, "In Situ Thermocouples in Micro-Components Fabricated Using SALD/SALDVI Techniques: I Thermochemical Modeling," *Materials and Manufacturing Processes*, Vol. 13, pp. 859-882.

Sun, L., Jakubenas, K.J., Crocker, J.E., Harrison, S., Shaw, L.L., and Marcus, H.L. 1998b, "In Situ Thermocouples in Micro-Components Fabricated Using SALD/SALDVI Techniques: II Evaluation of Processing Parameters," *Materials and Manufacturing Processes*, Vol. 13, pp. 883-907.

Sun, L., Jakubenas, K.J., Crocker, J.E., Harrison, S., Shaw, L.L., and Marcus, H.L. 1998c, "In Situ Thermocouples in Micro-Components Fabricated Using SALD/SALDVI Techniques: III Fabrication and Properties of the SiC/C Thermocouple Device," *Materials and Manufacturing Processes*, Vol. 13, pp. 909-919.

Taylor, C.A., Wayne, M.F., and Chiu, W.K.S., 2004, "Microstructural Characterization of Thin Carbon Films Deposited from Hydrocarbon Mixtures," *Surface and Coatings Technology,* Vol. 182, pp. 131-137.

Ueda, O., 1996, *Reliability and Degradation of III-V Optical Devices*, Artech House, Inc., Boston.

Van Doormaal, J.P., and Raithby, G.D., 1984, "Enhancements of the Simple Method for Predicting Incompressible Fluid Flows," *Numerical Heat Transfer*, Vol. **7**, pp. 147-163.

Williams, K., Maxwell, J., Larsson, K., and Boman, M., 1999, "Freeform Fabrication of Functional Microsolenoids, Electromagnets and Helical Springs Using High-Pressure Laser Chemical Vapor Deposition," *Proceedings of the* 1999 *12th IEEE International Conference on Micro Electro Mechanical Systems! MEMS,* pp. 232-237.

Zhang, Y., 2003, "Quasi-Steady State Natural Convection in Laser Chemical Vapor Deposition using a Moving Laser Beam," ASME *J. Heat Transfer*, Vol. 125, No. 3, pp. 429-437.

Zhang, Y., 2004a, "Effect of Natural Convection on Laser Chemical Vapor Deposition with a Stationary Laser Beam," *Int. J. Heat and Fluid Flow*, Vol. 25, No. 4, pp. 683-691.

Zhang, Y., 2004b, "A Simulation-Based Correlation of Cross-Sectional Area of the Thin Film Produced by Laser Chemical Vapor Deposition with a Moving Laser Beam," ASME *J. Manufacturing Science and Engineering,* Vol. 126, No. 4, pp. 796-800.

Zhang, Y., 2006, "Nonequilibrium Modeling of Heat Transfer in a Gas-Saturated Powder Layer Subject to Short-Pulsed Heat Source," *Numerical Heat Transfer*, Part A, Vol. 50, No. 6, pp. 509-524.

Zhang, Y., and Faghri, A., 2000, "Thermal Modeling of Selective Area Laser Deposition of Titanium Nitride on a Finite Slab with Stationary and Moving Laser Beams," *Int. J. Heat Mass Transfer*, Vol. 43, pp. 3835-3846.

In: Thermal Engineering Research Developments
Editors: J. Evgova and O. Kostadinov, pp. 317-344
ISBN: 978-1-60741-497-1

Chapter 7

THERMAL ANALYSIS OF INDUSTRIAL COOLING SYSTEMS

Martín Picón Núñez
Department of Chemical Engineering
University of Guanajuato, Mexico.

1. Introduction

Industrial cooling systems that operate under water recirculation schemes consist of a cooling tower, a pumping system and a set of heat exchangers. Cooling systems are required to remove the excess heat from process or product streams for its final temperatures to reach specified values. The set of heat exchangers where the heat removal process takes place is referred to as the cooling network. Depending on whether the heat transfer process that takes place within these exchangers is related to the removal of sensible or latent heat, they are further classified as coolers or condensers respectively.

The size and capacity of a cooling system varies from application to application, and they can comprise a single cooling tower for a single network of exchangers or a number of cooling towers servicing a network of heat exchangers each. With respect to the way cooling water is distributed amidst the network of exchangers, they are said to be arranged in parallel, series or a combination of both.

This chapter looks at the design, retrofit and the thermal performance analysis of industrial cooling systems. In design, the effect of the arrangement of the exchanger network upon total surface area and cost is analyzed. In retrofit, a targeting methodology is presented that allows for the determination of the minimum water flow rate that will accomplish the required heat duty considering the actual installed surface area. Energy savings are achieved by reducing the actual pumping power to the minimum required by the minimum targeted water flow rate.

Cooling systems are commonly subject to changes in ambient conditions and process changes that increase or reduce the heat load. From a thermal point of view, a cooling system is comprised of two main elements: a set of heat exchangers for indirect heat transfer (coolers and condensers) and a single heat exchanger for direct heat transfer (cooling tower). These two systems interact.

The various heat exchangers in a cooling network are connected to each other through the cooling media (cooling water), therefore any change to the water inlet temperature, will have an effect upon the performance of each unit and eventually as it reaches the cooling tower, will affect its performance. The set of "n" number of exchangers that make up the cooling network can be considered to be a complex assembly of heat exchangers, and as such, it is possible to derive a single expression to calculate its thermal response. The thermal response of the exchanger assembly can be calculated by means of the thermal effectiveness of the assembly.

A cooling tower, on the other hand, is a single heat exchanger where the cooling process takes place by the direct contact between air and water. The thermal effectiveness of these kinds of systems must include the effect of mass transfer in the cooling process. However, in this paper it is shown that for the case of variations in the temperature inlet conditions to the tower, the thermal effectiveness remains nearly constant such that a simplified model can be derived to determine its steady state response.

The assembly of heat exchangers and the cooling tower can now be combined into a single assembly for which an overall thermal effectiveness equation can be derived. The application of the overall cooling system thermal effectiveness model is demonstrated and the validity of the simplified cooling tower thermal effectiveness model is demonstrated by comparing the results with experimental data.

2. Effect of Network Arrangement on the Heat Transfer Area of Cooling Networks

Most industrial installations have cooling systems whose network of exchangers is arranged in parallel. From the operational point of view, in this arrangement, the cooling water enters every unit at the same temperature. There are cases, however, where not all exchangers require water at the coldest temperature; therefore, it is possible to generate different arrangements with positive results in terms of equipment size and therefore capital investment. In this section, various cooling network design approaches are analysed subject to restrictions such as the maximum permissible water return temperature and the allowable pressure drop between the supply and return points of the system.

Figure 1 shows an existing industrial cooling system; the facility consists of 50 exchangers and a cooling tower with a capacity to remove 69.3 MW of heat. The system re-circulates 144 m^3/min of water, and this flow is moved with five pumps operated by 500 hp motors each; the system operates between a cold water temperature of 29°C and a return temperature of 36°C (Picon et al., 2004). Most exchangers in the system are arranged in parallel; however, there are three areas of the plant where exchangers are set in series as in the case of exchangers: HX-19, HX-20 and HX-21.

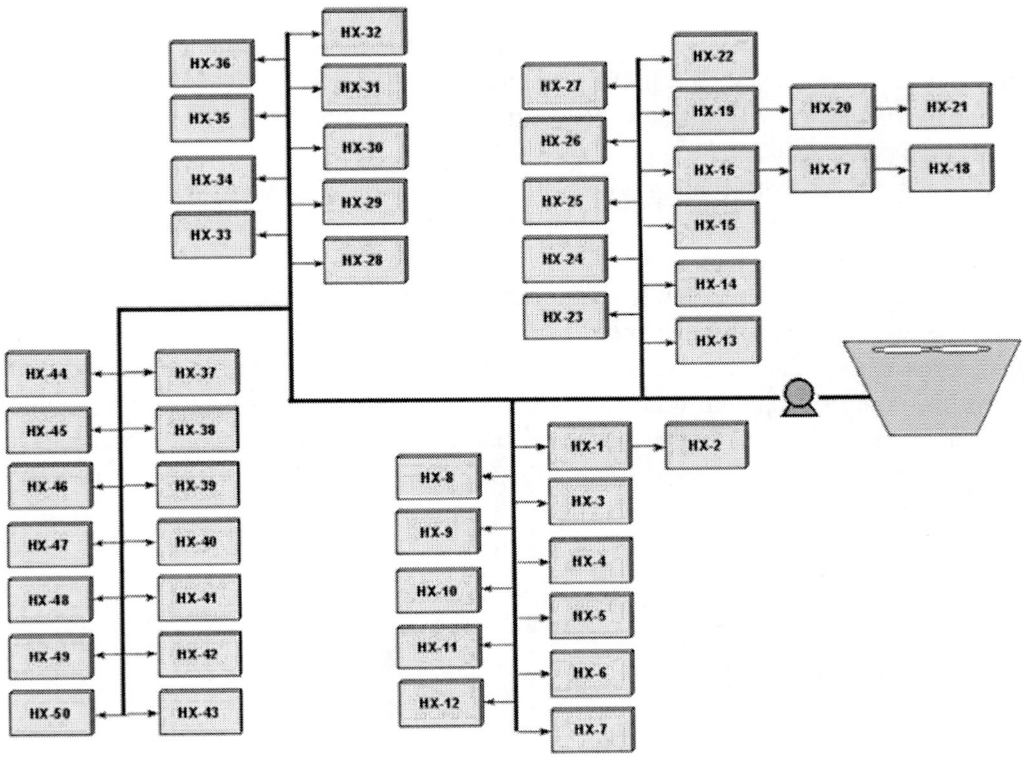

Figure 1. Industrial installation of a cooling network.

Cooling networks can be designed using various techniques. In this chapter the focus is on the following techniques:

1. The pinch design method based on vertical heat transfer (Linnhoff and Hindmarsh, 1983), which produces a network structure that combines a series-parallel arrangement.
2. The minimum water use design method (Kim and Smith, 2001), which produces a network structure that combines a series-parallel arrangement.
3. A design approach assuming that all exchangers are set in parallel.
4. A design approach assuming that all exchangers are set in series.

2.1. Coolng System Design Considerations

The technique used to design a cooling network determines the network structure and the number of heat exchangers. In turn, the way heat exchangers are arranged in a cooling network has an effect upon the total surface area. For this effect to be assessed, a detailed exchanger design must be carried out for each network. The design basis is the allowable pressure drop associated to losses due to friction through the exchangers which is assumed to be the same in each network structure and is distributed linearly upon single exchangers.

For a given cooling duty, there are water flow rate bounds within which the cooling system can be designed. These bounds are set by the minimum and the maximum permissible water flow rate. The minimum water flow rate is obtained when the maximum water outlet temperature is fixed; a practical condition that determines this maximum temperature is related to process constraints such as salt precipitation (Helalizadeh et al., 2000). On the other hand, the maximum water flow rate is determined by the maximum allowable pressure drop that the pump system can handle.

The relation between cooling tower dimensions, heat exchanger surface area and pumping power consumption for a given water flow rate can be described using figure 2. For instance, when the minimum water flow rate is used, the heat transfer surface area requirements are higher due to lower heat transfer coefficients on the water side; now since the cooling water reaches its maximum permissible temperature value, the height of the cooling tower tends to increase contrary to the cross sectional area and the pumping costs are low. For the case of maximum water flow rate, lower heat transfer surface area is required since heat transfer coefficients on the water side are higher; the pumping costs tend to increase and the cooling tower will be shorter in height but of larger cross sectional area since more water is circulated.

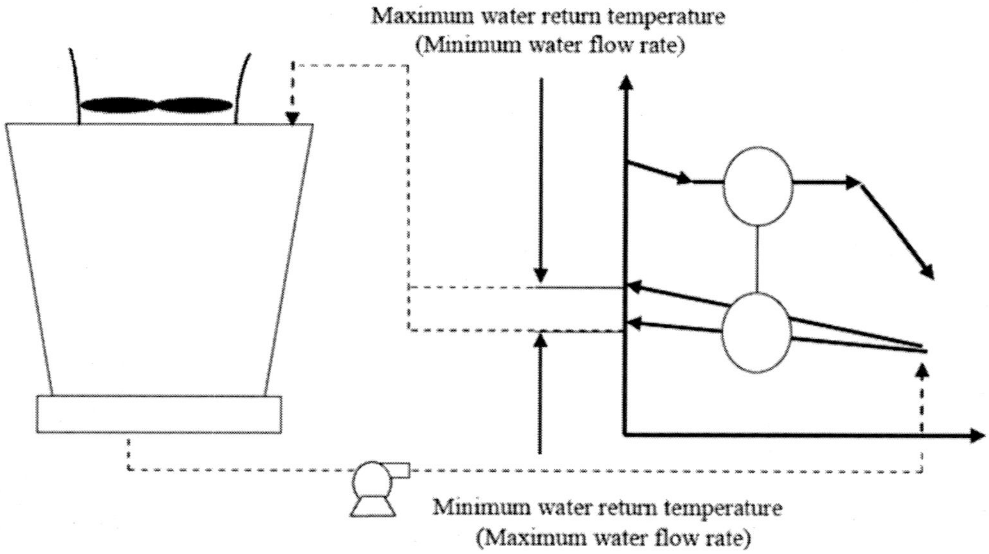

Figure 2. Effect of the choice of water flow rate upon the size of each of the components of a cooling system.

2.2. Heat Transfer Surface Area Requirements

A case study from the open literature (Kim and Smith, 2001) is used to determine the area requirements of various cooling networks capable of meeting the cooling duty of the process. The process has 4 streams with no phase change. Commercial software (B-Jac of Aspen Tech) is used for the detailed design of the exchangers that are assumed to be of the shell and tube type. Cold water is available at 20°C; the higher the return temperature to the cooling tower the better its performance (Bedekar , 1998) but this also increases the fouling problems in coolers due to the decrease of solubility of calcium and magnesium salts, so the maximum return temperature is fixed at 40°C. Process data for the design problem are given in table 1.

The overall pressure drop due to friction through a cooling network is made up of two main components: one is the pressure drop due to the flow of the fluid through pipes, fittings and valves, and the second one is the pressure drop across the exchangers. In this study only the pressure drop across the heat transfer units is considered and this is assumed to be 0.81 bar for all networks. Coolers arranged in parallel exhibit the same pressure drop, whereas in a series arrangement, the total pressure drop is the summation of the individual pressure drops of each unit. The costs of the exchangers are calculated using the cost models contained in the software.

Table 1. Process data for network cooling design

Stream	T_{in} [°C]	T_{out} [°C]	M C_p [kW/°C]	Q [kW]
1	50	30	20	400
2	50	40	100	1,000
3	85	40	40	1,800
4	85	65	10	200

Figures 3, 4 , 5 and 6 show the network structures for the parallel, series, series-parallel (vertical heat transfer), and series-parallel (minimum water flow rate) designs and tables 2, 3, 4 and 5 show the results of the detailed exchanger design.

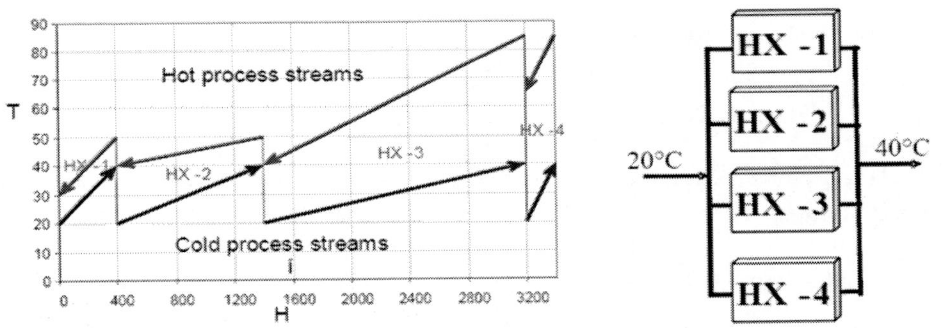

Figure 3. Cooling network in parallel arrangement.

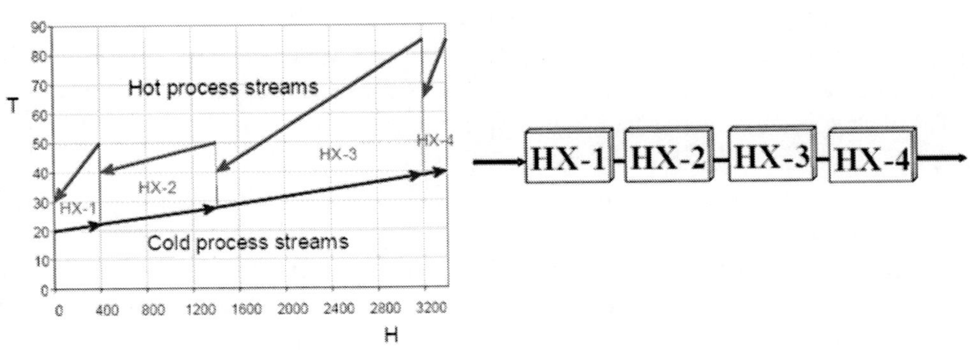

Figure 4. Cooling network in series arrangement.

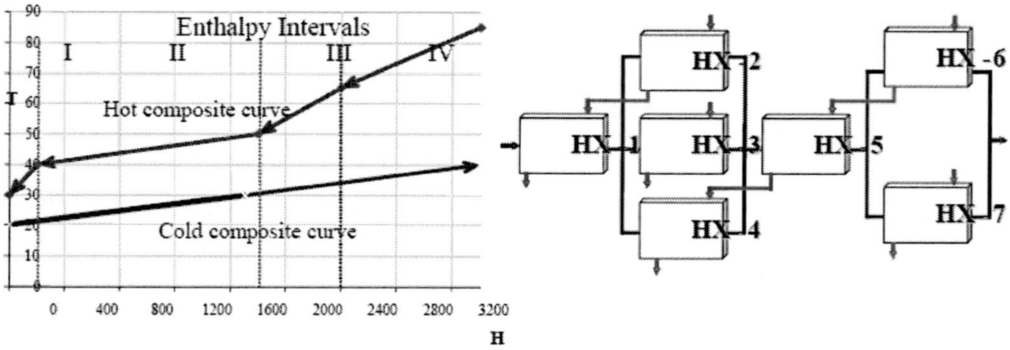

Figure 5. Cooling network designed with vertical heat transfer.

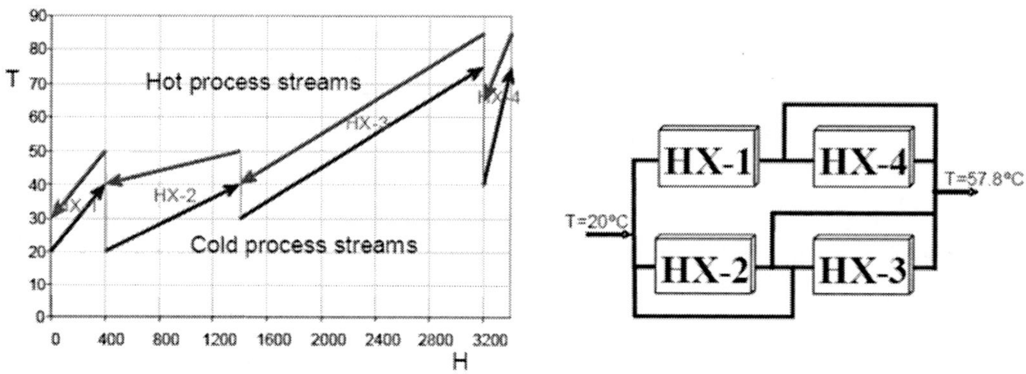

Figure 6. Cooling network designed for minimum water flow rate.

The cooling network for minimum water use has a total surfaces area greater than the other configurations. The reason for this can be explained using figure 6, where it is shown that all exchangers exhibit a minimum temperature approach of 10°C on the hot end; this leads to a more constrained design with a higher thermal effectiveness. The design of these exchangers results in the use of more than one shell, increasing the total cost. Another feature in this approach is that in order to achieve the minimum water flow rate target, the final water outlet temperature is relaxed to the point where its value reaches 57.8°C as shown in figure 6.

The design based on vertical heat transfer (figure 5) has the lowest overall heat transfer surface area but does not have the lower cost since the network structure requires more units compared to other designs. The series configuration is the structure that exhibits the minimum cost. The reason for this being that higher heat transfer coefficients on the water side are achieved due to the use of higher flow rates.

In practical designs, it is common to find cooling systems with a network structure of coolers set in parallel. This type of arrangement is convenient to build and operate since they easily accommodate to the layout features of the plant. However, there is a penalty associated to this type of arrangement to be paid for in terms of surface area. Whenever possible, subject to layout and pressure drop restrictions, the series arrangement must be preferred upon parallel arrangement (Picón, et al., 2007a).

Table 2. Detailed exchanger design and capital cost of network in parallel arrangement

HX	Hot Process Stream					Cooling Water					Heat Exchanger				
	T_{in}	T_{out}	ΔP	CP	M	T_{in}	T_{out}	ΔP	CP	M	Q	U	ΔT_{ML}	A	$
	[°C]	[°C]	[bar]	[kW/°C]	[kg/s]	[°C]	[°C]	[bar]	[kW/°C]	[kg/s]	[kW]	[W/m²°C]	[°C]	[m²]	[dls]
1	50	30	30.4	20	8.23	20	40	0.81	20	5.14	400	1,444	8.02	34.5	12,520
2	50	40	30.4	100	42.53	20	40	0.81	50	12.85	1,000	1,743	11.62	49.4	11,370
3	85	40	30.4	40	16.8	20	40	0.81	90	23.13	1,800	1,581	24.86	45.9	11,300
4	85	65	30.4	10	4.24	20	40	0.81	10	2.57	200	1,641	43.47	2.9	4,080
Total	---	---	---	170	---	20	40	0.81	170	43.69	3,400	6,409	---	132.7	39,270

Table 3. Detailed exchanger design and capital cost of network in series arrangement

HX	Hot Process Stream					Cooling Water					Heat Exchanger				
	T_{in}	T_{out}	ΔP	CP	m	T_{in}	T_{out}	ΔP	CP	M	Q	U	ΔT_{ML}	Area	$
	[°C]	[°C]	[bar]	[kW/°C]	[kg/s]	[°C]	[°C]	[bar]	[kW/°C]	[kg/s]	[kW]	[W/m²°C]	[°C]	[m²]	[dls]
1	50	30	30.4	20	8.23	20	22.35	0.1	170	43.69	400	1,687	17.34	14	6,430
2	50	40	30.4	100	42.53	20	28.24	0.24	170	43.69	1,000	2,057	19.12	25.4	9,150
3	85	40	30.4	40	16.8	20	38.82	0.42	170	43.69	1,800	1,540	21.15	56	10,610
4	85	65	30.4	10	4.24	20	40	0.05	170	43.69	200	1,144	34.73	5	6,270
Total	---	---	---	170	---	20	40	0.81	170	43.69	3,400	6,429	---	100.4	32,460

Table 4. Detailed exchanger design and capital cost of network with vertical heat transfer

HX	Hot Process Stream					Cooling Water					Heat Exchanger				
	T_{in}	T_{out}	ΔP	CP	m	T_{in}	T_{out}	ΔP	CP	M	Q	U	ΔT_{ML}	Area	$
	[°C]	[°C]	[bar]	[kW/°C]	[kg/s]	[°C]	[°C]	[bar]	[kW/°C]	[kg/s]	[kW]	[W/m^2°C]	[°C]	[m^2]	[dls]
1	40	30	15.2	20	8.24	20	21.18	0.05	170	43.69	200	1,631	13.95	8.8	6,420
2	50	40	15.2	20	8.24	21.18	30.59	0.38	21.3	5.49	200	1,587	18.36	6.9	4,860
3	50	40	30.4	100	42.53	21.18	30.59	0.38	106.3	27.46	1,000	2,153	18.27	25.4	9,250
4	50	40	7.1	40	16.81	21.18	30.59	0.38	42.5	10.74	400	1,595	18.27	14.5	7,380
5	65	50	10.1	40	16.81	30.59	34.12	0.14	170	43.69	600	1,334	24.34	18.6	8,050
6	85	65	13.2	40	16.81	34.12	40	0.24	136	32.85	800	1,753	37.03	14.6	8,140
7	85	65	30.4	10	4.24	34.12	40	0.24	34	10.84	200	1,506	36.97	3.6	4,330
Total	---	---	---	170	---	20	40	0.81	170	43.69	3,400	11,559	---	92.4	48,430

Table 5. Detailed exchanger design and capital cost of network for minimum water flow rate

HX	Hot Process Stream					Cooling Water					Heat Exchanger				
	T_{in}	T_{out}	ΔP	CP	m	T_{in}	T_{out}	ΔP	CP	M	Q	U	ΔT_{ML}	Area	$
	[°C]	[°C]	[bar]	[kW/°C]	[kg/s]	[°C]	[°C]	[bar]	[kW/°C]	[kg/s]	[kW]	[W/m^2°C]	[°C]	[m^2]	[dls]
1	50	30	30.4	20	8.23	20	40	0.33	20	4.77	400	1,177	7.39	47.3	15,020
2	50	40	30.4	100	42.53	20	40	0.33	50	11.93	1,000	1,633	11.84	55.3	11,220
3	85	40	30.4	40	16.8	30	75	0.48	40	9.55	1,800	1,097	7.32	228.1	50,840
4	85	65	30.4	10	4.24	40	75	0.48	5.7	1.37	200	1,303	14.37	11.2	9,660
Total	---	---	---	170	---	20	57.8	0.81	90	21.49	3,400	5,210	---	341.9	86,740

3. Retrofit for Minimum Power Consumption

In an existing cooling system, power is consumed to maintain the circulation of the cooling water in the circuit. Due to a number of reasons, the total amount of energy spent in pumping may be larger than it should be. For instance, the overdesign of the cooling tower and the heat exchangers generally is accompanied by a large water flow rate, a large pumping system and its corresponding power consumption. This section describes a practical approach for the reduction of energy consumption; the approach takes advantage of the installed network area and identifies situations where exchanger overdesign can be used to reduce water flow rate for the given process heat duty.

Contrary to the case of overdesign, is the case where the process cooling duty exceeds the actual capacity of the cooling system. In such situations a debottlenecking technique must be applied is the process is to continue to operate. For instance, Castro and Pinto (2000) conclude that in such cases, the most economical and practical way of de-bottlenecking a cooling system is the removal of hot water and its replacement with cold water. Other researches (Kim and Smith, 2003) have investigated the effect of techniques such as heat load redistribution amidst cooling towers, use of air coolers and the increase of air flow rate in the tower. However, the problem related to the excessive power consumption requires a different approach.

A common practice in the design of cooling systems in process plants is that its various components are sized by different companies. The likely result of this practice is the overdesign of the cooling tower, the heat exchangers and the flow system that ends up circulating large amounts of water thus giving rise to high operating costs due to pumping. In such cases, there is an opportunity to minimise water flow rate by taking advantage of the excess installed area. In doing so, the costs associated with pumping can be reduced. If the reduced flow rate is still within the tower practical operating region, further cooling will be achieved, thus bringing further benefits to the plant.

Power expenditure can be reduced through the implementation of a targeting methodology for determining the minimum water flow rate that for the installed heat transfer surface area can still provide the process cooling duty. Figure 2 can be used to describe the approach. Let us assume that the system has the largest possible water flow rate that together with the oversized exchanger provide more cooling than needed. Under these conditions, water flow rate can be reduced to the point where the process minimum cooling duty is reached. In doing so, the reduction in the heat transfer coefficient due to reduced flow rate is compensated for by excess heat transfer surface area. Thus excess water flow rate and heat transfer area are the key variables for implementing the minimum water flow rate targeting approach.

The first element of the targeting approach consists in the development of a simulation model that incorporates single phase and condensation heat transfer. Such a model requires the use of the most suitable expression for calculating the heat transfer coefficients that will allow the prediction of the thermal performance under variations in the water flow rate given the exchanger geometry.

In most practical applications, only water flow rate can be reduced, however, there are cases where even surface area can be removed; this is graphically shown in figure 7. A case like this would be the situation of overdesigned units that consist of two exchangers placed in parallel.

The cooling network structure shown in figure 1 is examined in this section. A simulation program implemented on a spreadsheet was used to simulate the performance of the units under changed water flow rate conditions. The program requires information such as physical properties of the fluids, exchanger geometry and heat transfer correlations. The maximum water outlet temperature is fixed at 40°C. The results of the simulations allow for the calculation of the target for minimum water flow rate of the system. Table 6 shows the results of the simulation of the most important units. The simulation was carried out for a scenario of maximum throughput. Under these conditions, the minimum water flow rate required by the 8 exchangers shown in table 6 is 28.14 m^3/min. Compared to the current flow rate of 57.43 m^3/min gives a difference of 29.29 m^3/min. A total of five pumps driven by 500 hp motors each deliver a flow rate of 144 m^3/min. Thus a single pump moves 28.8 m^3/min. This indicates that one pump can be taken out of operation. Considering an operating cost of \$0.055 USD/kWh, then a saving of \$167,200 USD a year is achieved.

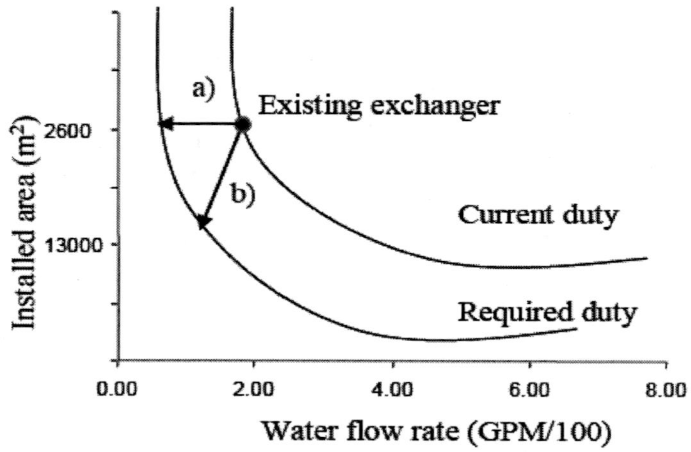

Figure 7. Minimisation of water flow rate in an overdesigned exchanger. a) Reduction of water flow rate with surface area unchanged, b) Reduction of water flow rate and surface area.

Table 6. Installed area and current and minimum water flow rate for most important exchangers in cooling network of figure 1

Exchanger No.	Installed Area (m^2)	Current water flow rate (m^3/min)	Minimum water flow rate (m^3/min)
E26	279.3	3.26	1.021
E31	913.1	13.98	5.632
E16	1,452.2	11.65	11.26
E2	275.3	20.81	6.68
E4	9.95	0.189	0.079
E3	27.8	2.0	0.685
E20	24	1.025	0.287
E19	54	4.511	2.494
Total	3,035.65	57.425	28.138

4. Steady State Response of Cooling Systems

This section looks at the propagation of temperature disturbances through cooling systems. Temperature disturbances are introduced into such systems in various ways, namely: changes in the wet bulb temperature, changes in the heat load and by the deterioration of the coolers performance due to fouling. The response of a cooling systems to temperature disturbances is shown to be a function of the overall system thermal effectiveness which in turn is a function of the cooling tower thermal effectiveness and of the network overall thermal effectiveness. The way the latter parameter is determined depends on the network structure.

4.1. Temperature Disturbances in Cooling Networks

As mentioned in the sections above, the design of cooling systems has traditionally been carried out on the basis of a parallel arrangement of the coolers with all of them taking water at the same temperature and discharging it at different temperature into a return header (figure 8). The temperature at which water leaves the coolers depends on the heat load and water flow rate. The water thus collected in the header is returned to the cooling tower where it is cooled down to be re-circulated back to the process. The ambient variables that determine the minimum water temperature attainable by a cooling tower are the wet bulb temperature and the air humidity of the location.

As with most process equipment, cooling towers and coolers are designed for fixed operating conditions, namely: fixed heat load, fixed water flow rate and fixed ambient conditions. During operation, it is common to see that all of these variables change over a range that may or may not have an effect upon the process. When the overall cooling system is overdesigned, as it is frequently the case, these variations are absorbed by the system itself and no effect upon the process is detected. However, overdesign is not always desirable since it leads to over expenditure of pumping power and capital.

The most common operating conditions subject to variation that have an effect upon the performance of a cooling system are: the wet bulb temperature and the process heat load. The latter may result from a change in the process throughput or a change in the overall heat transfer coefficient of the coolers. For instance, when the process heat load increases, the cooling water leaves the coolers at a higher temperature; this affects the performance of the cooling tower which will deliver the water at a temperature level above the cold temperature which was originally specified. The disturbance then propagates through the cooling network in cycles until a steady state is reached.

When the process plant is located in a place where ambient conditions change over the year, disturbances are introduced into the system as a result of the change in the wet bulb temperature. Again, when the wet bulb temperature increases, the cooling tower delivers the water at a higher temperature; this disturbance propagates through the network resulting in higher water temperature at the outlet of the coolers and thus a higher temperature reaching the top of the cooling tower. This disturbance enters the tower and its effect is added up to the original wet bulb temperature disturbance. If the steady state response of a cooling system is known then it is possible to design the system for flexible operation. This section presents the model to predict such behaviour.

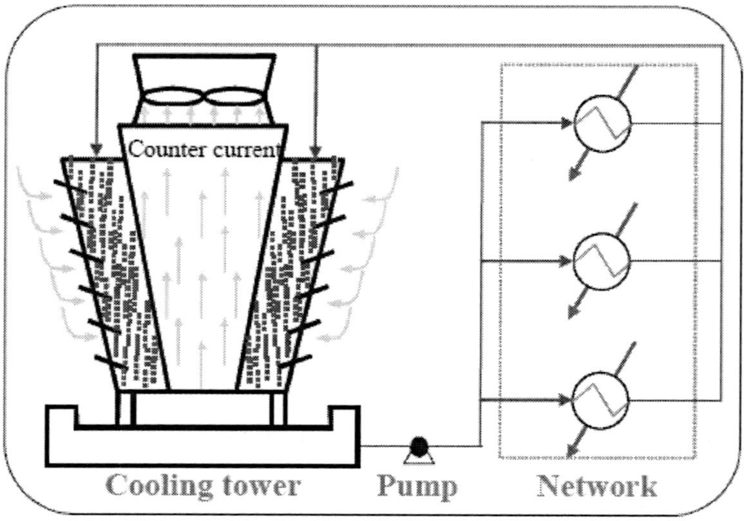

Figure 8. Typical cooling network with parallel arrangement.

The steady state response of heat exchangers to changes in operating conditions can easily be analysed and predicted using simple models derived from the thermal effectiveness (ε)-number of transfer units (NTU) relations. Picon and Polley (1995) introduced an approach to determine the response of heat recovery networks under variations in the operating conditions. The same approach is applied in this work to develop the concepts of cooling network overall thermal effectiveness and overall system effectiveness. The latter term is a function of the network overall effectiveness and the cooling tower thermal effectiveness. These models are developed in this section considering, for the case of the cooling network, only single phase heat exchangers.

4.2. Cooling Tower Thermal Effectiveness

Figure 9 shows, in an enthalpy vs. temperature diagram of a cooling process, the air saturation curve and the tower operating line. The lines indicated by f_R and f_{R+A} represent respectively, the slope of the saturated air enthalpy curve at the average temperature of the tower range (difference between tower inlet and outlet water temperature) and the slope of the saturated air enthalpy curve at the average temperature of the tower range plus the approach (difference between the water inlet temperature and the wet bulb temperature).

Using these parameters, El-Dessouky et al. (1997) defined a rigorous tower thermal effectiveness as:

$$\varepsilon_{\text{Tower,rigurous}} = \frac{f_R\left(T_{w1} - T_{w2}\right)}{f_{R+A}\left(T_{w1} - T_{wb}\right)} \tag{1}$$

where T_{w1} and T_{w2} are the tower inlet and outlet temperature respectively, and T_{wb} is the wet bulb temperature. The tower thermal effectiveness is a design parameter that remains approximately constant when temperature changes enter the system (Picón, et al., 2007b).

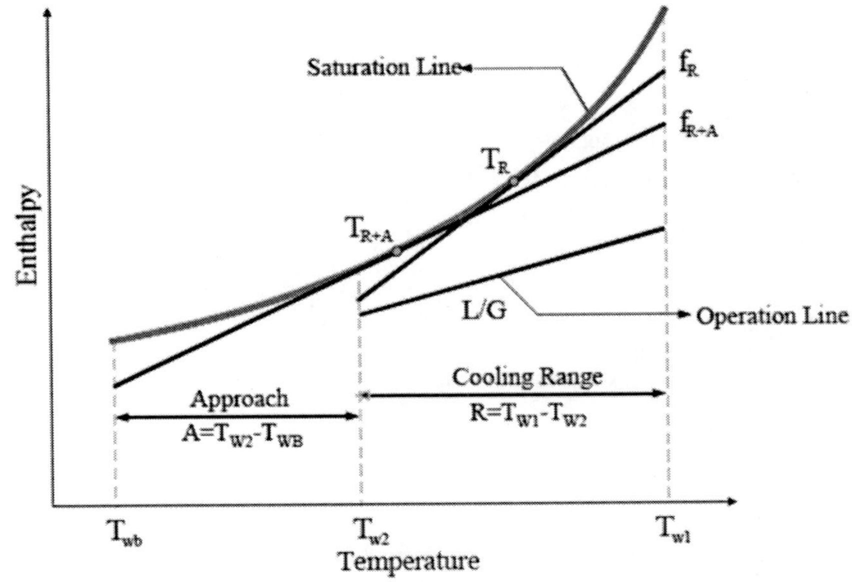

Figure 9. Graphical Representation of the f_R and f_{R+A} slopes.

For instance, when a wet bulb temperature change occurs the change in the f_{R+A} and f_R slopes counteract the change in the temperature difference ratio (Equation 1) maintaining the value of ε almost constant. In practical situations, unless the appropriate information is at hand, the computation of f_{R+A} and f_R can not be readily achieved. In such cases, a simplified tower thermal effectiveness is of great value. Such a parameter can be calculated form:

$$\varepsilon_{\text{Tower}} = \varepsilon_{\text{Tower,rigurous}} \frac{f_{R+A}}{f_R} = \frac{\left(T_{w1} - T_{w2}\right)}{\left(T_{w1} - T_{wb}\right)} \tag{2}$$

Equation (2) indicates that if the wet bulb temperature and the inlet and outlet water temperature are known, the tower effectiveness can be easily computed. This simplified tower effectiveness (or tower thermal effectiveness from now on) can be used as a quick means of evaluating the tower performance under changes in the process heat duty or changes in the wet bulb temperature.

From equation (2), the water outlet temperature T_{w2}, is calculated as:

$$T_{w2} = T_{w1}\left(1 - \varepsilon_{Tower}\right) + \varepsilon_{Tower}\, T_{wb} \tag{3}$$

when a wet bulb temperature disturbance ($\delta_{wb}=T^N{}_{wb}-T_{wb}$) enters the system, the new water outlet temperature ($T^N{}_{w2}$) can be calculated from equation (2) as:

$$T^N{}_{w2} = T_{w1}\left(1 - \varepsilon_{Tower}\right) + \varepsilon_{Tower}\, T^N{}_{wb} \tag{4}$$

Then the effect upon the water outlet temperature ($\hat{T}_{w2}$) is:

$$\hat{T}_{w2} = \delta_{wb} \ \varepsilon_{Tower} \tag{5}$$

where the term $\hat{T}_{w2}$, is the temperature change of the water at the outlet of the cooling tower. A similar analysis can be conducted to determine the effect upon the water outlet temperature when a change in the return temperature ($\delta_{w1}=T^N{}_{w1}-T_{w1}$) takes place. From equation (2) we have:

$$T^N{}_{w2} = T^N_{w1}\left(1-\varepsilon_{Tower}\right)+\varepsilon_{Tower}\ T_{wb} \tag{6}$$

Then the effect upon the water outlet temperature ($\hat{T}_{w2}$) is:

$$\hat{T}_{w2} = \delta_{w1}\ \left(1-\varepsilon_{Tower}\right) \tag{7}$$

For equation (2) to apply to the case above, the major assumption involved is that the tower thermal effectiveness remains approximately constant under changes in the temperature operating conditions. In order to verify the validity of this assumption, a rigorous cooling tower design method (El-Dessouky et al., 1997) is used; the method is further developed to operate in a rating mode. The validity of the approach is tested against the experimental data reported by Bernier (1994), where the experimentation was carried out with a wet bulb temperature between 15.8°C and 16°C. The results of the comparison between experimental data and predicted values for the outlet temperature are shown in table 7.

The close agreement between experimental data and the predicted values as shown in table 7, is an indication that the rating model can reproduce with enough accuracy the performance of a cooling tower under changed operating conditions. This result is an indication that the rating model can be used as a basis for assessing the accuracy of the predictions of the thermal effectiveness model.

Table 7. Comparison of the cooling tower rating model against experimental data (Bernier, 1994)

Cooling tower experimental data					
Experiment	**1**	**2**	**3**	**4**	
Water flowrate (kg/s)	0.2	0.3	0.398	0.495	
Air flowrate (°C)	0.67	0.656	0.664	0.658	
Tower inlet temperature (°C)	36.7	32	29.	27.9	
Tower outlet temperature (°C) (measured)	19.8	20.4	20.7	20.8	
Model rating results					
Tower outlet temperature (°C) (predicted)	19.20	20.39	20.85	21.20	

Given the operating conditions and design specification of a cooling tower, the rigorous rating model is used to compute the water outlet temperature for changes in the tower inlet temperature and for changes in the wet bulb temperature. In tables 8 and 9, the results of such an exercise are compared with those that are obtained using the thermal effectiveness model. The cooling tower is designed for a 35°C inlet temperature, a 25°C outlet temperature, 18°C wet bulb temperature, a water flow rate of 200 kg/s and L/G=1. The thermal effectiveness of the tower is 0.5882.

Table 8. Tower outlet temperatures with changes in inlet temperature

T_{W1} (°C)	29	31	33	35	37	39
T_{w2} , rigorous model (°C)	23.1	23.8	24.4	25	25.5	25.9
T_{w2} , effectiveness model, (°C)	22.8	23.6	24.3	25	25.6	26.1

Table 9. Tower outlet temperatures with changes in wet bulb temperature

T_{Wb} (°C)	15	16.5	18	19.5	21
T_{w2} , rigorous model (°C)	23.9	24.4	25	25.6	26.3
T_{w2} , effectiveness model, (°C)	23.2	24.1	25	25.9	26.7

The results of tables 8 and 9 show a very close approach between the rigorous and the effectiveness model. In table 8, changes in the tower inlet temperature vary in a range from 6°C below the specified point to 4°C above it. On the other hand, the changes in the wet bulb temperature vary from 3°C below the specified point to 3°C above it (table 9). The maximum error detected is of 3%. Thus the effectiveness model gives adequate results for them to be used in the prediction of the performance of the overall system.

4.3. Cooling Network Thermal Effectiveness

Cooling networks are, in practice, mostly designed in a parallel configuration. However, a series arrangement and a combination of both are, in some cases, desirable. Figure 10 shows various examples of these types of arrangements.

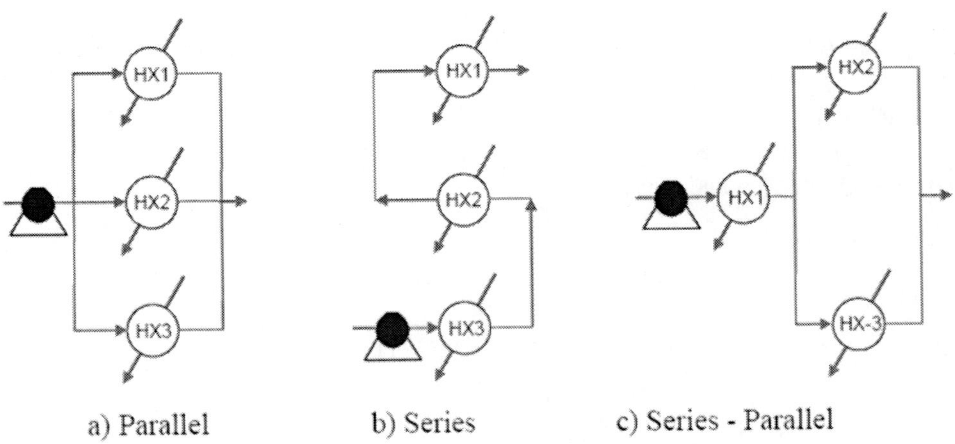

Figure 10. Examples of various cooling network arrangements.

For a single phase heat exchanger, where the cold fluid has a larger heat capacity-mass flow rate, it can be readily shown that the effect of a temperature disturbance (δ) upon the outlet temperature of the stream is calculated from:

$$\hat{T}_{out} = \varepsilon\left(1 - C\,\delta\right) \tag{8}$$

where $\hat{T}$ is the temperature change experienced by the cold fluid and C is the ratio of the hot fluid heat capacity-mass flow rate to that of the cold fluid (Picón and Polley, 1995). When a temperature disturbance enters a cooling network with all exchangers arranged in parallel, the temperature of the water after mixing in the return header can be calculated from a heat balance as:

$$\hat{T}_{w1}^{(1)} = \left[\frac{\sum_{1}^{n} m_n\left(1 - C_n\varepsilon_n\right)}{\sum_{1}^{n} m_n}\right]\hat{T}_{w2}^{(1)} \tag{9}$$

where m is the water mass flow rate. Equation (9) can be rewritten as:

$$\hat{T}_{w1}^{(1)} = \varepsilon_{\text{Network, parallel}}\,\hat{T}_{w2}^{(1)} \tag{10}$$

where $\varepsilon_{\text{Network,parallel}}$ represents the network overall thermal effectiveness for a parallel arrangement which is given by:

$$\varepsilon_{Network,parallel} = \frac{\sum_{1}^{n} m_n\left(1 - C_n\varepsilon_n\right)}{\sum_{1}^{n} m_n} \tag{11}$$

Assuming that these disturbances were originally caused by a wet bulb temperature change, the disturbance $\hat{T}_{w1}^{(1)}$ that enters the cooling tower adds to the original wet bulb disturbance. So, since the effects are additive, the new water outlet temperature can be calculated by combining equations (5) and (7) as:

$$\hat{T}_{w2}^{(2)} = \varepsilon_{Tower}\,\delta_{wb} + \hat{T}_{w1}^{(1)}\left(1 - \varepsilon_{Tower}\right) \tag{12}$$

Substitution of equation (5) and (10) into equation (12) gives:

$$\hat{T}_{w2}^{(2)} = \delta_{wb}\,\varepsilon_{Tower}\left(1 - \varepsilon_{System}\right) \tag{13}$$

where,

$$\varepsilon_{System} = \varepsilon_{Network,parallel}\left(1 - \varepsilon_{Tower}\right) \tag{14}$$

The propagation of the disturbance continues on until the steady state is reached. At this point we have that:

$$\hat{T}_{w2}^{(\infty)} = \delta_{wb}\, \varepsilon_{Tower} \left[1 + \varepsilon_{System} + \varepsilon_{System}^{2} + \varepsilon_{System}^{3} + \ldots\right] = \delta_{wb}\, \varepsilon_{Tower} \sum_{n=0}^{n} \varepsilon_{System}^{n} \qquad (15)$$

The summation of the ε_{System} terms can be approximated by the following expression:

$$\sum_{n=0}^{\infty} X^{n} = \left(\frac{1}{1-X}\right) \qquad \text{if } X \langle 1 \qquad (16)$$

Then equation (15) reduces to:

$$\hat{T}_{w2}^{(\infty)} = \delta_{wb}\, \varepsilon_{Tower} \left(\frac{1}{1-\varepsilon_{System}}\right) \qquad (17)$$

Equation (17) can be used to calculate the final change in the outlet temperature of the tower after a temperature disturbance has occurred.

For an arrangement in series, it can be readily shown that the network effectiveness is:

$$\varepsilon_{Network,series} = \prod_{n=1}^{\infty} \left(1 - C_n \varepsilon_n\right) \qquad (18)$$

For the combined arrangement of figure 10c, the expression for the network effectiveness is derived in various steps: exchangers HX2 and HX3 are in parallel (apply equation 11), and the resulting exchanger (HX2-3) is in series with exchanger HX1 (apply equation 18). The resultin expressions are:

$$\varepsilon_{2-3} = \frac{m_2 \left(1 - C_2 \varepsilon_2\right) + m_3 \left(1 - C_3 \varepsilon_3\right)}{m_2 + m_3} \qquad (19)$$

$$\varepsilon_{Network,combination} = \left(1\text{-}C_1 \varepsilon_1\right)\left(1 - C_{2-3}\, \varepsilon_{2-3}\right) \qquad (20)$$

where,

$$C_{2\ 3} = \frac{\sum_{n=2}^{n=3} \left(m_n C_p\right)_{process\ streams}}{\sum_{n=2}^{n=3} m_n} \qquad (21)$$

Similar expressions can be derived using the same approach for situations where the disturbance enters the system through a heat exchanger. In the following section the application of these models is demonstrated.

4.4. Steady State Response of Cooling Networks

To demonstrate the application of the network effectiveness approach, the three cooling networks shown in figure 10 are analyzed. Rigorous heat exchanger designs are carried out using commercial software; process information and physical properties are given in tables 10, 11 and 12. Tables 13, 14 and 15 show the results of the detailed exchanger design of each network.

Table 10. Process data for the design of HX1, HX2 and HX3 in parallel arrangement

HX1					
Fluid allocation	**Shell Side**		**Tube Side**		
Fluid name	**Process stream 1**		**Water**		
Mass flow rate	17.11		9.55		kg/s
Temperature (in/out)	50	30	20	41.98	°C
Density (in/out)	645.3	664.96	998.23	976.9	kg/m^3
Viscosity (in/out)	0.245	0.285	1.021	0.646	m-Pa-s
HX1					
Specific heat (in/out)	2.439	2.325	3.817	3.957	kJ/(kg-°C)
Thermal conductivity (in/out)	0.111	0.118	0.599	0.628	W/(m-°C)
Pressure drop	0.64		0.763		Bar
HX2					
Fluid allocation	**Shell Side**		**Tube Side**		
Fluid name	**Process stream 2**		**Water**		
Mass flow rate	14.66		11.94		kg/s
Temperature (in/out)	60	30.5	20	41.11	°C
Density (in/out)	675.78	700.74	998.23	977.75	kg/m^3
Viscosity (in/out)	0.35	0.476	1.021	0.657	m-Pa-s
Specific heat (in/out)	2.346	2.179	3.817	3.951	kJ/(kg-°C)
Thermal conductivity (in/out)	0.117	0.126	0.599	0.627	W/(m-°C)
Pressure drop	0.672		0.763		bar
HX3					
Fluid allocation	**Shell Side**		**Tube Side**		
Fluid name	**Process stream 3**		**Water**		
Mass flow rate	10.1		8.36		kg/s
Temperature (in/out)	70	40	20	40.94	°C
Density (in/out)	694.63	717.89	998.23	997.92	kg/m^3
Viscosity (in/out)	0.486	0.697	1.021	0.659	m-Pa-s
Specific heat (in/out)	2.326	2.158	3.817	3.95	kJ/(kg-°C)
Thermal conductivity (in/out)	0.121	0.128	0.599	0.627	W/(m-°C)
Pressure drop	0.59		0.763		bar

Table 11. Process data for the design of HX1, HX2 and HX3 in series arrangement

HX1					
Fluid allocation	**Shell Side**		**Tube Side**		
Fluid name	**Process stream 1**		**Water**		
Mass flow rate	17.11		29.84		kg/s
Temperature (in/out)	50	29.82	20	27.18	°C
Density (in/out)	645.3	665.12	998.23	991.32	kg/m^3
Viscosity (in/out)	0.242	0.291	1.016	0.868	m-Pa-s
Specific heat (in/out)	2.439	2.324	3.817	3.861	kJ/(kg-°C)
Thermal conductivity (in/out)	0.111	0.118	0.599	0.609	W/(m-°C)
Pressure drop	0.422		0.162		bar
HX2					
Fluid allocation	**Shell Side**		**Tube Side**		
Fluid name	**Process stream 2**		**Water**		
Mass flow rate	14.66		29.84		kg/s
Temperature (in/out)	60	30.5	27.18	35.74	°C
Density (in/out)	675.78	701.12	991.31	983.01	kg/m^3
Viscosity (in/out)	0.35	0.476	0.87	0.729	m-Pa-s
Specific heat (in/out)	2.327	2.223	4.191	4.189	kJ/(kg-°C)
Thermal conductivity (in/out)	0.117	0.126	0.609	0.621	W/(m-°C)
Pressure drop	0.856		0.269		bar
HX3					
Fluid allocation	**Shell Side**		**Tube Side**		
Fluid name	**Process stream 3**		**Water**		
Mass flow rate	10.1		29.84		kg/s
Temperature (in/out)	70	40.18	35.74	41.49	°C
Density (in/out)	694.63	717.74	983	977.37	kg/m^3
Viscosity (in/out)	0.486	0.695	0.729	0.652	m-Pa-s
HX1					
Specific heat (in/out)	2.326	2.158	3.916	3.954	kJ/(kg-°C)
Thermal conductivity (in/out)	0.121	0.128	0.621	0.628	W/(m-°C)
Pressure drop	0.843		0.156		bar

Table 12. Process data for the design of HX1, HX2 and HX3 in series-parallel arrangement

HX1					
Fluid allocation	**Shell Side**		**Tube Side**		
Fluid name	**Process stream 1**		**Water**		
Mass flow rate	17.11		29.84		kg/s
Temperature (in/out)	50	30	20	27.11	°C
Density (in/out)	645.3	664.93	998.23	991.38	kg/m^3
Viscosity (in/out)	0.245	0.285	1.021	0.872	m-Pa-s
Specific heat (in/out)	2.439	2.325	3.817	3.861	kJ/(kg-°C)
Thermal conductivity (in/out)	0.111	0.118	0.599	0.609	W/(m-°C)
Pressure drop	0.422		0.162		bar

Table 12. Continued

HX2					
Fluid allocation	**Shell Side**		**Tube Side**		
Fluid name	**Process stream 2**		**Water**		
Mass flow rate	14.66		17.55		kg/s
Temperature (in/out)	60	30.92	27.11	41.19	°C
Density (in/out)	675.78	700.39	991.38	977.67	kg/m^3
Viscosity (in/out)	0.35	0.474	0.872	0.656	m-Pa-s
Specific heat (in/out)	2.346	2.181	3.861	3.952	kJ/(kg-°C)
Thermal conductivity (in/out)	0.117	0.126	0.609	0.627	W/(m-°C)
Pressure drop	0.959		0.528		bar
HX3					
Fluid allocation	**Shell Side**		**Tube Side**		
Fluid name	**Process stream 3**		**Water**		
Mass flow rate	10.10		12.28		kg/s
Temperature (in/out)	70	41.48	27.11	40.59	°C
Density (in/out)	694.63	716.75	991.38	978.26	kg/m^3
Viscosity (in/out)	0.486	0.683	0.872	0.664	m-Pa-s
Specific heat (in/out)	2.326	2.166	3.861	3.948	kJ/(kg-°C)
Thermal conductivity (in/out)	0.121	0.128	0.609	0.627	W/(m-°C)
Pressure drop	0.616		0.274		bar

Table 13. Detailed design of the three heat exchangers in parallel arrangement

Unit	Area (m^2)	Q (kW)	$h_{shellside}$ (w/m^2°C)	$h_{tubeside}$ (W/m^2°C)	$\Delta P_{shellside}$ (Bar)	$\Delta P_{tubeside}$ (Bar)	Fouling (m^2°C/W)
HX1	89.3	815.65	2,318.3	6,080.3	0.636	0.752	0
HX2	47.7	978.44	2,631.7	7,387.8	0.672	0.667	0
HX3	21	679.2	6,205.4	7,965.4	0.59	0.412	0
	d_o (mm)	D_{shell} (mm)	No. Tubes	Shell passes	Tube passes	$MTD_{Corrected}$ (°C)	L (m)
HX1	20	508	248	2	8	6.2	5.77
HX2	20	457	182	2	6	12.0	4.21
HX3	16	219	72	1	2	18.9	5.84

Table 14. Detailed design of the three heat exchangers in series arrangement

Unit	Area (m^2)	Q (kW)	$h_{shellside}$ (w/m^2°C)	$h_{tubeside}$ (W/m^2°C)	$\Delta P_{shellside}$ (Bar)	$\Delta P_{tubeside}$ (Bar)	Fouling (m^2°C/W)
HX1	32.3	822.36	2,723.3	6,507.5	0.416	0.123	0
HX2	62.6	993.07	2,519.9	7,197.4	0.787	0.247	0
HX3	34.7	675.05	2,255.9	6,754.7	0.809	0.102	0
	d_o (mm)	D_{shell} (mm)	No. Tubes	Shell passes	Tube passes	$MTD_{Corrected}$ (°C)	L (m)
HX1	16	300	151	1	1	15.4	4.29
HX2	16	412	294	2	2	9.8	4.27
HX3	20	300	94	1	1	12.9	5.91

Table 15. Detailed design of the three heat exchangers in a combined series-parallel arrangement

Unit	Area (m^2)	Q (kW)	$h_{shellside}$ ($w/m^2{}^oC$)	$h_{tubeside}$ ($W/m^2{}^oC$)	$\Delta P_{shellside}$ (bar)	$\Delta P_{tubeside}$ (bar)	Fouling ($m^2{}^oC/W$)
HX1	31.5	814.4	2,661.6	6,907.1	0.342	0.147	0
HX2	95.9	964.9	2,384.5	5,345.0	0.956	0.454	0
HX3	26.4	646.6	2,042.4	6,845.4	0.128	0.627	0
	d_o (mm)	D_{shell} (mm)	No. Tubes	Shell passes	Tube passes	$MTD_{Corrected}$ (°C)	L (m)
HX1	16	356	140	1	1	15.6	4.52
HX2	20	508	280	2	4	6.9	5.49
HX3	20	290	72	1	2	17.4	5.87

Figures 11, 12 and 13 show the temperature profiles of the individual exchangers in the parallel, series and series-parallel arrangements. In table 16 simulation results when the network undergoes changes in the inlet temperature. Water enters the network at 20°C and changes go from 3°C below to 7°C above the original value. Commercial software (Aspen Plus linked to B-jac) is used to perform a rigorous simulation of the networks. The results include a comparison of the rigorous model ($T_{w,out,rigorous}$) against the effectiveness model ($T_{w,out,eff}$) for each heat exchanger and also for the network response ($T_{Network,out}$). Tables 17 and 18 show the same comparison for the series and series-parallel arrangements respectively.

Table 16. Cooling network temperature response to changes in the water inlet temperature. Thermal effectiveness of the parallel network: 0.4478

Parallel Configuration									
From Tower	HX1		HX2		HX3			δ	$T^N_{Network,out} = T^0_{Network} + \varepsilon\delta$
$T_{w,in}$	$T_{w,out}$	$T_{w,out,eff}$	$T_{w,out}$	$T_{w,out,eff}$	$T_{w,out}$	$T_{w,out,eff}$	$T_{Network,out,}$ rigurous	(°C)	$T_{Network,out,\ eff}$
17	41.2	41.2	39.7	40	39.2	39.9	40.34	-3	40.29
19	41.7	41.7	40.7	40.9	40.4	41	41.2	-1	41.18
20	42	42	41.1	41.4	40.9	41.59	41.63	0	41.63
21	42.3	42.3	41.6	41.8	41.5	42.2	42.06	1	42.08
23	42.8	42.8	42.6	42.8	42.7	43.3	42.92	3	42.97
25	43.3	43.3	43.5	43.7	43.9	44.4	43.78	5	43.87
27	43.9	43.9	44.5	44.6	45.1	45.6	44.65	7	44.77

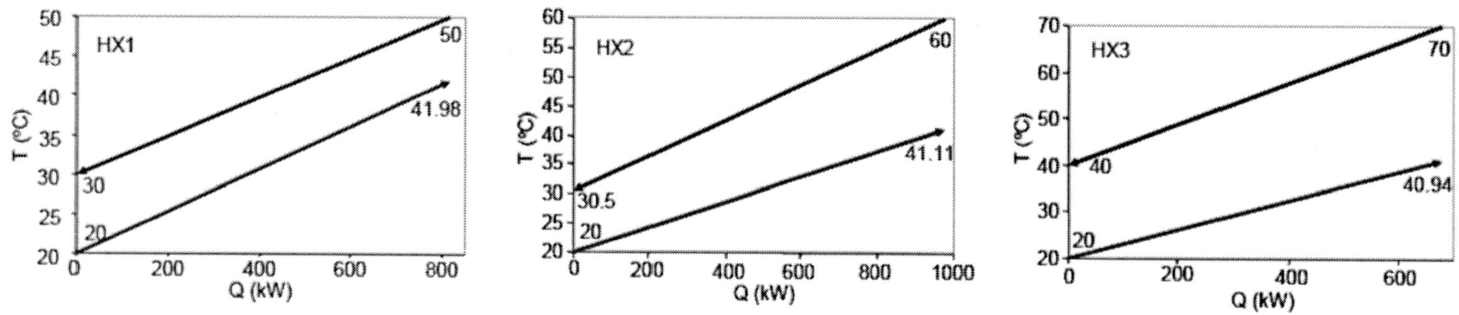

a) HX1 b) HX2 c) HX3

Figure 11. Temperature profiles of individual exchangers of cooling network with parallel arrangement.

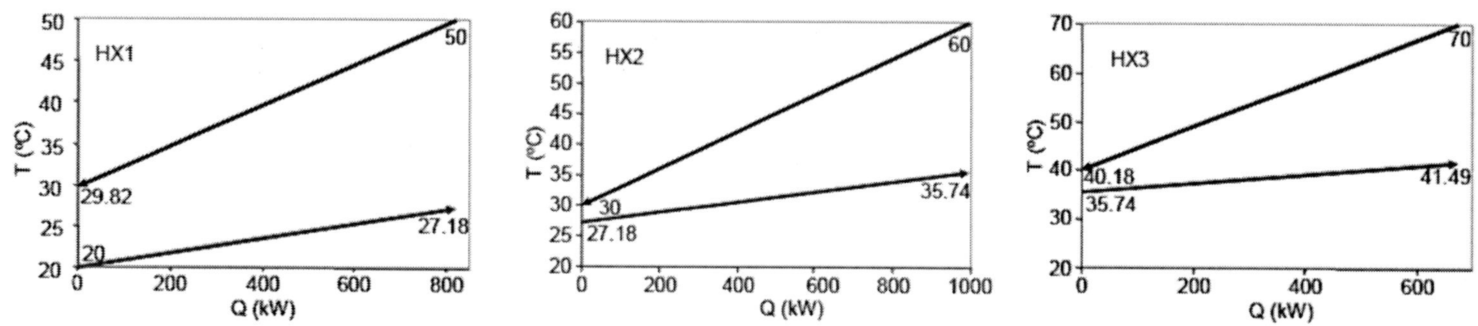

a) HX1 b) HX2 c) HX3

Figure 12. Temperature profiles of individual exchangers of cooling network with series arrangement.

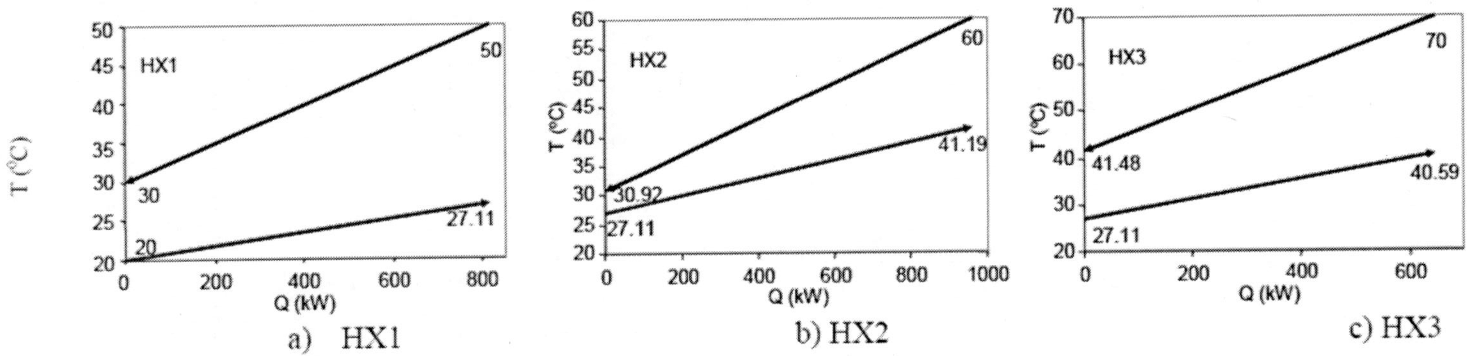

Figure 13. Temperature profiles of individual exchangers of cooling network with series-parallel arrangement.

Table 17. Cooling network temperature response to changes in the water inlet temperature. Thermal effectiveness of the series network: 0.4787

Series Configuration									
From Tower	HX1		HX2		HX3			δ	$T^N_{Network,out} = T^0_{Network} + \varepsilon\delta$
$T_{w,in}$	$T_{w,out}$	$T_{w,out,eff}$	$T_{w,out}$	$T_{w,out,eff}$	$T_{w,out}$	$T_{w,out,eff}$	$T_{Network,out,}$ rigurous	(°C)	$T_{Network,out,\ eff}$
17	24.2	24.9	33.5	32.9	39.6	39.6	39.6	-3	40.02
19	26.1	26.4	35	34.4	40.8	40.9	40.8	-1	40.98
20	27.1	27.2	35.7	35.1	41.5	41.46	41.5	0	41.46
21	28.1	27.9	36.4	35.9	42.1	42	42.1	1	41.94
23	30.1	29.5	37.9	37.4	43.3	43.3	43.3	3	42.89
25	32.1	31	39.4	38.9	44.5	44.5	44.5	5	43.85
27	34	32.5	40.8	40.3	45.7	45.7	45.7	7	44.81

Table 18. Cooling network temperature response to changes in the water inlet temperature. Thermal effectiveness of the series-parallel network: 0.4721

Series – Parallel configuration									
From Tower	HX1		HX2		HX3			δ	$T^N_{Network,out} = T^0_{Network} + \varepsilon\delta$
$T_{w,in}$	$T_{w,out}$	$T_{w,out,eff}$	$T_{w,out}$	$T_{w,out,eff}$	$T_{w,out}$	$T_{w,out,eff}$	$T_{Network,out,}$ rigurous	(°C)	$T_{Network,out,\ eff}$
17	24.8	24.8	39.9	39.9	39	39	39.53	-3	39.52
19	26.3	26.4	40.8	40.8	40.1	40.1	40.47	-1	40.47
20	27.1	27.1	41.2	41.2	40.6	40.59	40.94	0	40.94
21	27.9	27.9	41.6	41.6	41.1	41.1	41.42	1	41.41
23	29.4	29.4	42.5	42.5	42.2	42.2	42.36	3	42.35
25	30.9	30.9	43.4	43.4	43.2	43.2	43.30	5	43.30
27	32.5	32.5	44.3	44.2	44.3	44.3	44.25	7	44.24

In the case of exchanger HX1 in the parallel configuration the cooling water stream exhibits the lower heat capacity–mass flow rate (Figure 11, HX1). Thus, the water outlet temperature $T_{w,out,eff}$ is calculated from:

$$T_{water,out,eff} = \varepsilon\left(T_{process,in} \quad T_{water,in}\right) + T_{water,in} \tag{19}$$

For the rest of the exchangers in the same arrangement and other arrangements, water is the fluid with the larger value and $T_{water,\ out,eff}$ is calculated form:

$$T_{water,out,eff} = C\varepsilon\left(T_{process,in} \quad T_{water,in}\right) + T_{water,in} \tag{20}$$

In order to demonstrate the use of equation (17) to determine the steady state response of a cooling system to changes in wet bulb and return temperature, the following example is used: The cooling tower of a cooling system is designed for the following conditions: T_{w1}=40°C, T_{w2}=20°C, water flow rate=29.4 kg/s, L/G=0.8 and wet bulb temperature=17°C. The system is to be assessed assuming that the cooling network has the three arrangements shown in Figure 10. The system is simulated for a wet bulb temperature change of 2°C and a change of 2°C on the hot stream of exchanger HX3. The results are shown in tables 19 and 20 respectively.

Table 19. Steady state response of a cooling system to variations in wet bulb temperature

Wet bulb disturbance							
	T_{w1} (°C)	δ (°C)	ε_{tower} (°C)	$\varepsilon_{network}$ (°C)	ε_{system} (°C)	$\wedge^{(\infty)} T_{w2}$ (°C)	$T_{w2}^{(\infty)}$ (°C)
Parallel	40	2	0.8696	0.4478	0.0584	1.84	41.84
Series	40	2	0.8696	0.4787	0.0624	1.85	41.85
Series – Parallel	40	2	0.8696	0.4721	0.0616	1.85	41.85

Table 20. Steady state response of a cooling system to variations in water return temperature

Return water temperature disturbance							
	T_{w1} (°C)	δ (°C)	ε_{tower} (°C)	$\varepsilon_{network}$ (°C)	ε_{system} (°C)	$\wedge^{(\infty)} T_{w2}$ (°C)	$T_{w2}^{(\infty)}$ (°C)
Parallel	40	2	0.8696	0.4478	0.0584	0.27	40.27
Series	40	2	0.8696	0.4787	0.0624	0.27	40.27
Series – Parallel	40	3	0.8696	0.4721	0.0616	0.27	40.27

As can be seen from tables 19 and 20, the steady state response of a cooling system is independent of the network structure and changes in the web bulb temperature have a greater impact upon the performance of such these systems. The steady state response of cooling networks can be used to quickly assess its thermal performance under various operating conditions. The results have implications in design for flexible operation.

5. Conclusions

Three aspects related to cooling systems are covered in this work: design, retrofit for reduction of power consumption and thermal performance under variations in temperature conditions. The actual design of each of the various components of the system has an effect upon the power expenditure. For instance, overdesign is directly related to higher power consumption. The technique that is presented in this work, introduces a targeting methodology to determine the minimum water flow rate for the given installed heat transfer surface area and for the process heat duty. The approach is practical and can be implemented on any existing cooling facility.

In most practical applications, the network structure of a cooling network has to conform to the layout of the plant. This implies that in the majority of cases, the network of coolers and condensers are arranged in parallel. However, in this work it has been shown that the series arrangement reduces the surface area requirements.

From a thermal point of view, a cooling system can be considered as a complex assembly of heat exchangers. Such an assembly is made up of the following elements: a network of coolers and condensers (cooling network) on the one hand and a cooling tower on the other. A cooling network can have an arrangement of coolers in series, parallel or a combination of both. Whatever the arrangement, this work introduces a methodology to determine the network effectiveness. The cooling tower and the cooling network are set in series so that the overall system effectiveness is expressed as a function of these two terms.

The steady state response of a cooling system to changes in temperature operating conditions such as water return and wet bulb temperature is independent of the network structure. However, any change in wet bulb temperature has a bigger impact upon the cooling capacity of the system than a similar temperature change on the return temperature.

The thermal performance of counter current cooling towers can be determined with enough accuracy using the thermal effectiveness which can be computed only from the water inlet and outlet temperature and the wet bulb temperature. An important finding that allows for this simple calculation to be used in the prediction of the tower temperature response is that the effectiveness of a cooling tower remains approximately constant under temperature changes. The accuracy of the model has been tested against experimental data and rigorous theoretical models.

Nomenclature

A	Temperature approach (°C).
C	Hot to cold heat capacity-mass flow rate ratio.
C_p	Heat capacity (kJ/kg·K).
ΔP	Pressure drop (bar)
ΔT_{LM}	Logarithmic mean temperature difference (°C)
f	Slope.
G	Air mass flow rate (kg/s).
HX	Heat exchanger
h	Heat transfer coefficient, (kJ/m^2·K).
L	Water flow rate (kg/s).
m	Mass flow rate (kg/s).
NTU	Number of heat transfer units.
Q	Heat duty (kW)
R	Temperature range (°C)
T	Temperature (°C).
$\hat{T}$	Temperature change (°C)
U	Overall heat transfer coefficient (W/m^2°C)

Greek Symbols

δ	Temperature disturbance (°C).
ε	Thermal effectiveness.

Subscripts

1, in	Inlet condition.
2, out	Outlet condition.
eff	Effectiveness model
R	Cooling range.
R+A	Range plus approach.
w	Water.
Wb	Wet bulb.

Superscripts

(N)	New condition.
(O)	Original condition
(∞)	Steady state condition.

References

Bedekar, S., Nithiarasu, P., y Seetharamu, N., (1998). Experimental investigation of the performance of counter-flow, packed-bed mechanical cooling tower. *Energy,* Vol. 23, No. 11, pp. 943-947.

Bernier M. A., (1994). Cooling Tower Performance: *Theory and Experiments, ASHRAE Transactions: Research*, vol. 100, pp. 114-121.

Castro, M. M., Song T. W., and Pinto, J. M., (2000). Minimization of Operational Costs in Cooling Water Systems, *Trans IChemE,* vol. 78, Part A, pp. 192-201.

El-Dessouky, H. and Al-Juwayhel, F., (1997). Effectiveness of a Thermal Energy Storage System Using Phase-Change Materials, vol. 38, No. 6, pp. 601-617.

Helalizadeh, A., Muller-Steinhagen, H., Jamialahmadi, M., (2000), Mixed salt crystallisation fouling, *Chemical Engineering and Processing*, Vol. 39, pp. 29–43.

Kim, J. and Smith, R., (2001). Cooling Water System Design, *Chemical Engineering Science,* vol. 56, issue 12, pp. 3641-3658.

Kim, J. and Smith, R., (2003). Automated Retrofit Design of Cooling-water systems, *AIChe Journal,* vol. 49, no. 7, pp. 1712-1730.

Linnhoff, B. y Hindmarsh E., (1983), The pinch design method for heat exchanger networks, *Chem. Eng. Sci.,* 38 (5) Pp. 745-763.

Picón-Núñez, M, Morales-Fuentes, A. and Vázquez-Ramírez, E.E., (2007a). Effect of network arrangement on the heat transfer area of cooling networks, *Applied Thermal Engineering*. Vol. 27, No. 6, pp.2650-2656.

Picon-Nuñez, M., Nila-Gasca, C. and Gallegos-Muñoz, A., (2004). Energy Reduction through the Minimization of Water Flow rate in Water Cooling Networks. A case Study, *17th International Conference on Efficiency, Costs, Optimization, Simulation and Environmental Impact of Energy and Process Systems*, Guanajuato, Gto. México, July 7-9, pp. 1311-1319.

Picón-Núñez, M., Nila-Gasca, C. and Morales-Fuentes, A., (2007b). Simplified model for the determination of the steady state response of cooling systems, *Applied Thermal Engineering*. Vol. 27, pp. 1173-1181.

Picon-Nuñez, M. and Polley, G.T., (1995). Applying Basic Understanding of Heat Exchanger Network Behavior to the Problem of Plant Flexibility, Chemical Engineering Research and Design, *Trans IChemE,* vol. 73, pp. 941-952.

In: Thermal Engineering Research Developments
Editors: J. Evgova and O. Kostadinov, pp. 345-368
ISBN: 978-1-60741-497-1

Chapter 8

HEAT TRANSFER DURING CUTTING PROCESS: MODELLING AND PREDICTION TOWARDS REAL TIME CONTROL

M. Lazard[*]
GIP InSIC Institut Supérieur d'Ingénierie de la Conception
27 rue d'Hellieule, 88100 Saint Dié des Vosges, France

Abstract

Machining remains one of the most important manufacturing process and the need for a protection of the machine tool against wear and breakage has become a priority over the past few years. The damage of the tool insert is obviously strongly correlated with the maximum temperature obtained close to the tool's tip. The goal of this chapter is to propose an accurate methodology in two main steps in order to predict the heat transfer within the cutting tool.

The first part is dedicated to direct modelling of the cutting tool. After performing numerical simulations with the CFD code Fluent, an approach based on a quadrupole formulation is proposed in order to investigate thermal aspects of a cutting tool. The heat losses by convection and advection through the chip and the conduction in the tool holder are taken into account by heat transfer coefficients and a thermal contact resistance. The temperature evolution is obtained in the Laplace domain using Whittaker functions.

The second part concerns the inverse modelling in order to predict the temperature and the heat flux at the tool's tip. The model proposed which is not time consuming is implemented to inverse data. The difficulty in this problem comes from its ill-posed aspect: temperature and flux at the boundary (tool's tip) are determined through inner temperatures. If this model gives quite satisfactory results for temperature in the case of low noised data, this is not the case for heat flux. So the future time steps regularization technique is implemented to improve the estimations. The effects of particular in-time evolutions of the heat flux exhibiting step, ramp, slope and sinusoidal oscillations in the transient temperature determination are studied. The influence of the noise level, of the number of future time steps and also of the temperature profiles in the tool's tip vicinity (Cartesian, Cylindrical or Spherical) are also investigated.

Finally the conclusion presents some perspectives on a real-time tool's tip temperature and heat flux estimation adapted for turning process control.

[*] E-mail address: mlazard@insic.fr

I. Introduction

The need for a protection and for an adaptive control of the numerically controlled machine tool against wear has become a priority over the past few years [1].

The damage of the tool insert is obviously strongly correlated with the temperature [2, 3]. Indeed cutting requires mechanical energy that is quite completely converted into heat in the neighbourhood of the cutting zone. The maximum temperature is reached close to the tip of the insert. As a consequence, this parameter is of great importance since it governs the wear rate of the tool insert. That is the reason why several authors took interest in determining the cutting temperature. Measurements [4-16] and/or numerical simulations [17-22] are performed to determine the temperatures during cutting process (see figure 1).

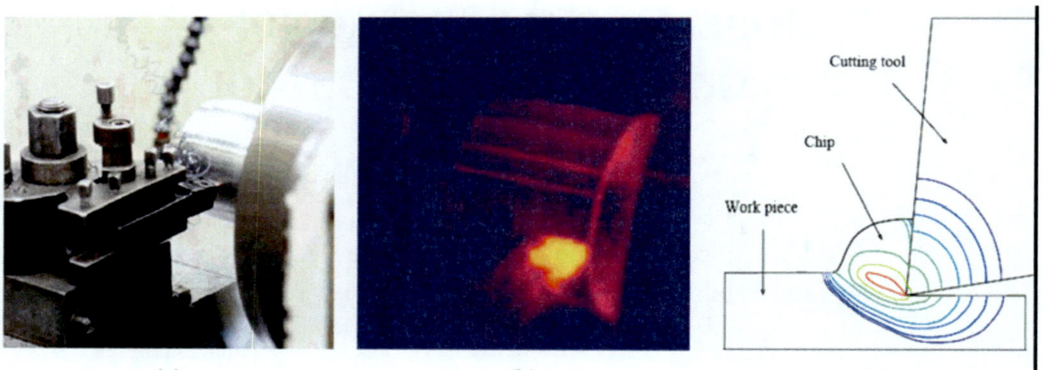

(a) (b) (c)

Figure 1. Illustrations of the cutting process : experiment and simulation.

Thermal measurements during cutting process should be performed essentially either with infrared cameras (see figure 1 (b)) or with thermocouples such as it is specified for instance in references [1, 2, 9, 10]. Main techniques used to evaluate temperatures during cutting process have been reviewed in those references and are also detailed in reference [11].

Some researchers measure steady-state tool-chip interface temperature with infrared camera such as [5, 8, 9]. But the use of infrared camera may raise some difficulties when a lubricant is used and it could be damaged for instance by chips. Moreover it is not "a low cost" set-up.

Others measure the temperature at the tool's tip with thermocouples such as [5, 14, 15] and use a non integer identified model to solve the inverse heat conduction problem [14]. Usually the thermocouple is located very close to the machining zone. This position permits to obtain directly the temperature of the insert's tip. However, it presents two major drawbacks: in the case of tool's wear, the thermocouple could be damaged. Moreover the thermal gradient is very high in this zone; as a consequence, the measurements are not very accurate.

Numerical simulations [17-22] could also be performed with numerical codes (see figure 1(c)). Some codes are general such as Ansys or Abaqus used for instance in reference [17] to model the cutting of Inconel 718. Others are especially dedicated to forming process such as for instance ThirdWave AdvantEdge used in reference [18] for differently coated tools or Deform used in references [19-20]. These codes are engineering softwares that enable designers to analyze metal forming on the computer and to provide solutions for optimizing traditional and state-of-the-art high speed machining process. Most of them are based on finite element method [17-21] using Lagrangian or Eulerian techniques, as well as a combination of both called Lagrangian-Eulerian formulation (denoted in the literature by ALE acronym) rather than finite difference method [22] or finite volume method [16].

The studies are focused not only on the tool but rather on the work piece or on the chip which is not our intention. Moreover some drawbacks due to these numerical codes have to be highlighted. In references [20-21], the authors raise some issues in the determination of the temperature during cutting process. They put the stress on the fact that due to the geometry and computational complexity, the formulation used in most FE-codes is not adequate because steady-state thermal conditions are not reached and often the simulation of the thermal phenomenon is ineffective. Moreover some important questions still remain such as how to define properly the boundaries conditions, on which surface the heat flux generated by the friction should be applied and so on. Furthermore these simulations are time consuming and as a consequence are not adequate to real time control.

As far as the attention is focused on the heat transfer during cutting process, one important question is what is the temperature field in the tool and what is the proportion of the heat flux going into the tool.

Some authors [23-26] give some ways to answer this question by introducing a heat partition coefficient which defines the percentage of the heat entering the moving chip and also it is possible to deduct the percentage of the dissipated energy going to the tool. Those approaches are essentially based on the theory of moving contact heat source and adimensional numbers such as Fourier or Nusselt numbers are involved. Then the temperature rise due to the frictional heat source is evaluated. It depends on a thermal number involving the chip sliding velocity, the natural contact length and also the thermal diffusivity of the work piece. The total temperature rise is calculated as the sum of the contribution due to the frictional source and of the plastic deformation [27]. This approach adopts a mechanical rather than a thermal point of view.

So, the aim of this chapter is not to simulate in a very precise way the whole cutting process involving the work piece, the chip and the tool because it is not an easy task even if numerical codes such as AdvantEdge or Deform are outstanding but to build a simple thermal model of the cutting tool in order to use it to inverse data. The approach used is based on a thermal analysis and on the implementation of a specific thermal model based on a quadrupole formulation. This model is dedicated to inversion in order to predict the temperature but also the heat flux in the tool and also especially at the tool's tip.

II. Direct Modelling of the Heat Transfer in the Cutting Tool

This part is dedicated to thermal simulations of a cutting tool. They are either numerical or semi-analytical. The use of a CFD code allows to model the heat transfer not only in the tool but also in the tool holder. Then the focus is put on the tool insert only. A specific thermal model involving Whittaker functions is built in order to inverse data for instance to identify the temperature or the heat flux at the tool's tip.

1. Numerical Simulations of the Heat Transfer in the Cutting Tool

In this section, numerical simulations have been performed with Fluent® in order to investigate the temperature and the heat flux within a cutting tool. The fluid and thermal point of view is considered, conduction and convection (by air or lubricant) are involved: as a consequence the finite volume method is then used to solve the governing equations whereas common studies took rather interest in the piece, treat the mechanical or thermo mechanical problem and use finite element method (use of commercial code for example Abaqus® [17], ThirdWave AdvantEdge [18] or Deform [19; 20]).

The cutting tool is presented in the figure 2. It is composed of two components: the tool insert (insert and tool's support) which is in contact with the work piece and the tool holder which is usually steel made. It corresponds to a real case chosen in the catalogue "Sandwik Coromant".

The tool is meshed with Gambit 2.0 and composed of 64 000 tetrahedral cells, whose volume is between $4.10^{-13}m^3$ and $10^{-10}m^3$. The grid is refined in the zones where the gradient temperature is high that is to say in the insert. The equiangleskew maximum (respectively the aspect ratio) is 0.78 (respectively 3.5). These values insure the quality of the simulations. Moreover, the independence of the results with the grid has been proved because simulations with a 120 000 cells grid have been made. The results obtained in terms of temperature distribution do not change more than 5% compared to those obtained with the first grid. So the 64 000 cells grid is well adapted for simulations.

The tool is essentially composed of Steel and Tungsten Carbide whose thermophysical properties usually used [28] are summarized in the table 1.

Table 1. Thermophysical Parameters used for simulations

	ρ (kgm^{-3})	c_p $(Jkg^{-1}K^{-1})$	k (Wm^{-1})
Steel	8030	502.5	26.3
Tungsten Carbide	19200	135	50
Cutting Oil	1000	4180	0.6

Figure 2. Continued on next page.

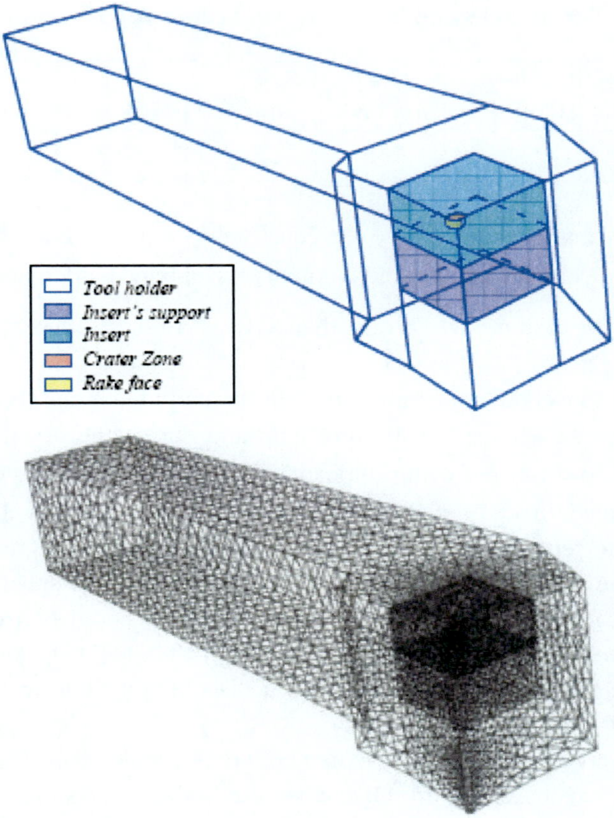

Figure 2. A cutting tool and its representations.

The attention is particularly focused on two processes commonly used in industrial applications. The first one is called "dry cutting": the tool and the piece are in contact without any lubricant and of course in this case the temperatures are more important. The second one is called "classical cutting" and a lubricant is used in order to cool the piece and the tool.

To simulate the heat transfer by conduction and convection, parameters and boundary conditions must be imposed (see table 2). The heat transfer coefficient h_{air} in the case of dry cutting (respectively h_{lub} in the case of classical cutting) is determined with the Colburn correlation from the speed of air or lubricant v, which corresponds to a cutting speed V_C. The backside of the tool holder is maintained at a constant temperature (condition *HW*). The crater zone (respectively the rake face) receives a heat flux of 70W (respectively 10 W) that corresponds to *HS1* (respectively to *HS2*). The value of these Heat Sources are close to those given by [28, 29]. The insert received a total heat flux of about 80W at the tip. The thermal contact between the different pieces (insert, support, tool holder) is not perfect. As a consequence, the heat transfer is locally modified. Thus, this contact thermal resistance (*TCR*) is among others things function of the surface state of the pieces and of the force exerted between the surfaces. Correlation obtained from experiences between the force and the thermal contact resistance is calculated. In the case studied here, the norm of the cutting forces is about $1500N$ (it is the same as in reference [30]). That leads to a thermal contact resistance (*TCR*).

Table 2. Parameters and boundary conditions used for simulations

v $(m \cdot s^{-1})$	V_C $(m \cdot min^{-1})$	h_{air} $(Wm^{-2}K^{-1})$	h_{lub} $(Wm^{-2}K^{-1})$	HW (K)	HS1 (MWm^{-2})	HS2 (MWm^{-2})	TCR (m^2KW^{-1})
2	120	40	2000	300	70	30	5 10^{-4}

The numerical scheme used to solve the equations is a power-law, well adapted to treat the heat transfer by conduction. The convergence is obtained after about $120\,s$ and the residuals on the temperature are lower than 10^{-7}. In each case, the global energy conservation is guaranteed.

The figure 3 (a) shows the temperature field in the tool holder and the insert. The highest temperature is of course obtained at the tool's tip and during the dry turning and is about $1600K$ (respectively 1400 K with lubricant). This value is close to the temperature measured or calculated in other papers [4, 31]. It also corroborates that the simulations performed are able to reproduce the real cutting conditions. The temperature decreases with the distance from the tool's tip. The isotherms in the insert have a spherical shape but as a first approximation, the heat transfer which is in reality 3D could be treated as a 2D heat transfer, and the isotherms as circular lines. Far from the tool's tip and near the contact between the insert and the tool holder, there is a discontinuity due to the thermal contact resistance.

The heat flux between the insert and its support is presented in the figure 3 (b). The exchange is maximal at the tool's tip. Of course, the higher values are obtained during dry cutting, because the heat transfer is dominated by conduction and the heat is evacuated from the tool by conduction. In the opposite, during cutting with lubricant, the heat is transferred by convection thanks to the fluid.

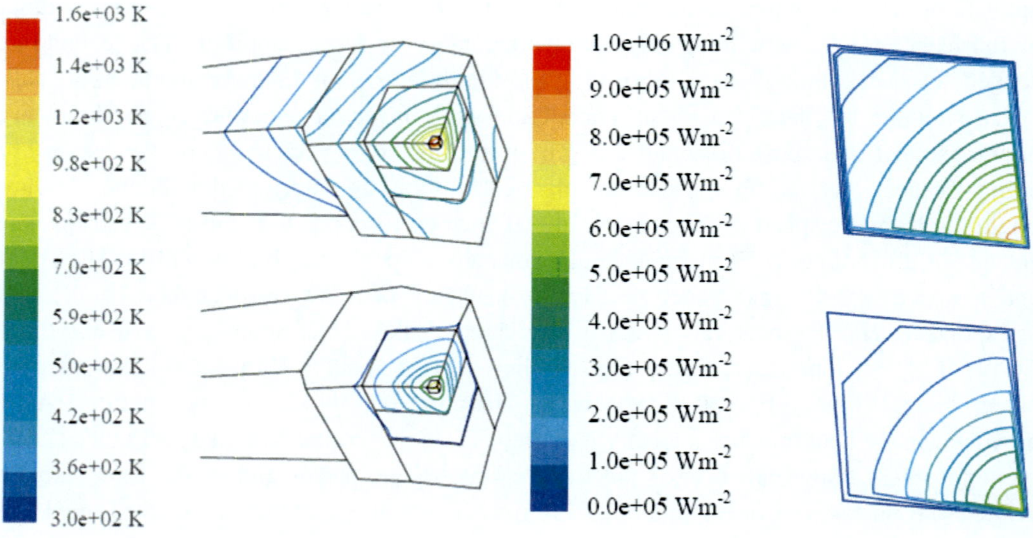

a b

Figure 3. Isolines of Temperature (a) and Heat Flux density (b) with air (top) and with lubricant (bottom).

The table 3 presents global heat fluxes for the insert and summarizes results in term of percents. It is quite surprising that the heat flux exchange in the case of cutting with oil is so important (close to 40%).

These simulations are in good agreement with experiments performed by others authors and are able to reproduce real cutting conditions. Therefore, it is highly likely that the quantities chosen for the simulations such as the heat fluxes, the thermal contact resistances and also the thermophysical properties of the materials are close to the real values of these parameters. The calculations performed with Fluent could then be used to test the specific thermal modelling based on quadrupole formulation presented below.

Table 3. Heat fluxes going through the insert

Cutting with air		
Convection	1 W	1.2%
Conduction		
• Insert/Support	45.5 W	56.9%
• Insert/Tool Holder	33.8 W	42.3%
Cutting with lubricant		
Convection		
• Insert upper face	27 W	33.7%
• Insert lateral face	22 W	27.5%
Conduction		
• Insert/Support	21.3 W	26.6%
• Insert/Tool Holder	10 W	12.5%

Moreover, those simulations put into light that quite only the tool insert is thermally affected (the tool holder is isothermal). As a consequence, the focus is now put only on the tool insert.

2. Specific Thermal Modelling

A complete model of the heat transfer in the cutting tool will be presented in this section. The aim is to predict the temperature and the heat flux within the tool with small computation times and then in the following section to use it as an inverse model to perform estimations.

A first simplified approach has already been proposed in [32]. But some assumptions were made such as very low heat losses. We proposed here to improve this model in order to build a more complete model which is closer to the physics reality of the tool during cutting process. The heat losses will be taken into account through heat transfer coefficients and also a thermal contact resistance.

Description of the Tool, Assumptions and Boundary Conditions

A schematic view of the insert is proposed on the figure 4. It is composed of a rounded tip (radius r_0). Cylindrical surfaces S_0, S_1, S_2 are corresponding to the radii r_0, r_1, r_2.

A heat flux Φ_0 goes through the surface S_0 for $r = r_0$ at the temperature T_0 and a heat flux Φ_1 goes through the surface S_1 for $r = r_1$ at the temperature T_1. The point N is located on the surface S_0 and corresponds to the insert's tip. The points $1, 2$ are located on isothermal surfaces where the measurements are made.

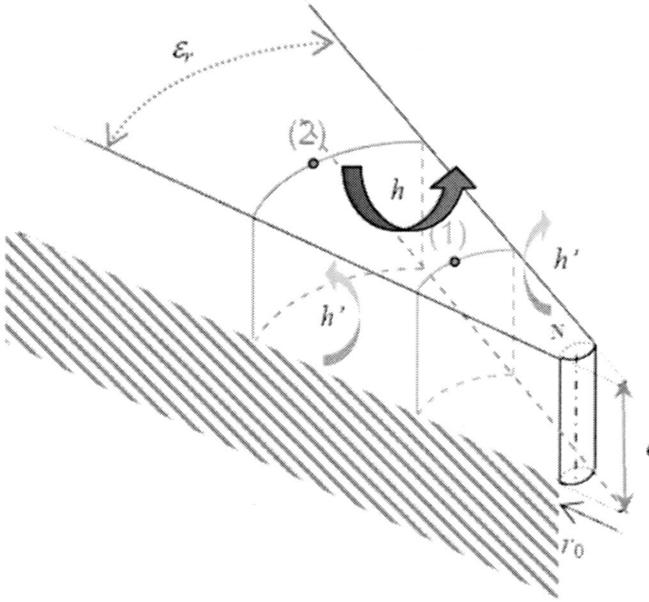

Figure 4. Geometry of the tool.

The heat losses due to convection (lateral faces and top face) and advection of the chip (top face) are taken into account by two heat transfer coefficients denoted h and h' respectively. The non-ideal contact between the tool's tip and the tool holder (bottom face) assumed as being a semi-infinite body with a temperature close to the room temperature T_{ext} is modelled by a thermal contact resistance R_c. The two thermocouples located in $r = r_1$ (1) and $r = r_2$ (2) will be used for the tool's tip temperature and heat flux estimation in N.

Governing Equations

The temperature field at the top of the tool's tip is spherical but becomes quickly cylindrical with space (it has been proved also with the numerical simulations performed with Fluent in the previous section). Thus, if the two thermocouples are in the vicinity but far enough from the tool's tip, the transient heat transfer in the tool can be described solving the following equation written in the Laplace domain:

$$\frac{\partial^2\overline{\theta}(r,p)}{\partial r^2}+\frac{1}{r}\frac{\partial\overline{\theta}(r,p)}{\partial r}+\frac{P}{\lambda}=\frac{p}{a}\overline{\theta}(r,p) \tag{1}$$

$$\text{with} \quad \overline{\theta}(r,p) = L(T - T_{ext}) \quad \text{and } p \text{ the Laplace variable} \tag{2}$$

P represents a heat source term defined by:

$$P.dV(r) = -[2h'dS'(r) + (h + 1/R_c)dS(r)]\overline{\theta}(r,p) \tag{3}$$

with

$$dV(r) = \varepsilon_r r.dr.e\,,\ dS'(r) = e.dr \text{ and } dS(r) = \varepsilon_r r.dr \tag{4}$$

Combining equations (3) and (4) leads to:

$$P = -\left(\frac{2h'}{\varepsilon_r r}\right)\overline{\theta}(r,p) - \left(\frac{(h + 1/R_c)}{e}\right)\overline{\theta}(r,p) \tag{5}$$

Using (5), the differential equation (1) becomes:

$$\frac{\partial^2 \overline{\theta}(r,p')}{\partial r^2} + \frac{1}{r}\frac{\partial \overline{\theta}(r,p')}{\partial r} - \frac{1}{r}\left(\frac{2h'}{\varepsilon_r \lambda}\right)\overline{\theta}(r,p') = \frac{p'}{a}\overline{\theta}(r,p) \tag{6}$$

$$\text{with} \quad p' = p + \frac{(h + 1/R_c)}{\rho C_p e} \tag{7}$$

The right-hand side term of equation (6) includes the heat losses on the upper face (by convection and advection by the chip) and on the lower face (by conduction in the tool holder).

The equation (6) is the same as:

$$x^2 \frac{d^2 f}{dx^2} + x\frac{df}{dx} - (x^2 + nx)f = 0 \tag{8}$$

with:

$$f(x) = \overline{\theta}(r,p'),\quad x^2 = (p'/a).r^2,\quad n = (2h'/\varepsilon_r\lambda).\sqrt{a/p'} \tag{9}$$

Solution of the Problem

The solution of the equations (8) involves either Bessel functions or Whittaker functions.

- If $n=0$ (i.e. $h'=0$), the solution is :

$$f(x) = C_1.I_0(x) + C_2.K_0(x) \tag{10}$$

with: I_0 and K_0 the Bessel functions.

In this case, if the following conditions are added: no convection on the top face and no advection by the chip, it is the same solution as in reference [32]

- If $n \neq 0$, it becomes:

$$f(x) = C_1 . \frac{W_M\left(-\frac{1}{2}n,0,2x\right)}{\sqrt{2x}} + C_2 . \frac{W_W\left(-\frac{1}{2}n,0,2x\right)}{\sqrt{2x}} \tag{11}$$

with W_M and W_W the Whittaker functions.

C_1 and C_2 are to be determined with the boundary conditions. But we will see in the section 2.5 that the determination of C_1 and C_2 could be skipped when the quadrupole formulation is used.

Let note:

$$W_M\left(-\tfrac{1}{2}n,0,2x\right) = M_{-\frac{1}{2}n,0}(2x) \quad \text{et} \quad W_W\left(-\tfrac{1}{2}n,0,2x\right) = W_{-\frac{1}{2}n,0}(2x) \tag{12}$$

The Bessel and Whittaker Functions

Some precisions must be given concerning the well known Bessel functions but the less known Whittaker functions. Below are presented some equations linking the Whittaker and the Bessel functions which could be obtained from relations given in the reference [33].

It allows in fact:

- to find again the result in the case of very low heat loss with the equations (13)-(14)
- to calculate the heat flux expression (which needs the derivation of the Whittaker functions) and to obtain the coefficient of the quadrupole in the section 2.5 with the equations (15)-(18)

- Relations between Whittaker and Bessel functions:

$$M_{0,\mu}(z) = 2^{2\mu}\Gamma(\mu+1)\sqrt{z}\,I_\mu\left(\frac{z}{2}\right) \qquad W_{0,\mu}(z) = \sqrt{\frac{z}{\pi}}K_\mu\left(\frac{z}{2}\right) \tag{13}$$

$$I_O(z) = \frac{M_{0,0}(2z)}{\sqrt{2z}} \qquad I_1(z) = \frac{M_{0,1}(2z)}{4\sqrt{2z}}$$

$$K_0(z) = \sqrt{\pi}.\frac{W_{0,0}(2z)}{\sqrt{2z}} \qquad K_1(z) = \sqrt{\pi}.\frac{W_{0,1}(2z)}{\sqrt{2z}} \tag{14}$$

- Derivation of the Whittaker functions:

$$\frac{d}{dz}\left(M_{\mu,\nu}(\alpha z)\right)=\alpha\left[\left(\frac{1}{2}-\frac{\mu}{\alpha z}\right)M_{\mu,\nu}(\alpha z)+\frac{(1/2+\nu+\mu)M_{\mu+1,\nu}(\alpha z)}{\alpha z}\right] \tag{15}$$

$$\frac{d}{dz}\left(W_{\mu,\nu}(\alpha z)\right)=\alpha\left[\left(\frac{1}{2}-\frac{\mu}{\alpha z}\right)W_{\mu,\nu}(\alpha z)-\frac{W_{\mu+1,\nu}(\alpha z)}{\alpha z}\right] \tag{16}$$

- Relations between the derivatives of Whittaker and Bessel functions:

$$I_1(z)=\frac{d}{dz}(I_O(z))=2.\frac{M'_{0,0}(2z)}{\sqrt{2z}}-\frac{M_{0,0}(2z)}{(2z)^{(3/2)}} \tag{17}$$

$$K_1(z)=-\frac{d}{dz}(K_0(z))=-2\sqrt{\pi}.\frac{W'_{0,0}(2z)}{\sqrt{2z}}+\sqrt{\pi}.\frac{W_{0,0}(2z)}{(2z)^{(3/2)}} \tag{18}$$

By taking the heat losses into account, the temperature does not involve Bessel functions I_0 and K_0 but the Whittaker functions M_0 and W_0. It is the same thing for the heat flux: the Whittaker functions take the place of I_1 and K_1. To apprehend the effect of the heat losses, it is interesting at first to plot and compare these functions.

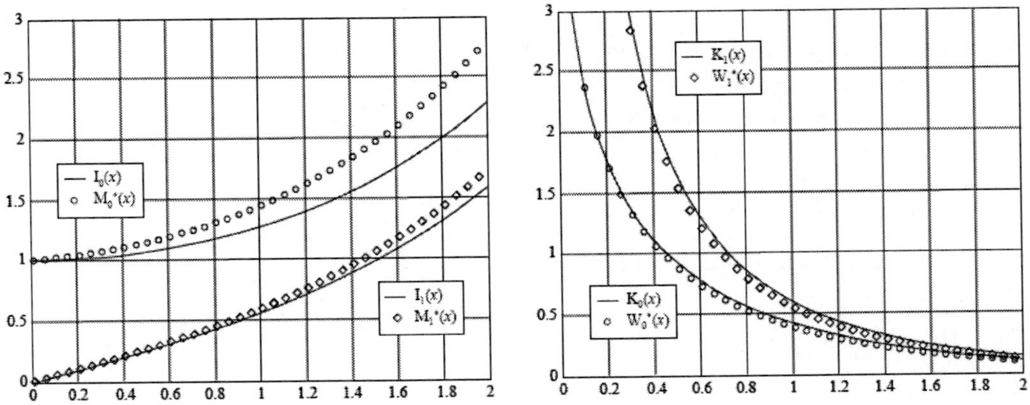

(a) (b)

Figure 5. Bessel and Whittaker functions.

The most important gap between Bessel and Whittaker functions is obtained between the functions I_0 and M^*_0. It occurs for x bigger than 1. For the other functions, the difference is very small. The curves representing Bessel I_0 and I_1 are always below the corresponding Whittaker functions whereas it is the opposite for K_0 and K_1. One can note that for x close to 0 (that is to say r close to 0), the values of the Bessel and Whittaker functions are the same. But if the measurement is made far from the tip, the results are very sensitive to the lateral heat losses and then the temperature becomes different from the temperature obtained with the ideal model without heat losses. Now, the attention will be focused on the quantities which are the core of our problem: the temperature and the heat flux which will be obtained through the quadrupole formulation.

Quadrupole Formulation of the Problem

- Quadrupole's history and basis :

From a mathematical point of view, the quadrupole method belongs to the class of analytical unified exact explicit method for solving linear partial differential equations in simple geometries. It relies on classical tools such as Laplace integral transforms (in time) or space integral transforms (Fourier for instance). Steady state or transient problems could be treated in one or multi dimensional heat transfer.

This approach was first presented by H.S. Carslaw in 1921 [34]. A precise name was not given to this method first used in engineering electrical applications but as it refers to an equation with four scales or poles then it was naturally called "quadrupole" method. The treatment of heat problems could follow in a similar way the modelling of electrical circuits once a boundary condition is known. The Laplace heat flux (respectively the Laplace temperature) is the analogue of the electrical current (respectively the electrical potential). Then the thermal solution can be expressed in terms of linear matrix relationships between transformed temperature flux vectors at the boundaries of the system. The intrinsic solutions obtained thanks to this formulation have a form which does not depend on these boundaries conditions.

Quadrupole formulation has been and is still widely used and developed to solve several different and complex thermal problems such as heat transfer in multilayer [35, 36] or in porous media [37], or coupled conductive radiative transient heat transfer in semi-transparent media with anisotropic scattering [38]. It is particularly well adapted to perform calculations quickly and to be used for instance to inverse data in order to remote to thermophysical parameters such as for instance the thermal diffusivity [39, 40].

- Quadrupole's formulation and representation :

So let consider the Laplace transform of the temperature called $\overline{\theta}_{in} = \overline{\theta}(r=0,p)$ (respectively $\overline{\theta}_{out} = \overline{\theta}(r=e,p)$) and the Laplace transform of the heat flux called $\overline{\varphi}_{in} = \overline{\varphi}(r=0,p)$ (respectively $\overline{\varphi}_{out} = \overline{\varphi}(r=e,p)$).

The quadrupole formulation of the heat transfer is given by the following equation:

$$\begin{bmatrix} \overline{\theta}_{in} \\ \overline{\varphi}_{in} \end{bmatrix} = \begin{bmatrix} A & B \\ C & D \end{bmatrix} \cdot \begin{bmatrix} \overline{\theta}_{out} \\ \overline{\varphi}_{out} \end{bmatrix} \tag{19}$$

The four coefficients of the matrix are either constant or functions of the Laplace variable. They involve the geometrical and thermophysical properties of the medium.

The quadrupole formulation presents the advantage to have a synthetic representation (figure 6).

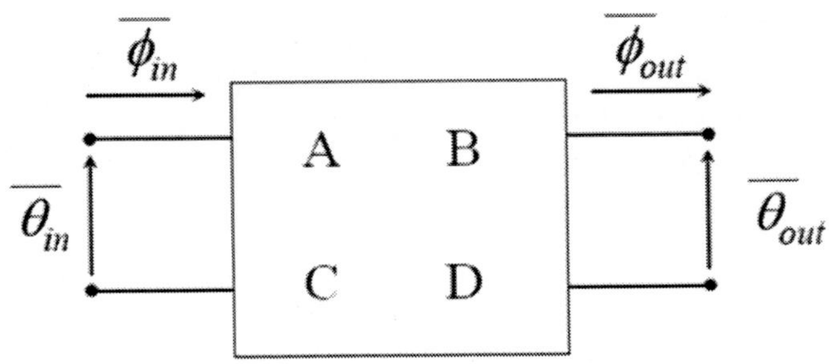

Figure 6. Quadrupole's representation.

- Quadrupole's formulation of the heat transfer in the cutting tool :

Using equation (11), the solution of the problem can be written in term of matrixes by introducing the inverse transfer matrix that linearly links in the Laplace domain the transformed temperature and flux at two different locations of the tool:

$$\begin{pmatrix} \overline{\theta}_0 \\ \overline{\varphi}_0 \end{pmatrix} = \begin{pmatrix} \alpha_0 & \beta_0 \\ \gamma_0 & \delta_0 \end{pmatrix} \begin{pmatrix} C_1 \\ C_2 \end{pmatrix} = \mathcal{M}_0 \begin{pmatrix} C_1 \\ C_2 \end{pmatrix} \text{ and } \begin{pmatrix} \overline{\theta}_1 \\ \overline{\varphi}_1 \end{pmatrix} = \begin{pmatrix} \alpha_1 & \beta_1 \\ \gamma_1 & \delta_1 \end{pmatrix} \begin{pmatrix} C_1 \\ C_2 \end{pmatrix} = \mathcal{M}_1 \begin{pmatrix} C_1 \\ C_2 \end{pmatrix} \tag{20}$$

with

$$\begin{aligned} &\alpha_i = M_{0,-\frac{1}{2}n}(2\,r_i)\Big/\sqrt{2\,r_i} \quad \gamma_i = -\lambda S_r(r_i)\Big[2M'_{0,-\frac{1}{2}n}(2\,r_i)\Big/\sqrt{2\,r_i} - M_{0,-\frac{1}{2}n}(2\,r_i)\Big/(2\,r_i)^{3/2}\Big] \\ &\beta_i = W_{0,-\frac{1}{2}n}(2\,r_i)\Big/\sqrt{2\,r_i} \quad \delta_i = -\lambda S_r(r_i)\Big[2W'_{0,-\frac{1}{2}n}(2\,r_i)\Big/\sqrt{2\,r_i} - W_{0,-\frac{1}{2}n}(2\,r_i)\Big/(2\,r_i)^{3/2}\Big] \end{aligned} \tag{21}$$

We can note that thanks to the two matrix equations (20), the determination of C_1 and C_2 involved in the equations (10) and (11) could be skipped as explained below.

Indeed the relations (20) links the Laplace transforms of the temperature and heat flux between the tool's tip N and the point 1 (C_1 and C_2 are eliminated):

$$\begin{pmatrix} \overline{\theta}_0 \\ \overline{\phi}_0 \end{pmatrix} = \begin{pmatrix} A_{01} & B_{01} \\ C_{01} & D_{01} \end{pmatrix} \begin{pmatrix} \overline{\theta}_1 \\ \overline{\phi}_1 \end{pmatrix} = \frac{1}{\alpha_1\delta_1 - \gamma_1\beta_1} \begin{pmatrix} \alpha_0\delta_1 - \gamma_1\beta_0 & -\alpha_0\beta_1 + \beta_0\alpha_1 \\ \delta_1\gamma_0 - \gamma_1\delta_0 & -\beta_1\gamma_0 + \alpha_1\delta_0 \end{pmatrix} \begin{pmatrix} \overline{\theta}_1 \\ \overline{\phi}_1 \end{pmatrix} = \mathcal{M}_0 \mathcal{M}_1^{-1} \begin{pmatrix} \overline{\theta}_1 \\ \overline{\phi}_1 \end{pmatrix} \tag{22}$$

Relation (22bis) links the temperatures and fluxes between points 1 and 2.

$$\begin{pmatrix} \overline{\theta}_1 \\ \overline{\phi}_1 \end{pmatrix} = \begin{pmatrix} A_{12} & B_{12} \\ C_{12} & D_{12} \end{pmatrix} \begin{pmatrix} \overline{\theta}_2 \\ \overline{\phi}_2 \end{pmatrix} = \frac{1}{\alpha_2\delta_2 - \gamma_2\beta_2} \begin{pmatrix} \alpha_1\delta_2 - \gamma_2\beta_1 & -\alpha_1\beta_2 + \beta_1\alpha_2 \\ \delta_2\gamma_1 - \gamma_2\delta_1 & -\beta_2\gamma_1 + \alpha_2\delta_1 \end{pmatrix} \begin{pmatrix} \overline{\theta}_2 \\ \overline{\phi}_2 \end{pmatrix} = \mathcal{M}_1 \mathcal{M}_2^{-1} \begin{pmatrix} \overline{\theta}_2 \\ \overline{\phi}_2 \end{pmatrix} \tag{22bis}$$

$$\begin{bmatrix}\theta_0\\ \varphi_0\end{bmatrix}=\begin{bmatrix}A_{01} & B_{01}\\ C_{01} & D_{01}\end{bmatrix}\cdot\begin{bmatrix}A_{12} & B_{12}\\ C_{12} & D_{12}\end{bmatrix}\cdot\begin{bmatrix}\theta_2\\ \varphi_2\end{bmatrix}$$

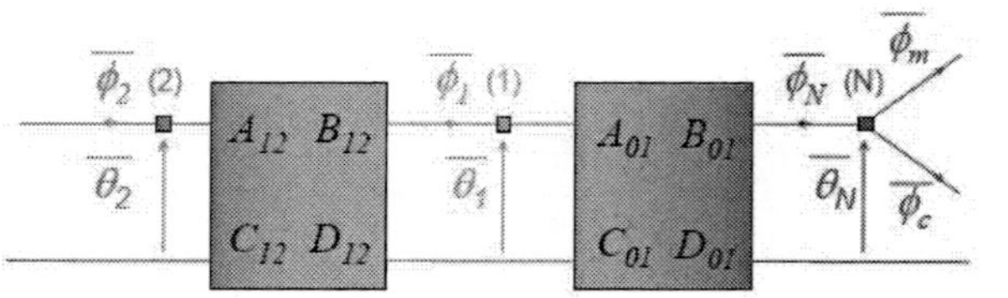

Figure 7. Quadrupole formulation and representation of the heat transfer in the cutting tool.

Combining (22) and (22bis) makes appear two linear relations between the temperature $\overline{\theta}_0$ and the heat flux $\overline{\phi}_0$ at the tool's tip as functions of the measured temperatures θ_1 and θ_2 in the cutting tool:

$$\begin{aligned}\overline{\theta}_0 &= \left(A_{01}B_{12}+B_{01}D_{12}\right)/B_{12}\cdot\overline{\theta}_1-\left(B_{01}/B_{12}\right)\overline{\theta}_2\\ \overline{\phi}_0 &= \left(C_{01}B_{12}+D_{01}D_{12}\right)/B_{12}\cdot\overline{\theta}_1-\left(D_{01}/B_{12}\right)\overline{\theta}_2\end{aligned}\tag{23}$$

The main interest of the double temperature measurement technique proposed is that temperature and heat flux at the tool's tip can be obtained without any particular assumptions on these quantities. In the following sections, $\overline{\theta}_2$ is assumed to be equal to zero that corresponds to an imposed temperature at r_2, close to the ambient temperature (it is true if r_2 is far from r_1). This assumption is not necessary since the method can also be implemented if $\theta_2 \neq 0$ but it is made here for simplifying calculations and showing some results in a more simple case.

III. Temperature and Heat Flux Estimation during Turning Process: Inverse Approach

Up to now, direct modelling has been proposed and as a consequence, the heat flux sources have been imposed as boundary conditions in the numerical simulations performed with Fluent. The values chosen were those proposed from experimental measurements made by authors. But unfortunately in real machining process, the heat flux generated at the boundary is unknown and strongly dependant on the cutting conditions and on the types of workpiece and cutting tool used.

To avoid these difficulties, inverse approach will be now built and applied. The specified thermal model based on quadrupole formulation will be used as an inverse model in order to estimate the temperature and the heat flux during cutting process.

Inverse methods have been widely applied and are the subject of many conferences and journals such as Inverse Problems in Science and Engineering. A state of art dealing with the inverse approach applied to machining is proposed in the reference [16]. Most of the works used very simplified models [41-44] most often for steady state conditions; the focus is put on the chip-tool interface temperature [41-42] or on the workpiece [43]. More appropriate models have been proposed by [45-48]. The aim of almost of these studies is to identify the heat flux flowing into the tool and the heat transfer coefficient between the environment during typical orthogonal cutting process. Indeed the effective determination of such quantities is necessary for instance for the numerical codes as input in order to predict the temperature field within the tool.

1. Temperature and Heat Flux Estimation: The Principle and the Procedure

In this section, the known quantities are the temperatures θ_1 and θ_2 (measured by thermocouples and far enough of the tool's tip in order not to damage the thermocouples). The four unknown are: $\theta_N, \phi_N, \phi_1, \phi_2$.

In order to evaluate the temperature and the heat flux at the tool's tip, the heat flux ϕ_1 is expressed as a function of θ_1 and θ_2 thanks to equation (22bis):

$$\phi_1 = C_{12}\theta_2 + D_{12}\phi_2 = C_{12}\theta_{12} + D_{12}\frac{\theta_1 - A_{12}\theta_2}{B_{12}} \tag{24}$$

It leads to the following expression of the temperature and the heat flux at the tool's tip as functions of the two measured temperatures θ_1 and θ_2:

$$\begin{pmatrix}\theta_N \\ \phi_N\end{pmatrix} = \frac{1}{B_{12}}\begin{pmatrix} A_{01}B_{12} + B_{01}D_{12} & -B_{01} \\ C_{01}B_{12} + D_{01}D_{12} & -D_{01}\end{pmatrix}\begin{pmatrix}\theta_1 \\ \theta_2\end{pmatrix} \tag{25}$$

This model gives the results in the Laplace domain. A numerical algorithm (Stehfest [49] or de Hoog [50]) permits to obtain the temperature and the heat flux as function of time.

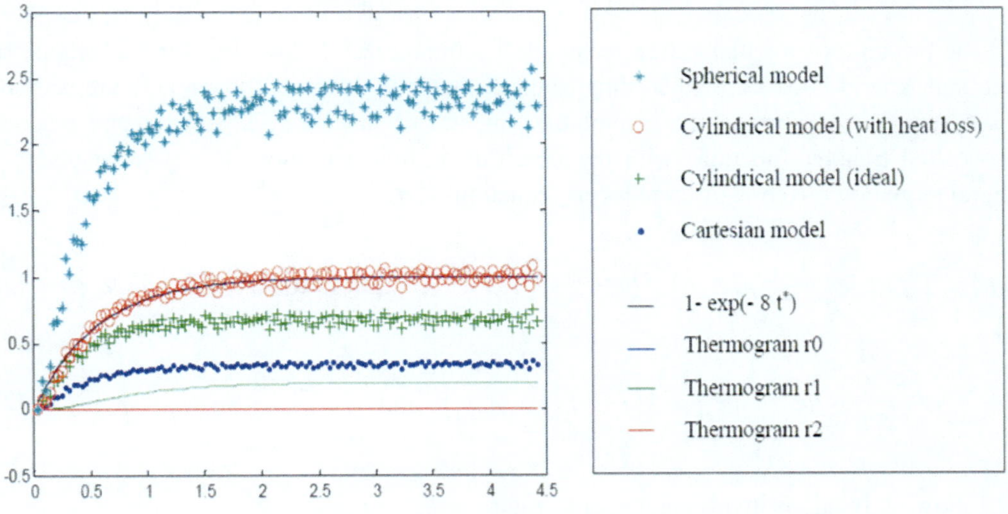

Figure 8. Inversion performed on noised data with different models.

In order to validate the methodology previously described, simulations have been performed to obtain thermograms at the three locations 0, 1 and 2. In order to test the analytical predictive models in the more realistic case, the thermal behaviour of the tool is noised : a random noise has been added on these thermograms. Indeed experimental measurements are always corrupted by noise. The causes are for instance perturbations at the thermocouples. In the case of thermal measurements during the turning process, perturbations are essentially due to the mechanical vibrations. Models based on quadrupole formulation (either cartesian, cylindrical, spherical or the specified thermal model -taking the heat losses into account- presented in the previous section) have been used to inverse data in order to obtain the temperature at the location 0. An example of such an inversion is presented in the figure 8.

The curves in full lines correspond to direct simulations and the curves with dashed lines correspond to the results obtained by inversion. The cylindrical model which takes the heat loss into account (that is to say the specified thermal model involving Whittaker functions) allows to find again the thermogram at the tool's tip.

If the specified thermal model allows a quite satisfactory estimation of the temperature even in the case of noised data, this is not the case for heat flux due to the ill-posed aspect of this problem as described by Beck et al [51]. As a consequence, the estimation of the heat flux at the tool's tip is difficult and requires the implementation of a regularization technique as described by Beck and Arnold [52] or Tikhonov and Arsénine [53].

The aim of the next section is to estimate the transient heat flux and temperature during cutting process by implementation of the future time steps method. The inversions are made on noised data for different in-time flux evolutions.

2. Improvement of the Estimations: Use of a Regularization Technique

Principle of the Future Time Steps Method

The principle of regularization technique is summarized below. For further details, see Beck and Arnold book [52]. According to the Duhamel's theorem, the temperature response of a system corresponds to the convolution integral of the impulsed temperature response (also called transfer function) with the heat flux in-time variation. It is written under an integral (equation (26) or a discrete forms (equations (26-27)):

$$T(t)=\int_0^t f(t)\varphi(t-\tau)d\tau=\int_0^t f(t-\tau)\varphi(\tau)d\tau \qquad \text{that is to say} \quad T=T_\delta \otimes \varphi \tag{26}$$

$$T_n=\sum_{i=1}^{n}\varphi_i \cdot f_{n-i+1}\cdot \Delta t \quad \text{with } f={}^t\{f_1 \quad f_2 \quad \dots \quad f_n\} \text{ and } \varphi={}^t\{\varphi_1 \quad \varphi_2 \quad \dots \quad \varphi_n\} \tag{27}$$

The following relations involving the temperatures are:

$$
\begin{aligned}
&T_1 = f_1 \cdot \varphi_1 \\
&T_2 = f_2 \cdot \varphi_1 + f_1 \cdot \varphi_2 \\
&\quad\vdots \\
&T_M = \underbrace{f_M \cdot \varphi_1 + f_{M-1} \cdot \varphi_2 + \ldots}_{\hat{T}_M} + \underbrace{f_1 \cdot \varphi_M}_{Z_1} \quad \rightarrow Y_M \\
&T_{M+1} = \underbrace{f_{M+1} \cdot \varphi_1 + f_M \cdot \varphi_2 + \ldots}_{\hat{T}_{M+1}} + \underbrace{(f_2 + f_1) \cdot \varphi_M}_{Z_2} \quad \rightarrow Y_{M+1} \\
&\quad\vdots \\
&T_{M+r-1} = \underbrace{f_{M+r-1} \cdot \varphi_1 + f_{M+r-2} \cdot \varphi_2 + \ldots + f_{r+1} \cdot \varphi_{M+r-22}}_{\hat{T}_{M+r-1}} + \underbrace{\sum_{k=0}^{r} f_k \cdot \varphi_M}_{Z_r} \quad \rightarrow Y_{M+r-1}
\end{aligned}
\tag{28}
$$

Then the aim is to estimate φ_M with the assumption that $Y_M, Y_{M+1}, \ldots, Y_{M+r-1}$ are known by minimizing $S_{\varphi_M} = \sum_{k=0}^{r-1} \left(Y_{M+k} - T_{M+k}\right)^2$.

T_{M+k} are changed into $T_{M+k} = \hat{T}_{M+k} + Z_{k+1} \cdot \varphi_M$. r represents the number of future time steps.

It comes:

$$
S_{\varphi_M} = \sum_{k=0}^{r-1} \left(Y_{M+k} - \hat{T}_{M+k} - Z_{k+1} \cdot \varphi_M\right)^2 \tag{29}
$$

$$
\begin{aligned}
&dS_{\varphi_M} / d\varphi_M = 0 \Rightarrow \hat{\varphi}_M = \sum_{k=0}^{r-1} Z_{k+1} \left(Y_{M+k} - \hat{T}_{M+k}\right) \Big/ \sum_{k=0}^{r-1} Z_{k+1}^2 \\
&\text{with } \hat{T}_{M+k} = \sum_{i=1}^{M+k-1} f_{M+r-i} \cdot \varphi_i, \quad Z_{k+1} = \sum_{i=1}^{k+1} f_i
\end{aligned}
\tag{30}
$$

Inversions

The inversions are made from the temperature evolution at r = r1. The estimations are performed either in the case of a Heavyside heat flux or in the case of a heat flux composed of different shapes such as pulse, ramp and sinusoidal variations. Results involving the heat flux (respectively the temperature) are presented in figure 9 (respectively figure 10). Several cases have been investigated: with or without noise, with or without regularization, with different levels of noise, with different numbers of time steps for the regularization and with cartesian, cylindrical and spherical models.

For temperature without noise (figure 10(a)), regularization is not needed because a classical inversion procedure allows an accurate estimation of the heat flux (figure 9(a)). But as soon as the data are corrupted by noise (figures 10(b) and 10(c)) and even if the level is low, a regularization technique must be used (the heat flux is not well estimated (figure 9(b)) whereas it is really better with regularization (figure 9(c))).

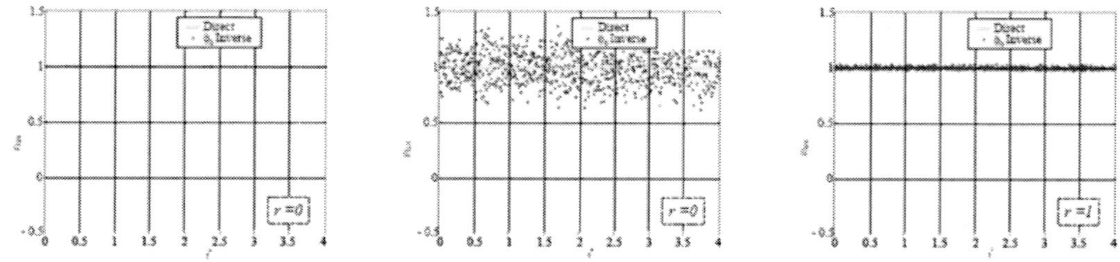

(a) No noise, no regularization

(b) Noise level :0.01% no regularization

(c) Noise level : 0.01% Regul. r=1

Influence of the noise and of the regularization

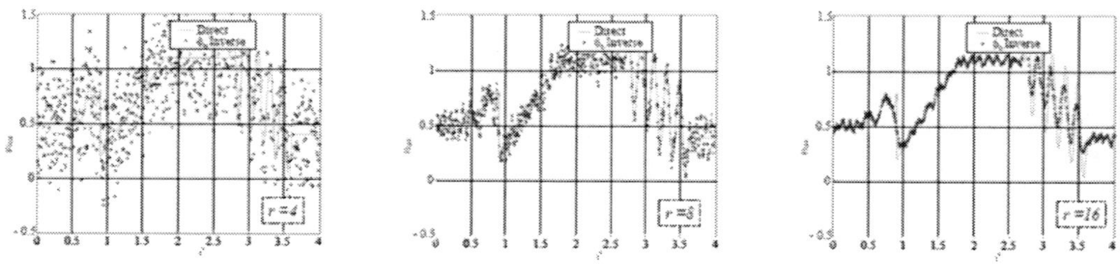

(d) Noise level : 2%, Regul. : r=4

(e) Noise level : 2%, Regul. : r=8

(f) Noise level : 2%, Regul. : r=16

Influence of the number of future time steps

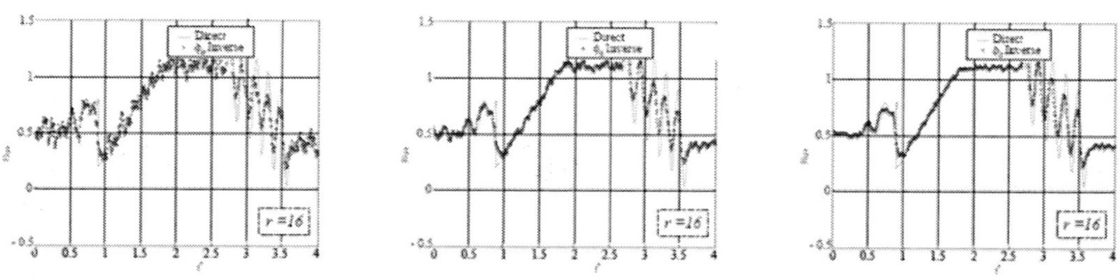

(g) Noise level : 2%, Regul. : r=16 Cartesian

(h) Noise level : 2%, Regul. : r=16 Cylindrical

(i) Noise level : 2%, Regul. : r=16 Spherical

Influence of the geometry

Figure 9. Heat flux estimation.

In this regularization method, an important parameter must be chosen: the number of the future time steps r. The quality of the estimation highly depends on this parameter (see figures 9(d), 9(e), 9(f), 10(d), 10(e) and 10(f)). For signal with a high level of noise, the number of time steps must be increased to obtain a good S/N ratio. Nevertheless, a particular attention must be paid to the choice of this regularization parameter, particularly for signal exhibiting high frequency sinusoidal oscillations (as for instance very quick variations of heat flux and/or temperature).

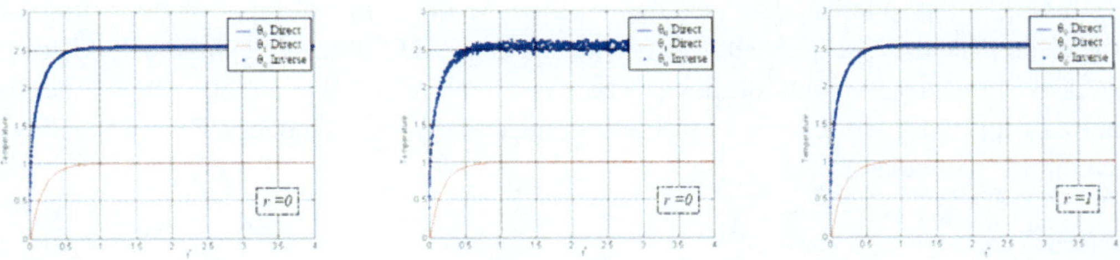

(a) No noise, no regularization

(b) Noise level :0.01% no regularization

(c) Noise level : 0.01% Regul. r=1

Influence of the noise and of the regularization

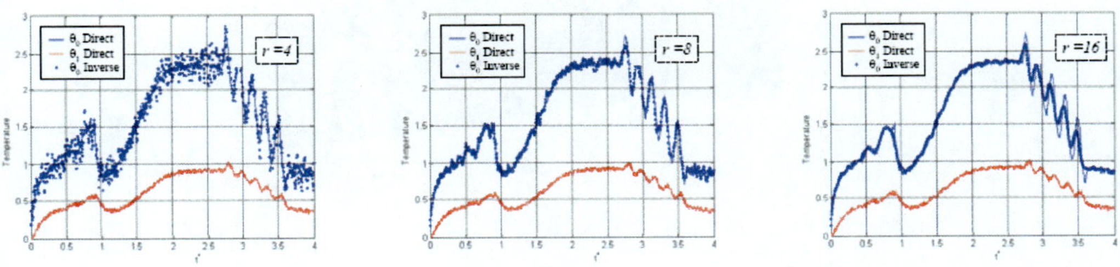

(d) Noise level : 2%, Regul. : r=4

(e) Noise level : 2%, Regul. : r=8

(f) Noise level : 2%, Regul. : r=16

Influence of the number of future time steps

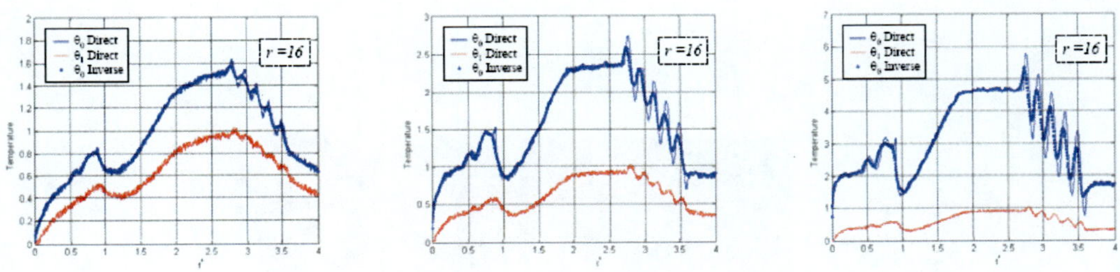

(g) Noise level : 2%, Regul. : r=16 Cartesian

(h) Noise level : 2%, Regul. : r=16 Cylindrical

(i) Noise level : 2%, Regul. : r=16 Spherical

Influence of the geometry

Figure 10. Thermograms.

Indeed, if r is too high then significant high frequencies components of this signal can be removed and not allow a precise estimation of its variations (figures 9(f) and 10(f)). The adapted choice thus results from a compromise solution between these two aspects.

Moreover the choice of the model (function of the geometry) is important for accurate temperature estimation. In the case presented here, the temperature at the tool's tip is 1.6 times higher than the temperature T1 measured at r=r1 (for cartesian model, see figure 10(g)) for the same input flux. It is 2.5 times higher in the cylindrical case (see figure 10(h)) and 5.25 higher in the spherical case (see figure 10(i)). So, the effect of the geometry can lead to large discrepancies. As explained before, if temperature field is mainly cylindrical, it becomes spherical in the close vicinity of the tip where temperature and flux are of interest. The main feature of this work will consist in the development of an hybrid model allowing to take into account this particular aspect.

Estimations

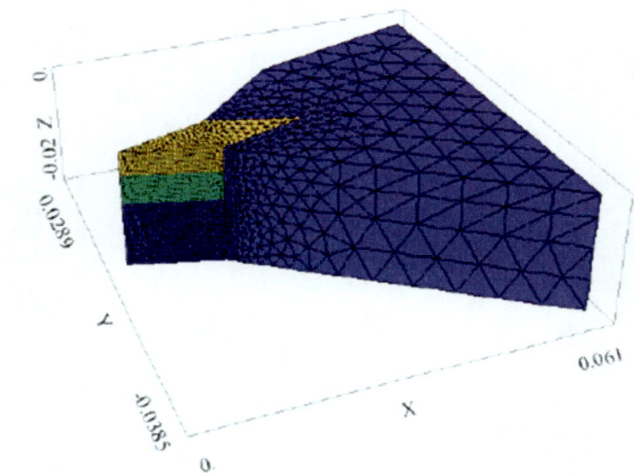

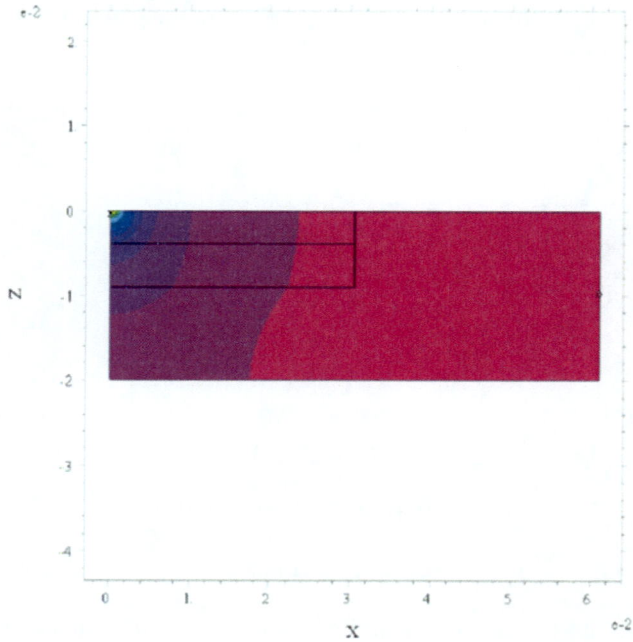

Figure 11. Continued on next page.

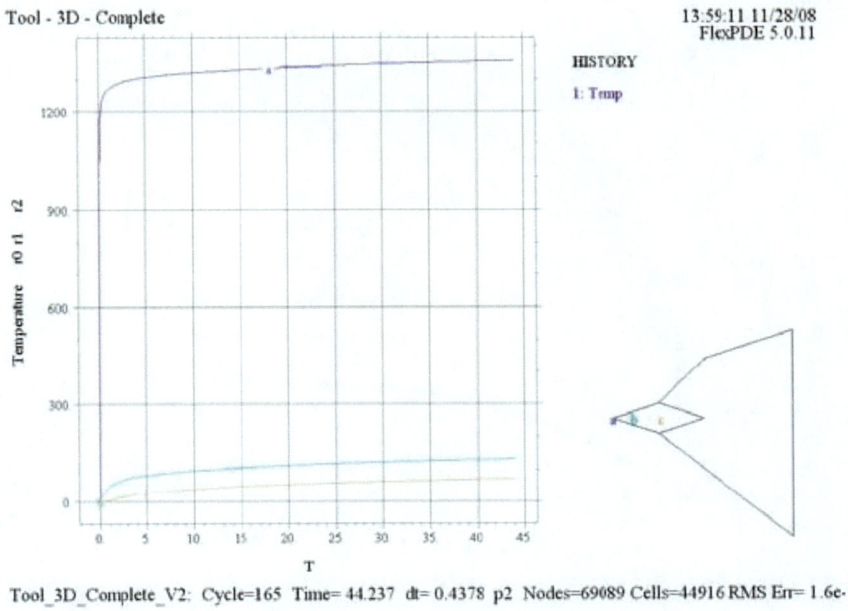

Figure 11. Simulations performed with FlexPDE –grid, temperature field and thermograms.

Simulations have also been performed using FlexPDE, a numerical computational code allowing to solve partial differential equations through a finite element method. An exponential variation of the temperature at the tool's tip (r=r0) has been considered and not only the tool but also the tool holder has been modelled (figure 11).

The corresponding heat flux variation is plotted in figure 11 as for the estimated heat flux and temperature obtained from the noised thermogram in r=r1 with or without regularization (noise level: 2%, regularization: r=16). The number of future time steps is r=30 for 2500 points in time.

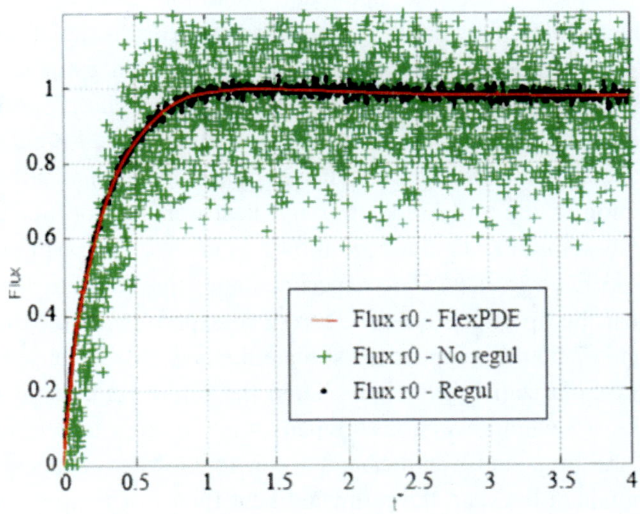

Figure 12. Continued on next page.

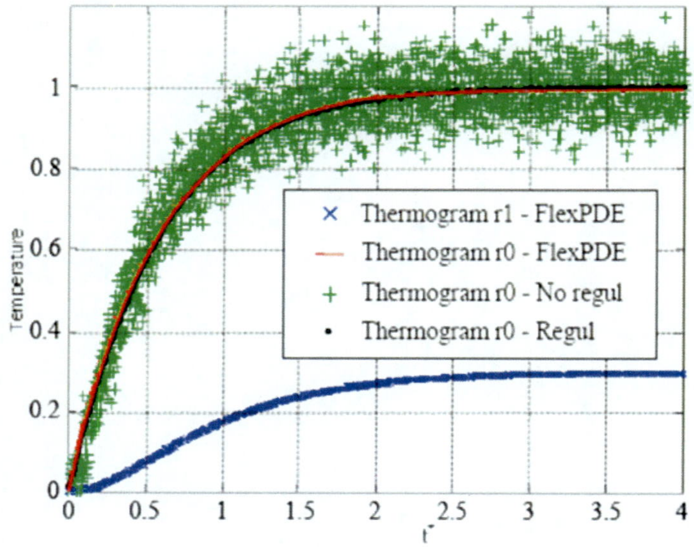

Figure 12. Heat flux estimation and Thermograms.

A significant improvement of the results due to the regularization is observed. The inverse estimated heat flux and temperature in r=r0 are in good agreement with the direct curves given by FlexPDE.

Conclusion

In this chapter, heat transfer in a cutting tool is investigated by numerical and semi-analytical simulations. The advantage of softwares such as for instance Fluent is the ability to reproduce in a very precise way the geometry of the cutting tool and to obtain accurate results as far as all the parameters and the boundary conditions are well known. But it is unfortunately not the case in the reality. Moreover numerical simulations are often time consuming and as a consequence are not adequate to perform estimations for real time control. That is the reason why, a specific thermal model is built and proposed to describe the transient heat transfer in the cutting tool. It is based on a quadrupole formulation and involves Whittaker functions. It takes into account heat losses by convection and advection through the chip and the conduction in the tool holder through heat transfer coefficients and a thermal contact resistance. This semi-analytical model allows an inverse approach in order to estimate data which are very difficult to determine experimentally: the transient temperature and the transient heat flux near the tip of the tool. To reproduce experimental measurements, data are corrupted by noise and difficulties appear to estimate the heat flux because the problem is ill-posed. The use of a regularization technique called the future time steps method allows to improve the estimations. Several inversions are performed in order to determine the influence of the noise's level, the time steps number and so on for different shapes of heat fluxes. A good agreement is obtained between the estimated heat fluxes or temperatures and the input data.

Some other studies concerning inverse methods applied to cutting process are underway. The technique will use a branch eigenmodes reduced model [54] in order to have an on-line temperature control at the tool chip workpiece interface.

References

[1] Trent, E. M.; Wright, P. K. *Metal Cutting;* Butterworth Heinemann: Woburn, MA, 2000.

[2] Boothroyd, G.; Knigh, W.A. *Fundamentals of Machining and Machine Tools;* Marcel Dekker: New York, UN, 1989.

[3] Ay, H.; Yang, W.J. *Exp. Heat Transfer.* 1994, 41, 203-216.

[4] Ay, H.; Yang W.J. *Int. J. Heat Mass Transfer.* 1998, 41, 613-623.

[5] Battaglia, J. L.; Batsale, J.C. *Inv. Prob. Eng.* 2000, 1-23.

[6] Jaspers, S.P.; Dautzenberg, J.H.; Taminiau, D.A. *Int. J. Adv. Manuf. Technology.* 1998, 7-12.

[7] Kwon, P.; Schiemann, T.; Kountanya, R. *Int. J. Mach. Tools Manuf.* 2001, 41, 1015-1030.

[8] M'Saoubi, R.; Lebrun, J.L.; Changeux, B. *Mach. Science Techn.* 1998, 2, 369-382.

[9] Davies, M. A.; Ueda, T.; M'Saoubi, R.; Mullany, B.; Cooke, A. L. In *CIRP Annals - Manufacturing Technology.* 2007, 56, 581–604.

[10] Komanduri, R; Hou, Z. B. *Tribology International.* 2001, 34, 653–682.

[11] O'Sullivan, D.; Cotterell, M. *J. Materials Processing Technology.* 2001, 118, 301-308.

[12] Grzesik, W. *Int. J. Mach. Tools Manuf.* 1999, 39, 355-369.

[13] Korkut, I.; Boy, M.; Karacan, I.; Seker, U. *Materials & Design*, 2007, 28, 2329-2335.

[14] Battaglia, J. L.; Cois, O.; Puigsegur, L.; Oustaloup, A. *Int. J. Heat Mass Transfer*, 2001, 44, 2671-2680.

[15] Lo Casto, S.; Lo Valvo, E.; Micari, F. *J. Mechanical Working Technology*, 1989, 20, 35-46.

[16] Carvalho, S.R.; Lima e Silva, S.M.M; Machadoa, A.R.; Guimaraes, G. *J. Materials Process. Technology.* 2006, 179, 97-104.

[17] Soo, S. L.; Aspinwall, D. K.; Dewes, R. C. *J. Materials Process. Technology.* 2004, 150, 116-123.

[18] Grzesik, W.; Bartoszuk, M.; Nieslony, P. *J. Materials Process. Technology.* 2005, 164, 1204-1211.

[19] Ceretti, E. ; Lazzaroni, C.; Menegardo, L. ; Altan, T. *J. Materials Process. Technology.* 2000, 98, 99-103.

[20] Filice, L. ; Umbrello, D.; Beccari, S.; Micari, F. *J. Materials Process. Technology.* 2006, 174, 286-292.

[21] Umbrello, D.; Filice, L.; Rizzuti, S.; Micari, F.; Settineri, L. *J. Materials Process. Technology*, 2007, 189, 284-291.

[22] Grzesik, W.; Bartoszuk, M.; Nieslony, P. *Int. J. Mach. Tools Manuf.* 2004, 44, 1451–1462.

[23] Shaw, M.C. *Metal Cutting Principles*, Clarendon Press: Oxford, 1989.

[24] Kato, T.; Fujii, H. *ASME Trans. J. Manuf. Sci. Eng.* 1999, 121, 393–398.

[25] Reznikov, A.N. *Thermophysical Aspects of Metal Cutting Processes; Mashinostroenie.* Moscow, 1981, (in Russian).
[26] Grzesik, W.; Nieslony, P. *Int. J. Mach. Tools Manuf.* 2004, 44, 889–901.
[27] Komanduri, R.; Hou, Z.B. *Int. J. Mechanical Sciences.* 2001, 43, 89-107.
[28] Grzesik, W. ; Nieslony, P. *Wear*, 2004, 256, 108-117.
[29] Rech, J. ; Kusiak, A.; Battaglia, J. L. *Surface Coatings Technology.* 2004, 186, 364-371.
[30] Huang, Y.; Liang, S. Y. *Int. J. Mach. Tools Manuf.* 2003, 43, 307-315.
[31] Liu, Y.R.; Liu, J.J.; Zhu, B.L.; Luo, Z.B.; Miao, H.Z. *Wear.* 1997, 210, 39-44.
[32] Lazard, M.; Corvisier, P. *Appl. Thermal Engineering.* 2004, 24, 839-849.
[33] Abramovitz, M.; Stegun, A. *Handbook of mathematical functions with formulas, graphs, and mathematical tables;* Dover, New York, 24 1964.
[34] Carslaw, H.S. *Introduction to the Mathematical Theory of Heat in Solids*; Macmillan, 1921.
[35] Degiovanni, A. *Int. J. Heat Mass Transfer.* 1988, 31, 553-557.
[36] Lazard, M. *J. Engineering Physics Thermophysics.* 2006, 79, 758-767.
[37] Barataud, F.; Moyne, C.; Breda, N.; Granier, A. *Plant Soil.* 1995, 172, 29-43.
[38] Lazard, M.; André, S.; Maillet, A. *J. Quant. Spect. Rad. Transfer.* 2001, 69, 23-33.
[39] Degiovanni, A. *High Temp. High Press.* 1985, 17, 683-689.
[40] Lazard, M.; André, S.; Maillet, D.; Degiovanni, A. *High Temp. High Press.* 2000, 32, 9-17.
[41] Yen, D.W.; Wright, P.K. *J. Eng. Ind.* 1986, 108, 252-263.
[42] Lin, J. *Int. J. Mach. Tools Manuf.* 1995, 35, 751-760.
[43] Ohadi, M.M.; Cheng, K.L. *J. Heat Transfer.* 1998, 115, 446-452.
[44] Stephenson, D.A. *J. Eng. Ind.* 1991, 113, 129-136.
[45] Lima, F.R.S.; Machado, A.R.; Guimaraes, G. In: *Proceedings of the Third International Conference on Metal Cutting and High Speed Machining*, Metz, 1, 2001, 25-34.
[46] Chen, W.C.; Tsao, C.C.; Liang, P.W. *Int. Commun. Heat Mass Transfer.* 1997, 24, 161-170.
[47] Yvonnet, J.; Umbrello, D. ; Chinesta, F.; Micari, F. *Int. J. Mach. Tools Manuf.* 2006, 46, 820-827.
[48] Lazard, M.; Corvisier P. In: *5th International Conference in Inverse Problems in Engineering: Theory and Practice* Cambridge 2005.
[49] Stehfest, H. *Com. A.C.M.* 624, 1970, 13, 47-49.
[50] De Hoog et al., *SIAM J. Sci. Stat. Comput.* 1982, 3, 357-366.
[51] Beck, J.V.; Blackwell, B.; St. Clair, C.R. *Inverse Heat Conduction Ill-posed Problems;* Wiley Interscience Publication: New York, 1985.
[52] Beck, J.V.;Arnold, K.J., *Parameter estimation in engineering and science;* John Wiley and Sons, New York, 1977.
[53] Tikhonov, A. ; Arsénine, V. *Méthodes de résolution des problèmes mal-posés;* Editions de Moscou, 1976.
[54] Videcoq, E. ; Lazard, M.; Quemener, O. ; Neveu, A. *Num. Heat Tranfer; Part A: Applications.* 2009, 55, 683-705.

In: Thermal Engineering Research Developments
Editors: J. Evgova and O. Kostadinov, pp. 369-391
ISBN: 978-1-60741-497-1

Chapter 9

PERFORMANCE EVALUATION OF A CONVENTIONAL AND A DOUBLE-GLASS COVER SOLAR STILLS

Sabah A. Abdul-Wahab**[*a]**, Rafiq A. Siddiqui**[a]**,
***Ali M. Al-Damkhi**[b]** and Hilal Al-Hinai**[a]*

[a] Mechanical & Industrial Engineering Department, College of Engineering, Sultan Qaboos University, P.O. Box 33, Al Khoud 123, Oman

[b]Department of Environmental Sciences,College of Health Sciences, Public Authority for Applied Education and Training (PAAET), State of Kuwait.

Abstract

The operating performance of a double-glass cover solar still with 10 mm spacing (i.e., solar still with cooling water flowing between a double-glass cover) was evaluated experimentally under the climatic conditions of Muscat, Oman. The base area of the still was 1 m^2 (i.e., 1 m length by 1 m width) and the cover of the still was made an angle of 23.4 degree with the horizontal optimized for 23.37 N latitude of Muscat. For comparison purposes, a conventional (single-glass cover) solar still with the same dimension was also designed and fabricated. The performances of the two stills were experimentally investigated under identical operational and weather conditions. The experimental study was conducted on some of the factors that affect the productivity of the solar stills. This included the water depth in the basin and the effect of the insulator. The tests were also performed using de-ionized water and sea water as feed. The productivities of the two stills and the temperatures of the water and the glass covers were determined. The quality of the water produced was tested. It was found that more yield of fresh water was obtained in case of a double-glass cover solar still as compared to the single-glass solar still. This is due to higher temperature difference between the water and inner glass cover. The chapter has also revealed that the water depth is inversely proportional to the productivity of the stills. Hence, the daily production of the two stills was increased by reducing the depth of the water in the basin.

Keywords: Conventional solar still; double glazing solar still; productivity; solar energy; Oman.

* E-mail address of corresponding author: sabah1@squ.edu.om (Sabah A. Abdul-Wahab). Tel.: +968-2414-1360; fax: +968-2414-1316.

Nomenclature

T_a	ambient temperature (°C)
T_{bS}	basin temperature of the single-glass cover solar still (°C)
T_{gS}	lower (inner) glass temperature of the single-glass cover solar still (°C)
T_{wS}	water temperature inside the single-glass cover solar still (°C)
T_{bD}	basin temperature of the double-glass cover solar still (°C)
T_{gl}	lower (inner) glass temperature of the double-glass cover solar still (°C)
T_{gu}	upper (outer) glass temperature of the double-glass cover solar still (°C)
T_{wD}	water temperature inside the double-glass cover solar still (°C)

Subscripts

a	ambient
b	basin
D	double-glass cover solar still
g	glass
l	lower
S	single-glass cover solar still
u	upper
w	water

Abbreviations

PH	the concentration of hydrogen ions
TDS	total dissolved salts (mg /L)
EC	Electrical Conductivity (μ s/cm)

1. Introduction

Desalination of saline water by using solar energy is simple and low cost technology. It can easily be adopted especially by local people living in remote areas of Oman that are affected by scarcity of fresh water resources. The most economic and practical way to accomplish solar desalination is by using a simple basin type solar still, where the heat collection and distillation processes take place in the same equipment. The documented use of solar stills began in the sixteenth century. According to Samee et al. (2007), the first system was built in 1872 to supply a mining community in Chile with drinking water. With its simplicity, the basin type solar still would be a suitable solution of the water shortage problem. However, the solar still output in terms of distilled water productivity is still very limited and unacceptable in some instances which, therefore offsets its advantages. This has directed many researchers towards improving the operating performance of basin type solar stills, to enhance the yield of water per solar collection area, through the development of more advanced concepts in its design, configuration, type and operational parameters. According to

Tiwari and Noor (1996), solar stills have been studied extensively and a number of new actions and methods have been proposed by several researchers in an attempt to improve their performances. According to Zaki et al. (1992), these methods include increasing the brine temperature, improving distillate collection process, injecting black dye in the seawater, reducing the heat conduction through basin walls and top cover, enhancing the evaporation process, or decreasing the condensation surface temperature.

On the basis of various modifications and mode of operations introduced in simple solar stills, these solar distillation systems are classified as passive and active type solar stills (Abu-Arabi et al., 2002; Al-Hinai et al., 2002; Singh and Tiwari, 2004; Nijmeh et al., 2005; Shukla and Sorayan, 2005; Murugavel et al., 2008). In active type of solar stills, an extra-thermal energy by external mode is fed into the basin of the solar still for faster evaporation (e.g., additional condensers or collectors). If no such external mode is used then that type of solar still is known as passive solar still. Therefore, in passive type still either simple modification has been done inside the still or some materials have been used in basin along with saline water (Murugavel et al., 2008). Passive methods include the use of dye or charcoal to increase the solar absorptivity of water (Sodha et al., 1980; Lawrence et al., 1988; Abu-Arabi et al., 2002; Al-Hayek and Badran, 2004), applying good insulation (Al-Hinai et al., 2002), lowering the water depth in the basin to lower its thermal capacity (Al-Hinai et al., 2002; Al-Hayek and Badran, 2004), ensuring vapor tightness, and using reflective side wall (Al-Hayek and Badran, 2004). According to Abu-Arabi et al. (2002), active type of solar stills include the use of solar collectors or waste heat to heat the basin water, the use of internal and external condensers or applying vacuum inside the solar still to enhance the evaporation/condensation processes, and cooling the glass cover to increase the temperature difference between the glass and the water in the basin and hence increase the rate of evaporation. According to Singh and Tiwari (2004), the work on passive solar distillation was reviewed by Malik et al. (1982) and further review was carried out by Tiwari (1992), which also includes work on active solar distillation. Further, Tiwari et al. (2003) carried out a study on the present status of research work on both passive and active solar distillation systems.

Several researchers have studied various aspects of solar stills with a view to increasing the daily total distillate. Their studies indicated that the distillate output of a solar still depends mainly on water-to-cover temperature differential. This is because the evaporation rate increases with increasing temperature difference between the water surface and the transparent glass cover. Hence, various scientists have made attempts to maximize the yield of the simple solar still either by passive methods or by active methods (Shukla and Sorayan, 2005). According to Al-Kharabsheh and Goswami (2003), a theoretical analysis by Dunkle (1961), and the relations that he derived for the heat and mass transfer within the still, formed the basis for many research efforts. Dunkle (1961) found out that the mass transfer rate depends upon the water and glass cover temperature difference. This temperature difference is very important as it controls the rate of evaporation from the basin and rate of condensation on the glass cover (Abu-Hijleh, 1996). This motivated the search for methods to enhance the still productivity by increasing temperature difference between the water surface and the glass cover. According to Tiwari et al. (1997), the temperature difference can be increased either by increasing the water temperature or by decreasing the glass cover (condensing cover) temperature or by having both conditions at the same time.

Several methods were proposed to increase the temperature difference between the water in the basin and the glass cover, such as coupling the solar still to an external flat plate solar collector (Yadav and Jha, 1989; Lawrence and Tiwari, 1990; Zaki et al., 1992; Yadav, 1993; Yadav and Prasad, 1995; Kumar and Tiwari, 1996; Tiwari et al., 1996; Boukar and Harmim, 2001). This arrangement was attempted in order to increase the basin water temperature. The results indicated that the still performance could improve significantly, but the system cost would increase (Al-kharabsheh and Goswami, 2003).

In order to increase this temperature difference, some researchers studied the effect of adding a condenser to the still. The results indicated that the still performance could increase significantly as the result of increasing of the heat sink capacity (Khalifa, 1985; Fath and Elsherbiny, 1993; El-Bahi and Inan, 1999; Al-Kharabsheh and Goswami, 2003). Another alternative cooling system for basin type solar stills was presented by Haddad et al. (2000), where an external condenser, constructed as a packed bed storage tank, was integrated with the still. The packed bed condenser was cooled during the night, using a radiative cooling panel by circulating water into the packed bed condenser and the radiative cooling panel.

Another way to increase the temperature difference is by reducing the temperature of the glass cover and/or utilizing the latent heat of condensation dissipated through the glass cover (i.e., among the active methods used). A lower glass temperature results in reducing convection and radiation energy loss to the ambient. The glass cover gains heat from water vapor condensation on the lower side and loses this heat to the ambient by convection and radiation. Normally the rates of heat transfer from the glass cover to the ambient by convection and radiation are small. This causes the glass temperature to remain high which adversely affects the still productivity. Cooling of the glass cover increases the temperature difference between water in the basin and the glass cover, which increases the rate of evaporation, giving higher still productivity (Abu-Arabi et al., 2002; Zurigat and Abu-Arabi, 2004).

According to Al-Kharabsheh and Goswami (2003), this idea of reducing the temperature of the glass cover was investigated by Kumar and Tiwari (1996) who considered the performance of a single and double effect solar still, with and without water flow over the still glass cover, and concluded that a still with water flows over the glass cover gives the maximum yield.

Abu-Hijleh (1996) focused on increasing the temperature difference between the water basin and glass cover by passing a low flow rate film of saline water on the top of the glass cover. He found that the presence of a water film with its intermediate index of refraction value (n=1.33), between the air (n=1), and the glass (n=1.52), would smooth the transmission of solar rays into the still. He also reported that the still efficiency was improved by utilizing part of the used cooling water as pre-heated makeup water to the still.

According to Abu-Arabi et al. (2002) and Zurigat and Abu-Arabi (2004), several theoretical and experimental studies were conducted on cooling the glass cover by flowing water film over the glass cover (Abu-Hijleh, 1996). Another technique used to cool the glass cover is by passing water between a double-glass cover (Prakash and Kavanthekar, 1986; Singh and Tiwari, 1993). There are different arrangements for the double glass cover. One arrangement leaves no space between the flowing water surface and the upper glass cover, thus no evaporation from the enclosed water takes place. According to Zurigat and Abu-Arabi (2004), this double glass channel arrangement was investigated by Abu-Arabi et al. (2002) who found significant improvement over the conventional still. In the second arrangement the spacing can also be made in such a way to have a second basin (or second effect) with and without flowing water. In this case evaporation takes place from the water surface on top of the lower glass cover and additional distillate can be collected from condensing the water vapor on the upper glass and at the same time the exiting heated water can be fed to the lower basin. This arrangement was investigated by Prakash and Kavanthekar (1986).

The cooling by flowing water between a double-glass cover was investigated by Prakash and Kavanthekar (1986). It was reported that significant improvement was achieved mainly at low water mass in the basin. Later, Singh and Tiwari (1993) numerically compared the thermal performance of a solar still having a double-glass cover cooling with and without a collector against a conventional solar still with and without a collector. They concluded that there is a significant enhancement in the still productivity due to the cooling effect, but no enhancement due to the availability of hot water from the collector (Abu-Arabi et al., 2002).

The parametric study of a conventional passive solar still by incorporating the effect of water flowing over the glass cover was studied by Tiwari and Rao (1984). Lawrence et al. (1990) analyzed the parametric study of a passive solar still in the presence and absence of dye in water mass in the basin with the effect of water flowing over the glass cover. Furthermore, the effect of heat capacity of FRP multiwick solar still with water flowing over the glass cover was analyzed by Singh and Tiwari (1992). Janarthanan et al. (2006) presented a simple transient performance of floating cum tilted-wick type solar still by incorporating the effects of water flowing over a glass cover, heat capacity of tilted-wick water surface and floating-wick water surface.

It may be concluded that there is a limited number of experimental studies published on the performance of double-glass cover cooling solar stills. In the present chapter the solar still with double-glass cover cooling was designed and tested experimentally under the climatic conditions of Oman. The objective of the chapter was to improve the quality and increase the productivity of fresh water for people living in remote desert areas in Oman. Experimental analysis was also undertaken to study the impact of some variables that affect on the solar still productivity and the system performance under the weather conditions of the Sultanate of Oman. The performance of a simple basin type solar still under similar climatic conditions was also investigated and compared with the performance of a double-glass cover cooling solar still. The two stills were designed, fabricated and tested side-by-side in order to allow direct comparison of their experimental performance.

2. Experimental Setup and Procedures

The experimental set up consists of conventional and double-glass cover cooling solar stills (figure 1). The two stills are identical and both are of single sloped type solar stills. They are designed and constructed to maintain the comparison under the same conditions. In the case of the conventional solar still, no water flows on top of the glass cover, which has only one glass cover. In case of the double-glass cover, there is cooling water flowing between a double-glass cover, which has two glass covers with 10 mm spacing. The photograph of actually fabricated double-glass cover solar still with cooling water flowing between its double-glass cover is shown in figure 2.

Figure 1. Photograph of the fabricated conventional and double-glass cover solar stills.

Figure 2. Photograph of a fabricated double-glass cover solar still with cooling water flowing between its double-glass cover.

Both solar stills are operated simultaneously and the experimental procedures are essentially the same for both. These identical units allow investigation of the effect of the cooling of the glass cover under the same operating and weather conditions. The schematic diagrams of conventional and double-glass cover cooling solar stills along with dimensions are shown in figures 3 and 4, respectively. It should be noted that the condensing surface in each unit is the glass cover only.

The solar still are fabricated and assembled from locally available materials. The two solar stills are made using galvanized iron sheet of 2 mm thickness. Each unit has an absorbing square basin area of 1 m^2 (i.e., 1 m length by 1 m width) and consists of a metallic box having four sides. The bottom basin area of the stills is painted black to enable maximum absorption of solar radiations (i.e., for greater absorptivity).

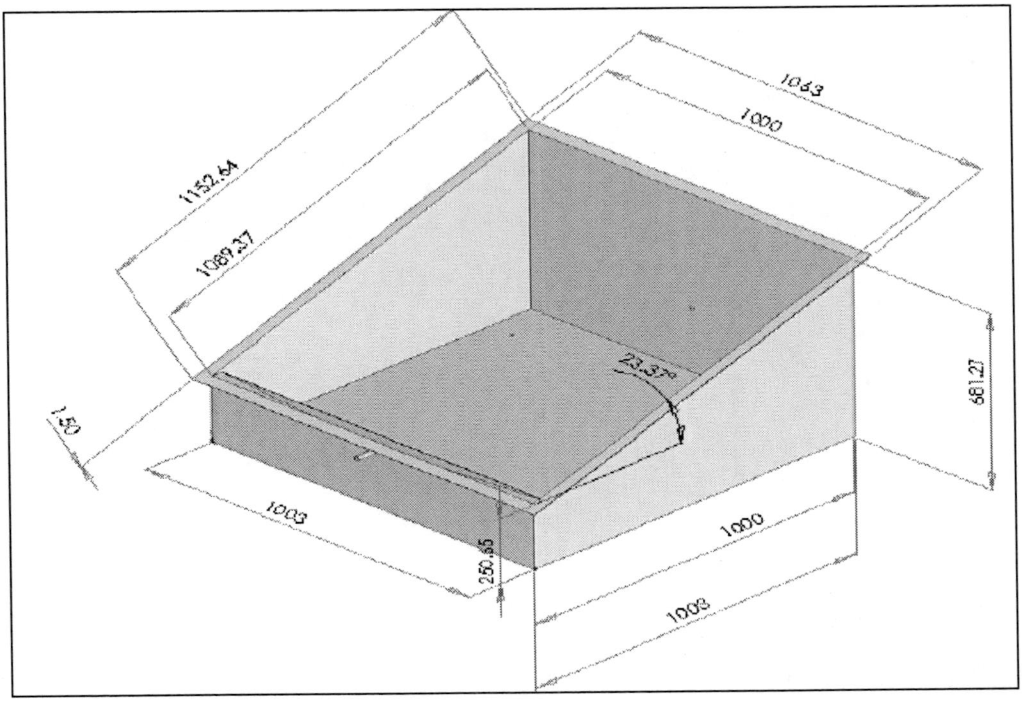

Figure 3. Schematic diagram of the conventional single-glass cover solar still (all dimensions are in mm).

The condensing covers of the stills are made up of simple window glass of 6 mm thick. They are adjusted on the edge of the rectangular sides of the metallic box. The whole unit is made air-tight using silicone rubber as sealant. A rubber gasket seal is also placed between the glass cover and the still box in order to make the entire assembly airtight. Once the condensing covers are fixed on the metallic boxes of the two stills, sets of tighten clamps were also used on the assemblies of both stills to ensure no leakage of vapor to the atmosphere and to make sure that the assemblies of both stills is air airtight. The condensing glass covers are tilted at an angle of 23.4° with horizontal optimized for 23.37 N latitude of Muscat (Al-Hinai et al., 2002). Each unit is placed on a stand facing to the south.

The feed water is fed through an inlet pipe while distillate and excess saline water is taken out through the respective outlets. Therefore, three holes in each solar still are made. The first hole is made at the backside of the solar still for feeding saline water through an inlet plastic pipe which is fixed at the rear hole. The second hole is made in the base for the drainage, whereas the third hole is made in the front wall of the still for collecting the distilled output. Further, a constant head tank is used to control the level of water inside each still by a float installed at the inlet pipe of each still (figure 1).

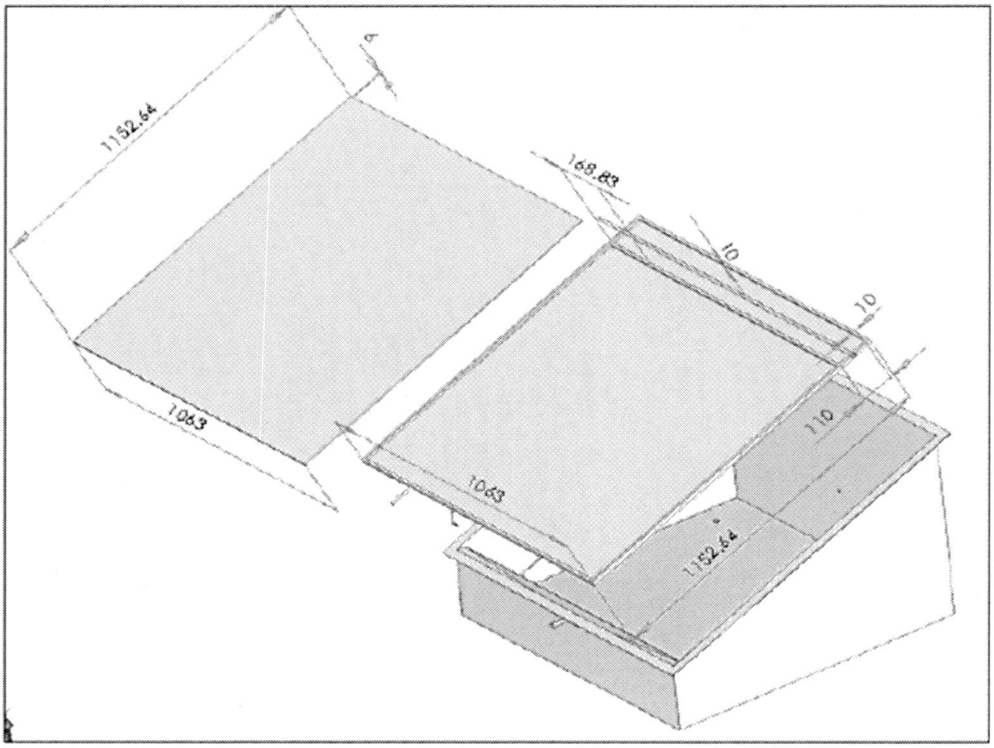

Figure 4. Schematic diagram of the double-glass cover solar still (all dimensions are in mm).

The produced fresh water is condensed on the inner surfaces of the inclined glass cover. The condensed water trickles down the lower edge of the inclined glass cover and is collected in the interior trough. This trough is fixed to the lower rectangular vertical side of the basin at the lower end of the glass cover. The output from the still is taken out of the trough through an outlet nozzle which drains the distillated water to an external storage bottle. The volume of the distillate produced is collected in a bottle and then the amount of distillate is measured by a graduated cylinder after every 2 hours.

Sets of thermocouples (K type) are used to measure temperatures of different parts of the still. These thermocouples are fixed in different positions of the still to measure the temperatures. This includes outer glass cover, inner glass cover, basin water, feed water and ambient temperatures. These thermocouples are further connected to a multi point data recorder (i.e., thermocouples reader) to record the temperatures after every one hour.

All experiments are conducted during 13-18 November, 2007 at Sultan Qaboos University, Oman. Experimentation is started at 8.00 a.m. local time and finished at 8.00 p.m. The runs have been conducted under various operating conditions, as reported in tables 1 and 2. The effect of the water depth on the solar still productivity for both solar stills is examined using three brine depths (5, 10, and 15 cm). In order to examine the effect of the insulation on the solar still productivity, the two stills are operated with and without the insulation. In order to reduce heat losses, the solar stills are insulated from outside as well as on the base of the units by 4 cm thick Styrofoam. All the tests are performed using sea water as feed as well as a lab-prepared de-ionized water samples.

Table 1. Operation conditions of the experiments investigated in this chapter

Operating parameter	Values
Water depth in the basin	5 cm 10 cm 15 cm
Existence of the Insulator	Without insulation Insulation with foam (Styrofoam), 4 cm thick
Type of feed water	Lab-prepared water samples Sea water samples

Table 2. Parameters used for the experimental tests investigated in this chapter

Experiment number	Date of test	Parameters
1	13 November 2007	Water depth = 5 cm. The outside walls and the base of the stills are insulated with foam (Styrofoam, 4 cm thick). Lab-prepared water sample is used as feed. TDS in the feed = 3500 mg/l
2	14 November 2007	Water depth = 10 cm. The outside walls and the base of the stills are insulated with foam (Styrofoam, 4 cm thick). Lab-prepared water sample is used as feed. TDS in the feed = 3500 mg/l
3	15 November 2007	Water depth = 15 cm. The outside walls and the base of the stills are insulated with foam (Styrofoam, 4 cm thick). Lab-prepared water sample is used as feed. TDS in the feed = 3500 mg/l
4	17 November 2007	Water depth = 5 cm. The outside walls and the base of the stills are insulated with foam (Styrofoam, 4 cm thick). Sea water is used as feed. TDS in the feed = 3800 mg/l
5	18 November 2007	Water depth = 5 cm. No insulation is used. Lab-prepared water sample is used as feed. TDS in the feed = 3500 mg/l

During experimentation, the temperatures are measured at different locations every hour for a period of 12 h for all operating conditions. This includes the outer glass temperature, the inner glass temperature, the water temperature, the basin temperature, and the ambient temperature. The productivity of each still is also measured simultaneously every two hours. At the end of each run, the total collected distillate outputs are measured to compare the relative performances of the two types of stills. In addition, the quality of the desalinated water produced is determined and is compared with that of the feed water. Three important parameters are measured in this regard such as total dissolved solids (TDS), pH and electrical conductivity.

3. Results and Discussion

3.1. Effect of Water Cooling of the Glass Cover on Cumulative Productivity

The effect of water cooling of the glass cover of the solar still on the cumulative productivity is shown graphically in figures 5a-5e. The figures show the cumulative productivity of the conventional and double-glass cover solar stills with time for the stated conditions as indicated in table 2. Each figure shows the variation of productivity obtained from both stills after every 2 hours. It is quite clear that the performance of the double-glass still is higher than the conventional one. The reason for this increase is because of the higher difference between the water and the glass temperature for the double-glass solar still than that of the conventional solar still. This represents the influence of the glass cover cooling on the productivity of the solar still. The increase in the still yield is due to the higher difference between the glass cover temperature and the water temperature which have increased the condensation process on the inner glass layer of the cover. This will be as a result of reduced convection and radiation energy losses to the ambient. The above conclusion regarding the performance of the double-glass still was seen for all the conducted experiments included in this chapter.

It is further to be noted that the hourly yields from both solar stills are significantly higher at night than during the day. This is attributed as we will indicate later on to the temperature differences between the water and glass which play the important rule on the productivity. As shown by figures 5a-5e, the accumulated water production starts very slowly due to warming of the stills and the somewhat low solar energy during the morning hours. A notable increase in the production rates are obtained at about 2.00 p.m. Furthermore, after 2.00 p.m., the production rates continue to decrease.

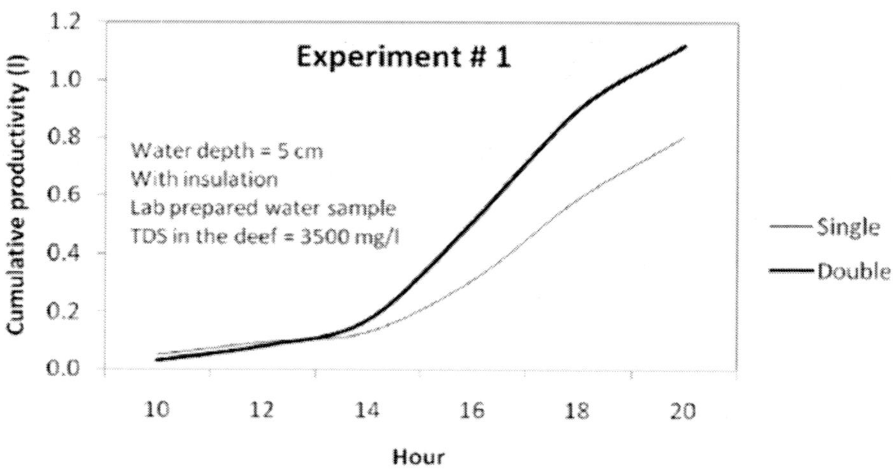

Figure 5a. Cumulative productivity of the two stills in 13 November 2007.

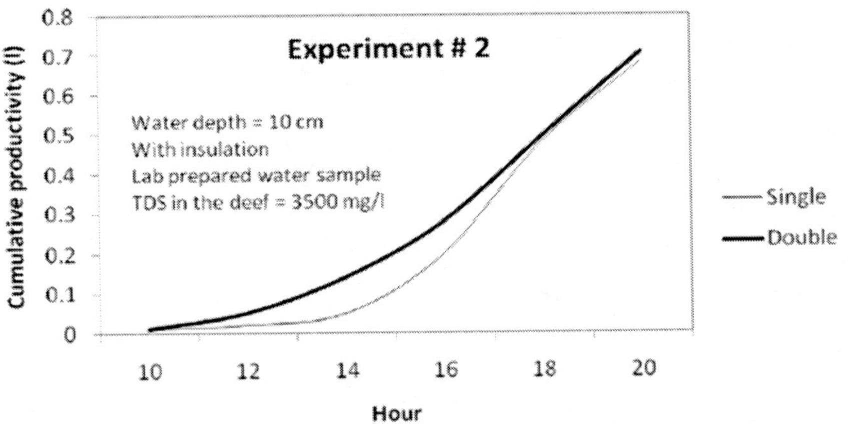

Figure 5b. Cumulative productivity of the two stills in 14 November 2007.

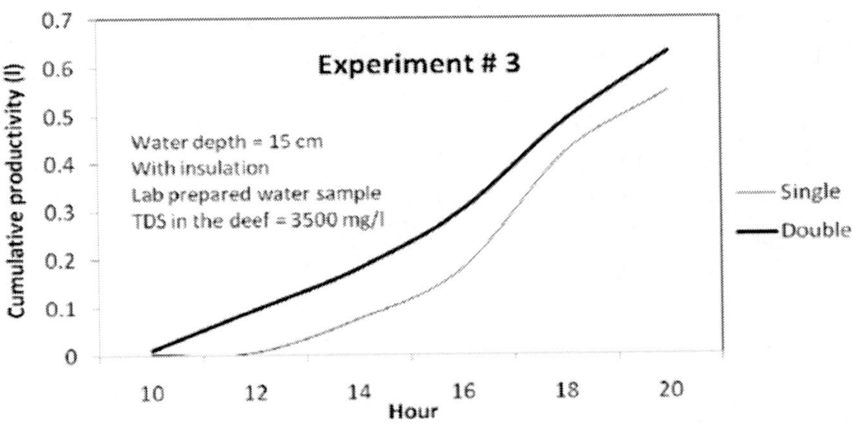

Figure 5c. Cumulative productivity of the two stills in15 November 2007.

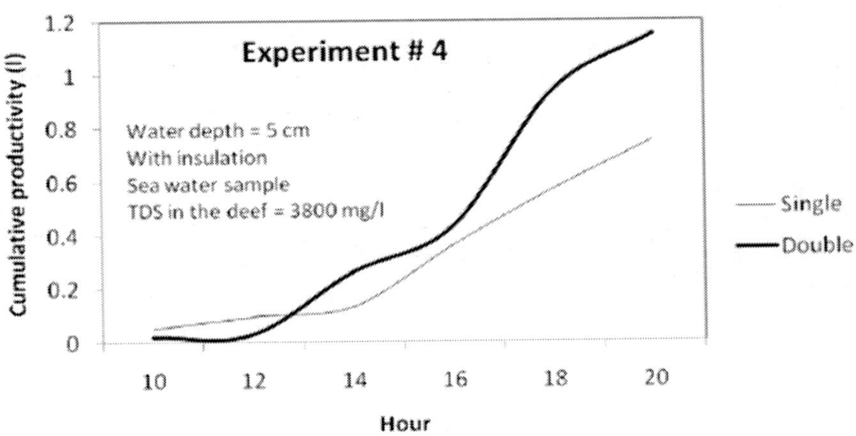

Figure 5d. Cumulative productivity of the two stills in 17 November 2007.

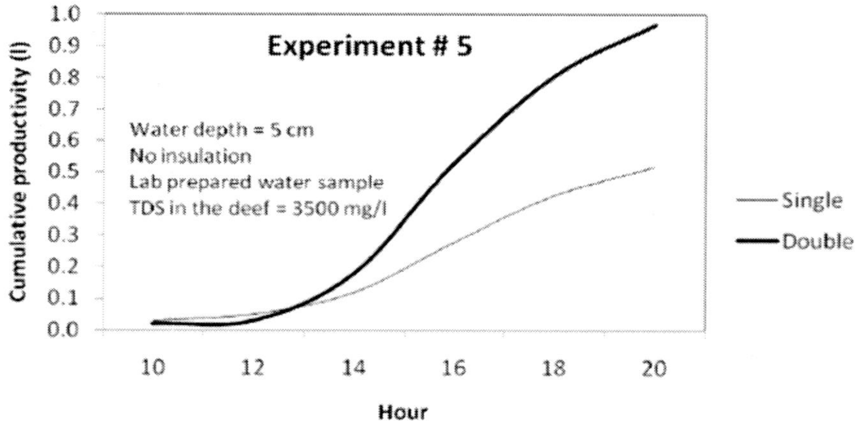

Figure 5e. Cumulative productivity of the two stills in 18 November 2007.

This finding is consistent with results reported by Murugavel et al. (2008) which indicated that the accumulated production is less during heating period and increases at higher rate during cooling period.

Figures 6a-6e show the variation of the hourly values of the various temperatures for the conducted experiments for the two stills. For the conventional solar still, the values of basin temperature *(T_{bS})*, lower glass temperature (T_{gS}) and water temperature (T_{wS}) are presented. For the double-glass cover solar still, the values of basin temperature *(T_{bD})*, lower glass temperature (T_{gl}), upper glass temperature (T_{gu}) and water temperature (T_{wD}) are illustrated.

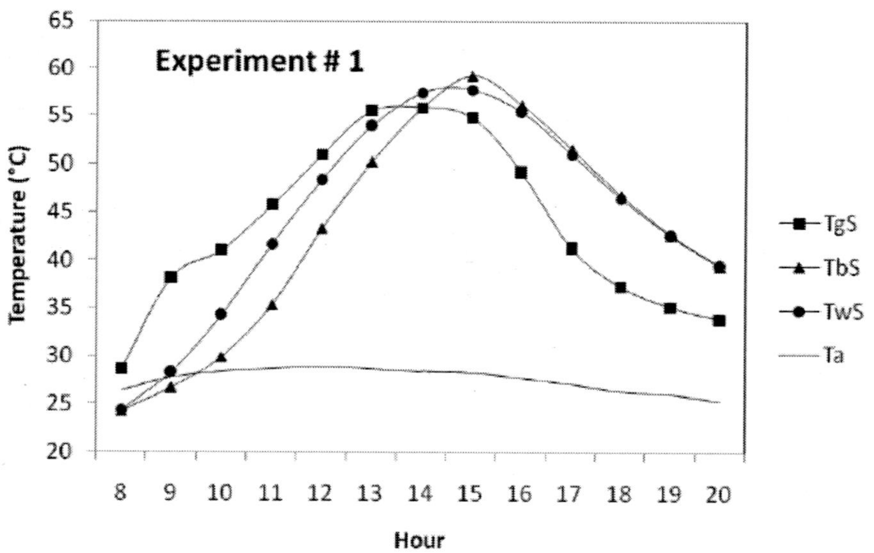

Figure 6a. Continued on next page.

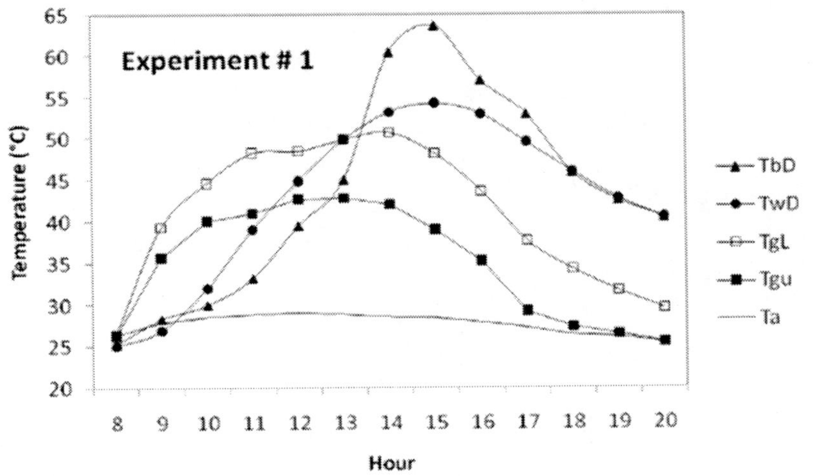

Figure 6a. Temperature variation of different parts of the two stills in 13 November 2007.

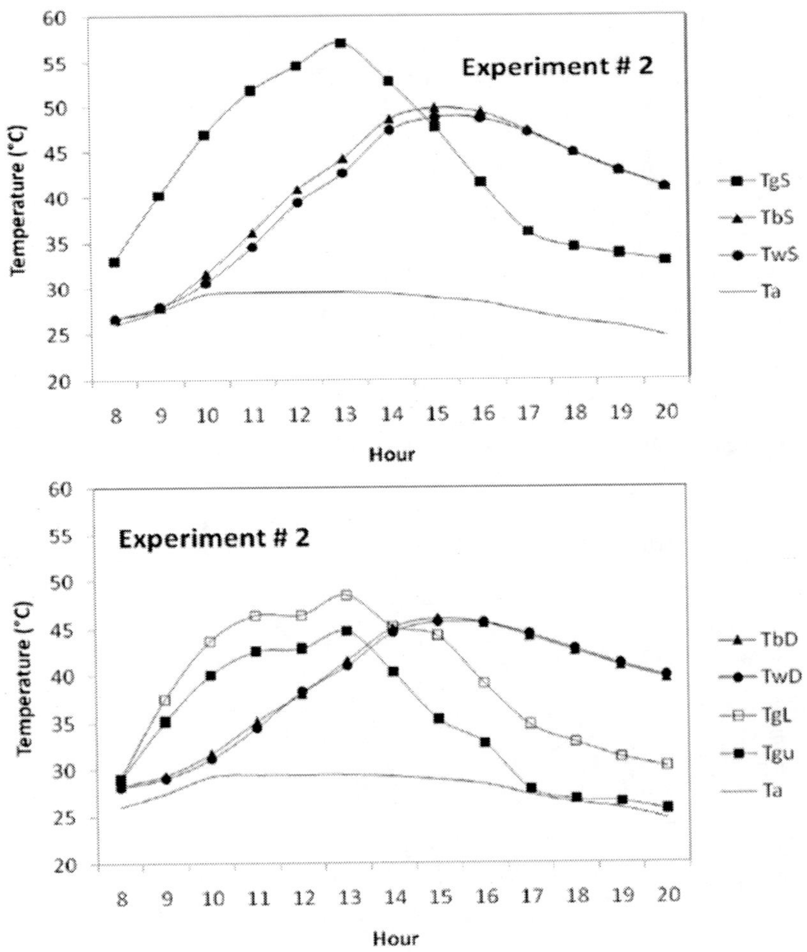

Figure 6b. Temperature variation of different parts of the two stills in 14 November 2007.

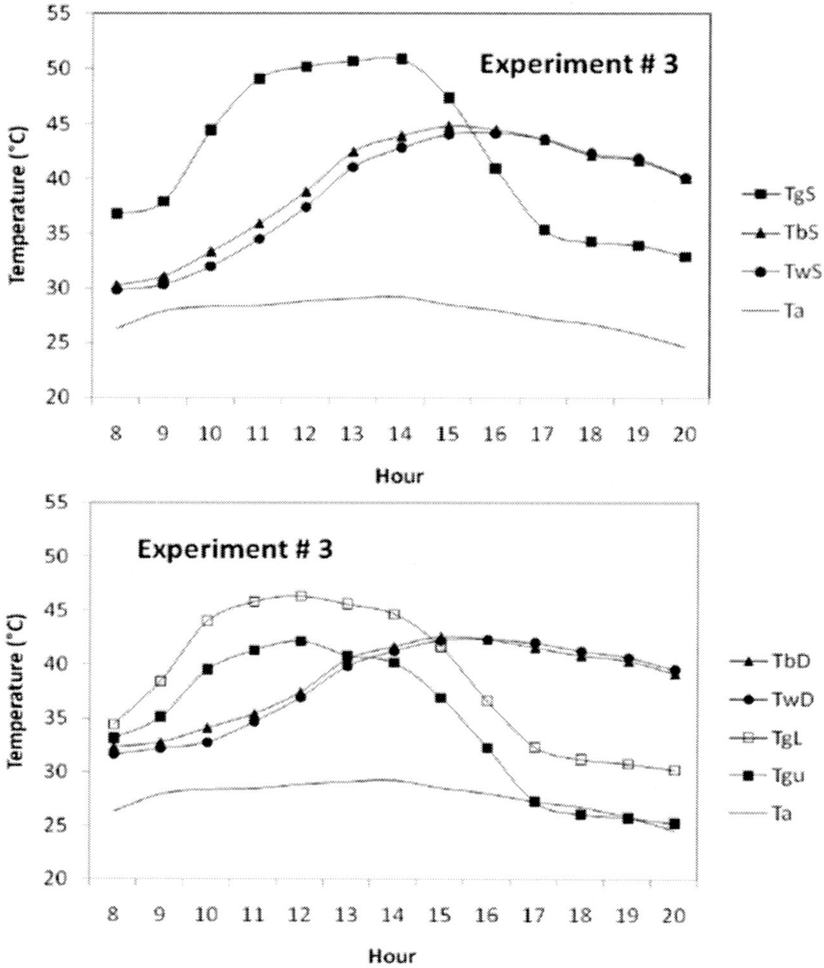

Figure 6c. Temperature variation of different parts of the two stills in 15 November 2007.

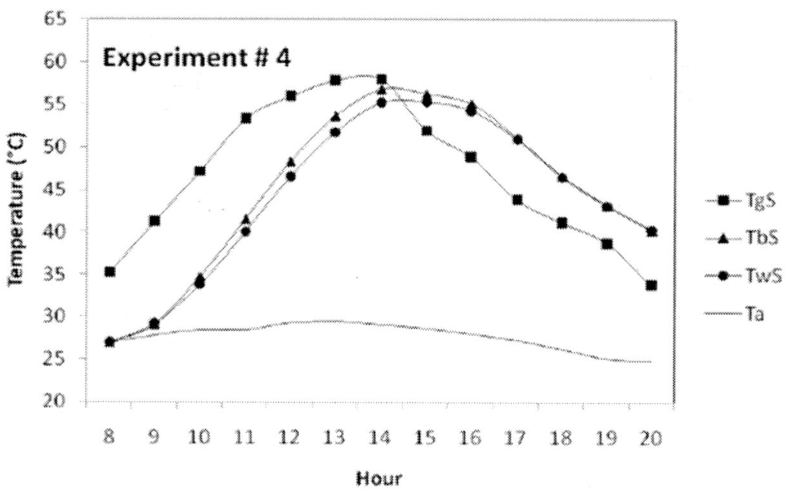

Figure 6d. Continued on next page.

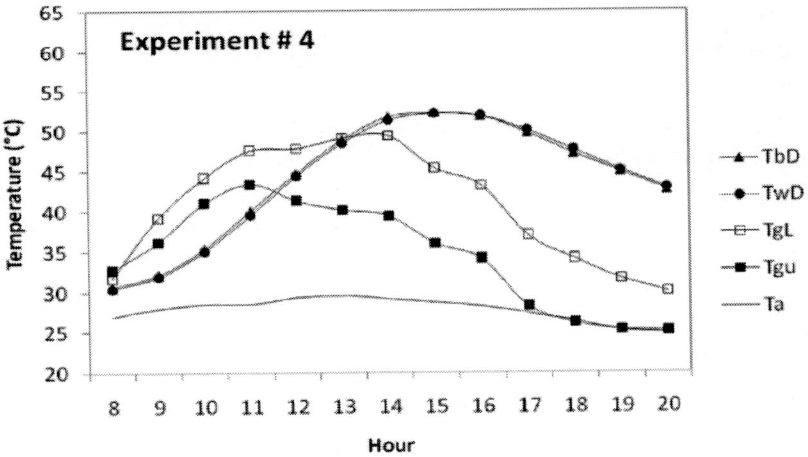

Figure 6d. Temperature variation of different parts of the two stills in 17 November 2007.

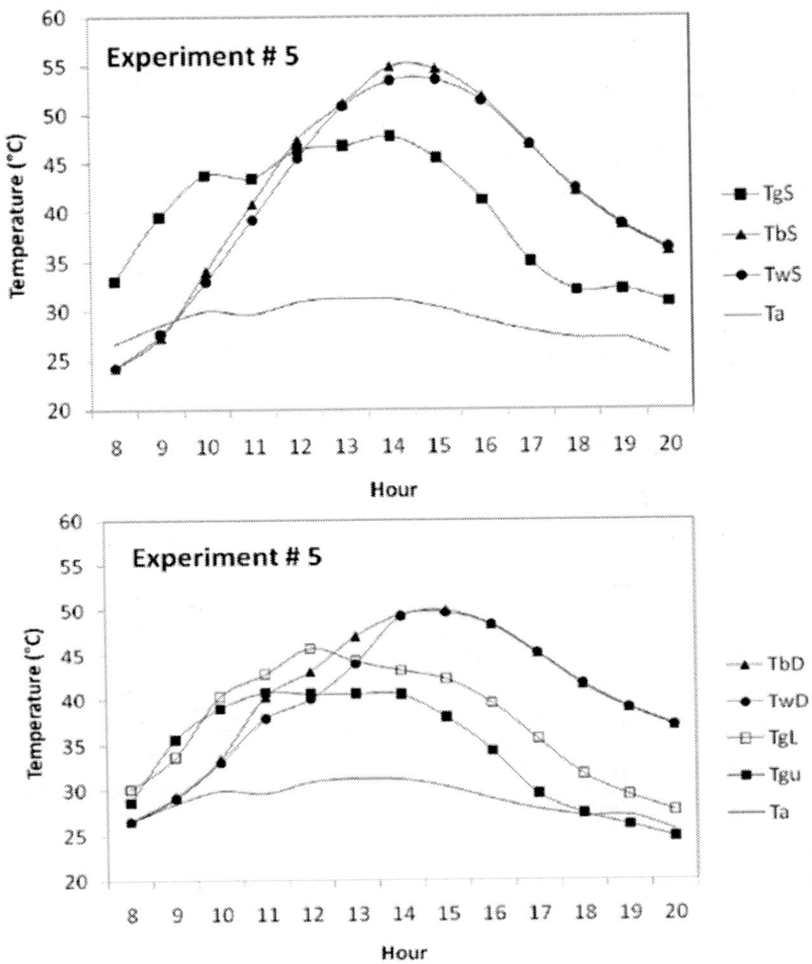

Figure 6e. Temperature variation of different parts of the two stills in 18 November 2007.

It can be seen from figures 6a-6e that similar trends are noticed for all the experiments (i.e., the temperature rises during the daytime up to the maximum value, and then it decreases). These results are consistent with the experimental findings that were reported by many investigators. Also, it is noticed that the water temperature in the base of the still is always the maximum due to incident rays there. This is followed by the temperatures of the inner glass where the condensation occurs and the outer glass which is in contact with the surrounding. The minimum temperature is always the ambient temperature. The values of the inner glass temperature (T_{gl}) are much higher than the condensing cover (T_{gu}) temperature which is due to the minimum heat loss from the inner glass cover to the surrounding due to the double glazed cover, as expected.

The flowing of cooling water between a double-glass cover solar still will lower its inner glass temperature and this will result in increasing the difference between water to glass temperatures. Hence, the temperature difference of the double-glass cover solar still is higher than that of the conventional type still. This is the reason behind that the production of the double-glass cover solar still is higher than that of the conventional type still as discussed before.

The difference between water and glass temperatures of the two types of stills is presented as shown in figure 7. It is also noticed that the temperature difference differ significantly in both type of stills over the collecting period. This is due to cooling of the condensing cover, the double-glass cover solar still is capable of producing higher temperature differences, as compared with the conventional solar still that is attained lower temperature difference. Hence, a greater temperature difference is obtained from the double-glass cover solar still than from the conventional.

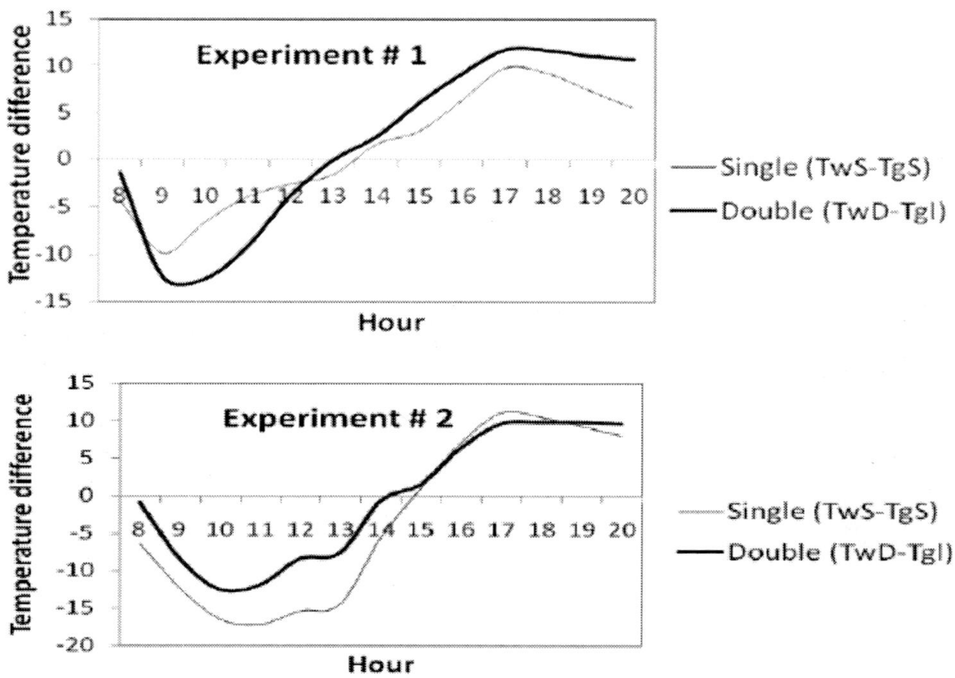

Figure 7. Continued on next page.

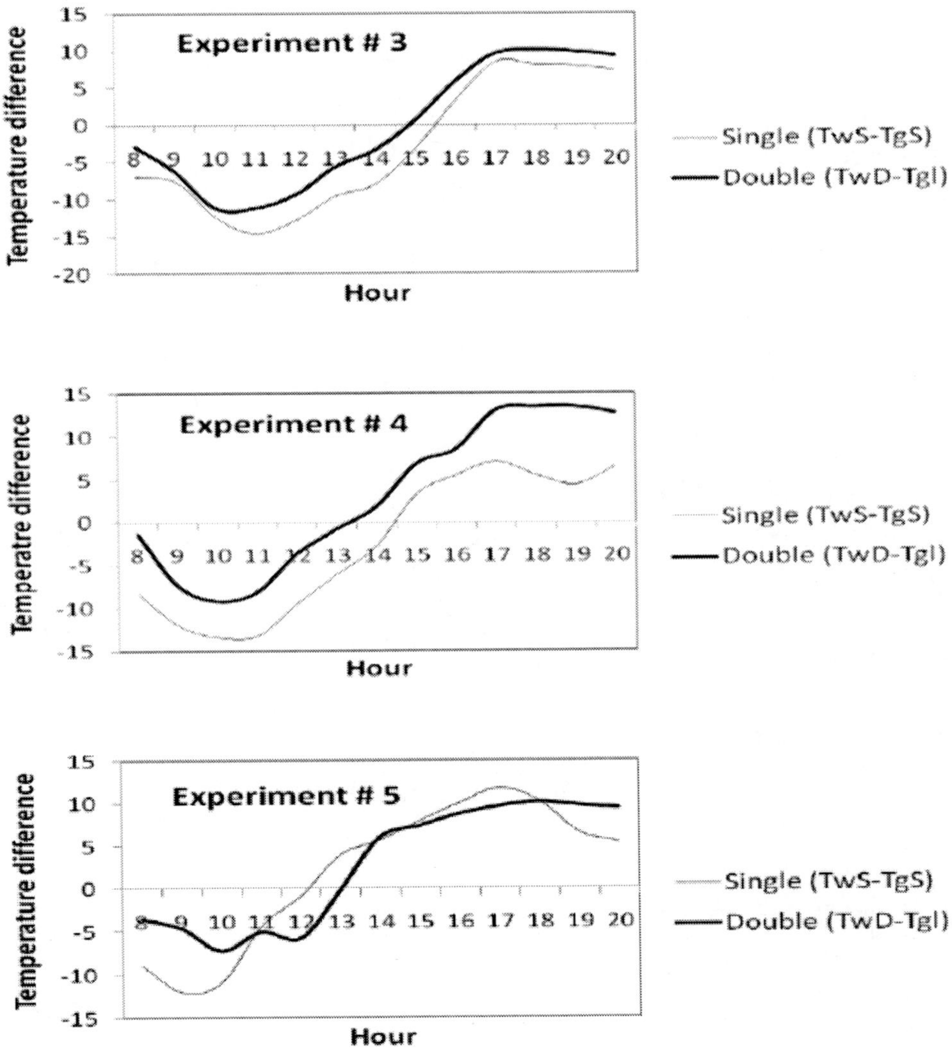

Figure 7. Difference between temperatures of water in the still and glass cover of the two stills for the experimental tests included in this chapter.

3.2. Effect of Water Depth on Cumulative Productivity

The effect of water depth in the still basin on the productivity is presented in figure 8. The data shows the cumulative amounts of distillates collected from the single-cover and double-cover solar stills, using different water depths (5, 10 and 15 cm). It can be seen that water depth has a significant effect on altering the productivity of the distilled water. As the water depth increases, the productivity will be decreased. This is attributed to the increase of the heat capacity of the water in the basin, results in lower water temperature in the basin leading to lower evaporation rate (figure 9.).

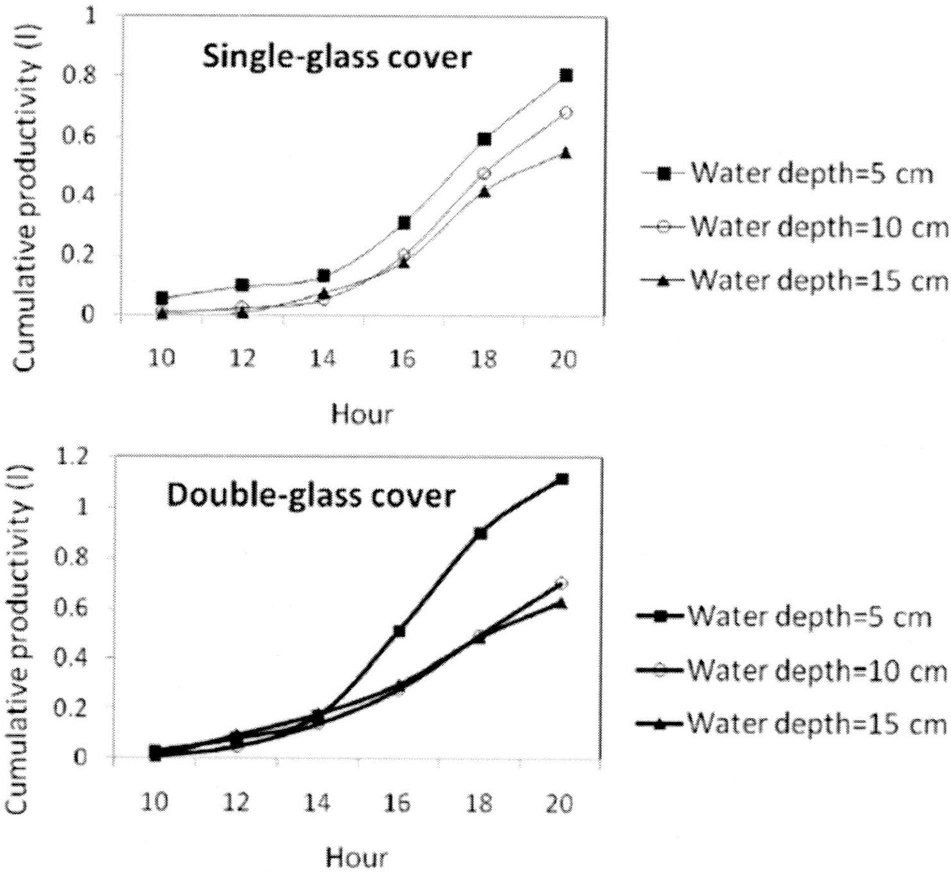

Figure 8. The effect of water depth on the cumulative productivity of the two stills.

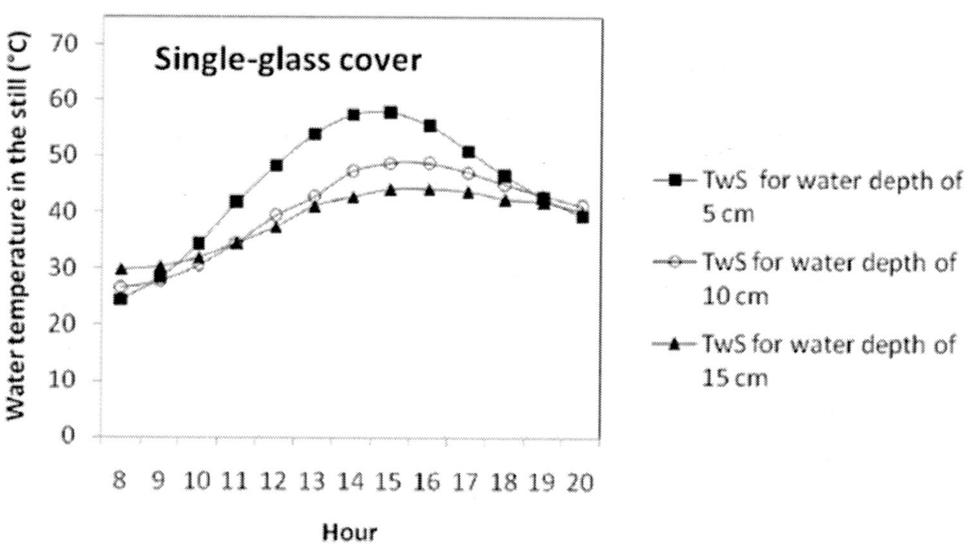

Figure 9. Continued on next page.

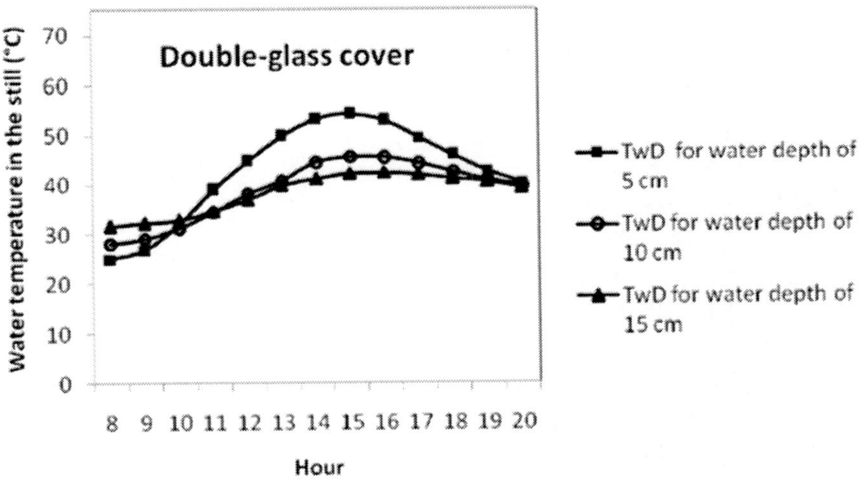

Figure 9. Water temperature in the basin at different water depths for the two stills.

3.3. Effect of Insulation on Cumulative Productivity

The effect of insulation on the cumulative productivity of the conventional and the double-glass cover solar stills, with and without insulation is reported in figure 10. It is interesting to see that when insulation is used, the cumulative productivity increases as the heat losses to the surrounding is reduced.

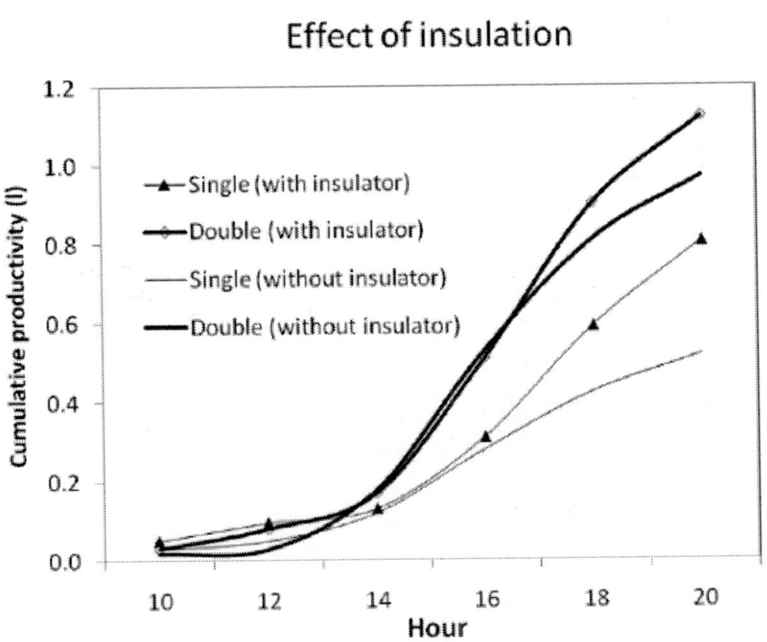

Figure 10. Effect of the insulation on the performance of the two stills.

Table 3. Quality parameters calculated before desalination and after desalination for both the conventional and double-glass cover solar stills

No.	Parameters	Solar still	Concentration TDS (mg /L)		PH		Electrical Conductivity (μ s/cm)	
			Before desalination	After desalination	Before desalination	After desalination	Before desalination	After desalination
1	Depth of water = 5 cm Insulation = 4 cm thick Lab-prepared water samples	Single	3.5×10^4	414.7	6.75	5.21	5×10^4	754
		Double	3.5×10^4	172.5	6.75	5.89	5×10^4	345
2	Depth of water = 10 cm Insulation = 4 cm thick Lab-prepared water sample	Single	3.5×10^4	61.5	6.75	5.68	5×10^4	123.1
		Double	3.5×10^4	34	6.75	5.76	5×10^4	68.1
3	Depth of water = 15 cm Insulation = 4 cm thick Lab-prepared water sample	Single	3.5×10^4	36.5	6.75	5.67	5×10^4	73
		Double	3.5×10^4	36	6.75	5.62	5×10^4	71.8
4	Depth of water = 5 cm Insulation = 4 cm thick Seawater samples	Single	3.8×10^4	55	7.31	5.90	5.43×10^4	109.4
		Double	3.8×10^4	38.5	7.31	6.14	5.43×10^4	76.6
5	Depth of water = 5 cm No insulation Lab-prepared water sample	Single	3.5×10^4	18	6.75	5.80	5×10^4	35.9
		Double	3.5×10^4	18	6.75	5.74	5×10^4	35.3

3.4. Quality of Desalinated Water

Table 3 shows the quality of the water before and after the desalination by the conventional and the double-cover solar stills. Three important properties are investigated in this regard: total dissolved solids (TDS), pH and electrical conductivity (EC). All these three parameters are determined for the samples at the start and end of each experiment period. It should be noted here that higher values of electrical conductivity indicate presence of more dissolved solids and hence more salinity.

The experimental tests indicate that the desalination has reduced the levels of TDS, pH and TC that present in the initial feed water samples. According to WHO guidelines for drinking-water Quality, the palatability of water with a TDS level of less than 600 mg/l is generally considered to be good. Drinking water becomes significantly and increasingly unpalatable at TDS levels greater than 1000 mg/l. Looking at table 3, it can be seen that the TDS values for all samples after desalination fall well within the 600 mg/l. The quantity of TDS after desalination ranges from 18.0 to 414.7 mg/liter, which is much below the values of WHO acceptable level.

It can also be seen that the pH values of the samples after desalination are also reduced when compared to the feed samples values before desalination. However, there is no health-based guideline for pH values that has been proposed by WHO that can be used for comparison purposes.

The effect of using the double-glass cover cooling solar still on both the TDS and conductivity of the samples is clearly seen in table 3. The values of TDS and conductivity of the samples that are desalinated by the double-glass cover solar still are found to be lower than those produced by the conventional single-cover solar still. This highlights an increase in the quality of the water produced by the double-glass cover solar still as compared with that of a conventional single-glass cover solar still under identical operation and weather conditions.

4. Summary and Conclusions

A double-glass cover solar still was fabricated to investigate and study its performance under the climatic conditions of Muscat, Sultanate of Oman. The conventional single-glass cover solar still was also fabricated and used to investigate its performance for comparison purposes. The two stills were tested and evaluated under the same operating and climatic conditions.

It was found that using a double-glass cover solar still, with cooling water flowing between its two glass covers, caused a significant improvement in water output per unit area of the solar still. Hence, the double-glass cover solar still was able to produce greater amounts of water than the conventional single-cover solar still over the same collection period.

The results indicated that the productivity of the produced water increased with decrease in the water depth. This was quite evident from the experimental results for both the conventional and double-glass cover cooling solar stills. It was also found that the performance of the two stills depend on whether they were insulated or not.

The results of this chapter could be extended and applied to other countries such as Kuwait and other Gulf States which have about the same climatic conditions.

Acknowledgments

The funding provided by Sultan Qaboos University (Project Number IG/ENG/MIED/08/05) is gratefully acknowledged.

References

Abu-Arabi, M., Zurigat, Y., Al-Hinai, H., & Al-Hiddabi, S. (2002). Modeling and performance analysis of a solar desalination unit with double-glass cover cooling. *Desalination*, **143**,173-182.

Abu-Hijleh, B.A.K. (1996). Enhanced solar still performance using water film cooling of the glass cover. *Desalination*, **107**, 235-244.

Al-Hayek, I., & Badran, O.O. (2004). The effect of using different designs of solar stills on water distillation. *Desalination*, **169**,121-127.

Al-Hinai, H., Al-Nassri, M.S., & Jubran, B.A. (2002). Effect of climatic, design and operational parameters on the yield of a simple solar still. *Energy Conversion & Management*, **43**, 1639-1650.

Al-Kharabsheh, S., & Goswami, D.Y. (2003). Experimental study of an innovative solar water desalination system utilizing a passive vacuum technique. *Solar Energy*, **75**, 395-401.

Boukar, M., & Harmim, A. (2001). Effect of climatic conditions on the performance of a simple solar still: a comparative study. *Desalination*, **137**, 15-22.

Dunkle, R.V. (1961). Solar water distillation: The roof type still and a multiple effect diffusion still. *Proceedings of the International Conference of Heat Transfer*, University of Colorado, Publication No. 108, pp. 895-902.

El-Bahi, A., & Inan, D. (1999). A solar still with minimum inclination, coupled to an outside condenser. *Desalination*, **123**, 79-83.

Fath, H.E.S., & Elsherbiny, S.M. (1993). Effect of adding a passive condenser on solar still performance. *Energy Conversion and Management*, **34**, 63-72.

Haddad, O.M., Al-Nimr, M.A., & Maqableh, A. (2000). Enhanced solar still performance using a radiative cooling system. *Renewable Energy*, **21**, 459-469.

Janarthanan, B., Chandrasekaran, J., & Kumar, S. (2006). Performance of floating cum tilted-wick type solar still with the effect of water flowing over the glass cover. *Desalination*, **190**, 51-62.

Khalifa, A.N. (1985). Evaluation and energy balance study of a solar still with an internal condenser. *Journal of Solar Energy Research*, **3**, 1-11.

Kumar, S., & Tiwari, G.N. (1996). Performance evaluation of an active solar distillation system. *Energy*, **21**, 805-808.

Lawrence, S., Gupta, S., & Tiwari, G. (1988). Experimental validation of thermal analysis of solar still with dye. *International Journal of Solar Energy*, **6**, 291-305.

Lawrence, S.A., Gupta, S.P., & Tiwari, G.N. (1990). Effect of heat capacity on the performance of solar still with water flow over the glass cover. *Energy Conversion and Management*, **30**, 277-285.

Lawrence, S.A., & Tiwari, G.N. (1990). Theoretical evaluation of solar distillation under natural circulation with heat exchanger. *Energy Conversion and Management*, **30**, 205-213.

Malik, M.A.S., Tiwari, G.N., Kumar, A., & Sodha, M.S. (1982). *Solar Distillation*, Pergamon Press, UK.

Murugavel, K.K., Chockalingam, K.N., & Srithar, K. (2008). Progresses in improving the effectiveness of the single basin passive solar still. *Desalination*, **220**,677-686.

Nijmeh, S., Odeh, S., & Akash, B. (2005). Experimental and theoretical study of a single-basin solar still in Jordan. *Heat and Mass Transfer*, **32**,565-572.

Prakash, J., & Kavanthekar, A.K. (1986). Performance prediction of a regenerative solar still. *Solar and Wind Technology*, **3,** 119-125.

Samee, M.A., Mirza, U.K., Majeed, T., & Ahmad, N. (2007). Design and performance of a simple single basin solar still. *Renewable & Sustainable Energy Reviews*, **11**, 543-549.

Shukla, S.K., & Sorayan, V.P.S. (2005). Thermal modeling of solar stills: an experimental validation. *Renewable Energy*, **30**,683-699.

Singh, A.K., & Tiwari, G.N. (1992). Performance study of double effect distillation in a multiwick solar still. *Energy Conversion and Management*, **33**, 207-214.

Singh, A.K., & Tiwari, G.N. (1993). Thermal evaluation of regenerative active solar distillation under thermosyphon mode. *Energy Conversion and Management*, **34**, 697-706.

Singh, H.N., & Tiwari, G.N. (2004). Monthly performance of passive and active solar stills for different Indian climatic conditions. *Desalination*, **168**, 145-150.

Sodha, M.S., Kumar, A., & Tiwari, G.N. (1980). Effects of dye on the performance of a solar still. *Applied Energy*, **7**,147-162.

Tiwari, G.N., & Rao, V.S. (1984). Transient performance of single basin solar still with water flowing over glass cover. *Desalination*, **49**, 231-239.

Tiwari, G.N. (1992). *Recent Advances in Solar Distillation*, Kamal, R., Maheshwari, K.P., & Sawhney, R.L., eds., Wiley Eastern, New Delhi, Chap. 2.

Tiwari, G.N., Kumar, S., Sharma, P.B., & Khan, M.E. (1996). Instantaneous thermal efficiency of an active solar still. *Applied Thermal Energy*, **16**,189-192.

Tiwari, G.N., & Noor, M.A. (1996). Characterization of solar stills. *International Journal of Solar Energy*, **18**, 147-171.

Tiwari, G.N., Kupfermann, A., & Aggarwal, S. (1997). A new design for a double-condensing chamber solar still. *Desalination*, **114**, 153-164.

Tiwari, G.N., Singh, H.N., & Tripathi, R. (2003). Present status of solar distillation. *Solar Energy*, **75**, 367-373.

Yadav, Y.P., & Jha, L.K. (1989). A double-basin solar still coupled to collector and operating in the thermosyphon mode. *Solar Energy*, **14**, 653-659.

Yadav, Y.P. (1993). Performance analysis of a solar still coupled to a heat exchanger. *Desalination*, **91**, 135-144.

Yadav, Y.P., & Prasad, A.S. (1995). Performance analysis of a high temperature solar distillation system. *Energy Conversion and Management*, **36**, 365-374.

Zaki, G.M., Al-Turki, A., & Al-Fatani, M. (1992). Experimental investigation on concentrator-assisted solar-stills. *International Journal of Solar Energy*, **11**, 193-199.

Zurigat, Y., & Abu-Arabi, M. (2004). Modeling and performance analysis of a regenerative solar desalination unit. *Applied Thermal Engineering*, **24**,1061-1072.

In: Thermal Engineering Research Developments
Editors: J. Evgova and O. Kostadinov, pp. 393-405
ISBN: 978-1-60741-497-1

Chapter 10

TRANSONIC AND TURBULENT FLOW AROUND GAS TURBINE BLADE

S. Djouimaa*
Physical department, sciences faculty,
University of El haj lakhdar, Batna, Algeria

Introduction

The very first gas turbine was developed at Gottingen University in the last of 1930. After some arrangements, the engine was used after for planes replacing the comparatively heavy piston engines in this time. The first gas turbines efficiency was relatively low, surrounding 20%, and since the first prototype is designed, it was a standard for the engineers to improve their performance.

Currently the efficiency of the gas turbines is in the surroundings from 35 to 50%, it is a very significant progress since the thirties and which evolves until our days.

1.1. Interest

The interest to simulate the gas turbine blade to blade flows relates to the comprehension of the induced phenomena, the internal flows characteristics determination and identification especially when the architecture of the blade is very complex. The transonic turbines constitute particularly a very interesting study case sight the complexity to model the internal flow which passes, through the blades, of the subsonic to the supersonic mode, from where the compressibility effects are non negligible.

* E-mail address: sihemdjouimaa@univ-Batna.dz; sdjouimaa@hotmail.com

1.2. Bibliographic Studies

Three methods were largely used in the analysis as well as the design of modern turbo machinery; experimental research, empirical analysis and numerical methods. The development of precise numerical methods and powerful computer made that the numerical simulation is more effective and sure. However, the questions of capacity and reliability are the key factors in the future development of the numerical analysis for the study and the design turbomachines.

1.2.1. Phenomena Induced by the Flow

The flows between blades are very complex. They are strongly three-dimensional, viscous with various secondary and swirling flows (horseshoe vortex, passage vortex, etc.) figure (1.1). Their blade interaction, the boundary layers, possibility of shock wave presence, shock wave /boundary layer interaction and wakes causes losses in general.

The transitory flows and high turbulence intensity generate additional complexities. Because of this nature, the analytical methods are very rare and non precise. The experimental investigations results are limited to a narrow range of the flow parameters in the turbomachines modeling.

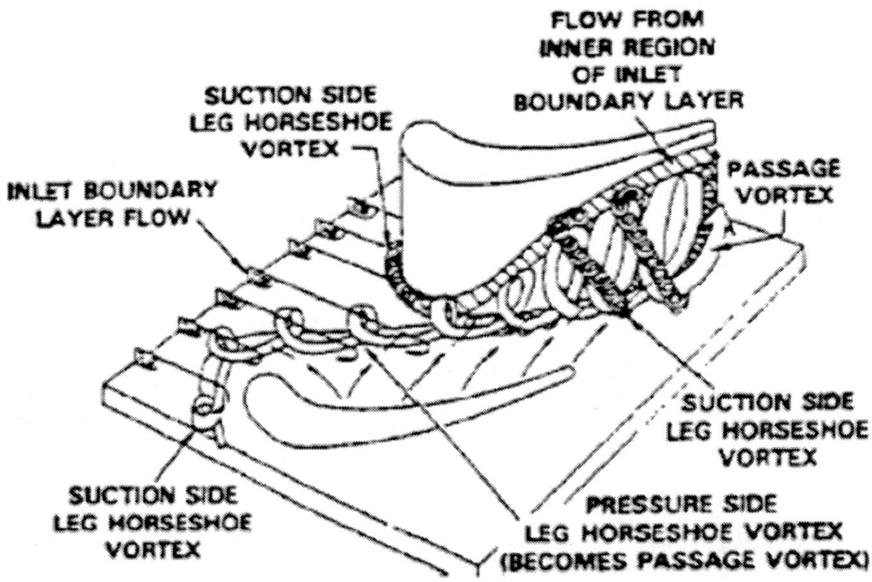

Figure 1.1. Secondary flows model by Sharma & Butler, (1987).

1.2.2. Experimental Works

For several decades, the interest of the transonic flows has not been neglected. So, many researchers were particularly interested in the experimental study in order to provide precise data for the numeric codes (CFD) validation. The latter are currently largely used by the industrial. Graham and Kost, (1979) evaluated the choc wave/boundary layer interaction in a transonic turbine cascade. Kiock et al. (1986) were rather interested in the experimental study through a plane turbine cascade. Mee et al. (1992) measured the losses caused by the transonic flow in a blades turbine cascade. Blair, (1994) proposed an experimental study of the heat transfer in a rotor turbine passage at large scale. Giel et al (1998) measured endwall heat transfer in linear cascade using the same method as Boyle and Russell (1990) in rotor cascade. Measurements are obtained for different Mach and Reynolds numbers with and without turbulence grid. Eight different flow conditions were investigated. The endwall heat transfer with the aerodynamic data presented by Giel et al. (1996) comprise a complete set of data suitable for CFD code and model validation Sveningsson, (2003). Personally, by using these data I approve as Vijay and Ameri (2001) that Giel et al ((1996, 1998) provided Benchmark quality data for CFD code and model verification. Also Vijay, (2001), was leaning on the study of the heat transfer.

1.2.3. Numerical Research

With efficiency and power increases of modern gas turbines, researchers tried continuously to increase the inlet temperature to the maximum. This can be done only with better blades cooling, great heat transfer comprehension and three-dimensional distribution of the temperature inside the turbine. To give a detailed cooling analysis as well as a good thermal structure of blades, several researchers treated numerically the three-dimensional flow which has very important effects, in particular the secondary flows, for the heat transfer in a real turbine Larsson, (1996); Giel et al., (1999); Laumert et al., (2000); and Vijay, (2001). On the other hand, much of authors Davidson, (1995) Larsson and Eriksson, (1995), Larsson, (1996); Yumin and Amano, (2000); Vogt and Fransson, (2002); and Bredberg, (2002); invested their efforts in the modeling and the two- dimensional flow comprehension.

Indeed, considering the blade complex form, the flow is characterized by regions with large continuous pressure gradients causing strong accelerations and decelerations. Moreover, the prediction transition region remains difficult to reach. The shock wave and the boundary layer interaction can cause the transition or the separation in a turbine stage.

Maciejewski and Moffat, (1992) and Thole and Bogard, (1995) showed that the strongly turbulent flow coming from the combustor causes an early boundary layer transition and increased heat transfer. Although Larsson and Eriksson, (1995) developed methods for the external heat transfer prediction in supersonic turbines, the major part of their article deals with details subsonic and transonic flow between blades. They showed that in the transonic case, the transition is not apparent on the pressure side and that a shock is localized on the suction side just on the trailing edge.

As for modeling the flow, very good results, especially for the transition, were obtained by Larsson and Eriksson, (1995) with Baldwin-Lomax algebraic model entrance turbulence intensities $Tu < 0.2$ %. But to model the higher intensities influence, only two equations models with were used (*Low-Re* k-ε et k-ω). On the suction side, only the Launder-Sharma model is able to capture laminar region. All other models predict an almost immediate transition and a fully turbulent boundary layer on the whole suction side. All models used predicted too high turbulence levels around the leading edge. According to Larsson, this excessive production is due to the high normal strains in the stagnation region. On the pressure side, all models give quite good results. In the same context, according to Durbin, (1995), the majority of turbulence models based on the Boussinesq approach give high levels and unreal values of the turbulent kinetic energy and turbulent viscosity at the stagnation region. Several researchers kalitzin, (1999), Sveningsson, (2003) used in the last few years the $\overline{v^2}$-f turbulent model originally suggested by Durbin, (1995), they proved that this model has an ability to correct the turbulence parameters damping without use damping functions. The $\overline{v^2}$-f model has also shown to be superior to other RANS methods in many fluid flows where complex flow features are present Sveningsson, (2003), the models modified by the additional terms at the ε equation were conceived to prevent turbulent viscosity to take nonphysical values. Sveningsson and Davidson, (2003) validated the theory of Durbin, (1995) by using the $\overline{v^2}$-f model in order to correct the jumps values of flow parameters at the leading edge and near wall. Concerning the trailing edge, Larsson, (1996) showed that all turbulence models give oscillations at separation point, thing which is very difficult to avoid. This was confirmed in experiments by Giel et al., (1996) and numerically by Djouimaa, (2007). Many researchers have previously used boundary layer and Navier-stokes solvers with algebraic turbulence models which gave good results, but these methods break down and are not good to predict correctly what occurs at the leading edge and often have difficulties at the trailing edge because of separation.

Furthermore, there are big problems with separated regions, shocks and high inlet turbulence levels. With a transport models, these problems can be solved or at least reduced. On the other hand some authors have successfully used the commercial code Fluent to simulate the flow in turbomachines Djouimaa, et al (2007).

1.3. Turbulence Models

Many losses in turbomachines components are due to viscous effects which are often concerned by turbulence. Nowadays, to model the flow region where these phenomena appear, we uses one equation turbulence algebraic models, two transport equations and the Reynolds model. Each kind of model take part of advantages and is characterized by a quite precise validity field. The problem is then to adopt not only one model poor in calculations and not complicated for simple flows, but also to avoid simulating complex phenomena with a limited model. Such compromises still appear difficult when one must take account of the type of space discretization which was adopted by the solver to model the flows. Thus, whereas the turbulence models are usually used on the structured grids, the implementation of certain very known turbulence models is far from being obvious when it is a question of using them with unstructured grids.

In order to understand the nature of a transonic flow in a turbine, the numerical tool can prove to be very useful. The complexity of such turbulent flows, containing strong gradients and regions where coexist very high and very low velocities, makes difficult any numerical simulation Alexis Bourgoing, (2002). The direct approaches which consist to solving the Navier-Stockes conservation equations are currently unrealizable because of the very small scales which we must solve to obtain precisely the fully flow dynamics.

The most often used technique actually for complex transonic flows calculations is based on the resolution of averaged Navier-Stockes equations in time when all the dynamics scales modeled.

1.3.1. Near Wall Flow (Passage from One Mode to Another)

1.3.1.1. Stagnation Zone (Khezar), (2005)

The stagnation region is a very critical in aerodynamic and heat transfer point of view. It is the region where the boundary layers start development on the suction and pressure sides. The heat transfer in this region can become very high and it is often the blade region where the maximum heat flow occurs. The leading edge heat transfer depends strongly on the inlet turbulence. Typically, a high turbulence level of the free flow (more than 10 %) can increase the stagnation point heat transfer by a ratio of 50 % compared with a laminar flow Larsson, (1996)

Not only the turbulence level has a significant influence on the stagnation region, but also the inlet length scale has importance. A small length scale as shown by several authors, gives usually a greater heat transfer increase. This increase is less by comparing it with the turbulence level influence Larsson, (1998)

1.3.1.2. Laminar Region

When the flow is laminar on the blade surface, transitory flows are often produced. The precise knowledge of the laminar boundary layer parameters is important for the good utilization of transition criteria Hu and. Fransson, (2000). Classically, the heat transfer in the laminar boundary layers region is independent of free flow turbulence; this is not the case for high turbulence levels. For this case, the swirls in the free flow enter in interaction with the laminar boundary layers and increase the heat transfer in these regions. This effect is greater close leading edge, particularly on the suction side. Downstream on the suction side, the more convex curve and the boundary layers thicker reduce the free flow turbulence effect. With a high inlet turbulence level, the laminar zone is smaller and the transition point advances. The influence of the high turbulence level of the free flow on the laminar heat transfer is usually larger on the blade pressure side than on the suction side, this influence extends also further with the downstream because of the concave curve of the pressure side Larsson, (1998).

1.3.1.3. Transition Region

The transition from laminar to turbulent region represents one of the more flows problems defying in the turbomachines field. The growth of complexities is due primarily to the transitory boundary layers along the blade surfaces which affect the aerodynamic and thermal performances of turbomachines, Simon and Ashpis, (1996).

The capacity to predict correctly the transition start and length is very important in the effective machines design.

Three kinds of transition arise; the first transition is called "natural" where the laminar boundary layer develops what is called the Tollmien-Schlichting waves followed by instabilities amplification and finally the fully turbulent flow figure (1.2). The natural transition usually occurs with small disturbances of the free flow.

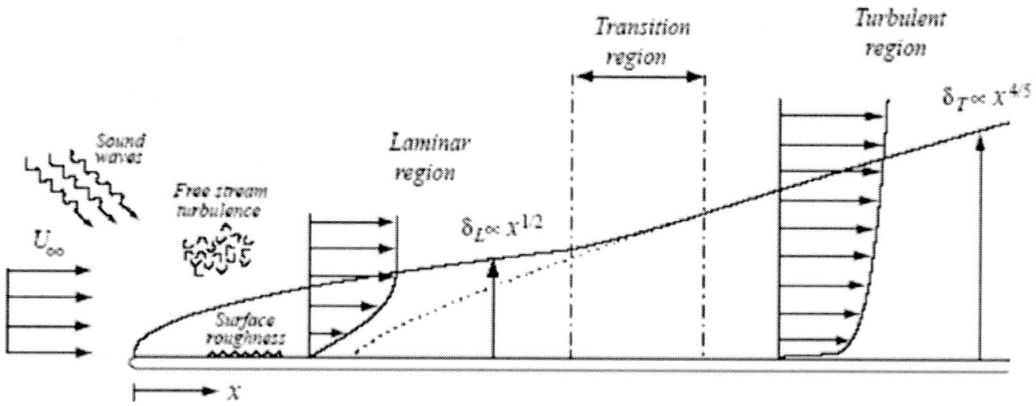

Figure 1.2. Induced Phenomenon by natural transition on a plaque plane with zero incidence degree.

The second transition type is caused by large disturbances of the free flow (apart of the boundary layer). This type is called "bypass" mode because the Tollmien-Schlichting instabilities are non-existent.

The third type is called transition «separated-flow " which occurs in the separation of the laminar boundary layer and which can or not generate the Tollmien-Schlichting waves Chernobrovkin and Lakshminarayana, (1999).

The turbulence level is very high in the average flow in the turbines, of this fact the transition mode is "bypass" type. The natural transition is non-existent in the practical cases and the third mode is specific to low pressure turbines.

The steady flow prediction is not satisfactory with turbulence models of high order this is with the insufficiency of the physical models used. These models are not in right proportion with the rotation, curves, heat transfer, compressibility, three-dimensional stress field, flow separation and instability effects. The Fluent *code* is sophisticated and most complete for the flows in the turbomachines. There is a need to evaluate its numerical techniques and to improve its efficiency Djouimaa, (2007).

Mayle, (1991) re-examined the transition phenomena in the gas turbines. Of a theoretical prospect, the first transition is seen as being an abrupt jump from the laminar to turbulent flow. The laminar flow is separated from the turbulent one by a line or a simple section. By the boundary layer properties modification, the two zones are connected. It is the approach which the majority of the numerical methods adopt Launder and Spalding, (1974); Schmidt and Patankar, (1991).

The experimental results showed that the transition is a no abrupt process. Emmons, (1951) was the first which proposed that the transition is a three-dimensional and an instationary process, which with region where the laminar and turbulent flows coexist. At a certain point of space, the flow can be laminar in a given time and turbulent at another one. This was described as being the intermittency phenomenon which it was the factor which had the most interest for much of theoretical investigations. Later, Mayle and Paxson, (1991) proposed a new theory, in their article they described the process of transition on the blade as being the total effect of two independent processes one "normal mode" and the other non stationary "wake-induced transition". For each mode they suggested models for the beginning of the transition and the distribution factor the intermittency downstream. Ameri and Arnone;

(1996) checked this theory by using the TRAF3D *" Transonic flow 3D"*code Arnone, (1994) which uses the approach of finite volumes with centered cells.

During the seven last years, in blade to blade flows, many researchers accentuated their work for the determination of the transition region.

Winoto et *al* (2000) treated the transition in the boundary layers on concave surfaces. Through its study, he compared the flow in the boundary layer for a plane plate with zero incidence angle and curved surface, by using the Goertler proposal which was the first to predict analytically that the laminar boundary layer developed along a curved surface will be the subject of centrifugal instabilities which appear as being a system of "streamwise counter-rotating vortices".

In the same context Wang, (2005) compared turbulent viscosities between two turbulence models, one with one equation [Spalart Allmaras], the other was the k-ω model. He remark that the entrance turbulent length scale has a small effect on the turbulence modeling properties Therefore the spalart allmaras model is an effective approach to model viscosities close to the wall with a developed flow along a concave surface.

Therefore Johnson, (2006) used the theory of Goertler Taylor (T-G). In his article, he confirms that the prediction of the traditional transition by the models developed for a plane plate failed in the transition prediction in the boundary layer on the blade pressure side of gas turbines. This because the centrifugal effects of the pressure gradients induced by the blade curve are destabilized which leads to the vorticities of Taylor Goertler formation. In order to determine the transition region, Mark W Johnson used the Transition criterion of Mayle, (1991) such as:

$$Re_{Tstart}=400\ T_u^{-0.625} \tag{1.1}$$

Koyabu et al., (2005) made an experimental study based on the waves induced by "bypass" transition from the boundary layers on a plane plate under adverse favorable pressure gradients. According to Koyabu, in the "bypass" transition we believes that the turbulent structures are create in the boundary layer like consequence of a rise in the turbulence level of the free flow for example. I.e. the conventional idea of two-dimensional instabilities of Tollmien-Schlichting is bypassed by this type of transitions. Since the flow field in turbomachines is characterized by the existence of intense external disturbances like free turbulence and periodic ondulations, it is easy to carry out that the "bypass" mode dominates the transition process in the boundary layers, the author also used the Mayle, (1991) transition criterion as well as the Abu-Ghannam and Shaw, (1980) Correlation to determine the transition region.

In their article Fransson et al., (2005) proposed that the most important source which induces the "bypass" transition in the boundary layer can be the free flow turbulence .In their study, they described the increase of the initial energy of the disturbances (streamwise-oriented) in the boundary layer origin of the presence of the F.S.T.(free-stream turbulence) for turbulence levels between 1.4% and 6.7%.

Suzen et al., (2003) used the transport equation for the intermittence factor to predict the transition flows in turbines with low pressure; the intermittence is taken into account near the transition flows and is incorporated in calculations by modifying turbulent viscosity with the intermittence factor.

The article presented by Stieger and Hodson, (2003) was the revelation and object of a new mechanism of transition resulting from an interaction from convective ondulations and of a separation bubble on the suction side part of a blade turbine at low pressure with strongly load.

Also, Walsh et al., (2004) presented in their article a new assumption saying that the sweeping force of the beginning transition of the boundary layer is the proper generation of the entropy rate; Walsh revealed that this assumption is speculative.

Skoda et al., (2007) used the realizable k-ε and the $\overline{v^2} - f$ models, to capture the transition effects on the suction side. The comparison with experimental data was made on the basis of another article where the pressure distribution and velocity profiles of boundary layer are considered.

1.3.1.4. Turbulent Zone

The major part of the flows in the turbomachines is turbulent with laminar region occurring close to blade leading edge. Turbulence influences the aerodynamic and thermodynamic analysis Keranen, (2003). Usually the turbulent region on the suction side is not fully influenced by the free flow turbulence. The blade convex curve can sometimes be important. However, a convex curve has a stabilizing effect on turbulence and reduces the fluctuations. The flow more strongly accelerates on the convex side and this gives a decrease in pressure and heat transfer. The heat transfer can be reduced from 15 to 20 % on a convex surface compared with the case of a plane plate Larsson, (1996)

1.4. Choc Wave Boundary Layer Interaction

In the transonic flows, the viscosity effects have a very important role and their influence on the flow field is very significant, in particular, under the supercritical conditions. The shock wave can be formed in the invicid region with strong interactions close to the shock root where results a quick thickness of the boundary layer, followed by a possible separation when the shock is strong.

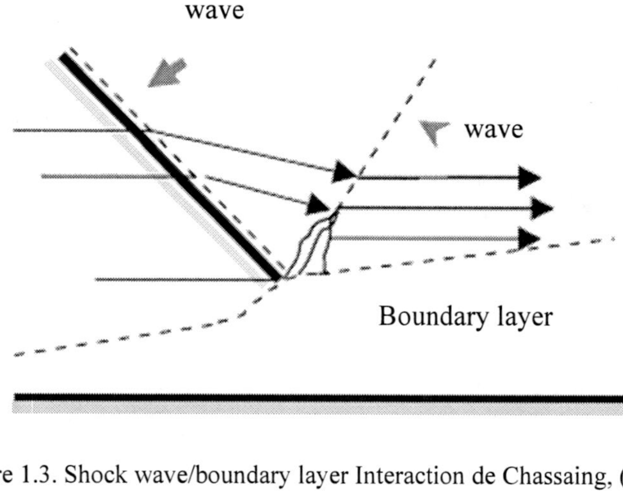

Figure 1.3. Shock wave/boundary layer Interaction de Chassaing, (2000).

The study of the shock wave/boundary layer interaction is as important as that of the transition. So Kozel et al., (2003) made a comparison between experimental and numerical results obtained by a study of a transonic flow through a blade cascade, for that he used the Baldwin-Lomax algebraic model, the one equation model Spallart Allmaras, the two equations model SST and the RNG *k-ε* implemented in *Fluent. He* found that the interferogram visualization of the shock wave/boundary layer interaction corresponds to that given numerically. Therefore appearance of a fine boundary layer without separation upstream the trailing edge which becomes thicker after the reflexion of the shock wave.

Figure 1.4. Shock wave/boundary layer Interaction. Interferogram Visualisation by Kozel et al, (2003).

In the same context, for the same geometry and the same conditions Mei and Guha, (2005) proposed moreover, one specific idea and which is traduced by a grid refinement in regions upstream the trailing edge in order to determine the thickness of the shock wave. This technique allowed to show that the shock wave becomes finer.

Quéré et al., (2003) have as an objective to improve the prediction of a model in the case of massive separations, this two transport equations model is based on the Boussinesq theory, the difference consists in modifying the expression of μ_t by the addition of a function of the turbulent fields .such as $\mu_t = C_\mu f_\mu (\rho k^2) / \varepsilon$. This being made, in order to study the correction effect of these *k-ε models* to compare the shock wave/boundary layer interaction for two modes transonic and supersonic.

Pagella et al., (2002) studied stability near shock wave/ laminar boundary layer interaction and the small amplitudes of the disturbances by a linear theory of stability and the direct numerical simulation (DNS). These results showed clearly the infiltration of the shock wave in the boundary layer by the flow parameters determination such as the density gradients.

Conclusion

The framework of the study was first specified. One is reminded that the comprehension of certain technological problems involved in the relaxed flows so complicated to model especially by an individual work, justifies us to presents the most of phenomenon induced by the flow in this domain. We also stuck so that it comes out from this work, an aspect more or less general I e collecting the most of researchers results and problems which they meted allowing the interested reader to easily implement the methodologies suggested, with a view to a possible technology transfer towards a maximum efficiency of transonic gas turbines.

References

Abu-Ghannam B. J., Shaw R., *1980* "Natural Transition of Boundary Layers-The Effect of Turbulence, Pressure Gradient, and Flow History", *IMechE,* Vol. 22, No. 5, pp. 213-288.

Alexis Bourgoing, 2002, "Instationnarité et Dissymétrie d'un Ecoulement Supersonique Décollé dans une Tuyère Plane", Thèse présentée pour l'obtention d'un Doctorat, Université Paris 6, Paris.

Ameri A. A., Arnone A., *April 1996* "Transition Modeling Effects on Turbine Rotor Blade Heat Transfer Predictions", *Journal of Turbomachinery*, Vol. 118 No. 1, pp. 307-313.

Arnone A., Pacciani R. et Sestini A., *1994* "Multigrid Computations of Unsteady Rotor-Stator Interaction Using the Navier-Stokes Equations", AD-Vol.40, ASME Winter Annual Meeting, Chicago, pp. 87-95.

Blair M. F, *1994* "An Experimental Study of Heat Transfer in a Large-Scale Turbine Rotor Passage", *ASME Journal of Turbomachinery*, Vol. 116 (1) pp. 1-13.

Bredberg J., *2002* "Turbulence Modeling for Internal Cooling of Gas-Turbine Blades", *Ph.D Thesis, Dep. Of Thermo and Fluid Dynamics, Chalmers University of Technology, Göteborg, Sweden.*

Boyle, R.& Russell, L., 1990 "Experimental determination of stator endwall heat transfer". *Journal of Turbomachinery.* **112**,547-558.

Chassaing P., *2000* "Mécanique des fluides, Eléments d'un premier parcours", *Collection Polytech de l'I.N.P. de Toulouse,* CEPADUES EDITIONS.

Chernobrovkin A., A. et Lakshminarayana B., July 1999 "Numerical Simulation of Complex Turbomachinery Flow", NASA / CR – 1999-209303 PSU CGTP 9901.

Davidson L., *1995* "An Efficient and Stable Solution Procedure of Compressible Turbulent Flow on General Unstructured Meshes Using Transport Turbulence Models", *AIAA 95-0342, 33rd Aerospace Sciences Meeting and Exhibit*, January 9-12, /Reno, NV.

Djouimaa S., Messaoudi L., Giel P.W., March 2007 " Transonic turbine blade loading calculations using different turbulence models-effects of reflecting and non-reflecting boundary conditions" *Science Direct Applied Thermal Engineering.* **27**,pp. 779-787

Djouimaa S., June (2007) " Simulation de l'écoulement transsonique turbulent autour d'une aube de turbine "thèse de doctorat d'état en génie mécanique, option énergétique, Universite de Batna, Algeria.

Durbin P., *1995* "On the *k-ε* stagnation point anomaly", *international journal of heat and fluid flow* **17**, *pp. 89-90.*

Emmons H.,W., *1951* "The Laminar-Turbulent Transition in Boundary Layer, Part 1", *J. of Aero. Sci.,* Vol. 18, pp. 490-498.

Fransson J. H. M., Matsubara M. et Alfredsson P.H., 2005 "Transition induced by free-stream turbulence", *Journal of Fluid Mechanics,* **527** Cambridge university press, pp.1-25.

Giel P. W., Thurman D. R., Lopez I., Boyle R. J., Van Fossen G. J. Jett T. A., Camperchioli. P. W. et La H., *1996* "Three-Dimensional Flow Field Measurements in a Transonic Turbine Blade Cascade", *ASME paper* 96-GT-113, presented at the *ASME International* Gas Turbine Conference, June 10-13, Birmingham, England.

Giel P. W., Van Fossen G. J., Boyle R., J., Thurman.D., R. et Civinskas K. C., *June 1999* "Blade Heat Transfer Measurements and Predictions in a Transonic Turbine Cascade", *ASME paper 99*-GT-125, presented at the ASME International Gas Turbine Conference, Indianapolis, IN, 7-10.

Graham C. G. et Kost F. H., *1979* "Shock Boundary Layer Interaction on High Turning Transonic Turbine Cascades", *ASME paper.* 79-GT-37.

Hu ,J and. Fransson, T.H., 2000, "Numerical Performance of Transition Models in Different Turbomachinery Flow Conditions*: A Comparative Study", ASME Paper* 2000-GT-520.

Johnson Mark W., *May 2006* "Predicting Transition On Concave Surfaces"*, Proceedings of GT ASME Turbo Expo 2006: Power for Land, Sea and Air Barcelona, Spain* 8-11,pp. 1-7

Kiock R., Lehthaus F., Baines N. C. et Sieverding C. H., 1986 "The Transonic Flow Through a Plane Turbine Cascade as Measured in Four European Wind Tunnels", *ASME J. of Engr. for Gas Turbines and Power,* Vol. 108, pp. 277-284.

Khezzar, R.(2005)" Simulation de l'écoulement transsonique aube à aube dans une turbine à gaz"

Thèse de Magistère en physique énergétique, Département de physique, Université ELHADJ LAKHDAR- Batna, Algéria

Koyabu Eitaro, Funazaki Ken-ichi et Kimura Manabu, *2005* "Experimental studies on wake induced bypass transition of flat plate Boundary layers under favorable and Adverse pressure gradients", *JSME International Journal.* Vol. 48, No. 3, pp. 579-588.

Larsson, J., Eriksson L. E. et Hall Ufl., *September 1995* "External Heat Transfer Predictions in a Supersonic Turbines Using The Reynolds Averaged Navier-Stokes Equations", *In Proc. 12th ISABE Conference, Melbourne,* Vol. 2, (Copies distributed by AIAA), pp. 1102-1112.

Larsson J., *1996* "Turbine Blade Heat Transfer Calculations Using Two-Equation Models", Department of Thermo and Fluid Dynamics, Chalmers University of Technology S-412 96 Gothenburg, Sweden, pp. 1-10.

Larsson J, 1998, "Numerical Simulation of Turbulent Flows for Turbine Blade Heat Transfer Applications", Department of Thermo and Fluid Dynamics Chalmers University of Technology S41296 Goteborg, Sweden, 1998.

Laumert Björn, Märtensson Hans Fransson, Torsten H., *May 2000* "Investigation of the Flow-Field in the Transonic VKI Brite EURAM Turbine Stage with 3D Steady and Ansteady NS Computations", *Proceedings of ASME TURBOEXPO 2000, Munich, Germany,* 8-11, pp. 1-11.

Launder B. E. and Spalding D. E., *1974* "The Numerical Computation of Turbulent Flows", *Computer Methods in Applied Mechanics and Engineering,* Vol. 3, pp.269-289.

Maciejewski P. K. et Moffat R., J., *1992* "Heat Transfer with Very High Free-Stream Turbulence: Part i and ii", *ASME Journal of Heat Transfer*, Vol. 114 pp. 227-239.

Mayle R. E., *1991* "The Role of Laminar-Turbulent Transition in Gas Turbine Engines", *J. of Turbomachinery*, Vol. 113, pp. 509-537.

Mayle R. E. et Paxson D. E., *1991* "A New Look at Transition", *J. of Fluid Mechanics.*

Mee D. J., Baines N. C., Oldfield M. L. G. et Dickens T. E., *1992a* "An Examination of the Contributions to Loss on a Transonic Turbine Blade in Cascade", *ASME Journal of Turbomachinery*, Vol. 114 No. 1 pp. 155-162.

Quéré Stéphane, Galli Arnaud, Benay Richard et Bur Reynald, *Septembre 2003* "Validation d'un modèle k-e faiblement non-linéaire", *16ème Congrès Français de Mécanique,* Nice, 1-5, pp. 1-6.

Schlichting H., *2000* "Boundary Layer Theory", *8th Revised and Enlarged Edition. Springer.*

Schmidt R. C. et Patankar S.V., *1991* "Simulating Boundary Layer Transition With Low-Reynolds-Number k-εTurbulence Models: Part 1 - An Evaluation of Prediction Characteristics", *Journal of Turbomachinery,* Vol. 113, January, pp. 1-26.

Sharma O. et Butler T., *1987* "Predictions of endwall losses and secondary flows in axial flow turbine cascades", *Journal of Turbomachinery* **109**, pp. 229-236.

Simon F.F. et Ashpis D. E., 1996 "Progress in Modeling of Laminar to Turbulent Transition on Turbine Vanes and Blades", *Int. J. of Heat Transfer and Fluid Flow* aussi publié à la NASA TM 107180.

Skoda Romuald, Schilling Rudolf et Schobeiri Meinhard T., *2007* "Numerical Simulation of the Transitional and Unsteady Flow Through a Low Pressure Turbine", *International Journal of Rotating Machinery,* Vol. 2007, Article ID 10940, pp.1-11.

Stieger R. D. et Hodson H. P., *2003* "The Transition Mechanism Of Highly-Loaded LP Turbine Blades", *Proceedings of ASME Turbo Expo power for land, Sea and Air Atlanta, Georgia*, USA, pp.1-10

Suzen Y. B., Huang P. G., Volino R. J., Corke T. C., Thomas F. O., Huang J., Lake J. P., King P. I., *June 2003* "A comprehensive CFD Study of Transitional Flows in Low-Pressure Turbines under a wide Range of operating Conditions", *33rd AIAA Fluid Dynamics Conference Orlando, Florida* 23-26, pp. 1-33

Sveningsson A. et Davidson L., *2003* "Assessment of Realizability Constraints and Boundary Conditions in v^2-f Turbulence Models", *Turbulence, Heat and Mass Transfer* **4**, Hanjalic,Y.Nagano and M. Tummers (Editors), Begell House, Inc, pp. 585-592.

Sveningsson A.,(2003) "Analysis of Performance of Different $\overline{v^2}$-f Turbulence models in a stator vane passage flow" thesis for the degree of Licentiate in Engineering, Department of Thermo and Fluid Dynamics, Chalmers University of Technology Goteborg, Sweden.

Thole K. A. et Bogard D. J., *1995* "Enhanced Heat Transfer Due to High Free-Stream Turbulence", *ASME Journal of Turbomachinery*, Vol. 117 pp. 418-424

Vijay K. Garg et Ameri Ali A., *June 2001* "Two-Equation Turbulence Models of Heat Transfer on a Transonic Turbine Blade", *Prepared for the 2001 Turbo Expo Sponsored by the American Society of Mechanical Engineers New Orleans, Louisiana,* pp.1-16.

Vogt D. M. et Fransson T. H., *Sep. 2002* "A New Turbine Cascade for Aerodynamical Testing", *The 16th Symposium on Measurement Techniques in Transonic and Supersonic Flow In Cascades and Turbomachines*, Cambridge, UK.

Walsh Ed J., Davies Mark R. D. et McEligot Donald M., *August 2004* "On use of Entropy to Predict Boundary Layer Stability", *Journal of Entropy,* **6**(4), pp. 375-387.

Wang Chi R., *2005* "Navier-Stokes Computations with One-Equation Turbulence Model for Flows Along Concave wall Surfaces", NASA/TM—213830, pp. 1-16.

Winoto S. H., Zhang D. H. et Chew Y. T., July-August.2000 "Transition in Boundary layers on a concave surface*", Journal of Propulsion and Power,* Vol. 16, N°. 4.

Yumin Xiao et Amano R. S., *May 2000* "Analysis of Flow and Heat Transfer in the End-wall Region of a Turbine Blade Passage", *Proceeding of ASME TURBOEXPO 2000*, 2000-GT-211, Munich, Germany.

In: Thermal Engineering Research Developments
Editors: J. Evgova and O. Kostadinov, pp. 407-427
ISBN: 978-1-60741-497-1

Chapter 11

OPTIMIZATION OF REFRIGERATION CARNOT CYCLE MACHINE. CASE STUDY

Camelia Petre, Michel Feidt, Monica Costea and Stoian Petrescu
LEMTA, University "Henri Poincare" Nancy I, UMR, Cedex France

Abstract

Reverse cycle machines are useful as refrigerating systems, heat pumps or air conditioning units. These machines generally imply phase change of working fluid in two heat exchangers (evaporator and condenser). Accordingly, the corresponding cycle is near of a reverse Carnot one. In this chapter, an optimization of such a machine is proposed and analysed by taking into account the internal irreversibility of the machine and generalized convective-like form of heat transfer laws, in case of limited size of heat exchangers through a parameter equivalent to a total NTU that is to be allocated to the two exchangers.

Various constraints are considered for the optimization; obtained results are discussed and compared. Perspectives are evoked for further developments.

Introduction

Let's start this chapter with a brief recall of the Irreversible Thermodynamics beginning. During the 60s, a new direction in Thermodynamics appeared, namely the analysis of internally reversible heat engines considering finite heat transfer irreversibilities [1], [2]. It was renewed during the 70s when an expression for the endoreversible efficiency at maximum power output was developed [3].

Since then, the research in the field was mostly intended for analyzing endoreversible thermal engines, the main irreversibility considered being the heat transfer at finite temperature difference between the sources of the studied system and the working fluid [4], [5]. Generally, the associated heat fluxes obeyed linear heat transfer laws (Newton's law of heat transfer) [5].

As inverse cycle systems have applications in several domains, numerous valid contributions are dedicated to their study and optimization, too. When applied to the study of heat pumps, the same method does not exhibit an optimum due to no theoretical limitation for gas temperature at the hot end [6]. Previously, an expression for the coefficient of performance of a heat pump has been obtained corresponding to minimum power consumption for a required heating load [7]. Also, some complementary results have been obtained for a similar case, considering a given heat exchanger area to be allocated [8].

It followed an extensive study on the theory of irreversible heat transfer refrigeration systems [8]. Formulae for the coefficient of performance and cooling rate of an endoreversible refrigerating machine were derived [9] and also for maximum specific cooling load [10]. All mentioned authors considered linear heat transfer laws.

A forward step was to introduce the internal irreversibilities, firstly represented by an irreversibility ratio [9], when maximum cooling load was determined for a required coefficient of performance.

A review has been done [12] providing a new approach regarding the internal irreversibilities, by introducing into the model the internal entropy generation term as parameter. Also non-linear convective-like heat transfer laws, $\dot{Q} \sim (\Delta T)^n$, were considered. Two main non-linear heat transfer laws, convective and radiative-like ones ($\dot{Q} \sim \left(T_S^n - T^n\right)$), have been explored extensively [13], the linear heat transfer law representing a particular case.

Other papers dealt with the optimum allocation of heat exchanger inventory of irreversible air refrigeration cycles [14] when a given heat exchanger area to be distributed was considered.

The thermoeconomic performance of refrigerators and heat pumps has been investigated [16] when combined effect of internal irreversibility and irreversible heat transfer were taken into account. Optimal performances and design parameters that maximize the objective functions (cooling load for the refrigerator and heating load for the heat pump per total unit cost) were derived.

In this chapter we are presenting an analytical model for predicting general performance characteristics of an irreversible Carnot cycle machine [17],[18],[19],[20] applying the model to a refrigeration cycle and considering a complete set of optimization procedures. Several objective functions could be considered, namely maximum useful effect with imposed coefficient of performance, minimum power consumption or minimum total dissipation. Operational and dimensional constraints were introduced in the model. Also, the non-linear convective-like heat transfer laws are applied, since in practice this case could represent natural convection or boiling.

Internal and external irreversibilities of the cycle are taken into account too, by introducing the two terms corresponding to internal and respectively, total entropy generation. The last mentioned one, which results by applying the Second Law of Thermodynamics to the whole system composed by the cycle and the two heat reservoirs, allows us to put emphasis on minimum dissipation operating regime. By applying this proposed model to a refrigeration cycle Carnot machine for which dimensional or performance constraints are imposed, one can notice the existence of different optimal operating regimes and also a limited range for the possible values of the system variables and parameters. Thus, the method has applicability to the design of the refrigerating machines in terms of selection, design and optimization of the main parameters.

Mathematical Approach for a Refrigeration Carnot Machine

Machine Description

In figure 1, the *T-s* diagram of an irreversible Carnot cycle machine is represented; therefore all processes are represented as irreversible ones. For the two adiabatic processes, the irreversibility is illustrated by entropy generation, while for the isothermal ones it is represented by finite temperature heat transfer processes, as well as entropy created along these irreversible isothermal.

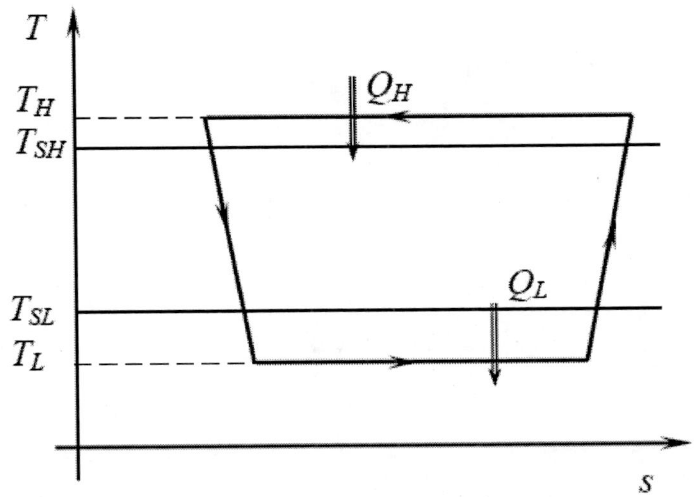

Figure 1. *T-s* diagram for an irreversible refrigeration Carnot cycle machine.

The internal irreversibility of the machine is represented by the corresponding internal entropy generation term, $\dot{S}_i$, as a model parameter. It takes into account the internal irreversibility generated on each cycle process, its value representing the sum of throttling losses, friction ones, etc. An external irreversibility is added, being generated by the finite temperature heat transfer process between the working fluid and the heat reservoirs.

We divide the mathematical development in two parts corresponding to two cases: the two heat reservoirs are considered to be thermostats (of constant temperatures T_{SH} and T_{SL}) or a second case when the two heat reservoirs are of finite thermal capacities. In this chapter we will detail the first case (thermostats). A steady state operating regime is considered.

Mathematical Equations

The proposed model is based on the most powerful and useful tools of Thermodynamics, namely the First and Second Laws of Thermodynamics, and also on the two heat transfer laws applied to the source and sink. For the considered hypotheses, the expression of the First Law of Thermodynamics is:

$$\dot{W} = \dot{Q}_H + \dot{Q}_L \tag{1}$$

where $\dot{W}$ represents the mechanical power [W]; $\dot{Q}$ is the heat flux [W] exchanged by the working fluid with the heat source (subscript *H*) and sink (subscript *L*), respectively.

The used sign convention is as follows: power consumption is considered negative and heat fluxes entering the cycle are considered positive, while leaving ones are negative.

The Second Law of Thermodynamics could be written in two different forms, function on the system to which it is applied. If it is applied to the endoirreversible machine, its expression is written taking into account the internal entropy generation term, $\dot{S}_i$ [W/K]:

$$\frac{\dot{Q}_H}{T_H} + \frac{\dot{Q}_L}{T_L} + \dot{S}_i = 0 \tag{2}$$

If it is applied to the whole system (cycle and the two heat reservoirs), it is written taking into account the total entropy generation term, $\dot{S}_T$ [W/K]:

$$\frac{\dot{Q}_H}{T_{SH}} + \frac{\dot{Q}_L}{T_{SL}} + \dot{S}_T = 0 \tag{3}$$

To these three above mentioned equations, the non-linear heat transfer laws between the heat reservoirs and working fluid are added. The general convective-like form is:

$$\dot{Q} = U_{gen} A \cdot (T_S - T)^n \tag{4}$$

where U_{gen} represents the "generalized" global heat transfer coefficient [$W/(m^2K^n)$] and *A* is the heat transfer surface [m^2].

Applied to the source, the heat transfer is written as:

$$\dot{Q}_H = U_{H,gen} A_H (T_{SH} - T_H)^{nH} (-1)^{nH-1} \tag{5}$$

while applied to the sink it becomes:

$$\dot{Q}_L = U_{L,gen} A_L (T_{SL} - T_L)^{nL} \tag{6}$$

As in this chapter we apply the model to refrigeration machines, the expression of the corresponding coefficient of performance, COP, is:

$$COP = \frac{\dot{Q}_L}{|\dot{W}|} \tag{7}$$

Combining the two expressions of the Second Law given by eq. (2) and (3), we get:

$$\dot{S}_T = \dot{S}_i + \dot{Q}_H\left(\frac{1}{T_H} - \frac{1}{T_{SH}}\right) + \dot{Q}_L\left(\frac{1}{T_L} - \frac{1}{T_{SL}}\right) \tag{8}$$

One can notice that the total entropy generation is a sum of three positive terms each of them taking into account one type of irreversibility: the first term – internal, the last two ones – external irreversibilities due to finite temperature heat transfer processes between the heat reservoirs and working fluid.

Generally, it is difficult to establish a variation law for the internal entropy generation. We have chosen to consider the following approaches [21]:

a) constant: $\dot{S}_i = const$ (9a)

b) linear variation law with temperature: $\dot{S}_i = const \cdot (T_H - T_L)$ (9b)

c) logarithmic variation law with temperature: $\dot{S}_i = const \cdot \ln\frac{T_H}{T_L}$ (9c)

Dimensionless Model Equations

In order to avoid cumbersome computations and to make the model easy to apply to any machine, it is preferred working non-dimensionly.

Since the working fluid is changing the phase (in between vapour and liquid) when passing through the evaporator and condenser, it is very important to make a proper choice for the reference state and the way of transforming dimensional parameters in dimensionless ones. Thus, we propose using the concept of Evaluation Criteria of Heat Transfer, denoted *CHT*, defined in a similar manner to the number of heat transfer units *NTU*, but taking as reference the product $\dot{m}c_p$ of the working fluid (refrigerant) in the superheated state (at the condenser inlet), denoted by $(\dot{m}c_p)_R$; the aim is to keep the same reference fluid for both heat echangers.

Since the heat transfer laws are non-linear, we transfer this non-linearity to the above introduced criteria *CHT*:

$$CHT_{gen} = \frac{(UA)T_{SH}^{n_H - 1}}{(\dot{m}c_p)_R} \tag{10}$$

The dimensionless procedure is chosen to be made relative to the Evaluation Criteria of Heat Transfer of the heat source, $CHT_{H,gen}$ and the heat source temperature (as environmental temperature), T_{SH}.

We use the following *dimensionless parameters and variables*:

- evaluation criteria of heat transfer corresponding to the two heat reservoirs:

$$CHT_{H,gen} = \frac{(UA)_H T_{SH}^{n_H - 1}}{(\dot{m}c_p)_R} ; \; CHT_{L,gen} = \frac{(UA)_L T_{SH}^{n_H - 1}}{(\dot{m}c_p)_R} \tag{11}$$

imposing a dimensionless constraint relative to a fixed heat exchanger area:

$$CHT_{T,gen} = CHT_{H,gen} + CHT_{L,gen} \tag{12}$$

which becomes a parameter of the model;

- temperature ratio with respect to the heat source temperature:

$$\theta_{SL} = \frac{T_{SL}}{T_{SH}} ; \; \theta_H = \frac{T_H}{T_{SH}} ; \; \theta_L = \frac{T_L}{T_{SH}} \tag{13}$$

- dimensionless heat fluxes:

at source

$$\tilde{q}_H = \frac{\dot{Q}_H}{(\dot{m}c_p)_R T_{SH}} = CHT_{H,gen}(1 - \theta_H)^{n_H}(-1)^{n_H - 1} \tag{14}$$

and sink:

$$\tilde{q}_L = \frac{\dot{Q}_L}{(\dot{m}c_p)_R T_{SH}} = CHT_{L,gen}(\theta_{SL} - \theta_L)^{n_L} = = (CHT_{T,gen} - CHT_{H,gen})(\theta_{SL} - \theta_L)^{n_L} \tag{15}$$

- dimensionless mechanical power:

$$\tilde{w} = \frac{\dot{W}}{(\dot{m}c_p)_R T_{SH}} = \frac{\tilde{q}_L}{COP} \tag{16}$$

- dimensionless internal and total entropy generation:

$$\tilde{s}_i = \frac{\dot{S}_i}{(\dot{m}c_p)_R} ; \; \tilde{s}_T = \frac{\dot{S}_T}{(\dot{m}c_p)_R} \tag{17}$$

Accordingly to the above mentioned sign convention and to the operating principle of any refrigerating system,

$$T_L < T_{SL} < T_{SH} < T_H ,$$

which implies the following constraints:

$$\theta_L < \theta_{SL} < \theta_{SH} < \theta_H \,;\; \theta_{SH} = 1 \,;$$

$$\widetilde{q}_H < 0 \;;\; \widetilde{q}_L > 0 \;;\; \widetilde{w} < 0 \;;\; \widetilde{s}_i > 0 \qquad (18)$$

These constraints are always true for any refrigerating system.

Optimisation under Constraints

The dimensionless variables of the model are chosen to be θ_L, θ_H and CHT_L, while the considered parameters are: θ_{SL}, CHT_T, $\widetilde{s}_i$ and the imposed performance characteristics (COP, refrigerating load, minimum dissipation or minimum power consumption).

In this chapter, we are presenting the case when a certain value for *COP* is required to exemplify the proposed methodology.

In the case of imposed *COP*, the objective functions could be: maximum refrigerating load, minimum power consumption or minimum total dissipation (minimum total entropy generation).

In the case of imposed refrigerating load, the objective functions could be: maximum *COP*, also corresponding to minimum power consumption or minimum total dissipation (minimum total entropy generation).

The analytical development for the optimization procedure is based on Lagrange Multipliers Method for which we wrote the Lagrangian function as:

$$L = FO + \lambda_1 C_1 + \lambda_2 C_2 \qquad (19)$$

where: *FO* is one of the considered objective functions stated above; λ_1 and λ_2 are the corresponding Lagrange multipliers; C_1 and C_2 are the problem constraints represented by the First Law of Thermodynamics including the imposed performance characteristic restrictions and Second Law of Thermodynamics applied to the cycle including the internal entropy generation restriction.

Imposed *COP* and Internal Entropy Generation $\widetilde{s}_i$

We will look for two objective functions: maximum refrigerating load and respectively, minimum total entropy generation. By applying the Lagrange Multipliers Method, we obtain a system of equations for each objective function. Solving this system of equations, we get the optimum values for the system variables.

LAGRANGE Multipliers Method

The Lagrangian associated to the problem is given by equation (19). The optimum values for the system variables are obtained by solving the following system of equations:

$$\left.\frac{\partial L}{\partial \, variable}\right)_{1->n\_variables} = 0\,;\ \frac{\partial L}{\partial \lambda_1} = 0\,;\ \frac{\partial L}{\partial \lambda_2} = 0 \tag{20}$$

Combining the First Law of Thermodynamics with equation (16) and taking into consideration the mathematical relations for the generalized heat exchange fluxes, one obtains:

$$CHT_{H,gen}(1-\theta_H)^{n_H}(-1)^{n_H-1} + CHT_{L,gen}(\theta_{SL}-\theta_L)^{n_L}\frac{1+COP}{COP} = 0 \tag{21}$$

One can generalize the above relation. It results the first constraint-equation (C1):

$$\boxed{CHT_{H,gen} f + CHT_{L,gen}\frac{1+COP}{COP}\xi = 0}\ (C_1) \tag{22}$$

where we have introduced the following auxiliary functions:

$$f = (1-\theta_H)^{nH}(-1)^{nH-1}$$
$$\xi = (\theta_{SL}-\theta_L)^{nL} \tag{23}$$

One can notice that f is a function of θ_H, while ξ is a function of θ_L.

In the above relation we take into account the dimensionless constraint relative to a fixed heat exchanger area, namely equation (12).

By introducing the mathematical relations for the generalized heat exchange fluxes in the mathematical expression of the Second Law of Thermodynamics applied to the cycle, it results:

$$\frac{CHT_{H,gen}(1-\theta_H)^{n_H}(-1)^{n_H-1}}{\theta_H} + \frac{CHT_{L,gen}(\theta_{SL}-\theta_L)^{n_L}}{\theta_L} + \tilde{s}_i = 0 \tag{24}$$

We use the auxiliary functions introduced above and obtain the second constraint-equation (C2):

$$\boxed{CHT_{H,gen}\frac{f}{\theta_H}+CHT_{L,gen}\frac{\xi}{\theta_L}+g=0}\ (C_2) \tag{25}$$

where g is a function of θ_H and θ_L according to the assumptions (9a-c).

Let us consider the first objective function – the refrigerating load:

$$FO=\tilde{q}_L=CHT_{L,gen}(\theta_{SL}-\theta_L)^{n_L}=CHT_{L,gen}\xi \tag{26}$$

The Lagrangian becomes:

$$\begin{aligned}&CHT_{L,gen}\xi+\\&+\lambda_1\left[(CHT_{T,gen}-CHT_{L,gen})f+CHT_{L,gen}\frac{1+COP}{COP}\xi\right]+\\&+\lambda_2\left[(CHT_{T,gen}-CHT_{L,gen})\frac{f}{\theta_H}+CHT_{L,gen}\frac{\xi}{\theta_L}+g\right]=0\end{aligned} \tag{27}$$

We compute the derivatives according to relations (20) for the considered variables of the system ($CHQ_{L,gen}$, θ_H, θ_L):

$$\frac{\partial L}{\partial CHT_{L,gen}}=0:\ \xi+\lambda_1\left[-f+\frac{1+COP}{COP}\xi\right]+\lambda_2\left[-\frac{f}{\theta_H}+\frac{\xi}{\theta_L}\right]=0 \tag{28}$$

$$\begin{aligned}&\frac{\partial L}{\partial\theta_H}=0:\\&\lambda_1(CHT_{T,gen}-CHT_{L,gen})f'+\lambda_2\left[(CHT_{T,gen}-CHT_{L,gen})\frac{\theta_H f'-f}{\theta_H^2}+g'_H\right]=0\end{aligned} \tag{29}$$

where g'_H represents the derivative of function g with respect to θ_H;

$$\begin{aligned}&\frac{\partial L}{\partial\theta_L}=0:\\&CHT_{L,gen}\xi'+\lambda_1 CHT_{L,gen}\frac{1+COP}{COP}\xi'+\lambda_2\left[CHT_{L,gen}\frac{\theta_L\xi'-\xi}{\theta_L^2}+g'_L\right]=0\end{aligned} \tag{30}$$

where g'_L represents the derivative of function g with respect to θ_L;

$$\frac{\partial L}{\partial \lambda_1} = 0 : \left(CHT_{T,gen} - CHT_{L,gen}\right)f + CHT_{L,gen}\frac{1+COP}{COP}\xi = 0 \quad (31)$$

$$\frac{\partial L}{\partial \lambda_2} = 0 : \left(CHT_{T,gen} - CHT_{L,gen}\right)\frac{f}{\theta_H} + CHT_{L,gen}\frac{\xi}{\theta_L} + g = 0 \quad (32)$$

Equations (28)-(32) give the system of equations to be solved in order to obtain the optimum values for the system variables. The procedure to solve this sytem of equations is the following:

- we divide equation (30) by $CHT_{L,gen}\xi'$;
- we divide equation (28) by ξ;
- we determine the ratio $\frac{\lambda_1}{\lambda_2}$ by substracting the above obtained relations;
- we also determine the ratio $\frac{\lambda_1}{\lambda_2}$ from equation (29);
- we equal these two expressions for $\frac{\lambda_1}{\lambda_2}$ and get an equation (the first from the below system of equations);
- the last step is to "assembly" the system of equations:

$$\boxed{\begin{array}{l} \dfrac{f^2}{\theta_H^2 f'} - \dfrac{fg'_H}{CHT_{H,gen}f'} = \dfrac{\xi^2}{\theta_L^2\xi'} - \dfrac{\xi g'_L}{CHT_{L,gen}\xi'} \\ CHT_{H,gen}f + \dfrac{1+COP}{COP}CHT_{L,gen}\xi = 0 \\ CHT_{H,gen}\dfrac{f}{\theta_H} + CHT_{L,gen}\dfrac{\xi}{\theta_L} + g = 0 \\ CHT_{H,gen} = CHT_{T,gen} - CHT_{L,gen} \end{array}} \quad (33)$$

The system is solved by Newton-Raphson method with respect to $CHT_{L,gen}$, θ_H, θ_L, for each of the three g functions given by equations (9a-c) in a dimensionless form.

Let us consider now the second objective function – the total entropy generation:

$$FO = \tilde{s}_T = -\tilde{q}_H - \frac{\tilde{q}_L}{\theta_{SL}} = -CHT_{H,gen}f - CHT_{L,gen}\frac{\xi}{\theta_{SL}} \quad (34)$$

The Lagrangian becomes:

$$\begin{aligned}&-\left(CHT_{T,gen}-CHT_{L,gen}\right)f-CHT_{L,gen}\frac{\xi}{\theta_{SL}}+\\&+\lambda_1\left[\left(CHT_{T,gen}-CHT_{L,gen}\right)f+CHT_{L,gen}\frac{1+COP}{COP}\xi\right]+\\&+\lambda_2\left[\left(CHT_{T,gen}-CHT_{L,gen}\right)\frac{f}{\theta_H}+CHT_{L,gen}\frac{\xi}{\theta_L}+g\right]=0\end{aligned}\tag{35}$$

We compute the derivatives according to relations (20) for the considered variables of the system ($CTQ_{L,gen}$, θ_H, θ_L):

$$\begin{aligned}&\frac{\partial L}{\partial CHT_{L,gen}}=0:\\&f-\frac{\xi}{\theta_{SL}}+\lambda_1\left[-f+\frac{1+COP}{COP}\xi\right]+\lambda_2\left[-\frac{f}{\theta_H}+\frac{\xi}{\theta_L}\right]=0\end{aligned}\tag{36}$$

$$\begin{aligned}&\frac{\partial L}{\partial\theta_H}=0:\\&\left(CHT_{T,gen}-CHT_{L,gen}\right)f'(\lambda_1-1)+\lambda_2\left[\left(CHT_{T,gen}-CHT_{L,gen}\right)\frac{\theta_H f'-f}{\theta_H^2}+g'_H\right]=0\end{aligned}\tag{37}$$

$$\begin{aligned}&\frac{\partial L}{\partial\theta_L}=0:\\&-\frac{CHT_{L,gen}\xi'}{\theta_{SL}}+\lambda_1 CHT_{L,gen}\frac{1+COP}{COP}\xi'+\lambda_2\left[CHT_{L,gen}\frac{\theta_L\xi'-\xi}{\theta_L^2}+g'_L\right]=0\end{aligned}\tag{38}$$

$$\frac{\partial L}{\partial\lambda_1}=0:\left(CHT_{T,gen}-CHT_{L,gen}\right)f+CHT_{L,gen}\frac{1+COP}{COP}\xi=0\tag{39}$$

$$\frac{\partial L}{\partial\lambda_2}=0:\left(CHT_{T,gen}-CHT_{L,gen}\right)\frac{f}{\theta_H}+CHT_{L,gen}\frac{\xi}{\theta_L}+g=0\tag{40}$$

Equations (36) – (40) give the system of equations to be solved in order to obtain the optimum values for the system variables. The procedure to solve this sytem of equations is the following:

- we divide equation (38) by $-\dfrac{CHT_{L,gen}\xi'}{\theta_{SL}}$;
- we divide equation (37) by $(CHT_{T,gen} - CHT_{L,gen})f'$;
- we determine the ratio $\dfrac{\lambda_1}{\lambda_2}$ by substracting the above obtained relations;
- we divide equation (36) by $f - \dfrac{\xi}{\theta_{SL}}$;
- we divide equation (38) by $-\dfrac{CHT_{L,gen}\xi'}{\theta_{SL}}$;
- we determine the ratio $\dfrac{\lambda_1}{\lambda_2}$ by substracting the above obtained relations;
- we equal these two expressions for $\dfrac{\lambda_1}{\lambda_2}$ and get an equation (the first from the below system of equations);
- the last step is to "assembly" the system of equations:

$$\boxed{\begin{aligned}
&\frac{f^2}{\theta_H^2 f'} - \frac{fg'_H}{CHT_{H,gen}f'} = \frac{\xi^2}{\theta_L^2\xi'} - \frac{\xi g'_L}{CHT_{L,gen}\xi'}\\
&CHT_{H,gen}f + \frac{1+COP}{COP}CHT_{L,gen}\xi = 0\\
&CHT_{H,gen}\frac{f}{\theta_H} + CHT_{L,gen}\frac{\xi}{\theta_L} + g = 0\\
&CHT_{H,gen} = CHT_{T,gen} - CHT_{L,gen}
\end{aligned}} \tag{41}$$

which is solved by Newton-Raphson method with respect to $CHT_{L,gen}$, θ_H, θ_L.

Let us compare the two systems of equations obtained for the two considered objective functions. One can notice that we obtained the same system of equations nomatter the considered objective function. This means that the extreme value of the refrigerating load (maximum or minimum) is obtained for the same optimum values of the system variables (obtained from solving the system of equations) as for achieving the extreme value (maximum or minimum) of the total entropy generation. So:

$$\boxed{Extrema(\tilde{q}_L) \xleftrightarrow{coresponds} Extrema(\tilde{s}_T)} \tag{42}$$

Similarly, one can prove that

$$\boxed{Extrema(\widetilde{q}_L) \xleftrightarrow{coresponds} Extrema(\widetilde{s}_T) \xleftrightarrow{coresponds} Extrema(\widetilde{w})} \quad (43)$$

The development of the second case when the two heat reservoirs are of finite thermal capacities was previously reported [19].

Results

Some of the obtained results are presented in what follows, in a graphical manner, thus easily to be interpreted. The case corresponding to imposed *COP* and $\widetilde{s}_i$ constant is illustrated here after.

Linear Form of the Heat Transfer Law

In figure 2, the variations of dimensionless temperature of the refrigerant are plotted for the two physical solutions of the system of equations. These two solutions determine two operating regimes: one corresponding to maximum refrigerating load, the other one corresponding to minimum total entropy generation. As it was analytically proved, the extrema of these two characteristics are obtained for the same (optimal) values for the system variables.

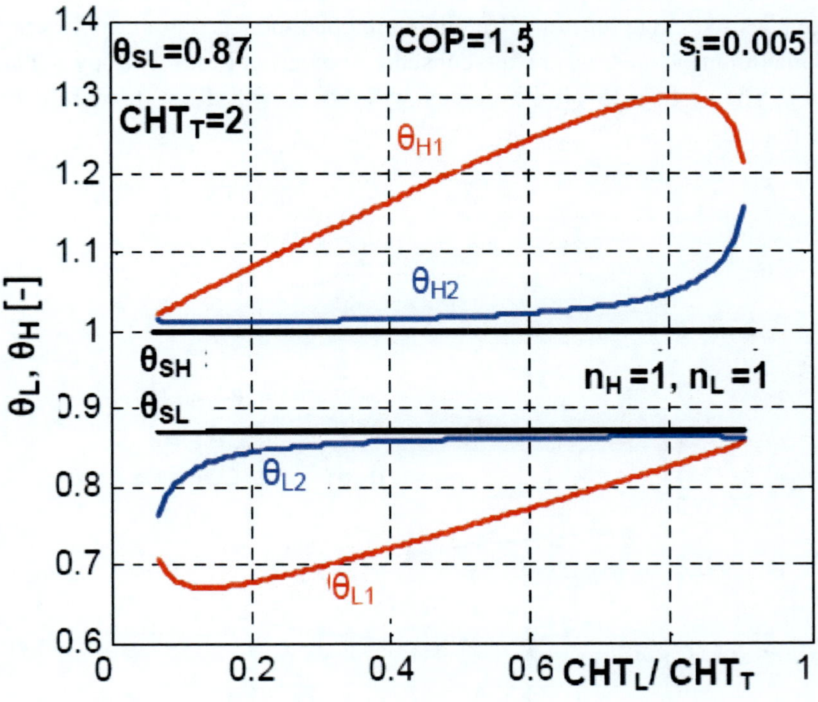

Figure 2. Dimensionless temperature of the refrigerant for the two solutions of the system of equations – case of imposed *COP*, $\widetilde{s}_i$ constant.

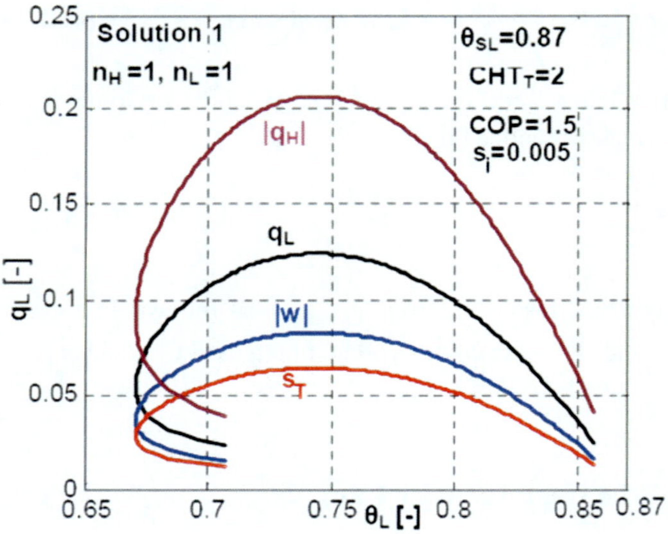

Figure 3. Dimensionless energetic fluxes and total entropy generation for the solution corresponding to maximum refrigerating load regime function of θ_L.

In figure 3, the dimensionless energetic fluxes and total entropy generation are plotted only for the solution corresponding to maximum refrigerating load regime, function of the system variable θ_L. One can draw very important information regarding the decision upon the values for the evaporation temperature, respectively evaporation pressure of the refrigerant. The operation desired range for low refrigerating temperatures is obviously the one in the left-side of the maximum values, where the consumed power and total entropy generation have lower values. The right-side of the curves will be preferred for moderate refrigerating temperature regimes.

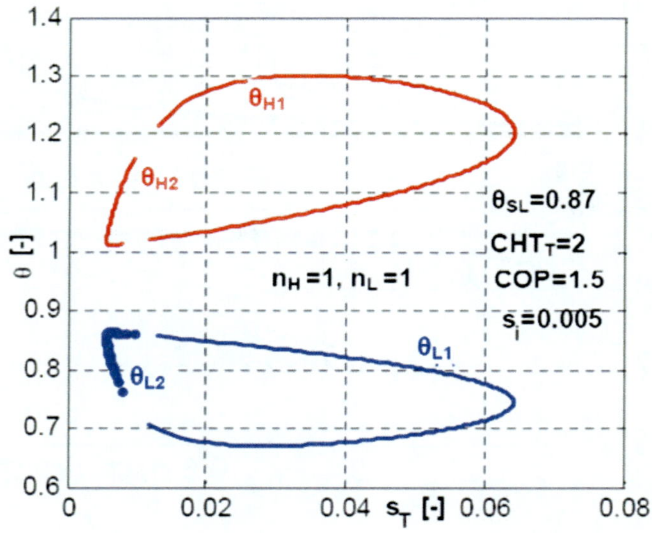

Figure 4. Refrigerant dimensionless temperatures for the two solutions of the equation system versus $\widetilde{s}_T$.

Figure 4 presents a comparison with the *T-s* representation of the cycle. One can notice the limited range of variation for the two refrigerant temperature levels (evaporation and condensation). Also, we notice that the second solution generates the least losses, but also the energetic fluxes have low values.

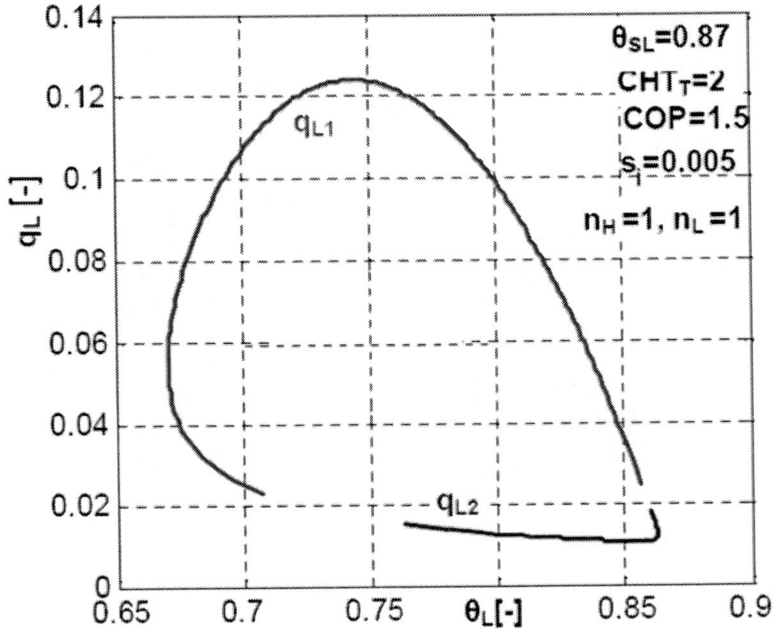

Figure 5. Dimensionless cooling load corresponding to the two operating regimes.

When the dimensionless cooling load is illustrated versus θ_L for both operating regimes, a loop shaped curve yields (see figure 5). It shows that for certain values of the dimensionless temperature θ_L, two values of the dimensionless cooling load correspond, a high one and a very small one. Also, the same value of the dimensionless cooling load is reached for a dimensionless temperature θ_L of the refrigerant closed to the reference temperature (environmental one) – the right side of the curves, as well as for a much lower value – the left side of the curves. Obviously, the operational regime is only one and it is determined by the values of other machine parameters and control variables (θ_H, CHT_L, etc).

A sensitivity study with respect to the machine coefficient of performance shows the effect of *COP* increasing on the variation range of system variables and performances (see figures 6 and 7). Thus, by imposing an operational regime with a high *COP* value, the cooling load decreases, and the same does the variation range of the refrigeration temperature or the ratio CHT_L/CHT_T.

Also, the internal entropy generation has a significant effect on the above mentioned performances and variables. The curves from figure 8 illustrate that higher value of the dimensionless internal entropy determine a less wide range of the refrigeration temperature and important decrease of the dimensionless cooling load values. However, the right side of the same curves that corresponds to regular refrigeration temperatures shows an insignificant effect of the internal entropy variation because of their superposition, while quite opposite effect is registered on the left side curves. It means that the internal entropy generation effect is not at all negligible in cryogenic applications.

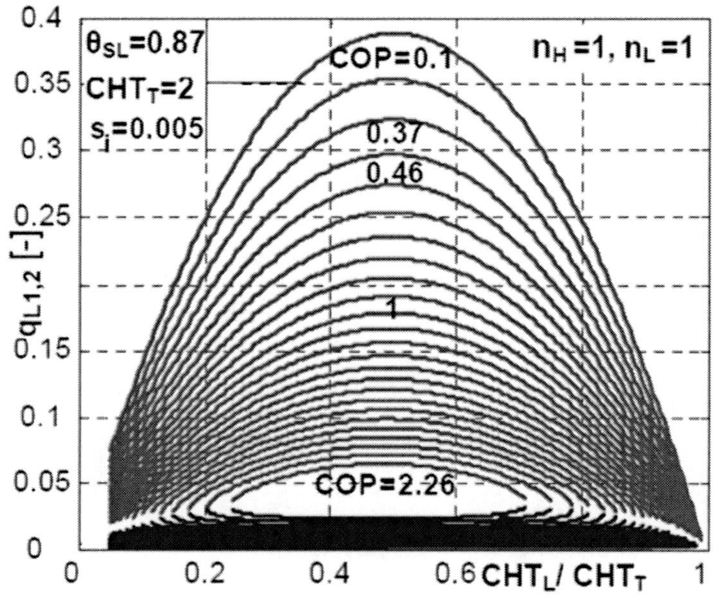

Figure 6. Dimensionless cooling load corresponding to the two operating regimes; sensitivity study with respect to *COP*.

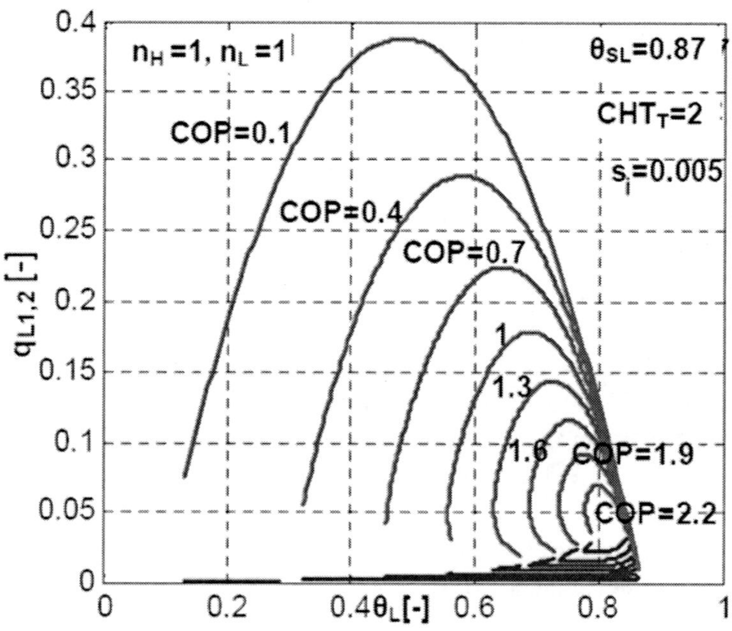

Figure 7. Dimensionless cooling load corresponding to the two operating regimes function on the control variable θ_L; sensitivity study with respect to *COP*.

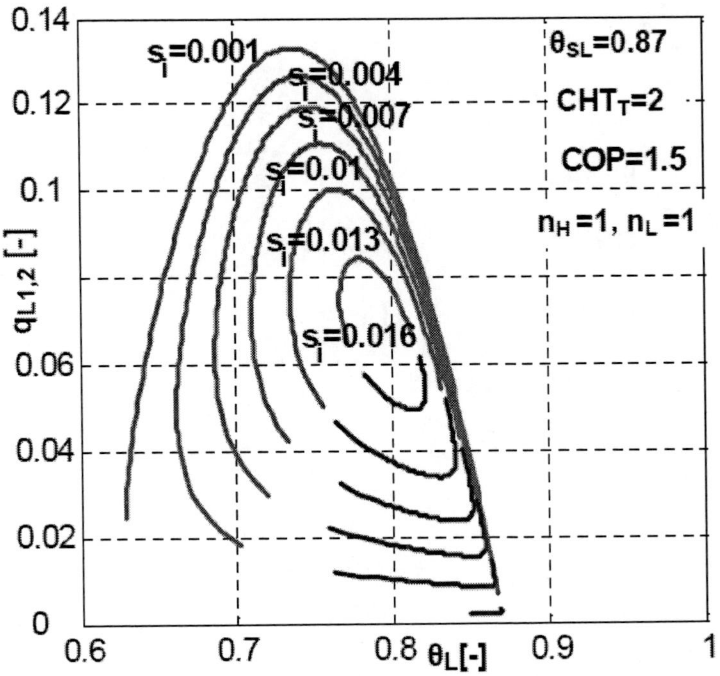

Figure 8. Dimensionless cooling load corresponding to the two operating regimes function on the control variable θ_L; sensitivity study for $\widetilde{S}_i$.

Non Linear Form of the Heat Transfer Law

Figures 9–12 illustrate same kind of results when non linear forms of the heat transfer laws in the machine heat exchangers are considered. Hence, the two curves from figure 9, one for each solution of the equation system, show the effect of the combination of linear (at the sink) with non linear (at the source) form of the heat transfer law on the extreme values of the performance correlated with the refrigeration temperature values. The comparison of the curves for *COP* = 1 from figure 7 and figure 9 indicates an important decrease of the refrigeration temperature (from 0.7 to 0.6 for solution 1, and from 0.82 to 0.67 for solution 2) corresponding to the maximum cooling load. Also the maximum value of the dimensionless cooling load decreases severely for the non linear heat transfer law form, namely from 0.18 in figure 7, to 0.026 in figure 9. One can say that this effect is more significant when the non linear heat transfer law is considered at the source heat exchanger (n_L = 3). Thus, the comparison of the maximum value of the dimensionless cooling load on the curves corresponding to *COP* = 0.46 in figures 10 and 11 clearly indicates its reduction from 0.28 to 0.095. Moreover, on the similar curve from figure 12, where symmetrical combination of the heat transfer laws appears, this maximum is only 0.05. Also, one notes that the part of the heat transfer area of the source heat exchanger (CHT_L/CHT_T) registered changes (see figures 6 and 10–12).

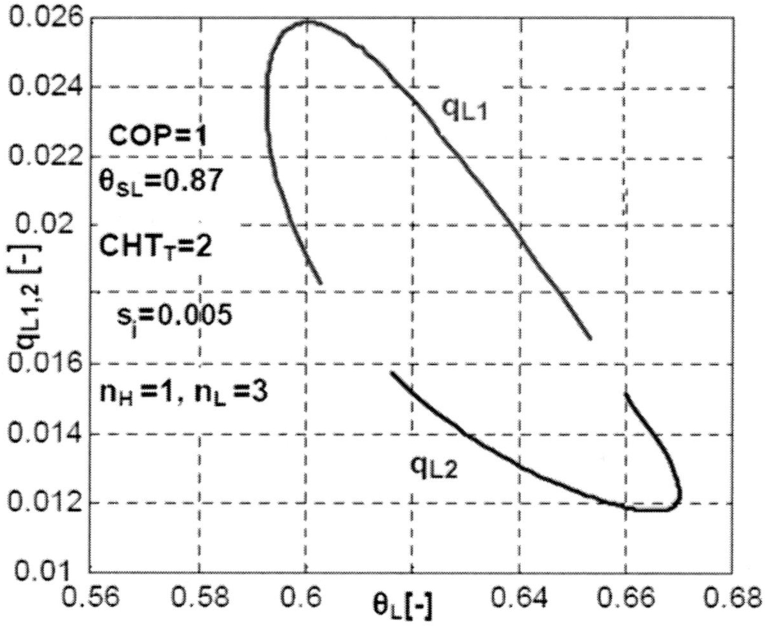

Figure 9. Dimensionless cooling load corresponding to the two operating regimes, non linear form of the heat transfer law in the sink heat exchanger.

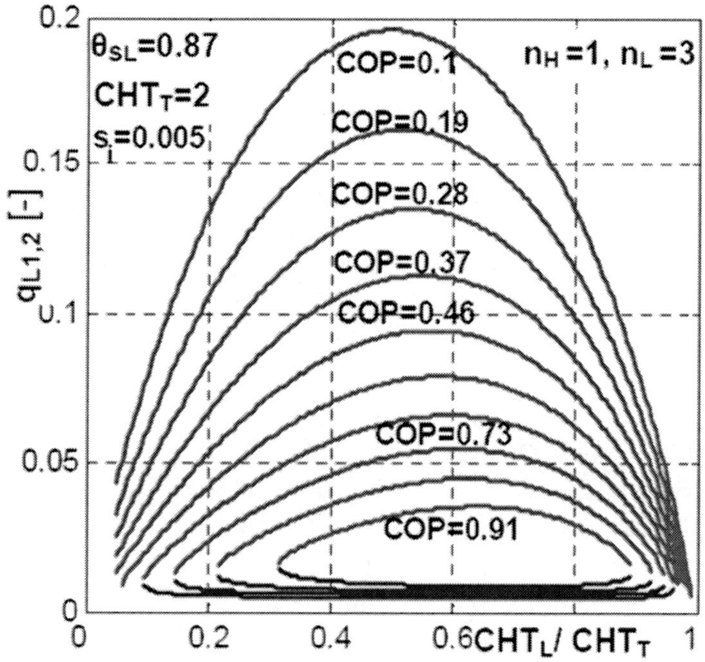

Figure 10. Dimensionless cooling load corresponding to the two operating regimes; sensitivity study with respect to *COP* and non linear form of the heat transfer law in the sink heat exchanger– case of imposed *COP*.

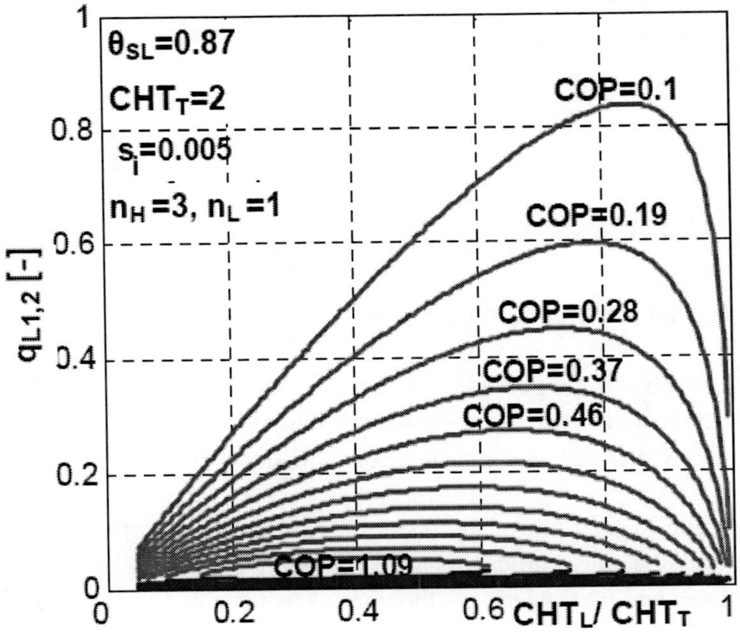

Figure 11. Dimensionless cooling load corresponding to the two operating regimes; sensitivity study with respect to *COP* and non linear form of the heat transfer law in the source heat exchanger – case of imposed *COP*.

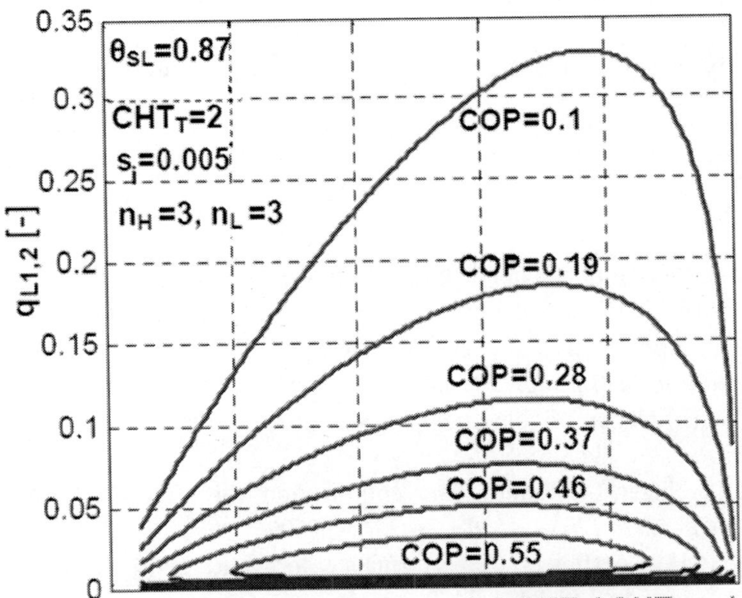

Figure 12. Dimensionless cooling load corresponding to the two operating regimes; sensitivity study with respect to *COP* and two non linear forms of the heat transfer laws.

The results of this study may have important consequences on the design of refrigerating machines, since the real operating regime in the two heat exchangers consists of phase change processes governed by non linear heat transfer laws.

Conclusions

An analytical model for the study and optimization of irreversible thermal machines applied here to Carnot refrigeration one is proposed. Two imposed constraints were considered, namely the variation law for internal entropy generation and the coefficient of performance.

The results have shown the possible variation range of model variables and parameters, as well as the heat exchanger surface correlation with machine performance and parameters helping the designer to make the appropriate choice in connection with space and economical constraints.

The effect of non linear form of the heat transfer laws in the machine heat exchangers was discussed. It pointed out significant changes of the machine performance correlated with those of the variables and parameters. As the non linear form of the heat transfer law comes close to the real operating conditions in the condenser and evaporator of a refrigerating machine, the conclusions of our study regarding the effects of the symmetrical and non symmetrical combinations of the heat transfer laws seems important.

The presented results confirm that the proposed model is a useful tool in optimization of real-operating machines in the design stage. Also, a similar model for the case when the two heat reservoirs are of finite thermal capacities was developed and the results were previously reported [18].

Presently extensions of the model for other operating regimes and different constraints are developed.

References

[1] Chambadal P. Évolution et applications du concept d'entropie. Paris: Dunod, 1963.

[2] Novikov I. The efficiency of atomic power stations. *Atommaya Energiya.* 1957;3:409.

[3] Curzon FL, Ahlborn B. Efficiency of a Carnot Engine at Maximum Power Output. *American Journal of Physics*. 1975;43:22-24.

[4] De Vos A. Efficiency of some heat engines at maximum power conditions. *American Journal of Physics.* 1985;53:570-573.

[5] Salamon P, Nitzan A. Finite Time optimizations of a Newton's law Carnot cycle. *Journal of Chemical Physics.* 1981;74:3546-3560.

[6] Blanchard CH. Coefficient of performance for finite speed heat pump. *Journal of Applied Physics.* 1980;51:2471-2472.

[7] Leff HS, Teeters WO. EER, COP and second law efficiency for air conditioners. *American Journal of Physics.* 1978; 46: 19-22.

[8] Bejan A. Theory of heat transfer irreversible refrigeration plant. *International Journal of Heat and Mass Transfer.* 1989; 32: 1631-1639.

[9] Agrawal DC, Menon VJ. Finite-time Carnot refrigerators with wall gain and product loads. *Journal of Applied Physics.* 1993; 74(4): 2153-2158.

[10] Wu C. *Maximum obtainable specific cooling load of a refrigerator. Energy Conversion and Management.* 1995; 36(1): 7-10.

[11] Assad M. Performance characteristics of an irreversible refrigerator. In: Recent Advanced in Finite Time Thermodynamics, Nova Science Publishers, New York, 2000.

[12] Feidt M. Advanced Thermodynamics of reverse cycle machine. In: Low Temperature and Cryogenic Refrigeration, Kluwer Academic Publishers, 2003.

[13] Feidt M. Thermodynamics and Optimization of Reverse Cycle Machines: Refrigeration, Heat Pumps, Air Conditioning, Cryogenics. In: Thermodynamics Optimization of Complex Energy Systems, Kluwer Academic Publishers, 1999.

[14] Luo J., Chen L., Sun F., Wu C. Optimum allocation of heat exchanger inventory of irreversible air refrigeration cycles. *Physica-Scripta.* 2002; 65(5): 410-415.

[15] Goth Y., Feidt M. Conditions optimales de fonctionnement des pompes à chaleur ou machine à froid associées à un cycle de CARNOT endoréversible. C.R. Acad. Sci. Paris 1986; t. 303, II(1) : 19-24.

[16] Kodal A., Sahin B., Yilmaz T. Effects of internal irreversibility and heat leakage on the finite time thermoeconomic performance of refrigerators and heat pumps. *Energy Conversion and Management.* 2000; 41(6):607-619.

[17] Feidt M., Petre C., Costea M., Petrescu S. General Model of Reversed Cycle Carnot Machine with Contraints and Non-Linear Heat Transfer Laws. In Proceedings of ECOS 2005, Trondheim, Norway, June 20-23, 2005, Vol. II, p. 529-536.

[18] Feidt M., Costea M., Petre C., Petrescu S. Optimization of Direct Carnot Cycle. *Applied Thermal Engineering.* 2007; 27:829-839.

[19] Petre C., Feidt M., Costea M., Petrescu S. A Model for Study and Optimization of Real-Operating Refrigeration Machines, International Journal of Energy Research (2008), published by John Wiley & Sons Ltd, ISSN (online): 1099-114X.

[20] Petre C., Utilisation de la Thermodynamique à vitesse finie pour l'etude et l'optimisation du cycle Carnot et des machines de Stirling. PhD Thesis 2007; University POLITEHNICA Bucharest and University HENRI POINCARE Nancy.

[21] Wijeysundera NE. Performance of three heat reservoirs absorption cycles with external and internal irreversibilities. *Applied Thermal Engineering.* 1997;17:1151-1161.

In: Thermal Engineering Research Developments
Editors: J. Evgova and O. Kostadinov, pp. 429-447
ISBN: 978-1-60741-497-1

Chapter 12

NUMERICAL COMPUTATION OF HEAT TRANSFER AND FLUID FLOW DURING FLOATING CRYSTAL GROWTH OF SILICON

***R. Younsi*[*1], *A. Harkati*[2], *M. Drir*[2] *and D.Ouadjaout*[3]**

[1] Applied Sciences Department, University of Quebec at Chicoutimi, G7H2B1 Chicoutimi (QC), Canada.
[2] Theoretical Physics Laboratory, Physics institute, USTHB, Algiers (Algeria).
[3] Simulation's Laboratory, UDTS, 2Bd Frantz Fanon, Algiers (Algeria).

Abstract

In this research article, we present numerical results concerning the simulation of semiconductor melts with free capillary surfaces, particularly silicon (Si) crystal growth by the floating zone method. The effect of fluid motion due to the gravity and surface tension is illustrated. The input power induced by the heat source is a Gaussian distribution and the surface tension is assumed dependent on temperature to models the heat experimental conditions. The mathematical model is a coupled system, which consists of axisymmetric heat equation and the Navier-Stokes equations in the melt with Marangoni boundary conditions and Boussinesq approximation. The partial differential equations obtained have been discretized by means of finite volume procedure, and the SIMPLER (Semi-Implicit Method for Pressure-Linked Equations Revised) algorithm is used to treat the coupled velocity-pressure equations. Different steady thermocapillary flows are obtained by varying the following significant parameters: temperature gradient conditions (Grashof number), surface tension gradient (Marangoni number), all of them relevant in practical crystal growth situations. Our results show that thermocapillary can have a quite significant effect on the stability of a primarily buoyancy-driven flow, as well as on the flow structures and dynamic behaviour.

Keywords: Heat transfer, Buoyancy-driven flow, Thermocapillary-driven flow, Crystal growth, Floating zone technique, Numerical simulation.

[*] E-mail address: ryounsi@uqac.ca, Tel: +1418-545-5011, Fax: +1418-545-5012

Nomenclature

A:	aspect ration of the floating zone
g:	gravity acceleration
Gr:	Grashof number (buoyancy forces on viscous diffusion ratio)
Ma:	Marangoni number (thermocapillary forces on thermal diffusion)
P:	dimensionless pressure
Pr:	Prandtl number (viscous on thermal diffusions ratio)
r, z:	radial and vertical coordinates
R, L :	radius and height of the floating zone
u, v:	dimensionless axial and radial velocities

Subscripts

m:	melt

Greek Symbols

ρ:	density
α:	thermal diffusivity
β:	expansion coefficient of the melt
ν:	kinematics viscosity
Θ:	dimensionless temperature
Ψ:	stream function
γ:	surface tension for interface

1. Introduction

The design and optimization of modern materials processing systems rely on the understanding of some basic mechanisms and their interactions. The most utilized crystal growth techniques from the melt are: the directional solidification (Bridgman technique); the crystal pulling technique named after Czochralski; the zone melting method, the floating zone method is one of variants of its vertical configuration. Figure 1 shows techniques commonly used for the crystal growth from the melt [1].

Until years 50, the main method used in crystal growth was Czochralski (CZ) method, making call for growing solid crystals from a melt [2-6]. However, the advent of semiconductors drove researchers to invent proceed which is adapted to the industrial needs. Within this context, of particular relevance is the Floating Zone (FZ) technique that seems to be one of the most promising candidates for obtaining such advanced materials. In comparison with float zone silicone, Czochralski silicone can be grown to very low resistivities. This is important for low and medium power applications, where the Czochralski silicone wafer serves as a carrier substrate for the actual components built on epitaxial material deposited on top of the Czochralski silicone wafer[7-9].

Floating zone technology for growing silicone monocrystals is by far the most pure method and results in silicone with unique properties as opposed to any other growth technology. Czochralski silicone does not have this high level of purity. For instance most of the very pure Silicon produced today is processed in this way. The main advantage of the float-zone process is that it is a container-less process and thus a main source of contamination is removed. Another advantage is the reduction of thermal stresses which are caused by differential thermal expansion between the crystal and the crucible.

Research, both experimental and theoretical, has been done in this field. The quantitative data from the experiments are limited, [10-13], because the experiments are difficult to design and the measurements are difficult to obtain (due to the opaqueness of semiconductor melts) and too expensive [14, 15]. Most theoretical [16] and numerical studies in this field, [17-21], are done for idealized geometries (the solid-melt interfaces and the free surface are assumed to be flat) and for rather limited operating conditions (the temperature difference across the zone is a prescribed quantity instead of the result of the heating mode).

Considerable attention has been paid in recent years to surface tension-driven convection in liquid floating zones and similar configurations under microgravity conditions. Kobayashi [2] computed thermocapillary convection under zero gravity conditions for two different Prandtl numbers, with and without crystal rotation while neglecting the deformation of the free surface. Kobayashi [6] extended the study to include buoyancy-induced convection. Young and Chait [22] has presented asymptotic solutions for the temperature, concentration and interface shapes in the limit of small aspect ratio and weak surface tension for a two-dimensional float-zone established in a vertical sheet. Neitzel et al. [23] have computed thermocapillary convection in a float-zone with a computed free surface shape. Chu and Chen [24] have obtained numerical solutions for float-zone growth of Molybdenum (Mo) crystals with an imposed heat input profile rather than an imposed temperature profile. Monti [25] investigated the onset of instability of Marangoni convection in space by using large liquid bridges with lengths of the order of several centimetres and in the absence of buoyancy effects. Recent experimental works have been performed by Petrov et al. [26], Muehlner et al. [27] and Schwabe et al. [28]. Lee et al.[29] have performed experiments to investigate oscillatory thermocapillary flow. They found that three-dimensional oscillatory flow appears when the Marangoni number exceeds a critical value.

In this method, a melted zone is obtained by melting, is maintained by its own superficial tension between two verticals solid and collinear. The superior cylinder is a polycrystalline solid, whereas the bottom one is a monocrystal, that growth as the melted zone recrystallise under the action of the vertical displacement of the lateral heating source. Two of the most important mechanisms are momentum transfer and heat transfer, which control such phenomena as fluid flow and solidification. The development of computational resources and numerical methods make computer modeling one of the major ways of studying this mechanism. In the industrial configuration of crystal preparation effected in soil, the melted phase is the seat of convective movement due to gravity and surface tension. The study of these coupled phenomena is revealed of big complexity. For this reason, we have examined every type of convection separately in order to understand the own dynamics of each.

Numerical simulation of heat and mass transfer in FZ crystal-growth configurations is a promising tool for investigating flow patterns and temperature fluctuations under different growth conditions.

The aim of this article is to characterize the flows of convection that could be observed. The numerical analysis of the flow in the melt is quite complex and simplified models are commonly used, where both the geometry and the thermal boundary conditions are simplified. A mathematical model for the flow and heat transfer during the floating-zone growth process is developed. The temperature and flow fields in the melt are obtained numerically. Natural and thermocapillary convection effects in the melt are included in the analysis.

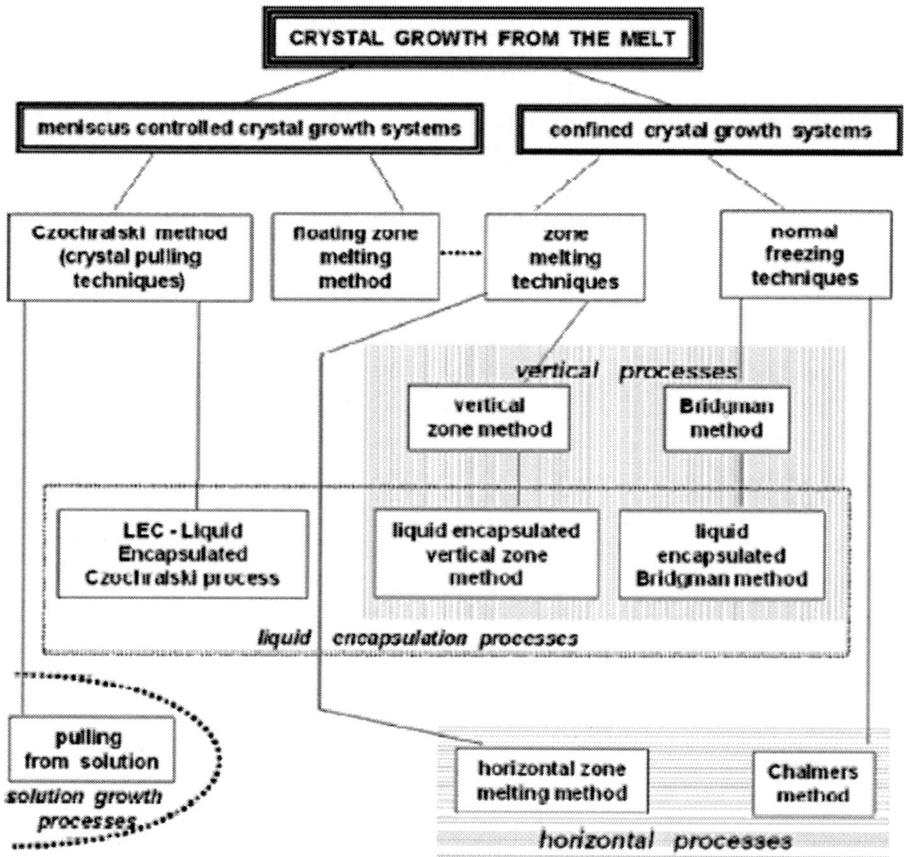

Figure 1. Main categories of crystal growth techniques from the melt [1].

2. Numerical Model and Method

The geometry of the problem is shown in figure 2. The floating zone is modeled as a cylindrical liquid bridge of radius R and length 2L, hold between two circular disks. A cylindrical coordinate system (r,Θ, z) with origin in the center of the zone is considered and is assumed to be axis-symmetric, $\partial/\partial\theta = 0$. The two disks that hold the melt are at temperature T_m. The known heat flux distribution on the free surface is supposed to be parabolic. This symmetric distribution is appropriate in the case of resistance furnaces, where the heating can be symmetric, but is not so in the case of monoellipsoidal mirror furnaces, where the heating is not symmetric. The simulation is based on the Boussinesq approximation. Thus, we study the motion of a laminar incompressible fluid in the presence of thermal buoyancy forces. The heat generated by viscous dissipation in the melt is neglected, the transport coefficients are assumed to be temperature independent.

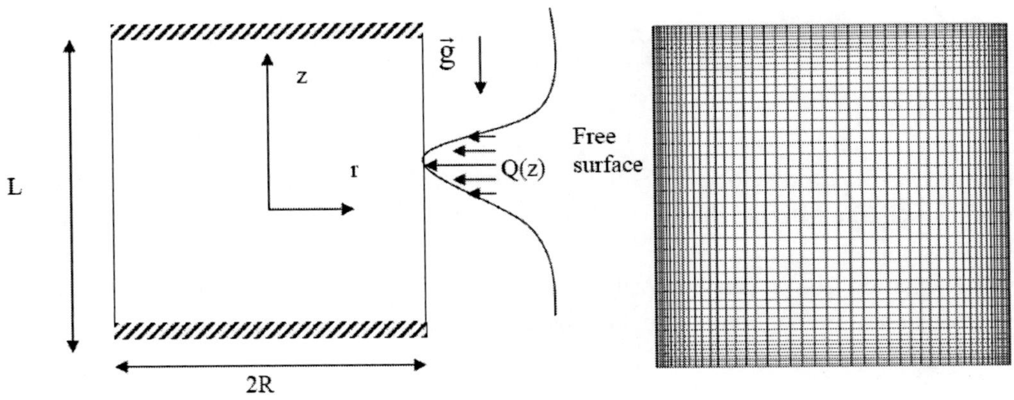

Figure 2. Schematic diagram of the floating zone system and mesh.

The non-dimensional quantities and parameters are defined as follows:

$$r^* = \frac{r}{2R},\ z^* = \frac{z}{2R},\ u^* = \frac{u}{\alpha/2R},\ v^* = \frac{v}{\alpha/2R}, \Theta = \frac{T - T_0}{\Delta T},\ \Delta T = Q2R/k,\ p^* = \frac{p}{(\alpha/2R)^2}$$

$$Gr = \frac{g\beta\Delta T(2R)^3}{\nu^2}, \Pr = \frac{\nu}{\alpha},\ A = \frac{L}{2R}$$

The superscript " * " denotes the non-dimensional quantities and will be omitted hereafter

The non-dimensional steady state, continuity, momentum, and energy equations governing the flow and heat transfer are given by:

Conservation of mass:

$$\frac{1}{r}\frac{\partial}{\partial r}(ru)\frac{\partial}{\partial z}(v) = 0 \quad (1)$$

Conservation of momentum:

r-momentum

$$\frac{1}{r}\frac{\partial}{\partial r}\left(ru^2\right) + \frac{\partial}{\partial z}(vu) = -\frac{\partial p}{\partial r} + \left[\frac{1}{r}\frac{\partial}{\partial r}\left(r\frac{\partial u}{\partial r}\right) + \frac{\partial^2 u}{\partial z^2} - \frac{u}{r^2}\right] \quad (2)$$

z-momentum

$$\frac{1}{r}\frac{\partial}{\partial r}(ruv)+\frac{\partial}{\partial z}\left(v^{2}\right)=-\frac{\partial p}{\partial z}+\left[\frac{1}{r}\frac{\partial}{\partial r}\left(r\frac{\partial v}{\partial r}\right)+\frac{\partial^{2}v}{\partial z^{2}}\right]+Gr.\Theta \tag{3}$$

Conservation of energy:

$$\frac{1}{r}\frac{\partial}{\partial r}(ru\Theta)+\frac{\partial}{\partial z}(v\Theta)=\frac{1}{\Pr}\left[\frac{1}{r}\frac{\partial}{\partial r}\left(r\frac{\partial\Theta}{\partial r}\right)+\frac{\partial^{2}\Theta}{\partial z^{2}}\right] \tag{4}$$

where (u, v) is the fluid velocity, P is the pressure and Θ is the temperature. Gr is the Grashof, number which characterise the intensity of convection, Pr is the Prandtl number, A is the aspect ratio

The equation above can be written in the following generalized from to facilitate the integration and discretization (symbol explanation is referred to nomenclature):

$$\frac{\partial\Phi}{\partial t}+\frac{\partial}{\partial z}(u\Phi)+\frac{1}{r}\frac{\partial}{\partial r}(rv\Phi)=\frac{\partial}{\partial z}\left(\Gamma_{\Phi}\frac{\partial\Phi}{\partial z}\right)+\frac{1}{r}\frac{\partial}{\partial r}\left(r\Gamma_{\Phi}\frac{\partial\Phi}{\partial r}\right)+S_{\Phi} \tag{5}$$

where Γ_{Φ} and S_{Φ} are shown in table 1.

Table 1. Governing differential equations

Φ	Γ_{Φ}	S_{Φ}
1	0	0
u	1	$-\frac{\partial P}{\partial z}+\frac{\partial}{\partial z}\left(\frac{\partial v}{\partial z}\right)+\frac{1}{r}\frac{\partial}{\partial r}\left(r\frac{\partial u}{\partial z}\right)+Gr\Theta$
v	1	$-\frac{\partial P}{\partial r}+\frac{\partial}{\partial z}\left(\frac{\partial V}{\partial r}\right)+\frac{1}{r}\frac{\partial}{\partial r}\left(r\frac{\partial u}{\partial r}\right)$
Θ	1/Pr	0

In order to complete the mathematical model, we still use, in addition to equ. 5, the appropriate boundary conditions. The non-dimensional boundary conditions are written as

- on the bottom surface : $\left(z=0;\ 0\leq \mathrm{r}<1\right)$

$$\mathrm{u{=}v{=}}\Theta=0 \tag{6}$$

- on the top surface : $\left(z=A;\ 0\leq \mathrm{r}<1\right)$

$$\mathrm{u{=}v{=}}\Theta=0 \tag{7}$$

- on the free surface (right): $\left(r=1\ ;\quad 0 \leq z \leq A\right)$

$$v=0\ ,\ \frac{\partial u}{\partial r}=-\frac{Ma}{Pr}\frac{\partial\Theta}{\partial z}.(1\text{-}(z\text{-}1)^2)^2,\ \ \frac{\partial\Theta}{\partial r}=(1\text{-}(z\text{-}1)^2)^2=0 \tag{8}$$

- on the symmetry axis (left): $\left(r=0\ ;\quad 0 \leq z \leq A\right)$

$$v=0\ ,\ \frac{\partial u}{\partial r}=0\ \ ,\ \ \frac{\partial\Theta}{\partial r}=0 \tag{9}$$

The dimensionless stream function is evaluated in order to ascertain the strength of the convective motion and is defined in the usual way:

$$v=-\frac{1}{r}\frac{\partial\psi}{\partial z}\ \text{and}\ u=\frac{1}{r}\frac{\partial\psi}{\partial r} \tag{10}$$

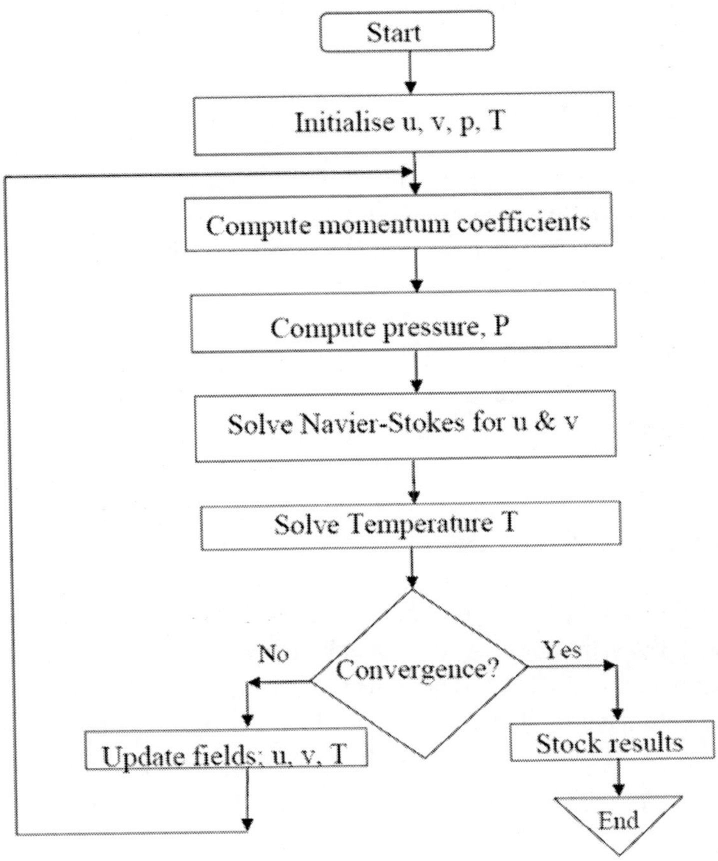

Figure 3. Simulation algorithm used for convection loop.

A control volume based discretization method, described by Patankar [30] and Nicolás et al. [31] has been applied for solving the system of equ.5, together with the boundary and interfacial conditions (6-9). The most attractive feature of the control volume based discretization scheme is that the resulting solution would imply that integral conservation of physical quantities such as mass, momentum and energy is exactly satisfied, over any group of control volumes and, of course, over the whole calculation domain. In evaluating convective and diffusive fluxes, a power-law scheme and the SIMPLER algorithm were considered. Use of 81x81 non-uniform grid (figure 2) permits the presence of several discretization points near the interface without introducing an excessive number of grid points, which would induce high CPU cost.

The choice of the convergence criterion is particularly important. The variable convergence criteria may be estimated as follows:

$$\left|\frac{u-u_{old}}{u_{old}}\right|_{\max} \le 1.0\times 10^{-6}, \left|\frac{v-v_{old}}{v_{old}}\right|_{\max} \le 1.0\times 10^{-6}, \left|\frac{\Theta-\Theta_{old}}{\Theta_{old}}\right|_{\max} \le 1.0\times 10^{-6} \tag{11}$$

where "old" denotes the previous iteration and "max" means the maximum value of the absolute value. Figure 3 shows the simulation algorithm used for convection loop.

3. Simulation Results

In this section we present numerical results for the case of silicon, which has a Prandtl number Pr= 0.015; the rest of its physical properties can be found in Cröll et al. [32]. With respect to the geometry, the case of sample radius R=5mm and aspect ratio A=2 is considered, which is characteristic of the floating-zone technique in mirror and resistance furnaces.

We analyse the character of the melt flow and the temperature field. Streamlines, velocity vector and isotherms are presented in the steady state case. The ranges of the other parameters studied herein are: $10^3 \le Gr \le 10^6$; $10 \le Ma \le 10^3$.

3.1. Effect of Buoyant Convection

We first consider the influence of natural convection on the flow when thermocapillary forces are absent by setting Ma = 0. Natural convection is induced by an adverse or destabilising temperature gradient. The effect of the gravity is represented through the Grashof number. We have reported in figures 3.1-3.4 the flow pattern for several values of Grashof number (Gr). For very low values of Gr (not presented here), no flow is detected in the melt (Ψ_{max}~0); the heat transfer is still dominated by conduction. As shown, one large vortex, which rotates counterclockwise in direction, dominates the flow field. It is caused by the ascending flow near the free surface the melt is directly heated by the outer heat source. Since the buoyancy grows larger as the increase of gravitational acceleration (Grashof number), the intensity of natural convection becomes highest at Gr=10^6 (Ψ_{max}=240).

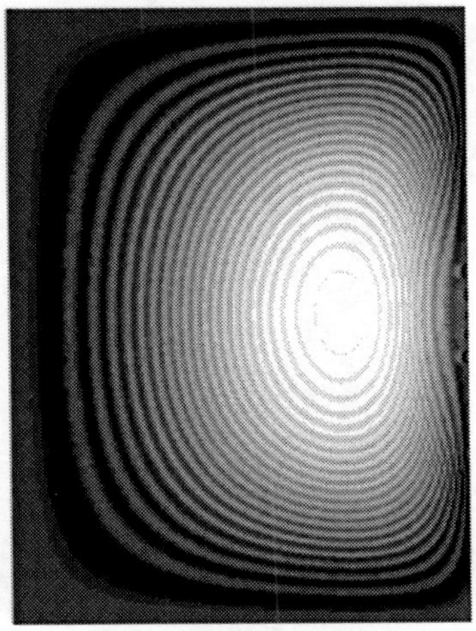

A) $\Psi_{max}=1.0$

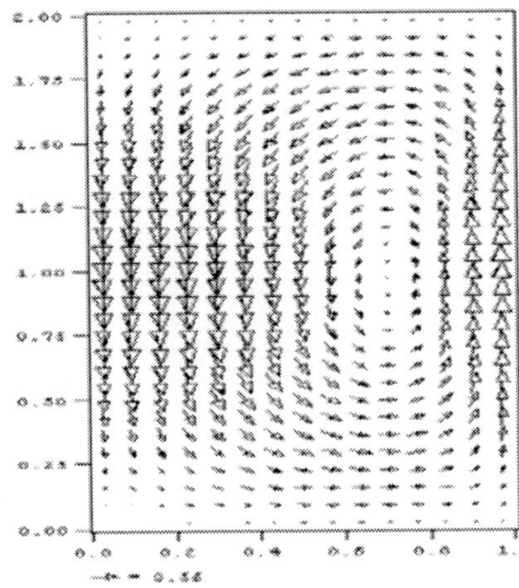

B)

C)

Figure 3.1. Effect of buoyant convection on: A) Streamlines, B) velocity vector, C) Isotherms. Pr=0.015, A=2, Ma=0, $Gr=10^3$.

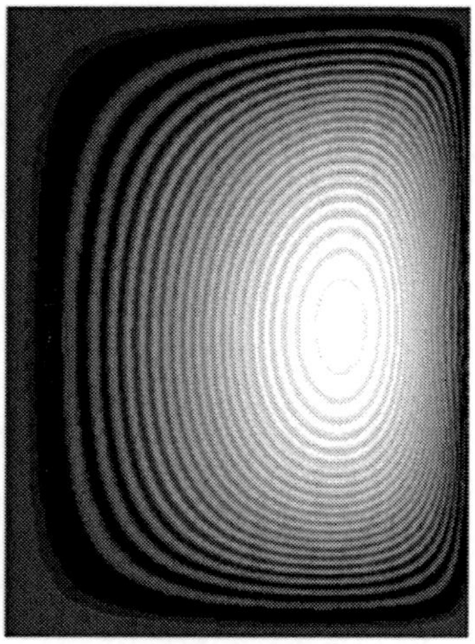

A) Ψ_{max}=8.5

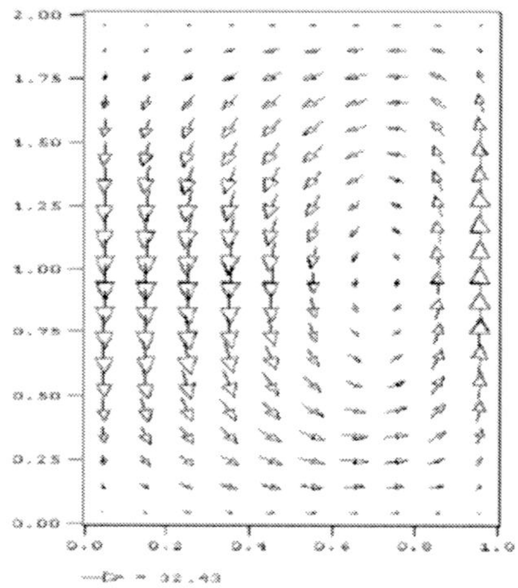

B)

C)

Figure 3.2. Effect of buoyant convection on: A) Streamlines, B) velocity vector, C) Isotherms. Pr=0.015, A=2, Ma=0, Gr=10^4.

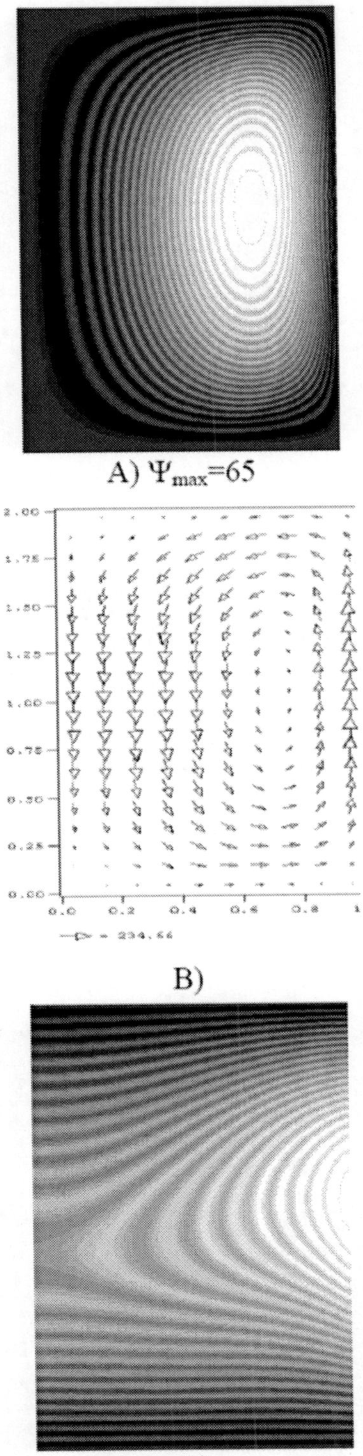

Figure 3.3. Effect of buoyant convection on: A) Streamlines, B) velocity vector, C) Isotherms. Pr=0.015, A=2, Ma=0, Gr=10^5.

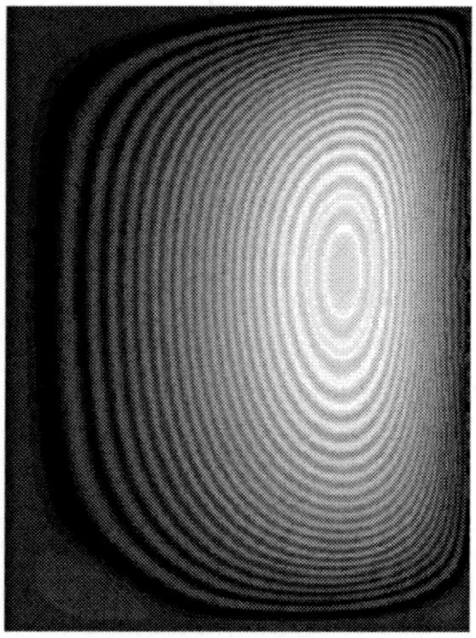

A) Ψ_{max}=240

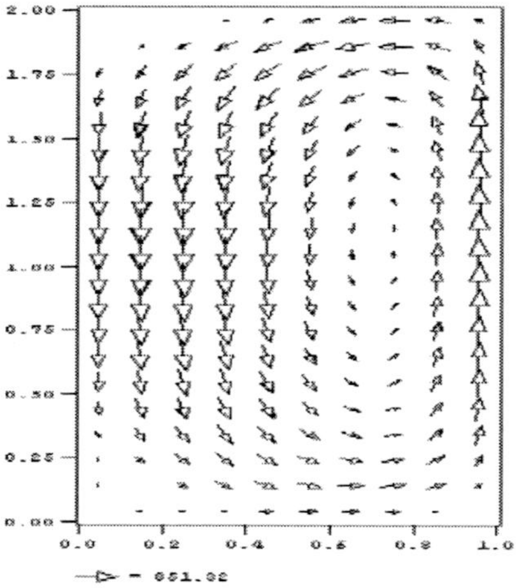

B)

C)

Figure 3.4. Effect of buoyant convection on: A) Streamlines, B) velocity vector, C) Isotherms. Pr=0.015, A=2, Ma=0, Gr=10^6.

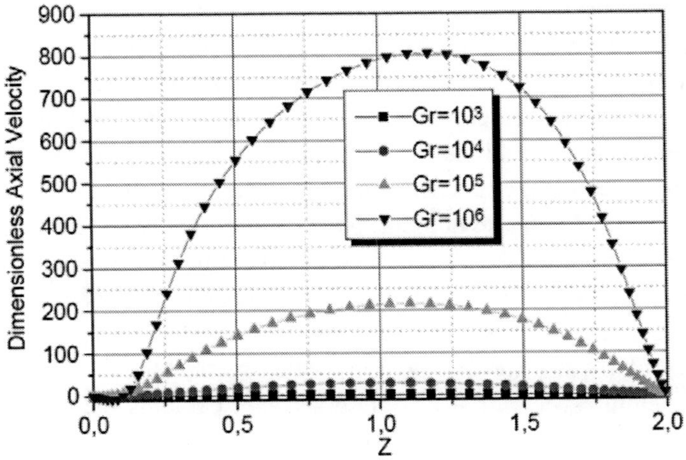

Figure 4. Profile of dimensionless axial velocity at the free surface versus Grashof number, Pr=0.015, A=2, Ma=0.

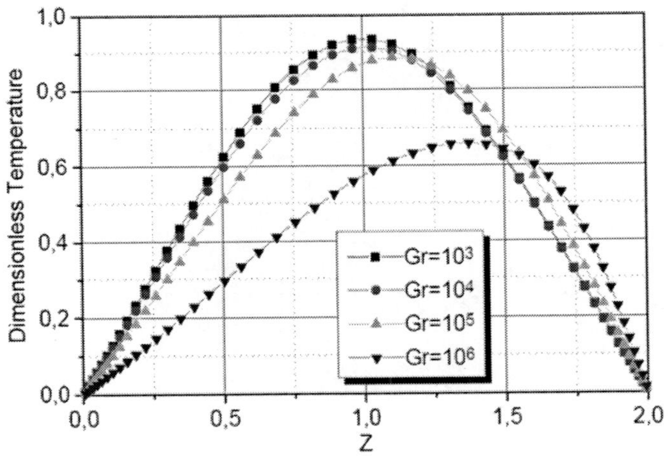

Figure 5. Profile of dimensionless temperature at the free surface versus Grashof number, Pr=0.015, A=2, Ma=0.

This is confirmed by the number of isolines in the stream function, which increases with the magnitude of Grashof number. On the contrary, the maximum temperature becomes lowest at $Gr=10^6$ due to a strong stir. The thermal convection destroys the symmetry with respect to the horizontal midplane. This observation is confirmed in figure 4 and figure 5, in which we have reported the axial velocity and temperature profiles at the free surface versus Grashof number. The maximum of temperature is slightly displaced to the top of the crystal.

3.2. Effect of Thermocapillary Convection

In this calculation, we consider only the effects of tension forces; a non-buoyant forces is imposed by setting Gr=0. This convection is obtained in experiences achieved in microgravity environment. Figures 6.1-6.3 illustrate typical streamlines, isotherms and velocity vector patterns for Gr=0 and Ma=10,10^2 and 10^3, respectively. Two vortices occupy the flow field as shown in the figures, and are symmetrical with respect to the central height of the melt zone due to no buoyancy force. The vortices on the top and the bottom, which rotate counterclockwise and clockwise, respectively, are formed by the Marangoni effect on the free surface. For the temperature field it doesn't present a difference with conductive field. As the Marangoni number increases, the flow filed intensifies and the convective strength that comes with the increases of the Marangoni number modifies slightly the temperature filed. Since the fluid with high temperature on the free surface is moved to the low-temperature regions by the gradient of surface tension, the temperature gradient becomes steeper in the vicinity of the Marangoni convection. As a result, the rims of the free surface are gradually moved to the solid phases (horizontal walls, top and bottom) which maintained at a constant temperature. The coupling temperature-velocity at r=1 (free surface) explains the evolution of the axial velocity that is more and more reduced near of these cold walls as depicted in figures 7-8.

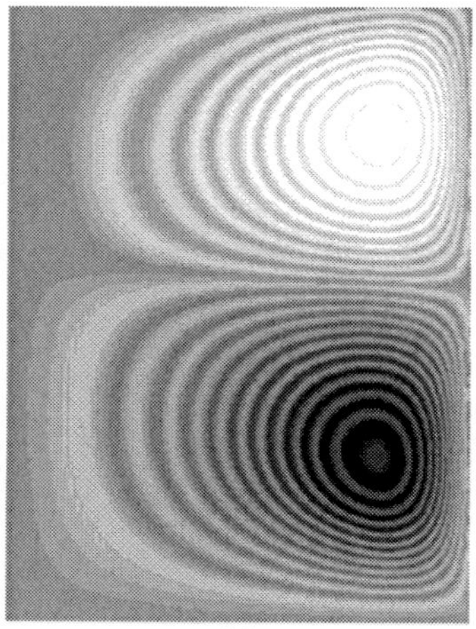

A)

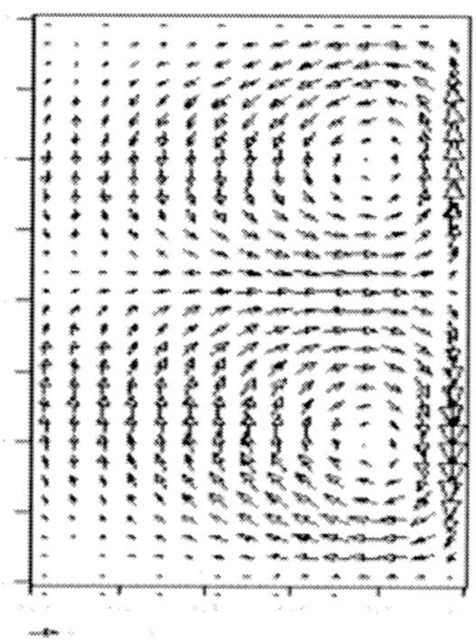

B)

Figure 6.1. Continued on next page.

C)

Figure 6.1. Effect of thermocapillary convection on: A) Streamlines, B) Velocity vector, C) Isotherms. Pr=0.015, A=2, Gr=0, Ma=10.

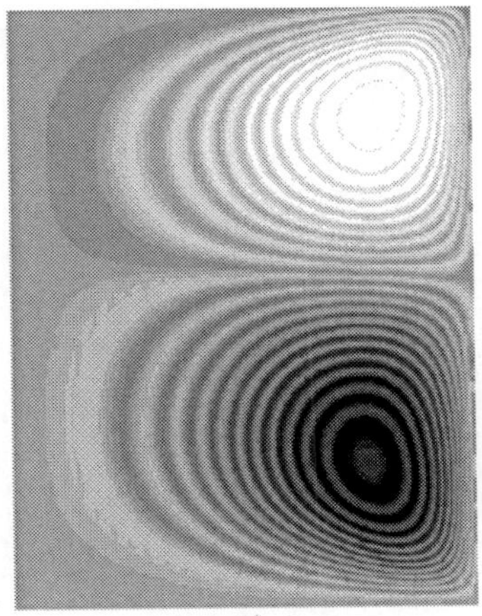

A)

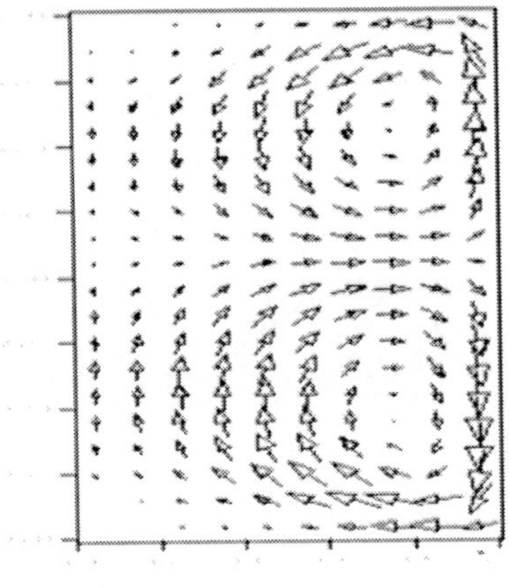

B)

Figure 6.2. Continued on next page.

C)

Figure 6.2. Effect of thermocapillary convection on: A) Streamlines, B) Velocity vector, C) Isotherms. Pr=0.015, A=2, Gr=0, Ma=10^2.

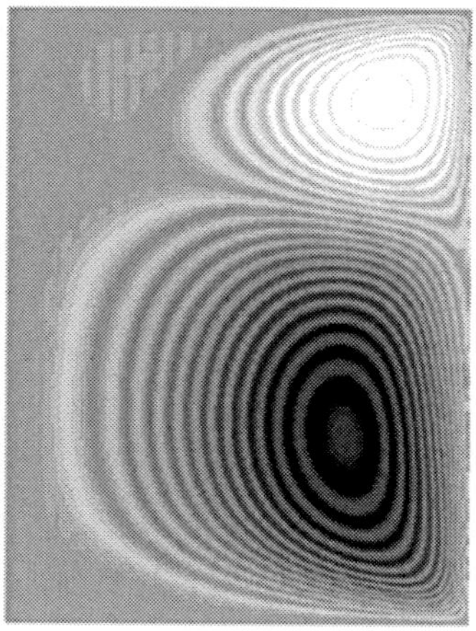

A)

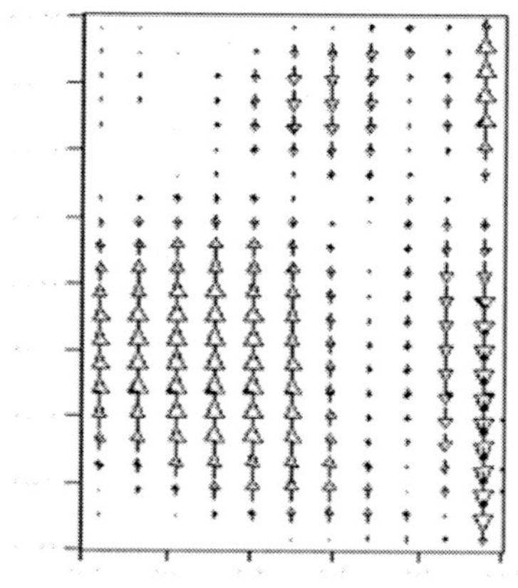

B)

Figure 6.3. Continued on next page.

C)

Figure 6.3. Effect of thermocapillary convection on: A) Streamlines, B) Velocity vector, C) Isotherms. Pr=0.015, A=2, Gr=0, Ma=10^3.

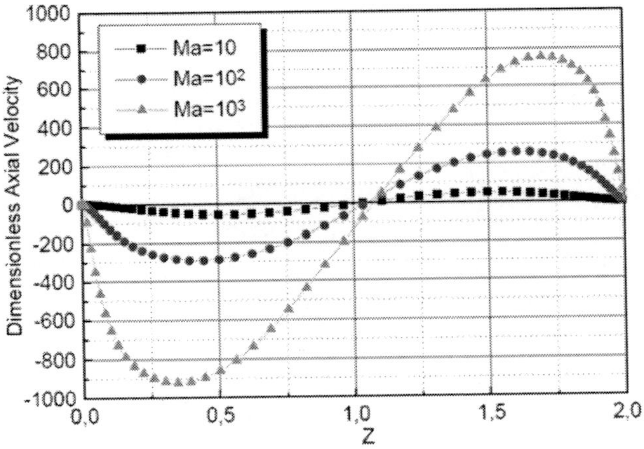

Figure 7. Profile of dimensionless axial velocity at the free surface versus Marangoni number, Pr=0.015, A=2, Gr=0.

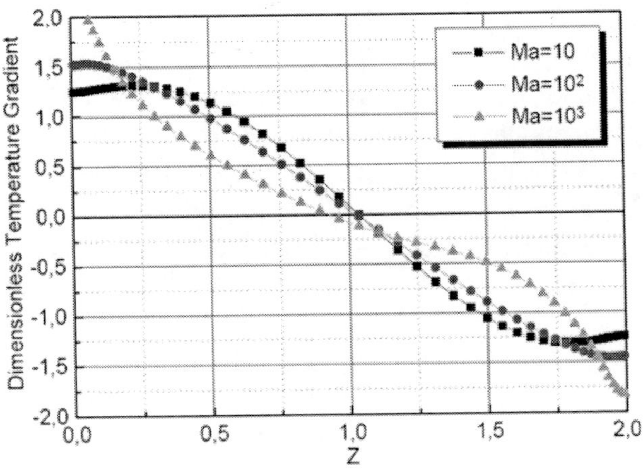

Figure 8. Profile of dimensionless temperature gradient at the free surface versus Marangoni number, Pr=0.015, A=2, Gr=0.

3. Conclusion

Flow and heat transfer in the melt, during floating zone growth of silicon, have been studied numerically. The model considered here, with a condition of flux and no temperature represents a new approach for the applied conditions of heating. We have analysed the structures of the flow of pure Marangoni convection and gravitational convection. The main conclusions of this study are summarized as follows:

- The structures of the flow are very dependent of the Marangoni number and where the temperature progresses slightly.
- In case of buoyant convection, the evolution of the flow with Grashof number is monotonous.
- The present analysis is focused on the influence of a limited number of dimensionless parameters. As an extension of this work, it is particularly relevant to take into account the effect Prandtl number (liquid melt) and combined buoyant and thermocapillary convections.

References

[1] Hiroyuki, O.;, Janusz S. S.; Toshio. T. In *Historical Evolution and Trends*; S. Molokov et al. ; Eds; Springer, Netherlands, 2007, 375–390.

[2] Kobayashi, N. *J. Crystal Growth.* 1978, 43, 357-367.

[3] Mihelcic, M.; Schroeck-Pauli, C.; Wingerath, K.; Wenzl, H.; Uelhoff, W.; Vander Hart A. *J. Crystal Growth.* 1981, 53, 337-354.

[4] Crochet, M. J.; Wouters, P. J.; Geyling F. T.; Jordan, A. S. *J. Crystal Growth.* 1983, 65, 153-165.

[5] Derby, J. J.; Atherton, L. J.; Thomas, P. D.; Brown, R. A. *J. Scientific Computing.* 1989, 2, 297-343.

[6] Kobayashi, N. *Japan J. Appl. Phys.* 1988, 27, 20-24.

[7] Ozoe, H. In *Modelling of Transport Phenomena in Crystal Growth;* Szmyd J.S.; Suzuki, K.; Eds; WITPress: SOUTHAMPTON, 2000, 201–237.

[8] Szmyd, J. S.; Suzuki, K. In *Modelling of Transport Phenomena in Crystal Growth;* Szmyd J.S.; Suzuki, K.; Eds; WITPress, Southampton, 2000 pp 279–321.

[9] Ivanova, L. D.; Granatkina, Yu. V. *Inorganic Materials*, 2007, 43(3), 295-300.

[10] Lind, M.D. AIAA-87-0618, AIAA 25th Aerospace Science Meeting, Jan.12-15, 1987

[11] Carlberg, T. *Acta Astronautica.* 13(11/12), 1986, 639-643.

[12] Kamotani, Y.; Ostrach, S.; Vargas, M. *J. Crystal Growth.* 1984, 83-90.

[13] Schwabe, D.; Scharmann, A.; Preisser, F.; Oeder, R. *J. Crystal Growth.* 1978, 43, 305-312.

[14] Kakimoto, K.; Eguchi, M.; Watanabe, H.; Hibiya, T. *J. Crystal Growth.* 1988,88, 365-370.

[15] Campbell, T.; Koster, J. *J. Crystal Growth* 1994, 140, 414-425.

[16] Okano, Y.; Hatano, A.; Hirata, A. *J. Chemical Engineering of Japan.* 1989, 22(4), 385-388.

[17] Murthy, J.Y.; Lee, P. *Int. J. Heat Mass Transfer* 1988, 31(9), 1932-1932.

[18] Murthy, J.Y. *J. Crystal Growth*. 1987, 83, 23-24.
[19] Smith, M.K. *J. Fluid Mech*. 1986, 166, 245-264.
[20] Kobayashi, N. *J. Crystal Growth*. 1984, 66, 63-72.
[21] Fu, B.I.; Ostrach, S. ASME PED-1983, HTD-(29)
[22] Young, G. W.; Chait, A. *J. Crystal Growth*. 1990, 106, 445-466.
[23] Neitzel, G. P.; Hyer, J. R.; Jankowski, D.F. *AIAA*. 1990, 90-0406.
[24] Chu, C. F.; Chen, J. C. *Int. J. Heat Mass transfer*. 1995, 38, 1841-1853.
[25] Monti, R. *Acta Astronautic,* 1987, 15, 557–560.
[26] Petrov, V.; Schatz, M. F.; Muehlner, K.A.; Van Hook, S.J.; McCormick, W.D.; Swift, J.B. *In Third Microgravity Fluid Physics Conference*, NASA. 1996, CP 3338, 487–492.
[27] Muehlner, K. A.; Schatz, M. F.; Petrov, V.; McCormick, W. D.; Swift, J. B.; Swinney, H. L. *Physics and Fluids*. 1997, 9(6), 1850–1852.
[28] Chwabe, D.; Hintz, P.; Frank, S. *Microgravity Science and Technology*. 1996, 9, 163–168.
[29] Lee, J.; Lee, D.J.; Lee, J. H. *J. Crystal Growth*, 1995, 152, 341-346.
[30] Patankar S.V. *Numerical Heat Transfer and Fluid Flow;* Hemisphere, Washington DC, USA, 1980.
[31] Nicolás, J. A.; Rivas, D.; Vega, J. M. *Math. Phys*. 1997, 48, 389–423.
[32] Cröll, A.; Müller-Sebert, W.; Benz, K.W.; Nitsche, R. *Microgravity Sci. Technol*. 1991, III, 204–215.

In: Thermal Engineering Research Developments ISBN 978-1-60741-497-1
Editors: J. Evgova and O. Kostadinov, pp. 449-528

Chapter 13

MODELING AND STABILITY ANALYSES OF NATURAL CIRCULATION BOILER

Walter Heimo*
Vienna University of Technology
Institute for Thermodynamics and Energy Conversion
Getreidemarkt 9, 1060 Vienna, Austria

PACS 05.45-a, 52.35.Mw, 96.50.Fm. **Keywords:** natural circulation, reverse flow, density wave oscillation, stability analysis, steam generator.
Key Words: natural circulation, reverse flow, density wave oscillation, stability analysis, steam generator AMS Subject Classification: 53D, 37C, 65P.

1. Introduction

Natural circulation boilers are used in many applications in the energy and chemical industry. The natural circulation steam generator belongs to the category of so-called water tube boilers. In this case the working medium of the steam generator flows inside and the flue gas circulates around the tubes. Natural circulation boilers, which are used in process facilities or power plants, usually have a water-steam system as a working medium. The simplest form of this circulation system is shown in figure 1 and contains a drum, a downcomer and a riser tube.

In natural circulation boilers the downcomer tubes can be designed as heated or unheated tubes. The circulation of the working medium is produced by the difference in density between the water-steam mixture in the riser tubes and the water in the downcomer. The density of the water-steam mixture in the heated riser tubes is lower than that of the water in the less or unheated downcomer tubes, which results in a lifting force in the riser system and in a circulation of the working medium. The water circulation is controlled autonomously and is based on the heat flux. With an increase in the operation pressure, the density difference between the water-steam mixture in the riser tubes and the water in the downcomer tubes decreases. The physical boundary of natural circulation systems is the

*E-mail address: heimo.walter@tuwien.ac.at

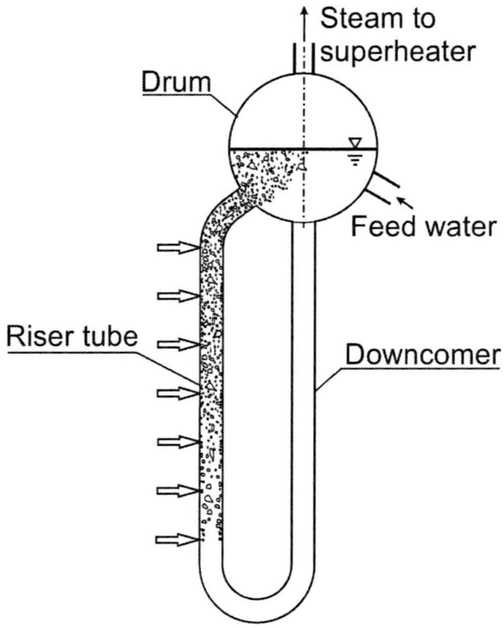

Figure 1. Simplest form of a natural circulation system.

critical point (no separation of the steam and the water is possible). But a natural circulation is also possible under supercritical pressure. Nevertheless, the application of natural circulation systems is usually limited to a maximum drum pressure of approximately 180 bars. This pressure is based on the fact that the latent heat decreases with an increase in the operation pressure. This results in a decrease in the percentage of the evaporator heating surface on the overall heating surface of the boiler. If the water-steam separation in the drum is done with the help of cyclones, then the live steam pressure in front of the steam turbine decreases to approximately 160 bars. This lower pressure is the result of the higher pressure loss in the cyclones.

Natural circulation boilers are provided with a large cylindrical pressure vessel, the steam drum in which the saturated steam is separated from the water steam mixture. These drums can be quite large, with diameters up to several meters. They are fabricated from thick plates that have been rolled into cylinders with hemispherical heads. The separation of the water-steam mixture which enters the drum through the riser tubes can be done in the simplest form by gravitation. A better separation of the two-phase mixture is given by using cyclones separators, which are applied at higher operation pressure of the boiler. The drum also represents the final evaporation point of the natural circulation system.

A important parameter for natural circulation systems is the so-called circulation ratio K_C. This factor is equal to the ratio between the mass flow $\dot{m}_{tota}$ fed to the heated tubes and the steam mass flow $\dot{m}_{st}$ actually generated in the evaporator and is defined by

$$K_C = \frac{\dot{m}_{tota}}{\dot{m}_{st}}. \tag{1}$$

The circulation ratio for a typically designed circulation system decreases with an increase in the pressure. For example, a usual value at 100 bars for K_C is approximately 10, or at 180 bars K_C is approximately 6.5. This behavior results in a higher steam quality at the outlet of the steam generator and a lower mass flux with increasing pressure [55]. Therefore, it is necessary to check the heat transfer conditions at higher operation pressure at steady state as well as under transient conditions. The design engineer has to prove that a sufficient distance to the boiling crisis DNB (departure of nucleate boiling) and DO (dryout) is given to avoid critical heat flux (CHF) conditions. The maximum heat flux just before the onset of the boiling crisis is called CHF (see section 2.5.2.). An oil-fired steam generator with a heat flux of approximately 550 kW/m^2 uses a higher circulation ratio as a coal-fired boiler with approximately 300 kW/m^2 because the critical vapor quality, which is a criterion to determine CHF, is a function of pressure p and heat flux $\dot{q}$ [16]. Under normal operation conditions the natural circulation steam generator does not have more than 30% steam quality. The usual objective in the designing of recirculation steam generators is to avoid CHF conditions, whereas in the case of one-through boilers the objective is to predict their location.

The preferred operation mode of a natural circulation steam generator is the operation with constant pressure, but variable pressure operation is also possible. Unfavorable for the natural circulation system is the limitation of the rate of pressure reduction, which is a result of the thermal stress of the heavy-walled components in the saturation region like the drum or evaporator headers. Therefore, a natural circulation boiler can be used only under specific conditions or under variable pressure operation because in this case the load change is connected with a pressure change and, consequently, with a change in the saturation temperature. A further disadvantage of natural circulation systems is the high sensitivity against a rapid pressure reduction. A rapid pressure reduction can lead to steam generation in the downcomer tubes of the boiler, which can result in failure of the water circulation. If this effect occurs in the evaporator tubes, the admissible tube temperature can be exceeded and the tube can be damaged. The allowable rate of pressure reduction for typical natural circulation boilers is up to approximately 6 to 8 bars per minute [71].

An advantage of the natural circulation is the high storage capacity of the boiler based on the high water mass in the tube network of the steam generator. In the case of a pressure reduction based on a sudden steam extraction, the latent heat of the liquid inside the tubes of the boiler will be delivered in the form of additionally produced steam. The relative storage capacity is approximately 2 to 2.5 kg vapor per % pressure reduction and kg/s steam output [54]. Compared to other types of steam generators, a further advantage of natural circulation systems is the reduced investment, maintenance and operation cost due to the reduced power requirement of the circulation pump. Under normal circumstances, natural circulation systems do not need special start-up devices.

The experience shows that the most critical operation mode of a natural circulation boiler is the hot start-up or a heavy load change. In these cases, stagnation and/or reverse flow ([95], [101], [99], [97] or [100]) as well as dynamic instabilities e. g. density wave oscillations ([98], [92], [90] or [102]) or pressure drop oscillations of the working fluid can occur. Therefore the boiling flow instability is a problem of great current interest in many heat transfer applications such as nuclear reactors, refrigeration systems or conventional steam generators. Operation conditions where such instability can occur can lead to a restricted operation or a damage of the facility. It becomes important in designing of boiling flow systems to predict the conditions under which a system will perform without any instability.

In this chapter some modeling and stability aspects of natural circulation boilers will be discussed.

2. Models for Simulating a Natural Circulation Boiler

In the following sections selected mathematical models for the numerical simulation of a natural circulation boiler, which are included in the simulation program DBS (Dynamic Boiler Simulation), will be presented. This program was developed at the Vienna University of Technology, Institute for Thermodynamic and Energy Conversion. The section starts with the formulation and discretization of the balance equations for the fluid flow in the tube and is followed by the mathematical formulation of the most important components of a natural circulation boiler the drum and the header. The section will be completed by a description of some heat transfer aspects between fluid flow and wall.

2.1. Model of the Fluid Flow in the Tube

The mass flow in the tubes of a steam generator can be assumed to be one-dimensional, as the length of the tubes is much greater compared to their diameter. The present mathematical model [86] for the working medium is one-dimensional in flow direction and uses a homogeneous equilibrium model for the two-phase flow and applies a correction factor Φ^2_{fric} for the two phase pressure drop due to friction according to Friedel [28]. The formula derived by Friedel for a vertical upward or a horizontal fluid flow is:

$$\Phi^2_{fric} = C_1 + \frac{3.24\, C_2\, C_3}{Fr_{2ph}^{0.045}\, We_{2ph}^{0.035}} \tag{2}$$

with the coefficients

$$\begin{aligned} C_1 &= (1-\dot{x})^2 + \dot{x}^2 \left(\frac{\rho_l}{\rho_g} \frac{\zeta_{g0,fric}}{\zeta_{l0,fric}} \right) \\ C_2 &= \left(\frac{\rho_l}{\rho_g} \right)^{0.91} \left(\frac{\eta_g}{\eta_l} \right)^{0.19} \left(1 - \frac{\eta_g}{\eta_l} \right)^{0.7} \quad \text{and} \\ C_3 &= \dot{x}^{0.78} (1-\dot{x})^{0.224} . \end{aligned}$$

For a vertical downward flow direction of the fluid the following equation for Φ^2_{fric} was suggested by Friedel

$$\Phi^2_{fric} = C_1 + \frac{48.6\, C_2\, C_3 Fr_{2ph}^{0.03}}{We_{2ph}^{0.12}} \tag{3}$$

with the coefficients

$$\begin{aligned} C_1 &= (1-\dot{x})^2 + \dot{x}^2 \left(\frac{\rho_l}{\rho_g} \frac{\zeta_{g0,fric}}{\zeta_{l0,fric}} \right) \\ C_2 &= \left(\frac{\rho_l}{\rho_g} \right)^{0.9} \left(\frac{\eta_g}{\eta_l} \right)^{0.73} \left(1 - \frac{\eta_g}{\eta_l} \right)^{7.4} \quad \text{and} \\ C_3 &= \dot{x}^{0.8} (1-\dot{x})^{0.29} . \end{aligned}$$

The Froude- and Weber-number for the two-phase flow is defined by

$$Fr_{2ph} = \frac{\dot{m}^2_{flux,tota}}{g\, d_i\, \rho^2} \tag{4}$$

$$We_{2ph} = \frac{d_i \dot{m}^2_{flux,tota}}{\sigma \rho}. \tag{5}$$

For calculating the pressure drop coefficient ζ Friedel uses the equations derived by Techo [76].

$$\zeta_{j,fric} = \frac{64}{Re} \qquad \text{for} \quad Re_j \leq 1055$$
$$\zeta_{j,fric} = \left[0.86859 \ln\left(\frac{Re_j}{1.964 \ln(Re_j) - 3.8215}\right)\right]^{-2} \qquad \text{for} \quad Re_j > 1055 \tag{6}$$

with $j = l0, g0$.

For a straight tube with constant cross section the governing equations in flow direction are written for the mass and momentum conservation of the fluid:

$$\frac{\partial \rho}{\partial t} + \frac{\partial \rho w}{\partial x} = 0. \tag{7}$$

$$\frac{\partial}{\partial t}(\rho w) + \frac{\partial}{\partial x}(\rho w w + p) = -\rho g_x - \left(\frac{\partial p}{\partial x}\right)_{fric} \tag{8}$$

The density ρ and the velocity w are averaged values over the cross section of the tube. Considering the fluid flow in steam generators, the thermal energy is much higher than the kinetic and the potential energy as well as the expansion work. Therefore, the balance equation for the thermal energy can be simplified with the help of

$$e = h - \frac{p}{\rho} \tag{9}$$

to:

$$\frac{\partial}{\partial t}(\rho h) + \frac{\partial}{\partial x}(w \rho h) = \dot{q}\frac{U}{A} + \frac{\partial p}{\partial t}. \tag{10}$$

The heat exchange between the fluid and the tube wall is governed by Newton's law of cooling and the heat transfer through the wall is assumed to be in the radial direction only.

The one-dimensional partial differential equations for the conservation of momentum, mass and energy in Cartesian coordinates can be written in general form:

$$\frac{\partial}{\partial t}(\rho \Phi) + \frac{\partial}{\partial x}(\rho w \Phi) = \frac{\partial}{\partial x}\left(\Gamma \frac{\partial \Phi}{\partial x}\right) + S_\Phi. \tag{11}$$

where Φ and Γ is shown in Table 1.

The volumetric source term can be approximated by the linear source relation:

$$S_\Phi = S_c + S_P \Phi \tag{12}$$

Table 1. Relation between the variables Φ and Γ of the general form of the balance equation

Balance equation	dependent variable Φ	diffusion coefficient Γ
Mass	1	0
Momentum	w	η
Energy	h	η/Pr

The discretization of the partial differential equations for the conservation laws was done with the aid of the finite-volume-method. In this case the calculation domain is subdivided into a number of control volumes, each is associated with a grid point. The scalar variables and the pressure are stored at the grid points, whereas the velocities are stored at staggered locations at the interface between two control volumes. This so called staggered grid is used to prevent checkerboard pressure fields (for more information about the checkerboard pressure field see [65]). The algebraic equations for mass and energy are obtained by integrating the general partial differential equation (11) for the control volume i shown in figure 2(a), whereas for the momentum balance equation (11) is integrated for the control volume $i+1/2$ shown in figure 2(b).

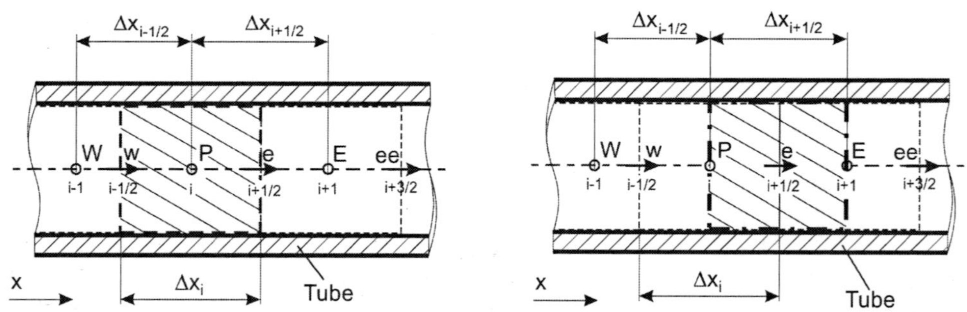

(a) Control volume for mass and energy balance

(b) Control volume for momentum balance

Figure 2. Arangement of the control volumes for the one dimensional tube flow.

In case of the control volume i the discretized equation for the one dimensional fluid flow in a tube with constant cross section area can be written in the following form:

$$a_{Pi}\Phi_i = a_{Wi}\Phi_{i-1} + a_{Ei}\Phi_{i+1} + b_i = \sum_{nb} a_{nb}\Phi_{nb} + b_i \tag{13}$$

with the coefficients:

$$a_{Pi} = a_{Wi} + a_{Ei} + a^0_{Pi} - S_{pi}A_i\Delta x_i \tag{14}$$

$$a_{Wi} = \left(\frac{\Gamma_{i-\frac{1}{2}}}{\Delta x_{i-\frac{1}{2}}} - [[(\rho w)_{i-\frac{1}{2}}, 0]] \right) A_{i-\frac{1}{2}}, \tag{15}$$

$$a_{Ei} = \left(\frac{\Gamma_{i+\frac{1}{2}}}{\Delta x_{i+\frac{1}{2}}} + [[(-\rho w)_{i+\frac{1}{2}}, 0]] \right) A_{i+\frac{1}{2}}, \tag{16}$$

$$b_i = S_{ci}A_i\Delta x_i + a^0_{Pi}\Phi^0_i \quad \text{and} \tag{17}$$

$$a^0_{Pi} = \frac{\rho^0_i A_i \Delta x_i}{\Delta t}. \tag{18}$$

The special symbol $[[A,B]]$, which is equivalent to the FORTRAN command DMAX(A,B), stands for the largest of the quantities contained within. A detailed derivation of equation (13) can be found amongst others by [86], [65] or [26]. Equation (13) must be solved for each grid point within the calculation domain. For the one dimensional fluid flow in a tube the resulting matrix can be solved with the TDMA (Tri-Diagonal Matrix Algorithm) solver. A detailed description of the TDMA solver is given e.g. in [65]. It can be seen on equation (15) and (16) that for the convective term of the general equation the UPWIND scheme is used which was first put forward by [18].

2.1.1. Pressure Correction Algorithm

If the finite-volume method is used on the basis of the so called primitive variables pressure and velocity, then mass and momentum balance must be fulfilled at the same time. In this case the balance equations represent a elliptic-parabolic system of equations. The main difficulty in the calculation of the velocity field lies in the unknown pressure field. In the set of balance equations there is no obvious equation for obtaining pressure. Regardless of this circumstance the pressure field is determined indirectly by the velocity field because the velocity must fulfill simultaneous both the momentum and mass balance.

The difficulty associated with the determination of the pressure field has led to different methods that eliminates the pressure from the governing equations. For a problem in two dimensions the pressure can be eliminated from the two momentum balances by cross differentiation. This leads to a so called vorticity transport equation. In combination with the definition of a stream function for steady two-dimensional problem is the basis for the well known "stream function/vorticity method". The feature of this method is, that we must not solve the continuity and the two momentum balances. Only the two equations to obtain the stream function and the vorticity. A disadvantage of the method is, that the value of the vorticity at the wall is difficult to specify. This causes in some cases difficulties in getting a converged solution. Also the pressure must be extracted from the vorticity, if it is needed for the calculation of the fluid properties. In this case the computing savings become lost. A expansion of the method to three dimensions results in the use of six dependent variables - three components for the vorticity and three components for the stream function vector. Thus, the complexity to determine this six dependent variables is greater than that of treating the velocity components and the pressure [8].

Therefore other solution methods are developed which uses the so called primitive variables namely the pressure and the velocity. In these methods a equation for the pressure is used which is a result by a combination of the continuity and the momentum balance. These methods can be used in principle for compressible as well as for incompressible fluid flow problems [42]. One of these methods is the pressure correction algorithm.

In the following, the pressure correction algorithm SIMPLE (Semi Implicit Method for Pressure Linked Equations), which was first proposed by [65], are presented for a one dimensional fluid flow through a tube with constant cross section area. Figures 2(a) and 2(b) shows the grid structure and the storage locations for the dependent variables. A typical control volume for the mass and energy balance is denoted by the shaded area of figure 2(a) and for the momentum balance by the shaded area shown in figure 2(b). The velocities, which are located at the interface between two control volumes, e. g. cell i and $i+1$, are denoted with the index $i+1/2$.

SIMPLE algorithm The SIMPLE algorithm will be presented so that necessary equations are available later, when the models for the numerical simulation of a natural circulation boiler will be discussed. The momentum equation can be deduced from equation (11) by specifying Φ to the velocity w and Γ to the dynamic viscosity η. The pressure gradient of the source term b_i expressed separately, equation (13) becomes:

$$a_{ei} w_{i+\frac{1}{2}} = \sum_{nb} a_{nb} w_{nb} + (p_i - p_{i+1}) A_{i+\frac{1}{2}} + b_{ei} \tag{19}$$

For a guessed pressure field p^* the velocity w^* satisfies

$$a_{ei} w^*_{i+\frac{1}{2}} = \sum_{nb} a_{nb} w^*_{nb} + \left(p^*_i - p^*_{i+1}\right) A_{i+\frac{1}{2}} + b_{ei} \tag{20}$$

The momentum equation (20) supplies a velocity field. If the velocity field does not satisfy the continuity equation, the pressure field is incorrect. The pressure field can be corrected using the following term:

$$p = p^* + p' \tag{21}$$

where p' is called the pressure correction. The velocities are corrected in the same way:

$$w = w^* + w'. \tag{22}$$

The relation between the pressure corrections and the velocity corrections can be developed by subtracting equation (20) from (19) to get

$$a_{ei} w'_{i+\frac{1}{2}} = \sum_{nb} a_{nb} w'_{nb} + \left(p'_i - p'_{i+1}\right) A_{i+\frac{1}{2}}. \tag{23}$$

In the SIMPLE algorithm the velocity correction equations are obtained by omitting $\sum_{nb} a_{nb} w'_{nb}$ in equation (23). Substitution of the velocity correction w' in the simplified equation (23) by help of equation (22) yields in the velocity correction equation:

$$w_{i+\frac{1}{2}} = w^*_{i+\frac{1}{2}} + d_{ei} \left(p'_i - p'_{i+1}\right) \tag{24}$$

with

$$d_{ei} = \frac{A_{i+\frac{1}{2}}}{a_{ei}}. \tag{25}$$

Equation (24) shows how the velocities react to the pressure corrections. The pressure correction equation is developed from the discretized mass balance by substituting the velocities at the boundaries of the control volumes through the velocity correction.

$$a_{Pi} p'_i = \sum_{nb} a_{nb} p'_{nb} + b_i \tag{26}$$

So the corrected velocities in equation (24) satisfy the continuity constraint. The SIMPLE procedure is implemented by executing the following sequence of steps:

1. Guess a pressure field p^* (e. g. a homogeneous pressure distribution over the whole grid).
2. Calculate the coefficients of the momentum balance (20) and solve to obtain the velocity field w^*.
3. Evaluate the coefficients of the pressure correction equation (26) and solve for p'.
4. Correct the velocity field, using the equations (24) and (25), and the pressure field using equation (21).
5. Solve other Φ equations (e. g. energy balance).
6. Update properties, coefficients, etc.
7. Using the pressure field found in step 4 as the new guessed p^* field; check convergence.
8. If convergence is not achieved return to step 2.

2.2. Model of the Header

The header is one of the important structural component of every steam generator and is used to connect a low number of connection tubes with a large diameter and the high number of tubes of a bundle heating surface or a combustion wall with a small tube diameter. In figure 3 a evaporator wall with a header on the upper and lower wall end can be shown. The evaporator wall is carried out as a membrane wall. This membrane wall is covered with a higher number of tubes, which are connected by flat steel elements along the tube axis. This combination of tubes and bars is called membrane wall. The working medium enters the evaporator wall through the header at the bottom (small header diameter) and leaves the wall at the top (large header diameter). The different diameter of both headers is a result of the different fluid volume between inlet and outlet of the evaporator wall. The cross section area of the header depends also on the highest allowed fluid velocity inside the header. The high number of tubes which are connected to the header causes a reduction of the cross section and results in combination with the high operation pressure in a heavy-walled structural component of the power plant.

Figure 3. Membrane wall with headers.

Figure 4 shows for a better understanding of the header model a sketch of a header with only four connected tubes. These tubes should be representative for all other tubes which are connected to the header. The variables of the header are denoted with the index h; j represent values at the header entrance and k the values at the outlet. V_h is the fluid volume inside the header and A the cross sectional area of the connected tubes.

For the computation of such a header, the following simplification assumptions may be made:

- Assuming that the distribution of the thermodynamic state of the header is homogeneous, the collector can be seen as a single control volume for the calculation.
- The gravity distribution of density and pressure of the fluid inside the header can be neglected because the vertical dimension of the header is small compared to that of the remaining tube system.
- The huge difference in the cross section area of the header and the header connected tubes causes strong turbulence, avoiding thus a segregation of the fluid in the header.
- Because the headers are assumed to be a single control volume, the equations for the mass and energy balance are ordinary differential equations with time t as independent variable:

$$\frac{\mathrm{d}}{\mathrm{d}t}(\rho_h V_h) = \sum_j \rho_j w_j A_j - \sum_k \rho_k w_k A_k \tag{27}$$

$$\frac{\mathrm{d}}{\mathrm{d}t}(\rho_h h_h V_h) = \sum_j \rho_j w_j h_j A_j - \sum_k \rho_k w_k h_k A_k \tag{28}$$

- Similar to the treatment of the fluid flow in the tube, the kinetic energy as well as the expansion work is neglected in the energy balance for the header.

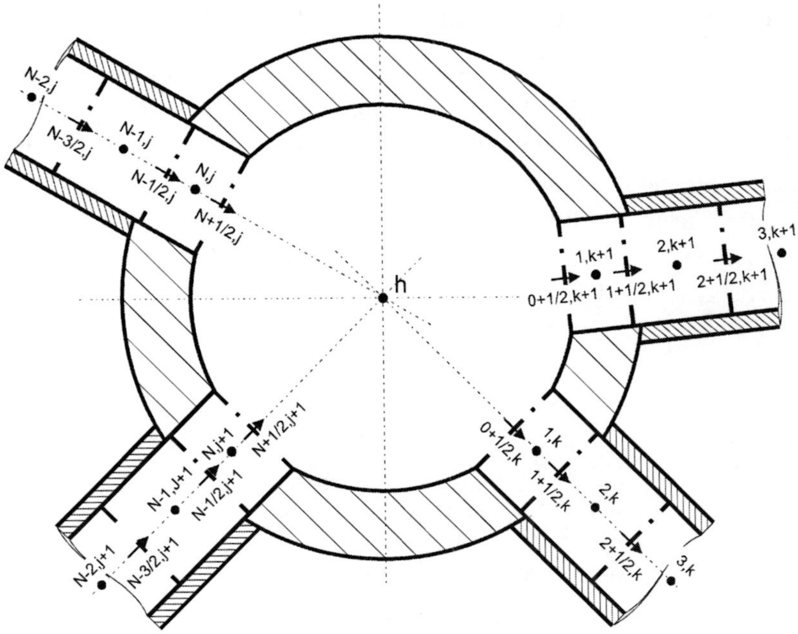

Figure 4. Discretization of the header and the header connected tubes.

- The velocity of the fluid in the header is rather small compared to that inside the tubes. So it can be assumed that the momentum of the fluid will be lost at the inlet of the header and has to be rebuilt at the outlet. Based on this assumption, the momentum balance of the header reduces to a pressure balance. The changes of the momentum at the inlet and the outlet can be taken into account by a pressure loss coefficient ζ:

$$p_h = p_j - \frac{\zeta_j}{2}\rho_j w_j \left|w_j\right| \tag{29}$$

$$p_h = p_k + \frac{\zeta_k}{2}\rho_k w_k \left|w_k\right| \tag{30}$$

  ζ_k takes into account the sum of the pressure loss due to the acceleration of the fluid inside the header from approximately zero to the tube inlet velocity and the frictional pressure loss across the sudden contraction. The pressure loss for e. g. a sharp-edged entrance can be found in [67] or [27] and is identical 0.5. ζ_j represents the frictional pressure loss of the fluid flowing from the tube outlet cell into the header. Based on the model assumptions for the header the recovery of the pressure energy from the kinetic energy of the fluid is set to zero.

In the following the discretized equations for the header and header connecting tubes will be presented. This set on equations is based on SIMPLE algorithm which was proposed by Patankar [65]. The algorithm will not be discussed here in detail. For further information about the SIMPLE algorithm see amongst others [86], [65], [26] or [63]. All indices in the next two sections correspond with the indices shown in figure 4.

2.2.1. Balance Equations for the Header

Taking into consideration the above described tube-header-model the dissipation and regeneration of the momentum leads to a separation of the momentum balance for the header and the connected tubes. At the formulation of the momentum equations for the single control volumes of the tube network the control volume of the header can be neglected. Therefore the tri-diagonal structure of the matrix can be preserved. The calculation of the header pressure is not independent from the connected tubes and therefore the pressure correction equation for the header must be written in the following form:

$$a_{mhP}\, p'_h = \sum_j a_{mhW,j}\, p'_{N,j} + \sum_k a_{mhE,k}\, p'_{1,k} + b_h \tag{31}$$

with the coefficients:

$$a_{mhP} = \sum_j a_{mhW,j} + \sum_k a_{mhE,k}, \tag{32}$$

$$a_{mhW,j} = (\rho A)_{N+\frac{1}{2},j}\, d_{wN,j}, \tag{33}$$

$$a_{mhE,k} = (\rho A)_{0+\frac{1}{2},k}\, d_{e1,k}, \quad \text{and} \tag{34}$$

$$b_h = \frac{(\rho_h^0 - \rho_h)V_h}{\Delta t} + \sum_j (\rho w^* A)_{N+\frac{1}{2},j} - \sum_k (\rho w^* A)_{0+\frac{1}{2},k}. \tag{35}$$

The spec. enthalpy of the fluid inside the header can be calculated with

$$a_{hP}\, h_h = \sum_j a_{hW,j}\, h_{N,j} + \sum_k a_{hE,k}\, h_{1,k} + b_h \tag{36}$$

with the coefficients:

$$a_{hP} = \sum_j a_{hW,j} + \sum_k a_{hE,k} + a_{hP}^0 - S_{ph} V_h, \tag{37}$$

$$a_{hW,j} = [[(\rho w)_{N+\frac{1}{2},j}, 0]] A_{N+\frac{1}{2},j}, \tag{38}$$

$$a_{hE,k} = [[-(\rho w)_{0+\frac{1}{2},k}, 0]] A_{0+\frac{1}{2},k}, \tag{39}$$

$$b_h = S_{ch} V_h + a_{hP}^0 h_h^0 \quad \text{and} \tag{40}$$

$$a_{hP}^0 = \frac{\rho_h^0 V_h}{\Delta t}. \tag{41}$$

2.2.2. Balance Equations for the Header Connected Control Volumes

In this section, the discretized balance equations for the header connected control volumes will be presented. The number of control volumes which are connected to the header are not restricted by the model. The numbering of the header connected control volumes can be seen in figure 4. All equations are independent from the flow direction.

With the following equations the pressure, velocity and the spec. enthalpy for the control volume $1,k$ at the tube inlet can be calculated. The discretized momentum balance for computation of the velocity at the staggered grid can be written as:

$$a_{e0,k} w_{0+\frac{1}{2},k} = a_{ee0,k} w_{1+\frac{1}{2},k} + (p_h - p_{1,k}) A_{0+\frac{1}{2},k} + b_{0,k} \tag{42}$$

with the coefficients:

$$a_{e0,k} = a_{ee0,k} + a^0_{e0,k} - S_{p0,k} A_{0+\frac{1}{2},k} \Delta x_{0+\frac{1}{2},k} \tag{43}$$

$$a_{ee0,k} = \left(\frac{\eta_{1+\frac{1}{2},k}}{\Delta x_{1+\frac{1}{2},k}} + \left[\left[-(\rho w)_{1+\frac{1}{2},k}, 0 \right]\right] \right) A_{1+\frac{1}{2},k}, \tag{44}$$

$$b_{0,k} = S_{c0,k} A_{0+\frac{1}{2},k} \Delta x_{0+\frac{1}{2},k} + a^0_{e0,k} w^0_{0+\frac{1}{2},k} \quad \text{and} \tag{45}$$

$$a^0_{e0,k} = \frac{\rho^0_{0+\frac{1}{2},k} A_{0+\frac{1}{2},k} \Delta x_{0+\frac{1}{2},k}}{\Delta t}. \tag{46}$$

The proportional part of the source term leads to:

$$S_{p0,k} \Delta x_{0+\frac{1}{2},k} = -\frac{|\Delta p_{fric0+\frac{1}{2},k}|}{|w_{0+\frac{1}{2},k}|} - (1+\zeta_{in}) \frac{\rho_{0+\frac{1}{2},k} w_{0+\frac{1}{2},k} [[w_{0+\frac{1}{2},k}, 0]]}{2 \left| w_{0+\frac{1}{2},k} \right|}. \tag{47}$$

With the factor ζ_{in} the pressure loss due to the acceleration of the fluid inside the header from approximately zero to the tube inlet velocity is considered. For a sharp-edged entrance ζ_{in} is identical to 0.5 according to [67]. In case of reverse flow at the tube inlet the pressure loss due to acceleration will be identical to zero with the help of the operator $[[A, B]]$.

The pressure correction formula for the control volume $1,k$ can be written independent from the direction of the fluid in the channel in the following form:

$$a_{mP1,k} p'_{1,k} = a_{mW1,k} p'_h + a_{mE1,k} p'_{2,k} + b_{1,k} \tag{48}$$

where

$$a_{mP1,k} = a_{mW1,k} + a_{mE1,k}, \tag{49}$$

$$a_{mW1,k} = (\rho A)_{0+\frac{1}{2},k} d_{w1,k}, \tag{50}$$

$$a_{mE1,k} = (\rho A)_{1+\frac{1}{2},k} d_{e1,k} \quad \text{and} \tag{51}$$

$$b_{1,k} = \frac{(\rho^0_{1,k} - \rho_{1,k}) A_{1,k} \Delta x_{1,k}}{\Delta t} + (\rho w^* A)_{0+\frac{1}{2},k} - (\rho w^* A)_{1+\frac{1}{2},k}. \tag{52}$$

The energy balance for the control volume $1,k$ is given with:

$$a_{P1,k} h_{1,k} = a_{W1,k} h_h + a_{E1,k} h_{2,k} + b_{1,k} \tag{53}$$

with the coefficients:

$$a_{P1,k} = a_{W1,k} + a_{E1,k} + a^0_{P1,k} - (S_p A \Delta x)_{1,k} \tag{54}$$

$$a_{W1,k} = [[(\rho w)_{0+\frac{1}{2},k}, 0]] A_{0+\frac{1}{2},k}, \tag{55}$$

$$a_{E1,k} = [[-(\rho w)_{1+\frac{1}{2},k}, 0]] A_{1+\frac{1}{2},k} \tag{56}$$

$$b_{1,k} = (S_c A \Delta x)_{1,k} + (a^0_P h^0)_{1,k} \quad \text{and} \tag{57}$$

$$a^0_{P1,k} = \frac{\rho^0_{1,k} A_{1,k} \Delta x_{1,k}}{\Delta t}. \tag{58}$$

The equations for a control volume at the tube outlet (subscript N) are similar to the equations presented above. Therefore they are not described here in detail. This set of equations can be derived from the equations (42) to (58).

It can be seen that the equations for the pressure correction and the spec. enthalpy of the header and the header connected control volumes are not independent from each other. Therefore the tri-diagonal structure of the matrix will be lost and the resulting band width of the sparse matrix is a function of numbering the header connected control volumes. It is well known that for a sparse coefficient matrix a optimization of the band width should be done.

For the pressure correction algorithms SIMPLER (Semi Implicit Method for Pressure Linked Equations Revised) and PISO (Pressure-Implicit with Splitting of Operators), which is a time-marching procedure in which during each time step one predictor step followed by one or more corrector steps, the discretized equations of the above presented tube-header-model can be found in [86], [96], [91] or [105].

2.3. Model of the Drum

Natural circulation boilers are provided with a large cylindrical pressure vessel, the steam drum. These drums can be quite large (up to approximately 30 m) with diameters up to several meters. They are fabricated from thick plates that have been rolled into cylinders with hemispherical heads.

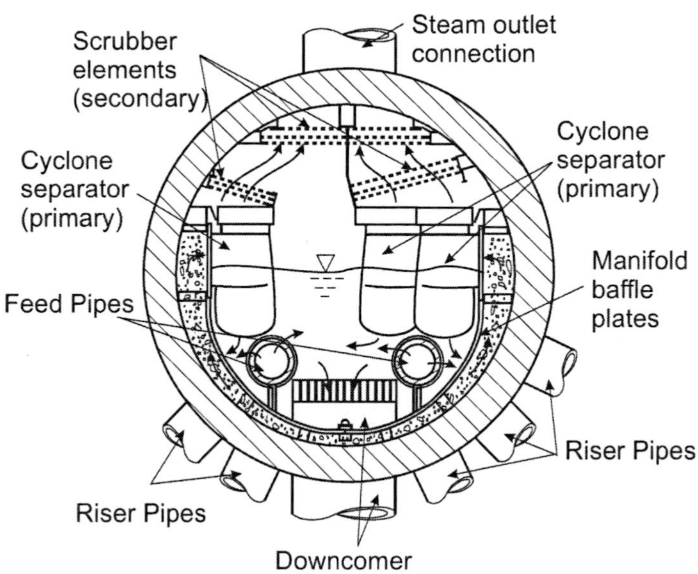

Figure 5. Drum with internals.

The primary function of the drum is to link the downcomer and riser tubes, which are necessary for the circulation of the working fluid through the evaporator, with the feed water and the steam outlet tubes as well as for the separation of the water-steam mixture into saturated steam and saturated liquid (see figure 5). For a better separation of the two-phase flow mixture different devices will be included such as baffle plates or cyclone separators.

The saturated steam leaves the drum at the drum vertex in direction to the superheater. The drum is also used for

- Mixing the feed water with the saturated liquid after steam separation,
- Containing limited water storage to accommodate some changes in the boiler level,
- Removing a portion of the boiler water to control the boiler water chemistry (i.e. blowdown) and
- Mixing the chemicals added to control corrosion.

Figure 5 shows a drum with their typical arranged internals. In high-pressure boilers the cyclones are used to separate the most of the water from the steam. Because the low difference in density between the water and the steam makes other methods uneconomical [50]. Scrubber or dryer elements are then used to remove the remaining moisture from the steam before leaving the drum. The feed water tubes are arranged along the drum length to provide uniform mixing of the preheated feed water with the saturated liquid inside the drum. The feed water tubes should be arranged in such a way that the feed water do not enter directly into the downcomer tubes because in some cases the feed water cannot preheated up to the saturation enthalpy at drum pressure. If the feed water enters in this case directly into the downcomer tubes, then the working medium in the downcomer is subcooled and not saturated.

In figure 6 a sketch of the drum with all for the derivation of the drum balance equations necessary quantities is shown. In the present drum model it is possible to connect k feed water, n downcomer and j riser tubes. For the blow down as well as the steam mass flow which leaves the drum one tube is provided in the model. Corresponding to the header model it will be assumed also at the drum model that the momentum of the fluid will be lost at the inlet of the drum and has to be rebuilt at the outlet. Therefore the momentum balance for the drum pressure is reduced to the pressure balance (69). Based on the nomenclature shown in figure 6 the mass and energy balance of the drum can be written in the following form:

mass balance:

$$\frac{\mathrm{d}\,M_{l,D}}{\mathrm{d}t}+\frac{\mathrm{d}\,M_{st,D}}{\mathrm{d}t}=\sum_{i=1}^{k}\dot{m}_{feed_i}+\sum_{i=1}^{j}\dot{m}_{R_i}-\sum_{i=1}^{n}\dot{m}_{Do_i}-\dot{m}_{Bl}-\dot{m}_{st,D,out} \tag{59}$$

energy balance:

$$\frac{\mathrm{d}\,M_{l,D}\,h_{l,D}}{\mathrm{d}t}+\frac{\mathrm{d}\,M_{st,D}\,h_{st,D}}{\mathrm{d}t}=\sum_{i=1}^{k}\dot{m}_{feed_i}\,h_{feed_i}+\sum_{i=1}^{j}\dot{m}_{R_i}\,h_{R_i}-$$
$$\sum_{i=1}^{n}\dot{m}_{Do_i}\,h_{Do_i}-\dot{m}_{Bl}\,h_{Bl}-\dot{m}_{st,D,out}\,h_{st,D,out}+V_D\frac{\mathrm{d}\,p_D}{\mathrm{d}t} \tag{60}$$

with the mass of the liquid

$$M_{l,D}=\rho_{l,D}A_{l,D}l_D \tag{61}$$

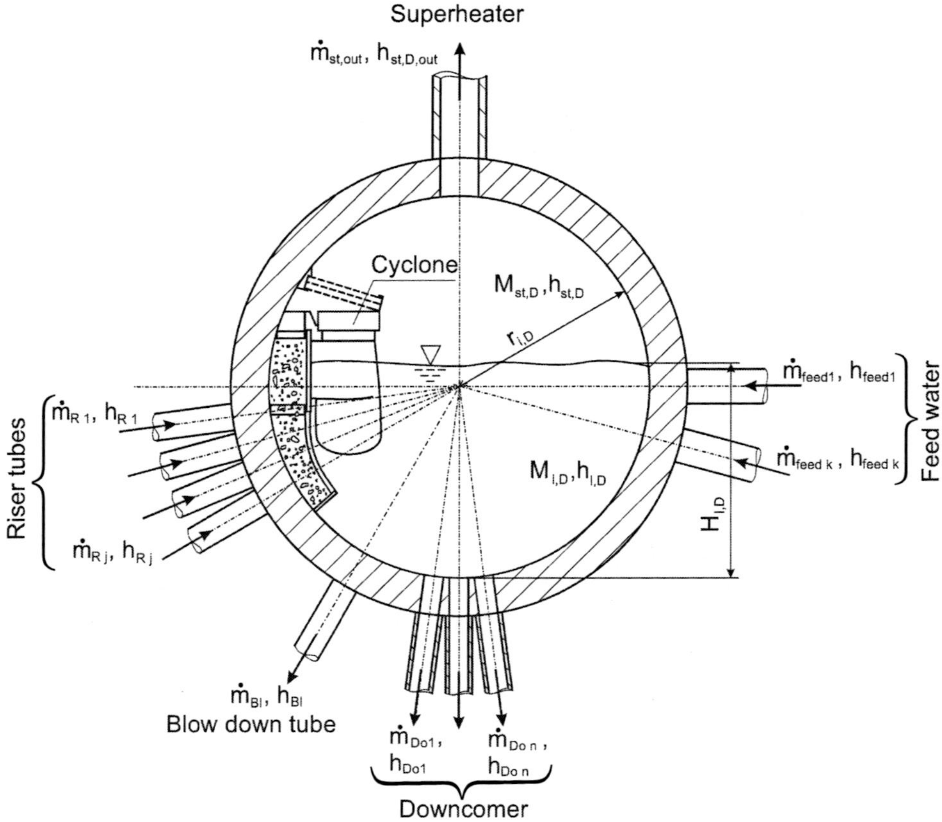

Figure 6. Drum model with connected tubes.

and the steam inside the drum:

$$M_{st,D} = \rho_{st,D}\,(A_{D,tota} - A_{l,D})\,l_D, \tag{62}$$

The distinction between the spec. enthalpy of the steam inside the drum $h_{st,D}$ and the spec. enthalpy of the steam which leaves the drum $h_{st,D,out}$ is based on the circumstance, that the steam inside the drum and leaving the drum can have a different portion of liquid.

The derivatives in the energy and mass balance will be substituted by backward finite differences and results for the first term on the left hand side of equation (59) to:

$$\frac{\mathrm{d}\,M_{l,D}}{\mathrm{d}t} = \frac{M_{l,D} - M^0_{l,D}}{\Delta t} \tag{63}$$

with the liquid mass inside the drum at the old time step $M^0_{l,D}$. The other derivatives in the energy and mass balance of the drum can be substituted by a backward finite difference in the same way. A combination of the continuity balance for the drum (59) and the steam mass inside the drum (62) with the drum energy balance results in a equation for the cross section area for the liquid inside the drum.

$$A_{l,D} = \frac{\dfrac{\Delta t}{l_D} Z_1 + Z_2}{\rho_{l,D}\,h_{l,D} - \rho_{st,D}\,h_{st,D} + h_{st,D,out}\,(\rho_{st,D} - \rho_{l,D})} \tag{64}$$

with the coefficient

$$
\begin{aligned}
Z_1 &= \Sigma_{i=1}^{k} \dot{m}_{feed_i}\, h_{feed_i} + \Sigma_{i=1}^{j} \dot{m}_{R_i}\, h_{R_i} - \Sigma_{i=1}^{n} \dot{m}_{Do_i}\, h_{Do_i} - \dot{m}_{Bl}\, h_{Bl} \\
&\quad - h_{st,D,out} \left(\Sigma_{i=1}^{k} \dot{m}_{feed_i} + \Sigma_{i=1}^{j} \dot{m}_{R_i} - \Sigma_{i=1}^{n} \dot{m}_{Do_i} - \dot{m}_{Bl} \right)
\end{aligned}
\tag{65}
$$

and

$$
\begin{aligned}
Z_2 &= \rho_{st,D}\, A_{D,tota} \left(h_{st,D,out} - h_{st,D} \right) + \left(\rho_{l,D}\, A_{l,D}\, h_{l,D} \right)^0 \\
&\quad - h_{st,D,out} \left\{ \left(\rho_{l,D}\, A_{l,D} \right)^0 + \left[\left(A_{D,tota} - A_{l,D} \right) \rho_{st,D} \right]^0 \right\} \\
&\quad + \left[\left(A_{D,tota} - A_{l,D} \right) \rho_{st,D}\, h_{st,D} \right]^0 + A_{D,tota} \left(p_D - p_D^0 \right).
\end{aligned}
\tag{66}
$$

The superscript 0 specifies the value at the old time step.

With the help of the drum geometry a second equation for the cross section area of the liquid inside the drum as a function of the water level $H_{l,D}$ and the inner drum radius $r_{i,D}$ is given.

$$
A_{l,D} = r_{i,D}^2 \arccos\left(1 - \frac{H_{l,D}}{r_{i,D}} \right) + \left(H_{l,D} - r_{i,D} \right) \sqrt{H_{l,D} \left(2 r_{i,D} - H_{l,D} \right)} \tag{67}
$$

for $H_{l,D} \leq r_{i,D}$ and

$$
A_{l,D} = r_{i,D}^2 \pi \left[1 - \frac{1}{\pi} \arccos\left(\frac{H_{l,D}}{r_{i,D}} - 1 \right) \right] + \left(H_{l,D} - r_{i,D} \right) \sqrt{H_{l,D} \left(2 r_{i,D} - H_{l,D} \right)}
$$

for $H_{l,D} \geq r_{i,D}$.

The water level of the drum can be decided from equation (67) or (68) with the aid of the Newton algorithm and the formula (64) for the cross section area. The steam mass flow which leaves the drum is a result of the continuity equation and results to:

$$
\begin{aligned}
\dot{m}_{st,D,out} &= -\frac{\Delta t}{l_D} \Big\{ A_{l,D} \left(\rho_{l,D} - \rho_{st,D} \right) + \rho_{st,D} A_{D,tota} \\
&\quad - \left[\left(\rho_{l,D} A_{l,D} \right)^0 + \left(\left(A_{D,tota} - A_{l,D} \right) \rho_{st,D} \right)^0 \right] \Big\} \\
&\quad + \sum_{i=1}^{k} \dot{m}_{feed_i} + \sum_{i=1}^{j} \dot{m}_{R_i} - \sum_{i=1}^{n} \dot{m}_{Do_i} - \dot{m}_{Bl}
\end{aligned}
\tag{68}
$$

2.3.1. Integration of the Drum Model into the SIMPLE Algorithm

As presented in the section before, the mass and energy balance of the drum was used to calculate the water level of the drum and the steam mass flow which leaves the drum. Therefore no additional balance equation for the drum is available which can be included in the SIMPLE algorithm. To solve this problem, the calculation domain for the whole steam generator is subdivided into four sub-domains. This sub-domains include the following tube networks of the steam generator:

- evaporator,

- feed water system,
- drum and
- superheater system

and can be seen in figure 7.

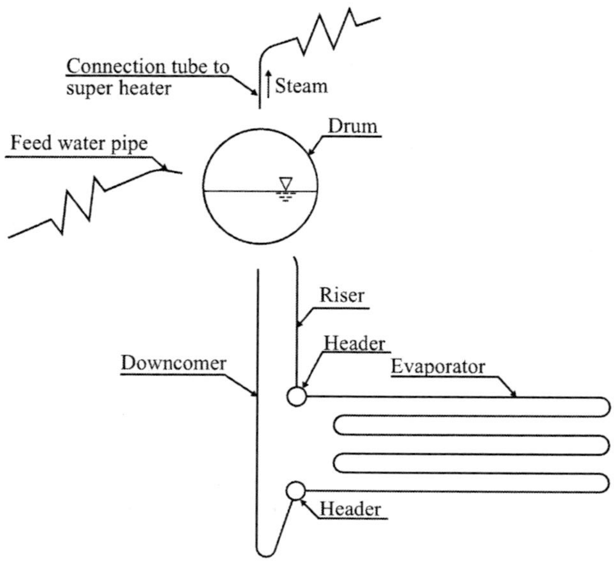

Figure 7. Sub-systems of the steam generator.

Figure 7 shows the different sub-domains for the numerical simulation at a simplified sketch of a natural circulation heat recovery steam generator (HRSG). At the connection points of the sub-systems evaporator, feed water and superheater to the drum it is necessary to define internal boundary conditions for the calculation. This internal boundary conditions must be defined in such a way that they are physically correct. For the feed water system and the evaporator with his downcomer and riser tubes the pressure must be defined. For the evaporator system as second condition the spec. enthalpy must be given at that tubes where the mass flow enters the downcomer or, in case of a reverse flow, the riser tubes. For the feed water tubes the second boundary condition is needed only in case of a reverse flow. For the tube where the steam leaves the drum the mass flow $\dot{m}_{st,D,out}$, which is calculated with equation (68), and the spec. enthalpy of the steam must be defined.

The drum pressure can be calculated with following formula:

$$p_D = p_{1,k} - \zeta_{out} \frac{\dot{m}^2_{st,D,out}}{2\,\rho_{st,D,out}\,A^2_{1,k}} \tag{69}$$

with the pressure loss coefficient at the tube where the steam mass flow leaves the drum ζ_{out}. The nomenclature for equation (69) is shown in figure 8. It should be noted at this point that a additional iteration over the sub-domains is necessary to calculate the drum pressure. During the iteration over the sub-systems the drum pressure does not change.

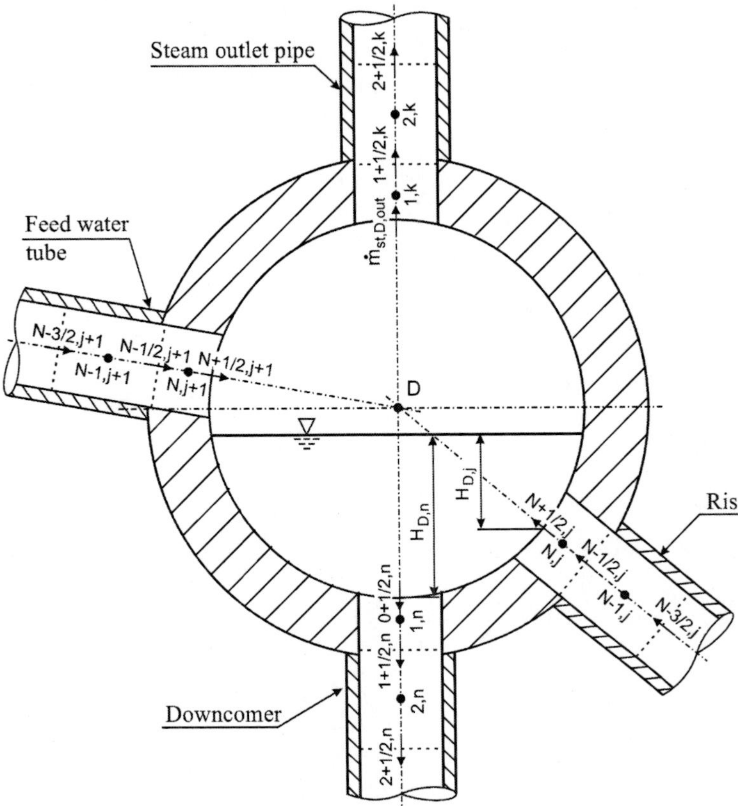

Figure 8. Control volume of the drum and drum connected tubes.

Figure 8 shows a sketch with the control volume of the drum and the drum connected tubes. These tubes should be representative for all other tubes which are connected to the drum. The variables of the drum are denoted with the index D; j represents values at the drum entrance and k and n the values at the outlet. A is the cross sectional area of the connected tubes.

In the following the equations necessary to calculate the drum connected control volumes will be presented.

At first the equations for the control volume where the steam leaves the drum are described. The discretized momentum balance for computation of the velocity at the staggered grid for the control volume $1+\frac{1}{2},k$ can be written as:

$$a_{e1,k}w_{1+\frac{1}{2},k} = a_{w1,k}\frac{\dot{m}_{st,D,out}}{\rho_{st,D,out}A_{1+\frac{1}{2},k}} + (p_{1,k} - p_{2,k})A_{1+\frac{1}{2},k} + b_{1,k} \tag{70}$$

with the coefficients:

$$a_{e1,k} = a_{w1,k} + a^0_{e1,k} - S_{p1,k} A_{1+\frac{1}{2},k} \Delta x_{1+\frac{1}{2},k} \tag{71}$$
$$a_{w1,k} = \dot{m}_{st,D,out}, \tag{72}$$
$$b_{1,k} = S_{c1,k} A_{1+\frac{1}{2},k} \Delta x_{1+\frac{1}{2},k} + a^0_{e1,k} w^0_{1+\frac{1}{2},k} \quad \text{and} \tag{73}$$
$$a^0_{e1,k} = \frac{\rho^0_{1+\frac{1}{2},k} A_{1+\frac{1}{2},k} \Delta x_{1+\frac{1}{2},k}}{\Delta t}. \tag{74}$$

The pressure correction equation results to:

$$a_{mP1,k} p'_{1,k} = a_{mE1,k} p'_{2,k} + b_{1,k} \tag{75}$$

where

$$a_{mP1,k} = a_{mE1,k}, \tag{76}$$
$$a_{mE1,k} = (\rho A)_{1+\frac{1}{2},k} d_{e1,k} \quad \text{and} \tag{77}$$
$$b_{m1,k} = \frac{(\rho^0_{1,k} - \rho_{1,k}) A_{1,k} \Delta x_{1,k}}{\Delta t} + \dot{m}_{st,D,out} - (\rho A)_{1+\frac{1}{2},k} w^*_{1+\frac{1}{2},k} \tag{78}$$

with

$$d_{e1,k} = \frac{A_{1,k}}{a_{e1,k}}.$$

The algebraic equation for the energy balance of the drum connected control volume where the steam mass flow leaves the drum is reduced to

$$a_{P1,k} h_{1,k} = b_{1,k}. \tag{79}$$

with the coefficients:

$$a_{P1,k} = 1 \quad \text{and} \tag{80}$$
$$b_{1,k} = h_{st,D,out}. \tag{81}$$

For the remaining drum connected tubes with outgoing fluid (downcomer tubes and in case of reverse flow the riser or the feed water tubes) the following balance equations must be used.

The momentum balance for the tubes with the index n (downcomer) is given as follows:

$$a_{e0,n} w_{0+\frac{1}{2},n} = a_{ee0,n} w_{1+\frac{1}{2},n} + (p_D - p_{1,n}) A_{1+\frac{1}{2},n} + b_{0,n} \tag{82}$$

with

$$a_{e0,n} = a_{ee0,n} + a^0_{e0,n} - S_{P0,n} A_{0+\frac{1}{2},n} \Delta x_{0+\frac{1}{2},n}, \tag{83}$$
$$a_{ee0,n} = \left(\frac{\eta_{1+\frac{1}{2},k}}{\Delta x_{1+\frac{1}{2},k}} + [[-(\rho w)_{1+\frac{1}{2},n}, 0]] \right) A_{1+\frac{1}{2},n}, \tag{84}$$
$$b_{e0,n} = S_{c0,n} A_{0+\frac{1}{2},n} \Delta x_{0+\frac{1}{2},n} + a^0_{e0,n} w^0_{0+\frac{1}{2},n} \quad \text{and} \tag{85}$$
$$a^0_{e0,n} = \frac{\rho^0_{0+\frac{1}{2},n} A_{0+\frac{1}{2},n} \Delta x_{0+\frac{1}{2},n}}{\Delta t}. \tag{86}$$

Analogous to the header model the proportional part of the source term for equation (85) takes also into consideration the pressure loss due to acceleration ζ_{in} and leads to:

$$S_{p0,n}\Delta x_{0+\frac{1}{2},n} = -\frac{|\Delta p_{fric0+\frac{1}{2},n}|}{|w_{0+\frac{1}{2},n}|} - (1+\zeta_{in})\frac{\rho_{0+\frac{1}{2},n} w_{0+\frac{1}{2},n}[[w_{0+\frac{1}{2},n}, 0]]}{2\left|w_{0+\frac{1}{2},n}\right|}. \quad (87)$$

Values for ζ_{in} are defined e.g. in [67] or [27]. The static hight which is a function of the difference in hight between the actual water level of the drum and the center line of the pipe n must be taken into account in the constant part of the linearized source term (see figure 8).

$$S_{c0,n}\Delta x_{0+\frac{1}{2},n} = -\rho_{0+\frac{1}{2},n} g H_{D,n} \quad (88)$$

The pressure correction equation of the control volume $1,n$ is given as follows:

$$a_{mP1,n}\, p'_{1,n} = a_{mE1,n}\, p'_{2,n} + b_{1,n} \quad (89)$$

where

$$a_{mP1,n} = a_{mE1,n}, \quad (90)$$

$$a_{mE1,n} = (\rho A)_{1+\frac{1}{2},n} d_{e1,n} \quad \text{and} \quad (91)$$

$$b_{m1,n} = \frac{(\rho^0_{1,n} - \rho_{1,n}) A_{1,n}\Delta x_{1,n}}{\Delta t} + (\rho A)_{0+\frac{1}{2},n} w^*_{0+\frac{1}{2},n} - (\rho A)_{1+\frac{1}{2},n} w^*_{1+\frac{1}{2},n} \quad (92)$$

with

$$d_{e1,n} = \frac{A_{1,n}}{a_{e1,n}}.$$

The conditional equation for the spec. enthalpy $h_{1,n}$ of the tube n, which is independent from the flow direction of the fluid, is given with:

$$a_{P1,n}\, h_{1,n} = a_{W1,n}\, h_{l,D} + a_{E1,n}\, h_{2,n} + b_{1,n} \quad (93)$$

with the coefficients

$$a_{P1,n} = a_{W1,n} + a_{E1,n} + a^0_{P1,n} - (S_P A\Delta x)_{1,n}, \quad (94)$$

$$a_{W1,n} = [[(\rho w)_{0+\frac{1}{2},n}, 0]] A_{0+\frac{1}{2},n}, \quad (95)$$

$$a_{E1,n} = [[-(\rho w)_{1+\frac{1}{2},n}, 0]] A_{1+\frac{1}{2},n}, \quad (96)$$

$$b_{1,n} = (S_c A\Delta x)_{1,n} + (a^0_{hP} h^0)_{1,n} \quad \text{and} \quad (97)$$

$$a^0_{P1,n} = \frac{\rho^0_{1,n} A_{1,n}\Delta x_{1,n}}{\Delta t}. \quad (98)$$

The equations for a control volume at the tube outlet (subscript j) are similar to the equations presented above. Therefore they are not described here in detail. This set of equations can be derived from the equations (82) to (98).

For the pressure correction algorithm SIMPLER the discretized equations for the drum model are presented in detail in [86].

The model for calculating the heat transfer through the thick-walled drum wall will not described here in detail. It should be referred to the literature [86] or [104].

2.3.2. Subcooling Model for the Downcomer Mass Flow

As mentioned above the mass flow which enters the downcomer can be subcooled. This behavior is a result of the drum design. Therefore the liquid mass flow which leaves the drum through the downcomer can have the following two extreme positions:

- No feed water enters directly the downcomer pipe. In this case the feed water is completely mixed with the saturated liquid inside the drum. Only saturated water leaves the drum through the downcomer tubes.
- The feed water which enters the drum is supplied directly beside the downcomer entrance. Based on this model assumption a mixture of the feed water and the saturated liquid will appear and leave the drum through the downcomer.

The possibility of a subcooling of the downcomer mass flow will be allowed for by a so called subcooling factor f_U. This factor must be designed in such a way that the two extreme positions for subcooling the downcomer mass flow can be calculated. With this factor the fraction of feed water mass flow is defined which enters directly into the downcomer. The subcooling factor used in the current model is defined with

$$f_U = \frac{\Delta h_U}{\Delta h_{U,max}} \tag{99}$$

with the difference of the spec. enthalpy for subcooling:

$$\Delta h_U = h_{sat} - h_l \tag{100}$$

where $\Delta h_{U,max}$ is the highest value for subcooling which can be reached.

Figure 9 shows the model used for subcooling the downcomer mass flow. In this model it is possible to connect k feed water and n downcomer pipes. ξ_U describes the mass flow ratio of feed water which enters directly into the downcomer. In the current model a averaged value for the feed water enthalpy can be calculated with following equation:

$$\overline{h}_{feed} = \frac{\sum\limits_{i=1}^{k} \dot{m}_{feed_i} h_{feed_i}}{\dot{m}_{feed}}. \tag{101}$$

with $\dot{m}_{feed} = \sum_{i=1}^{k} \dot{m}_{feed_i}$.

The total feed water mass flow $\dot{m}_{feed}$ which enters the drum will be distributed in the following way:

- the feed water mass flow $\dot{m}_{feed}(1-\xi_U)$ is mixed with the saturated liquid in the drum and
- the feed water mass flow $\dot{m}_{feed}\xi_U$ enters directly into the downcomer pipes. The feed water and the mass flow from the drum are mixed homogeneous. It is assumed that the feed water mass flow $\dot{m}_{feed}\xi_U$ is evenly distributed to the downcomer tubes.

The spec. enthalpy at the inlet of the downcomer tube i is given by following formula:

$$h_{Do_i} = \frac{\dot{m}_{Do_i} h_{l,Do} + \frac{\xi_U}{n_{Do}} (\overline{h}_{feed} - h_{l,Do}) \dot{m}_{feed}}{\dot{m}_{Do_i}} \tag{102}$$

with the number of downcomer tubes n_{Do}.

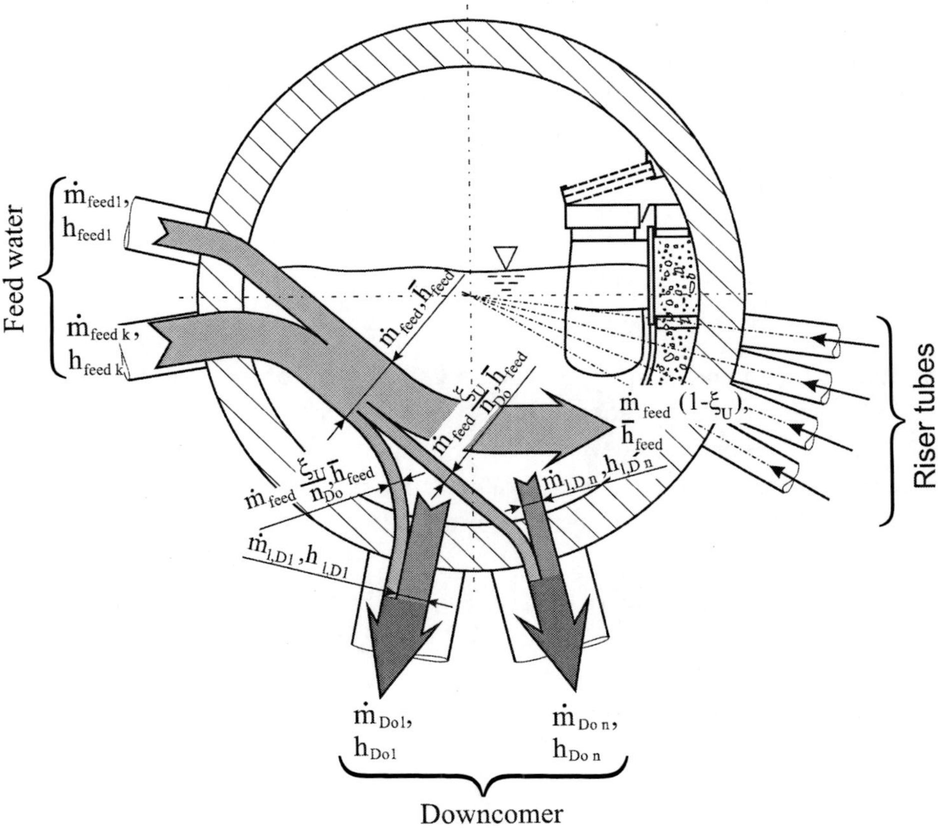

Figure 9. Model for subcooling downcomer mass flow.

2.3.3. Controller for the Water Level

During the operation, especially during load changes and start up, the water level inside the drum oscillates as well as changes. Therefore it is necessary to control the water level because the water level should not exceed the variation limits (high and low water level) defined by the boiler manufacturer.

The widely-used controllers have according to [37] a P-, PI- or PID behavior. In the literature, e.g. [37], [81] or [39], the idealized differential equation for the PID-controller is given with:

$$u_{out}(t) = K\left[u_{in}(t) + \frac{1}{T_I}\int_0^t u_{in}(t)\mathrm{d}t + T_{Der}\frac{\mathrm{d}u_{in}(t)}{\mathrm{d}t}\right]. \tag{103}$$

Based on the different nature of the I- and D-part of the PID controller the approximation must be done with different methods. According to [81] it is favorable to discretize the differential quotient of equation (103) with a backward finite difference scheme of first order whereas for the continuous integration of the I-part two methods will be suggested in the literature, e. g. [37], [81] or [39], for the approximation. The first method approximates the integral with the sum of rectangles while for the second method the so called Tustin-method (bilinear transformation or trapezium integration) is used. With the Tustin-method a better approximation should be given [39]. With the help of the backward finite difference scheme of first order for the differential quotient and the Tustin-Method for the integration part of the controller equation (103) can be written in the following form:

$$u_{out}(t) = K\left[u_{in}(t) + \frac{t_0}{T_I}\left(\frac{u_{in}(0) - u_{in}(t)}{2} + \sum_{i=1}^{t-1} u_{in}(i)\right) + \frac{T_{Der}}{t_0}\left(u_{in}(t) - u_{in}(t-\Delta t)\right)\right], \quad (104)$$

where t_0 denotes the sampling time. This non-recursive form of the controller algorithm is not suitable for numerical applications because for the generation of the output variable of the controler $u_{out}(t)$ all previous deviations in regulation $u_{in}(t)$ must be stored.

Subtraction of equation (104) for the last deviation in regulation $u_{in}(t-\Delta t)$ from equation (104) results in the recursive algorithm for the PID-controller.

$$u_{out}(t) = u_{out}(t-\Delta t) + v_0 u_{in}(t) + v_1 u_{in}(t-\Delta t) + v_2 u_{in}(t-2\Delta t) \quad (105)$$

with the parameters

$$v_0 = K\left(1 + \frac{T_{Der}}{t_0} + \frac{t_0}{2T_I}\right), \quad (106)$$

$$v_1 = -K\left(1 + 2\frac{T_{Der}}{t_0} - \frac{t_0}{2T_I}\right) \quad \text{and} \quad (107)$$

$$v_2 = K\frac{T_{Der}}{t_0}. \quad (108)$$

Equation (105) is a more useful formula for the PID-controller included in numerical simulation applications.

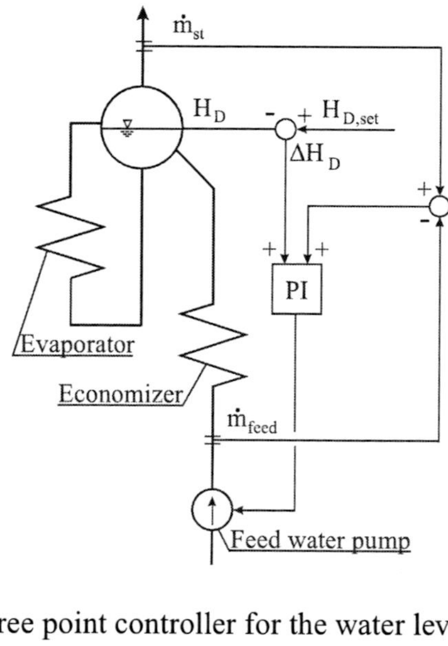

Figure 10. Three point controller for the water level of the drum.

In figure 10 a so called three point controller for the water level of the drum is shown. The actual water level of the liquid in the drum H_D will be subtracted from the set point for the water level $H_{D,set}$. The resulting difference must be converted into a value for the mass difference between the actual and the set point water level with the following formula:

$$\Delta \dot{m}_{l,D} = \frac{\Delta M_{l,D}}{\Delta t} = \frac{[A_{l,D,set} - A_{l,D}(H_D)]l_D}{\Delta t} \rho_{l,D} \tag{109}$$

with the cross section area of the liquid in the drum at the set point $A_{l,D,set}$. $\Delta \dot{m}_{l,D}$ is one of the input values for the PI-controller shown in figure 10. The second input value for the PI-controller is the difference between the steam mass flow which leaves the drum and the feed water mass flow. The output of the PI-controller is the correction value for the feed water mass flow $\Delta \dot{m}_{feed}$.

The advantage of this controller concept is the possibility to charge the feed water mass flow during a load change in a very smooth way. This is a result of adding the difference between the steam and the feed water mass flow as second input value for the PI-controller.

Beside the controller for the drum water level, a second control loop for the numerical simulation as well as for the operation of the drum is necessary. This second control loop is used to check the upper limit for the water level and can be seen in figure 11.

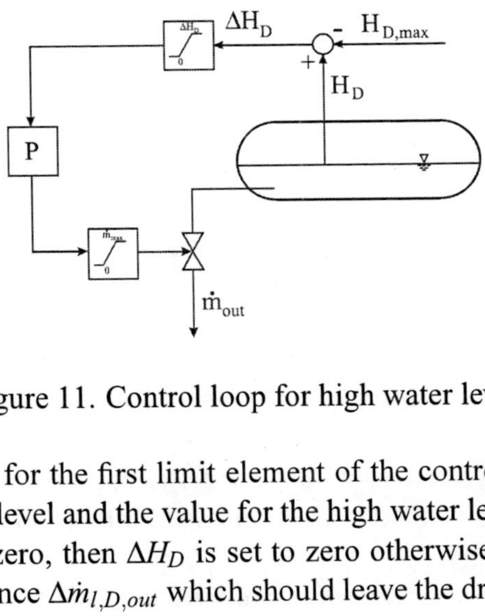

Figure 11. Control loop for high water level.

The input value ΔH_D for the first limit element of the controller loop is the difference between the actual water level and the value for the high water level $H_{D,max}$. If the resulting difference is lower than zero, then ΔH_D is set to zero otherwise ΔH_D will be recalculated into the mass flow difference $\Delta \dot{m}_{l,D,out}$ which should leave the drum. This mass flow can be calculated with

$$\Delta \dot{m}_{l,D,out} = \frac{[A_{l,D}(H_D) - A_{D,max}]l_D}{\Delta t} \rho_{l,D} \tag{110}$$

and is used for the P-controller as input value. The output value of the P-controller passes a second limiter and the resulting value is the actuating variable for the shut-of valve.

2.4. Model for the Flue Gas Pass

Corresponding to the fluid flow in the tubes of the boiler the structure for the flue gas side was conceived in the same way. That means, that the flue gas pass is also build up as a network.

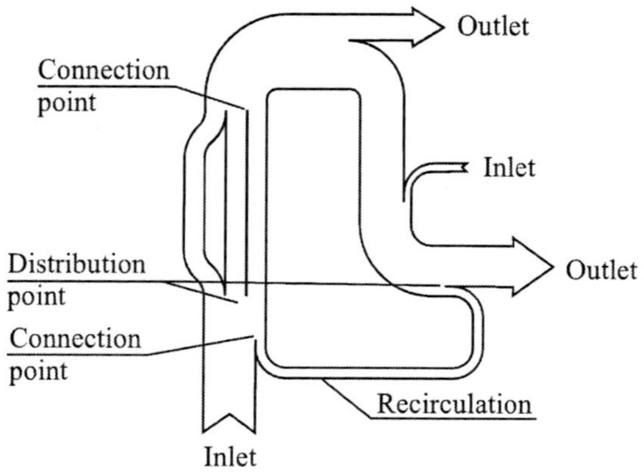

Figure 12. Structure elements for the flue gas pass.

The structure elements for the flue gas pass – connection and distribution points, recirculation or inlets and outlets – which are necessary to build up the flue gas passes used in the energy and chemical industry is presented in figure 12. The number of flue gas mass flow inlet or outlet points is not limited in the present model. In dependence on the design of the steam generator the following boundary conditions for the flue gas network must be realized:

- the flue gas mass flow and temperature is given or
- the fuel and air mass flow as well as their temperatures must be given.

Both boundary conditions are Dirichlet-conditions. With the boundary condition presented in point 1 it is possible to connect the flue gas pass of the boiler with every other process, if the time history of the discharged mass flow and temperature or spec. enthalpy is known. In this case it is possible e.g. to simulate the dynamic behavior of a heat recovery steam generator (HRSG) arranged downstream of a e.g. gas turbine or blast furnace if the time rate of change of the blast furnace or gas turbine exhaust flue gas mass flow and temperature is known. In this case the flue gas composition must be also known for calculating the flue gas properties. The arrangement of a combustion chamber downstream of the flue gas pass results on the other hand to a situation where the fuel and air mass flow as well as their temperatures must be given.

As shown in figure 12 it is possible in the presented model to subdivide and mix the flue gas mass flow of a single flue gas pass. With this possibility to divide the flue gas mass flow into different streams the simulation of a non-uniform mass flow distribution can be done. This different flue gas streams can be a result of e.g. the shut down of coal mills or a unfavorable design of the flue gas pass. A heat transfer between the flue gas streams is not included in the presented model. In the next sections the mathematical models for the flue gas will be presented.

2.4.1. Balance Equations for the Flue Gas Pass

Corresponding to the tube network of the working medium the fluid flow of the flue gas is also modeled one dimensional. In the present model the transition between two different steady states conditions should be simulated but it was out of interest to calculate the pressure drop or the pressure variation in the flue gas column in case of a burner failure and so on. Therefore the momentum balance was neglect for the flue gas.

The flue gas mass flow was modeled quasi-stationary and the mass flow partitioning in the distributor elements was done equivalent to the ratio of the cross section area between the different flue gas channels.

Contrary to the momentum and continuity balance the energy balance is solved unsteady. The discretization of the energy balance will be presented in the next section.

Energy Balance for the Flue Gas Figure 13 shows the grid structure and the storage locations for the dependent variables. The symbol • located in the middle of every control volume defines the calculation point for the cell. A typical control volume for the energy balance is denoted by the shaded area of figure 13. The velocities, which are located at the interface between two control volumes, e. g. cell i and $i+1$, are denoted with the index $i+1/2$.

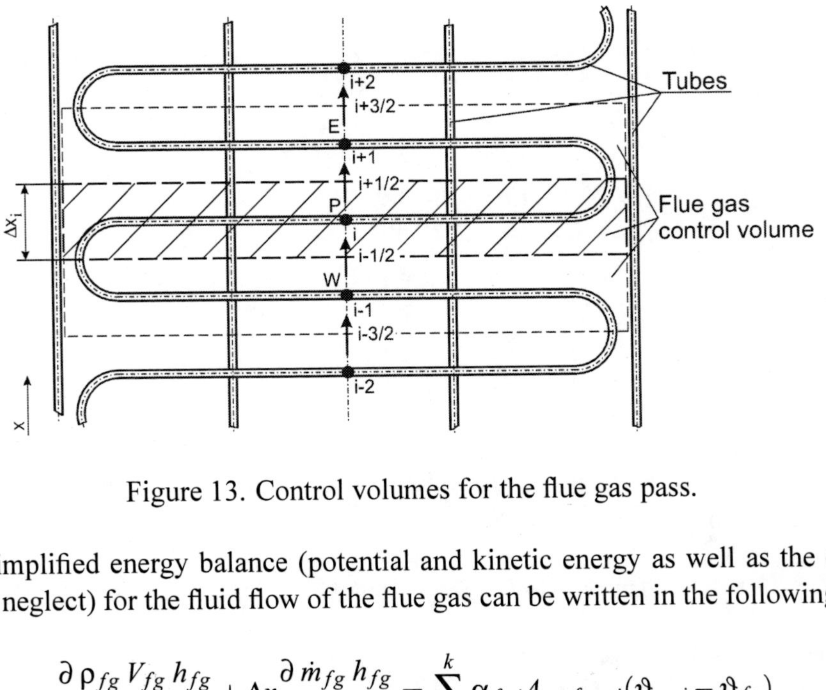

Figure 13. Control volumes for the flue gas pass.

The simplified energy balance (potential and kinetic energy as well as the expansion work was neglect) for the fluid flow of the flue gas can be written in the following form:

$$\frac{\partial \rho_{fg} V_{fg} h_{fg}}{\partial t} + \Delta x \frac{\partial \dot{m}_{fg} h_{fg}}{\partial x} = \sum_{j=1}^{k} \alpha_{fgj} A_{surf,waj} (\vartheta_{waj} - \vartheta_{fg}) \tag{111}$$

with the spec. enthalpy h_{fg} of the flue gas, the flue gas mass M_{fg}, the flue gas mass flow $\dot{m}_{fg}$, the surface area $A_{surf,waj}$ of the different heating surfaces j, the heat transfer coefficient between flue gas and tube wall α_{fgj}, the flue gas temperature ϑ_{fg} and the averaged outer tube wall temperature ϑ_{waj} of the heating surface.

In figure 14 a more detailed sketch of the control volume i presented in figure 13 is shown. Besides the inlet and outlet flue gas mass flows also the averaged outer tube wall temperatures $\vartheta_{wai,j}$ and the associated heat transfer coefficients $\alpha_{fgi,j}$ of the different heating surfaces j can be seen.

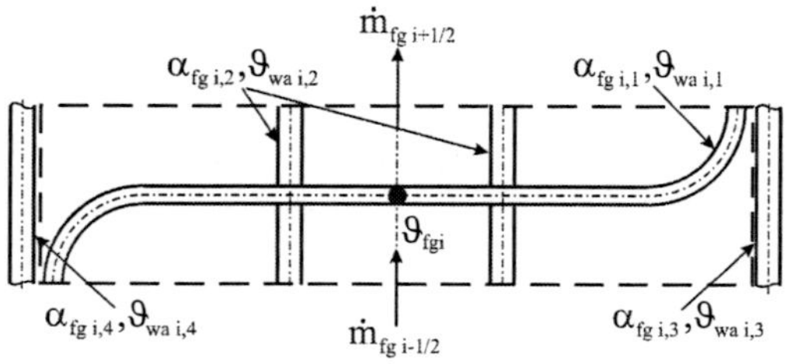

Figure 14. General control volume of the flue gas pass.

The calculation of the averaged outer tube wall temperature of a heating surface j is done in such a way that a weighted arithmetic average over the different outer tube wall temperatures of the tube control volumes was accomplished. The weighting was done with the help of the ratio between the outer surface area of the tube control volume and the outer surface area of the whole heating surface inside the control volume of the flue gas.

Integration of the balance equation for the energy (111) over the control volume results in

$$a_{Pi}h_{fgi} = a_{Wi}h_{fgi-1} + a_{Ei}h_{fgi+1} + b_{fgi} + \sum_{j=1}^{k} \alpha_{fgi,j}A_{surf,wai,j}(\vartheta_{wai,j} - \vartheta_{fgi})$$

with the coefficients

$$a_{Pi} = a_{Wi} + a_{Ei} + a^0_{Pi} - S_{Pi}A_{fgi}\Delta x_i, \tag{112}$$

$$a_{Wi} = [[\dot{m}_{fg\,i-\frac{1}{2}}, 0]], \tag{113}$$

$$a_{Ei} = [[-\dot{m}_{fg\,i+\frac{1}{2}}, 0]], \tag{114}$$

$$b_i = S_{ci}A_{fgi}\Delta x_i + a^0_{Pi}h^0_{fgi} \quad \text{and} \tag{115}$$

$$a^0_{Pi} = \frac{\rho^0_{fgi}\, V^0_{fgi}}{\Delta t}. \tag{116}$$

Substitution of the spec. enthalpy h_{fgi} in equation (112) with $\overline{c}_{p\,fgi}\vartheta_{fgi}$ results in a equation for the flue gas temperature ϑ_{fgi}:

$$a_{Pi}\vartheta_{fgi} = a_{Wi}\overline{c}_{p\,fgi-1}\vartheta_{fgi-1} + a_{Ei}\overline{c}_{p\,fgi+1}\vartheta_{fgi+1} + b_i \tag{117}$$

with the coefficients for the diskretized energy balance of the flue gas

$$a_{Pi} = \overline{c}_{p\,fgi}\left(a_{Wi} + a_{Ei} + a^0_{Pi} - S_{Pi}A_{fgi}\Delta x_i\right) + \sum_{j=1}^{k} \alpha_{fgi,j}A_{surf,wai,j}, \tag{118}$$

$$a_{Wi} = [[\dot{m}_{fg\,i-\frac{1}{2}}, 0]], \tag{119}$$

$$a_{Ei} = [[-\dot{m}_{fg\,i+\frac{1}{2}}, 0]], \tag{120}$$

$$b_i = S_{ci}A_{fgi}\Delta x_i + a^0_{Pi}\overline{c}^0_{p\,fgi}\vartheta^0_{fgi} + \sum_{j=1}^{k} \alpha_{fgi,j}A_{surf,wai,j}\vartheta_{wai,j} \quad \text{and} \tag{121}$$

$$a^0_{Pi} = \frac{\rho^0_{fgi}\, V^0_{fgi}}{\Delta t}. \tag{122}$$

With the source term S_{ci} and S_{Pi} it is possible to take into account an additional heat source or sink in the control volume i. The additional heat source must be included in the source term S_{ci} or S_{Pi} in the following form:

$$S_{ci} A_{fgi} \Delta x_i = [[\dot{Q}, 0]] \tag{123}$$

or

$$S_{pi} A_{fgi} \Delta x_i = -\frac{[[-\dot{Q}, 0]]}{\overline{c}_{p\,fgi} \vartheta_{fgi}}. \tag{124}$$

For the one dimensional energy balance of the flue gas the resulting matrix is tri-diagonal and can be solved with the TDMA (Tri-Diagonal Matrix Algorithm) solver.

Energy balance for the connection and distribution control volumes With the help of equation (117) the energy balance for a flue gas pass without distributions and connections can be solved. But in case of e.g. a distribution of the flue gas mass flow a another flue gas energy balance must be solved.

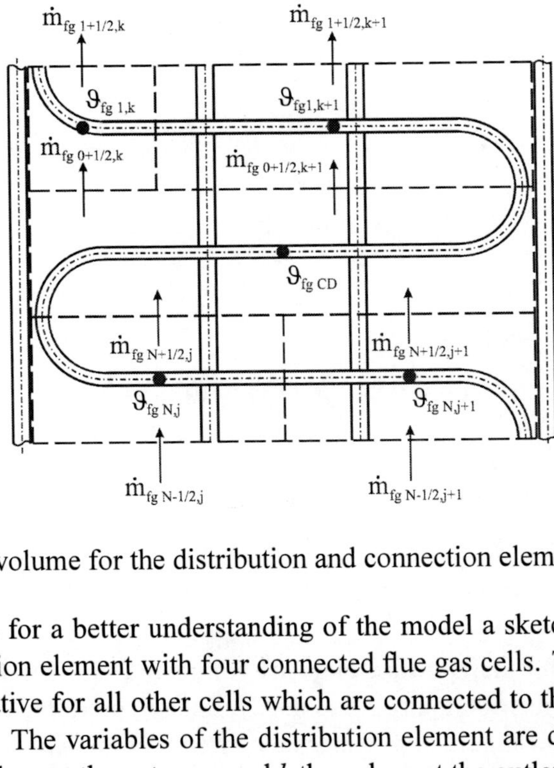

Figure 15. Control volume for the distribution and connection element of a flue gas pass.

Figure 15 shows for a better understanding of the model a sketch of a distribution respectively a connection element with four connected flue gas cells. These control volumes should be representative for all other cells which are connected to the distribution element of the flue gas pass. The variables of the distribution element are denoted with the index *CD*; *N* represents values at the entrance and *k* the values at the outlet.

In figure 16 a more detailed sketch of the control volume i presented in figure 15 is shown. The averaged outer tube wall temperatures ϑ_{wa} and the associated heat transfer coefficients α_{fg} of the different heating surfaces can be seen.

The flue gas temperature of a distribution or connection element is not independent from the neighbor control volumes and is given with:

$$a_{PCD} \vartheta_{fgCD} = \sum_j a_{W\,CD,j} \overline{c}_{p\,fgN,j} \vartheta_{fgN,j} + \sum_k a_{E\,CD,k} \overline{c}_{p\,fg1,k} \vartheta_{fg1,k} + b_{CD} \tag{125}$$

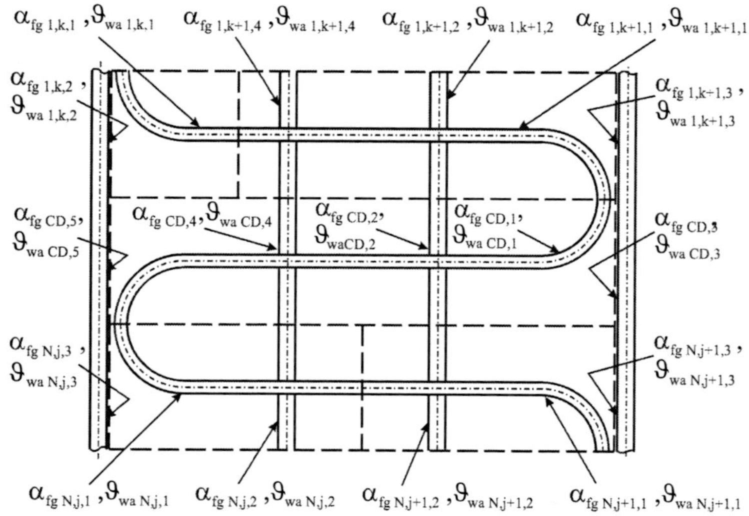

Figure 16. Arrangement of the heating surfaces in the control volume for the distribution and connection element of the flue gas pass.

where

$$a_{PCD} = \overline{c}_{p\,fgCD}\left(\sum_j a_{W\,CD,j} + \sum_k a_{E\,CD,k} + a^0_{PCD} - S_{PCD}A_{fgCD}\Delta x_{CD}\right) + \sum_i \alpha_{fgCD,i}A_{surf,waCD,i}, \quad (126)$$

$$a_{W\,CD,j} = [[\dot{m}_{fg\,N+\frac{1}{2},j}, 0]], \quad (127)$$

$$a_{E\,CD,k} = [[-\dot{m}_{fg\,0+\frac{1}{2},k}, 0]], \quad (128)$$

$$b_{CD} = S_{c\,CD}A_{fgCD}\Delta x_{CD} + a^0_{PCD}\overline{c}^0_{p\,fgCD}\vartheta^0_{fgCD} + \sum_i \alpha_{fgCD,i}A_{surf,waCD,i}\vartheta_{waCD,j} \quad \text{and} \quad (129)$$

$$a^0_{PCD} = \frac{\rho^0_{fgCD}V^0_{fgCD}}{\Delta t}. \quad (130)$$

The tri-diagonal form of the coefficient matrix will be lost for the distribution element.

Energy balance for the distribution element connected control volumes Figure 16 shows additional to the distribution cell also the distributor connected control volumes. The counting direction for the distributor connected control volumes corresponds with the flow direction of the flue gas mass flow. If the flue gas mass flow leaves the distributor control volume, then the counting will start with the value 1 otherwise the counting will be stopped.

The mass flow which enters the distribution control volume through the flue gas pass N is indicated by the index $N+\frac{1}{2}, j$ whereas the flue gas mass flow which leaves the distribution cell through the flue gas pass designated with k gets the index $0+\frac{1}{2}, k$.

The algebraic energy balance for the control volume N of the flue gas pass j is given with:

$$a_{PN,j}\vartheta_{fgN,j} = a_{W\,N,j}\overline{c}_{p\,fgN-1,j}\vartheta_{fgN-1,j} + a_{E\,N,j}\overline{c}_{p\,fgCD}\vartheta_{fgCD} + b_{N,j} \quad (131)$$

with the coefficients

$$a_{PN,j} = \bar{c}_{pfgN,j}\left(a_{WN,j} + a_{EN,j} + a^0_{PN,j} - S_{PN,j}A_{fgN,j}\Delta x_{N,j}\right) + \sum_i \alpha_{fgN,j,i}A_{surf,waN,j,i}, \tag{132}$$

$$a^0_{PN,j} = \frac{\rho^0_{fgN,j}\, V^0_{fgN,j}}{\Delta t}, \tag{133}$$

$$a_{WN,j} = [[\dot{m}_{fg\,N-\frac{1}{2},j}, 0]], \tag{134}$$

$$a_{EN,j} = [[-\dot{m}_{fg\,N+\frac{1}{2},j}, 0]] \quad \text{and} \tag{135}$$

$$b_{N,j} = S_{cN,j}A_{fgN,j}\Delta x_{N,j} + a^0_{PN,j}\bar{c}^0_{pfgN,j}\vartheta^0_{fgN,j} + \sum_j \alpha_{fgN,j,i}A_{surf,waN,j,i}\vartheta_{waN,j,i}. \tag{136}$$

The equations for a control volume at the flue gas pass inlet (subscript k) are similar to the equations presented above. Therefore they are not described here in detail. This set of equations can be derived from the equations (131) to (136).

2.5. Heat Transfer Model for the Fluid Flow in Tubes

For the numerical simulation of steam generators it is also necessary to include models for the heat transfer between the fluid and the fluid surrounding wall. In this section the heat transfer model used in the simulation program DBS will be described.

As mentioned above the natural circulation steam generator works under sub-critical conditions up to approximately 180 bars (maximum operation pressure for the drum). Therefore it is not necessary to include in the model for the heat transfer equations to describe the super-critical region as well as the region around the critical point of the working medium where pseudo-critical boiling occur. Correlations for this region are presented in the literature among others [1], [21], [20], [74], [60], [73], [111], [40], [34] or [32].

The sub-critical region of the working medium water must be subdivided for the heat transfer model into a single- and two-phase flow region.

2.5.1. Convective Heat Transfer for the Single-Phase Flow

For the convective heat transfer of the fully developed single-phase flow the equation according to [30] as well as [31]

$$Nu_{tur} = \frac{\zeta_{fric}}{8}\frac{(Re-1000)\,Pr_{wm}}{1+12.7\left(Pr_{wm}^{2/3}-1\right)\sqrt{\frac{\zeta_{fric}}{8}}}K_1, \tag{137}$$

can be used. This equation is valid for both the single-phase flow water and steam. For the pressure loss coefficient due to friction ζ_{fric} the equation according to [51] must be used.

$$\zeta_{fric} = (1.8\log Re - 1.5)^{-2} \tag{138}$$

The correction factor K_1 is used to take into consideration the temperature dependence of the properties of water. Whereby in particular the Prandtl-number is used to take into consideration the high dependence of the water viscosity from the temperature.

$$K_1 = \left(\frac{Pr_{wm}}{Pr_{wa}}\right)^{0.11} \tag{139}$$

Pr_{wm} represents the Prandtl-number calculated at the averaged fluid temperature and Pr_{wa} represents the Prandtl-number calculated at the averaged wall temperature.
Gnielinski [31] recommends for the correction factor K_1 in case of a single-phase steam flow the following formula:

$$K_1 = \left(\frac{T_{wm}}{T_{wa}}\right)^{-0.18}, \tag{140}$$

with the averaged temperature of the working medium T_{wm} and the averaged wall temperature T_{wa}.

The definition region for equation (137) is given with

$$\begin{array}{lll} 2300 & \leq Re \leq & 10^6 \\ 0.6 & \leq Pr \leq & 10^5 \quad \text{and} \\ & d_i/l \leq & 1. \end{array}$$

For the laminar region of the fluid flow in a tube the correlation

$$Nu_{lam} = 4.364 \tag{141}$$

according to [82] is used, whereby the definition region is given for Reynold-numbers $Re <$ 2300.

In the transition region between laminar and turbulent flow, $2300 \leq Re < 10^4$, a linear interpolation between the two equations (137) and (141) will be done.

$$Nu = \frac{Nu_{tur}(Re - 2300) + Nu_{lam}(10000 - Re)}{7700} \tag{142}$$

2.5.2. Convective Heat Transfer for the Two-Phase Flow

The description of the heat transfer in the two-phase flow region is more complex then for the single phase region. Therefore the model for the two-phase flow heat transfer will be subdivided into the following areas: subcooled and boiling region, critical heat flux and post dryout region.

Subcooled boiling and boiling region In the subcooled region the wall temperature is a small amount higher than the saturation temperature and approximately constant. At the same time the core temperature of the fluid is smaller than the saturation temperature and increases. For a subcooled vertical upward flow the correlation of Jens and Lottes [38]

$$\vartheta_{wa} - \vartheta_{sat} = 0.79057\dot{q}^{0.25}e^{-\left(p/62.05\cdot 10^5\right)} \tag{143}$$

was used to calculate the temperature difference between the wall and the saturation temperature. This equation is valid for higher operation pressure and higher heat flux. For the pressure range of $p < 50$ bars and a heat flux lower than 300 kW/m^2 the modified correlation of Thom *et al.* [78]

$$\vartheta_{wa} - \vartheta_{sat} = 0.02265\dot{q}^{0.5}e^{-(p/87\cdot 10^5)} \tag{144}$$

gives the better values.

The heat transfer coefficient α for the subcooled region – from single-phase flow up to saturated boiling – of a vertical upward flow can be described with:

$$\alpha = \frac{\dot{q}}{\vartheta_{sat} - \vartheta_{bal} + (\vartheta_{wa} - \vartheta_{sat})}. \tag{145}$$

For the temperature difference $(\vartheta_{wa} - \vartheta_{sat})$ the values calculated with equation (143) respectively (144) must be used. The fluid temperature ϑ_{bal} must be calculated with the help of the energy balance.

The region for nucleate boiling of the saturated fluid follows the region of subcooled boiling. Under the assumption of a thermodynamic equilibrium the starting point for nucleate boiling is defined as the one who the mass vapor quality is identical to zero. In this case the mechanism of the heat transfer is independent from the mass flow and the local enthalpy and therefore identical to subcooled boiling. Equation (145) is also valid for nucleate boiling, if it is taken into account that the fluid has reached the saturation temperature. In this case equation (145) can be simplified to the following formula:

$$\alpha = \frac{\dot{q}}{(\vartheta_{wa} - \vartheta_{sat})}. \tag{146}$$

The temperature difference $(\vartheta_{wa} - \vartheta_{sat})$ must be also calculated with equation (143) respectively (144).

In the most cases the heat transfer correlations are valid for a vertical tube with upward mass flow. Therefore it is of great importance to check the used equations for their validity for horizontal and inclined tubes. In comparison to vertical tubes it should be not possible for horizontal and inclined tubes to design a heat transfer model which is independent from the dimensions and the properties of the tube [36].

As mentioned above for the vertical tube, the wall temperature in the subcoled region is approximately constant. This was also found for the subcooled region for horizontal and inclined tubes by [13]. This behavior was also confirmed by [46]. A good consistency of the correlations from Jens and Lottes [38] with experimental data was found from [46]. Therefore this equations are used also for the horizontal and inclined tubes.

The location for the boiling crisis is appeared, if the wetting of the heated channel wall cannot maintained after exceeding a certain value for the mass flow quality. With the boiling crisis the heat transfer respectively the heat transfer coefficient between the tube wall and the fluid decreases. This results in systems with impressed heat flux, e.g. radiation or electrical or nuclear heating, to a discontinuous increase of the wall temperature. At systems with a impressed wall temperature, e.g. condenser or heat exchanger, the heat flux decreases discontinuous after exceeding the boiling crisis.

In principle there are two different types of boiling crisis namely

- film boiling or departure of nucleate boiling (DNB) and
- dryout (DO)

are described in the literature. The critical heat flux (CHF) is defined as the maximum heat flux just before the onset of the boiling crisis is given. In the following paragraph the correlations to calculate the critical heat flux will be presented.

Critical heat flux Under normal operation conditions the natural circulation steam generator do not have more than 30% steam quality. But in case of a flow instability, namely reverse flow, the flow direction of the fluid in the tube changes. During this process a stagnation of the mass flow occurs and as a consequence the fluid velocity is identical to zero for a certain time. Therefore it is necessary to include the region between the onset of the boiling crisis and the single-phase steam into the heat transfer model.

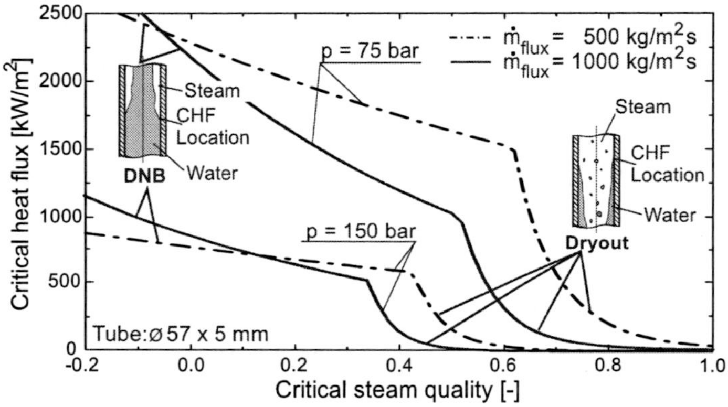

Figure 17. Critical heat flux for vertical tube with upward flow rate at constant pressure and mass flux.

Figure 17 shows the critical heat flux for departure of nucleate boiling (DNB) and dryout (DO) as a function of the critical steam quality at different constant pressures and mass fluxes. It can be seen, that with increasing operation pressure the critical heat flux decreases. In case of DNB the continues phase is the liquid of the two-phase mixture (see sketch for DNB in Figure 17). Under slightly subcooled and low steam conditions a bubble layer which is formed at the channel wall separates the liquid from the wall. With detaching of a bubble layer from the wall a stagnation of the liquid at the heating surface occurs for a short time period. This stagnation causes at high heat flux a very intense boiling and results to a steam film at the heated wall. Based on the poor thermal conduction of the vapor, the heat transfer coefficient decreases rapidly. The lower the steam quality the higher the CHF. Dryout can be observed in a two phase-flow region with high void fraction (annular flow). The DO is caused by the disruption or the dry out of the liquid layer from the channel wall. In this case the vapor is the continuous phase. In comparison to the DNB the heat transfer coefficient at DO is higher. This higher heat transfer coefficient is based on the higher cooling effect by convection which is a result of the higher steam mass flow fraction. In the literature there are many empirical correlations available to estimate the CHF conditions in smooth bore tubes with uniform heating, e. g. [12], [68], [69] or [45]. But many of these correlations can be used only in a small parameter range. Based on the different mechanism which leads to the boiling crisis it can be a advantage to use different equations for DNB and DO. Drescher and Khler recommend in their study [8] after an extensive comparison with experimental data the correlation from Kon'kov [52] and Doroshchuk [22] for the calculation of the boiling crisis DNB and DO. A comparison of [23] to other instructions to calculate DNB has shown a small loss of accuracy of the correlation according to [22]. However, Drescher and Koehler recommend that for the practical use to calculate DNB the correlation of Doroshchuk should be the preferred mathematical method.

Doroshchuk's correlation for vertical tubes takes the form:

$$\dot{x}_{crit} = \frac{\ln\left(\frac{\dot{m}_{flux}}{1000}\right)\left(0.68\frac{p}{p_{crit}} - 0.3\right) - \ln(\dot{q}_{crit}) + \ln(b)}{1.2\ln\left(\frac{\dot{m}_{flux}}{1000}\right) + 1.5} \tag{147}$$

where $\dot{x}_{crit}$ is the critical mass vapor quality. The critical heat flux can be calculated from

$$\dot{q}_{krit} = b\left(\frac{\dot{m}_{flux}}{1000}\right)^{0.68\left(\frac{p}{p_{crit}}\right) - 1.2\dot{x} - 0.3} e^{-1.5\dot{x}} \tag{148}$$

with the coefficient

$$b = 10^3\left[10.3 - 17.5\left(\frac{p}{p_{crit}}\right) + 8\left(\frac{p}{p_{crit}}\right)^2\right]\left(\frac{8\cdot 10^{-3}}{d_i}\right)^{0.5}$$

and q_{crit} in kW/m^2. The definition region for the correlation (147) and (148) is given by

$$\begin{array}{rcccl} 29\ \text{bar} & \leq & p & \leq & 200\ \text{bar} \\ 500\ \text{kg/(m}^2\ \text{s)} & \leq & \dot{m}_{flux} & \leq & 5000\ \text{kg/(m}^2\ \text{s)} \\ 4\ \text{mm} & \leq & d_i & \leq & 25\ \text{mm} \end{array}$$

The correlation for the departure of nucleate boiling losses their validity at low heat flux. The critical mass vapor quality or the critical heat flux for dryout in a vertical tube with upward flow can be calculated using the correlations of Kon'kov [52]:

$$\dot{x}_{crit} = 10.795\,\dot{q}^{-0.125}\,\dot{m}_{flux}^{-0.333}\,(1000d_i)^{-0.07}\,e^{0.017150\,p} \tag{149}$$

4.9 bar to 29.4 bar

$$\dot{x}_{crit} = 19.398\,\dot{q}^{-0.125}\,\dot{m}_{flux}^{-0.333}\,(1000d_i)^{-0.07}\,e^{-0.00255\,p} \tag{150}$$

29.4 bar to 98 bar

$$\dot{x}_{crit} = 32.302\,\dot{q}^{-0.125}\,\dot{m}_{flux}^{-0.333}\,(1000d_i)^{-0.07}\,e^{-0.00795\,p} \tag{151}$$

98 bar to 196 bar

or

$$\dot{q}_{crit} = 1.8447 \cdot 10^{8}\, \dot{x}^{-8} \dot{m}_{flux}^{-2.664} (1000\, d_i)^{-0.56}\, e^{0.1372\, p} \tag{152}$$

4.9 bar to 29.4 bar

$$\dot{q}_{crit} = 2.0048 \cdot 10^{10}\, \dot{x}^{-8} \dot{m}_{flux}^{-2.664} (1000\, d_i)^{-0.56}\, e^{-0.0204\, p} \tag{153}$$

29.4 bar to 98 bar

$$\dot{q}_{crit} = 1.1853 \cdot 10^{12}\, \dot{x}^{-8} \dot{m}_{flux}^{-2.664} (1000\, d_i)^{-0.56}\, e^{-0.0636\, p} \tag{154}$$

98 bar to 196 bar

with the heat flux $\dot{q}$ in kW/m^2 and the pressure p in bar. The definition region for the correlation (149) to (154) is given by

$$\begin{array}{rcccl} 200\ \text{kg/(m}^2\ \text{s)} & \leq & \dot{m}_{flux} & \leq & 5000\ \text{kg/(m}^2\ \text{s)} \\ 4\ \text{mm} & \leq & d_i & \leq & 32\ \text{mm}. \end{array}$$

In some cases it is not possible to define the nature of the boiling crisis. Therefore, Drescher and Koehler [23] have suggested that both values (the value for DNB and DO) should be calculated. The lower of the two values should be used for further calculations. The intersection point between the equations according to Doroshchuk and Kon'kov is outside of their definition area. Koehler [49] has recommended, that a exceeding of the definition area cannot be avoided because otherwise a closed calculation is not possible.

In the parameter range of

$$\begin{array}{rcccl} & & p & < & 5\ \text{bar} \\ & & \dot{m}_{flux} & < & 300\ \text{kg/(m}^2\ \text{s)} \\ 10 & < & l/d_i & < & 100 \\ \text{-0,2} & < & \dot{x}_{in} & < & 0 \end{array}$$

the critical heat flux can be calculated with the equation according to [79]:

$$\dot{q}_{crit} = b \left(\frac{\dot{m}_{flux}}{l/d_i} \right)^{0.8} (1 - 2\dot{x}_{in}). \tag{155}$$

For water the coefficient b should be set to 460 [84].

The above presented models for calculating the boiling crisis can be used only for vertical tubes. Based on the phase separation due to gravitation the boiling crisis will be initiate earlier in a horizontal or inclined tube in comparison to the vertical tube. The location of the boiling crisis can be found mostly at the upper side of the tubes while the bottom side of the tube is wetted with a liquid film. The stratification effects of the two-phase flow in a horizontal tube can be described according to [84] by the dimensionless Froude-number Fr. This number is defined as the ratio of inertia to gravity forces in the flow. The formulation derived by [84] for horizontal tubes was extended by [47] taking into account only those gravity force components that act perpendicular to the tube wall. The effect of the tube orientation on the stratification of the two-phase flow can be described as

$$Fr = \frac{\dfrac{\dot{x}_{crit}\, \dot{m}_{flux}}{\sqrt{\rho_g}}}{\sqrt{g\, d_i\, (\rho_l - \rho_g) \cos\varphi}} \tag{156}$$

with the tube inclination angle φ between the evaporator tube and the horizontal. The reference parameter x_{crit} is the critical mean steam quality in the transition region calculated for upward flow in a vertical tube. This allows to use correlations developed for vertical evaporator tubes, such as that of Kon'kov [52] for horizontal and inclined tubes also. For $Fr \geq 10$, the tube orientation has no influence on the boiling crisis. If the Froude number is smaller 3, then the influence of the tube inclination on the boiling crisis is very high.

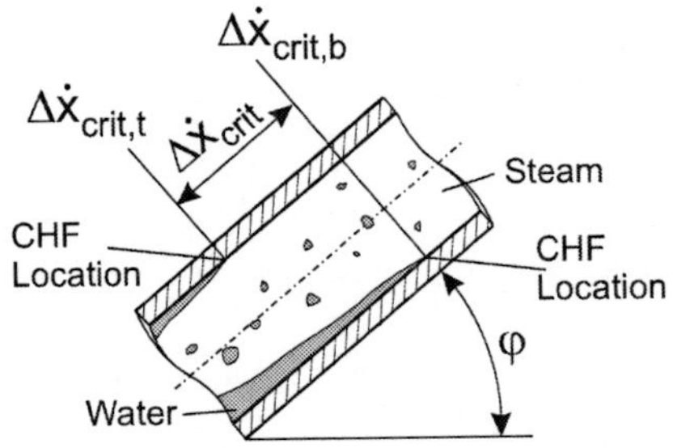

Figure 18. Location of the boiling crisis at the top and bottom of a inclined tube with upward mass flow.

Figure 18 shows the different locations for the boiling crisis at the top and the bottom of the inclined tube with upward mass flow. The boiling crisis extends from the top to the bottom in the form of a ellipse over the circumference of the tube and along a certain, in some cases over several meters, tube length. The difference between the critical steam quality at the top and the bottom $\Delta\dot{x}_{crit}$ of the inclined and horizontal tube can be calculated by the empirical equation according to [47] with:

$$\Delta\dot{x}_{crit} = \frac{16}{(2+Fr)^2}. \tag{157}$$

The parameter range of this equation is:

$$\begin{aligned} 25\ \text{bar} &\leq & p &\leq 200\ \text{bar} \\ 500\ \text{kg/(m}^2\ \text{s)} &\leq & \dot{m}_{flux} &\leq 2500\ \text{kg/(m}^2\ \text{s)} \\ 12.5\ \text{mm}^2 &\leq & d_i &\leq 24.3\ \text{mm} \\ 200\ \text{kW/m}^2 &\leq & \dot{q} &\leq 600\ \text{kW/m}^2. \end{aligned}$$

The steam qualities at which dryout at the top and the bottom occurs can be calculated with:

$$\dot{x}_{crit,tb} = \dot{x}_{crit} \mp \frac{\Delta\dot{x}_{crit}}{2} \tag{158}$$

If $\dot{x}_{crit,tb} > 1$, then $\dot{x}_{crit,tb}$ must be set to 1.

For the design of natural circulation boilers with horizontal evaporator tubes it is most important to check the Froude-number Fr as well as the flow pattern of the fluid flow. The Froude-number should be great enough so that the tube orientation has no influence on the boiling crisis and the fluid flow should not have a stratified flow. This check should also done for the DBN.

Post-dryout region After exceeding the location for onset of the boiling crisis (dryout), the steam is the continuous phase whereas the water droplets represents the disperse phase. The heat transfer in this region is called post-dryout. The model used for this region was developed by Koehler [49] and is valid for a vertical upward flow. The model is based on the thermodynamic imbalance between steam and the water droplets and shows a good consistency with the experimental data.

The height of the thermal imbalance, which is necessary to determine the temperature difference between steam and the water droplet $\Delta\vartheta_{imb}$, was identified by Koehler [49] with the help of the energy balance.

$$\Delta\vartheta_{imb} = \vartheta_{st} - \vartheta_{sat} = \frac{\Delta h_v}{2c_{pst}} \left(\sqrt{1 + \frac{4\, c_{pst}\, \dot{q}}{\Delta h_v \left(A_{surf}\alpha\right)_{drop}}} - 1 \right) \tag{159}$$

The heat transfer coefficient between the water droplet and the steam as well as the surface of the vaporizing water droplet can be calculated with the formula:

$$\begin{aligned} \left(A_{surf}\alpha\right)_{drop} &= 1.473 \cdot 10^{-4} \left(\frac{\dot{m}_{flux}}{b_{Lapl}}\right)^{1.33} \quad \text{for} \quad \frac{\dot{m}_{flux}}{b_{Lapl}} \leq 1767 \cdot 10^3 \\ \left(A_{surf}\alpha\right)_{drop} &= 3.078 \cdot 10^{-21} \left(\frac{\dot{m}_{flux}}{b_{Lapl}}\right)^{4} \quad \text{for} \quad \frac{\dot{m}_{flux}}{b_{Lapl}} > 1767 \cdot 10^3 \end{aligned} \tag{160}$$

with $\left(A_{surf}\alpha\right)_{drop}$ in W/m^{2}K and the Laplace-constante

$$b_{Lapl} = \sqrt{\frac{\sigma}{g\left(\rho_l - \rho_{st}\right)}}. \tag{161}$$

The effective mass vapor quality $\dot{x}_{eff}$ according Koehler [49] is given by:

$$\dot{x}_{eff} = \frac{h - h_l}{\Delta h_v + c_{pst}\Delta\vartheta_{imb}}. \tag{162}$$

The spec. enthalpy of the fluid h can be calculated with the aid of the energy balance:

$$h = \frac{\dot{q}\, A_{surf,out}}{\dot{m}_{flux} A} + h_{wm}. \tag{163}$$

If steam and water droplet are in thermal equilibrium, then equation (162) passes over to a constitutional equation for the mass vapor quality.

$$\dot{x}_{bal} = \frac{h - h_l}{\Delta h_v} \tag{164}$$

For the calculation of the two-phase flow heat transfer coefficient the correlation (137) according to Gnielinski must be used and the correction factor K_1 in formula (137) must be neglect [49]. The two-phase Reynolds-number Re_{2ph} must be calculated with the averaged fluid velocity under neglecting the slip between the two phases.

$$Re_{2ph} = \frac{\dot{m}_{flux} d_{hyd}}{\eta_g} \left[\dot{x}_{eff} + \left(1 - \dot{x}_{eff}\right) \frac{\rho_g}{\rho_l} \right] \tag{165}$$

The reference temperature for the properties of state used in equation (137) is the averaged boundary layer temperature. This reference temperature is equivalent to the arithmetical mean temperature between wall and steam. The density ρ_{st} used in (165) is applied to the effective steam temperature.

The model for the post dryout region described above is only valid in case of a full developed thermal imbalance. That means, that the location of the lowest heat transfer $\dot{x}_{\alpha_{min}}$ is reached.

$$\dot{x}_{\alpha_{min}} = \dot{x}_{crit} + \frac{\dot{x}_{crit} c_{pst} \Delta\vartheta_{imb}}{\Delta h_v} \tag{166}$$

A linear interpolation will be done between the location of the second boiling crisis and the point of minimal heat transfer. If the thermodynamic steam quality is greater then

$$\dot{x}_{lim} = 0.7 + 0.002 \cdot 10^{-5} p, \tag{167}$$

then the upper definition boundary for the model according to [49] is reached. For a mass vapor quality $\dot{x} > \dot{x}_{lim}$ the equations for the single-phase heat transfer coefficient are valid.

The calculation of the heat transfer coefficient in the post dryout region for horizontal and inclined tubes can be done according to [82] with the aid of the Froude-number (156) and the critical steam quality at the upper and the lower tube side (157) and (158). With these values the calculation of the heat transfer coefficient can be done in the following way:

- The heat transfer coefficient is independent from the tube orientation if $Fr \geq 10$ or $\dot{x}_{crit,b} - \dot{x}_{crit,t} \leq 0.1$, whereas for $\dot{x} > \dot{x}_{crit,b}$ a drop flow exist. In both cases the method for calculating the heat transfer coefficient described above for the vertical tube can be used [82].

- The calculation of the heat transport in the region of $Fr < 10$ and $\dot{x}_{crit,t} \leq \dot{x} \leq \dot{x}_{crit,b}$ is according to Kefer [46] and Hein *et al.* [36] not independent from the wall thickness, the wall properties and the wetting of the inner tube wall. Kefer *et al.* [47] suggested for this region extensive calculations with the help of the two dimensional Fourier differential equation for thermal conduction. This method to determine the heat transfer coefficient is possible for a steady state calculation or for a detailed analysis of a single pipe or a group of pipes. But this method is not applicable for transient calculations. Therefore the calculation of the heat transfer coefficient was done in such a way that the limiting values for $Fr = 10$ and $\dot{x}_{crit,b} - \dot{x}_{crit,t} = 0.1$ for a vertical tube will be calculated and the lower value of both results are used for the heat transfer coefficient.

For further details about the heat transfer model see [25] or [105]

3. Stability Analysis of Natural Circulation Steam Generators

It is well known and described in the literature that the multiphase flow has a stronger tendency towards unstable behavior than single phase flow. These so-called fluid-dynamic instabilities can manifest themselves in form of pressure drop or mass flow fluctuations, a periodical change of the flow pattern of the two phase mixture, or fluctuations of the wall temperatures. The instabilities can cause vibrations of plant components or can lead to a non-uniform mass flow distribution in the tube bank. They can accelerate the initiation of a boiling crisis, which is characterized by a sudden rise of the surface temperature due to the drop of the heat transfer coefficient, a non-uniform heat flux or a difficult measurement and control of system parameters. Thermo-hydraulic oscillations and flow instabilities can influence the operation and the control of the plant and are therefore undesirable.

In the following sections a short overview about the different types of two-phase flow instabilities and some examples about the stability analysis of natural circulation steam generators will be presented.

3.1. Generals about Two-Phase Flow Instabilities

The first pioneering paper about instabilities in two-phase flow systems was presented in the last century by Ledinegg [53]. In this work Ledinegg studied the conditions leading to flow instabilities in heated parallel boiler tubes between two connecting headers. Ledinegg presented in his article criteria for the prediction of a so called static instability. A higher attention on the phenomenon of two-phase flow instabilities was started approximately 25 years after publishing the paper of Ledinegg with the development of high-power-density boilers and boiling water reactors.

Since Ledinegg, many works have been done by researchers on flow instabilities and thermo-hydraulic oscillations. A large part of these studies are focused on forced circulation (e. g. [3], [4], [43], [75], [106] or [61]) and only a relative smaller part has been done on natural circulation systems (e. g. [33], [62], [95], [98] or [112]). A main classification of the flow instabilities and thermo-hydraulic oscillations were presented by e. g. Bour *et al.* [15], Bergles [10] or Yadigaroglu [110].

3.1.1. Classification of Two-Phase Flow Instabilities

The main classification, which is presented in figure 19, was done by Bour *et al.* [15] and is based on the distinction between the static and dynamic character of the conservation laws needed to explain the dynamics of unstable equilibrium states.

As shown in figure 19 the two-phase flow instabilities are subdivided into two groups: the static and the dynamic instabilities. The static instability can be predicted from steady-state consideration only, while the triggering of dynamic instabilities involves transient inertia and dynamic feedback effects. According to [80] an instability is referred to *compound* when several elementary mechanism interact in the process simultaneously and cannot be studied separately. It is *fundamental* in the opposite sense. The *compound instability as secondary phenomenon* occurs if the oscillation is a result of a static instability as primary phenomenon.

In the following a short description of the different types of instabilities will be presented:

Static instability:

- **Ledinegg instability:** The flow excursion involves a sudden change of the mass flow rate to a lower value. It occurs when the internal characteristic of the channel becomes

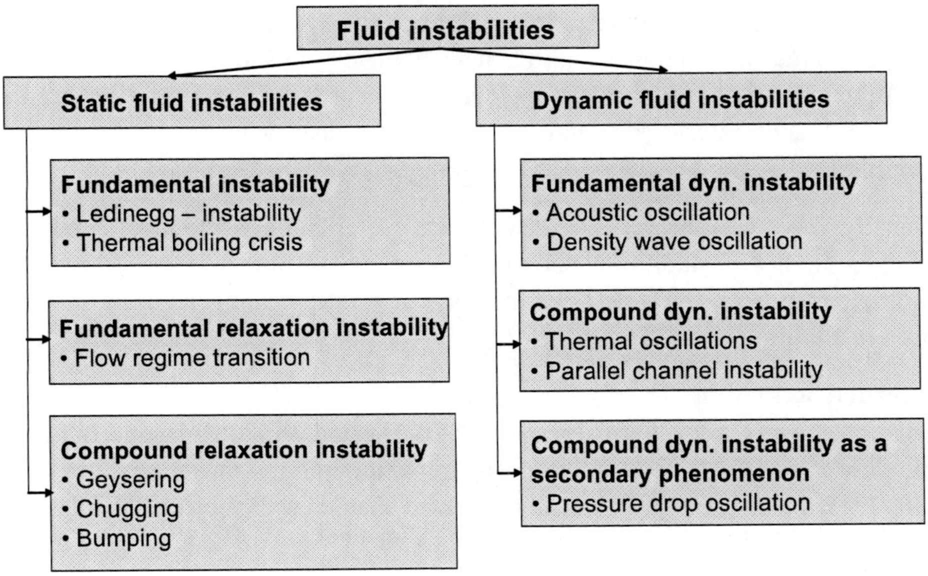

Figure 19. Classification of two-phase flow instabilities.

algebraically smaller than the external characteristic.

- **Thermal boiling crisis:** is caused by a substantial change of the heat transfer coefficient which results in a sudden rise of the wall temperature with a possible flow oscillation. A final experimental verification of this phenomenon is not given [80].

- **Flow regime transition:** This form of instability occurs mainly when the flow conditions are close to the point of transition between bubbly- and annular flow. A small disturbance of the flow rate results in a temporary increase of the bubbly population in the bubbly-slug flow which results in a change of the flow pattern to annular flow. The bubbly flow has less void but a higher pressure drop then the annular flow. The lower pressure drop will speed up the mass flow momentarily. The higher mass flow reduces the vapor generation and the flow pattern reverts to bubbly-slug flow. This cycle can be repeated.

- **Geysering, chugging, bumping:** These instabilities are different forms of the compound relaxation instability and they involve a static phenomena that are coupled so as produced a repetitive behavior which is not necessarily periodic. When the cycles are irregular, each flow excursion can be considered hydraulically independent of others. Geysering can be observed in natural-, e.g. Baars [9] or Aritomi *et al.* [7], [5], as well as one-through systems, e.g. Aritomi *et al.* [6]. Bumping is exhibited in boiling of the alkali metals at low pressure. Chugging is a periodic phenomenon, which can be characterized by a periodic expulsion of coolant from the flow channel. A overview to this form of instability can be found in [35].

Dynamic instability:

- **Acoustic oscillation:** This instability is characterized by a high frequency (10 to 100 Hz). The period has approximately the same order of magnitude as the time required for a pressure wave to travel through the system. This form of oscillation was observed in subcooled boiling, bulk boiling, and film boiling. According to [11] the acoustic oscillation can have a significant influence on the system in case of a decreasing interference and response of the disturbance and a increase of the amplitude.

- **Density wave oscillation:** The density wave oscillation is the most common type of dynamic flow instability encountered in two-phase flow systems. It is a result of multiple feedback effects between the mass flow rate, steam generation and pressure drop in a boiling channel.

- **Thermal oscillation:** This form of instability was identified by [70]. The thermal oscillation appears as the thermal response of the heated wall after dryout. In this region the flow can oscillate between film- and transition boiling, thus producing large amplitudes temperature oscillations in the heated channel wall subjected to constant heat flux. The periodic time of oscillation is 2 to 80 s [15]. The density wave oscillation is apparently required for the thermal oscillation.

- **Parallel channel instability:** A heat exchanger is designed with a higher number of parallel and heated tubes with the same geometry (tube length and diameter) which are connected with headers. Under stable operation conditions the same pressure difference between the inlet- and outlet headers is given for every channel. This results in the same mass flow for every tube. A parallel channel instability is given when a small disturbance in one of these tubes result in a interaction between the other tubes of the heat exchanger. This form of instability can be found in natural- as well as one-through systems.

- **Pressure drop oscillation:** Pressure drop oscillation occurs in systems having a compressible volume upstream of, or within the heated section [15]. The pressure drop oscillation is associated with operation on the negative sloping portion of the characteristic curve for a tube (pressure loss - mass flow curve).

For further details on the different forms of instability see among others [15], [110], [80] or [104].

The main types of instability encountered in natural circulation systems are the Ledinegg instability (in form of reverse flow) and the density wave oscillation. In the following paragraphs some examples will be given for the stability analyses of natural circulation steam generators.

3.2. Ledinegg Instability

Figure 20 shows a simple system composed of two pressure reservoirs p_1 and p_3, a pump, a uniform heated channel and their connecting pipes. If we make a stability analysis for the system presented in figure 20 around a steady state position so it can be shown that the following criterion must be fulfilled to get a stable system:

$$\left(\frac{\partial \Delta p}{\partial w}\right)_{ext} < \left(\frac{\partial \Delta p}{\partial w}\right)_{int} . \tag{168}$$

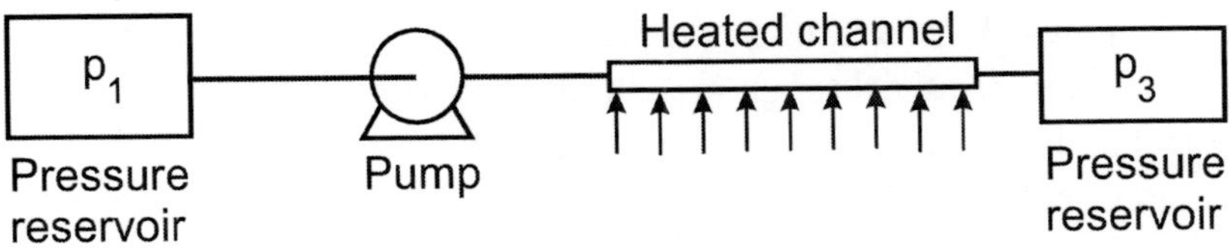

Figure 20. Simple system for a two-phase flow system.

The index *int* stays for the internal and the index *ext* for the external characteristic of the system. This behavior requires that the channel characteristic exhibit a region where the pressure drop decreases with increasing flow. In a two-phase flow system this situation does exist where the sum of the component terms (friction, momentum and gravity terms) increases with increasing mass flow. This dependence of the mass flow respectively velocity from the pressure drop was described the first time by Ledinegg [53]. A derivation of the criterion can be found among others in [104] or [110].

A decreasing of the subcooling at the inlet of a heated channel, a higher operation pressure or the assembling of a throttle at the channel inlet (in the region of single phase flow) can help to stabilize (decrease) the aperiodic instability. The increase of the pressure drop is a disadvantage of the throttle. Because the throttle must be designed for the unfavorable operation mode and results in many cases in a high pressure drop under full load operation for the steam generator. Further articles to the Ledinegg instability are e.g. [64], [66], [77], [101], [98], [95], [97], [87], [89] or [107].

3.2.1. Reverse Flow in Natural Circulation Systems

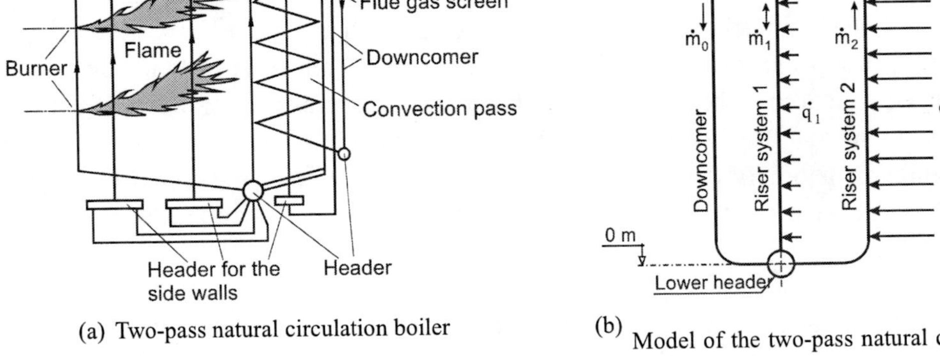

(a) Two-pass natural circulation boiler

(b) Model of the two-pass natural circulation boiler

Figure 21. Two-pass natural circulation boiler. Printed with permission of the publisher from [87].

Figure 21(a) shows the sketch of a two-pass natural circulation steam generator. The fuel enters the steam generator through the burners and will be burned with the air inside the combustion chamber. The resulting flue gas leaves the combustion chamber through the flue gas screen in direction to the convective pass, the so called second pass. In the second pass the flue gas flows downstream over the heating surfaces and leaves the boiler at the bottom of the second pass in direction to the chimney. The main heat transfer mechanism in the combustion chamber of the boiler is the radiation (flame and gas radiation) while in the second pass the forced convection is the main heat transfer mechanism. The combustion chamber as well as the second pass are enclosed with evaporator tubes which consists of a bar tube combination, the so called membrane wall (for the calculation of the temperature distribution in membrane walls see e.g. [93]). These evaporator tubes are connected with the unheated downcomer tubes through headers.

A unequal heating of the evaporator tubes of the boiler can be the result of e.g.:

- different heat transfer mechanism in the combustion chamber and the convective pass,
- the fouling of the heating surfaces is not homogeneous,
- covering of some parts of the heating surface,
- in many cases the side walls are heated lower,
- unevenly heat flux along the combustion chamber walls,
- decrease of the flue gas temperature along the flue gas pass ...

It can be seen in figure 21(a) that the lower and the higher heated evaporator tubes are connected with headers with the unheated downcomer tubes. Such a combination of the heating surfaces can be result under unfavorable conditions to unstable behavior which should be analyzed during the design of natural circulation boilers.

Figure 21(b) shows a simplified model of the two-pass natural circulation steam generator. The higher heated riser system 2 corresponds to the tubes of the combustion chamber and the lower heated riser system 1 represents the evaporator tubes of the wall in the second pass. The large diameter downcomers are unheated. All evaporator tubes are subdivided into a heated and unheated part. The unheated part of both riser systems are connected directly with the drum. The dimension for the downcomer was defined with ø108 x 6.3 mm and with a tube length of 22.25 m. For the riser system 1 and 2 the heated tube length was 18 m while the length of the unheated part was 3 m. The tube diameter and wall thickness for both riser systems was ø57 x 4.5 mm. The difference in height can be taken from figure 21(b).

It is well known that the number of possible solutions for a system depends on the number of bifurcations. Such a system can have stable as well as unstable solutions. The numerical methods to analyze a unstable system can only find stable solutions. Therefore analytical or graphical methods must be used to find also the unstable solutions. In the following some methods – graphical as well as numerical – will be presented to analyze such natural circulation systems.

Graphical method The graphical analysis is used to study the mass flow distribution under steady state conditions. For the determination of the mass flow rate in the individual tubes of the boiler it is necessary to combine the different tube characteristics. The characteristic curve for a tube can be obtained by plotting the total pressure difference between the drum and the lower header $\Delta p_{Dh,tota}$ against the mass flow rate (see figure 22). The total pressure difference $\Delta p_{Dh,tota}$ can be calculated by

$$\Delta p_{Dh,tota} = \Delta p_{fric} + \Delta p_H + \Delta p_{accel}. \tag{169}$$

Δp_{fric} includes the pressure loss to friction as well as the pressure loss to e. g. in- and outflow and bends. The pressure difference due to acceleration Δp_{accel} is identical to zero in case of an unheated tube. Δp_H describes the pressure difference of the static head.

For the graphical analysis the total pressure difference $\Delta p_{Dh,tota}$ for every individual tube must be equal (equation (170)) and also the mass and energy balance for the lower header (equation (171) and (172)) must be fulfilled.

$$\Delta p_{Dh,tota,0} = \Delta p_{Dh,tota,1} = \Delta p_{Dh,tota,2}, \tag{170}$$

$$\dot{m}_0 = \pm|\dot{m}_2| + \dot{m}_2 \quad \text{and} \tag{171}$$

$$\dot{m}_0 h_0 = \pm|\dot{m}_2| h_1 + \dot{m}_2 h_2 \tag{172}$$

The negative sign in equation (171) and (172) must be used if the mass flow in the lower heated tube is directed downward. The counterclockwise flow direction is counted positive.

In equation (172) the potential and kinetic energy of the mass flow at the inlet and at the outlet of the header is neglected. The header is considered as adiabatic. The spec. enthalpy at the riser inlet, which is used for the calculation of the characteristic curve for the riser systems, is identical with the spec. fluid enthalpy of the header. The spec. fluid enthalpy at the outlet of the drum must be calculated with equation (102).

The exact determination of the spec. enthalpy from the balance equations for the lower header and the drum results in a more complex procedure for the graphical solution. According to [72] it is not necessary to make an exact calculation of the spec. enthalpy, because the resulting difference between the exact and an approximately calculated characteristic curves is very small. Therefore the influence on the results of the graphical solution is also small. In figure 21(b) the possible flow directions for the higher and the lower heated riser systems is included. In such a system the flow direction in the higher heated riser system is always directed upward.

Figure 22 should help to explain the procedure for the graphical determination of the mass flow distribution in the tubes of the boiler shown in figure 21(b). At the horizontal axis of figure 22 the mass flow and at the vertical axis $\Delta p_{Dh,tota}$ is outlined.

The characteristic curve for the downcomer is illustrated in figure 22 by the full line and that for the riser system 1 and 2 by the doted lines. Dependent on the flow direction of riser system 1 the mass flow for the unequally heated riser systems 1 and 2 must be added or subtracted at constant pressure difference $\Delta p_{Dh,tota}$ (In case of reverse flow in the lower heated riser system the characteristic curve for $\dot{m}_1 < 0$ must be subtracted (see arrows in figure 22) from the characteristic $\dot{m}_2 > 0$). This results in circulation characteristics (broken lines) dependent on the flow direction of riser system 1.

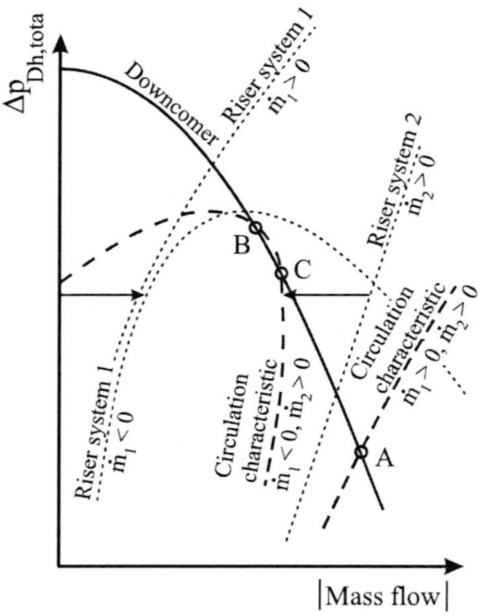

Figure 22. Graphical solution method. Printed with permission of the publisher from [100].

It can be seen that the circulation characteristic for upward flow direction in both riser systems (signed with $\dot{m}_1 > 0$ and $\dot{m}_2 > 0$) has only one intersection point with the characteristic of the downcomer (point A). In this intersection point equations (170), (171) and (172) are fulfilled. The associated circulation mass flow $\dot{m}_0$, $\dot{m}_1$ and $\dot{m}_2$ as well as $\Delta p_{Dh,tota}$ can be read from the axis. In case of reverse flow in riser system 1 the circulation characteristic (signed with $\dot{m}_1 < 0$ and $\dot{m}_2 > 0$) has two intersection points with the graph for the downcomer (point B and C). Only solution B is stable.

Higher heat absorption at the lower heated system 1 moves the circulation characteristic for reverse flow in direction of lower mass flow rate and $\Delta p_{Dh,tota}$. This leads to the consequence that the solutions B and C move together. The point where B and C are merged represents the limit of univocal flow direction in the riser system with the lower heat absorption. For further increase of the heat flow $\dot{q}_1$ the downcomer characteristic does not cross the graph for the circulation characteristic with the mass flow $\dot{m}_1 < 0$ and $\dot{m}_2 > 0$. In this case only upward flow is possible in both riser tubes.

The graphical solution is very illustrative but also very time-consuming. Therefore a numerical simulation is preferable to find design criteria to prevent reverse flow. In the following paragraphs the results of such numerical analysis will be presented.

Numerical method As mentioned above the hot start-up as well as the heavy load change are the most critical operation modes for natural circulation steam generators. The start-up behavior of the boiler depends on the time difference between the shutdown and restart of the boiler. For the start-up from the cold condition (cold start, the system pressure in the drum is equal to the atmospheric pressure) or part load condition (warm start) the admissible rate of temperature (pressure) change in the natural circulation systems is mostly conditioned by the thermal stresses in the large diameter components in particular headers and drums. This limitation applies both to load increase and reduction. A load change linked to a pressure change is accompanied by a change of the saturation temperature. Therefore variable pressure operation of a natural circulation system is only acceptable under certain conditions. Especially in the low pressure region the rate of change of the saturation temperature as a function of pressure is high. Tables with acceptable start-up time for drums, cyclones and headers for different design pressures and wall thickness can be seen in [55].

Therefore a hot start-up was simulated for the present example. In this case the time difference between shutdown and restart of the two-pass boiler is up to approximately one day (overnight standstill) and the operation pressure is supposed unchanged during this time. Therefore the limitation of the pressure (temperature) change is not taken into account and the boiler can start in a faster way [91]. Gradients in superheater headers are not within the scope of this analysis.

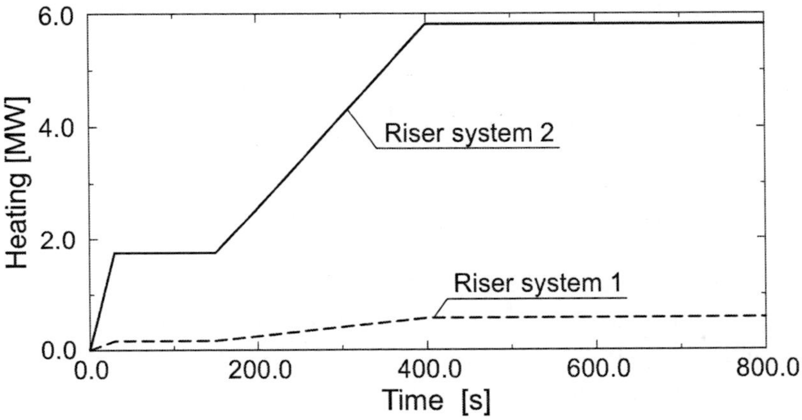

Figure 23. Heating ramp. Printed with permission of the publisher from [100].

The heating ramp for the simulated hot start-up was selected according to [56] and is shown in figure 23. It can be seen that in the first period of the simulation the heat flow increases up to 30% of full load. Between 30 s and 150 s after the ignition of the burner the load is constant. During the time period between 150 s and 400 s the ramp-like heat input reaches the full load. The total time for the simulation was 10000 s. The heat flux over the tube length was supposed as uniform.

The total heat absorption (i.e. the heat absorption for both riser systems) was varied in steps between 8 MW and 0.8 MW. The value of 8 MW was chosen as upper extreme. For industrial furnaces with oil firing the corresponding heat absorption can be assumed between 5 MW - 6 MW. The value of 0.8 MW can represent the condition for low load operation or special equipment in process engineering (e. g. refuse incineration plant).

The following initial conditions are used for the dynamic simulations:

- the steam generators are filled with water near boiling condition
- the pressure distribution of the fluid is due to gravity and the velocity of the working fluid in the tube network of the evaporator is zero and
- the fluid temperature in the evaporator of the boiler is equivalent to the boiling temperature at drum pressure.

As boundary condition it was defined that the drum pressure of the steam generator is constant during the whole simulation. The heat fluxes to the two riser systems of the two-pass boiler are also boundary conditions for the simulation. The water level was controlled at the drum centerline.

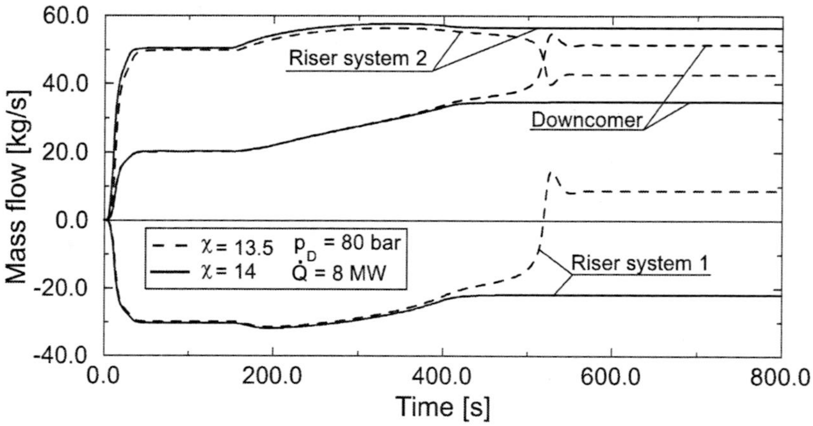

Figure 24. Mass flow distribution at a heat absorption ratio of $\chi = 13.5$ and $\chi = 14$.

Figure 24 shows as a result of the dynamic simulation the mass flow in the different tubes of the boiler. The simulation was done for a operation pressure of 80 bars, a total heat flow of 8 MW and a heat absorption ratio of $\chi = 14$ (full line) and $\chi = 13.5$ (broken line). The heat absorption ratio, which was hold constant during the simulation, is defined as

$$\chi = \frac{\dot{q}_2}{\dot{q}_1}. \tag{173}$$

In the following the results for the heat absorption ratio of $\chi = 13.5$ will be described in detail:

It can be seen that the lower heated riser system 1 starts in the downward direction with a relatively high mass flow rate. The higher heated riser system 2 starts in the upward direction with approximately the double mass flow rate compared to riser system 1. With the beginning of the steam production in the lower heated riser system 1, the mass flow starts to slow down. Approximately 500 seconds after the start of the simulation the flow direction of the fluid in riser system 1 changes rapidly to upward flow. Following the change of the flow direction, the mass flow tends after a short peak to a much lower mass flow. After about 600 seconds the mass flows in the downcomer and both riser systems have achieved the steady state values.

The situation for the circulation mass flow after a moderate change of the heat absorption ratio from $\chi = 13.5$ to $\chi = 14$ is different. The mass flow in the tubes of the boiler at $\chi = 14$ starts similar to that shown for a heat absorption ratio of $\chi = 13.5$. The mass flow of the working fluid in riser system 1 starts also in the downward direction too. With the beginning of the steam production in the lower heated riser system 1, the mass flow starts to slow down. But, compared to the case with $\chi = 13.5$ the heat flux to the riser system 1 is not high enough to change the flow direction. Therefore the flow direction remains different in both riser systems at full load.

The heat absorption ratio of 13.5 represents the critical value χ_{crit} between a stable upward flow in both riser systems and a possible operation condition with reverse flow in the lower heated riser system 1. This can be seen also in figure 25

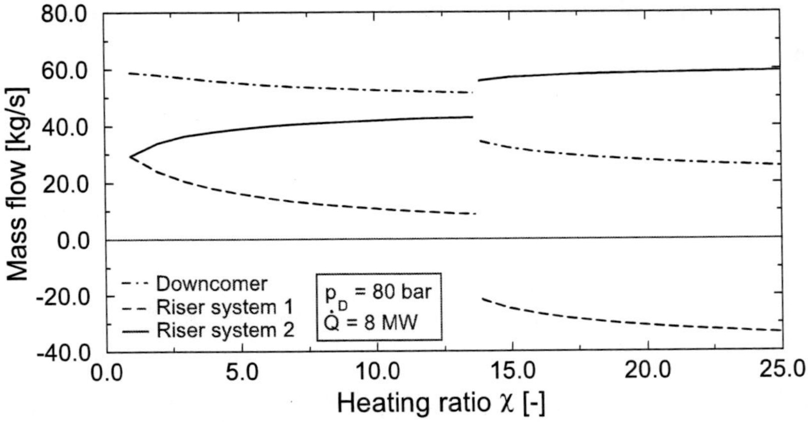

Figure 25. Mass flow distribution at different heat absorption ratios χ. Printed with permission of the publisher from [87].

Figure 25 shows the mass flow in the tubes of the two-pass boiler at full load for different heat absorption ratios χ. The calculation was done for a operation pressure of 80 bars and a total heat flow of 8 MW. Figure 25 contains for every heat absorption ratio one of the three possible solutions (see chapter 3.2.1. graphical method). With increasing heat absorption ratio the difference between the mass flow in the different riser systems growth, whereas the total circulation mass flow decreases slightly (see downcomer mass flow). With achieving the critical heat absorption ratio the mass flow in the lower heated riser system 1 changes the flow direction from upward to downward flow. This can be seen in figure 25 by the jump in the graphs for the mass flows. The lower heated riser system 1 works after crossing χ_{crit} as second downcomer (negative mass flow in figure 25). As a consequence of the change of the flow direction the mass flow in the downcomer decreases. With a further increasing of the heat absorption ratio the mass flow in the downcomer decreases and in the riser system 1 increases. This is a result of the lower heat absorption of riser system 1. The physical circumstance for this behavior can be seen in the time evolution of the pressure differences in the single tubes.

Figure 26 shows the chronological evolution of the total pressure difference between the drum and the lower header $\Delta p_{Dh,tota}$ and the pressure difference due to height Δp_H (static head) for both riser systems and the downcomer. The graphs for the static head of the downcomer for $\chi = 13.5$ and $\chi = 14$ are identical. The difference between the graphs of $\Delta p_{Dh,tota}$ and the static heads Δp_H is the sum of the pressure difference due to acceleration Δp_{accel} and friction Δp_{fric}. At first the results for the heat absorption ratio of $\chi = 13.5$ (broken line) will be discussed:

During the first phase of the hot start-up, the total pressure difference between the drum and the lower header $\Delta p_{Dh,tota}$ is smaller than the static head Δp_H of the lower heated riser system and therefore the mass flow in riser system 1 is directed downward. With the beginning of the steam production in the lower heated riser system 1 the static head

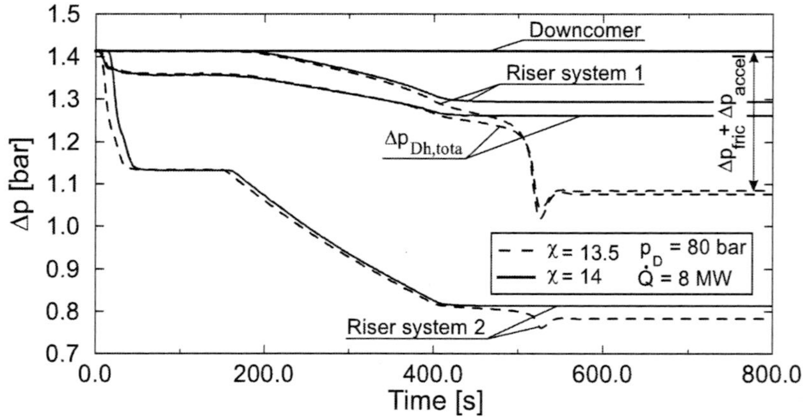

Figure 26. Development of the pressure difference for the simulated hot start-up.

decreases. During the time difference of approximately 500 s and 540 s the increasing steam production reduces the static head and the curve for Δp_H crosses the graph of $\Delta p_{Dh,tota}$. In this period the mass flow in the lower heated system changes the flow direction from downward to upward.

The pressure difference due to height Δp_H is always higher than $\Delta p_{Dh,tota}$ in case of a heat absorption ratio of $\chi = 14$ (see figure 26). Therefore the mass flow is directed downward in the lower heated riser system 1 during the whole startup and also at full load.

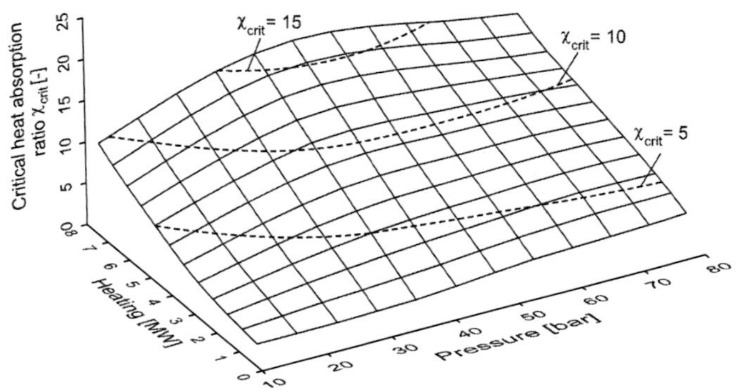

Figure 27. Critical heat absorption ratios χ_{crit} at different pressure and heat load. Printed with permission of the publisher from [87].

Figure 27 shows the critical heat absorption ratio χ_{crit} for the boiler at different system pressures and heat loads. A heat absorption ratio of $\chi_{crit} \geq 5$ is assumed to be a design criterion for the two-pass boiler for stable operation at all heat loads. A curve with the constant critical heat absorption ratio of $\chi_{crit} = 5$ is included in figure 27. It can be seen that the system tends to flow reversal for low heat loads even at 80 bars. This leads to the circumstance that the part load condition for the steam generator is restricted. The maximum of the critical heat absorption ratio is given in the area of approximately 35 to 45 bars with $\chi_{crit} \approx 17$. The surface for χ_{crit} decreases rapidly in direction to lower operation pressures.

Beside the overall circulation ratio the fluid velocities are also of important interest for the boiler design. In a certain temperature (pressure) range (approximately 140 up to 250 saturation temperature) high velocities can be a reason for erosion-corrosion (see e.g. [58], [44] [48], [14] or [24]). It should be noted at this point that besides the fluid velocity and the fluid temperature also the selection of the material and the fluid chemistry (pH-value of the water) have a strong influence to the erosion-corrosion.

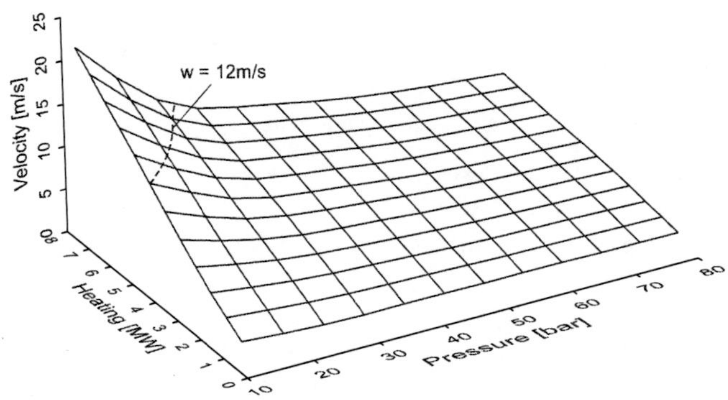

Figure 28. Velocity of the working medium in the higher heated riser system 2. Printed with permission of the publisher from [87].

Figure 28 shows the velocity in the higher heated tubes. It can be seen that the velocity corresponds over a wide pressure and heat load range with the usual design rules (The velocity of the steam-water mixture should be approximately ≤ 12 m/s and the velocity in the downcomer $\leq$ 4m/s [101]). A curve with the constant velocity of 12 m/s is included in figure 28.

The analysis of the current boiler design has shown that the steam generator should not operate in the pressure range ¡ 30 bars and in the low load region, even at 80 bars. Practical experience and this investigation clearly indicate that natural circulation systems with unequally heated risers and a common downcomer should be only designed that way if they can be operated well below the critical heat absorption ratio χ_{crit}. χ_{crit} must be determined for every new system configuration and heating condition. Furthermore it can be stated that downward flow in riser systems should be avoided under any circumstances. Therefore modifications on the system geometry are necessary. One of the most successful methods to get a higher stability is to modify the thermo-hydraulic behavior of the circulation system. This can be done by reducing the pressure loss in the downcomer or increasing the pressure loss in the riser systems ([57], [101], [95], [100], [97], [99]).

There are a number of papers available which deals with the influence of different system parameters on the stability of a boiling system presented in figure 21(b). The most important parameters can be summarized as follows:

- **flow resistance:** a higher stability of the boiler can be achieved by changing the thermo-hydraulic behavior of the boiler. This can be done by increasing the pressure loss in the riser systems. But the method to improve the boiler stability by increasing the pressure loss in the riser systems is restricted by the highest allowed fluid velocity in the riser tubes.

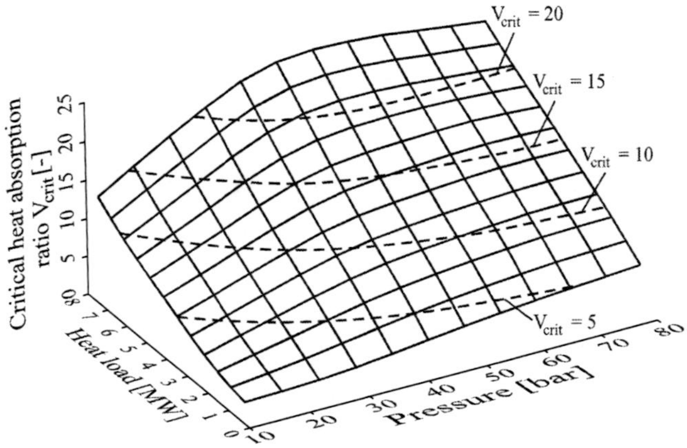

Figure 29. Critical heat absorption ratios χ_{crit} at different pressure and heat load for a redesigned two-pass boiler.

Figure 29 shows the 3-dimensional surface of the critical heat absorption ratio χ_{crit} at different pressure and heat load for a redesigned two-pass boiler. For this investigation the unheated part of both riser systems was replaced by an intermediate header and two riser relief tubes for the lower heated system and four riser relief tubes for the higher heated riser system 2. The modification of the thermo-hydraulic behavior by increasing the pressure loss in the riser system results in higher values for χ_{crit} at all operation pressures and heat loads. Compared with the results presented in figure 27 the region with χ_{crit} ¡ 5 is moved towards lower heat loads and operation pressure (see [89], [87]).

- **nonuniform heating profile over the tube length:** a higher heat flux at the upper end of the lower heated tubes (the end where the fluid leaves the tube in case of upward flow direction) improves the stability whereas a higher heat flux at the lower end of the lower heated tube system results in a more unstable behavior of the boiler (see [99], [95], [88]).

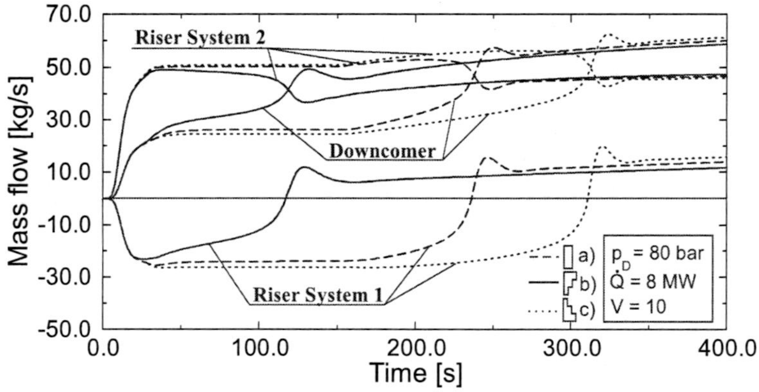

Figure 30. Mass flow at nonuniform heating profile over the tube length.

Figure 30 shows as an example a comparison of the mass flow in the tubes of the two-pass boiler at different heating profiles. The simulation was done at an operation pressure of 80 bars, different heating profiles and a heat absorption ration of V = 10. The full lined curves correspond to the heating profile b), the broken lines to the uniform heating profile a) and the dotted lines to the heating profile c). It can be seen that the development of the mass flow in the individual tubes is different for the three heating profiles. The result for the heating profile b) show that during the first phase of the hot startup the mass flow in both riser systems is lower and in the downcomer higher compared to the uniform heating profile a). The results for the heating profile c) shows an opposite behavior compared to profile b). In this case the mass flow in both riser systems is higher and in the downcomer lower compared to the profile a) (see [89], [87]).

- **operation pressure:** the flow stability of the boiler increases with increasing operation pressure by a given heat flux and geometry (see [101], [89]).
- **downcomer diameter:** a larger downcomer tube diameter tends to a more stable system. The larger downcomer diameter decreases the single phase pressure drop due to friction which results in a higher pressure in the lower header. The higher pressure in the lower header leads to a more stable system ([87]).

Further investigations have shown that all modifications which lead to a higher pressure in the lower header results in a higher stability (see e. g. [101], [95], [100], [97], [99]).

The above presented parameters can improve the stability of the boiler but the possibility of a reverse flow in the lower heated riser system cannot be completely eliminated with this solutions. The best way to solve the problem of reverse flow in natural circulation systems where a unheated downcomer is connected with a header to a lower and a higher heated riser system is to separate the different heated systems. This can be done by e. g. installing a downcomer tube for every riser system.

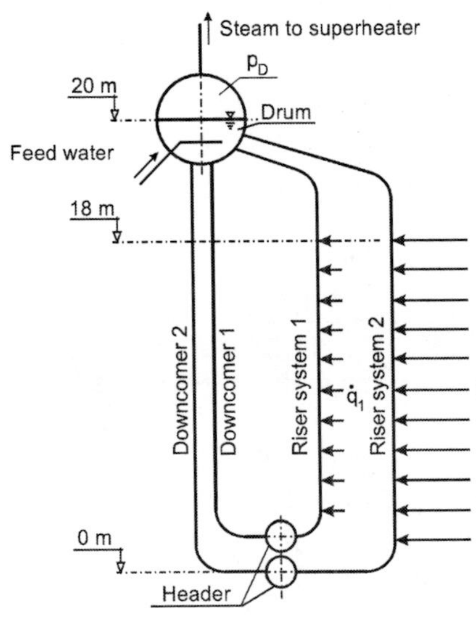

Figure 31. Model for the two-pass boiler with separated systems. Printed with permission of the publisher from [87].

In figure 31 a sketch of the model where both riser systems have their own downcomer tube is presented. All geometry data as well as boundary and initial conditions for the numerical simulation are the same as for the model presented in figure 21(b).

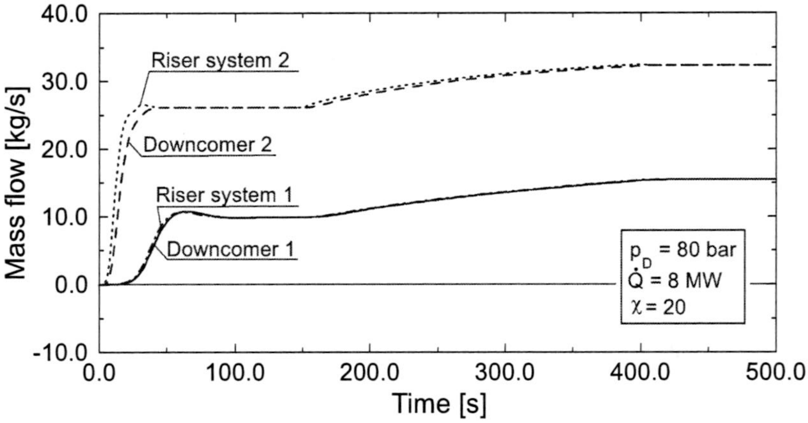

Figure 32. Mass flow distribution for the boiler design with separate downcomers for both riser systems. Printed with permission of the publisher from [87].

The evolution of the mass flow in the different tubes of the boiler presented in figure 31 is shown in figure 32. The simulation was done for an operation pressure of 80 bars, a total heat flow to both riser systems of 8 MW and a heat absorption ratio of 20. It should be mentioned here that the heat absorption ratio of 20 is higher than χ_{crit} for the boiler design presented in figure 21(b). It can be seen that the mass flow for both riser systems increase during the whole start-up without reverse flow.

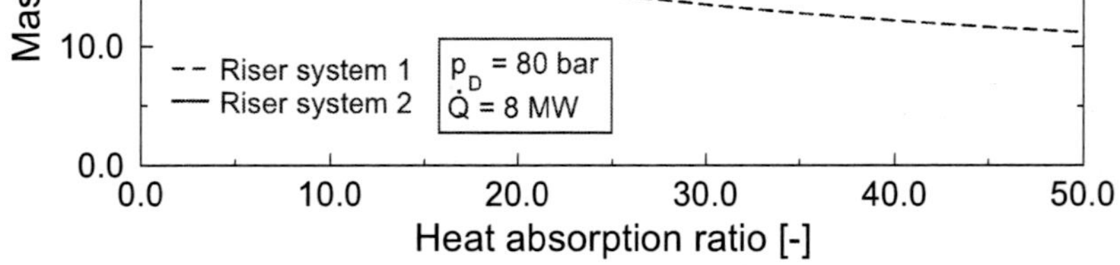

Figure 33. Mass flow at different heat absorption ratios χ for the boiler design with separate downcomers for both riser systems. Printed with permission of the publisher from [87].

Figure 33 presents the mass flow distribution at full load for different heat absorption ratios χ at an operation pressure of 80 bars and a total heat flow to both riser systems of 8 MW. The mass flow in the downcomer tubes at full load are identical to the mass flow in the corresponding riser systems (see figure 32).

With increasing heat absorption ratio the mass flow in the lower heated system 1 decreases while the mass flow in the higher heated system 2 is approximately constant. This is a result of the decreasing heat flux to riser system 1. A reverse flow was not detected for all analyzed conditions. A comparison with the results in figure 25 shows that the system with a separation of the lower and higher heated system results in a stable behavior of the boiler.

A second example to the Ledinegg instability should show that also a internal circulation of the working medium can exist in the tube network of the evaporator of a natural circulation boiler.

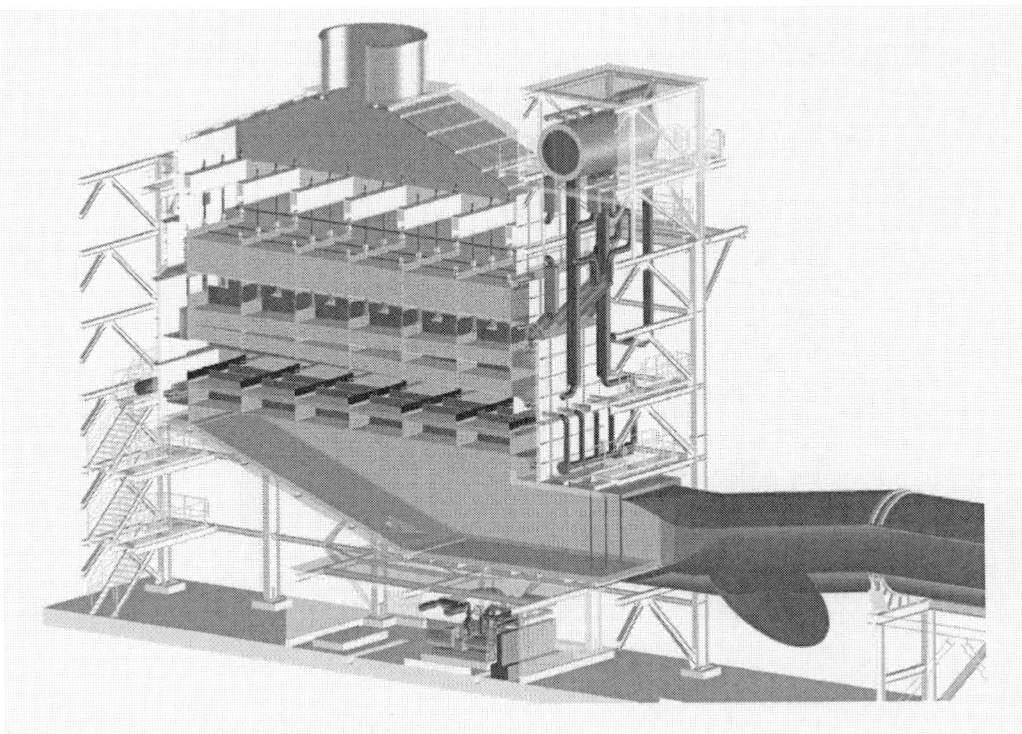

Figure 34. Sketch of a vertical type natural circulation heat recovery steam generator. Printed with permission of BERTSCHenergy Kessel- und Energietechnik Apparatebau, Austria.

Figure 34 shows the sketch of a vertical type Heat Recovery Steam Generator (HRSG). Many of these boilers use the waste heat of gas turbines. In these so called combined power cycles the steam generator is arranged downstream the gas turbine. Modern gas turbines for combined cycles are highly flexible in their mode of operation, concerning start up, load change and shut down. Heat recovery steam generators are forced to operate in such a way, that the gas turbine operation is not restricted by them. Therefore, they should be designed for a high cycling capability. The design of modern gas turbines are standardized and therefore it is necessary for the boiler manufacturer to design the HRSG in such a way that an optimal use of the gas turbine waste heat is given. This results in a boiler design up to three pressure stages. For an economic operation of the HRSG a individual pinch- and approach point is realized for every pressure stage. Today a pinch point of approximately 5 to 10 K (the higher value is used for low pressure systems whereas the lower value is used for high pressure systems) and a approach point between 0 and 5 K are usual. Advantages of HRSG's with natural circulation evaporators are the reduced investment, maintenance and operation costs due to the absence of a circulation pump. The bundle heating surfaces of the vertical HRSG's are arranged horizontally and the flue gas enters the boiler in the most cases at the bottom and leaves it on the top. The typical tube length per layer in such a boiler configuration is up to 20 m. In natural circulation systems with horizontal arranged evaporators the water steam mixture has no preferred flow direction during start-up. Therefore the vertical type natural circulation HRSG is more unstable than a system with vertical evaporator tubes (horizontal type HRSG). Especially vaporizers at low operation pressure show tendency to flow instabilities (see e.g. [103], [87], [92], [90], [98], [102]).

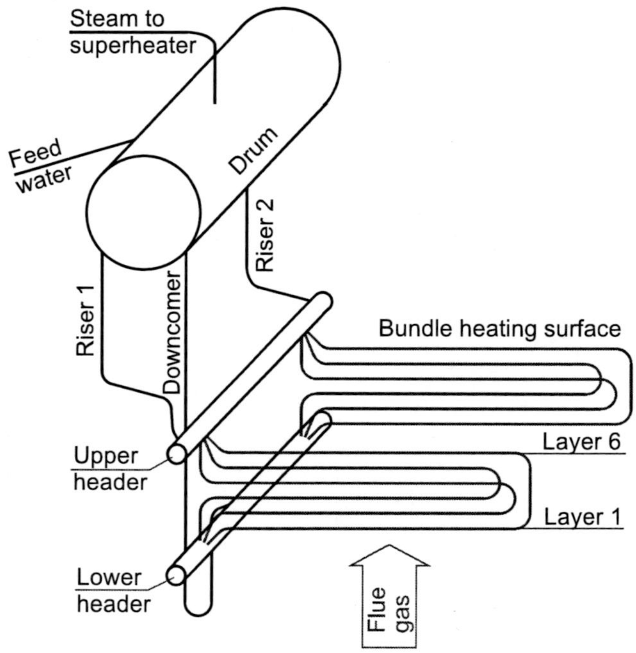

Figure 35. Intermediate pressure evaporator of a three stage heat recovery steam generator.

Figure 35 presents the simulation model of an intermediate pressure evaporator of a vertical type natural circulation heat recovery steam generator. The evaporator system shown in figure 35 consists of a tube bank with three parallel tube paths arranged in 6 layers. The three tube paths are connected to a lower and an upper header. The upper header is connected with the drum by two riser relief tubes with different geometry. The geometry of both riser tubes is different in the followings points:

- different tube length (riser tube 2 is approximately 2.5 m longer compared to riser tube 1).
- riser system 1 has one 90^0 bend less and two 45^0 bends more then riser system 2.

- the spatial arrangement of the bends differs between both riser tubes.

The working fluid enters the drum subcooled through the feed water tube and leaves the drum saturated through the downcomer. The downcomer is feeding the lower header with boiling water at drum pressure. The saturated water leaves the upper header through the riser tubes in direction to the drum.

The initial conditions used for the numerical simulation of the HRSG are the same as for the simulation of the two-pass boiler. The operation pressure of the intermediate pressure evaporator was 30 bars and constant during the whole simulation. The development of the gas turbine exhaust flue gas temperature and mass flow over the time, which is also a boundary condition for the simulation of the HRSG, is shown in figure 36. The flue gas temperature presented in figure 36 is the temperature in front of the analyzed bundle heating surface of the evaporator.

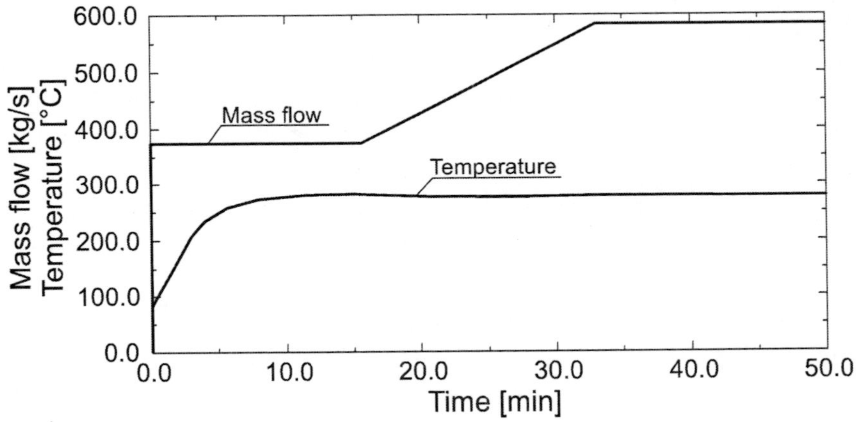

Figure 36. Development of the gas turbine exhaust flue gas temperature and mass flow over the time.

The exhaust flue gas mass flow enters the flue gas channel with a step up to approximately 64% of full load. This mass flow is constant for approximtaely 15 min followed by a ramp. After approximately 33 min full load is reached. The whole simulation time was 50 min. The flue gas temperature at simulation start is approximately 80 . In the first 3 min after simulation start the temperature increases linearly up to ≈ 234 . In the following the temperature gradient decreases and is identical to zero after approximately 15 min. At this time the gas turbine exhaust flue gas temperature has obtained the maximum value of ≈ 280 . In the following the flue gas temperature decreases slightly down to the steady state value of ≈ 277 at simulation end.

In the following the results for the investigated intermediate pressure evaporator will be presented.

The following agreement for the sign of mass flow is used in all figures:

- A positive sign of the mass flow in the riser tubes describes a flow direction from the upper header to the drum.
- A positive sign of the mass flow in the downcomer tubes indicates a flow direction from the drum to the lower header.

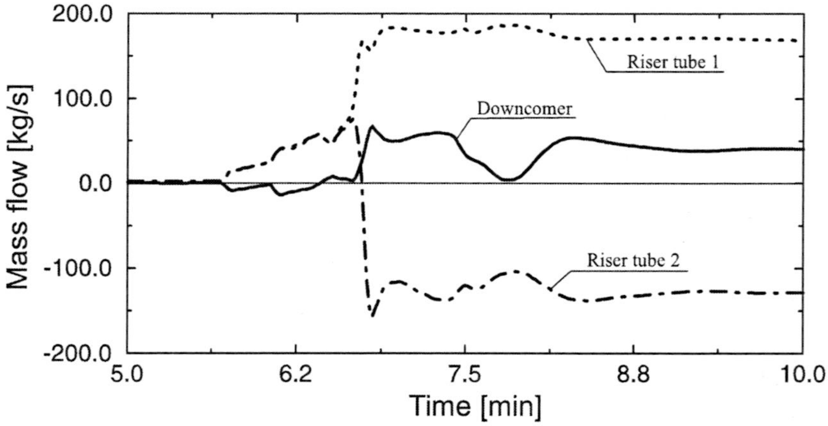

Figure 37. Mass flow in the downcomer and both riser tubes.

Figure 37 presents as one of the simulation results the mass flow in the downcomer (full line), in riser system 1 (dot-dashed line) and in riser system 2 (dotted line). With the beginning of the steam production in the most heated layer 1, which is not presented in figure 37, the mass flow of the working fluid leaves the tubes of layer 1 in direction of both, the downcomer and the riser tubes. This is a result of the increasing fluid volume which is based on the steam production. Therefore the mass flow in the downcomer and both riser tubes flows from the bundle heating surface in direction to the drum (reverse flow in the downcomer tube). The mass flow in the other two tube paths will be accelerated in direction from the lower header to the drum.

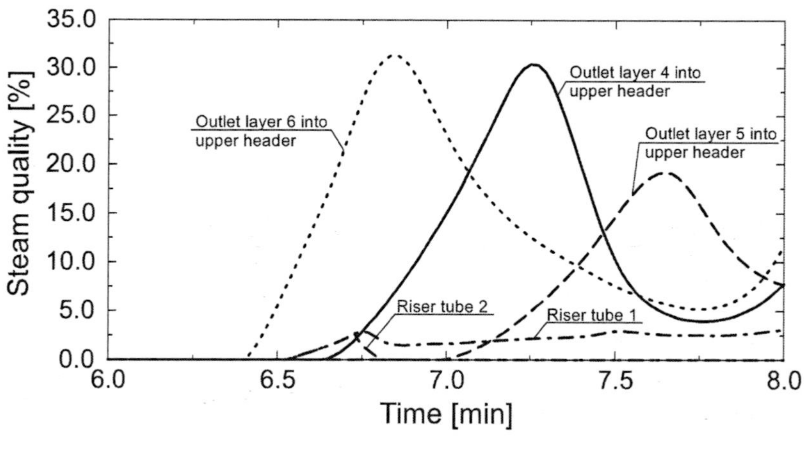

Figure 38. Steam quality at the inlet to the upper header.

The steam production in the residual tube layers start with different time lags and is shown in figure 38. This results also in a higher fluid mass flow in direction to the drum (see figure 37). During this period the absolute values of the mass flow in both riser tubes is identical. Approximately 6.5 min after simulation start vapor enters the first time both riser tubes (see figure 38). This leads to a change of the pressure distribution in both riser tubes 1 and 2. Based on the different riser geometry the two-phase flow pressure loss in both riser systems is also different. This results in a change of the flow direction in riser 2 from upward to downward flow (see figure 37). Riser tube 2 works in this period of the hot start-up as a further downcomer. The increase of the pressure drop in both riser tubes is different in spite of the same mass flow change in the riser tubes 1 and 2 and be responsible for this circumstance. The different changes of the pressure drop in both riser is a consequence of the different total flow resistance. The entrance of the vapor in the riser tubes increases this inequality. After the change of the flow direction in riser tube 2 the mass flow in the tubes of the investigated evaporator oscillates with decreasing amplitude and tend to the steady state values. The reverse flow in riser tube 2 is also given at steady state.

Different possibilities can be done to get an water circulation for the natural circulation evaporator without an internal circulation between riser tube 1 and 2. The changes can be done by a change of the heating ramp or a change of the boiler geometry.

While the change of the heating ramp is subject to the gas turbine operation or/and an installed by-pass duct the change of the boiler geometry depends in first of all on the status of boiler design (is the boiler erection finished or in the design phase). In case of an erected boiler the structural alteration of the geometry is more restricted in comparison to a boiler in design phase, because the costs for rebuilding an erected boiler are high.

However, in the following a number of simple to realize geometry modifications will be presented:

1. implementation of two geometrically identical riser tubes (case 1),
2. splitting of the upper header into two equivalent parts. This can be done by the implementation of two upper headers with the same geometry or the implementation of an plate in the middle of the upper header.
3. implementation of two geometrically identical riser tubes and splitting of the upper header into two equivalent parts.

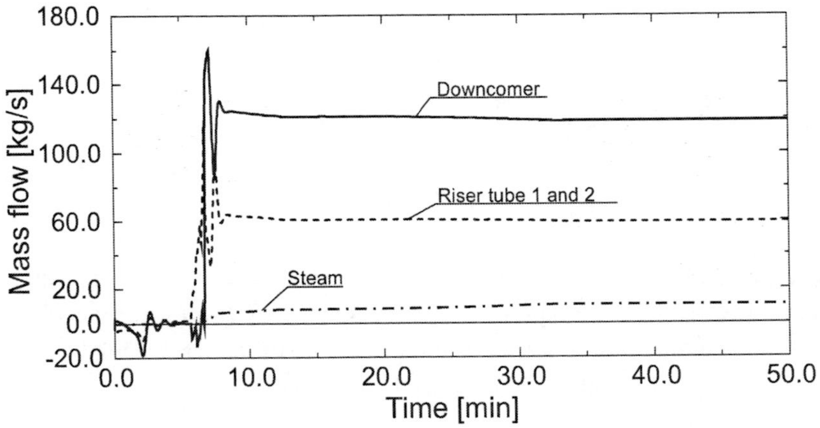

Figure 39. Mass flow in the downcomer and both riser tubes after modification of riser tube 2.

Figure 39 shows the mass flow in the downcomer and both riser tubes as well as the steam mass flow which leaves the drum for the evaporator design two geometrically identical riser tubes. During the first period of the warm start-up (up to the entering of the first vapor into both riser tubes) the development of the mass flow in the different tubes are approximately the same as shown in figure 37. In contrary to the base design of the boiler the further development of the mass flow in both riser systems of the modified boiler design shows no different behavior. This is a result of the corresponding total flow resistance for both riser tubes also in case of a two-phase flow. In the following the mass flow in both riser systems do not change the flow direction also at full load.

A comparison of the results presented in figure 37 and 39 shows that the mass flow in the riser tubes at full load is different. The modified design has a higher mass flow in the downcomer and a lower in the riser tubes. This is also a result of the changed pressure distribution in the evaporator.

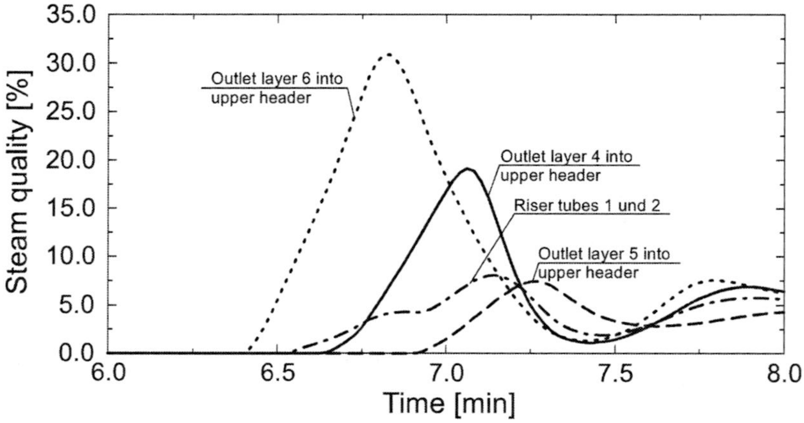

Figure 40. Steam quality at the inlet to the upper header after modification of riser tube 2.

The development of the steam quality at the inlet to the upper header can be seen in figure 40. In comparison to the base design (figure 38) the curves for the steam quality at the outlet of layer 4 and 5 shows a clear lower gradient. This is a result of the higher circulation mass flow through the bundle heating surface.

The simulation results for the other two above suggested simple to realize geometry modifications (e.g. splitting of the upper header into two equivalent parts) shows approximately the same transient behavior than the presented modified design case 1. In all cases no reverse flow in the riser tubes of the analyzed natural circulation evaporator at full load was detected.

The above presented examples to the Ledinegg instability in natural circulation boilers should show that such boilers respond in a very sensitive way to perturbations. Further analyzed parameters which influence the static instability of natural circulation can be found amongst other in [95], [96], [98], [101], [100], [99], [97], [104] or [87].

3.3. Density Wave Instability

Beside the static instability (Ledinegg instability) a natural circulation boiler must be analyzed also for the occurrence of the most common type of dynamic instabilities, namely the density wave instability (DWO).

In the following a short description of the physical mechanisms of the density wave oscillation and a example for the occurrence of the DWO in the evaporator of a heat recovery steam generator should be presented.

3.3.1. Physical Mechanisms of the Density Wave Oscillation

Generally, the DWO is the most common type of dynamic instability encountered in two-phase flow systems [110]. The instability is a result of the multiple feedback effects in relationship between the flow rate, steam generation and pressure drop in a boiling channel. An inlet flow fluctuation creates a perturbation of the enthalpy in the single phase region. When the enthalpy perturbation reaches the boiling region they are transformed in void fraction and quality perturbations. These fluctuations travel with the flow along the channel. This disturbance affects the pressure drop as well as the heat transfer behavior. The combination of the void fraction and flow perturbation and the change in the lengths of the two-phase area creates a pressure-drop oscillation in the two-phase region. If the overall pressure drop of the channel is imposed by the external characteristic of the channel, then the two-phase pressure-drop fluctuations will create a feedback pressure perturbation in the single phase region with the opposite sign. This pressure perturbation in the single phase region can attenuate or re-enforce the imposed oscillation by creating a feedback inlet flow perturbation. For certain combinations of the geometry arrangements, boundary conditions and operating conditions the pressure-drop perturbation can acquire a 180^0 out of phase pressure fluctuation at the exit which leads to a self-sustaining of the flow at the inlet. Therefore a unstable transient behavior exist. Further details to the physical mechanism of the DWO instability are described in e. g. [109], [110] and [80].

The DWO is characterized by large amplitudes and a nearly sinusoidal period (see figure 45). DWO is a low frequency oscillation ($\approx$1 Hz) related to a period of one or two times which is necessary for a fluid particle to travel along the channel [18]. The transportation delays in the channels are of paramount importance for the stability of the system. The transportation delays in the channels are of high importance for the stability of the system. The oscillation of pressure and mass flux are in phase. Fukuda and Kobori [29] have shown analytically and experimentally that two different types of DWO (Type 1 and 2) exist. For a heated channel with a subcooled mass flow at the inlet a dimensionless phase change number

$$N_{pch} = \frac{\dot{Q}}{A\,w_{in}\,\rho_l\,\Delta h_v}\frac{\rho_l - \rho_g}{\rho_g}. \tag{174}$$

and a dimensionless subcooling number

$$N_{sub} = \frac{h_l - h_{in}}{\Delta h_v}\frac{\rho_l - \rho_g}{\rho_g}. \tag{175}$$

can be calculated. With the help of these numbers a stability diagram can be developed for the heated channel as shown schematically in figure 41(a).

The type 1 instability is dominant when the steam quality is low at channel outlet, while the type 2 instability is relevant at high steam quality. The type 2 is well known as the typical density wave oscillation. But the mechanisms of both DWO - types are approximately equal.

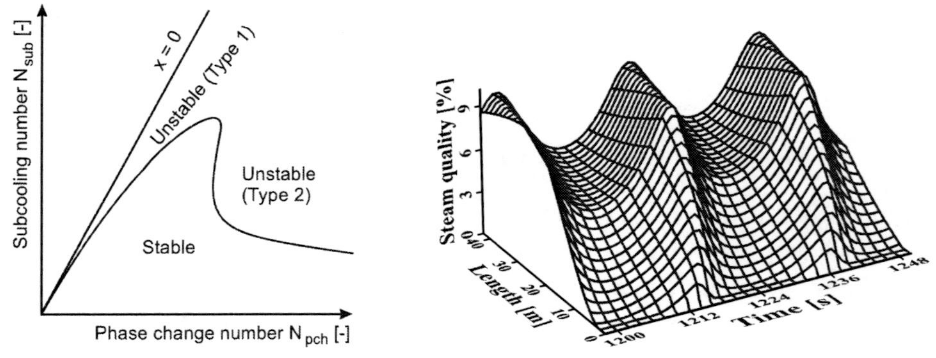

(a) Schematically stability diagram for DWO

(b) Steam quality along a tube path of the bundle heating surface of the HRSG

Figure 41. Stability diagram and oscillation cycles of the DWO.

In figure 41(b) the time dependent distribution of the steam quality along a tube path of the bundle heating surface during the hot start-up of a HRSG is shown. The figure shows the time period for approximately two oscillation cycles of the DWO. The fast increase of the steam quality in first part of the tube (layer 1) is given by the higher heat flux to the tube compared to the second part of the tube (layer 7). The second part of the tube path is arranged in the flue gas pass downstream of layer 1 (see figure 42).

3.3.2. Example of Density Wave Oscillations in a Natural Circulation HRSG

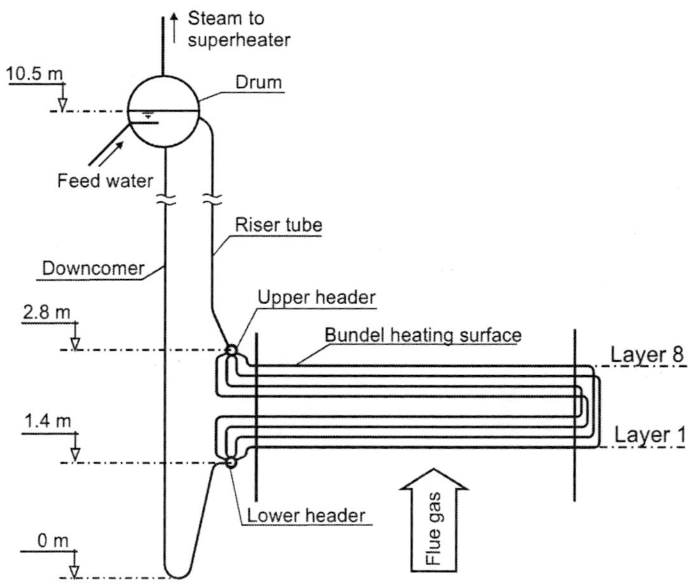

Figure 42. Sketch of the analyzed evaporator of a vertical type HRSG.

A model of the investigated evaporator of a vertical type HRSG is shown in figure 42. The model includes a drum, a downcomer with siphon, a lower and a upper header, a heated tube bank with four horizontal parallel tube paths (the heated length of the tube per layer is 20 m) and a riser pipe.

For the bundle heating surface helically coiled finned tubes with segments are used which are arranged in a staggered way. The dimension of the evaporator tubes for the analyses was defined with ø48.3 x 3.2 mm, that of the downcomer ø168.3 x 4.5 mm and that of the riser tubes ø219.1 x 6.3 mm. The number of fins per m length was 236, the fin height 12.7 mm and the fin spacing 4.5 mm. The averaged fin thickness was 1 mm. The number of tubes per layer was 70. The difference in height can be taken from figure 42.

The working medium enters the drum subcooled through the feed water tube and leaves the drum saturated through the downcomer. The water steam mixture, which is produced in the tube bank of the evaporator, enters the drum through the riser tubes. The two-phase flow will be separated in the drum. The steam leaves the drum at the top in direction to the superheater.

The present example was done for a hot start of the boiler. The following initial conditions are used for the dynamic simulations:

- the steam generator is filled with water near boiling conditions,
- the pressure distribution of the fluid is due to gravity,
- the velocity of the working fluid in the tube network of the evaporator is zero and
- the fluid temperature in the evaporator of the boiler is equivalent to the boiling temperature at drum pressure.

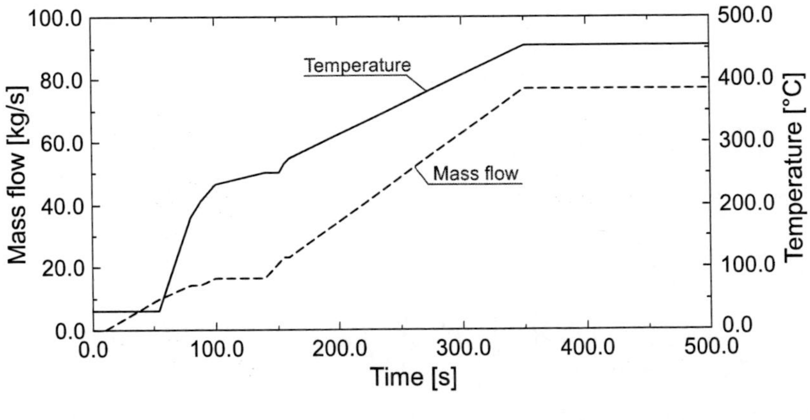

Figure 43. Flue gas mass flow and temperature in front of the bundle heating surface. Printed with permission of the publisher from [92].

The drum pressure, which is constant during the whole simulation, and the gas turbine exhaust flue gas temperature and mass flow, which are given as a function of time, are boundary conditions for the simulation. Figure 43 shows the time evolution of the gas turbine exhaust flue gas mass flow (broken line) and temperature (full line) for the hot start-up of the steam generator. The values for the flue gas temperature in figure 43 are located in front of the bundle heating surface. The gas turbine reaches the steady state approximately

350 s after the start of simulation. 350 s after simulation start the gas turbine works at full operation load. The total time for the simulation was 2000 s.

During the whole simulation the water level of the drum was controlled at the drum centerline.

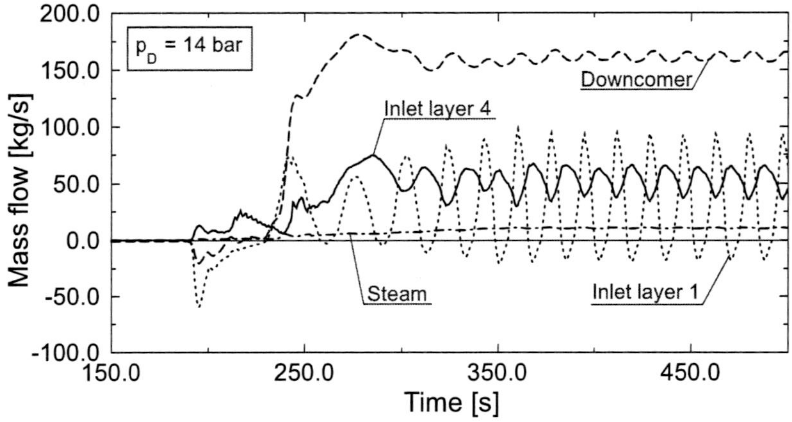

Figure 44. Mass flow in selected tubes of the HRSG. Printed with permission of the publisher from [87].

Figure 44 shows as a result of the numerical simulation the development of the mass flow in selected tubes of the natural circulation HRSG over the time at a drum pressure of 14 bars. In this case the overall steady state circulation ratio was approximately 14.8.

It can be seen that the mass flow in the four tube paths starts approximately 200 s after start of simulation to oscillate. In the following the mass flow in the downcomer and the riser as well as in the tubes of the bundle heating surface oscillate. This detected oscillation does not decay and also exists at steady state. The oscillation is identified as a Type 2 density wave oscillation according to [29].

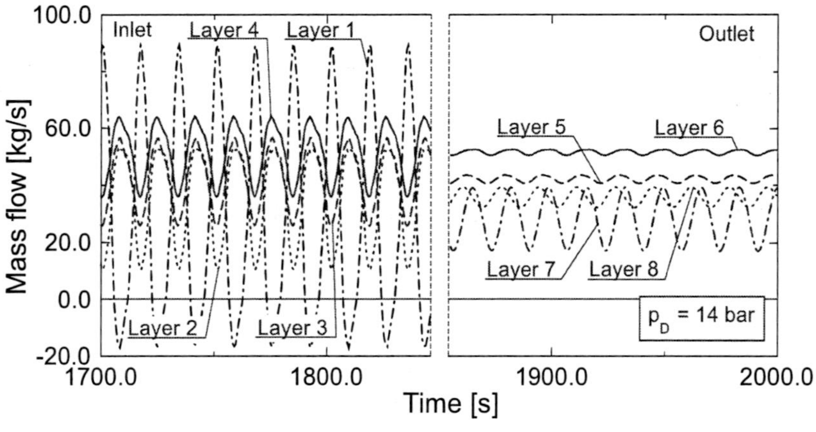

Figure 45. Mass flow in the tubes of the bundle heating surface. Printed with permission of the publisher from [87].

The mass flow at the inlet of the layers 1 to 4 and at the outlet of the layers 6 to 8 is shown in figure 45. The phase displacement of the oscillation in the four tube paths of the evaporator is shifted in time. The oscillation amplitude at the inlet of the first four layers is larger than at to the outlet of the four lower heated layers. Therefore the oscillation is damped during its way through the tube paths from e. g. the inlet of layer 1 to the outlet of the associated layer 7. This oscillation behavior of the DWO results in a low influence on the circulation mass flow in the downcomer and riser as well as on the steam rate which leaves the drum (see in figure 44).

At the inlet of the layers 1 to 4 the amplitude of the oscillation is high compared to the outlet, and therefore reverse flow of the working medium can occur inside these tubes for short time periods during the oscillation (negative mass flow in the most heated layer 1 in figure 45). As a result of the interaction of the fluid flow in the four tube paths in the upper header the oscillation of the mass flow in all other tubes (e. g. the downcomer or riser relief tubes) is in some cases not sinusoidal as described above (see figure 44). This behavior increases with decreasing system pressure at given power and geometry (figure 46). For further details see [90], [92] or [103].

In the literature a high number of analytical and experimental studies exist which have analyzed the influence of different system parameters on the density wave oscillation. The influence of the most important parameters can be summarized as follows:

- **operation pressure:** with increasing system pressure at given heat input the steam quality at the outlet of a heated channel decreases, which results in a lower two-phase pressure drop and mass flow amplitude. A damping of the DWO is the consequence of the raised system pressure, which can be seen in figure 46 and 47 for the analyzed boiler shown in figure 42 in the decrease of the oscillation amplitude.

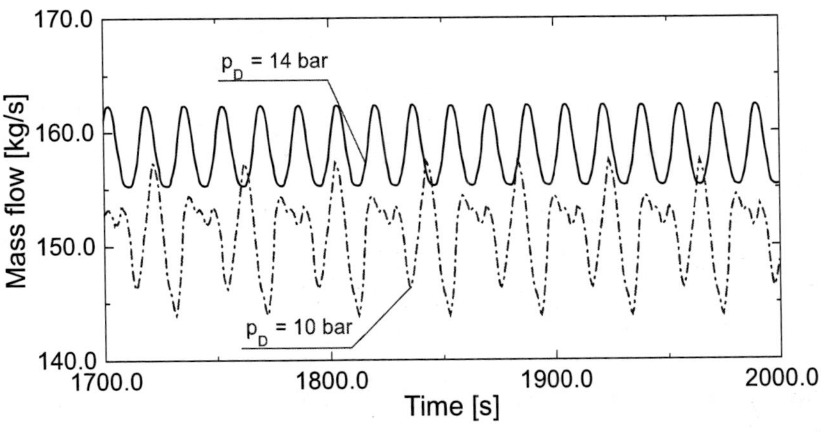

Figure 46. Mass flow in the downcomer of the HRSG at a drum pressure of 14 and 10 bars.

This effect is equal to a reduction of the heat absorption or a increase of the mass flow of the working medium. For further details see amongst others [106], [92], [103], [90], [102], [80] or [15].

The variation of the operation pressure of the analyzed HRSG evaporator presented in figure 42 has shown that a system pressure of 18 bars works very close to the stability boundary. The oscillation amplitude of the DWO decreases very slowly (see figure 47). Approximately 5000 s after the simulation start the oscillation disappears.

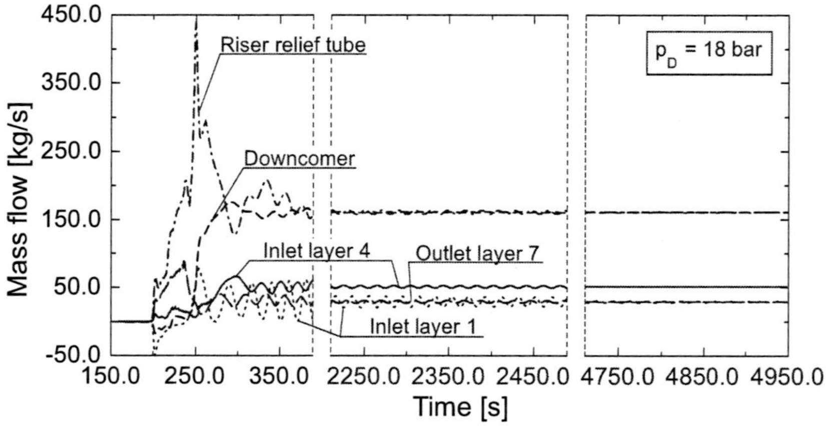

Figure 47. Mass flow in selected tubes of the HRSG at a drum pressure of 18 bars. Printed with permission of the publisher from [92].

- **additional flow resistance at the tube inlet or outlet:** the thermo-hydraulic behavior depends on the flow resistance in the individual sections of the boiler. It is well known that additional flow resistance can improve the stability of heated circulation systems (see, e.g., [15], [97]). The implementation of a orifice at the tube inlet where a single phase flow enters the channel reduces the amplitude of the DWO. A additional flow resistance at the inlet of the heated channel increases the single phase pressure drop, which is in phase with the mass flow at the tube inlet. This improves the stability of the fluid flow in the channel.
  A additional flow resistance implemented at the outlet of the heated tube increases the two-phase flow pressure drop, which is not in phase with the inlet mass flow. This leads to a more unstable behavior of the system ([2], [15], [85], [94], [97], [98] or [103]).

- **fluid velocity and heat flux:** a increase of the fluid velocity or a decrease of the heat flux at a given geometry results in a more stable circulation system. Both measure leads to a lower steam quality at the outlet of the heated channel which results also to a lower two-phase flow pressure drop at the channel outlet.

- **subcooling at the channel inlet:** a high subcooling of the single phase flow at the channel inlet stabilizes the system whereas a lower subcooling leads to a more unstable behavior of the system. A higher subcooling of the single phase flow leads to a lower void fraction at the tube outlet and a increase of the single phase region in the heated channel ([106] or [109]).

Further investigations are done for example to the influence of the gravity ([59]), the length of the pipe ([19], [59]), a non-uniform heating profile over the bundle heating surface ([87], [90], [102], [103]), the bypass ratio of parallel tubes ([17], [41], [83]) or of higher-mode DWO which is connected with a sudden change of the period and the flow pattern ([108], [109], [110]).

4. Conclusion

This chapter dealt with some aspects of modeling and stability analysis of the natural circulation steam generator.

In the introduction a short description of the phenomenon of natural circulation and of some important design criteria as well as the advantages and disadvantages of natural circulation steam generators were presented.

A detailed description of the mathematical models for the most important components of water tube boilers—drum, header and tubes—was given. The modeling of the fluid flow in the tubes of the steam generator was done in one dimension. This one-dimensional mathematical model for the fluid flow includes a homogeneous equilibrium model for the two-phase flow and applies a correction factor for the two-phase pressure loss. The headers and drums, respectively, are assumed to be single control volumes. The equations for the mass and energy balance of the header and the drum are ordinary differential equations. The steam, which enters the header, is evenly distributed to the outlet tubes.

The discretization of the partial differential equations for the fluid flow in the tubes as well as for the flue gas were done with the help of the finite volume method. The pressure-velocity coupling and overall solution procedure are based on the SIMPLE algorithm. To prevent checkerboard pressure fields a staggered grid is employed and for the convective term the UPWIND scheme is used.

The heat transfer model between the fluid flow of the working medium and the tube wall was also presented.

After a short overview of the different types of two-phase flow instabilities a more detailed description was given for the two most common instabilities in natural circulation steam generators, namely, the static instability reverse flow (in particular, the case of the Ledinegg instability) and the dynamic instability density wave oscillation.

For the determination of the static instability, different methods (a graphical and a numerical method) were presented. Different operation pressures and heat loads were investigated. The examples showed that for every system pressure, heat load and boiler configuration, a special critical heat absorption ratio χ_{crit} exists. This critical heat absorption ratio must be determined for every configuration of such a boiler. The critical heat absorption ratio depends also on the thermo-hydraulic behavior of the circulation system. Heating conditions that exceed this critical value can lead to unstable conditions including the possibility of flow stagnation and or reverse flow. Further analysis has shown that all modifications that lead to a higher pressure in the lower header results in a higher stability.

To prevent reverse flow in natural circulation systems where an unheated downcomer is connected with a header to a lower and a higher heated riser system, the unequally heated riser systems are separated. This can be done by installing a downcomer tube for every riser system.

The second example of static instability showed that natural circulation boilers respond in a very sensitive way to perturbations in the pressure drop.

The example of dynamic instability—namely, the density wave oscillation—in natural circulation systems was done for a vertical type HRSG under hot start-up conditions. The horizontal bundle heating surface was designed with four horizontal parallel tube paths. The example showed that the oscillation amplitude was damped during its way through the heated channel. The mass flows inside the four tube paths of the evaporator interact in the upper header of the HRSG. Superposition of the mass flow oscillation can lead to a reduction or amplification of the resulting mass flow amplitude as well as to a non-sinusoidal oscillation amplitude.

The influence of the most important system parameters on the DWO were discussed.

5. Nomenclature

Latin Symbol

Variable	Description
a	discretization coefficient [kg/s]
a_{Ei}	SIMPLE discretization coefficient for the east neighbor of the cell i [kg/s]
a_{hE}	SIMPLE coefficient for the energy balance of the east neighbor cell of the header [kg/s]
a_{hP}	SIMPLE coefficient for the energy balance of the header [kg/s]
a_{hW}	SIMPLE coefficient for the energy balance of the west neighbor cell of the header [kg/s]
a_{mhE}	SIMPLE pressure correction discretization coefficient of the east neighbor cell of the header [m s]
a_{mhP}	SIMPLE pressure correction discretization coefficient of the header [m s]
a_{mhW}	SIMPLE pressure correction discretization coefficient of the west neighbor cell of the header [m s]
a_{Pi}	SIMPLE discretization coefficient of the cell i [kg/s]
a_{Wi}	SIMPLE discretization coefficient for the west neighbor of the cell i [kg/s]
A	cross sectional area [m^2]
b	coefficient [-]
b_i	constant term in discretization equation of the cell i [-]
b_h	constant term in discretization equation for the header [-]
b_{Lapl}	Laplace constant [m]
$\overline{c}_p$	integral spec. isobaric heat capacity [J/kg K]
C_1	coefficient [-]
C_2	coefficient [-]
C_3	coefficient [-]
d	diameter [m]
e	spec. internal energy [J/kg]
f_U	subcooling factor [-]
Fr	Froude-number [-]
g	gravitational acceleration [m/s^2]

Variable	Description
h	spec. enthalpy [J/kg]
Δh_U	spec. enthalpy difference for subcooling [J/kg]
Δh_v	latent heat [J/kg]
H	height [m]
K	amplification factor of the controller [-]
K_C	circulation ratio [-]
K_1	correction factor for heat transfer in single phase flow [-]
l	length [m]
$\dot{m}$	mass flow [kg/s]
$\dot{m}_{flux}$	mass flux [kg/m^{2}s]
M	mass [kg]
n_{Do}	number of downcomer tubes [-]
N_{pch}	phase change number [-]
N_{sub}	subcooling number [-]
Nu	Nusselt-number [-]
p	pressure [Pa]
p'	pressure correction [Pa]
Δp	pressure difference [Pa]
Pr	Prandtl-number [-]
$\dot{q}$	heat flux [W/m^2]
$\dot{Q}$	heat flow [W]
r	radius [m]
Re	Reynolds-number [-]
S_Φ	source term
S_c	constant part of the linearized source term S_Φ
S_P	proportional term of the linearized source term S_Φ
t	time [s]
t_0	sampling time of the controller [s]
T_I	integral action time of the controller [-]
T_{Der}	derivative action time of the controller [-]
u_{in}	input variable of the controller [-]
u_{out}	output variable of the controller [-]
U	perimeter [m]
v_0	controller parameter [-]
v_1	controller parameter [-]
v_2	controller parameter [-]
V	volume [m^3]
V_h	volume of the fluid inside the header [m^3]
w	velocity [m/s]
We	Weber-number [-]
x	length [m]
x	steam quality [-]

Variable	Description
$\dot{x}$	mass vapor quality [-]
Z_1	coefficient for the cross section area of the drum liquid [J/s K]
Z_2	coefficient for the cross section area of the drum liquid [J/m K]

Greek Symbol

Variable	Description
α	heat transfer coefficient [W/m^2 K]
Γ	diffusion coefficient [kg/m s]
η	dynamic viscosity [Pa s]
ρ	density [kg/m^3]
φ	tube inclination angle [-]
Φ	dependend variable of the general balance equation
Φ^2_{fric}	two-phase flow multiplier [-]
ϑ	temperature [ ]
ζ	pressure loss coefficient [-]
ξ_U	mass flow ratio [-]
χ	heating ratio [-]
σ	surface tension [N/m]

Subscript

Variable	Description
$accel$	acceleration
b	bottom
bal	balance
Bl	blow down
$crit$	critical
CD	connection, distribution
$drop$	droplet
D	drum
Der	derivative
Dh	drum-header
Do	downcomer
e	east grid point in momentum balance
eff	effective
ext	external
E	east grid point in mass and energy balance

Variable	Description
feed	feed
fg	flue gas
fric	friction
flux	flux
g	gas
*g*0	gas only
h	header
hyd	hydraulic
H	height
i	inner, cell number
I	integration
imb	imbalance
in	inlet
int	internal
j	index, counter
k	index, counter
l	liquid
*l*0	liquid only
lam	laminar
leav	leaving
lim	limit
Lapl	Laplace
m	mass
max	maximum
n	index, counter
nb	neighbor grid points
N	index, counter
P	grid point in mass and energy balance
out	outlet
pch	phase change
R	riser
sat	saturated
st	steam
sub	subcooling
surf	surface
t	top
tota	total
tur	turbulent
U	subcooled
w	west grid point in the momentum balance
wa	wall
wm	working medium

Variable	Description
W	west grid point in mass and energy balance
α_{min}	lowest value for the heat transfer
$2ph$	two phase

Superscript

Variable	Description
0	value at the old time step
*	guessed value
'	correction value
-	averaged value

References

[1] J. W. Ackerman. Pseudoboiling Heat Transfer to Supercritical Pressure Water in Smooth and Ribbed Tubes. *Trans. of ASME, Ser. C, Journal of Heat Transfer*, **91**:1–8, 1969.

[2] R. P. Anderson, L. T. Bryant, J. C. Carter, and J. F. Marchaterre. Transient Analysis of Two-Phase Natural Circulation Systems. Argonne National Laboratory USAEC Report ANL-6653, Argonne National Laboratory, 1962.

[3] M. Aritomi, Sh. Aoki, and A. Inoue. Instabilities in Parallel Channel of Forced-Convection Boiling Upflow System, (I) Mathematical Model. *Journal of Nuclear Science and Technology*, **14**(1):22–30, 1977.

[4] M. Aritomi, Sh. Aoki, and A. Inoue. Instabilities in Parallel Channel of Forced-Convection Boiling Upflow System, (II) Experimental Results. *Journal of Nuclear Science and Technology*, **14**(2):88–96, 1977.

[5] M. Aritomi, J. H. Chiang, and M. Mori. Fundamental Studies of Safty-Related Thermo-Hydraulics of Natural Circulation Boiling Parallel Channel Flow Systems Under Startup Conditions (Mechanism of Geysering in Parallel Channels). *Nuclear Safety*, **33**(2):170–182, 1992.

[6] M. Aritomi, J. H. Chiang, and M. Mori. Geysering in parallel boiling channels. *Journal of Nuclear Engineering and Design*, **141**:111–121, 1992.

[7] M. Aritomi, J. H. Chiang, T. Nakahashi, M. Wataru, and M. Mori. Fundamental Study on Thermo-Hydraulics during Start-Up in Natural Circulation Boiling Water Reactors, (I) Thermo-Hydraulic Instabilities. *Journal of Nuclear Science and Technology*, **29**(7):631–641, 1992.

[8] J. D. Aziz, K. und Hellums. Numerical Solutions of the Three-Dimensional Equation of Motion for Laminar Natural Convection. *The Physics of Fluid*, **10**:314, 1967.

[9] A. Baars. Stationaere und instationaere Betriebsbedingungen eines Naturumlaufverdampfers. *Fortschr.-Bericht VDI* **779**, VDI Verlag, Duesseldorf, 2003.

[10] A. E. Bergles. *Two-Phase Flows and Heat Transfer*, chapter Review of Instabilities in Two-Phase Systems. Hemisphere Pub. Corp., Washington, 1st edition, 1976.

[11] A. E. Bergles, P. Goldberg, and J. P. Maulbetsch. Acoustic Oszillations in High Pressure Single Channel Boiling Systems. In *Euratom Symposium on Two-Phase Flow Dynamics*, volume 1, pages 525–550. EURATOM, 1967.

[12] L. Biasi, G. C. Clerici, R. Sala, and A. Tozzi. A Theoretical Approach to the Analysis of an Adiabatic Two-Phase Annular Dispersed Flow. *Energia Nucleare*, **15**(6):394–405, 1968.

[13] K. Bier, D. Goetz, and D. Gorenflo. Zum Einfluss des Umfangswinkels auf den Waermeuebergang beim Blasensieden an horizontalen Rohren. *Waerme- und Stoffuebertragung*, **15**:159–169, 1981.

[14] G. Bohnsack. Das Verhalten des Eisen-II-hydroxides bei hoeheren Temperaturen. *Mitteilung der VGB*, **51**(4):328–338, 1971.

[15] J. A. Bouré, A. E. Bergles, and L. S. Tong. Review of Two-Phase Flow Instability. *Nuclear Engineering and Design*, **25**:165–192, 1973.

[16] D. Brockel, G. von der Kammer, W. Rettemeier, and H. Weber. *Jahrbuch der Dampferzeugertechnik*, volume 1, chapter Gro"se Naturumlaufdampferzeuger, pages 362–383. Vulkan Verlag, Essen, 5th edition, 1985.

[17] M. B. Carver. Effect of By-Pass Characteristics on Parallel-Channel Flow Instabilities. In *Proceedings of the Institution of Mechanical Engineers*, volume 184, pages 84–92, 1969.

[18] R. Courant, E. Isaacson, and M. Rees. On the Solution of Non-Linear Hyperbolic Differential Equations by Finite Differences. *Comm. Pure Appl. Math.*, **5**:243, 1952.

[19] C. J. Crowley, C. Deane, and S. W. Gouse Jr. Two-Phase Flow Oscillations in Vertical, Parallel Heated Channels. EURATOM Rep., Proc. Symp. on Two-Phase Flow Dynamics, Eindhoven, 1967.

[20] N. L. Dickinson and C. P. Welch. Heat Transfer to Supercritical Water. *Transactions of the American Scociety of Mechanical Engineers*, **80**(3):746–752, 1958.

[21] G. Domin. Wrmebergang in kritischen und berkritischen Bereichen von Wasser in Rohren. *BWK*, **15**(11):527–532, 1963.

[22] V. E. Doroshchuk, L. L. Levitan, and F. P. Lantsman. Recommendations for Calculating Burnout in a Round Tube with Uniform Heat Release. *Teploenergetika*, **22**(12):66–70, 1975.

[23] G. Drescher and W. Khler. Die Ermittlung kritischer Siedezustaende im gesamten Dampfgehaltsbereich fuer innendurchstroemte Rohre. *BWK*, **33**(10):416–422, 1981.

[24] P. H. Effertz, P. Forchhammer, and A. Heinz. Korrosion und Erosion in Speisewasservorwaermern – Ursachen und Verhuetung. *Der Maschinenschaden*, **51**(4):154–161, 1978.

[25] B. Epple, R. Leithner, W. Linzer, and H. Walter. *Simulation von Kraftwerken und wrmetechnischen Anlagen*. Springer Verlag, 1st edition, 2009.

[26] J. H. Ferziger and M. Perić. *Computational Methods for Fluid Dynamic*. Springer Verlag, Berlin, Heidelberg, New York, 2nd edition, 1999.

[27] E. Fried and I. E. Idelchik. *Flow Resistance: A Design Guide for Engineers*. Hemisphere Publishing Corporation, New York, Washington, Philadelphia, London, 1989.

[28] L. Friedel. Improved Friction Pressure Drop Correlations for Horizontal and Vertical Two-Phase Pipe Flow. In *European Two Phase Flow Group Meeting*, pages 1–25, Ispra, Italien, 1979.

[29] K. Fukuda and T. Kobori. Classification of Two-Phase Flow Instability by Density Wave Oscillation Model. *Journal of Nuclear Science and Technology*, **16**(2):95–108, 1979.

[30] V. Gnielinski. Neue Gleichungen fr den Wrme- und den Stoffbergang in turbulent durchstrmten Rohren und Kanlen. *Forschung im Ingenieurwesen*, **41**(1):8–16, 1975.

[31] V. Gnielinski. Ein neues Berechnungsverfahren fr die Wrmebertragung im bergangsbereich zwischen laminarer und turbulenter Rohrstrmung. *Forschung im Ingenieurwesen*, **61**(9):240–248, 1995.

[32] H. Griem. A new Procedure for the Prediction of Forced Convection Heat Transfer at Near- and Supercritical Pressure. *International Journal of Heat and Mass Transfer: Wrme und Stoffbertragung*, **31**:301–305, 1996.

[33] S. Guanghui, J. Dounan, K. Fukuda, and G. Yujun. Theoretical and Experimentel Study on Density Wave Oscillation of Two-Phase Natural Circulation of Low Equilibrium Quality. *Nuclear Engineering and Design*, **215**:187–198, 2002.

[34] W. B. Hall and J. D. Jackson. Heat Transfer near the Critical Point. In *Proceedings of the 6th International Heat Transfer Conference*, volume 6, pages 377–392, Toronto, Canada, 7.-11. August 1978.

[35] P. Hawtin. Chugging Flow. AERE-R 6661, 1970.

[36] D. Hein, W. Kastner, and W. Koehler. Einfluss der Rohrlage auf den Waermeuebergang in einem Verdampferrohr. *BWK*, **34**(11):489–493, 1982.

[37] R. Isermann. *Grundlagen, Deterministische Regelung*, volume 1 of *Digitale Regelsysteme*. Springer Verlag, Berlin, 2nd edition, 1988.

[38] W. H. Jens and P. A. Lottes. Analysis of Heat Transfer, Burnout, Pressure Drop and Density Data for High Pressure Water. Argonne National Laboratory USAEC Report ANL-4627, Argonne National Laboratory, 1951.

[39] H. P. Jrgl. *Repititorium Regelungstechnik*, volume 2. R. Oldenburg Verlag, Vienna, Munich, 1994.

[40] S. Kakaç, R. K. Shah, and W. Aung. *Handbook of Single-Phase Convective Heat Transfer*. John Wiley & Sons, New York, 1987.

[41] S. Kakaç, T. N. Veziroglu, K. Akyuzlu, and O. Berkol. Sustained and Transient Boiling Flow Instabilities in a Cross-Connected Four-Parallel-Channel Upflow System. In *Proceedings of the* 5th *International Heat Transfer Conference*, volume 4., pages 235–239, Tokyo, Japan, 3.-7. September 1974.

[42] S. V. Karki, K. C. und Patankar. Pressure based calculation procedure for viscous flows at all speeds in arbitrary configurations. *AIAA-Journal*, **27**(9):1167–1174, 1989.

[43] S. Karsli, M. Yilmaz, and O. Comakli. The effect of internal surface modification on flow instabilities in forced convection boiling in a horizontal tube. *International Journal of Heat and Fluid Flow*, **23**:776–791, 2002.

[44] W. Kastner, K. Riedle, and H. Tratz. Experimentelle Untersuchungen zum Materialabtrag durch Erosionskorrosion. *VGB Kraftwerkstechnik*, **64**(5):452–465, 1984.

[45] Y. Katto and H. Ohno. An Improved Version of the Generalized Correlation of Critical Heat Flux for the Forced Convective Boiling in Uniformly Heated Vertical Tubes. *International Journal of Heat and Mass Transfer*, **27**(9):1641–1648, 1984.

[46] V. Kefer. *Stroemungsformen und Waermeuebergang in Verdampferrohren unterschiedlicher Neigung*. PhD thesis, 1989.

[47] V. Kefer, W. Khler, and W. Kastner. Critical Heat Flux (CHF) and Post-CHF Heat Transfer in Horizontal and Inclined Evaporator Tubes. *International Journal of Multiphase Flow*, **15**(3):385–392, 1989.

[48] H. Keller. Erosionskorrosion an Nassdampfturbinen. *VGB Kraftwerkstechnik*, **54**(5):292–295, 1974.

[49] W. Khler. *Einfluss des Benetzungszustandes der Heizflaeche auf Waermeuebergang und Druckverlust in einem Verdampferrohr*. PhD thesis, Technichal University Munich, 1984.

[50] J. B. Kitto Jr. and M. J. Albrecht. *Once-Through Boilers*, chapter Fossil-Fuel-Fired Boiler: Fundamentals and Elements, pages 179–275. John Wiley & Sons, Inc., New York, Cichester, Brisbane, 1991.

[51] P. K. Konakov. Eine neue Formel fr den Reibungskoeffizienten glatter Rohre (Orig. russ.). *Berichte der Akademie der Wissenschaften der UDSSR*, **51**(7):503–506, 1954.

[52] A. S. Kon'kov. Experimental Study of the Conditions under which Heat Exchange Deteriorates when a Steam-Water Mixture Flows in Heated Tubes. *Teploenergetika*, **13**(12):77, 1965.

[53] M. Ledinegg. Unstabilitaet der Stroemung bei natuerlichem und Zwangumlauf. *Die Waerme*, **61**(48):891–898, 1938.

[54] H. Lehmann. *Handbuch der Dampferzeugerpraxis*. Resch-Verlag, Munich, 2.nd edition, 1990.

[55] R. Leithner. *Once-Through Boilers*, chapter Thermohydraulic Design of Fossil-Fuel-Fired Boiler Components, pages 277–362. John Wiley & Sons, Inc., New York, Cichester, Brisbane, 1991.

[56] V. Linzer. Das Ausstroemen von Siedewasser und Sattdampf aus Behaeltern. *BWK*, **22**(10):470–476, 1970.

[57] W. Linzer and H. Walter. Flow reversal in natural circulation systems. *Applied Thermal Engineering*, **23**(18):2363–2372, 2003.

[58] C. Loos and E. Heitz. Zum Mechanismus der Erosionskorrosion in schnell stroemenden Fluessigkeiten. *Werkstoff und Korrosion*, **24**(1):38–48, 1973.

[59] R. P. Mathisen. Out of Pile Channel Instability in the Loop Sk"alvan. In *International Symposium on Two-Phase Dynamics*, Eindhoven, Niederlande, September 1967.

[60] W. H. McAdams, W. E. Kennel, and J. N. Addoms. Heat Transfer of Superheated Steam at High Pressures. *Transactions of the American Scociety of Mechanical Engineers*, **72**:421–428, 1950.

[61] H. C. nal. The Period of Density-Wave Oscillations in Forced Convection Steam Generator Tubes. *Int. J. Heat Mass Transfer*, **25**(3):419–422, 1982.

[62] A. K. Nayak, P. K. Vijayan, D. Saha, V. V. Raj, and M. Aritomi. Analytical Study of Nuclear-Coupled Density-Wave Instability in a Natural Circulation Pressure Tube Type Boiling Water Reactor. *Nuclear Engineering and Design*, **195**:27–44, 2000.

[63] B. Noll. *Numerische Stroemungsmechanik: Grundlagen*. Springer Verlag, Berlin, Heidelberg, New York, 1st edition, 1993.

[64] M. M. Padki, K. Palmer, S. Kakaç, and T. N. Veziroglu. Bifurcation analysis of pressure-drop oscillations and the Ledinegg instability. *International Journal of Heat and Mass Transfer*, **35**(2):525–532, 1992.

[65] S. V. Patankar. *Numerical Heat Transfer and Fluid Flow*. Series in Computational Methods in Mechanics and Thermal Sciences. Hemisphere Publ. Corp., Washington, New York, London, 1980.

[66] P. Profos. Die Stabilisierung der Durchflussverteilung in Zwanglaufheizflaechen. *Energie*, **11**(6):241–247, 1959.

[67] H. Richter. *Rohrhydraulik*. Springer Verlag, Berlin, Heidelberg, 4th edition, 1962.

[68] M. M. Shah. A Generalized Graphical Method for Predicting CHF in Uniformly Heated Vertical Tubes. *International Journal of Heat and Mass Transfer*, **22**:557–568, 1979.

[69] M. M. Shah. A General Correlation for Critical Heat Flux in Annuli. *International Journal of Heat and Mass Transfer*, **23**:225–234, 1980.

[70] A. H. Stenning and T. N. Veziroglu. Flow Oscillations Modes in Forced Convection Boiling. In *Proc. 1965 Heat Transfer and Fluid Mech. Inst., Stanfort Univ. Press*, pages 301–316, 1965.

[71] K. Strausz. *Kraftwerkstechnik: zur Nutzung fossiler, regenerativer und nuklearer Energiequellen*. Springer Verlag, Berlin, Heidelberg, New York, 1992.

[72] A. Stribersky and W. Linzer. Ein Beitrag zum Problem der Stabilit"at beim Naturumlauf. *Fortschr.-Ber. VDI* **154**, VDI Verlag, D"usseldorf, 1984.

[73] M. A. Styrikowitsch, S. L. Miropolsky, and M. E. Schitzman. Wrmebergang im kritischen Druckgebiet bei erzwungener Strmung des Arbeitsmediums. *VGB Kraftwerkstechnik*, **61**:288–294, 1959.

[74] H. S. Swenson, J. R. Carver, and C. R. Kakarala. Heat Transfer to Supercritical Water in Smooth-Bore Tubes. *Trans. of ASME, Ser. C, Journal of Heat Transfer*, **87**:477–484, 1965.

[75] K. Takitani and T. Takemura. Density Wave Instability in Once-Through Boiling Flow System, (I) Experiment. *Journal of Nuclear Science and Technology*, **15**(5):355–364, 1978.

[76] R. Techo, R. R. Tickner, and R. E. James. An Accurate Equation for the Computation of the Friction Factor for Smooth Pipes from the Reynolds Number. *Journal of Appl. Mech.*, **6**, 1965.

[77] F. Thelen. Stroemungsstabilitaet in Verdampfern von Zwangsdurchlaufdampferzeugern. *VGB Kraftwerkstechnik*, **61**(5):357–367, 1981.

[78] J. R. S. Thom, W. M. Walker, T. A. Fallon, and G. F. S. Reising. Boiling in Subcooled Water During Flow Up Heated Tubes or Annuli. In *Symp. Inst. Mech. Eng.*, London, 15.-16. September 1965.

[79] B. Thompson and R. Macbeth. Boiling Water Heat Transfer Burnout in Uniformly Heated Round Tubes. *Technical Report AAEW-R* **356**, 1964.

[80] L. S. Tong and Y. S. Tang. *Boiling Heat Transfer and Two-Phase Flow*. Series in Chemical and Mechanical Engineering. Taylor & Francis, 2nd edition, 1997.

[81] H. Unbehauen. *Regelungstechnik II – Zustandsregelungen, digitale und nichtlineare Regelsysteme*. Vieweg Verlag, Braunschweig, 6th edition, 1993.

[82] VDI–Waermeatlas. *VDI–Waermeatlas - Berechnungsblaetter fuer den Waermeuebergang*. VDI–Verlag GmbH., Duesseldorf, 8th edition, 1996.

[83] T. N. Veziroglu and S. S. Lee. Boiling-Flow Instabilities in a Cross-Connected Parallel Channel Upflow System. In *National Heat Transfer Conference, ASME Paper 71-HT-12, ASME*, New York, 1971.

[84] G. B. Wallis. *One-dimensional Two-phase-Flow*. McGraw - Hill Book Company, New York, St. Louis, San Francisco, London, 1969.

[85] G. G. Wallis and J. H. Heasley. Oscillations in Two-Phase Flow Systems. *Trans. of ASME, Ser. C, Journal of Heat Transfer*, **83**:363–369, 1961.

[86] H. Walter. Modellbildung und numerische Simulation von Naturumlaufdampferzeugern. *Fortschr.-Bericht VDI* **457**, VDI Verlag, Duesseldorf, 2001.

[87] H. Walter. Ein Beitrag zur statischen und dynamischen Stabilitaet von Naturumlaufdampferzeugern. *Fortschr.-Ber. VDI* **546**, VDI Verlag, Duesseldorf, 2006.

[88] H. Walter. Numerical Stability Analysis of a Natural Circulation Steam Generator with a Non-uniform Heating Profile over the tube length. In *Proceedings of the* 4th *IASME / WSEAS International Conference on Heat Transfer, Thermal Engineering and Environment*, pages 187–193. WSEAS, 21.-23. August, Elounda, Greece 2006.

[89] H. Walter. Theoretical Stability Analysis of a Natural Circulation Two-Pass Steam Generator: Influence of the Heating Profile and Operation Pressure. *WSEAS Transactions on Heat and Mass Transfer*, **1**(3):274–282, 2006.

[90] H. Walter. Dynamic Flow Instability of Natural Circulation Heat Recovery Steam Generators. *International Journal of Transport Phenomena*, **9**(2):119–132, 2007.

[91] H. Walter. Dynamic simulation of natural circulation steam generators with the use of finite-volume-algorithms – A comparison of four algorithms. *Simulation Modelling Practice and Theory*, **15**:565–588, 2007.

[92] H. Walter. Numerical Analysis of Density Wave Oszillations in the Horizontal Parallel Tube Paths of the Evaporator of a Natural Circulation Heat Rocovery Steam Generator. In S. H. Sorab, H. J. Catrakis, N. Kobasko, and S. Necasova, editors, *Proceedings of the* 5th *IASME/WSEAS International Conference on HEAT TRANSFER, THERMAL ENGINEERING and ENVIRONMENT (HTE07)*, pages 172–179, Vouliagmeni, Athens, Greece, 25. - 27. August 2007. WSEAS, WSEAS Press.

[93] H. Walter. Development of Nomograms for the Calculation of the Temperature Distribution in a Membrane Wall. In *Proceedings of the 2008 ASME International Mechanical Engineering Congress and Exposition*, number IMECE2008-66361, pages 1–10. American Society of Mechanical Engineers, October 31-November 6, Boston, Massachusetts, USA 2008.

[94] H. Walter and W. Linzer. Flow Stability of Heat Recovery Steam Generators. In *Proceedings of the ASME Turbo EXPO, Power for Land, Sea and Air 2004*, number GT2004-53040, pages 1–9. American Society of Mechanical Engineers, 14. -17. June, Vienna, Austria 2004.

[95] H. Walter and W. Linzer. Investigations to the Stability of a Natural Circulation Two-Pass Boiler. In A. E. Bergles, I. Golobic, Ch. H. Amon, and A. Bejan, editors, *Proceedings of the ASME - ZSIS International Thermal Science Seminar II*, pages 469–474. American Society of Mechanical Engineering, 13.-16. June, Bled, Slowenia 2004.

[96] H. Walter and W. Linzer. Numerical Simulation of a Three Stage Natural Circulation Heat Recovery Steam Generator. *IASME Transactions*, **2**(8):1343–1349, 2005.

[97] H. Walter and W. Linzer. Flow Reversal in a Horizontal Type Natural Circulation Heat Recovery Steam Generator. In *Proceedings of the ASME Turbo EXPO, Power for Land, Sea and Air 2006*, number GT2006-90689, pages 1–9. American Society of Mechanical Engineers, 8.-11. May, Barcelona, Spain 2006.

[98] H. Walter and W. Linzer. Flow Stability of Heat Recovery Steam Generators. *Trans. of ASME, Ser. D, Journal of Engineering for Gas Turbines and Power*, **128**:840–848, 2006.

[99] H. Walter and W. Linzer. Investigation of the Stability of a Natural Circulation Two-Pass Boiler. *Heat and Mass Transfer*, **42**(6):562–568, 2006.

[100] H. Walter and W. Linzer. Stability Analysis of Natural Circulation Systems. In C. A. Long, S. H. Sohrab, H. Catrakis, A. G. Fedorov, F. Sotiropoulos, A. C. Benim, G. Wang, and T. Pham, editors, *Proceedings of the 2006 IASME/WSEAS International Conference on HEAT and MASS TRANSFER*, pages 62–68, 18.-20. January, Miami, Florida, USA, 2006. WSEAS.

[101] H. Walter and W. Linzer. The influence of the operating pressure on the stability of natural circulation systems. *Applied Thermal Engineering*, **26**(7/8):892–897, 2006.

[102] H. Walter and W. Linzer. Density Wave Oscillation in a Vertical Type Natural Circulation Heat Recovery Steam Generator - A Numerical Study. In *Proceedings of the ASME Turbo EXPO, Power for Land, Sea and Air 2008*, number GT2008-50035, pages 1–11. American Society of Mechanical Engineers, 9.-13. June, Berlin, Germany 2008.

[103] H. Walter, W. Linzer, and Th. Schmid. Dynamic Flow Instability of Natural Circulation Heat Recovery Steam Generators. In *ISTP-16, Proceedings of the 16th International Symposium on Transport Phenomena*, pages 1–11. Pacific Center of Thermal-Fluids Engineering, 29. August - 1. September, Prague, Czech Republic 2005.

[104] H. Walter and K. Ponweiser. *Simulation von Kraftwerken und wrmetechnischen Anlagen*, chapter Dampferzeugersimulation - Simulation der Wasser- und Dampfstrmung. Springer Verlag, 1st edition, 2009.

[105] H. Walter and A. Weichselbraun. Ein Vergleich unterschiedlicher Finite-Volumen-Verfahren zur dynamischen Simulation beheizter Rohrnetzwerke. *Fortschr.-Bericht VDI* **477**, VDI Verlag, Duesseldorf, 2002.

[106] Q. Wang, X. J. Chen, S. Kakaç, and Y. Ding. An experimental investigation of density-wave-type oscillation in a convective boiling upflow system. *International Journal of Heat and Fluid Flow*, **15**(3):241–246, 1994.

[107] K. Woerrlein. Instabilitaeten bei der Durchflussverteilung in beheizten Rohrstraengen von Dampferzeugern. *VGB Kraftwerkstechnik*, **55**(8):513–518, 1975.

[108] G. Yadigaroglu and A. E. Bergles. An Experimental and Theoretical Study of Density-Wave Oscillation in Two-Phase Flow. MIT Report DSR 74629-3, Massachusetts Institute of Technology, Cambridge, MA, 1969.

[109] G. Yadigaroglu and A. E. Bergles. Fundamental and Higher-Mode Density-Wave Oscillations in Two-Phase Flow. *Trans. of ASME, Ser. C, Journal of Heat Transfer*, **94**:189–195, 1972.

[110] R. Yadigaroglu. *Thermohydraulics of Two-Phase Flow Systems for Industrial Design and Nuclear Engineering*, chapter Two-Phase Flow Instabilities and Propagation Phenomena, pages 353–404. Series in Thermal and Fluids Engineering. McGraw-Hill Book Company, New York, 1981.

[111] K. Yamagata, K. Nishikawa, S. Hasegawa, T. Fuji, and S. Yoshida. Forced Convective Heat Transfer to Supercritical Water Flowing in Tubes. *International Journal of Heat and Mass Transfer*, **15**:2575–2593, 1972.

[112] G. Yun, G. H. Su, J. Q. Wang, W. X. Tian, S. Z. Qiu, D. N. Jia, and J. W. Zhang. Two-phase instability analysis in natural circulation loops of China advanced research reactor. *Annals of Nuclear Energy*, **32**:379–397, 2005.

INDEX

C

D

E

F

G

H

I

N

O

P

Q

R

S

T

U

V

W

Y

Z